The Clinical Toxicology Laboratory

The Clinical Toxicology Laboratory
Contemporary Practice of Poisoning Evaluation

Editor-in-Chief
Leslie M. Shaw, Ph.D., DABCC

Co-Editor-in-Chief
Tai C. Kwong, Ph.D., DABCC

Associate Editors
Thomas G. Rosano, Ph.D., DABCC
Paul J. Orsulak, Ph.D., M.B.A.
Bryan A. Wolf, M.D., Ph.D.
Barbarajean Magnani, Ph.D., M.D.

2101 L Street, NW, Suite 202
Washington, DC 20037-1558

©2001 American Association for Clinical Chemistry, Inc. All rights reserved. No part of this publication may be reproduced, stored in a retrieval systems, or transmitted in any form by electronic, mechanical, photocopying, or any other means without written permission of the publisher.

1 2 3 4 5 6 7 8 9 0   PCP   03 02 01

Printed in the United States of America

Library of Congress Cataloging-in-Publication Data

The clinical toxicology laboratory : contemporary practice of poisoning evaluation / edited by Leslie M. Shaw . . . [et al.].
   p. ; cm.
   Includes bibliographical references and index.
   ISBN 1-890883-53-0 (alk. paper)
   1. Toxicology. 2. Poisoning. I. Shaw, Leslie M., 1941–
   [DNLM: 1. Toxicology—methods. 2. Chemistry, Analytical—methods. 3. Laboratory Techniques and Procedures. 4. Poisons—analysis. 5. Psychotropic Drugs—poisoning. QV 602 C641 2001]
   RA1211 .C588 2001
   615.9—dc21   2001037325

Contents

Preface		vii
Editors		ix
Contributors		xi
1.	Epidemiology of Poisoning	1
2.	Toxicokinetics: Principles and Practical Applications	11
3.	Pharmacokinetics	17
4.	Clinical Approach to the Poisoned Patient	27
5.	Marijuana	43
6.	Opioids	73
7.	Cocaine	97
8.	Amphetamines	113
9.	γ-Hydroxybutyrate	127
10.	Point-of-Care Testing for Drugs of Abuse	145
11.	Urine Adulteration before Testing for Drugs of Abuse	157
12.	Volatile Alcohols: Ethanol, Methanol, and Isopropanol	173
	Addendum: Point-of-Care Testing for Alcohol	190
13.	Glycols	197
14.	Psychotrophic Agents: The Benzodiazepines	211
15.	Antidepressant Drugs	223
16.	Agents for the Treatment of Bipolar Disorder	237
17.	Antiepileptic Drugs	247
18.	Digoxin and Other Cardiac Glycosides	263
19.	Calcium Channel Blockers: An Overview	291
20.	Biological Monitoring of Chemical Exposure	329

21.	Carbon Monoxide Poisoning	345
22.	Organophosphate and Carbamate Pesticide Poisoning	359
23.	Lead Testing	369
24.	Arsenic, Mercury, and Cadmium	383
25.	Acute Iron Poisoning	401
26.	Alternative Samples: Oral Fluid (Saliva), Sweat, Hair, and Meconium	411
27.	Advanced Analytical Techniques	423
28.	Fundamentals of Pharmacogenetics	437
29.	The Toxicology Laboratory	455

Appendix A: Answers to Self-Assessment Questions 467

Appendix B: Various Toxins Associated with Vital Sign Abnormalities 477

Appendix C: Classic Toxidromes: Clinical Manifestations and
Agents Commonly Involved 479

Appendix D: Concentrations of Compounds that Produce Positive Results 481

Appendix E: Methods for Salicylate and Acetaminophen Measurement 499

Appendix F: Selected Book Reviews 503

Appendix G: Abbreviations Used 507

Index 513

Preface

This exciting new textbook, *The Clinical Toxicology Laboratory: Contemporary Practice of Poisoning Evaluation*, is based on the course notes of the successful AACC Review Course—Contemporary Practice in Clinical Toxicology. All the chapters have been updated and in some cases expanded. Four new chapters have been added to reflect the evolving needs and interests of the laboratory and medical toxicology communities: "Urine Adulteration before Testing for Drugs of Abuse," "Biological Monitoring of Chemical Exposure," "Carbon Monoxide Poisoning," and "Fundamentals of Pharmacogenetics."

The contributors to this unique textbook are leading experts in laboratory medicine, emergency medicine, and toxicology who work at poison control centers and university, governmental, and private laboratories. The wide range of disciplines represented points to the very essence of the contemporary practice of toxicology: the dramatic advances in the care of the poisoned patient due to improved standards in the specialty of emergency medicine, the advent of regional poison centers over the past 25 years, and the success of laboratory toxicologists in adapting cost-effective technologies to support the changing diagnostic and treatment modalities used. Critical to the development of an effective toxicology laboratory service is good communication between the members of all of these disciplines, a key value on which this book is based. It is our hope this book will contribute to the development of clinically effective and cost-efficient toxicology laboratory services in the new millennium.

Major topics in this textbook include:

- Epidemiology of poisoning and substance abuse
- Pharmacokinetic and toxicokinetic behavior of the most prevalent drugs of abuse, medications, and environmental poisons
- Clinical manifestations of poisons and approaches for diagnosis
- Critical appraisal of laboratory and point-of-care testing methods for drugs of abuse and ethanol, sample types, and tests for sample adulteration
- Biological monitoring of chemical exposures
- Pharmacogenetic principles and toxicity risk assessment
- Principles and applications of advanced analytical techniques, especially high-performance liquid chromatography–mass spectrometry
- A timely assessment of the provision of clinical toxicology laboratory service that is based on examining clinical needs, technology requirements, and financial justifications
- For each of the toxins, important information is presented including:
  - Epidemiology
  - A patient case integrated into the chapter to illustrate the symptomatology and the interpretation of toxicologic and other laboratory test data
  - Chemical, pharmacologic, and toxicologic properties
  - Toxicokinetics

- Methodology, including discussion of the comparative data of available measurement methods
- Guidelines for interpretation of concentration data
- A discussion of current controversies and questions

We would like to express our thanks to the many colleagues who have offered encouragement, support, and advice for this book. Our deepest appreciation goes to Joanna Grimes and Shauna Roberts for their careful, professional, and tireless efforts at copyediting and book production.

This book is dedicated to our wives and children: Mary, Margaret, Jim, Michael, Teresa, and James; and Joanie, David, and Michael, who are continuing sources of inspiration and support.

Leslie M. Shaw, Ph.D., Editor-in-Chief
Tai C. Kwong, Ph.D., Co-Editor-in-Chief

Editors

EDITOR-IN-CHIEF

Leslie M. Shaw, Ph.D., DABCC
Professor, Pathology and
 Laboratory Medicine
University of Pennsylvania
 School of Medicine
Director, Clinical Toxicology Laboratory
Department of Pathology and
 Laboratory Medicine
Hospital of the University of Pennsylvania
Philadelphia, Pennsylvania

CO-EDITOR-IN-CHIEF

Tai C. Kwong, Ph.D., DABCC
Professor, Pathology and
 Laboratory Medicine
University of Rochester School of Medicine
 and Dentistry
Director, Toxicology Laboratory
Department of Pathology and
 Laboratory Medicine
Strong Memorial Hospital
Rochester, New York

ASSOCIATE EDITORS

Thomas G. Rosano, Ph.D., DABCC, DABFT
Professor, Pathology and
 Laboratory Medicine
Albany Medical College
Director of Forensic Toxicology and
 Clinical Chemistry
Albany Medical Center Hospital
Albany, New York

Paul J. Orsulak, Ph.D., M.B.A.
Professor of Psychiatry and Pathology
University of Texas Southwestern
 Medical School
Dallas, Texas

Bryan A. Wolf, M.D., Ph.D.
Professor
University of Pennsylvania
 School of Medicine
Pathologist-in-Chief, Pathology and
 Laboratory Medicine
The Children's Hospital of Philadelphia
Philadelphia, Pennsylvania

Barbarajean Magnani, Ph.D., M.D.
Director, Clinical Chemistry
Associate Chief, Department of
 Laboratory Medicine
Boston Medical Center
Boston University School of Medicine
Boston, Massachusetts

Contributors

Linda C. Akers, M.S.
Research Chemist
Mental Health Service
Veterans Administration
North Texas Health Care System
Dallas, Texas

Lee M. Blum, Ph.D., DABFT
Director, Occupational and
 Environmental Toxicology
National Medical Services, Inc.
Willow Grove, Pennsylvania

Larry D. Bowers, Ph.D., DABCC, FACB
Senior Managing Director,
 Technical/Information Resources
United States Anti-Doping Agency
Colorado Springs, Colorado

Daniel J. Cobaugh, Pharm.D., DABAT
Clinical Associate Professor
Department of Emergency Medicine
The George Washington University
Associate Director
American Association of
 Poison Control Centers
Washington, D.C.

Edward J. Cone, Ph.D.
ConeChem Research
Severna Park, Maryland

Francis J. De Roos, M.D.
Assistant Professor, Emergency Medicine
University of Pennsylvania
 School of Medicine
Residency Program Director,
 Emergency Medicine
Hospital of the University of Pennsylvania
Philadelphia, Pennsylvania

Henry R. Drott, Ph.D.
Clinical Assistant Professor of Chemistry
 in Pediatrics
University of Pennsylvania
 School of Medicine
Director of Clinical Chemistry
Department of Pathology and
 Laboratory Medicine
The Children's Hospital of Philadelphia
Philadelphia, Pennsylvania

Anne F. Eder, M.D., Ph.D.
Assistant Professor, Pathology and
 Laboratory Medicine
University of Pennsylvania
 School of Medicine
Medical Director, Hematology Laboratory
Assistant Medical Director, Blood Bank
 and Apheresis Service
Department of Pathology and
 Laboratory Medicine
The Children's Hospital of Philadelphia
Philadelphia, Pennsylvania

Edward A. Emmett, M.D., M.S.
Professor and Director, Academic
 Programs in Occupational Medicine
Department of Emergency Medicine
University of Pennsylvania
 School of Medicine
Philadelphia, Pennsylvania

Contributors

Albert D. Fraser, Ph.D., FCACB, DABCC, DABFT
Professor, Department of Pathology
Dalhousie University
Clinical and Forensic Toxicologist
Department of Pathology and
    Laboratory Medicine
Queen Elizabeth II Health Science Centre
Halifax, Nova Scotia, Canada

Bruce A. Goldberger, Ph.D., DABFT
Director of Toxicology and
    Clinical Associate Professor
Department of Pathology, Immunology and
    Laboratory Medicine
University of Florida College of Medicine
Gainesville, Florida

Fred M. Henretig, M.D.
Professor of Pediatrics
University of Pennsylvania
    School of Medicine
Director, Section of Clinical Toxicology
Division of Emergency Medicine
The Children's Hospital of Philadelphia
Medical Director
The Poison Control Center
Philadelphia, Pennsylvania

Marilyn A. Huestis, Ph.D., M.S., B.A.
Adjunct Associate Professor
Department of Epidemiology
University of Maryland Medical School
Acting Chief, Department of Chemistry and
    Drug Metabolism
National Institute on Drug Abuse
    Intramural Research Program
National Institutes of Health
Baltimore, Maryland

Daniel Isenschmid, Ph.D., DABFT
Chief Toxicologist
Wayne County Medical Examiner's Office
Detroit, Michigan

Sarah Kerrigan, Ph.D.
Clinical Professor
Department of Pathology
University of New Mexico
Bureau Chief, Department of Toxicology
New Mexico Department of Health,
    Scientific Laboratory Division
Albuquerque, New Mexico

Gabor Komaromy-Hiller, Ph.D., DABCC
Technical Director, Special Chemistry
Specialty Laboratories
Santa Monica, California

Tai C. Kwong, Ph.D., DABCC
Professor, Pathology and
    Laboratory Medicine
University of Rochester School of Medicine
    and Dentistry
Director, Toxicology Laboratory
Department of Pathology and
    Laboratory Medicine
Strong Memorial Hospital
Rochester, New York

Barry Levine, Ph.D., DABFT, DABCC
Toxicologist
Office of the Chief Medical Examiner
Baltimore, Maryland
Director, Forensic Toxicology Laboratory
Armed Forces Medical Examiner
Rockville, Maryland

Mark W. Linder, Ph.D., FACB, DABCC
Assistant Professor, Pathology and
    Laboratory Medicine
Associate Director,
    Pharmacogenetics Laboratory
University of Louisville School of Medicine
Assistant Director, Clinical Chemistry and
    Toxicology Laboratory
University of Louisville Hospital
Louisville, Kentucky

Pei-Ke Liu, M.D.
Research Fellow, Psychiatry
University of Texas Southwestern
    Medical School
Dallas, Texas

Barbarajean Magnani, Ph.D., M.D.
Director, Clinical Chemistry
Associate Chief, Department of
    Laboratory Medicine
Boston Medical Center
Boston University School of Medicine
Boston, Massachusetts

Karla A. Moore, D.V.M., Ph.D., DABFT
Deputy Toxicologist
Department of Toxicology
Office of the Chief Medical Examiner
Baltimore, Maryland

Paul J. Orsulak, Ph.D., M.B.A.
Professor of Psychiatry and Pathology
University of Texas Southwestern
   Medical School
Dallas, Texas

Kevin C. Osterhoudt, M.D.
Assistant Professor of Pediatrics
University of Pennsylvania
   School of Medicine
Consultant Toxicologist
The Children's Hospital of Philadelphia
Associate Medical Director
The Poison Control Center
Philadelphia, Pennsylvania

Frank P. Paloucek, B.S., Pharm.D., DABAT
Clinical Associate Professor
Director, Residency Programs
Pharmacy Practice
University of Illinois
Chicago, Illinois

K. Michael Parker, Ph.D., DABCC
Professor and Vice Chair
Department of Pathology
The University of Oklahoma
   Health Sciences Center
Director, Department of Toxicology
The University of Oklahoma Medical Center
Oklahoma City, Oklahoma

Jeanmarie Perrone, M.D., FACEP
Assistant Professor, Emergency Medicine
University of Pennsylvania
   School of Medicine
Attending Physician, Emergency Medicine
Hospital of the University of Pennsylvania
Philadelphia, Pennsylvania

Petrie M. Rainey, M.D., Ph.D., DABMT
Professor, Laboratory Medicine
University of Washington
Director, Clinical Chemistry
University of Washington Medical Center
Seattle, Washington

Jeri D. Ropero-Miller, Ph.D.
Assistant Clinical Professor
Department of Pathology and
   Laboratory Medicine
University of North Carolina–Chapel Hill
Deputy Chief Toxicologist
Office of the Chief Medical Examiner
Chapel Hill, North Carolina

Thomas G. Rosano, Ph.D., DABCC, DABFT
Professor, Pathology and
   Laboratory Medicine
Albany Medical College
Director of Forensic Toxicology and
   Clinical Chemistry
Albany Medical Center Hospital
Albany, New York

Arti N. Shah, M.S., M.D.
Internal Medicine Resident/
   Cardiovascular Fellow
Department of Internal Medicine
Hahnamann University
Philadelphia, Pennsylvania

Sandy Shah, D.O.
Post-Doctoral Cardiovascular Fellow
Department of Cardiovascular Medicine
Yale University School of Medicine
New Haven, Connecticut
and
Cardiovascular Fellow
Department of Cardiovascular Medicine
Robert Wood Johnson Medical Center
University of Medicine and Dentistry
   of New Jersey
New Brunswick, New Jersey

Leslie M. Shaw, Ph.D., DABCC
Professor, Pathology and
   Laboratory Medicine
University of Pennsylvania
   School of Medicine
Director, Clinical Toxicology Laboratory
Department of Pathology and
   Laboratory Medicine
Hospital of the University of Pennsylvania
Philadelphia, Pennsylvania

Offie Porat Soldin, Ph.D., M.B.A.
Research Scientist
Consultants in Epidemiology and
 Occupational Health
Washington, DC

Steven John Soldin, Ph.D., FACB, FCACB
Professor, Pediatrics and Pathology
The George Washington University
 School of Medicine
Director, Clinical Chemistry
Department of Laboratory Medicine
Children's National Medical Center
Washington, DC

Roland Valdes Jr., Ph.D., FACB
Professor, Pathology and
 Laboratory Medicine
Director, Pharmacogenetics Laboratory
University of Louisville School of Medicine
Chief, Clinical Chemistry and
 Toxicology Laboratory
University of Louisville Hospital
Louisville, Kentucky

Jo-Anne Vergilio, M.D.
Fellow
Department of Pathology and
 Laboratory Medicine
University of Pennsylvania
 School of Medicine
Philadelphia, Pennsylvania

Andrew Volosov, Ph.D.
Research Associate
Department of Laboratory Medicine
Children's National Medical Center
Washington, DC

Philip D. Walson, M.D.
Professor, Department of Pediatrics
University of Cincinnati
Director, Clinical Pharmacology Division
Clinical Trials Office
Children's Hospital Medical Center
Cincinnati, Ohio

Bryan Wolf, M.D., Ph.D.
Professor
University of Pennsylvania
 School of Medicine
Pathologist-in-Chief, Pathology and
 Laboratory Medicine
The Children's Hospital of Philadelphia
Philadelphia, Pennsylvania

Alan H.B. Wu, Ph.D., DABCC
Professor, Pathology and Laboratory
 Medicine
University of Connecticut
Director, Clinical Chemistry
Department of Pathology and
 Laboratory Medicine
Hartford Hospital
Hartford, Connecticut

CHAPTER 1

Epidemiology of Poisoning

Jeanmarie Perrone, M.D., FACEP

LEARNING OBJECTIVES

After completing this chapter, the reader should be able to:

1. Describe the various routes by which poisoning occurs.
2. List new substances of abuse.
3. Identify data resources for tracking poisoning epidemiology.
4. Specify common substances implicated in poisoning and their associated toxicity.
5. List the more common drugs implicated in pediatric and adult poisonings.

OVERVIEW

Despite measures at prevention, unintentional and accidental poisonings continue to occur. Last year, 2.2 million poisonings and 873 deaths were reported to the American Association of Poison Control Centers (AAPCC) (1) (Table 1–1). Although most adult poisonings are intentional, 53% of these exposures were in children under the age of 6 and were most likely preventable. New substances of abuse and new pharmaceuticals have also resulted in increased toxicity. The expansion of alternative therapies, including herbals and nontraditional medications, has created a spectrum of toxic effects. In addition, the ubiquity of the Internet as an information source continues to contribute to sales of unregulated products and the dissemination of information on synthesis of specific abused substances (2,3). The challenge for the toxicology laboratory is to be both contemporary and timely. Affiliation with clinical staff in emergency medicine, psychiatry, and medical toxicology and poison control is important for understanding the spectrum of poisonings in your region.

Epidemiology

Although carbon monoxide is responsible for the greatest number of poisoning deaths annually (>5000/year), cyclic antidepressants have the highest case fatality rate for ingestions. Salicylates and acetaminophen (aspirin and Tylenol) account for many fatalities (340/year) but are so ubiquitous in over-the-counter (OTC) and prescription formulations that the number of exposures is very large (74,602 adult exposures in 1999; see Table 1–2), and the case fatality rate therefore is relatively low.

Routes of Exposure

Poisoning may occur in various settings. The most common unintentional poisoning reported to poison control centers is accidental pediatric exposure (see Table 1–3). For example, a mother tells her young children everyday to take vitamins to "grow strong"; the older child enthusiastically feeds the younger sibling a bottle of vitamins, with resultant severe iron poisoning.

More serious outcomes result from intentional adult suicide attempts. The irony is that many depressed patients overdose on

TABLE 1–1. Categories with largest numbers of deaths, 1999

Substance	Deaths per year
Analgesics	340
Antidepressants	153
Cardiovascular drugs	127
Stimulants and street drugs	121
Sedative-hypnotics	110
Alcohols	97
Gases and fumes	45
Chemicals	43
Antihistamines	28
Muscle relaxants	18
Automotive products	15
Pesticides and insecticides	15
Anticonvulsants	14
Hormones and hormone antagonists	13
Cleaning substances	13

Adapted from Litovitz TL, Klein-Schwartz W, White S, et al: 1999 Annual Report of the American Association of Poison Control Centers Toxic Exposure Surveillance System. *Am J Emerg Med* 2000;18:517–574. Note: Carbon monoxide exposures are the leading cause of poisoning death, but are not tabulated in this database.

TABLE 1–2. Top 15 substances most frequently involved in adult (>19 years old) exposures, 1999

Substance	Cases per year
Analgesics	74,602
Cleaning substances	64,691
Bites and envenomations	52,349
Sedative-hypnotics	50,311
Antidepressants	42,983
Food poisoning	36,924
Cosmetics and personal care	30,029
Chemicals	29,441
Alcohols	28,020
Hydrocarbons	25,676
Fumes, gases, and vapors	25,056
Pesticides and insecticides	24,510
Cardiovascular drugs	22,947
Plants	16,764
Cough and cold preparations	15,868

Adapted from Litovitz TL, Klein-Schwartz W, White S, et al: 1999 Annual Report of the American Association of Poison Control Centers Toxic Exposure Surveillance System. *Am J Emerg Med* 2000;18:517–574.

TABLE 1–3. Top 10 substances most frequently involved in pediatric (<6 years old) exposures, 1999

Substance	Cases per year
Cosmetics and personal care preparations	153,057
Cleaning substances	123,575
Analgesics	87,471
Plants	79,287
Foreign bodies	76,268
Topical agents	65,561
Cough and cold preparations	63,951
Pesticides and insecticides	43,107
Vitamins	38,651
Gastrointestinal preparations	36,597

Adapted from Litovitz TL, Klein-Schwartz W, White S, et al: 1999 Annual Report of the American Association of Poison Control Centers Toxic Exposure Surveillance System. *Am J Emerg Med* 2000;18:517–574.

their antidepressant medications, many of which can be lethal (cyclic antidepressants, monoamine oxidase inhibitors).

Although federal agencies such as the Occupational Safety and Health Administration (OSHA) have improved occupational safety, many significant exposures continue to occur on the job. The people most at risk are the self-employed, employees at smaller companies that are not subject to OSHA inspections, and temporary workers who do not get appropriate safety orientation because of the transient nature of their employment. Approximately 3% of poisonings occur in the workplace. Iatrogenic toxicity is also a significant cause of preventable poisonings. Approximately 6% of cases of drug toxicity result from administration errors, decimal point misreading, prescription illegibility, or inadequate patient instructions.

Additionally, patients chronically on medication, especially the elderly, may develop renal or hepatic insufficiency, decreasing clearance and excretion of the drug. This leads to accumulation of the drug and toxic effects. The most common example is that of an elderly patient on digoxin for heart failure and atrial fibrillation. Progressive renal dysfunction leads to digoxin toxicity and potential cardiac dysrhythmia.

POISONING EXPOSURES

Factors

There are four factors involved in poisoning exposure: age, geography, season, and setting. Different substances are popular among different age groups, which reflects access to substances. In young adolescents and teenagers, solvent or inhalant abuse may occur due to its extensive availability. The Internet has also become a resource for teen experimentation, including recipes for the synthesis of γ-hydroxybutyrate (GHB) (3) and dextromethorphan abuse.

Geographic variation in drug use is especially prominent in border states such as Texas and California, where the illegal importation of drugs, such as benzodiazepines sold without a prescription in Mexico, is prevalent. Methamphetamine, which is synthesized in elaborate home laboratories, plagues certain regions of the Western United States, probably due to abuse in certain indigent populations. Heroin abuse is more common in the East, especially in urban port cities such as Baltimore, New York, and Philadelphia.

Season may influence the availability of plant poisonings such as Jimson weed anticholinergic poisoning in the fall and carbon monoxide poisoning in the fall and winter months, when accessory heat sources are used.

The setting, urban vs. rural, also influences the types of poisonings seen. Organophosphate poisoning or nicotine poisoning from wet tobacco harvests may be a common problem in rural North Carolina, while exceedingly rare in urban areas, where lead poisoning in the cities and concerns for "arsenic" in playground equipment made with wood pressure-treated with copper-chromium-arsenic are more of an issue.

Substance Abuse

Although alcohol and cigarettes continue to lead the numbers of new substance users and abusers, marijuana, hallucinogens, and inhalants are also prominent. However, when databases such as the Drug Abuse Warning Network (DAWN), which tallies emergency department visits related to use of an illicit drug, are used, cocaine and heroin are most commonly implicated.

DAWN data are limited in their sampling scope in that the focus is on emergency department hospital visits; patients may be treated for substance use in other areas of hospitals or in an outpatient setting and are not detected in these numbers.

The following long-term trends have been observed from recent (1999) DAWN data comparing 1990 and 1999 (http://www.samhsa.gov/oas/p0000018.htm). The total number of drug-related emergency department visits reported to DAWN increased 49% from 1990 to 1999. During this period, reported emergency department visits for heroin abuse increased the most (149%), followed by cocaine (110%) and methamphetamine 100%. The vast majority of estimated emergency department mentions of methamphetamine or speed in 1999 came from five cities in the United States: Los Angeles, San Diego, San Francisco, Seattle, and Phoenix. Since 1988, heroin and cocaine visits in patients ≥35 years old have tripled, reflecting an aging population of substance users. Marijuana-related visits have increased, especially in white males age 18–25, often in combination with other drugs.

Additional epidemiologic data focusing on first-time substance users are collected through the National Center for Alcohol and Drug Abuse Interventions (NCADI). Recent estimates include more than half a million new first-time cocaine users per year.

Incidence of Substance Abuse

There are several government agencies that track data on substance use. The National Household Survey on Drug Abuse estimated >13.6 million users of illicit drugs in 1998. However, this number is considered a gross underestimation, considering the survey only contacts people with a home and a telephone. DAWN (discussed above) is limited by including only patients who need medical attention. The National Institute of Justice Drug Use Forecasting (DUF) collects data on the results of drug screening in the arrestee population. Although this population has a high incidence of substance abuse, relative changes from year to year do underscore trends in drug use. Not surprisingly, 48–62% of arrestees tested positive for at least one drug in 1994 (cocaine, marijuana, and opioid had the highest incidence) (4). These three data sources can be followed for an approximation of trends in substance use.

NEW, UNUSUAL, OR EMERGING SUBSTANCES OF ABUSE

The AAPCC annually tabulates exposures that are reported to regional poison control centers. Although this data set is incomplete and reflects reporting selection bias, many new trends of poisoning or substance abuse have been revealed. Through the poison center network, public health alerts can be issued in real time, potentially diminishing future exposures. Various trends and epidemics have been followed, and the network is often the best source of regional substance abuse data.

Case 1: Scopolamine-Tainted Heroin

In March 1995, eight patients presented to a New York City hospital with altered mental status. The patients were brought by ambulance from a party where people were known to be using heroin. Physical examination revealed pulse 130–140/min, blood pressure 170/100 mmHg, temperature 99–101 °F (37.2–38.3 °C), and respiratory rate 18–26/min. The patients were noted to have markedly dilated pupils that were minimally reactive to light, as well as dry mouth and dry axilla. Diminished bowel sounds and evidence of urinary retention were noted when Foley catheters were placed. Their skin was flushed, and they were described as in an agitated delirium. The patients were restrained and sedated with benzodiazepines as needed; some underwent computed tomography (CT) scanning and lumbar punctures for altered mental status and fever. All were normal. Over 24–72 hours in the hospital, the patients improved to normal mental status but admitted only to using heroin and/or cocaine. Urine drug screens by enzyme-multiplied immuno technique (EMIT) detected morphine and/or cocaine metabolites.

Throughout March and April, as many as 50 patients were seen in New York City emergency departments with a similar presentation. Some were treated with physostigmine for presumed anticholinergic toxidrome with improvement; however, drug screening for anticholinergic compounds was negative. Ultimately, a sample of the heroin was obtained and analyzed by gas chromatography–mass spectrometry (GC/MS) and found to contain scopolamine. Subsequently, urine and blood from affected patients were also shown to contain scopolamine.

The epidemic has been noted in several Northeastern cities (5). It is still not clear why scopolamine was present in some heroin samples. Scopolamine contamination of cocaine was previously reported in early 1990 in Spain (6). It is likely an intentional adulterant to potentiate the effects of heroin in the central nervous system (CNS). The incidence has now tapered off, but cases continue to occur sporadically. Scopolamine is not detected in routine toxicology laboratory screening.

Case 2: Health-Food Fad Leads to Toxicity

A 26-year-old bodybuilder was brought to the emergency department after having a seizure in a bar. Physical examination revealed a vigorous young male, somnolent with sonorous respirations and the following vital signs: pulse 70/min, respiratory rate 10/min, blood pressure 120/70 mmHg, temperature 99 °F (37.2 °C). The patient's pupils were 3 mm and reactive. He had moist mucous membranes and a decreased gag reflex. His lungs were clear, and his skin was warm, dry, and not flushed. His neurologic exam was notable for flaccid coma.

The patient was intubated to protect his

airway and assist ventilation. He was given activated charcoal (50 g) via nasogastric tube. Intravenous fluids were administered for an hour, and when the patient awakened he extubated himself. Friends then admitted "slipping" something in his drink, GHB.

γ-Hydroxybutyrate

The use and abuse of GHB is increasing in the United States. GHB emerged as a popular diet fad in bodybuilders in the early 1990s and was purported to release growth hormone and "burn fat into muscle while you sleep." Although it has no medical indications in the United States, it has been investigated in Europe as an agent to treat alcohol withdrawal and as an anesthetic. It was initially sold in health-food stores. However, reports of toxicity and abuse prompted many states to prohibit sales, and it is now largely obtained illicitly and via mail order. Apparently, GHB is synthesized easily by adding sodium hydroxide to γ-hydroxybutyrolactone (GBL), a common industrial solvent (see Figure 1–1). Caustic burns due to sodium hydroxide in these homemade mixtures have been reported (7). Additionally, intoxication from GBL alone is now reported (8).

Clinical Toxicity of GHB and GHL

Hundreds of cases of GHB and GBL toxicity have been reported to the Centers for Disease Control and Prevention (9,10). Typical manifestations of GHB intoxication include CNS and respiratory depression and can occur after recommended doses, which vary from 1/2 to 3 teaspoons (2.5 to 14.8 mL) at bedtime. Approximately 10–15% of emergency department cases of GHB intoxication require intubation, highlighting the potency of the CNS depressant effects of this agent (11,12). Concurrent ethanol use, which is common, appears to exacerbate the effects. A case series of adverse events including one death was recently reported in nine patients with GBL abuse (13). Although several fatalities have been reported, polysubstance abuse may be contributory. Treatment is largely supportive, and one of the hallmarks of GHB intoxication is the relatively rapid recovery of consciousness after severe obtundation.

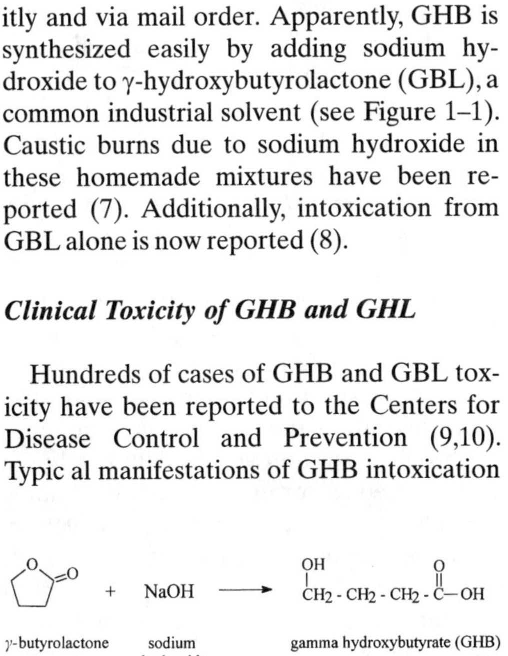

Herbals

Ephedrine

Ephedrine, like GHB, has become a popular health-food supplement. Known by its Chinese herbal nomenclature, ma huang, or simply as ephedra, it is alleged to have various properties. Ephedrine is advertised as an energy supplement and as a dieting aid. It has become known as "herbal ecstasy," allegedly a safer, plant-derived "herbal" form of ephedrine. A recent report described >500 adverse events associated with ephedrine-containing dietary supplements over a 2-year period (14). Adverse effects included palpitations, tachycardia, syncope, hypertension, psychoses, convulsions, coronary vasospasm, chest pain, acute myocardial infarction, and cerebrovascular ischemia (14,15). Eight deaths have been reported due to cardiovascular or cerebrovascular complications. It has been proposed that most incidents occurred because the public did not perceive the potential for adverse reactions from a natural or herbal product.

FIG. 1–1. GHB synthesis. GBL + NaOH yields GHB. However, γ-butyrolactone, a solvent that is readily available, appears to be rapidly hydrolyzed to active GHB in the stomach, obviating the need for any presynthesis steps.

Case 3: Concerned College Coed

A 19-year-old college freshman came to the emergency department early one Sunday morning concerned that she had been "drugged." She reported that she was in a bar the night before and did not remember anything after that. The following morning she awakened in her dormitory room but did not know how she got home. She had heard about other students getting "drugged." Her physical examination revealed normal vital signs and was unremarkable. In addition, there was no evidence of trauma or sexual assault. This case illustrates the common concern of patients seeking emergency care due to fear they have been "poisoned." The absence of signs and symptoms and a "negative" drug screen is typical. This patient's history is consistent with her having been a victim of a "date-rape" drug, although toxicology lab analytical data to confirm this hypothesis is lacking.

Malicious Tampering and "Knockout" Drops

The concept of rendering someone unconscious or delirious as an endeavor of chemical persuasion has been present since antiquity. Nearly 100 years of references to various "knockout" drops in literature, film, and medical history have been described. The term "slipped him a Mickey," or "Mickey Finn," emerged in the early 20th century and is thought to refer to chloral hydrate. The potency and toxicity of chloral hydrate were enhanced in the presence of ethanol. Historically, scopolamine has also been implicated in cases of chemical submission. In an epidemic in New York City, scopolamine was used to intoxicate wealthy businessmen, who were then stripped of their personal belongings, including clothing, watches, jewelry, and money. A similar series of cases occurred in Florida when jewelry store employees were poisoned when they were asked to participate in an orange juice survey.

Flunitrazepam, the intermediate-acting oral benzodiazepine, is the current drug being used for such purposes. Also known as "roofies," flunitrazepam's abuse among teenagers in Florida has become problematic.

Rohypnol® (Hoffmann–La Roche, Nutley, NJ) is not approved for use in the United States; however, it is available by prescription in Europe and without a prescription in South America. Nasal insufflation (snorting or sniffing) of flunitrazepam is a common form of substance abuse in Chile.

Although flunitrazepam is a benzodiazepine, its chemical structure is slightly different so that it produces more hypnotic and amnesic effects than diazepam. Additionally, it is more potent, and thus clinically significant doses may not be detected on many benzodiazepine screens because of lower threshold limits of detectability for the urinary metabolite. Some hospitals have integrated an extensive "send-out" serum and urine drug testing program for all victims of sexual assault to specifically address this issue.

Succumbing to public pressures, Hoffmann–La Roche has "volunteered" to formulate a new Rohypnol pill that releases a blue dye and flakes of white material in the process of solubilization. GHB reportedly has also been used in a similar mode.

Case 4: Solvent Abuse—A Surprising Case in Philadelphia

In February 1999, a devastating car crash left five bright high school girls dead at the scene. The crash occurred at 1500 h on a cold, clear sunny afternoon on dry road and without provocation. Investigation into the cause of the crash was unrevealing. Ultimately, the county medical examiner reported finding "significant" concentrations of a solvent, a computer keyboard cleaner, in the blood of the driver and in several of the passengers of the car. Concentrations were too high to be associated with ambient exposure and were alleged to be related to intentional inhalation (huffing). Solvent abuse occurs in younger people because it is easily available and an inexpensive alternative to alcohol and other substances. A recent poison-center study described the epidemiology of inhalant abuse in its region (16). Although the mean age of the patients reported was 16 years, the age ranged from 4 to 45 years. Spray paint and gasoline were the most common substances involved. Abuse was described in all settings, but more commonly reported in rural environments.

CLASSIC POISONING CASES

Case 5: Common Poisoning Exposures—Emptying the Medicine Cabinet

A 23-year-old female student was brought to the emergency department with altered mental status several hours after overdosing on everything in her medicine cabinet. She was agitated and was not cooperative. She was flushed and awake with the following vital signs: pulse 140/min, respiratory rate 20/min, blood pressure 150/95 mmHg, temperature 99.8 °F (37.7 °C). Her pupils were 8 mm dilated, and she had dry mouth and oral mucosa. Her lungs were clear, and a normal cardiovascular exam was noted. She had decreased bowel sounds, and her skin was noted to be dry and flushed, without axillary sweat. She was unable to follow commands; however, she appeared to be moving all extremities equally. Her reflexes were brisk.

The patient was presumed to have an anticholinergic poisoning due to ingestion of an antihistamine such as doxylamine or diphenhydramine found in many OTC cold remedies. The patient received activated charcoal and sedation with lorazepam. Her mental status improved over 24 hours in the hospital. The following day, however, she complained of right upper quadrant pain and was noted to have an aspartate transaminase of 2150 U/L.

Toxicity of Over-the-Counter Medications

Acetaminophen is the most commonly used medication in the world. Acetaminophen and salicylates, because of their prevalence in numerous OTC preparations, result in many cases of serious toxicity as well as fatalities annually. Although patients may not expect to get too sick by overdosing on nonprescription pills, many fatalities are reported annually from OTC medications such as acetaminophen, salicylates, and cold medications and decongestants (especially phenylpropanolamine and diphenhydramine).

Although generally safe in therapeutic dosing, intentional overdose of acetaminophen can be fatal (1). Early symptoms after overdose are vague and may be absent; therefore, recognizing acetaminophen poisoning by clinical signs and symptoms may be impossible. Furthermore, patients and physicians may not recognize that acetaminophen was co-ingested when a more clinically symptomatic poisoning is present as in the case above in a hospital that had very limited toxicology laboratory support. In addition, early recognition of acetaminophen toxicity and treatment with the antidote N-acetylcysteine prevents hepatotoxicity and death (17). Therefore, a serum acetaminophen concentration measurement has been recommended in most patients with intentional ingestions to rule out the presence of clinically silent acetaminophen toxicity (18).

Phenylpropanolamine is a potent decongestant with amphetamine-like properties. It can precipitate severe hypertension, often accompanied by a reflex bradycardia. Severe headaches and, rarely, intraparenchymal hemorrhages have been reported (19). Due to a recent case-control series, phenylpropanolamine containing products were voluntarily recalled in December 2000 (20).

Dextromethorphan, a common ingredient in cough suppressants, is an opioid derivative with some abuse potential. Because of dose and weight considerations, dextromethorphan toxicity occurs mostly in toddlers. Antihistamines and decongestants, present in many cough and cold preparations, may result in anticholinergic toxicity or acute hypertensive episodes, respectively. Visine® contains tetrahydrozoline, a clonidine-like imidazolidine that causes CNS depression, hypotension, and bradycardia. Airway management, fluids, and atropine may be required.

Case 6: Rose Spray

A 22-year-old male was brought to the emergency department after collapsing in his parents' garage. He was noted to be pale and diaphoretic, complaining of shortness of breath.

His vital signs were pulse 55/min, respiratory rate 24/min, blood pressure 130/98 mmHg, temperature 98 °F (36.7 °C). The physical examination revealed a lethargic male with small pupils and tearing bilaterally, very moist oral mucosa and rhinorrhea, chest with rales bilaterally, and evidence of incontinence of urine and stool. His skin was markedly diaphoretic and pale, and he had gurgling sonorous respirations when he was supine.

He was intubated and ventilated and given atropine because of bradycardia and suspicion of organophosphate overdose. After standard doses of atropine (0.5–1.0 mg) failed to improve his condition, increased atropine doses of 2–3 mg at a time were administered, with subsequent improvement in his oxygenation and heart rate. Pralidoxime was initiated when the family brought an empty bottle of chlorpyrifos, which the father used to control garden pests.

Organophosphate and carbamate exposures are common. Last year, the AAPCC reported 21,000 exposures and 9 fatalities due to unintentional or intentional poisoning (1). Although specific antidotal therapy exists (atropine and pralidoxime), outcome depends on diagnosis and airway support initially. Cholinesterase concentrations can be helpful in patients who are chronically exposed or in whom the diagnosis is in question. However, when suspected poisoning occurs, antidotal therapy should be instituted immediately and not await laboratory detection of the poison or of depression of cholinesterase values. Treatment with pralidoxime normalizes erythrocyte cholinesterase concentrations, so blood specimens should be obtained for laboratory assays before initiating treatment. Most toxicology laboratories do not measure the organophosphate or carbamate insecticides, except for reference-level toxicology laboratories.

Typical clinical manifestations of cholinesterase inhibitor poisoning are illustrated by this case. Patients demonstrate excess cholinergic stimulation, resulting in parasympathomimetic findings such as salivation, lacrimation, increased gastrointestinal motility with diarrhea and urinary incontinence, bronchorrhea, and pulmonary edema. Nicotinic findings such as fasciculations, weakness, respiratory depression, and seizures are more common with organophosphates and thus organophosphates are considered more toxic than carbamates.

CONCLUSIONS

Although less common, multiperson poisoning exposures do occur, especially when contaminated drugs of abuse or natural products (herbals) are shared. Despite negative initial drug screens in the scopolamine heroin epidemic, persistent investigation using GC/MS methodology revealed the etiology of the poisoning. Along with scopolamine-tainted heroin, both GHB and flunitrazepam have become new drugs of abuse. Of more concern, however, is the use of GHB and flunitrazepam for date rape and other malicious acts. Confirming the presence of these agents in a patient with suspected poisoning will become increasingly important to clinicians.

Although patients may appear with fulminant anticholinergic findings after an overdose of OTC medications, screening for co-ingestion of acetaminophen is always important because an antidote is readily available and early treatment prevents severe hepatotoxicity. Conversely, cholinergic toxicity may result in patients with occupational exposure to organophosphates or carbamates or in patients who intentionally ingest these agents. The antidotes atropine and pralidoxime should be promptly administered when classic signs of cholinergic toxicity are manifest.

REFERENCES

1. Litovitz TL, Klein-Schwartz W, White S, et al: 1999 Annual Report of the American Association of

1. Poison Control Centers Toxic Exposure Surveillance System. *Am J Emerg Med* 2000;18:517–574.
2. Weisbord SD, Soule JB, Kimmel PL: Poison on line: acute renal failure caused by oil of wormwood purchased through the Internet. *N Engl J Med* 1997; 337:825–827.
3. Henretig F, Vassalluzo C, Osterhoudt K, et al: "Rave by Net" gamma-hydroxybutyrate (GHB) toxicity from kits sold to minors via the Internet [Abstract]. *J Toxicol Clin Toxicol* 1998;36:503.
4. Rouse BA: Epidemiology of illicit and abused drugs in the general population, emergency department drug related episodes, and arrestees. *Clin Chem* 1996;42:1330–1336.
5. Perrone J, Hamilton R, Nelson L, et al: Scopolamine poisoning among heroin users—New York City, Newark, Philadelphia and Baltimore, 1995 and 1996. *MMWR Morb Mortal Wkly Rep* 1996; 45:457–460.
6. Nogue S, Sanz P, Munne P, et al: Acute scopolamine poisoning after sniffing adulterated cocaine. *Drug Alcohol Depend* 1991;27:115–116.
7. Dyer JE, Reed JH: Alkali burns from illicit manufacture of GHB [Abstract]. *Clin Toxicol* 1997;35: 553.
8. Lo Vecchio F, Curry SC, Bagnasco T: Butyrolactone-induced central nervous system depression after ingestion of RenewTrient, a "dietary supplement." *N Engl J Med* 1998;339:847–848.
9. Carter J, Mofenson H, Caraccio T, et al: Gamma hydroxy butyrate use—New York and Texas, 1995–1996. *MMWR Morb Mortal Wkly Rep* 1997;46:281–283.
10. Adverse events associated with ingestion of gamma butyrolactone-Minnesota, New Mexico and Texas 1998–1999. *MMWR* 1999; 48:137–140.
11. Chin RL, Sporer KA, Cullison B, et a: Clinical course of gamma hydroxybutyrate overdose. *Ann Emerg Med* 1998;31:716–722.
12. Li J, Stokes SA, Woeckener A: A tale of novel intoxication: a review of the effects of gamma-hydroxybutyric acid with recommendations for management. *Ann Emerg Med* 1998;31:729–736.
13. Zvosec DL, Smith SW, McCutcheon JR, et al: Adverse events including death associated with the use of 1,4 butanediol. *N Engl J Med* 2001;344:87–94.
14. Perrotta DM, Coody G, Culmo C: Adverse events associated with ephedrine-containing products—Texas, December 1993-September 1995. *MMWR Morb Mortal Wkly Rep* 1996;45:689–693.
15. Doyle H, Kargin M: Herbal stimulant containing ephedrine has also caused psychosis. *BMJ* 1996; 313:756.
16. Spiller HA, Krenzelok EP: Epidemiology of inhalant abuse reported to two regional poison centers. *J Toxicol Clin Toxicol* 1997;35:167–173.
17. Smilkstein MJ, Knapp GL, Kulig KW, et al: Efficacy of oral N-acetylcysteine in the treatment of acetaminophen overdose. *N Engl J Med* 1988;319:1557–1562.
18. Goldfrank LR, Flomenbaum NE, Lewin NA, et al, eds: *Goldfrank's toxicologic emergencies*, 6th ed. Stamford. CT: Appleton and Lange, 1998.
19. Lake CR, Gallant S, Masson E, et al: Adverse drug effects attributed to phenylpropanolamine: a review of 142 case reports. *Am J Med* 1990;89:195–208.
20. Kernan WN, Viscoli CM, Brass LM, et al: Phenylpropanolamine and the risk of hemorrhagic stroke. *N Engl J Med* 2000;343:1826–1832.

SELF-ASSESSMENT QUESTIONS

1. Which are the most common poisoning exposures reported to the American Association of Poison Control Centers?
   a. adult, intentional
   b. adult, unintentional
   c. pediatric, intentional
   d. pediatric, unintentional

2. From which of the following do the greatest number of poisoning deaths result?
   a. acetaminophen (Tylenol)
   b. aspirin
   c. Prozac
   d. carbon monoxide

3. To what does the most recently implicated drug in malicious chemical submission, "knockout" drops or "date rape" drug, refer?
   a. heroin
   b. cocaine
   c. flunitrazepam
   d. carbamazepine

4. For what should all adult patients with intentional ingestions be screened?
   a. cocaine abuse
   b. heroin abuse
   c. acetaminophen toxicity
   d. phenytoin toxicity

5. Which is not a toxic over-the-counter medication?
   a. acetaminophen
   b. salicylates
   c. tetrahydrozoline (Visine?)
   d. ibuprofen

CHAPTER 2

Toxicokinetics: Principles and Practical Applications

Philip D. Walson, M.D.

LEARNING OBJECTIVES

After completing this chapter, the reader should be able to:

1. Define the volume of distribution (V_d).
2. Estimate the amount of drug in the body if given the plasma concentration and the V_d of the drug.
3. List at least three reasons for delayed or multiple "peak" concentrations after an ingestion.
4. Explain the distribution phase.
5. Give an example of an error in interpretation of a value as a result of sampling during the distribution phase.
6. List potential problems that could occur from sampling during the distribution phase.

INTRODUCTION

Several basic pharmacokinetic principles must be understood to interpret drug concentrations in overdosed patients. Most important is the volume of distribution (V_d). The V_d is defined as the constant that, when multiplied by the plasma concentration, gives the amount of drug in the body (the "body burden"). V_d's are drug specific, but are altered in specific patients. There is much less variability in V_d than there are in half-lives or clearances, but there are multiple distribution volumes. It is important to realize that the V_d of the central compartment is different from the postdistributional V_d at steady state.

Volume of distribution at steady state (V_{dss}) is what will be used for the calculations in this chapter. For example, if a patient is found to have a 22 mg/L (122 µmol/L) concentration of theophylline at steady state, by definition, the amount of theophylline in the body is this concentration times the V_d for theophylline. Because the concentration is 22 mg/L (122 µmol/L), all that needs to be known is the V_d of theophylline to calculate the amount of drug in the body as long as the sample is drawn after distribution (see below). The V_d of theophylline ranges from 0.3 to 1.0 L/kg; the larger volumes are found in fluid-overloaded and premature neonatal patients, for example. However, the average V_d is very close to 0.5 L/kg. Therefore, the amount in the body of a patient with 22 mg/L (122 µmol/L) is somewhere between 6.6 and 22 mg/kg, and in most patients or in the "average patient" it will be 11 mg/kg. The ability to do this simple calculation is critical for anyone interpreting drug concentrations in overdosed patients.

"PEAK" CONCENTRATIONS

The peak or maximal concentration of drug in the body after a dose occurs when the amount of drug going into the body equals the amount of drug going out. It does not occur when absorption is complete. This simple fact is often misunderstood. Even if the amount of drug being cleared from the body is fixed, there can be delayed or multiple peak concentrations. After a single therapeutic dose, the amount of drug being absorbed into the body starts at a maximal rate that then decreases exponentially over time (so called first-order absorption). Obviously, this is not always the case in an overdosed patient, especially when the product taken is an extended-release preparation. In overdosed patients there may be concretions of tablets or capsules (bezoars) that produce variable amounts of drug that are absorbed with time as the concretion is broken up and dissolves in the intestine. Many overdosed patients also develop delayed gastric emptying or even an ileus from the effects of the ingested substances. Ingestion of an extended-release product, presence of paralytic ileus, and delayed gastric emptying can all cause delayed time of peak concentrations, as well as multiple peaks.

DISTRIBUTION

After a drug is administered, there is a period during which the drug in the circulation comes into equilibrium with drug in the rest of the body. This is called the distribution phase. During this distribution phase, concentrations of drug in the body are not easily predictive of tissue concentrations. Only after distribution is complete can the plasma concentration be multiplied by the steady-state V_d to calculate how much drug is in the body. Before the end of the distribution phase, this is not true. Drug concentrations drawn during distribution can be used to calculate the body burden, but only with more sophisticated mathematical models. For some drugs, such as digoxin, distribution can be quite prolonged.

Digoxin has a prolonged distribution phase after both oral and intravenous administration. The distribution phase can last 6–8 hours after digoxin administration and even longer in patients with congestive heart failure or other causes of poor cardiac output or decreased tissue perfusion. Digoxin samples drawn during the distribution phase overestimate the amount of drug in the body. This overestimate can lead to misinterpretation and unnecessary antidotal therapy.

There are also some additional practical analytical issues with digoxin that must be considered in the overdosed patient. Skin contamination can be a major problem if samples are collected by skin puncture. If a patient has even touched a digoxin preparation, there can be significant, artifactually elevated concentrations. Interference from skin contamination occurs because concentrations in the blood are measured in nanograms per milliliter, yet there may be microgram amounts of drug on the skin. For similar reasons, blood should never be drawn from lines used to administer digoxin. There may also be significant interference from digoxin-like immunoreactive substances, which may be elevated in newborns, pregnant women, and patients with renal or hepatic disease. There are also other drugs (for example, spironolactone and digitoxin) that interfere with digoxin immunoassays.

Finally, after digoxin Fab antibodies (the antidote for digoxin poisoning) are administered, very high total digoxin concentrations are found. Unbound digoxin concentrations, but not total digoxin concentrations, can be used to predict effects because after using Fab antibodies total concentrations do not correlate well with the amount in the body. All of these facts must be considered when interpreting a digoxin result.

ASSUMPTIONS

These simple calculations may be inaccurate for drugs that exhibit multicompartment kinetics but work whether the drug has first-order or zero-order (saturation or Michaelis-Menten constant) kinetics.

AVAILABILITY OF CONSULTANTS

A few drugs will be reviewed to provide the reader an opportunity to understand some basic principles of toxicokinetic analysis. It is important to recognize, however, that there are always toxicologists available at Regional Poison Centers who are trained in the proper evaluation, collection, and interpretation of drug concentrations in poisoned patients. These specialists include physician toxicologists, analytical chemists, clinical pharmacists, and other poison center staff who are available by phone 24 hours a day, 7 days a week. All health-care professionals who collect, suggest, or analyze samples from poisoned patients should be aware of their closest nationally certified Regional Poison Center.

There are no "normal" drug concentrations for poisoned patients. Standard reference ranges are seldom useful for the overdosed patient. Although a number of references give advice on when and how many samples should be drawn, this advice can be wrong or even dangerous in treating individual patients. Decisions about when to measure concentrations and how many concentration measurements to do depend on the individual clinical situation. Often, these decisions require consultation with an experienced toxicologist or Regional Poison Center.

BASIC PRINCIPLES

Duration

The toxicity of many substances is influenced by the duration of exposure. Acute, acute-on-chronic, and chronic toxicity often differ. Examples include lithium, salicylate, theophylline, and digoxin.

Individual Susceptibility

There are both pharmacokinetic and pharmacodynamic reasons for individual differences in therapeutic and toxic responses to the same plasma concentration. These include distribution effects, concomitant medications, differences in receptor numbers or affinity, metabolite concentrations, disease state, and age.

PHARMACOKINETIC PRINCIPLES

Definitions

Definitions include:

- C_{peak} or C_p: The maximal concentration
- Volume of distribution (V_d): The number that, when multiplied by the concentration, gives the amount in the body. It has units of a "volume," but is not a physical volume.
- K_a: The rate of drug absorption
- K_e: The rate of elimination
- Half-life ($t_{1/2}$): The time it takes for the amount in the body to fall in half (for substances eliminated by first-order kinetics)

Absorption

Both rate and extent are important. The peak occurs when rate in equals rate out, not when absorption is complete. Anticipate prolonged or delayed peaks with bezoars, ileus, or extended-release products.

Distribution

Distribution can be prolonged. Concentrations drawn during distribution may not reflect tissue concentrations or effects. Examples include digoxin and aminoglycosides. V_d is used to calculate dose from peak or peak from dose, and V_d and concentration are used to interpret history.

Factors to consider in distribution include:

- Calculation of amount of drug in the body and assumptions
- Variability of V_d
- One-compartment and multicompartment drugs

Elimination

The rate of the elimination of a compound is usually proportional to the concentration of the compound present; the higher the concentration, the faster the rate. This relation is called a "first-order" process.

However, if the concentration of the compound gets high enough, then the rate of elimination no longer depends on concentration. It becomes a fixed, constant rate. When this loss of dependence on concentration occurs, elimination is "zero order." All compounds can become zero order when or if they "saturate" their elimination processes and go from being first-order eliminated to zero order. This change can occur at "therapeutic" or even "subtherapeutic" concentrations, as it does with ethanol, phenytoin, or salicylates. However, zero-order elimination can be seen with any compound if concentrations get high enough, as they often do after overdose.

It is important to realize that when zero-order elimination occurs, the simple first-order calculations of volume and half-life are no longer valid.

Analytical Considerations

These include interferences, metabolites, assay linearity, nomograms, and units.

SUMMARY

An appreciation of basic toxicokinetic principles is necessary for the proper interpretation of toxicology test results. In Chapter 3, these principles are illustrated with five patient case scenarios.

SELF-ASSESSMENT QUESTIONS

1. A 1-year-old patient given a single 110-mg dose of theophylline (weight recorded as 22 kg on emergency department sheet) develops signs of toxicity and is found to have a theophylline concentration of 22 mg/L (122 µmol/L). Provide an interpretation.

2. Interpret a digoxin concentration of 3 ng/mL (3.8 nmol/L) drawn 1 hour after a dose in an asymptomatic patient taking digoxin chronically.

3. Interpret a digoxin concentration of 3 ng/mL drawn 1 hour after a dose in an asymptomatic young child found with an open bottle of his grandmother's Lanoxin.

CHAPTER 3

Pharmacokinetics

Frank P. Paloucek, B.S., Pharm.D., DABAT

LEARNING OBJECTIVES

After completing this chapter, the reader should be able to:

1. Calculate, with one of three formulas, a serum toxin concentration based on a specific history of exposure.
2. List three reasons for either under- or overpredicting an initial toxin concentration.
3. Calculate a worst-case highest potential serum toxin concentration.
4. Explain the rationale for differentiating acute, acute-on-chronic, and chronic intoxications.
5. Define at least two "volume of distribution" terms relevant to the interpretation of serum toxin concentrations in a poisoning.

COMMON PHARMACOKINETIC CALCULATIONS

A common no-calculation approach in the assessment of the poisoned patient is to obtain a history of the dose ingested (assuming the maximum value possible), convert it to a weight-based value (mg/kg), and compare it with published standards for either prognosis or treatment decisions (1–4). For a limited number of agents including but not limited to acetaminophen, salicylate, theophylline, lithium, digoxin, and phenytoin, the serum concentration at the time of presentation, or subsequent peak value, can be similarly and often preferentially used.

Predictive pharmacokinetic calculations, based on historical data and/or comparative analysis of predicted vs. observed data, can further enhance the assessment and need-to-treat decisions (4). The use of these formulas should not provide sole or absolute criteria in patient management issues but should be combined with overall clinical assessments, preferably with the assistance of poison control centers or clinical toxicologists.

Case 1

An 8-day-old (33-weeks gestation) infant was in a neonatal intensive care unit. One morning the infant was noted to have poor feeding, "jitteriness," and tachycardia. The nursing staff recognized the possibility of a drug reaction. The neonate had been receiving theophylline 1.1 mg/kg intravenous (i.v.) bolus every 8 hours for 4 days. A total of 12 doses had been administered, the last dose 8 hours earlier. The morning dose was held pending a stat theophylline concentration determination. The theophylline serum concentration was found to be 57 mg/L (316 µmol/L). A regional poison control center was contacted, which forwarded the case to the medical toxicology service on call. The medical toxicology fellow evaluated the case over the telephone.

The patient's birthweight was 1.5 kg and had dropped to 1.35 kg during the past 8 days.

On recalculation with the current body weight, the service evaluated the case as chronic toxicity due to actual 2 mg/kg dosing. Supportive therapy was started, but the child's age prevented the initiation of oral charcoal for fear of necrotizing enterocolitis. Per nephrology consult, the patient's size and the absolute magnitude of the serum concentration precluded use of extracorporeal elimination therapies, such as hemodialysis or hemoperfusion.

Worst-Case Scenario

The following formula can be used to predict the highest concentration for a specified dose and patient:

$$C_{peak} = dose/[V_d \times weight (kg)] \quad (1)$$

C_{peak} is the highest observed serum concentration, dose is the greatest possible ingested dose in milligrams, and V_d is the population value for the volume of distribution (in L/kg) of the ingested agent.

Observed values will vary from predicted on the basis of the accuracy of dose ingested, time between ingestion and serum sampling, and absorption and distribution processes. The formula assumes complete bioavailability (F = 1.0) and ignores any reductions in toxin dose due to the presence of a salt form (s = 1.0). Additionally, any intervening symptoms and/or treatments (vomiting or gastric decontamination) will also affect predictive performance.

Applying this formula to Case #1:

$$C_{peak} = (12 \times 2 \text{ mg/kg})/0.6 \text{ L/kg}$$

$$C_{peak} = 40 \text{ mg/L } (222 \text{ μmol/L})$$

The V_d used in the calculation was 0.6 L/kg (it ranges from 0.3 to 1.0 L/kg in neonates), the value commonly used to dose theophylline in this population. The calculation significantly underpredicts the observed concentration of 57 mg/L (8 hours after the last dose). The evaluation of chronic intoxication couldn't have realistically resulted in the observed peak concentration (C_p), and theophylline in excess of the history had to have been provided.

This is a common occurrence in neonatology, usually manifested as a 10-fold dilution error in dosage calculation or drug preparation (5–6). Recalculation assuming the last dose was a tenfold error resulted in a significant underprediction, when the assumption of absolutely of no drug elimination occurring is considered:

$$C_{peak} = [(11 \times 2) + (1 \times 20)]/0.6$$

$$C_{peak} = 70 \text{ mg/L } (389 \text{ μmol/L})$$

Recalculating assuming all doses were a tenfold error resulted in a significant overprediction:

$$C_{peak} = (22 \times 2)/0.6$$

$$C_{peak} = 400 \text{ mg/L } (2220 \text{ μmol/L})$$

Impaired metabolism or enzyme saturation (nonlinear) kinetics was considered but not supported by three serial C_p's drawn over 8 hours. The calculated [using the three concentrations plotted as concentration vs. time on semi-log paper (3)] half-life of 24 hours was consistent with reported values for the age group.

Accounting for the passage of time, the following formula is preferred to the formula above for a more precise prediction:

$$C_p = \{dose/[V_d \times weight (kg)]\} \times e^{(-Ke \times t)} \quad (2)$$

or

$$C_p = C_{peak} \times [e^{(-Ke \times t)}] \quad (3)$$

where C_p is the serum concentration at any time (t, in hours) after an ingestion, e is the inverse natural logarithm, Ke is the population value for the elimination rate of the toxin (h^{-1}), and dose and V_d are defined as above.

Formula 3 can be generalized to:

$$C_{p2} = C_{p1} \times [e^{(-Ke \times t)}] \quad (4)$$

where C_{p1} and C_{p2} are any two consecutive serum concentrations with time interval (t) between them. This allows the ability to anticipate future sampling times and/or potential windows of threat, opportunity, or resolution.

Patient-Specific Data

Formula 4 can be rearranged to:

$$Ke = [\ln(C_{p1}/C_{p2})]/t \quad (5)$$

where ln is the natural logarithm, allowing calculation of a patient-specific elimination rate from observed data. This can be compared with a known population value to assess the competency of the patient to eliminate the toxin and provide an etiology for the presentation, especially with chronic overdoses. It can also be used to assess the effect of intervening treatment strategies and/or the need to add or continue elimination-enhancing modalities.

Elimination rates are best determined graphically with at least three values to determine a linear plot. The use of formula 5 and its frequent adaptation to calculate half-life (formula 6 below) assumes that the absorption and distribution processes are essentially complete, which is frequently not true of initial samplings in the evaluation of a poisoning. Falsely low (absorption-phase sampling) or falsely high (distribution-phase sampling) values can occur and have to be considered.

Elimination half-life is calculated using Ke by the following formula:

$$T_{1/2} = 0.693/Ke \quad (6)$$

The toxicokinetic analyses from the data in case 1 suggested that alternative causes had to exist for the observed serum concentrations. Careful hands-on chart review identified a 10-fold dosing admixture error by a temporarily assigned health-care provider, a float person, on day 4, affecting the last three doses. Calculations supported this as an acceptable cause. The calculation sums the residual contribution of every dose administered within the last three half-lives (24 hours), or 72 hours, before the initial sampling. This calculation is necessary because the patient was not at steady-state on a stable consistent dose to allow the use of calculation by steady-state formulas. Three half-lives was chosen both for simplicity (fewer calculations) and because any doses administered prior to 72 hours before sampling would contribute less than 12% to the final sum. Thus, Formula 2 for nine different times and two different doses (2 mg/kg and 20 mg/kg, the latter for the three most recent):

$$C_p = C_{p8} + C_{p16} + C_{p24} + C_{p32} + C_{p40} + C_{p48} + C_{p56} + C_{p64} + C_{p72}$$

$$C_p = 0.41 + 0.52 + 0.66 + 0.82 + 1.04 + 1.32 + 16.6 + 21 + 26.4$$

$$C_p = 68.8 \, (382 \, \mu mol/L)$$

A preventable error was uncovered, and corrective action and specific staff education were instituted.

Case 2

A 12-year-old boy began vomiting severely on his school bus. When examined by a school nurse, he admitted to a suicide attempt at 0700 h with multiple medications and was transported to the emergency department by 0930 h.

The patient admitted to ingesting five tetracycline 250 mg, eight Theo-Dur 100 mg SR, 10 Extra-Strength Tylenol, a half-bottle bottle of Nyquil elixir, and 1 teaspoonful of Pepto-Bismol.

Nyquil contains 10 mg of pseudoephedrine, 2.1 mg of doxylamine, 167 mg of acetaminophen, and 5 mg of dextromethorphan in every 15 mL (10% ethanol solution). The greatest initial concerns were with the calculated acetaminophen ingestion of 286 mg/kg and a theophylline ingestion of 23 mg/kg.

On physical examination, this was a well-developed, well-nourished 1.7-m tall, 35-kg male. His vital signs were blood pressure 110/84 mmHg, heart rate 100/min, and respiration rate 22/min. He was afebrile. Significant physical findings included vomiting and abdominal pain; otherwise the examination was noncontributory.

Routine labs were normal except for a potassium of 3.1 mEq/L (3.1 mmol/L). Serum toxin concentrations at presentation (2 hours after ingestion) were acetaminophen of 84.4 mg/L (559 μmol/L) and theophylline of 63.0 mg/L (350 μmol/L). Ethanol was undetectable. The acetaminophen concentration was drawn before the minimum 4-hour time frame required by the Rumack-Matthew nomogram (Figure 3–1) for assessing toxicity and/or antidotal indication.

An orogastric tube was placed and the patient was lavaged with 3 L of normal saline ("pumping the stomach"). Lavage return yielded no pill fragments. Activated charcoal of 35 g was ordered; a single bottle of 25 g was given. Because of the history of a massive acetaminophen ingestion (>140 mg/kg), Mucomyst 140 mg/kg antidote was ordered (7).

Toxicokinetic analysis using the worst-case scenario would yield an acetaminophen C_{peak} of 336 mg/L (286 mg/kg divided by 0.85 L/kg) and a theophylline C_{peak} of 46 mg/L (23 mg/kg divided by 0.5 mg/kg). Acetaminophen was overpredicted by fourfold, and theophylline was underpredicted by ~33%. These predictions need to be tempered by several variables: bioavailability, time since oral ingestion, distribution, and dosage form (sustained-release theophylline).

Accounting for the Absorption Process

The formulas above assume essentially complete and instantaneous absorption and hold limited value (consistently overpredicting) for predicting concentrations for overdoses from sustained-release formulations (4,8). Although consistent with the worst-case approach to evaluating patients, accounting for the absorption process can improve accuracy of predictions and may refine evaluation and need-to-treat decisions.

The following formula provides the potentially most accurate prediction of a serum concentration at any time for any toxin for which the parameters V_d, Ka (absorption rate), Ke, salt form (s), and bioavailability (F) are available.

$$C_p = [\text{dose} \times s \times F)/(V_d \times wt)]$$
$$\times [Ka/(Ka - Ke)]$$
$$\times [e^{(-Ke \times t)} - e^{(-Ka \times t)}] \quad (7)$$

Unfortunately, Ka's are rarely found in the common biomedical literature, limiting the application of this calculation and evaluation of its potential benefit in use.

Application of the last formula, with known population pharmacokinetic parameters (including Ka's), to the patient in Case 2 above yields a predicted acetaminophen concentration for the specific interval at 2 hours after ingestion calculated using mean reported population values of Ka = 1.4 h^{-1} and Ke = 0.17325 h^{-1}.

$$C_p = [(\text{dose} \times F)/(V_d \times wt)]$$
$$\times [Ka/(Ka - Ke)]$$
$$\times [e^{(-Ke \times t)} - e^{(-Ka \times t)}]$$

$$C_p = [(286 \times 0.89)/0.85]$$
$$\times [1.4/(1.4 - 0.17325)]$$
$$\times [e^{(-0.17325 \times 2)} - e^{(-1.4 \times 2)}]$$

$$C_p = 220.2 \text{ mg/L} (1458 \text{ μmol/L})$$

Similarly calculated for theophylline:

$$C_p = [\text{dose}/(V_d \times wt)]$$
$$\times [Ka/(Ka - Ke)]$$
$$\times [e^{(-Ke \times t)} - e^{(-Ka \times t)}]$$

$$C_p = [23/0.5] \times [0.18/(0.18 - 0.091)]$$
$$\times [e^{(-0.091 \times 2)} - e^{(-0.18 \times 2)}]$$

$$C_p = 11.7 \text{ mg/L} (65 \text{ μmol/L})$$

The net result is an improvement in overprediction of the acetaminophen and a worsening of the underprediction for theophylline. This result seems consistent with

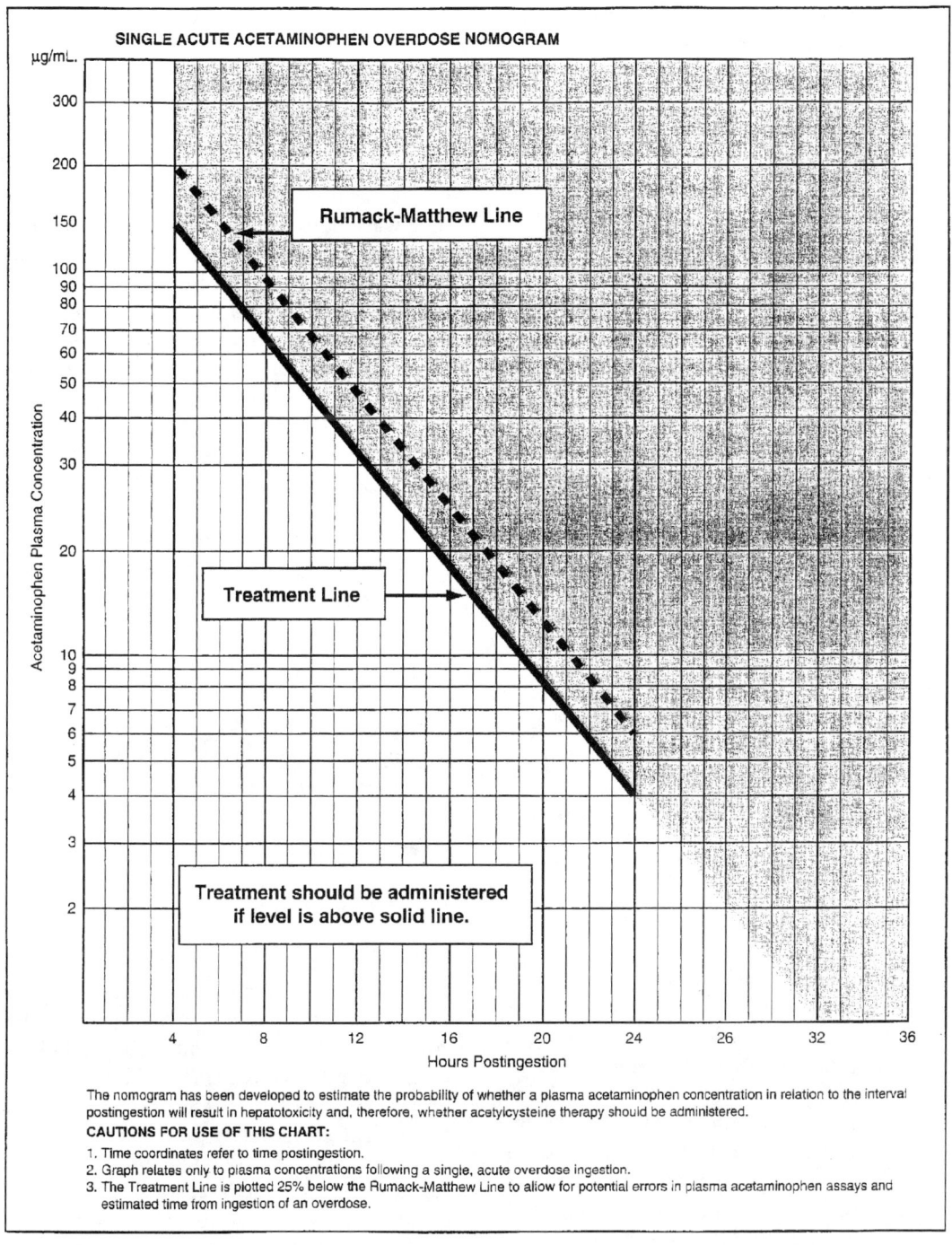

FIG. 3–1. The Rumack-Matthew nomogram. Courtesy McNeil Pharmaceuticals.

the ability to account for absorption (much slower with sustained-release theophylline) and elimination rates over the 2-hour period as well as bioavailability. These "enhanced" calculations suggest an inaccuracy in the actual dose ingested, the time of ingestion, or both, or a marked deviation in the normal pharmacokinetics of either drug because of the overdose setting.

A reasonable approach would be to assume an inaccuracy of dose. Back-calculation using the same formula and the observed concentration to determine an "actual" dose ingestion can be performed. Repeat calculations can be made for future serum concentration determinations based on the calculated "actual" dose ingested and assuming the time of ingestion history was accurate. An alternative would be to apply the degradation formula, Equation 4, to the observed values, but only if the absorption and distribution phases can be assumed to be completed. In this patient, that situation would be true only for the acetaminophen, not the theophylline. The back-calculation method was used to predict a 4-hour acetaminophen concentration (which was being obtained to allow application to the Rumack-Matthew nomogram) and for an 8-hour theophylline concentration.

The observed and predicted 4-hour acetaminophen C_p's were 82.1 and 81.4 mg/L (544 and 539 μmol/L), respectively.

$$C_p = [dose/(V_d \times wt)] \times [Ka/(Ka - Ke)] \\ \times [e^{(-Ke \times t)} - e^{(-Ka \times t)}]$$

$$C_p = [143/0.85] \times [1.4/(1.4 - 0.17325)] \\ \times [e^{(-0.17325 \times 4)} - e^{(-1.4 \times 4)}]$$

$$C_p = 81.4 \text{ mg/L } (539 \text{ μmol/L})$$

Note that the dose used was half the historical dose based on the observation that the ethanol concentration was undetectable, suggesting that no Nyquil was ingested (Nyquil accounted for half the assumed ingestion of acetaminophen). This result suggests that the back-calculated actual dose ingested was reasonable and reliable. Less acetaminophen was ingested than reported and/or substantially less was absorbed due to vomiting.

The observed and predicted 8-hour theophylline C_p's were 78.2 and 94.5 mg/L (434 and 524 μmol/L), respectively:

$$C_p = [dose/(V_d \times wt)] \times [Ka/(Ka - Ke)] \\ \times [e^{(-Ke \times t)} - e^{(-Ka \times t)}]$$

$$C_p = [124/0.5] \times [0.18/(0.18 - 0.091)] \\ \times [e^{(-0.091 \times 8)} - e^{(-0.18 \times 8)}]$$

$$C_p = 94.5 \text{ mg/L } (524 \text{ μmol/L})$$

Although not as accurate a prediction as was seen with the acetaminophen, the magnitude was felt to be clinically comparable and the actual-dose-ingested back-calculation was deemed reasonable and reliable. That allowed prediction, and prospective consideration, of the potential maximal theophylline concentration for the calculated dose ingested to occur 6 hours before its predicted occurrence. This provides ample time to alert or caution individuals about the probability of potentially future significant toxicities. It was known that the theophylline C_p could approach 100 mg/L (555 μmol/L), a value considered an indication for hemodialysis or hemoperfusion (9–12). There was time to alert nephrology to the potential need for their services.

EXPLANATIONS FOR VARIATIONS BETWEEN PREDICTED AND OBSERVED VALUES

Overpredictions

For Initial Concentrations

In instances in which the calculated initial concentration exceeds the observed initial determination, several explanations must be considered. The most common and likely is an inaccurate assessment of the dose ingested. This can be either as a consequence of the patient exaggerating the amount ingested or the conservative practice, the worst-case scenario, of using the largest value of a range of potential doses ingested. If the patient has vomited or undergone gastric decontamination before sampling, a diminishment of the dose absorbed will result. Similarly, the reported or estimated time at which the ingestion occurred could be incorrect. The time could be much more recent than estimated and absorption is not as complete as calculated.

Incomplete absorption assessment can also occur with concretions or bezoars (13–16). These are masses of incompletely dissolved drug or toxin, presumably secondary to saturation of solubility characteristics of the ingestant. Also sustained-release or enteric coated dosage forms need to be considered. In this case, the patient could be expected to develop symptoms in the near future. Alternatively, it could have been much longer ago than estimated and significantly more elimination of the ingestant than calculated has occurred and the patient should be at lower risk for the development of any dose-related toxicities.

For Subsequent Concentrations

Overpredictions of subsequent concentrations also occur. A likely reason to consider for this occurrence would be that the patient remains in an absorption state or that the patient's underlying metabolic or elimination capacity exceeds population estimates. Another possibility is the inadvertent or improper use of the V_d for the central compartment rather than the post-distribution or V_d steady-state.

Case 3

A 31-year-old African-American woman presented to the emergency room after supposedly taking "2 handfuls" of 325-mg immediate-release aspirin tablets 2 hours before. The estimated ingested dose based on bottle counts was 225 mg/kg, which can be assessed as probably toxic. The patient denied any complaints and was given activated charcoal for gastric decontamination in the emergency department. The initial aspirin serum C_p was 24.3 mg/L (0.135 mmol/L; units were verified). Serial serum aspirin C_p's every 2 hours, as recommended by the toxicology service, were ordered. The serum C_p's were 26.6 mg/L, 24.3 mg/L, 27.4 mg/L, and 28.3 mg/L (0.148, 0.135, 0.152, and 0.157 mmol/L) at 2, 4, 7, and 11 hours, respectively.

The continually rising, or minimally changed, concentrations are indicative of prolonged and continuous absorption. Bezoar formation as noted above and/or inadequate gastric decontamination needed to be considered. An abdominal X-ray (kidneys, ureter, and bladder) was read as negative, reducing the possibility of a bezoar. Chart review found only that one of four scheduled activated charcoal doses had been administered. The incomplete and inadequate gastric decontamination was corrected by administering two doses at 12 and 14 hours, respectively. The aspirin C_p at 16 hours after ingestion was 20.8 mg/L (0.116 mmol/L). This may reflect improved gastric decontamination, cessation of tablet dissolution and absorption, or both processes.

Underpredictions

For Initial Concentrations

In instances in which the calculated initial concentration is lower than the observed initial determination, several explanations must be considered. The reasons to consider are often the opposite of those listed above for overpredictions. Inaccuracies in the calculated dose ingested may be due to the patient understating the dose ingested or the estimated time of ingestion was too short while the actual absorption is complete. Also, if the calculations were performed assuming a sustained-released dosage form when in fact an immediate-release dosage form was used, an underprediction will result. This can also happen if a sustained-release dosage form is chewed or tampered with, thus destroying its sustained-release characteristics. Finally, inadvertent or improper use of the V_d pharmacokinetic parameter would occur with the use of the postdistribution V_d or steady-state V_d during the peridistribution phase. During the peridistribution phase, the central compartment V_d is a more appropriate parameter to use. The underprediction of subsequent concentrations occurs.

Underprediction also occurs when an elimination rate is calculated using two of the patient's observed values (using formula 5) when one or both concentrations were drawn during the distribution phase. Remember, in this phase, both distribution and elimination are occurring, leading to a much faster drop in concentrations than elimination alone. This event is unlikely to occur with a sustained-release dosage form.

Another regrettable reason for underpredicting would be new administration or ingestion of the toxin. Examples of this would be the patient who smuggled in additional ingestant, the patient given additional ingestant by another individual, or even the new administration of the toxin as a therapeutic agent. This is not uncommonly seen with acetaminophen, where the newly admitted acetaminophen overdose patient is given orders to treat headache or fever with acetaminophen.

Importantly, the occurrence of underpredictions of subsequent concentrations can be used to suggest ineffective or incomplete decontamination or elimination-enhancement modalities (such as hemodialysis).

It is also very important to bear in mind that all of the population parameters, V_d or K_e, used in the calculations are mean values from sample population surveys and have an associated range. Interpatient variability, reflecting the distribution of values around the population mean, will result in variations between calculated and observed values.

Case 4

A 25-year-old woman presented 2 hours after ingesting "handfuls" of Tylenol ER and Nyquil elixir. Additional maintenance medications, whose ingestion was denied, included paroxetine, thyroxine, thioridazine, clonidine, and omeprazole. Vital signs, physical examination, and most laboratory findings were normal. Serial acetaminophen concentrations at 2, 6, and 10 hours were 172.8, 69.8, and 50.5 mg/L (1144, 462, and 334 μmol/L), respectively. All these concentrations fall below the lower toxic ingestion–treatment indication line on the Rumack-Matthew acetaminophen nomogram (Figure 3–1). A fourth acetaminophen C_p obtained at 14 hours was 161 mg/L (1066 μmol/L), well above the toxic ingestion–treatment indication line. This result is a demonstration of either altered dissolution characteristics in an overdose or altered absorption rate processes due to overdosage or multidrug ingestion.

Case 5

A 25-year-old homeless man was found unresponsive behind a pharmacy. In his pockets were found several empty vitamin bottles and an empty Tylenol bottle. Initial examination was performed without significant physical findings except sedation. The patient was worked up for toxin ingestion as well as physical pathologies. Conventionally, acetaminophen C_p's are of limited value in this setting because nomograms cannot be used without a specific time of ingestion. Serial acetaminophen C_p's would be useful to determine whether absorption is continuing, which would provide an indication for aggressive decontamination. Also, if hepatotoxicity from overdosage occurs, then the acetaminophen elimination rate will decrease due to the impaired metabolism, and the antidote would clearly be indicated. One would hope for declining concentrations at a rate reflective of the normal half-life.

The initial and repeat acetaminophen C_p's at presentation and 2 hours later were 56 mg/L and 32 mg/L (371 and 212 μmol/L), respectively. These declining values suggest a postabsorption state, either the distribution or elimination phase. Calculation of the elimination rate (K_e) as $(C_{p2}/C_{p1})/t$ was 0.28 h^{-1}, which converts to a half-life of 2.5 hours (normal). Confirmation of a "normal" half-life and normal liver function tests greatly decreased the possibility of acetaminophen toxicity. Antidotal therapy was held, and the patient had no adverse outcomes.

POTENTIAL BENEFITS IN THE EVALUATION AND/OR MANAGEMENT OF THE OVERDOSE PATIENT

These formulas and principles can be of value in clinical toxicology for specific reasons. First, they can be used to differentiate the acute or acute-on-chronic overdose from a chronic intoxication. Acute-on-chronic ingestions are defined as an acute ingestion of a drug in a patient receiving the drug therapeutically and who is presumably at steady-state conditions (5,8,11,13,16,17). Acute or acute-on-chronic ingestions generally require higher serum concentrations to occur before significant target organ damage is manifest. Early identification or confirmation of a potentially serious toxic ingestion would allow for earlier and possibly more aggressive treatment measures to prevent or limit observed toxicity. Acute ingestions and acute-on-chronic ingestions can be identified by significantly larger than calculated (Formulas 1, 3, or 7) increases in concentrations than would be seen with a normal therapeutic dose. This is only observed within the immediate absorption phase. Two or more consecutive unchanged or minimally changed concentrations are another indicator of an acute ingestion. Chronic ingestions can be identified by the observation of serially decreasing concentrations with a calculated elimination half-life (Formula 5) within the range of values reported for the relevant population. This issue is most relevant for ingestions of digoxin, lithium, salicylates, and theophylline and theoretically would seem applicable to all ingestions.

Another common application of these principles is in the clarification or confirmation of the patient's history. Comparing predicted with observed values can lead to revised etiologies, which may reflect a reversible process, as in the neonatal theophylline exposure described in Case 1. They can assist in the prognosis of an exposure such as the acetaminophen exposure without any history of time or dose of ingestion described in Case 5.

Finally, the formulas can be used to calculate future serum concentration changes, both their relative direction and size. This could allow for the prospective use of treatment measures vs. traditional reactive implementation (for projected significant increases in concentration) or for limiting or minimizing empiric therapies of a reported ingestion for projected decreasing, or relatively minor increases in, concentrations (18).

REFERENCES

1. Ellenhorn MJ: Toxicokinetics. In: Ellenhorn MJ, Schonwald S, Ordog G, et al: *Ellenhorn's medical toxicology*, 2nd ed. Baltimore: Williams & Wilkins,1997:128–148.
2. Weisman RS, Howland MA, Reynolds JR, et al: Pharmacokinetic and toxicokinetic principles. In: Goldfrank LR, Flomembaum NE, Lewin NA, et al, eds: *Goldfrank's toxicologic emergencies*, 5th ed. Norwalk, CT: Appleton & Lange, 1994:85–98.
3. Watson WA, Rose SR: Pharmacokinetics and toxicokinetics. In: Ford MD, Delaney KA, Ling LJ, et al, eds: *Clinical toxicology*. Philadelphia: WB Saunders, 2001:73–78.
4. Paloucek FP: Clinical toxicokinetics. *J Pharm Practice* 1997;10:271–277.
5. Shannon M: Effect of acute versus chronic intoxication on clinical features of theophylline poisoning in children. *J Pediatr* 1992;121:125–130.
6. Osborn HH, Henry G, Wax P, et al: Theophylline toxicity in a premature neonate—elimination kinetics of exchange transfusion. *J Toxicol Clin Toxicol* 1993;4:639–644.
7. Lewis RK, Paloucek FP: Assessment and treatment of acetaminophen overdose. *Clin Pharm* 1991;10: 765–774.
8. Paloucek FP: Theophylline toxicokinetics. *J Pharm Practice* 1993;6:57–62.
9. Ellenhorn MJ: Elimination enhancement. In: Ellenhorn MJ, Schonwald S, Ordog G, et al: *Ellenhorn's medical toxicology*, 2nd ed. Baltimore: Williams & Wilkins, 1997:79–88.
10. Pond SM: Techniques to enhance elimination of toxic compounds. In: Goldfrank LR, Flomembaum NE, Lewin NA, et al, eds: *Goldfrank's toxicologic emergencies*, 5th ed. Norwalk, CT: Appleton & Lange, 1994:78–84.
11. Paloucek FP, Rodvold KA: Evaluation of theophylline overdoses and toxicities. *Ann Emerg Med* 1988;17:135–144.
12. Shannon M: Predictors of major toxicity after theophylline overdose. *Ann Intern Med* 1993;119: 1161–1167.
13. Dupuis RE, Cooper AA, Rosamond LJ, et al: Multiple delayed peak lithium concentrations following acute intoxication with an extended-release product. *Ann Pharmacother* 1996;30:356–360.
14. Buckley NA, Dawson AH, Reith DA: Controlled release drugs in overdose. Clinical consideration. *Drug Saf* 1995;12:73–84.
15. Albertson TE, Fisher CJ Jr, Shragg TA, et al: A prolonged severe intoxication after ingestion of phenytoin and phenobarbital. *West J Med* 1981;135:418–422.
16. Sue YJ, Shannon M: Pharmacokinetics of drugs in overdose. *Clin Pharmacokinet* 1992;23:93–105.
17. Temple AR: Acute and chronic effects of aspirin toxicity and their treatment. *Arch Intern Med* 1981; 141:364–369.
18. Dolgin JG, Nix DE, Sanchez J, et al: Pharmacokinetic simulation of the effect of multiple dose activated charcoal in phenytoin poisoning—report of two pediatric cases. *DICP* 1991;25:646–649.

SELF-ASSESSMENT QUESTIONS

1. What is the worst-case scenario formula for predicting the peak serum concentration after an ingestion?

2. What is the relative ranking for the acute, acute-on-chronic, and chronic intoxications for potential morbidity and mortality (from most to least)?

3. What is the formula for calculating an elimination half-life with two separate serum concentrations?

4. How is elimination rate converted to elimination half-life?

5. Which is the larger volume of distribution, central or steady state?

CHAPTER 4

Clinical Approach to the Poisoned Patient

Francis J. De Roos, M.D.

LEARNING OBJECTIVES

After completing this chapter, the reader should be able to:

1. Develop an organized and effective management strategy for acutely overdosed patients.
2. Identify vital sign changes and toxidromes in poisoned patients and use them to predict potential intoxicants.
3. Use the anion gap metabolic acidosis as a tool for assessing overdosed patients and designing strategies to narrow this differential.
4. Understand and describe the role of the clinical toxicologist in the management of overdosed and poisoned patients.

INTRODUCTION

Most acutely poisoned patients are managed within an emergency department, and many of the management decisions must be based primarily on the exposure history and the patients' presenting signs and symptoms. In fact, in most cases, these clinical parameters are all the information needed to narrow the differential diagnosis to a small list of potential toxins and to design an effective treatment strategy.

This chapter will describe some of the useful physical examination findings and the toxins most likely to produce those physiologic responses. Specific consideration will be given to vital signs, the concept of toxidromes, several common presentations, and unique patient populations. How these toxidromes can improve the utilization of the toxicology laboratory will also be described.

IDENTIFICATION OF THE TOXIN

Although much of emergent decision making is based on the history of the exposure (when, how much, and what type of medications ingested), it is important to understand that the history may not be reliable in acute-exposure situations. Specific examples include suicidal patients who may not want people to know what they ingested or a toddler who is accidentally exposed to a household product, particularly if the child has symptoms such as respiratory distress, vomiting, or crying. This is very distressing for parents, who are likely to rush their child to a health-care facility (an appropriate response) rather than call the regional poison control center serving their locale.

Sometimes product names are very similar and may cause confusion. This is sometimes referred to in poison centers as the "name game." The similarity among clozapine, clonapin, and clonidine is an excellent example of this confusion. Sometimes exposures are misidentified because the pills are not stored in the original prescription bottle or are simply loose. And finally, remember that illicit drugs and herbal medications are in no way regulated, so adulterants and contaminants are a real concern.

Sometimes patients may not be aware of their exposure, particularly in environmental or occupational exposures, therapeutic misadventures, or attempted homicides. Finally, clues in the history may identify stressors or signs of depression that may place the patient at risk for suicide.

Identification of the Pill, Product, or Plant Involved

With all these potential pitfalls in the history, what are some strategies and resources that are effective in identifying exposures? First, and most important, is direct observation by the emergency physician of the product. In addition, if the patient has a history of hypertension, then the chances for ingesting an antihypertensive are great. This method of using the patient's or the family's medical history may be extremely enlightening in developing a list of possible exposures. The poison center is an invaluable resource not only in assisting in the clinical management of poisoned patients but also in identifying products.

A large database called Poisondex contains the trade names of thousands of products, their active and toxic ingredients, manufacturer contacts, and recommendations for treatment. In addition, if only a tablet is available, this program may be able to identify the product given the shape, color, and scoring of the tablet.

For plants and mushrooms, most poison centers either work with a clinical toxicologist who is interested in these or have relationships with local botanists, horticulturists, and mycologists. Finally, the Web has some very interesting information specifically when it comes to drugs of abuse. Two sites with excellent listings of slang terms for illicit drugs are:

- http://www.drugs.indiana.edu/slang/home.html
- http://www.iec.org/drugsearch/documents/slangterms.html

When identification of the toxin is not successful, the clinician must rely on physical examination findings to direct the search both from the patient's or other people's history and from the analytical laboratory. Patients who are seriously suicidal typically will not give specifics when asked "What did you take?" because that question implies that the clinician does not know what's going on. On the other hand, if the physical examination reveals a dramatic physical finding that greatly narrows the differential diagnosis, and a few historical clues focus on one or a few toxins, most patients will tell you the specific pills when they realize you have figured it out.

Toxin-Induced Changes in Vital Signs

Often, clinical toxicologists focus on one profoundly abnormal vital sign to develop a differential diagnosis and treatment strategy. This strategy is highlighted in the next two cases.

Case 1

The patient was a 17-year-old male brought to the emergency department by his mother for complaints of feeling weak for a few hours. The patient felt well at school that day until 2 hours earlier, when he was lightheaded on standing. He denied a history of depression or depressive symptoms, and he denied any overdose. He had no past medical history and was not taking medications, including over-the-counter (OTC) products and herbals. He denied any drug or ethanol use. His physical examination was remarkable only for a heart rate of 50 beats/min with blood pressure of 110/70 mmHg.

What's going on and how do we solve this case?

Step 1. Identify the most significant physical finding and vital sign abnormality and generate a differential diagnosis

In this patient, the heart rate of 50 beats/min is the most remarkable physical examination finding. His rhythm is best evaluated with an electrocardiogram (ECG). In this case, formation of a differential diagnosis for bradycardia is done including the potential causative agents. The differential diagnosis for toxin-induced bradycardia includes the following agents:

- Cardiac glycosides
- Calcium channel antagonists
- β-Adrenergic antagonists
- Centrally acting α-agonists (imidazolines including clonidine)
- Potent α-agonists (phenylpropanolamine)
- Organophosphates
- Antidysrhythmics
- Sedative-hypnotics, opioids (?)

In order to understand this differential, it's valuable to review how the heart rate is modulated. A balance between the adrenergic and cholinergic hormonal tone establishes the heart rate. Bradycardia is commonly caused by direct myocardial toxins including digoxin, β-adrenergic receptor and calcium channel antagonists, and antidysrhythmics or indirect toxins that increase cholinergic effects, including organophosphates (decreased acetylcholine metabolism) and phenylpropanolamine (increased vagus tone causing reflex bradycardia from severe peripheral hypertension).

Tachycardia can be caused by sympathomimetic agents including cocaine and theophylline, anticholinergic agents including tricyclic antidepressants and antihistamines, or increased adrenergic effects due to withdrawal from agents such as ethanol and other sedative hypnotics, pain produced by caustic burns, or hypovolemia induced by toxins including arsenic, iron, organophosphates, and ricin.

Appendix A includes a listing of the most common agents associated with various vital sign abnormalities. This list is not meant to be exhaustive but rather a useful list of the most common and significant toxins involved.

Step 2. Refine the differential diagnosis

Knowing other significant clinical effects that each agent can produce allows one to refine and narrow the differential diagnosis. For example, organophosphates produce the SLUDGE syndrome:

- **S**alivation
- **L**acrimation
- **U**rination
- **D**efecation
- **G**astrointestinal upset
- **E**mesis

In this case, reevaluation of the physical examination eliminates organophosphates (no other cholinergic signs), sedative-hypnotics (the patient is not obtunded), and phenylpropanolamine (there is no elevation in blood pressure causing a reflex bradycardia).

Step 3. Develop an initial treatment strategy

At this point, even if you do not have a specific history the physician can, should, and sometimes must initiate therapy. Treatment may be more aggressive depending on the history. For example, massive ingestions and ingestions involving a highly toxic agent or one with a potentially delayed onset of toxicity such as a sustained-release calcium channel blocker or monoamine oxidase inhibitor would typically be aggressively treated.

This patient was kept in a lying position (his symptoms were most likely related to hypoperfusion) and monitored. An external pacer was placed but not activated, and atropine (two 1-mg doses) was intravenously (i.v.) administered, without improvement.

Step 4. Identify the toxin by clarifying the history, by following the clinical course, or by performing focused analytical testing

Even though the patient's treatment has begun, it is often desirable to identify the intoxicating agent to anticipate and prevent any further potential effects. The laboratory can also assist in this step with good communication between the clinician and the laboratory personnel.

After the patient's friends were questioned further, it turned out that, unknown to the patient, the friends had added a few drops into his soft drink to "get him high." The only agent on the list that is liquid is an imidazoline, tetrahydrazoline (Visine), which in fact was the product the patient's friends brought to the emergency department. The patient was closely monitored. Sixteen hours later his symptoms and the conduction delays seen on his ECG had resolved.

For further information regarding imidazoline poisoning see reference 1.

Case 2

The patient was a 26-year-old female who was brought in by paramedics after being found confused and disoriented on a university campus. Physical examination revealed a lethargic and combative female with obvious deep respirations (hyperpneic). Her vitals signs included a heart rate of 130 beats/min, blood pressure of 100/60 mmHg, and a respiratory rate of 28/min (tachypneic).

Step 1. Generate a differential diagnosis for the most significant physical finding or vital sign abnormality

In this case, a dramatic increased ventilatory effort was seen.

A differential for toxin-induced hyperpnea vs. dyspnea was needed. Toxin-induced increased respiratory effort (hyperpnea or tachypnea) has several general physiologic precipitants, and most etiologic agents can be classified in this manner (see Table 4–1 for a summary). The central nervous system can be

TABLE 4–1. *Agents that can cause clinically significant increased respiratory effort*

Respiratory stimulants
- Salicylates
- Amphetamines and cocaine
- Methylxanthines (theophylline, caffeine)

Acidosis (MUDPILES)
- **M**ethanol
- **U**remia
- **D**iabetic ketoacidosis, alcoholic ketoacidosis
- **P**araldehyde, phenformin, metformin
- **I**soniazid, iron
- **L**actate (cyanide, carbon monoxide, seizures, shock)
- **E**thylene glycol
- **S**alicylates
- Hydrogen sulfide
- Sodium monofluoroacetate

Hypoxia
- Carbon monoxide
- Methemoglobin inducers
- Simple asphyxiants

Pulmonary injury (decreased oxygen diffusion)
- Hydrocarbons
- Paraquat

Increased metabolic needs or carbon dioxide production
- Dinitrophenol
- Thyroid hormone

directly stimulated to increase respirations (as with amphetamines or methylxanthines), it can be indirectly stimulated by an increase in carbon dioxide or a decrease in pH [as with many agents, including salicylates, cyanide, and toxic alcohols (the MUDPILES mnemonic contained in Table 4–1)], or indirectly stimulated by a decrease in oxygen, including carbon monoxide or methemoglobin. In addition, any increase in oxygen consumption or carbon dioxide production via increasing metabolic activity (as with dinitrophenol) may indirectly stimulate increased respiratory effort. Finally, if the lungs themselves are directly injured (as with hydrocarbon aspiration or paraquat) and their ability to exchange gas is impaired, tachypnea results.

2. Refine this differential diagnosis knowing other significant clinical effects each agent can also produce

Further examination of the patient revealed an obtunded Japanese female with

5-mm pupils, normal skin without cyanosis, supple neck, clear lungs, and not following commands but withdrawing to painful stimuli. There was no stimulant toxidrome and no primary pulmonary process (clear lung examination).

3. Develop an initial evaluation and treatment strategy

One of the most difficult aspects of emergency medicine is caring for the critically ill patient when no history is available. In this patient, given her age and where she was found, the most likely things included multiple seizures, an acute intracranial event, or an overdose. Her treatment focused on acute stabilization, monitoring, i.v. access, high-flow oxygen, chest X-ray, and frequent reassessments. Early airway management with orotracheal intubation should always be considered. Then the evaluation and diagnostic process should target our differential diagnosis including:

- Evaluation of the altered mental status by testing for hypoglycemia and performance of a head computed tomography scan and treatment according to the findings
- Evaluation for possible metabolic acidosis with arterial blood gas analyses (co-oximeter including carbon monoxide and methemoglobin), basic electrolytes tests, renal function tests, and the serum lactate procedure
- Assessment of a possible overdose including a urine ferric chloride test
- Consideration given to the use of activated charcoal administration by oral gavage

The following initial laboratory test values were obtained at this point: pH = 7.37, pCO_2 = 17 mmHg, pO_2 = 126 mmHg, and serum HCO_3 = 8 mEq/L (8 mmol/L), and the anion gap was determined from the electrolyte results to be 42 mEq/L (42 mmol/L). The urine ferric chloride test was done and found to be positive.

The ferric chloride or Trinder reaction is a rapid bedside qualitative test used to assess patients' exposure to salicylic acid (aspirin). It is performed by adding 3–4 drops of a premixed solution of 10% ferric chloride (Trinder solution) into 2–3 mL of urine and then observing the mixture for an immediate color change as follows:

- Positive test: blue or purple color change
- Negative test: no change
- False-positive test: brown if phenothiazines present

Clinically, this test is fairly accurate, with a reported sensitivity of 94% and specificity of 75% in identifying patients with a salicylate concentration >30 mg/dL (2.17 mmol/L) (2).

After some consideration of the most likely etiologies, aspirin remained at the top of the list, and we initiated empiric therapy with several doses of sodium bicarbonate i.v. and began a sodium bicarbonate drip. This was to alkalinize the patient's urine to "ion trap" the salicylic acid in the urine.

When initiating empiric therapies on patients, one must always consider the risks and benefits. If this had been an intracerebral hemorrhage, then the hyperventilation would be a sign of increased intracranial pressure, and the most rapid method to reduce this would be by orotracheal intubation and mechanical hyperventilation. However, if this had been an aspirin overdose, the patient's altered mental status would be a sign of severe poisoning and the patient's hyperventilation would keep her serum alkalemic, which would hold the acidic aspirin from entering the more acidic brain tissue. During endotracheal intubation, it is extremely difficult to maintain such effective hyperventilation, and patients invariably have an elevation in pCO_2 and become acidic. This process can in essence drive salicylic acid into the brain and other tissues (3).

Step 4. Identify the toxin by clarifying the history, by following the clinical course, or by performing focused analytical testing

The use of ancillary history or sources (for example, searching clothing and bags, calling phone numbers in the patient's wallet) is often very helpful.

A search of the patient's book bag revealed three bottles (one empty and two half full) of OTC analgesics containing aspirin and acetaminophen. Her initial ferric chloride urine screen was positive, which confirmed that she had ingested at least some aspirin recently. However, during the first 60 minutes, while waiting for the quantitative aspirin concentration, the patient's mental status and vital signs improved. She was now awake and able to answer some questions. Her heart rate was 100 beats/min and her respiratory rate was 22/min. This is not the typical course for a severely aspirin-poisoned patient because her mental status improved much too rapidly.

A further search of her belongings revealed a vial of white crystals that her father, who had just arrived and was a geneticist on campus, immediately identified as potassium cyanide from his laboratory. This discovery coincided with the reporting of a nondetectable serum aspirin concentration from the toxicology laboratory. She was treated with supportive care alone and in 4 hours was sitting upright, alert, awake, and interacting with her family and the psychiatrist.

Toxidromes

Often it is not just one vital sign but several findings on physical examination that aid the clinician. A toxidrome is defined as a constellation of signs and symptoms that are typically produced by particular toxins (4,5). Appendix B lists the most common toxidromes, the typical clinical manifestations, and the causative agents. Cases 3 and 4 are examples of classic toxidromes.

Case 3

The patient was a 2-year-old, previously healthy child who was brought to the emergency department after being found obtunded in her crib. At that time she was being watched by her grandmother, who takes two unidentified pills, one for diabetes and one for hypertension. Physical examination revealed an obtunded child who had poor response to painful stimuli with vital signs as follows: Heart rate = 100 beats/min, blood pressure = 70/p mmHg, temperature, 96 °F (35.6 °C), and respiratory rate = 20/min and shallow.

She had 1- to 2-mm pupils that were poorly reactive, cool and dry skin, and normal heart, lung, and abdominal examinations.

Step 1. Generate a differential diagnosis for the most significant physical findings

In this case, hypoglycemia is number one on the list because any patient with altered mental status could be hypoglycemic and there are known oral hypoglycemic agents in the home. In addition, the depressed mental status, the reduced respiratory effort, and the miotic or pinpoint pupils are a classic description or "triad" for opioid toxicity. However, a few other agents must be considered.

The differential diagnosis of toxin-induced pinpoint pupils (miosis) associated with decreased mental status and respiratory effort included:

- Hypoglycemia
- Opioids
- Centrally acting α_2-adrenergic agonists (clonidine, imidazolines)
- Phenothiazines
- Cholinergic agents such as organophosphates

Step 2. Refine this differential diagnosis knowing other significant clinical effects that each agent should also produce, and identify the toxidrome

This patient's rapid glucose check was 105 mg/dL (5.8 mmol/L), eliminating hypoglycemia. Cholinergic agents could be eliminated because no other cholinergic effects (see SLUDGE list in Case 1) such as nausea, vomiting, diarrhea, profuse sweating, or pulmonary signs from bronchorrhea and bronchospasm were seen.

Step 3. Develop an initial evaluation and treatment strategy

It is always important to remember that the most successful treatments in overdoses are basic supportive care and monitoring. Therefore, this unconscious patient should be given oxygen therapy and placed on a continuous cardiac monitor, and a rapid assessment of the respiratory status should be made to determine whether assisted ventilation is needed.

In addition, any patient with this triad and significant respiratory depression should receive an empiric dose of naloxone, a short-acting opioid antagonist (6), because opioids are by far the most likely etiology for this triad, response is rapid, and side effects are minimal. The only exception to the usual minimal side effects is the potential to precipitate significant withdrawal in a patient who is opioid dependent (7). In patients with clues for this including a history of substance abuse, physical evidence of track marks, drug paraphernalia, or a methadone clinic card, very small doses of naloxone (0.1–0.2 mg) should be used and titrated to the desired effect of improved ventilatory status. This dosing is much smaller than the "standard" dose of 0.4–2.0 mg i.v. (8).

This child was administered 0.4 mg of naloxone i.v., resulting in slight arousal (within 10–15 seconds). After 0.4 mg more, the patient became alert with good ventilatory effort.

Step 4. Refine or clarify the history to identify the toxin

The positive response to naloxone eliminates phenothiazines, but both opioids and centrally acting α-adrenergic agonists such as clonidine still remain on the list. Send a serum and urine specimen to the toxicology lab and place a phone call so that the laboratory understands what you are looking for. This communication between the analytical laboratory and the clinician is so helpful and yet so often overlooked. In addition, return to the family members and question them about any analgesics or drug use or dependence they may not admit to initially.

In this case, the grandmother confessed that she was in a methadone maintenance program and that she had smuggled one of her doses home from the clinic and stored it in the refrigerator in a small baby food jar "just in case." The child was placed on a naloxone drip, treated, and observed for >3 days in the intensive care unit and did well.

For most patients with an acute overdose or poisoning, a focused and particular physical examination is critical to evaluating for potential toxidromes. This focused exam (summarized in Table 4–2) is no different than what an ophthalmologist or a cardiologist does: Evaluating the parts of the body and its physiology that are potentially involved in their area of expertise.

Classic toxidromes include the opioid, anticholinergic, cholinergic, sympathomimetic, and sedative-hypnotic. See Appendix B for further details.

Case 4

The patient was a 15-year-old male with no known prior medical history who was brought to the emergency department by his family for "not thinking clearly." The physical examination revealed a well-developed male in no distress randomly talking about Jesus, cars, and the curtains in the room. Assessment of his vital signs revealed a heart rate of 110 beats/min, a respiratory rate of 16/min, blood pressure of 130/90 mmHg, and a temperature of 100.5 °F (38 °C). A neurologic evaluation revealed that the patient was alert, awake, hypervigilant, and talkative; he was able to follow simple com-

TABLE 4–2. *The clinical toxicologist's focused physical examination*

Vital signs (the most important part)
Neurologic examination (mental status, reflexes, muscle tone)
Pupils examined for size and response to light
Skin and mucous membranes assessed for color, moisture, and markings
Bowel sounds

mands but not prolonged conversation; his speech was filled with perseverations and descriptions of illusions (the dinosaurs in the curtains are dancing); he had good motor and sensory strength with normal tone and reflexes. The patient's pupils were 7 mm and poorly reactive to light. His skin was warm and dry. An examination of his abdomen revealed few bowel sounds.

Step 1. Generate a differential diagnosis for the most significant physical finding and vital sign abnormality and identify the toxidrome

This case is a classic presentation of an anticholinergic toxidrome. The patient has an increased heart rate and temperature, acute delirium, dilated pupils, dry, flushed skin, and decreased bowel sounds.

A mnemonic for the clinical presentation of the anticholinergic syndrome is "Hot as Hades, mad as a hatter, blind as a bat, dry as a bone, and red as a beet." Symptoms can be subtle if the overdose involves a mildly anticholinergic agent such as a tricyclic antidepressant.

The list of agents that can produce anticholinergic findings is extensive:

- Atropine, homatropine, scopolamine, benztropine
- Many plants such as Jimson weed (*Datura stramonium*), henbane (*Hyoscyamus niger*), and Solanaceae family members including the deadly nightshade (*Atropa belladonna*)
- Antihistamines (diphenhydramine, hydroxyzine)
- Tricyclic antidepressants
- Carbamazepine
- Antipsychotics

Step 2. Develop an initial evaluation and treatment strategy

In patients with isolated and severe anticholinergic poisoning, including manifestations of severe hyperthermia, seizures, or dysrhythmias, the use of physostigmine (an acetylcholinesterase inhibitor) can be effective.

Our patient had no severe manifestations, so he received a benzodiazepine to sedate him slightly and allow him to sleep.

Step 3. Refine or clarify the history to identify the toxin

In further questioning, the patient was asked several short, directed questions about specific drugs of abuse. He admitted drinking a tea that was brewed from the seeds of the locoweed that morning with four friends, but had no recollection of the day's events. He was observed overnight and had returned to his normal mental status by morning.

For more information about Jimson weed intoxication see references 9 and 10.

UNIQUE SITUATIONS IN MANAGEMENT OF THE POISONED PATIENT

The first four cases were used to highlight the use of the physical examination and toxidromes to refine differential diagnoses and focus the treatment. Other less palpable clues to poisoning may be symptoms such as vomiting or dyspnea. One very significant and not uncommon finding is a metabolic acidosis. In addition, specific drug groups, such as those with delayed toxicity, or specific patient populations, such as pregnant patients and those with underlying psychiatric disease, are ones whose care raises very specific considerations. The next cases will highlight these issues.

Metabolic Acidosis

Case 5

The patient was a 57-year-old male, newly diagnosed with the human immunodeficiency virus, complaining of abdominal pain. He admits to drinking a "cupful" of a fluid from a gasoline can at his home. His physical examination is notable for normal vital signs, an abdomen with some mild to moderate tenderness in the lower quadrants, a severely depressed affect with intermittent bursts of tears, and psychomotor retardation with slowed speech. His pupils were 4 mm and reactive, his skin was warm without excessive dryness, and he had good bowel sounds.

Step 1. Generate a differential diagnosis for the most significant physical finding and vital sign abnormality

This physical examination was notable only for the patient's abdominal tenderness and the features of his major depression.

Step 2. Develop an initial evaluation and treatment strategy

If the patient's physical examination does not allow one to narrow the differential diagnosis, sometimes clues from the history can. Often the patient's occupation or hobby or where the patient was found can suggest specific exposures or ingestions, for example if the patient were a gardener, consider pesticides or fungicides; if a jeweler, consider cyanide or hydrofluoric acid; if a health-care employee, consider potassium, high-potency opioids, or barbiturates; if from a drug house, high on the list should be heroin or crack cocaine.

In this case the patient's history of ingesting something from a gasoline can suggests an exposure to a product used in the garage or for cars. These can include solvents, fuels, brake fluid, antifreeze, dry gas, and windshield washer fluid. Some of these products contain methanol and ethylene glycol. One method of screening for these potentially fatal ingestions is to evaluate the patient for an acidosis and an osmolal gap (Table 4–3 and references 11 and 12).

In this patient the arterial blood gas results were pH = 7.27, pCO_2 = 26 mmHg, and pO_2 = 102 mmHg. His electrolytes included sodium = 137 mEq/L (137 mmol/L), potassium = 4.6 mEq/L (4.6 mmol/L), and chloride = 107 mEq/L (107 mmol/L), and an anion gap of 17 mEq/L [17 mmol/L; the reference range was 1–15 mEq/L (1–15 mmol/L)]. The blood urea nitrogen (BUN) and creatinine values were 10 and 1.2 mg/dL (3.57 mmol/L and 106 μmol/L), respectively. The patient's glucose was 113 mg/dL (6.3 mmol/L).

Thus, our patient has an anion gap metabolic acidosis.

The mnemonic MUDPILES helps us remember the causes of an anion gap metabolic acidosis (13) (see Table 4–1).

Using other historical information, physical findings, and uncomplicated laboratory studies, we can greatly narrow the list for this patient:

- For anion gap metabolic acidosis due to methanol poisoning, visual symptoms occur late and eye findings occur very late after an ingestion. A stat methanol serum concentration can be very helpful in this circumstance to rule methanol poisoning in or out as a possible cause of the metabolic acidosis.
- The creatinine and BUN tests provide a rapid assessment for the possibility that uremia was a cause of the acidosis.
- For evaluation of a possible diabetic ketoacidosis, glucose and ketones in serum provide this assessment.
- The possibility of alcoholic ketoacidosis as the cause is assessed effectively by knowing of a history of ethanol dependence involving typically a history of recent vomiting or abdominal pain.

TABLE 4–3. *Laboratory tests for evaluation of an acidosis and an osmolal gap*

Arterial blood gas analysis including pH, pCO_2, and pO_2
Serum electrolytes including Na^+, K^+, Cl^-, and HCO_3^-
Renal function assessment, blood urea nitrogen, and creatinine
Glucose
Assessment of the anion gap:
$Na^+ - (Cl^- + HCO_3^-)$

- A rare potential cause of metabolic acidosis is paraldehyde, which is used rarely to treat severe pediatric seizures.
- The possibility that a biguanide oral hypoglycemic agent–induced lactic acidosis associated with an overdose of one of these agents can be suggested by a history of oral hypoglycemic agent usage and a current elevated creatinine. The latter situation favors the accumulation of the oral hypoglycemic agent and the development of serious lactic acidosis.
- An acute overdose ingestion of isoniazid and a presentation with seizures is associated with a profound metabolic acidosis.
- Metabolic acidosis due to an acute overdose of iron associated with vomiting, shock, and an altered sensorium is another consideration.
- Lactic acidosis can be evaluated by ordering a stat serum lactate concentration.
- The possibility that ethylene glycol caused the metabolic acidosis can be evaluated by checking for calcium oxalate crystals in urine, fluorescence in the urine, an elevated creatinine, and a stat serum ethylene glycol concentration.
- For the assessment of salicylate as a cause of the metabolic acidosis, evaluation for a respiratory alkalosis, measurement of salicylate in urine with the quick ferric chloride test, and a stat serum salicylate concentration should be done.

Our patient had a normal creatinine and BUN, thereby eliminating uremia; normal glucose and no ketones eliminate the possibility of diabetic ketoacidosis; no history of oral hypoglycemic use probably eliminates metformin; no history of seizures, vomiting, or evidence of hemodynamic compromise eliminates isoniazid and iron as a cause in this patient; a stat lactate concentration of 1 mg/dL (0.1 mmol/L) eliminates lactic acidosis; and no detectable salicylate eliminates salicylates as a cause in this patient.

Eliminating these possibilities left us with methanol, alcoholic ketoacidosis, and ethylene glycol as the possible causes. Because of the patient's ingestion history, we loaded him with ethanol to block the metabolism of ethylene glycol and methanol into their toxic metabolites. Over the next hour the patient received 1.5 L of 5% dextrose normal saline, and an ethanol infusion was initiated. An available alternate for use in this situation is fomepizole (4-methylpyrizole) as an extremely effective and safe, although expensive, alternative (14).

Repeat laboratory evaluations were unchanged, and no ketones were seen by repeat urinalysis. These results essentially eliminate alcoholic ketoacidosis because no ketones were detected, and this entity resolves rapidly when the patient is given glucose and fluids, which did not happen in our patient.

Unfortunately, because this case occurred in a small community hospital and the methanol and ethylene glycol determinations were sent to a larger reference laboratory, the methanol and ethylene glycol concentrations could not be obtained until the morning. However the history was concerning enough that hemodialysis was initiated to remove any potential toxic alcohols from the patient's serum. After 4 hours of hemodialysis, the patient's acidosis had resolved. He was kept on an ethanol drip until the results of his serum methanol and ethylene glycol analysis were reported. The initial ethylene glycol and methanol values were, respectively, 0 and 53 mg/dL (0 and 1.65 mmol/L). A postdialysis methanol concentration was 25 mg/dL (0.78 mmol/L). With the concentration of methanol still 25 mg/dL (0.78 mmol/L) after dialysis, the patient was again hemodialyzed, and the final concentration was 0 mg/dL (0 mmol/L).

Typically, severe metabolic acidosis (pH <7.0) is produced by hypovolemia, sepsis, cardiac arrest, and other causes of shock (15). In these situations the patient appears moribund, with a depressed concentration of consciousness. If a patient has a relatively intact mental status with a severe acidosis, then the cause of the acidosis is almost certainly methanol, ethylene glycol, or metformin.

In addition to unique clinical manifestations, there are specific drug classes, such as those with delayed toxicity, or patient populations, such as pregnant patients, whose care raises specific considerations.

Drugs with Delayed Onset and Peak of Toxicity

Typically, a patient who has overdosed on a medication receives an initial evaluation and basic treatment. If the patient remains asymptomatic or has mild symptoms that resolve, the patient can be discharged to an appropriate psychiatric care plan after 4–6 hours. However, several drugs require much more prolonged observation because of their potential for delayed toxicity. If any of these are ingested by history, then hospital admission to a monitored setting is required. These drugs include:

- Sustained-release formulations (calcium channel antagonists, β-adrenergic antagonists, lithium)
- Monoamine oxidase inhibitors
- Oral hypoglycemics (glipizide, glyburide)
- Colchicines
- Some chemotherapeutic agents
- Acetaminophen (because this is such a prevalent drug, all overdose patients should be routinely screened for its presence)

The Pregnant Patient

The pregnant patient who overdoses is often difficult to care for because of a variety of factors, including physiologic and psychosocial, but there are several basic principles and epidemiological trends that one should be aware of. These patients are often involved in impulsive ingestions during emotional and stressful periods. Readily available medications that are ingested include acetaminophen and salicylates, cough and cold remedies, prenatal vitamins with iron, and iron supplements; if the pregnancy is not desired, patients may ingest abortifacients. A prospective observational study (16) recommended pregnancy screening of all females 12–30 years old who presented with self-poisoning. Of 1142 patients, 371 (32%) got pregnancy tests and 43 of those patients were pregnant. Of these 43 intentional ingestions, five were with known abortifacients, including methotrexate, misoprostol, methylergonovine, quinine, and birth control pills. Other abortifacients include essential oils, particularly pennyroyal oil, and many herbal products, including blue cohosh.

The Patient with Psychiatric Disease

One population of patients that is poorly understood and yet represents some of the most severe self-poisonings is patients with psychiatric disease. Often their underlying medical condition, whether it be schizophrenia or depression, makes it difficult for physicians to interact, obtain a history, and sometimes effectively care for these patients (17,18).

There are many reasons patients with psychiatric disease are at high risk for poisoning, including underlying disease, depression, irrational thoughts, hallucinations, poor impulse control, opportunity for the use of and easy access to highly toxic agents [some of the most highly toxic therapeutics, namely, monoamine oxidase inhibitors and tricyclic antidepressants, are prescribed to patients with major depression who are at high risk for suicide (19)], societal consequences of their disease, substance abuse (for example, ethanol or cocaine) (20), self-treatment for hallucinations or depression, marginalization in society (some are homeless or migrate from one family member's home to the next, so odd ingestions are common), ingestion of other people's medications either to get high or for self-treatment, and accidental exposures to chemicals and waste in garbage and abandoned homes.

When these patients come to the emergency department, they have an altered cognitive function by definition, but frequently no one knows what their baseline is. The clinician is faced with having to decide whether this is a manifestation of their underlying disease or an intoxication. Screening for drugs of abuse, in addition to taking a probing substance abuse history, in this population can improve patient care (21). Patients who are crack cocaine users may benefit from detox treatment and/or referral to a 10-step program. The results of drug screening may also explain the exacerbation of their depression or schizophrenia.

EFFECTIVE CLINICAL USE OF THE ANALYTICAL LABORATORY

The clinical tools discussed above are essential when providing care for poisoned patients because in many smaller hospitals, the analytical laboratory does not have the resources to perform every assay possible and certainly not in a timely or useful manner. Thus, many uncommon tests are sent to referral laboratories (22). A telephone survey to all hospitals that had emergency medicine residency training programs found the following (based on 95 completed surveys of 115 emergency departments surveyed) (23):

- Methanol
  - Availability, 37/95 (39%)
  - Turn-around time, 1 hour (at site) vs. 36 hours (sent out)
- Ethylene glycol
  - Availability, 24/95 (25%)
  - Turn-around time, 1.5 hours (at site) vs. 42 hours (sent out)

In addition, many patients do not require specific drug identification or quantification because few treatment strategies revolve around that data. Many exposures, >80% of poison center calls, are simply managed by certified specialists in poison information over the phone, with the patient staying at home. There are, however, a few relatively common and relatively toxic drugs in which the laboratory is vital to designing an effective treatment strategy. Table 4–4 includes my "essentials" list when it comes to quantitative determinations of serum drug concentrations (24). The tests and the reasons for

TABLE 4–4. *A summary of the rationale for requiring rapid testing of the following oxin concentrations*

Initiation of specific antidotal therapy
Acetaminophen (*N*-acetylcysteine)
Digoxin (digoxin-specific antibody fragments)
Iron (deferoxamine), although this one is debatable
Initiation of hemoperfusion or hemodialysis
Theophylline
Aspirin
Methanol
Ethylene glycol
Lithium
Titrating therapy
Ethanol drip
Anticonvulsants (reloading patients in the emergency department after seizures)

my requiring them in a rapid or "stat" manner are summarized in Table 4–1.

Case 6

The patient was a 16-year-old female who presented to the emergency department after ingesting acetaminophen. The patient had had an altercation with her boyfriend after discovering that she was pregnant and locked herself in the bathroom, only to return 20 minutes later with an empty bottle of acetaminophen. She stated that she had ingested the entire bottle. On arrival at the emergency department, the patient was without any complaints and stated that she took only "a few pills" and was just trying to scare her boyfriend. Her physical examination was entirely unremarkable. She was placed on a cardiac monitor and treated with 50 g of activated charcoal orally. Her ECG was unremarkable, and her urine pregnancy test was positive.

In this case, an appropriately timed acetaminophen plasma concentration is critical in determining the patient's ultimate treatment —home after psychiatric evaluation or 3 days of hospitalization for 17 doses of oral *N*-acetylcysteine (NAC). This is based on the Rumack-Matthew nomogram (25), which plots plasma acetaminophen concentration (y-axis) vs. time (x-axis), and this practice was validated and supported by landmark investigations done by Smilkstein et al. (26).

If the patient's acetaminophen concentration falls above the nomogram's possible toxicity line, which begins at 150 µg/mL (993 µmol/L) at 4 hours and drops to <5 µg/mL (<33 µmol/L) at 24 hours, then NAC therapy is indicated. Proper use of the nomogram requires both an accurate or very conservatively estimated time of ingestion and an accurate serum specimen collection time. The nomogram was not designed to interpret acetaminophen serum concentrations in patients with multiple, serial ingestions of acetaminophen, patients with chronic acetaminophen toxicity, or patients who present >24 hours after their ingestion. These "atypical" cases are best managed on a case-by-case basis with consultation with a clinical toxicologist or poison control center.

In this case the patient's 4-hour acetaminophen concentration was 138 µg/mL (914 µmol/L). She was discharged home with her family after a psychiatric evaluation determined her to be at low risk of suicide.

Case 7

The patient was a 56-year-old female with a history of hypertension, angina, diabetes, chronic obstructive pulmonary disease, and depression who was brought in by paramedics after her daughter found her vomiting in the bathroom. Her medications include enalapril, hydrochlorothiazide, isosorbide dinitrate, SL nitroglycerin, enteric-coated aspirin, glyburide, theophylline, albuterol metered-dose inhaler, sertraline, and lorazepam. She admitted to overdosing on several handfuls of her theophylline in a suicide attempt and complained of vomiting and crampy abdominal pain. On physical examination she was noted to be an elderly female in mild distress with an emesis basin at the bedside. Vital signs included a heart rate of 120 beats/min, respiratory rate of 20/min, blood pressure of 165/90 mmHg, and a temperature of 99.3 °F (37.4 °C). Neurologically she was awake and slightly somnolent but responded appropriately to verbal stimuli. Her motor function was intact with symmetrical reflexes, and she had a resting tremor of her hands. Pupils were 4 mm and reactive, her skin was warm and moist, and her bowel sounds were present.

Step 1. Generate a differential diagnosis for the most significant physical finding and vital sign abnormality

This physical examination was notable for the vomiting, the tachycardia, and the moist skin. These findings are somewhat nonspecific but are very consistent with theophylline-induced sympathomimetic effects.

Step 2. Develop an initial evaluation and treatment strategy

All patients with toxic exposures should be placed on cardiac monitors, and i.v. access should be obtained. Because of the history of theophylline overdose and the physical examination finding consistent with that, a serum theophylline concentration, in addition to an acetaminophen screen, was sent. The patient was treated with 100 g of activated charcoal orally.

One of the most important concepts in clinical toxicology is that patients' clinical conditions are dynamic and therefore quantitative concentrations must be interpreted in the context of an evolving situation. A serum theophylline concentration of 50 µg/dL (278 µmol/L) 10 hours after an acute overdose is of much less concern than if it occurs only 2 hours after an acute ingestion or in a patient who has been chronically poisoned (27). In most cases, trends in serum concentrations are much more valuable than an absolute number. Concentrations can determine treatment based on an absolute value or a rising trend despite appropriate therapy, or can be used to modify treatments. One helpful technique is to "stack" concentrations. What this means is to send a serum sample every 1–2 hours regardless of whether the results from the first determination have returned.

At 1 hour, the patient's heart rate had increased to 140 beats/min, and she was having difficulty tolerating her activated charcoal even though it was being administered slowly via nasogastric tube and aggressive antiemetic therapy with prochlorperazine and ondansetron was being used. Shortly thereafter, her initial theophylline concentration was reported as 55 µg/dL (305 µmol/L).

The patient at this point was demonstrating signs of more significant toxicity, and although the initial concentration was "only" 55 µg/dL (305 µmol/L), clearly it was rising. How rapid this slope is rising will determine whether or not she will require hemoperfusion or hemodialysis [most textbooks recommend hemodialysis at 100 µg/dL (555 µmol/L)]. However, with the patient's clinical status worsening and her first concentration of 55 µg/dL (305 µmol/L), the nephrologist was contacted and notified that if her concentration was significantly elevated above the first one, then she would require hemoperfusion.

The patient's 1-hour serum theophylline concentration was 86 µg/dL (477 µmol/L). She underwent 4 hours of hemoperfusion with continued antiemetics and multiple-dose activated charcoal. She ultimately did well, with her peak serum theophylline concentration on initiation of hemodialysis at 97 µg/dL (538 µmol/L).

REFERENCES

1. Higgins GL, Cambell B, Wallace K, et al: Pediatric poisoning from over-the-counter imidazoline-containing products. *Ann Emerg Med* 1991;20:655–658.
2. Ford M, Tomaszewski C, Kerns W, et al: Bedside ferric chloride urine test to rule out salicylate intoxication [Abstract]. *Vet Hum Toxicol* 1994;36:364.
3. Fries J: Toward an understanding of NSAID-related adverse events: the contribution of longitudinal data. *Scand J Rheumatol* 1996;102 [Suppl]:3–8.
4. Goldfrank LR, Flomenbaum NE, Lewin NA, et al: Vital signs and toxic syndromes. In: Goldfrank LR, Flomenbaum NE, Lewin NA, et al, eds: *Goldfrank's toxicologic emergencies*, 6th ed. Norwalk, CT: Appleton and Lange, 1998:277–283.
5. Olson KR, Pentel PR, Kelley MT: Physical assessment and differential diagnosis of the poisoned patient. *Med Toxicol* 1987;2;52–63.
6. Hoffman RS, Goldfrank LR: The poisoned patient with altered consciousness. Controversies in the use of a "coma cocktail." *JAMA* 1995;274:562–569.
7. Tornabhene VW: Narcotic withdrawal syndrome caused by naltrexone. *Ann Intern Med* 1974;81:785–787.
8. Howland MA: Opioid antagonists. In: Goldfrank LR, Flomenbaum NE, Lewin NA, et al, eds: *Goldfrank's toxicologic emergencies*, 6th ed. Norwalk, CT: Appleton and Lange, 1998:996–1000.
9. Gowdy JM: Stramonium intoxication. *JAMA* 1972;221:585–587.
10. Shervette RE, Schydlower M, Lampe RM, et al: Jimson "loco" weed abuse in adolescents. *Pediatrics* 1979;63:520–523.
11. Jacobsen D, Bredesen JE, Eide I, et al: Anion and osmolal gaps in the diagnosis of methanol and ethylene glycol poisoning. *Acta Med Scand* 1982;212:17–20.
12. Glaser DS: Utility of the serum osmol gap in the diagnosis of methanol or ethylene glycol ingestion. *Ann Emerg Med* 1996;27:343–346.
13. Hoffman RS: Fluid, electrolyte, and acid-base principles. In: Goldfrank LR, Flomenbaum NE, Lewin NA, et al, eds: *Goldfrank's toxicologic emergencies*, 6th ed. Norwalk, CT: Appleton and Lange, 1998:244–260.
14. Brent J, McMartin K, Phillips S, et al: Fomepizole for the treatment of ethylene glycol poisoning. *N Engl J Med* 1999;340:832–838.
15. Gabow PA, Kachny WD, Fennessey PV, et al: Diagnostic importance of an increased serum anion gap. *N Engl J Med* 1980;303:354–361.
16. Perrone J, Hoffman RS: Toxic ingestions in pregnancy: abortifacient use in a case series of pregnant overdose patients. *Acad Emerg Med* 1997;4:206–209.
17. Wright N: An assessment of the unreliability of the history given by self-poisoned patients. *Clin Toxicol* 1980;16:381–387.
18. Robins E, Murphy GE, Wilkinson RH, et al: Some clinical considerations in the prevention of suicide based on a study of 134 successful suicides. *Am J Public Health* 1959;49:888–889.
19. Henry JA: A fatal toxicity index for antidepressant poisoning. *Acta Psychiatr Scand* 1989;354:37–45.
20. Kandal DB: Epidemiological trends and implications for understanding the nature of addiction. In: O'Brien CP, Jaffe JH. eds: *Addictive states*. New York: Raven, 1992:23–30.
21. Perrone J, Hollander JE, Jayaraman S, et al: History versus drug screening in detecting substance abuse in ED psychiatric patients. *Am J Emerg Med* 2001;19:49–51.
22. Nice A, Leikin JB, Maturen A: Toxidrome recognition to improved efficiency of emergency urine drug screens. *Ann Emerg Med* 1988;17:676–681.
23. Kearney J, Rees S, Chiang W: Availability of serum methanol and ethylene glycol levels: a national survey [Abstract #68]. *J Toxicol Clin Toxicol* 1997;5:509.
24. Brett AS, Rothschild N, Gray R, et al: Predicting the clinical course in intentional drug overdose: implications for use of the intensive care unit. *Arch Intern Med* 1987;147:133–137.
25. Rumack BH, Matthew H: Acetaminophen poisoning and toxicity. *Pediatrics* 1975;55:871–876.
26. Smilkstein MJ, Knapp GL, Kulig KW, et al: Efficacy of oral N-acetylcysteine in the treatment of acetaminophen overdose: analysis of the national multicenter study (1979–1985). *N Engl J Med* 1988;319:1557–1562.
27. Olson KR, Benowitz NL, Woo OF, et al: Theophylline overdose: acute single ingestion versus chronic repeated overmedication. *Am J Emerg Med* 1985;3:386–394.

SELF-ASSESSMENT QUESTIONS

1. Which of the following agents is not associated with bradycardia?
   a. clonidine
   b. verapamil
   c. aspirin
   d. digoxin

2. Which of the following agents is not associated with an increased respiratory effort?
   a. salicylates
   b. paraquat
   c. carbon monoxide
   d. methadone

3. Which of the following toxins is not associated with the formation of an anion gap metabolic acidosis?
   a. methanol
   b. phenothiazines
   c. metformin
   d. isoniazid

4. Which of the following agents is not associated with the formation of an anticholinergic toxidrome?
   a. lithium
   b. Jimson weed
   c. tricyclic antidepressants
   d. benztropine

CHAPTER 5

Marijuana

Marilyn A. Huestis, Ph.D., M.S., B.A.*

LEARNING OBJECTIVES

After completing this chapter, the reader should be able to:

1. *List physiologic and behavioral effects of marijuana.*
2. *Describe the pharmacokinetics of tetrahydrocannabinol (THC), the primary psychoactive compound in marijuana.*
3. *Normalize urinary cannabinoid concentrations to creatinine concentrations and describe how to use these data to predict new marijuana use.*
4. *Describe newly identified cannabinoid receptors, endogenous ligands, and antagonists.*
5. *Explain counterclockwise hysteresis found in marijuana concentration-effect curves.*
6. *List several medicinal uses of marijuana or synthetic THC.*

INTRODUCTION

Marijuana (MJ), hashish, sinsemilla, and other psychoactive products obtained from the complex *Cannabis sativa* plant are the most widely used illicit drugs in the world. Cannabis has been used for its euphoric effects for more than 4000 years. The stout, aromatic annual herb originated in Central Asia and is cultivated widely in the North Temperate Zone. A canelike variety, devoid of psychoactive effects, provides an important source of hemp fiber. The Assyrians incorporated cannabis into their religious rites and used the drug as medicine for neurological and psychiatric diseases. Cannabis was recently identified in plant ash on the skeleton of a young woman who died during childbirth in the 4th century A.D. Additional evidence of the use of cannabis during birthing comes from information documented in Egyptian papyri. Medicinal properties of the plant were recognized in China 2700 years ago for relief of pain, muscle spasms, convulsions, epilepsy, asthma, and rheumatism. O'Shaughnessy, an Irish surgeon, introduced cannabis to Europe in 1842. However, similar to other herbal preparations, its potency was unreliable, which contributed to the decline of its therapeutic use. In 1845, the French psychiatrist de Tour Moreau described cannabis's effects as "a gradual weakening of the power to direct thoughts at will" (1).

Mexican laborers introduced the drug into the South and Southwest around 1910. New Orleans was the site of the first urban concentration of cannabis users. Alcohol prohibition in the 1920s increased the spread of drug use to other urban centers (2). In the United States, annual use reached a high of 51% in 1979 and has since decreased over time. MJ continues to be the most commonly used illicit drug in the United States. It is estimated that there are 18–20 million regular MJ users, usually young adults, in the United States; the average age of first use is 13. Nearly 14% of high school students reported use of MJ

---

*Updated from Huestis MA: Marijuana. *DRE* 1994;6: 2–23.

within the previous 30 days in a national survey in 1991 (3). In 1988, 7.3% of military service applicants screened positive for cannabinoid metabolites (4), and approximately 20% of Chicago Police Department recruits also were positive for cannabinoids (5).

MJ is self-administered for its mood-altering properties and has been described as an addictive dependence-producing drug due to the production of euphoria, the presence of reversible psychological impairment, an abstinence syndrome, and tolerance (6). A mixture of depressant and stimulant effects is noted at low doses; MJ acts as a central nervous system (CNS) depressant at high doses. Cannabinoids share effects with other psychoactive drugs, yet possess a distinct pattern of effects that distinguish this unique pharmacological drug class.

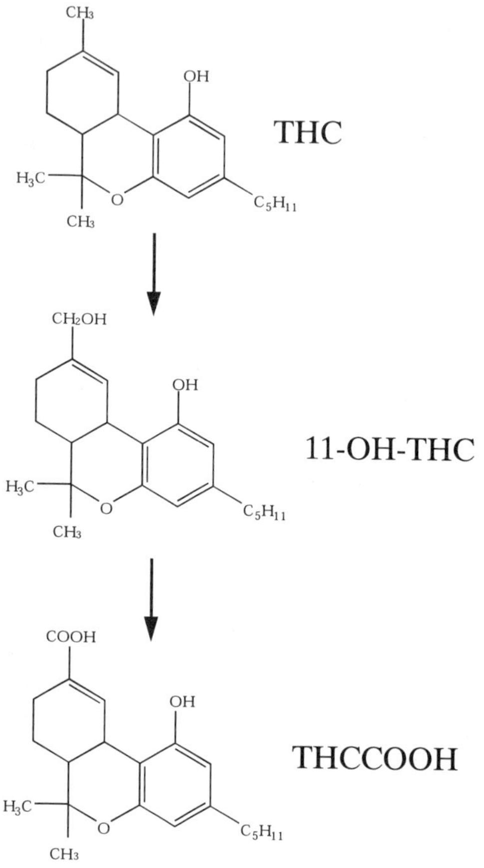

CHEMISTRY

Source

Cannabis preparations include loose MJ, kilobricks (the classical Mexican-produced material), buds, sinsemilla, Thai sticks, hashish (cannabis resin), and hash oil. Different parts of the plant vary in chemical composition. Sinsemilla, a seedless and more potent form of MJ produced from the unfertilized flowering tops of female cannabis plants, first appeared in 1977 and is usually produced in the United States. Δ^9-Tetrahydrocannabinol (THC), the primary psychoactive analyte, is found in the plant's flowering or fruity tops, leaves, and resin. The structure of THC is shown in Figure 5–1. Cannabis plant material is far more complex than pure THC and may produce different effects due to the presence of additional cannabinoids and other chemicals. The ratio of cannabinoids in MJ depends on the age of the sample, its geographic origin, and plant strain. Cannabinol is essentially a chemical degradation product and its relative abundance increases as samples age. The potency of a preparation is described by its THC concentration, usually as the % THC per dry weight of material. Potency has been steadily increasing over the

FIG. 5–1. Major metabolic route for tetrahydrocannabinol (THC) including the primary active metabolite, 11-nor-Δ^9-tetrahydrocannabinol (11-OH-THC), and the primary inactive metabolite, 11-nor-$\Delta^9$9-tetrahydrocannabinol (THCCOOH).

years through selective cultivation. In the early 1990s, the average percentages of THC in MJ, hashish, and hash oil seizures in the United States were 2.9, 3.4, and 16.5%, respectively (MA ElSohly, personal communication, 1993). The highest THC concentrations in seized materials were MJ (including buds, sinsemilla, etc.) 29.9%, hashish 27.7%, and hash oil 43.2%. Approximately 30% of all seizures were of domestic origin.

NOMENCLATURE

MJ contains more than 421 different chemical compounds, including 61 cannabinoids. During smoking, more than 2000 compounds may be produced through pyrolysis. Eighteen different classes of chemicals, including nitrogenous compounds, amino acids, hydrocarbons, sugars, terpenes, and simple and fatty acids contribute to MJ's known pharmacological and toxicological properties. Other cannabinoids include cannabinol, which is approximately 10% as psychoactive as THC, and cannabidiol, which is not a mood-altering agent.

The structure of THC was elucidated in 1964 (7). THC contains a tricyclic 21-carbon structure and is a volatile, viscous oil, insoluble in water with high lipid solubility. The pK_a of THC is 10.6. Two different numbering systems, the dibenzopyran, or Δ^9, system and the monoterpene, or Δ^1, system, are used in the literature to describe THC. The dibenzopyran system will be used throughout this chapter. THC contains no nitrogen and has two chiral centers in trans-configurations.

Structure-Activity Relationships

Early studies using impure materials suggested that cannabinoid actions were not highly stereospecific. However, more recent studies have reported highly stereospecific cannabinoid effects. These strict structural requirements indicate that THC interacts with a chiral biochemical entity, such as an enzyme or receptor. Cannabinoids are natural (–)-enantiomers with (3R,4R) stereochemistry. Structure-activity studies have demonstrated that the tricyclic structure, aromatic ring, and phenolic group are required for central activity. Animal studies have shown that a C5 hydroxy group confers potent peripheral activity (8).

Chemical Stability

THC is usually present in cannabis plant materials as a mixture of monocarboxylic acids that readily decarboxylate upon heating. THC decomposes when exposed to air, heat, or light; exposure to acid can oxidize the compound to cannabinol. THC binds readily to glass and plastic. THC adsorption can be minimized with storage of solutions in amber silylated glassware and maintenance of the compound in a basic solution or organic solvent.

MECHANISM OF ACTION

Two hypotheses for THC's mechanisms of action were proposed for many years. One hypothesis suggested that THC exerted its effects through nonspecific interactions of the drug with cellular and organelle membranes. The other hypothesis suggested that THC interacted with specific cannabinoid receptors. Delineating mechanisms of action was difficult due to demonstrated THC activity at many sites, including the opioid and benzodiazepine receptors, and noted effects on prostaglandin synthesis, DNA, RNA, and protein metabolism. Cannabinoids inhibit macromolecular metabolism in a dose-related manner and have a wide range of effects on enzyme systems, hormone secretion, and neurotransmitters. These numerous and diffuse effects lent support to the nonspecific interaction hypothesis.

However, in the past 10 years our knowledge of cannabinoid pharmacology has increased tremendously. Central (CB1) and peripheral (CB2) cannabinoid receptors have been characterized, endogenous ligands (such as anandamide) have been identified, and specific CB1 and CB2 receptor antagonists have been synthesized. CB1 receptors in the brain were mapped and indicated that distribution of the high-affinity, stereoselective, and pharmacologically distinct brain receptors is anatomically selective. Dense binding has been documented in the striatum, cerebral cortex, and hippocampus. Cannabinoid receptors belong to the G protein class of receptors and are allosterically regulated. The distinct peripheral cannabinoid receptor may play an immunomodulatory role.

The endogenous cannabinoid ligand, anandamide or arachidonylethanolamide, is an arachidonic acid derivative (9). It mimics THC in binding and pharmacodynamic activity studies. SR141716 is the first CB1 specific antagonist and is an effective antagonist to THC in preclinical studies. The first clinical studies of the pharmacokinetic and pharmacodynamic effects of SR141716 and MJ are ongoing in my research unit now. These significant advances have opened new frontiers in MJ research and should lead to a better understanding of MJ's effects and the role of the endogenous ligand in human pharmacology. It is suggested that THC may act as a direct or indirect dopamine agonist to stimulate electrical brain reward circuits. Stimulation of the brain's reward circuits is an essential characteristic of drugs of abuse. MJ produces substantial changes in human behavior that are linked to physiologic and biochemical changes. Subjective responses and performance effects are interrelated with other bodily functions; behavior is the highest level of human response. It is the integration of all of MJ's pharmacological effects that results in behavioral modification.

EFFECTS

MJ's behavioral and physiological effects have been well described. Behavioral effects include feelings of euphoria and relaxation, altered time perception, lack of concentration, impaired learning and memory, and mood changes such as panic reactions and paranoia. MJ's spectrum of behavioral effects is unique, preventing classification of the drug as a stimulant, sedative, tranquilizer, or hallucinogen. Subjective effects of MJ, such as drug "liking" and "feel drug" may appear after the first or second puff of a cigarette (10). MJ smoking was also found to produce rapid changes in some physiological effects including heart rate and diastolic blood pressure. The most frequent physiological effects include increased heart rate, conjunctival suffusion, dry mouth and throat, increased appetite, vasodilation, and decreased respiratory rate.

Most behavioral and physiological effects of THC return to base-line concentrations within 3–6 hours after exposure (11,12), although some investigators have demonstrated residual effects in specific behaviors as long as 24 hours after drug use (13,14). More research is needed to define the onset, magnitude, and duration of MJ's behavioral effects, especially after long-term, frequent use of the drug. Other effects of MJ use have been reported including cell metabolism alterations, immune system and endocrine transformations, lung damage, electroencephalogram alterations, and psychomotor impairment.

Effect vs. time curves for heart rate and subjective "high" display a counterclockwise hysteresis, indicating a delay between effects and plasma concentrations (15). A counterclockwise hysteresis is generally indicative of a prominent distribution phase, perhaps due to redistribution of drug from the vascular compartment to the drug's site of action, the brain. The subjective "high" effect was found to be directly proportional to the mean plasma concentration of THC from approximately 1 to 4 hours after MJ smoking (16). Domino et al. (17) determined that THC plasma concentrations at 50% maximal effect were 7.2 and 16.8 ng/mL (22.9 and 53.4 nmol/L) in light and heavy MJ users, respectively. More recently, initial changes in blood concentrations were reported to be out of phase (hysteresis) with physiological and behavioral changes (18). However, after blood-tissue equilibrium was established, a direct correlation of THC blood concentration and effect was observed.

Acute toxic effects of MJ include behavioral effects, such as panic attacks and psychosis, increased heart rates, and CNS depression. The National Institute on Drug Abuse's Drug Abuse Warning Network (DAWN) reported 20,703 MJ-related hospital emergency room visits in 1989, less than 5% of the total number of drug-related visits (19). This number decreased by almost 25% in 1990. Serious CNS depression has been observed after inadvertent ingestion of MJ by children. However, with supportive care, these cases have resolved successfully with few residual effects. MJ may be a contributing factor to fatalities, such as car accidents and multiple drug overdoses.

There are conflicting reports on MJ's chronic toxic effects in the literature. This may be due in part to different experimental protocols, type and potency of MJ materials, schedule and length of exposure, subject characteristics and defined end point of effect. Impaired health, including lung damage, behavioral changes, and reproductive, cardiovascular, and immunology effects, has been associated with MJ use. MJ smoke condensate yield, including potential mutagens, was more than 50% higher than tobacco cigarette smoke yield (20). Cannabinoids readily cross placental membranes and expose the developing fetus. Cannabinoids can be embryocidal, affect gestational length and labor, and induce maturational delays, and may have effects on behavioral parameters in the human neonate (21). There are indications that MJ inhibits the human immune system. Some of the most reproducible findings suggest that THC inhibits the progression of responsive macrophages to full activation by limiting their capacity to respond to immunogenic signals (22).

THERAPEUTIC USES

Potential therapeutic uses of cannabinoids have been actively pursued despite removal of THC from the British and United States Pharmacopoeias in the 1930s and 1940s due to central hallucinogenic actions of the drug. In 1996 two states in the United States, California and Arizona, passed referenda to permit physicians to write prescriptions for the therapeutic use of cannabinoids. Synthetic analogs, and MJ itself, have been used as antiemetic agents after chemotherapy, as antihypertensives in the management of glaucoma, as antispasmodics in multiple sclerosis, and as appetite-enhancing drugs for individuals with acquired immunodeficiency syndrome (AIDS). Dronabinol (Marinol®), a synthetic THC, has been available in the United States since 1986. Dronabinol is licensed for the treatment of nausea and vomiting associated with cancer chemotherapy. Some oncologists have indicated that smoked MJ is more effective than the synthetic oral medication due to dronabinol's low and less reliable bioavailability, some patients' inability to tolerate the oral medication, and the absence of other active compounds found in MJ plant material. Other studies have focused on cannabinoids' analgesic, anti-inflammatory, antitumor, and antiepileptic effects.

MJ would not be considered the first drug of choice for any of the conditions described above; however, proponents of MJ's therapeutic potential believe that the drug can provide help for individuals resistant to the primary therapeutic choices. Attempts to synthesize cannabinoid analogs that produce desired therapeutic responses with minimal undesirable side effects, such as the psychoactive effects, have been relatively unsuccessful. These preparations are chemically very different from the complex plant material and have different therapeutic efficacy. A small number of cancer, glaucoma, and AIDS patients have been supplied with MJ by the government and have been permitted to smoke MJ legally. Physicians have indicated a willingness to prescribe MJ for therapeutic purposes. A survey of oncologists indicated that 48% would prescribe MJ for the control of emesis if it were legal (23). The Public Health Service had begun to accept new applications for supply of therapeutic MJ in response to the human immunodeficiency virus epidemic. However, this policy was reversed, and no new applications were approved (24). The possibility of lung cancer, dysfunction of the immune system, and behavioral effects of the drug were listed as reasons for the policy reversal. In 1997, the National Institutes of Health approved and encouraged additional clinical studies on the therapeutic uses of MJ.

PHARMACOKINETICS

Absorption

The smoking route, the principal means of MJ administration, provides a rapid and highly efficient method of drug delivery, although oral use is not uncommon. Approximately 30% of THC is estimated to be destroyed by pyrolysis during smoking. Smoked drugs are highly abused, in part due to the efficiency and speed of delivery of drug from the lungs to the brain. Intense pleasurable and strongly reinforcing effects may be produced due to almost immediate drug exposure to the CNS. Drug delivery during MJ smoking is characterized by rapid absorption of THC with slightly lower peak concentrations than those found after intravenous administration.

Bioavailability after the smoking route has been reported to be 18–50%, due in part to the intra- and intersubject variability in smoking dynamics that contribute to the uncertainty in dose delivery (25,26). The number, duration, and spacing of puffs; hold time; and inhalation volume greatly influence the degree of drug exposure. THC can be measured in the plasma within seconds after inhalation of the first puff of MJ smoke (27). Mean ± SD THC concentrations of 7.0 ± 8.1 ng/mL (22.3 ± 25.8 nmol/L) and 18.1 ± 12.0 ng/mL (57.6 ± 38.2 nmol/L) were observed after the first inhalation of a low (1.75% THC) or high (3.55% THC) dose cigarette, respectively. Concentrations continued to increase rapidly and the peak concentrations occurred at 9.0 minutes, prior to initiation of the last puff sequence at 9.8 minutes.

Absorption is slower after the oral route of administration, with lower, more delayed peak THC concentrations. Bioavailability is reduced to 6–18% after oral use (28), due in part to degradation of drug in the stomach. Also, there is significant first-pass metabolism to active and inactive metabolites. The pharmacodynamic effects of MJ taken orally are due both to THC and its equipotent metabolite, 11-hydroxy-tetrahydrocannabinol (11-OH-THC).

Distribution

THC has a large volume of distribution, 10 L/kg, and is 97–99% protein bound in plasma, primarily to lipoprotein (29). Highly perfused organs, including the brain, are rapidly exposed to the drug. Less highly perfused tissues accumulate drug more slowly as THC redistributes from the vascular compartment to tissue. THC's high lipid solubility results in concentration and prolonged retention of the drug in fat (30). Low concentrations of THC in brain tissue have been documented in distribution studies in rats (31). Slow release of drug from fat and significant enterohepatic recirculation contribute to THC's long drug half-life in plasma, most recently reported as more than 4.1 days (32). Isotopically labeled THC and sensitive analytical procedures were used to obtain this estimate of drug half-life. Less sensitive methods have resulted in much lower estimates of the terminal half-life and a more simplified description of the drug's pharmacokinetics.

Metabolism

Hydroxylation of THC by the hepatic cytochrome P450 enzyme system leads to production of the active metabolite, 11-OH-THC, believed by early investigators to be the true active analyte. Concentrations of this analyte are low after MJ smoking, with metabolite concentrations attaining less than 10% of the THC concentration. Concentrations of 11-OH-THC may be as high as 50% of THC concentrations after MJ ingestion. Other tissues, including lung, may contribute to the metabolism of THC, although alternate hydroxylation pathways may be more prominent. Further metabolism to di- and tri-hydroxy compounds, ketones, aldehydes, and carboxylic acids has been documented. Oxidation of active 11-OH-THC produces the inactive metabolite 11-nor-9-carboxy-Δ^9-tetrahydrocannabinol (THCCOOH). The inactive THCCOOH metabolite and its glucuronide conjugate have been identified as the major end products of biotransformation in most species including humans (33). Renal clearance of these polar metabolites is low due to extensive protein binding (29).

THCCOOH concentrations gradually increase and surpass concurrent THC concentrations shortly after completion of smoking, due to the precipitous drop in THC concentrations during this period. The time course of detection of THCCOOH is much longer than either of the active analytes. The redistribution of THC from tissue to blood has been demonstrated to be the rate-limiting step in its metabolism. Fifteen to 20 percent of a

THC dose is eliminated as acidic urinary metabolites, with approximately 20% of these conjugated and unconjugated THCCOOH. No significant difference in metabolism between men and women has been reported (28). Figure 5–1 illustrates the common metabolic profile of THC.

Elimination

More than 65% of the drug is excreted in the feces, with approximately 20% excreted in the urine. A total of 80–90% of the drug is excreted within 5 days, mostly as hydroxylated and carboxylated metabolites. Many of these metabolites are conjugated with glucuronic acid, which increases the compounds' water solubility. The primary urinary metabolite is the acid-linked THCCOOH glucuronide conjugate, while 11-OH-THC predominates in the feces. The excretion half-life in humans has been estimated to be 3–4 days.

Plasma Concentrations

Huestis et al. (27) characterized the absorption of THC and formation of 11-OH-THC and THCCOOH during MJ smoking (Figure 5–2) and followed the time course of detection of these compounds over 7 days. THC was detected in the plasma immediately after the first MJ puff. Concentrations continued to increase rapidly. After smoking of one 1.75% or 3.55% THC cigarette, peak THC concentrations ranged from 50 to 129 ng/mL (159.0 to 410.2 nmol/L) [mean of 84.3 ng/mL (268.1 nmol/L)] and from 76 to 267 ng/mL (241.7 to 849.0 nmol/L) [mean of 162.2 ng/mL (515.8)], respectively. Mean THC concentrations were approximately 60% and 20% of peak concentrations 15 and 30 minutes after smoking, respectively. Within 2 hours, THC concentrations were at or below 5 ng/mL (15.9 nmol/L). The time of detection of THC (GC/MS LOD = 0.5 ng/mL) varied from 3 to 12 hours after the low-dose and from 6 to 27 hours after the high-dose MJ cigarette.

11-OH-THC plasma concentrations were approximately 6–10% of the concurrent THC concentrations for as long as 45 min after the start of smoking. Peak 11-OH-THC concentrations were noted 13.5 min (range 9.0–22.8) after the start of smoking. Mean peak concentrations of 6.7 ng/mL (20.3 nmol/L) [range 3.3–10.4 ng/mL (10.0–31.5 nmol/L)] and 7.5 ng/mL (22.7 nmol/L) [range 3.8–16 ng/mL (11.5–48.4 nmol/L)] were measured after smoking of

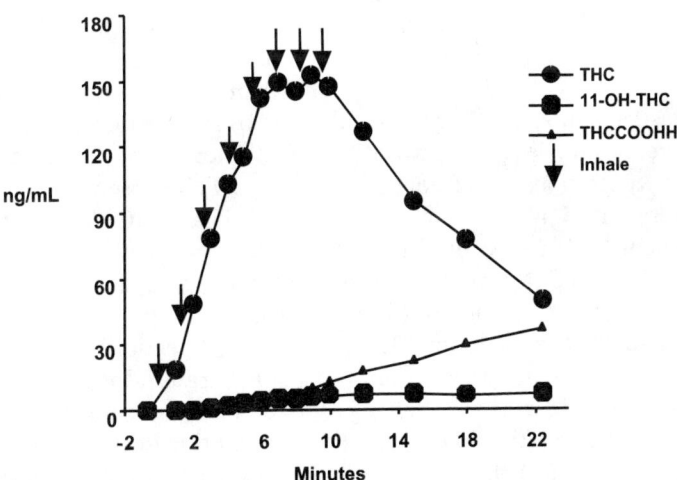

FIG. 5–2. Mean (n = 6) tetrahydrocannabinol (THC), 11-nor-Δ^9-tetrahydrocannabinol (11-OH-THC), and 11-nor-Δ^9-tetrahydrocannabinol (THCCOOH) concentrations during smoking of a 1.75% THC cigarette. Each arrow represents one inhalation or puff on the marijuana cigarette.

one MJ cigarette (1.75% or 3.55% THC), respectively. 11-OH-THC concentrations decreased gradually, with mean detection times of 4.5 hours and 11.2 hours after the two doses. THCCOOH concentrations in plasma increased slowly and plateaued for as long as 4 hours. This inactive metabolite was detected in all subjects' plasma by 8 minutes after the start of MJ smoking. Peak concentrations were consistently lower than peak THC concentrations, but were higher than 11-OH-THC peak concentrations. Peak THCCOOH concentrations were 24.5 ng/mL (71.1 nmol/L) [range 15–54 ng/mL (43.5–156.8 nmol/L)] and 54.0 ng/mL (156.8 nmol/L) [range 22–101 ng/mL (63.9–293.2 nmol/L)] after the low- and high-dose MJ cigarettes, respectively. After smoking of a 1.75% THC cigarette, THCCOOH was detected from 48 to 168 hours, with a mean of 84 hours. Detection times ranged from 72 to 168 h with a mean of 152 hours after smoking of the 3.55% THC cigarette.

Peak plasma concentrations after oral administration were reported after 2–3 hours and reached concentrations of 4.4 to 11 ng/mL (12.8–31.9 nmol/L) after a 20-mg dose (34). Law et al. (35) reported peak THC concentrations of less than 10 ng/mL (31.8 nmol/L) at 4–6 hours after ingestion of 20 mg THC. Cone et al. (36) reported increases in subjective behavioral measures after the ingestion of MJ-laced brownies. Peak effects occurred 2.5–3.5 hours after dosing.

Most THC plasma data have been collected after acute exposure; less is known of plasma THC concentrations in frequent users. Peat (38) reported plasma concentrations in frequent MJ users of THC, 11-OH-THC, and THCCOOH of 0.86 ± 0.22, 0.46 ± 0.17, and 45.8 ±13.1 ng/mL (2.7 ± 0.7, 1.4 ± 0.5, and 133.0 ± 38.0 nmol/L), respectively, a minimum of 12 hours after smoking. Furthermore, THC was detectable for as long as 6 days after smoking MJ in frequent users, but for less than 1 day in infrequent users. There were no differences in terminal half-lives in frequent or infrequent users.

Urine Concentration

Interpretation of positive urine tests requires an understanding of the excretion pattern of metabolites in humans. However, limited urinary excretion data from controlled clinical studies of MJ use are available to aid interpretation. Substantial intersubject variability occurs among subjects and among doses in patterns of THCCOOH excretion. THCCOOH concentration in the first specimen after smoking is indicative of how rapidly the metabolite appears in urine. Mean first urine THCCOOH concentrations were 47 ± 22.3 ng/mL (136.4 ± 64.7 nmol/L) and 75.3 ± 48.9 ng/mL (218.6 ±142.0 nmol/L) after smoking one 1.75 or 3.55% THC cigarette, respectively. Fifty percent of the subjects' first urine specimens after the low dose and 83% of the first urine specimens after the high dose were positive by gas chromatography–mass spectrometry (GC/MS) [15 ng/mL (43.5 nmol/L) THCCOOH cutoff concentration]. THCCOOH concentrations in the first urine specimen depend on the relative potency of the cigarette, the elapsed time after drug administration, smoking efficiency, and individual differences in drug metabolism and excretion. Mean peak urine THCCOOH concentrations averaged 89.8 ± 31.9 ng/mL (260.7 ± 92.6 nmol/L) [range 20.6–234.2 (59.8–679.9 nmol/L)] and 153.4 ± 49.2 ng/mL (445.3 ± 142.8 nmol/L) [range 29.9–355.2 (86.8–1031.2 nmol/L)] after smoking of approximately 15.8 mg and 33.8 mg of THC, respectively. The mean times of peak urine concentration were 7.7 ± 0.8 hours after the 1.75% THC dose and 13.9 ± 3.5 hours after the 3.55% THC dose. Although peak concentrations appeared to be dose related, there was a 12-fold variation among individuals.

Drug detection time, or the time after drug administration that an individual tests positive, is an important factor in the interpretation of urine drug results. Detection time depends on pharmacological factors (such as drug dose, route of administration, and rates of metabolism and excretion) and analytical factors (such as assay sensitivity, specificity, and accuracy). Mean detection times in urine after smoking vary considerably among subjects even in controlled smoking studies in which MJ dosing is standardized and smoking is computer-paced. During the terminal elimination phase, consecutive urine specimens may fluctuate between positive and negative as THCCOOH concentrations approach the cutoff concentration. After smoking a 1.75% THC cigarette, three of six subjects had additional positive urine samples interspersed between negative urine samples. This had the effect of producing much longer detection times for the last positive specimen. Using the 15 ng/mL cutoff for THCCOOH currently required by the Department of Health and Human Services, Department of Transportation, and Department of Defense for regulated urine drug testing, the mean GC/MS THCCOOH detection times for the last positive urine sample after the smoking of a single 1.75% or 3.55% THC cigarette were 33.7 ± 9.2 hours (range 8–68.5 hours) and 88.6 ± 23.2 hours (range 57–122.3 hours). GC/MS detection times were shorter than those obtained with the less-specific immunoassays.

Significant differences exist among available immunoassay products that affect the efficiency of detection of cannabinoid use. Knowledge of the sensitivity and specificity of immunoassays is essential for their proper use. Reports of prolonged drug excretion have provided the basis for the common assumption that cannabinoid metabolites may be detected in urine for a week or longer. The accuracy, sensitivity, and specificity of immunoassays for the detection of cannabinoids and metabolites are unique for a specific assay and may change over time. It is important that individuals who select assays and those who interpret test results be aware of qualitative and quantitative changes that occur. In general, detection times of cannabinoid metabolites in urine have decreased over time as the specificity of immunoassays has increased. Periodically, this variability should be assessed by determining individual cannabinoid excretion profiles with available cannabinoid immunoassays.

Urinary cannabinoid detection times vary substantially across assays, subjects, doses, and cutoff concentrations. Mean detection times across nine commercial cannabinoid assays were found to range from a maximum of one-half day after smoking one 1.75% THC cigarette to as long as a day and a half after one 3.55% THC cigarette using a 100 ng/mL immunoassay cutoff. Monitoring acute usage with a commercial cannabinoid immunoassay with a 50 ng/mL cutoff concentration provides only a narrow window of detection of 1–2 days. Mean detection times were less than 1 day after the low-dose and less than 2 days after the high-dose exposure with the 50 ng/mL cutoff. Mean detection times were longer, 1–6 days after smoking, with a 20 ng/mL cutoff immunoassay. GC/MS detection times were approximately twice as long as mean detection times with an immunoassay cutoff of 50 ng/mL. Consecutive urine voids may produce either positive or negative results when drug concentrations approach the cutoff concentration. Quantitative results, not adjusted for creatinine concentration, are subject to normal variation in excretion of cannabinoids due to differences in water content of the urine specimens. Cannabinoid concentration or dilution in the urine has an important effect on the detection time of the last positive urine and in the interpretation of urine test results.

An average of 93.9 ± 24.5 μg of THCCOOH (range 34.6–171.6 μg) was measured over a 7-day period after smoking of a single 1.75% THC cigarette. The average amount of THCCOOH excreted in the same time period after the high dose was 197.4 ± 33.6 μg (range 107.5–305.0 μg). This represented an average of only 0.54 ± 0.14% and 0.53 ± 0.09% of the original amount of THC in the low- and high-dose cigarettes, respectively. The small percentage of the total dose found in the urine as THCCOOH is not surprising considering the many factors that influence THCCOOH excretion after smoking. Before harvesting, cannabis plant material contains little active THC. When smoked, THC carboxylic acids spontaneously decarboxylate to produce THC with nearly complete conversion upon heating. Pyrolysis of THC during smoking destroys additional drug. Drug availability is further reduced by loss of drug in the sidestream smoke. It is estimated that the systemic availability of smoked THC is approximately 18% with a range of 10–14% for light users and 23–27% for heavy users.

The major route of excretion of THC and metabolites is in the feces (65%), rather than in the urine (20%). In addition, numerous cannabinoid metabolites are produced through human metabolism of THC. THC bioavailability is reduced due to the combined effect of these factors; the actual available dose is much lower than the amount of THC and THC precursor present in the cigarette.

ANALYSIS

Screening

Selection of an appropriate testing methodology requires knowledge of the type of available specimen, the analyte's metabolic profile, and the assay's characteristics. The nature and abundance of specific metabolites in different body fluids must be known. Initial testing or screening methodologies for cannabinoids in body fluids include immunoassays and thin-layer chromatography (TLC). A wide variety of immunoassays are available including enzyme immunoassays (EIA), fluorescence polarization immunoassays (FPIA), radioimmunoassays (RIA), cloned enzyme donor immunoassay (CEDIA), and kinetic interaction of microparticles in solution (KIMS). Urine is usually tested without specimen preparation due to high concentrations of drug and/or metabolites and low concentrations of other interfering components such as proteins and lipids.

The cross-reactivity of the immunoassay's antibodies to drug metabolites, including glucuronides, is important in method selection. Most of the immunoassays contain antibodies directed toward THCCOOH, the primary urinary metabolite, although cross-reactivities of the different immunoassays vary considerably toward other THC metabolites. This specificity toward THCCOOH makes these assays appropriate for analysis of urine because little to no parent THC or 11-OH-THC metabolite is present in urine. A high percentage of cannabinoid metabolites are excreted in the urine as glucuronides; however, the combined cross-reactivities of the glucuronide metabolites and the abundance of free carboxy metabolites have provided adequate sensitivity to forego hydrolysis of the urine during screening. Initial test cutoffs for screening cannabinoids in urine include 20, 50, and 100 ng/mL.

In April 1988, the Health and Human Services (HHS) guidelines for testing urine for the federal sector established the initial test (screen) cutoff for cannabinoids at 100 ng/mL. However, a more-sensitive 50 or 20 ng/mL cutoff is used in many other testing environments. HHS subsequently lowered the required initial test cutoff for cannabinoids to 50 ng/mL.

Saliva also is a suitable specimen for screening for the presence of cannabinoids. However, the immunoassay should be directed toward detection of the parent THC compound, rather than the 11-OH-THC or THCCOOH metabolites. The oral mucosa serves as a depot for THC after MJ smoking. Only minor amounts of drug and metabolites diffuse from the plasma into the saliva. No measurable (limit of quantification 0.5 ng/mL) 11-OH-THC or THCCOOH was found in the saliva for 7 days after MJ smoking when a sensitive GC/MS procedure was used for analysis. Specific RIAs targeted for the parent THC analyte are available for testing saliva and require little to no specimen preparation. Depending on the assay, a protein precipitation step may be required to improve assay performance.

Whole blood, plasma, and tissues may also be analyzed for THC, 11-OH-THC, THCCOOH, and other metabolites. Whole blood cannabinoid concentrations are approximately one-half the concentrations found in plasma specimens due to the low partition coefficient of drug into erythrocytes. Simple preparation steps, such as protein precipitation or single-step solvent extraction, may be required with some methodologies; more extensive extraction and concentration schemes may be required for the use of other immunoassays. Many different immunoassays have been adapted for testing cannabinoids in blood, plasma, or tissue, including RIAs and EIAs. Some immunoassay reagents contain antibodies that cross-react with the parent THC compound; other reagents contain antibodies to the inactive THCCOOH metabolite. THC cross-reacts poorly with antibodies found in most of the commercially available reagents, and a small number of cases containing high THC concentrations and low THCCOOH concentrations, such as specimens collected immediately after smoking, may produce negative results when assayed with THCCOOH-specific reagents. However, THCCOOH concentrations begin to rise during smoking, increase over time, and have a much longer time course of detection than the parent compound.

TLC is also used as a sensitive screening method for cannabinoids in urine. Early studies of THC metabolism used classical TLC methods. Commercial applications of these systems are available for the identification of THCCOOH in urine, and although the time required for testing and the hands-on technologist time may be greater than with immunoassays, several specimens may be tested simultaneously and at a reduced cost in some instances. TLC methods are specific for the THCCOOH metabolite and include alkaline hydrolysis, extraction, concentration, separation, and visualization steps. Fast Blue BB staining reagent is the reagent of choice for visualization; sensitivity limits of 5–10 ng/mL (14.5–29.0 nmol/L) THCCOOH have been achieved.

Confirmation

A chemical technique that is based on a different scientific principle from the chemical technique used in the initial test is required for confirmation of cannabinoid results, for example, immunoassay for the initial test and chromatography for confirmation. Confirmation methodologies include TLC, gas chromatography (GC), high performance liquid chromatography (HPLC), and GC/MS. Ideally, the sensitivity of the confirmation assay should be equal to or greater than the sensitivity of the initial test method. Most forensic toxicology applications confirm results by GC/MS. Selected ion monitoring, full-scan ion-monitoring, chemical-ionization methods, and direct probe insertion GC/MS have been used for the confirmation of cannabinoids in body tissues.

Cannabinoids are difficult to analyze due to their high lipophilicity and low concentration in body fluids. Complex specimen matrices may require multistep extractions to separate cannabinoids from lipids and proteins. Cannabinoid extraction techniques include liquid-liquid extractions and solid-phase extraction with bonded silica, XAD-2, and anion-exchange columns. Recovery may decrease due to the high affinity of cannabinoids for glass and plastic containers. Specimen preparation for urine cannabinoid testing includes either enzymatic hydrolysis with β-glucuronidase or, more commonly, alkaline hydrolysis to free THCCOOH from the glucuronide conjugate. Most GC/MS confirmation procedures in urine use a 15 ng/mL cutoff and are specific for THCCOOH. Some procedures have also used a hydrolysis step in the analysis of cannabinoids in blood.

The importance of glucuronide derivatives of cannabinoids in blood remains a contested issue. The efficiency of glucuronide hydrolysis in cannabinoid extraction methods should be routinely evaluated by inclusion of a THCCOOH glucuronide quality-control sample. This sample can be prepared from a pool of cannabinoid-positive specimens or from THCCOOH-glucuronide–spiked specimens.

Cannabinoids have been detected with reverse-phase HPLC, with ultraviolet, electrochemical, mass spectrometric, and RIA detection. High specificity has been achieved with adequate resolution of cannabinoid analytes and separation from interfering endogenous compounds; however, in many procedures sensitivity has been a limiting factor.

Several GC methods for both packed and capillary columns are available, although for forensic purposes, GC/MS has supplanted GC analyses. One of the most widely used GC methods for THCCOOH in urine uses flame ionization detection of the methyl ether-methyl ester derivative [detection limit 20 ng/mL (58.1 nmol/L)]. Improved sensitivity has been obtained with electron capture detection of the pentafluoro derivative of THCCOOH.

GC/MS confirmation of THC, 11-OH-THC, and THCCOOH provides adequate sensitivity and specificity in a wide variety of biological tissues. Several deuterated THCCOOH materials are available as internal standards and are recommended for the highest accuracy of results. Compounds with multiple deuterium ions are available and provide adequate resolution between the deuterated and native ions for use in full-scan ion-monitoring techniques. Carboxyl and hydroxyl groups on the THC, 11-OH-THC, and THCCOOH molecules require derivatization to increase volatility and improve chromatographic performance. Trimethyl silyl and methyl derivatizing reagents are frequently used to achieve acceptable chromatography and sensitivity. Electron-impact mass spectrometry (MS) of cannabinoids after alkaline hydrolysis and derivatization with one of a wide variety of derivatization reagents was made achievable for most forensic toxicology laboratories due to the availability of low-cost benchtop GC/MS instruments. The increased selectivity of chemical-ionization techniques is also readily available with the release of several new benchtop instruments. One of the most sensitive methods for detection of THCCOOH uses negative ion chemical-ionization MS of the methyl ester trifluoroacetate derivative. Sensitivity limits of 0.1 ng/mL (0.29 nmol/L) THCCOOH in blood, plasma, or urine can be achieved due to the high ionization efficiency and selectivity. Metastable ion detection techniques are also available for applications that require highly sensitive methods, for example, pharmacokinetic studies.

RIA, enzyme-multiplied immunoassay, and fluorescence polarization immunoassay screening methods have been used with sensitive GC/MS or GC/MS/MS confirmation methodologies in the analysis of cannabinoids in hair. Most studies have focused on identifying the most abundant cannabinoid analyte in hair, the more neutral and lipophilic parent compound, THC. However, other methods target THCCOOH, which is present in much lower concentrations in the hair. An advantage of measuring THCCOOH in hair is the reduced possibility of external contamination of hair from this metabolite.

Analysis of cannabinoids in hair has challenged the sensitivity limits of immunoassay and confirmation assays; GC/MS/MS has been required in most cases to increase the confirmation rate of presumptive-positive results.

INTERPRETATION

Urine

Detection of cannabinoids in urine is indicative of prior MJ exposure, but the long excretion half-life of THC in the body, especially in chronic MJ users, makes it difficult to predict the timing of past drug use. In a single extreme case, one individual's urine was positive at a concentration greater than 20 ng/mL (63.6 nmol/L) by immunoassay as long as 67 days after the last drug exposure (39). This individual had used MJ heavily for more than 10 years. However, a naive user's urine may be found negative by immunoassay only a few hours after the smoking of a single MJ cigarette. Assay cutoff concentrations affect drug-detection times. Another component to the detection of MJ use is the sensitivity and specificity of the immunoassay system used. In general, immunoassay reagents have become more specific for THCCOOH over time. This has resulted in improved correlation among immunoassays, and also among immunoassays using 50 ng/mL (145.2 nmol/L) initial test cutoffs and confirmation procedures with a 15 ng/mL (43.5 nmol/L) cutoff concentration. However, increased specificity also has contributed to the noted decrease in the time course of detection of MJ use. Nevertheless, a positive urine test for cannabinoids indicates only that drug exposure has occurred. The result does not provide information on the route of administration, the amount of drug exposure, when drug exposure occurred, or the degree of impairment.

Normalization of the cannabinoid drug concentration to the urine creatinine concentration aids in the differentiation of new from prior drug use. Due to the long half-life of drug in the body, especially in chronic MJ users, toxicologists and practitioners are frequently asked to determine whether a positive urine test represents a new episode of drug use or represents continued excretion of residual drug. Random urine specimens contain varying amounts of creatinine depending on the degree of concentration of the urine. An increase of more than 50% in the creatinine-normalized cannabinoid concentration above the previous specimen is considered to be indicative of a new episode of drug exposure in individuals who are infrequent users.

Recently, it has been proposed that the amount of THC and 11-OH-THC measured in urine samples can be related to the elapsed time after MJ smoking. Hydrolysis of urinary cannabinoid glucuronides with β-glucuronidase from *Escherichia coli* (bacteria) before GC/MS analysis resulted in higher concentrations of THC and 11-OH-THC than with β-glucuronidase from *Helix pomatia*. Further research is necessary to determine the validity of estimating time of MJ use from THC and 11-OH-THC concentrations in urine.

Saliva

Detection times of cannabinoids in saliva are shorter than in urine and more indicative of recent MJ use. Saliva THC concentrations correlate temporally with plasma cannabinoid concentrations and behavioral and physiological effects, but wide intra- and interindividual variation precludes the use of saliva concentrations as indicators of drug impairment. THC may be detected at low concentrations by RIA for as long as 24 hours after use. Saliva THC concentrations greater than 10 ng/mL (31.8 nmol/L) may be useful indicators of recent MJ exposure.

Blood, Plasma, and Serum

Scientific advances have improved our ability to identify and quantify cannabinoids in body fluids; however, the interpretation of results remains a difficult task. Forensic scientists receive frequent requests to interpret the significance of cannabinoid concentrations in blood specimens from individuals involved in accidents, criminal investigations, and traffic violations. Relevant facts, such as the amount of drug used, route of administration, and history of use, generally are unknown. To date, a practical presumptive concentration of blood THC cannot be related to a measurable concentration of impairment as is possible with blood ethanol concentrations (39). Due to chemical and pharmacokinetic differences between MJ and ethanol, we cannot use ethanol as a model for relating drug concentrations to effects. Consequently, the pattern of distribution to and elimination from active sites of these two molecules are quite different. Pharmacokinetic and pharmacodynamic models that account for the dispositional differences of THC may be more successful in defining blood concentrations that can be associated with the psychoactive effects of THC (18).

Although there continues to be controversy in the interpretation of blood cannabinoid results, some general concepts have wide support. A dose-response relationship has been demonstrated for smoked THC and THC plasma concentrations (40). It is well established that plasma THC concentrations begin to decline before the time of peak effects, although it has recently been shown that THC effects appear rapidly after initiation of smoking (10). Individual drug concentrations and ratios of cannabinoid metabolite to parent drug concentration have been suggested as potentially useful indicators of recent drug use (35,41). The ratio of plasma THCCOOH to THC was found to exceed one, 45 minutes after MJ smoking (42). This is in agreement with results reported by Mason and McBay (43) and Huestis et al. (27), who found that peak effects occurred when THC and THCCOOH concentrations reached equivalency, within 30 to 45 minutes after initiation of smoking. Measurement of cannabinoid analytes with short time courses of detection, such as 8β,11-dihydroxy-tetrahydrocannabinol, as markers of recent exposure has not found widespread use (44). Recent exposure (6 to 8 hours) and possible impairment have been linked to plasma THC concentrations in excess of 2–3 ng/mL (27,44,45). Gjerde and Kinn (46) suggested that 1.6 ng/mL (5.1 nmol/L) THC in whole blood may indicate possible impairment. This correlates well with the suggested concentration of plasma THC, due to the fact that THC in hemolyzed blood is approximately one-half the concentration of plasma THC (43). Interpretation is further complicated by residual THC and THCCOOH concentrations found in blood of frequent MJ users. In general, it is suggested that chronic MJ smokers may have residual plasma THC concentrations of less than 2 ng/mL (6.4 nmol/L) 12 hours after smoking the last MJ cigarette (37). Significantly higher residual concentrations of THCCOOH may be found.

Accurate prediction of the time of MJ exposure would provide valuable information in establishing the role of MJ as a contributing factor in events under investigation. Recently, two mathematical models have been described for the prediction of time of MJ use from the analysis of a single plasma specimen for cannabinoids (47). Model I was based on THC concentrations and Model II was based on the ratio of THCCOOH to THC in plasma. Model I and Model II correctly predicted the time of exposure within the 95% confidence interval for more than 90% of the specimens evaluated. Plasma THC and THCCOOH concentrations were evaluated after oral and smoked MJ exposure, in frequent and infrequent MJ smokers, and with measurements obtained by a wide variety of methods, including RIA and GC/MS. Plasma THC concentrations less than 2.0 ng/mL (6.4 nmol/L) were excluded from use in both models due to the possibility of residual THC concentrations in frequent smokers. These models may be used for the prediction of time of MJ use from cannabinoid plasma concentrations within defined confidence limits and may be beneficial to forensic scientists in the interpretation of cannabinoid plasma concentrations.

In Vitro Stability

One concern in the analysis of THC and metabolites in biological specimens is the stability of these compounds over time. THC is a lipophilic molecule and binds to hydrophobic surfaces. This can potentially cause a reduction in concentration depending on the type of storage container. It is recommended that blood or plasma specimens be stored in glass tubes rather than plastic tubes, because THC may adsorb to plastic tubes, thus reducing the measured concentration.

However, stability studies in blood or plasma have found that, in general, stability of concentration has not been a problem. THC remained stable in blood stored at 41, 23, and -4 °F (5, -5, and -20 °C) for as long as 17 weeks (48). Degradation began to occur after this time; the drug was not detected by 23 weeks. Temperature conditions did not influence recovery of the drug, nor did repetitive freeze-thaw cycles. The authors hypothesized that decreasing THC concentrations may have been due to unsuccessful extractions of the drug due to irretrievable binding to degrading proteins.

Another study on the stability of THC, 11-OH-THC, and THCCOOH in blood and plasma found no significant changes in concentrations for the first month of storage at room temperature, 39 °F (4 °C), or 14 °F (-10 °C) (49). However, THC and 11-OH-THC concentrations decreased significantly by 2 months with room-temperature storage. Blood stored at 39 °F (4 °C) showed no significant changes for 4 months, but extractions after 6 months demonstrated poor precision and inefficient extraction. Cannabinoid concentrations in blood and plasma stored at 14 °F (-10 °C) and plasma stored at 39 °F (4 °C) were stable for as long as 6 months. Furthermore, studies evaluating contact between the specimen and the collection tube's rubber stopper noted that the size of the blood specimen, storage temperature, and extent of contact with the rubber stopper did not affect cannabinoid concentration for as long as 24 hours.

THCCOOH was found to be stable for as long as 30 days in blood-collection tubes containing ethylenediaminetetraacetic acid, heparin, or sodium fluoride and in tubes with no anticoagulant at refrigerated and room temperatures (50). In addition, repetitive freeze-thaw cycles did not contribute to loss of this analyte. THC, 11-OH-THC, and THCCOOH concentrations in blood and plasma are stable for at least 1 month when stored at room, refrigerated, or frozen temperatures. After this time, reductions may be seen if the blood is stored at room temperature. This stability in blood or plasma may be attributed to extensive protein binding, which limits loss via adsorption to container surfaces.

Romberg (51) evaluated the stability of THCCOOH-spiked urine specimens stored at room temperature, under nitrogen, in the dark, or exposed to light. THCCOOH concentrations were within 10% of the initial concentration after 4 weeks of storage and within 20% after 7 weeks. In another study by Romberg and Past (52), THCCOOH concentrations in frozen routine urine specimens stored over variable time periods were noted to decrease an average of 24%, although considerable variability was observed. The percent change in concentration ranged from an increase of 28% to a decrease of 80%. There was no correlation between the decrease in concentration and the time in frozen storage. The stabilities of in-house prepared frozen control solutions also were evaluated at pH 6.0, 7.5, and 9.0. THCCOOH concentrations were stable over the 9-month evaluation period, except for the pH 9.0 control that decreased by 30%. Thawing and refreezing did not affect the stability of the drug in the control solutions. Other studies have indicated that THCCOOH is stable for at least 6 months at $-4\,°F\,(-20\,°C)$ (53). The most consistent results were obtained by thoroughly thawing the urine and sonicating before extraction. THCCOOH can concentrate in urine sediment and in foam formed upon shaking.

Passive Inhalation

Environmental exposure to MJ smoke can occur through passive inhalation by nonusers of sidestream and exhaled smoke. Several research studies have indicated that it is possible to produce detectable concentrations of MJ metabolites in the urine and plasma after passive inhalation of MJ smoke. It is generally agreed that passive exposure is not a valid explanation for a positive urine test (54). In studies in which positive urine drug tests were obtained, exposure conditions have been characterized as unrealistic (55). In one case, the smoke was so intense that the subjects required goggles to protect their eyes. Lowering the urine cannabinoid initial test cutoff from 100 to 50 ng/mL significantly increases the identification of true-positive specimens and increases drug detection time after MJ exposure.

IMPAIRMENT

It would be helpful if a drug concentration in a driver's blood could be linked to impairment of driving performance. However, interpretation of the data is complicated by factors such as drug interactions and tolerance, age and health of the driver, driving experience, and road and weather conditions. Evaluation of MJ's effects on driving has taken three approaches: epidemiological studies of MJ use and accident or fatality rates; performance studies of cognitive or psychomotor impairment; and driving or flying simulator or closed- or open-course driving tests. Each of these approaches offers a different perspective on MJ's effects on driving performance.

Epidemiological Studies

Epidemiological studies of MJ use and motor vehicle fatalities or injuries, driving under the influence of drugs (DUID), random safety checks, and occupational fatalities are included in Table 5–1. Most statistical studies that examine the frequency of accidents or fatalities in individuals who have consumed drugs lack proper control groups

TABLE 5-1. *Epidemiological studies*

Study population	Findings (percent positive for various drugs)	Blood concentrations
Fatalities in Ontario, Canada[a] (n = 401)	3.7% THC; 0.2% only THC; usually THC with ethanol; 10.7% positive for urine cannabinoids, but no THC in blood	13/15 blood THC <5 ng/mL
Single-vehicle fatalities in North Carolina[b] (n = 600)	7.8% THC; 0.7% only THC	RIA and EMIT screen >3 ng/mL (range 3.1–37 ng/mL)
Fatalities in young male California drivers[c] (n = 440)	37% THC or metabolite; 4.3% THC only drug; MJ's role in crash responsibility undetermined	THC ng/mL: 38% 0.2–0.9; 22% 1.0–1.9; 26% 2.0–4.9; 14% >5.0
Fatalities in Ontario, Canada[d] (n = 1169)	10.9% THC; 1.7% only THC	RIA and GC/MS
Fatalities in Los Angeles County[e]	19% THC; 41.4% ethanol	—
Motor vehicle fatalities in Rhode Island[f]	59% ethanol; 13% alcohol and drugs; 6% drugs only	
Motor vehicle fatalities in Norway[g]	5% THC, most common drug; 20% impaired by alcohol; 8% by alcohol and drugs	
Road accidents in Tasmania[h]	6% THC	RIA
Trauma patients in Maryland[i] (n = 1023)	37.4% THC; 16.5% alcohol and THC; 18.3% THC alone; similar percentage in vehicular and nonvehicular trauma patients	RIA ≥2 ng/mL, some GC/MS confirm; THC ng/mL: 39.7% 2–4.9; 25.6% 5–9.9; 34.6% ≥10.0
Trauma center admissions; injured drivers in Knoxville, Tennessee[j]	BAC 37%; other drugs 40%; >50% of BAC-positive also positive for other drugs	—
Motor vehicle injuries in Toronto, Canada[k]	13.9% THC; 40.5% positive for drug besides ethanol	—
DUID in California[l]	14.4% THC; 23% of ethanol negative; all THC positives failed roadside sobriety test	Blood THC ≥5.5 ng/mL, range 5.5–23 ng/mL, RIA
DUID in St. Louis, Missouri[m] Negative ethanol	47% of positive drug cases THC; impaired driving resulted in police stop	Urine cannabinoids
DUID in Norway[n]	26% suspected alcohol positive for cannabinoids; 13% confirmed THC; 43% DUID positive for THC; 15–20% with BAC <0.05% influenced by drugs	—

DUID negative for ethanol[o]	56% THC; 82% positive for more than one drug	EMIT and GC/MS, 50% <5 ng/mL THC
DUID in Pennsylvania and Washington, DC[p]	39% and 50% THC; 50% THC metabolite; GC/MS; 3.1% THC only; also THC and PCP, THC and cocaine; blood sampling >1 h after arrest; lower THC-positive rate	RIA cutoff 10 ng/mL; GC/MS confirm
Tractor-trailer drivers[q]	15% cannabinoids; 3% blood THC; ethanol <1%	Blood or serum THC 2.5–12 ng/mL
Occupational fatalities in Alberta, Canada[r]	8.5% urines positive cannabinoids; only illicit drug	—
Occupational fatalities in Ontario, Canada[s]	17% of those tested THC positive	—

[a] Cimbura G, Lucas DM, Bennett RC, et al: Incidence and toxicological aspects of drugs detected in 484 fatally injured drivers and pedestrians in Ontario. *J Forensic Sci* 1982;27:855–867.
[b] Mason AP, McBay AJ: Ethanol, marijuana, and other drug use in 600 drivers killed in single-vehicle crashes in North Carolina, 1978–1981. *J Forensic Sci* 1984;29:987–1026.
[c] Williams AF, Peat MA, Crouch DJ, et al: Drugs in fatally injured young male drivers. *Public Health Rep* 1985;100:19–25.
[d] Cimbura G, Lucas DM, Bennett RC, et al: Incidence and toxicological aspects of cannabis and ethanol detected in 1394 fatally injured drivers and pedestrians in Ontario (1982–1984). *J Forensic Sci* 1982;35:1035–1041.
[e] Budd RD, Muto JJ, Wong JK: Drugs of abuse found in fatally injured drivers in Los Angeles County. *Drug Alcohol Depend* 1989;23:153–158.
[f] Haley NR, Iwuc PS, Ogilvie LM, et al: Motor vehicle fatalities in Rhode Island (FY 1990–1991): a report on driver impairment. *RI Med J* 1992;75:397–400.
[g] Gjerde H, Behlich KM, Morland J: Incidence of alcohol and drugs in fatally injured car drivers in Norway. *Accid Anal Prev* 1993;25:479–483.
[h] McLean S, Parsons RS, Chesterman RB, et al: Drugs, alcohol and road accidents in Tasmania. *Med J Aust* 1987;147:6–11.
[i] Soderstrom CA, Triffillis AL, Shankar BS, et al: Marijuana and alcohol use among 1023 trauma patients. *Arch Surg* 1988;123:733–737.
[j] Kirby JM, Maull KI, Fain W: Comparability of alcohol and drug use in injured drivers. *South Med J* 1992;85:800–802.
[k] Stoduto G, Vingilis E, Kapur BM, et al: Alcohol and drug use among motor vehicle collision victims admitted to a regional trauma unit: demographic, injury, and crash characteristics. *Accid Anal Prev* 1993;25:411–420.
[l] Zimmermann EG, Yeager EP, Soares JR, et al: Measurement of Δ9-tetrahydrocannabinol in whole blood samples from impaired motorists. *J Forensic Sci* 1983;28:957–962.
[m] Poklis A, Maginn D, Barr JL: Drug findings in driving under the influence of drugs cases: a problem of illicit drug use. *Drug Alcohol Depend* 1987;20:57–62.
[n] Christophersen AS, Gjerde H, Bjorneboe A, et al: Screening for drug use among Norwegian drivers suspected of driving under influence of alcohol or drugs. *Forensic Sci Int* 1990;45:5–14.
[o] Gjerde H, Kinn G: Impairment in drivers due to cannabis in combination with other drugs. *Forensic Sci Int* 1991;50:57–60.
[p] Sutton LR, Paegle I: The drug impaired driver, detection and forensic specimen analysis. *Blutalkohol* 1992;29:134–138.
[q] Lund AK, Preusser DF, Blomberg RD, et al: Drug use by tractor-trailer drivers. *J Forensic Sci* 1988;33:648–661.
[r] Alleyne BC, Stuart P, Copes R: Alcohol and other drug use in occupational fatalities. *J Occup Med* 1991;33:496–500.
[s] Shannon HS, Hope L, Griffith L, et al: Fatal occupational accidents in Ontario, 1986–1989. *Am J Ind Med* 1993;23:253–264.

from the population at risk. The value of these studies in predicting possible impairment and linking drug use to increased accidents or fatalities is weakened because incidence rates of MJ use in control populations are unknown.

However, data are available on MJ usage rates in the general population and on usage rates shortly before or while driving. Twenty-nine percent of Massachusetts teenagers drove after using MJ within the previous month in 1979 and 21% in 1981, as measured by self-report (56). In Boston, 43% of teenagers reported that they were the driver or a passenger in a car in which the driver was under the influence of MJ in 1981–1982 (57). Furthermore, 6.3% of University of Illinois students operated a motor vehicle on a weekly or daily basis shortly after or while using MJ (58).

Other difficulties in interpreting epidemiological data relate to the selection of assay methods, cutoff concentrations, and specimens to be tested for drugs. Some studies only evaluated specimens for drugs if the blood alcohol concentration (BAC) was less than the legal limit. In some cases, blood was not obtained until after negative results were obtained on a breath alcohol test. Furthermore, some studies measured cannabinoids in urine, others in whole blood or plasma, and some used confirmatory methods, whereas others did not. Assay detection limits also varied considerably.

Despite these limitations, a number of conclusions can be drawn from the data presented in Table 5–1. THC was found in the blood of 3.7–37% of fatalities; THC was the only drug found in 0.2–4.3% of the cases. Most THC concentrations were low, <5 ng/mL (<15.9 nmol/L). Sutton (59) suggested this result was due to the rapid decline of THC concentrations after smoking and the time required to obtain blood specimens. THC was found in combination with ethanol or other drugs in a high percentage of the cases. In motor vehicle injuries, 40–50% of casualties were positive for at least one drug other than ethanol; THC was the most common drug found. In a prospective study of 1023 trauma cases in Maryland, 34.7% of subjects had greater than 2 ng/mL (6.4 nmol/L) THC in their serum; more than one-third of these subjects had serum THC concentrations of 10 ng/mL (31.8 nmol/L) or more (60). A high percentage of DUID cases were positive for THC, especially in specimens negative for BAC. Driving impairment was evident in THC-positive cases, as demonstrated by failures in roadside sobriety checks (61) and police stops for erratic driving (62). Random stops of tractor-trailer drivers indicated a 15% positive rate for cannabinoids, with 3% positive for blood THC ≥2.5 ng/mL (≥7.9 nmol/L) (64). MJ is not the drug of choice for professional drivers due to its sedative effects; therefore, this rate may not reflect general population use. In two studies of occupational fatalities in Canada (64,65), cannabinoids were found in 8.5–17% of those tested; MJ use in the general population was judged to equal or exceed these rates. Furthermore, these incident rates were based in part on urine cannabinoid studies. MJ's contribution to the cause of the fatalities was judged to be limited.

Drugs other than ethanol must be recognized as important contributors to driving impairment. MJ was the most frequently identified drug, other than ethanol, in studies of drivers involved in unsafe driving practices. Frequently, MJ and ethanol are found together, and the effects of both drugs must be considered in evaluating possible impairment. Critical skills needed for the safe operation of motor vehicles and other forms of transport can be impaired after MJ use. Research studies have indicated that MJ may decrease accuracy in cognitive and psychomotor tests, impair time estimation, and reduce learning and memory functions. MJ concentrations decline rapidly, requiring rapid collection of blood specimens to evaluate the drug's contribution to driving impairment.

Performance Studies

Controlled clinical studies of cognitive and psychomotor performance after MJ use are included in Table 5–2. Relating MJ-induced

TABLE 5–2. *Performance effects*

Task	Findings	Dose	Blood THC
Pursuit meter[a]	Significant impairment	5 mg smoke	Unable to detect in TLC procedure
Delayed auditory feedback[a]	5/9 significant increased errors	0, 2.5, 5 mg THC smoke	Not applicable
Pursuit meter[b]	Impairment both doses, but poor dose-response	0, 0.13 g/kg ethanol	Not applicable
Delayed auditory feedback[b]	Additive effect with ethanol	0, 12.5, 25, 50, 100 μg/kg smoke	Not applicable
Pursuit meter (hand-eye coordination)[c]	Significant linear dose-dependent relationship		
Delayed auditory feedback[c]	4/9 linear dose-dependent decrements; 5/9 no effect		Not applicable
Stability, wobble board[c]	Decreased stability with dose	25 μg/kg smoke	Not applicable
Recall of narrative material[d]	Significant impairment; addition of irrelevant material		Not applicable
Visual autokinetic phenomenon (motion of stationary light source)[e]	Apparent movement increased with two high doses	0, 50, 100, 200 μg/kg smoke	Not applicable
Information processing[f]	Disrupted short-term memory	20, 40, 60 mg oral	Not applicable
Information processing[g]	Dose-related impaired peripheral vision detection	0, 50, 100, 200 μg/kg smoke	Not applicable
Broad battery of psychomotor and cognitive tasks[h]	Attention, concentration, and motor function impairment; reduced adaptability in stressful situations; effects noticeable 8–10 hours after smoking	0, 350, 400, 450 μg/kg oral	Not applicable
Pursuit meter[i]	Significant impairment	0, 6, 12, 18 μg/kg smoke	Not applicable
Delayed auditory feedback[i]	No consistent change		
Stability, wobble board[i]	Decreased stability with dose		
Pursuit meter[j]	Impaired performance with marijuana	0, 25 mg THC smoke; 0, 10 mg/70 kg amphetamine	Not applicable
Wobble board[j]	No change in effects when amphetamine added		
Delayed auditory feedback[j]			
Psychomotor tracking[k]	Dose-related errors in simple and complex tracking; THC and ethanol additive effects on tracking; no effects on observer-rated car-handling ability	0, 1.6, 6.8 mg smoke; 0.3, 0.7% blood ethanol	Not applicable
Information processing[l]	Inconsistent effects in simple and complex tasks; slowed with high dose only	0, 10, 20 mg smoke	Not applicable
Reaction time[l]			
Standing steadiness, pursuit rotor[m]	Significant impairment THC in all measures; ethanol and THC additive effects	215 μg/kg oral; 0.54 g/kg ethanol	Not applicable
Simple and complex reaction times[m]			
Vienna Determination Apparatus[m]			
Information processing[n]	Impaired speed of visual information processing	6 mg smoke	Not applicable
Roadside sobriety tests[o]	Failures if THC >25 ng/mL 94% failed at 90 min and 60% at 150 min after smoking	Ad lib smoke	49.1% negative THC by RIA 150 min after smoking
Coordination tests[o]			

(Table continued on next page.)

TABLE 5-2. Performance effects (Continued)

Task	Findings	Dose	Blood THC
Memory[p]	Dose-related decreases in immediate and delayed free recall; increase in intrusion errors; no reduction inrecognition memory	0, 5, 10, 14 mg smoke	Not applicable
Concentrated attention[q]	Significant and prolonged drug impairment	0, 50, 100, 200 μg/kg smoke	Not applicable
Divided attention[q]	Significant impairment		
Divided attention, tracking errors[r]	Significant linear correlation 5–25 ng/mL for 2 h	100, 200, 250 μg/kg smoke	Collected same conditions but different times
Critical tracking breakpoint[r]	Sigmoid relation with log THC over 2–25 ng/mL for 7 h		
Pursuit meter[s]	No significant difference between doses, decrement with both doses; impairment with THC >5 ng/mL	37.5, 75 μg/kg smoke	RIA, 0–120 min, THC and THCCOOH
Body sway, pursuit rotor[t] Simple and choice reaction time[t] Addition and subtraction[t]	At these dosages and with these tests, potency ranged as follows: smoked MJ > oral MJ > ethanol; 0.05% BAC ≈ 8.5 mg oral MJ ≈ 1.5 mg smoked MJ	0, 5, 10, 15, 20 mg THC oral and smoke 0, 0.75, 1.0 g/kg ethanol	Not applicable
Perceptuomotor tasks[u]	Heavy cannabis users reacted more slowly	Not applicable	Not applicable
Intelligence and memory[u] Memory[v]	No differences	Not applicable	Not applicable
	Short-term memory deficits in cannabis-dependent adolescents for >6 weeks		
Circular lights[w] DSST[w]	Low and high dose MJ decreased response speed but not accuracy on DSST; high-dose ethanol impaired all tasks	0, 1.3, 2.7% THC smoke; 0, 0.6, 1.2 g/kg ethanol	Not applicable
Automated tracking task[w]			
Digit span (immediate recall)[x] Divided attention[x] DSST[x]	Inconsistent results for two doses on digit span; no effect on divided attention task; impairment on DSST with high dose only	0, 1.3, 2.7% THC smoke	Not applicable
Information processing[y]	Long-term cannabis users decreased ability to filter out irrelevant material and to focus attention	Not applicable	Not applicable
Cognitive tasks: serial addition and subtraction, logical reasoning, manikin[z] Short-term memory and attention: matrix and serial addition and subtraction[z] Time estimation: time wall[z] Psychomotor: DSST[z]	No MJ effects on serial addition and subtraction, matrix, manikin, DSST, and time wall; decreased accuracy on the logical reasoning task	0. 1.75, 3.55% THC smoke	Comprehensive GC/MS data during absorption to 7 d postdrug
Repeated acquisition task[aa] Simple and choice reaction times[aa] DSST[aa]	Combination high cocaine and MJ increased errors; prolonged subjective effects	0, 2.7% THC smoke; 0, 16, 32 mg intravenous cocaine	Not applicable

[a] Manno JE, Kiplinger GF, Haine SE, et al: Comparative effects of smoking marihuana or placebo on human motor and mental performance. Clin Pharmacol Ther 1970;11:808–815.
[b] Manno JE, Kiplinger GF, Scholz N, et al: The influence of alcohol and marihuana on motor and mental performance. Clin Pharmacol Ther 1971;12:202–211.
[c] Kiplinger GF, Manno JE, Rodda BE, et al: Dose-response analysis of the effects of tetrahydrocannabinol in man. Clin Pharmacol Ther 1971;12:650–657.
[d] Drew WG, Kiplinger GF, Miller FF, et al: Effects of propranolol on marihuana-induced cognitive dysfunctioning. Clin Pharmacol Ther 1972;13:526–533.
[e] Sharma S, Moskowitz H: Effect of marihuana on the visual autokinetic phenomenon. Percept Mot Skills 1972;35:891–894.
[f] Tinklenberg JR, Kopell BS, Melges FT, et al: Marijuana and alcohol: time production and memory functions. Arch Gen Psychiatry 1972;27:812–815.
[g] Moskowitz H, Sharma S, McGlothlin W: Effect of marihuana upon peripheral vision as a function of the information processing demands in central vision. Percept Mot Skills 1972;35:875–882.
[h] Kielholz P, Hobi V, Ladewig D, et al: An experimental investigation about the effect of cannabis on car driving behaviour. Pharmakopsychiatria 1973;6:91–103.
[i] Evans MA, Martz R, Brown DJ, et al: Impairment of performance with low doses of marihuana. Clin Pharmacol Ther 1973;14:936–939.
[j] Forney R, Martz R, Lemberger L, et al: The combined effect of marihuana and dextroamphetamine. Ann N Y Acad Sci 1976;281:162–170.
[k] Hansteen RW, Miller RD, Lonero L: Effects of cannabis and alcohol on automobile driving and psychomotor tracking. Ann N Y Acad Sci 1976;282:240–256.
[l] Schaefer CF, Gunn CG, Dubowski KM: Dose-related heart-rate, perceptual, and decisional changes in man following marihuana smoking. Percept Mot Skills 1977;44:3–16.
[m] Bird KD, Boleyn T, Chesher GB, et al: Intercannabinoid and cannabinoid-ethanol interactions and their effects on human performance. Psychopharmacology (Berl) 1980;71:181–188.
[n] Braff DL, Silverton L, Saccuzzo DP, et al: Impaired speed of visual information processing in marijuana intoxication. Am J Psychiatry 1981;138:613–617.
[o] Reeve VC, Grant JD, Robertson W, et al: Plasma concentrations of Δ9-tetrahydrocannabinol and impaired motor function. Drug Alcohol Depend 1983;11:167–175.
[p] Miller LL: Marijuana: acute effects on human memory. In: Agurell S, Dewey W, Willette R, eds: The cannabinoids: chemical pharmacologic, and therapeutic aspects. New York: Academic Press, 1984:21–46.
[q] Moskowitz H: Attention tasks as skills performance measures of drug effects. Br J Clin Pharmacol 1984;18:51S–61S.
[r] Barnett G, Licko V, Thompson T: Behavioral pharmacokinetics of marijuana. Psychopharmacology (Berl) 1985;85:51–56.
[s] Manno JE, Ferslew KE, Franklin LS, et al: Human pursuit tracking performance (PTP) and plasma concentrations of delta-9-tetrahydrocannabinol (THC) and 11-nor-delta-9-tetrahydrocannabinol-9-carboxylic acid (Nor-COOH THC) after smoking marijuana. In: Harvey DJ, ed: Marihuana '84. Oxford, UK: IRL Press, 1985:605–612.
[t] Chesher GB, Bird KD, Stramarcos A, et al: A comparative study of the dose-response relationship of alcohol and cannabis on human skills performance. In: Harvey DJ, Paton W, Nahas GG, eds: Marihuana '84. Oxford, UK: IRL Press, 1985:621–627.
[u] Varma VK, Malhotra AK, Dang R, et al: Cannabis and cognitive functions: a prospective study. Drug Alcohol Depend 1988;21:147–152.
[v] Schwartz RH, Gruenewald PJ, Klitzner M, et al: Short-term memory impairment in cannabis-dependent adolescents. Am J Dis Child 1989;143:1214–1219.
[w] Heishman SJ, Stitzer ML, Bigelow GE: Alcohol and marijuana; comparative dose effect profiles in humans. Pharmacol Biochem Behav 1989;31:649–655.
[x] Heishman SJ, Stitzer ML, Hingling JE: Effects of tetrahydrocannabinol content on marijuana smoking behavior, subjective reports, and performance. Pharmacol Biochem Behav 1989;34:173–179.
[y] Solowij N, Michie PT, Fox AM: Effects of long-term cannabis use on selective attention: an event-related potential study. Pharmacol Biochem Behav 1991;40:683–688.
[z] Huestis MA, Sampson AH, Holicky BJ, et al: Characterization of the absorption phase of marijuana smoke. Clin Pharmacol Ther 1992;52:31–41.
[aa] Foltin RW, Fischman MW, Pippen PA, et al: Behavioral effects of cocaine alone and in combination with ethanol or marijuana in humans. Drug Alcohol Depend 1993;32:93–106.

impairment to actual decrements in performance during driving, flying, or other complex behavior is difficult. Translating the importance of observed performance effects in a controlled laboratory study to impairment of specific driving skills is complex. In addition, some results have not been replicated in other laboratories. Laboratory performance studies help us to understand drug effects and to predict possible impairment in performance of complicated tasks. These results also provide the basis for additional studies using driving and flying simulators and closed- and open-course driving trials that more closely reflect actual driving conditions. Review of the literature requires careful evaluation of study design including:

- Inclusion of appropriate placebo controls
- Double-blind dose presentation
- Dose and route of drug administration
- Inclusion of more than one dose to demonstrate a dose-response effect
- Performance task difficulty
- Subjects' experience with the task

Generally, the more complex the task, the more sensitive the task to MJ impairment. Also, tasks that are well practiced, such as driving, tend to be more resistant to drug effects. Therefore, laboratory tests that require divided attention and response to stressful, demanding situations are more likely to demonstrate performance impairment. Another key factor in evaluating reported effects is the availability of simultaneously collected blood THC concentrations. Few performance studies have included measurement of drug concentrations in blood. Drug concentration measurements are important due to the difficulty in delivering a specified THC dose by the smoking route. Clinical studies including strict controls on smoking dynamics, such as number of puffs, time between puffs, hold time, and inhalation time, have reported wide intra- and intersubject variability in blood THC concentrations due to the subjects' ability to titrate drug dose (27).

In general, laboratory performance studies (Table 5–2) indicate that sensory functions are not highly impaired, but perceptual functions are significantly affected. Perceptual errors are the most frequently cited errors leading to driving accidents when people are under the influence of ethanol and could be significant factors in MJ impairment (66). Furthermore, epidemiological studies indicate that MJ is usually combined with ethanol and sometimes with other licit or illicit drugs. MJ's and ethanol's effects have been documented to be additive (67–69). In fact, one of the difficulties in interpreting epidemiological data is defining impairment attributable to MJ alone, due to the presence of other psychoactive agents. This problem accentuates the need for laboratory performance studies evaluating the effect of MJ alone and with ethanol and other drugs.

The user's ability to concentrate and maintain attention may be decreased during MJ intoxication. Perceptual motor skills, decision-making, and car-handling skills may also be reduced. Overall, impairment was noted in studies using the pursuit meter or other tracking devices; impairment increased with dose except under very low dose conditions (70, 71). However, Heishman et al. (72) did not report impairment in an automated tracking task after smoking of a single 2.7% THC cigarette. The author indicated that the tracking task might have been less difficult than tasks in other tracking studies and therefore less sensitive to MJ's effects.

The brain's centers for coordination of movement are the cerebellum and basal ganglia, two areas rich in cannabinoid receptors. Impairment of hand-eye coordination is dose-related over a wide range of MJ dosages and has been shown to be additive with MJ and ethanol exposure (69). Coordination and body sway decrements have been documented. Stability of stance was decreased in a dose-related manner after MJ use (73,74). Reeve et al. (75) found that after ad lib smoking of MJ, 94% of subjects failed roadside sobriety and coordination tests 90 minutes after smoking; there was a 60% failure rate after 60 minutes.

Deficits in information processing have been documented after MJ exposure (1,76). Impairment was observed in immediate and delayed free recall, but not in recognition memory. Impairment was not demonstrated in the digit span immediate recall and divided attention tasks after MJ smoking in a study by Heishman et al. (77).

The most consistently reported cognitive effect after MJ use is a disruption of short-term memory (78). Huestis et al. (10) demonstrated significant impairment in only one of six performance tasks after MJ smoking; decreased accuracy in the logical reasoning task was reported after smoking a single 1.75% or 3.55% THC cigarette. Dose-related impairment of peripheral vision and an increase in the visual autokinetic phenomenon could contribute to poor driving performance (79–81). Significant effects were noticeable 8–10 hours after smoking in a battery of psychomotor and cognitive tasks after ingestion of MJ (82).

Barnett et al. (83) found a linear correlation between THC concentration and divided attention performance over the range of 5–25 ng/mL (15.9–79.5 nmol/L) THC. In addition, a sigmoid relation between the critical tracking breakpoint and the log of the THC concentration over the range of 2–25 ng/mL (6.4–79.5 nmol/L) THC was observed. Effects were noted for as long as 7 hours. Prolonged impairment in both concentrated and divided attention tasks were reported by Moskowitz (66).

A study of perceptuomotor tasks with heavy MJ users found a slower reaction time in these tasks; however, no differences were found in intelligence or memory functions (84). Long-term MJ users also were reported to possess a decreased ability to filter out irrelevant material and in focusing attention on the required task (86). Forney et al. (86) observed MJ impairment on psychomotor tests; the combination of amphetamine and MJ did not reduce the observed psychomotor impairment. In a recent study of MJ's and cocaine's effects on performance of a repeated acquisition task, including simple and choice reaction time tests and the digit symbol substitution test, high-dose cocaine and MJ exposure increased the number of errors over that with either drug alone and also prolonged subjective effects (87).

Driving and Flying Simulator and Closed- and Open-Course Driving Studies

Results of driving simulator and closed- and open-course driving studies are found in Table 5–3. In general, MJ use impaired performance on driving simulator tasks and on open and closed driving courses. MJ decreased car handling performance (68,88) and reduced reaction times (89,90). MJ impaired time and distance estimation in a dose-related manner and affected decision-making that relied upon these skills (91,92). However, improvement in driving performance was observed in some subjects in these low-dose MJ studies. MJ reduced the number of attempts at passing other cars and risky behavior, in contrast to ethanol, which increased risk taking. As previously described, the more difficult the task, the more apparent the impairment.

Sutton (59) described no significant effects in a closed-course task when only ethanol or MJ was used; however, significant effects were noted after combined ethanol and MJ exposure. The author suggested that driving is a well-practiced task, and the course did not significantly challenge subjects' ability to perform.

Results of flying simulator studies are found in Table 5–4. Impairment for as long as 24 hours has been reported in flying simulator studies after MJ smoking (93,94). Short-term memory, attention, and concentration were affected. Impairment was also found to relate to the difficulty of the task and the age of the pilot (95). One of the most important aspects of the studies was the lack of pilot awareness of decreased performance or impairment (13). The study designs of some of these studies have been criticized due to the lack of placebo controls, double-blind conditions, and concurrent THC blood concentrations.

Laboratory performance measures and driving behavior are altered after MJ use; however, there are no unequivocal studies linking increased accident rates with MJ use (96). It is difficult to determine whether the observed decrements in performance are large enough to account for actual driving

TABLE 5–3. Driving and simulator studies

Subject	Findings	Dose
Driving simulator[a]	Cannabis and ethanol increased brake and start time; ethanol increased and cannabis decreased gear changes; cannabis impaired estimation of time and distance in dose-related manner	8, 12, 16 mg THC oral; 70 g ethanol
Driving and psychomotor tracking studies[b]	Impaired car-handling performance after ethanol and high-dose MJ when measured with analytical instruments; observers in the vehicle were unable to document impairment after MJ	0, 1.4, 5.9 mg THC smoke; ethanol 0.07%
Perceptual-motor tasks[c]	Impaired at highest dose	0, 100, 200 µg/kg THC smoke; 0.425, 0.68 g/kg ethanol
Perceptual tasks[c]	Inconsistent results, simplified task	
Decision making, passing[c]	MJ reduced number of attempts	
Decision making, rapid stop[c]	Reduced by MJ, significantly impaired by combination	
Driving simulator[d]	MJ reduced accuracy and prolonged reaction time; ethanol impairment enhanced by MJ in additive or synergistic manner	One 2.4% THC cigarette; 0, 0.42, 0.85 g/kg ethanol 0, 4.9, 8.4 mg THC smoke
Closed courses and city streets[e]	Impaired judgment, concentration closed course; amplified city streets; improved performance in some subjects	
Closed course[f]	Car-handling performance impaired; THC plasma concentrations ranged from 8.5 to 21.2 ng/mL (75 µg/kg) and 12.9 to 1.2 ng/mL (150 µg/kg); drivers did not weave on road or brake or accelerate erratically	75, 150 µg/kg THC smoke; 0.63, 1.25 g/L ethanol
Closed course[g]	Significant effects under combination condition; no significant effects on a simple task when THC or ethanol alone; serum THC, BAC provided	0, 2% THC smoke; 0, 0.06% BAC

[a] Rafaelsen OJ, Bech P, Rafaelsen L: Simulated car driving influenced by cannabis and alcohol. *Pharmakopsychiatria* 1973;6:71–83.
[b] Hansteen RW, Miller RD, Lonero L: Effects of cannabis and alcohol on automobile driving and psychomotor tracking. *Ann N Y Acad Sci* 1976;282:240–256.
[c] Smiley A, Moskowitz H, Ziedman K: *Effects of drugs on driving.* [U.S. Department of Health and Human Services, Publication no. ADM85-1386.] Washington, DC: U.S. Government Printing Office, 1985.
[d] Perez-Reyes M, Hicks RE, Blumberry J, et al: Interaction between marihuana and ethanol: effects of psychomotor performance. *Alcohol Clin Exp Res* 1988;1201:268–276.
[e] Klonoff H: Marijuana and driving in real-life situations. *Science* 1974;186:317–324.
[f] Attwood D, Williams R, McBurney L, et al: Cannabis, alcohol and driving: effects on selected closed-course tasks. In: Goldberg L III, ed: *Alcohol, drugs and traffic safety.* Stockholm: Almquist and Wiksell International, 1981:938–968.
[g] Sutton LR: The effects of alcohol, marihuana and their combination on driving ability. *J Stud Alcohol* 1983;44:438–445.

TABLE 5–4. *Flight simulator studies*

Findings	THC concentrations	Dose
Impairment short-term memory, attention, and concentration up to 4 h[a]	Not applicable	0, 0.9 mg/kg THC smoke
Performance decrements up to 24 h; pilots not aware of impairment[b]	Not applicable	19 mg THC smoke
Impairment with turbulent flight after high dose. Impairment increases as difficulty or complexity of task increases and by pilot age[c]	Serum THC 7 ng/mL at 1 h after high-dose smoke	0, 10, 20 mg THC smoke
24-hour impairment; pilots not aware of decreased performance[d]	Not applicable	0, 20 mg THC smoke

[a] Janowsky DS, Meacham MP, Blaine JD, et al: Marijuana effects on simulated flying ability. *Am J Psychiatry* 1976;133:384–388.
[b] Yesavage JA, Leirer VO, Denari M, et al: Carry-over effects of marijuana intoxication on aircraft pilot performance: a preliminary report. *Am J Psychiatry* 1985;142:1325–1329.
[c] Leirer VO, Yesavage JA, Morrow DG: Marijuana, aging, and task difficulty effects on pilot performance. *Aviat Space Environ Med* 1989;60:1145–1151.
[d] Leirer VO, Yesavage JA, Morrow DG: Marijuana carry-over effects on aircraft pilot performance. *Aviat Space Environ Med* 1991;62:221–227.

impairment. Impaired functioning of psychomotor activities including measures of coordination, tracking, and vigilance and cognitive behavior including memory, learning, attention, information processing, decision-making, and perception have been reported. Most effects have returned to base-line concentrations within 3–4 hours, although some complex divided-attention tasks have indicated decrements in performance as long as 24 hours after MJ use (13,14). Additional performance data are required after repetitive and multiple drug use. Performance impairment after MJ exposure remains a major safety issue.

REFERENCES

1. Miller LL: Marijuana: acute effects on human memory. In: Agurell S, Dewey W, Willette R, eds: *The cannabinoids: chemical, pharmacologic, and therapeutic aspects.* New York: Academic Press, 1984:21–46.
2. Roman PM, Trice HM: Deviance and work: the influence of alcohol and drugs on job behaviors. *Rev Environ Health* 1972;1:9–58.
3. Current tobacco, alcohol, marijuana, and cocaine use among high school students—United States, 1990. *MMWR Morb Mortal Wkly Rep* 1991;40:659–663.
4. Vogl WF, Peterson MR, Jewell JS: Prevalence of drug use among applicants for military service—United States, June-December 1988. *MMWR Morb Mortal Wkly Rep* 1989;38:580–583.
5. Ostrov E, Cavanaugh JL: Validation of police officer recruit candidates' self-reported drug use. *J Forensic Sci* 1987;32:496–502.
6. Nahas GG: Cannabis: toxicological properties and epidemiological aspects. *Med J Aust* 1986;145:82–87.
7. Mechoulam R: Marihuana chemistry. *Science* 1970;168:1159.
8. Evans FJ: Cannabinoids: the separation of central from peripheral effects on a structural basis. *Planta Med* 1991;57:560–567.
9. Devane WA, Hanus L, Breuer A, et al: Isolation and structure of a brain constituent that binds to the cannabinoid receptor. *Science* 1992;258:1946–1949.
10. Huestis MA, Sampson AH, Holicky BJ, et al: Characterization of the absorption phase of marijuana smoke. *Clin Pharmacol Ther* 1992;52:31–41.
11. Perez-Reyes M, Owens SM, Di Guiseppi S: The clinical pharmacology and dynamics of marihuana cigarette smoking. *J Clin Pharmacol* 1981;21;201S–207S.
12. Hollister LE, Gillespie HK, Ohlsson A, et al: Do plasma concentrations of delta-9-tetrahydrocannabinol reflect the degree of intoxication? *J Clin Pharmacol* 1981;21:171S–177S.
13. Leirer VO, Yesavage JA, Morrow DG: Marijuana carry-over effects on aircraft pilot performance. *Aviat Space Environ Med* 1991;62:221–227.
14. Heishman SJ, Huestis MA, Henningfield JE, et al: Acute and residual effects of marijuana: profiles of plasma THC levels, physiological, subjective, and performance measures. *Pharmacol Biochem Behav* 1990;37:561–565.
15. Barnett G, Chiang CN, Perez-Reyes M, et al: Kinetic study of smoking marijuana. *J Pharmacokinet Biopharm* 1982;10:495–506.
16. Chiang NC, Barnett G: Marijuana effect and delta-9-tetrahydrocannabinol plasma level. *Clin Pharmacol Ther* 1984;36:234–238.
17. Domino LE, Konimo SE, Domino EF: Relation of plasma delta-9-THC concentrations to subjective "high" in marijuana users: a review and reanalysis.

In: Agurell S, Dewey WL, Willette RE, eds: *The cannabinoids: chemical, pharmacologic and therapeutic agents.* Orlando, FL: Academic Press, 1984: 245–261.
18. Cone EJ, Huestis MA: Relating blood concentrations of tetrahydrocannabinol and metabolites to pharmacologic effects and time of marijuana usage. *Ther Drug Monit* 1993;15:527–532.
19. Drug-related visits to hospital emergency rooms decline. *NIDA Notes* 1990;Summer/Fall:31. [DHHS Publication No. (ADM)91-1488.]
20. Wehner FC, Van Rensburg SJ, Thiel PG: Mutagenicity of marijuana and Transkei tobacco smoke condensates in the Salmonella/microsome assay. *Mutat Res* 1980;77:135–142.
21. Dalterio SL: Cannabinoid exposure: effects on development. *Neurobehav Toxicol Teratol* 1986;8:345–352.
22. Cabral GA, Mishkin EM: Delta-9-tetrahydrocannabinol inhibits macrophage protein expression in response to bacterial immunomodulators. *J Toxicol Environ Health* 1989;26:175–182.
23. Doblin R, Kleiman MAR: Medical use of marijuana. *Ann Intern Med* 1991;114:809–810.
24. Bowersox J: PHS cancels availability of medicinal marijuana. *J Natl Cancer Inst* 1992;84:475–476.
25. Ohlsson A, Lindgren JE, Wahlen A, et al: Single dose kinetics of deuterium labelled Δ9-tetrahydrocannabinol in heavy and light cannabis users. *Biomed Mass Spectrom* 1982;9:6–10.
26. Agurell S, Halldin M, Lindgren: Pharmacokinetics and metabolism of delta 1 tetrahydrocannabinol and other cannabinoids with emphasis on man. *Pharmcol Rev* 1986;38:21–43.
27. Huestis MA, Henningfield JE, Cone EJ: Blood cannabinoids. I. Absorption of THC and formation of 11-OH-THC and THCCOOH during and after smoking marijuana. *J Anal Toxicol* 1992;16:276–282.
28. Wall ME, Sadler BM, Brine D, et al: Metabolism, disposition, and kinetics of delta-9-tetrahydrocannabinol in men and women. *Clin Pharmacol Ther* 1983;34:352–363.
29. Hunt CA, Jones RT: Tolerance and disposition of tetrahydrocannabinol in man. *J Pharmacol Exp Ther* 1980;215:35–44.
30. Johansson E, Noren K, Sjovall J, et al: Determination of Δ1-tetrahydrocannabinol in human fat biopsies from marihuana users by gas chromatography-mass spectrometry. *Biomed Chromatogr* 1989;3:35–38.
31. Kreuz DS, Axelrod J: Delta-9-tetrahydrocannabinol: localization in body fat. *Science* 1973;179:391–393.
32. Johansson E, Agurell S, Hollister L, et al: Prolonged apparent half-life of Δ1-tetrahydrocannabinol in plasma of chronic marijuana users. *J Pharm Pharmacol* 1988;40:374–375.
33. Harvey DJ, Paton WDM: Metabolism of the cannabinoids. *Rev Biochem Toxicol* 1986;6:221–264.
34. Ohlsson A. Lindgren JE, Wahlen A, et al: Plasma delta-9-tetrahydrocannabinol concentrations and clinical effects after oral and intravenous administration and smoking. *Clin Pharmacol Ther* 1980;28:409–416.
35. Law B, Mason PA, Moffat AC, et al: Forensic aspects of the metabolism and excretion of cannabinoids following oral ingestion of cannabis resin. *J Pharm Pharmacol* 1984;36:289–294.
36. Cone EJ, Johnson RE, Paul BD, et al: Marijuana laced brownies: behavioral effects, physiologic effects, and urinalysis in humans following ingestion. *J Anal Toxicol* 1988;12:169–175.
37. Peat MA: Distribution of Δ9-tetrahydrocannabinol and its metabolites. In: Baselt RC, ed: *Advances in analytical toxicology*, vol II. Chicago: Yearbook Medical Publishers, 1989:186–212.
38. Ellis GM, Mann MA, Judson BA, et al: Excretion patterns of cannabinoid metabolites after last use in a group of chronic users. *Clin Pharmacol Ther* 1985;38:572–578.
39. Hawks RL: Introduction and overview. In: Hawks RL, ed: *The analysis of cannabinoids in biological fluids.* [NIDA Research Monograph 42.] Washington, DC: US Government Printing Office, 1982:1–6.
40. Perez-Reyes M, DiGuiseppi S, Davis KH, et al: Comparison of effects of marihuana cigarettes of three different potencies. *Clin Pharmacol Ther* 1982;31:617–624.
41. Hanson VW, Buonarati MH, Baselt RC, et al: Comparison of 3H- and 125I-radioimmunoassay and gas chromatography/mass spectrometry for the determination of Δ9-tetrahydrocannabinol and cannabinoids in blood and serum. *J Anal Toxicol* 1983;7:96–102.
42. Kelly P, Jones RT: Metabolism of tetrahydrocannabinol in frequent and infrequent marijuana users. *J Anal Toxicol* 1992;16:228–235.
43. Mason AP, McBay AJ: Cannabis: pharmacology and interpretation of effects. *J Forensic Sci* 1985;30:615–631.
44. McBurney LJ, Bobbie BA, Sepp LA: GC/MS and EMIT analysis for Δ9-tetrahydrocannabinol metabolites in plasma and urine of human subjects. *J Anal Toxicol* 1986;10:56–64.
45. Barnett G, Willette RE: Feasibility of chemical testing for drug-impaired performance. In: Baselt RC, ed: *Advances in analytical toxicology*, vol II. Chicago: Yearbook Medical Publishers, Inc., 1989:218–250.
46. Gjerde H, Kinn G: Impairment in drivers due to cannabis in combination with other drugs. *Forensic Sci Int* 1991;50:57–60.
47. Huestis MA, Henningfield JE, Cone EJ: Blood cannabinoids. II. Models for the prediction of time of marijuana exposure from plasma concentrations of delta-9-tetrahydrocannabinol (THC) and 11-nor-9-carboxy-delta-9-tetrahydrocannabinol (THCOOH). *J Anal Toxicol* 1992;16:283–290.
48. Wong AS, Orbanosky MW, Reeve VC, et al: Stability of delta-9-tetrahydrocannabinol in stored blood and serum. In: Hawks RL, ed: *The analysis of cannabinoids in biological fluids.* [NIDA Research Monograph 42.] Washington, DC: U.S. Government Printing Office, 1982:119–124.
49. Johnson JR, Jennison TA, Peat MA, et al: Stability of Δ9-tetrahydrocannabinol (THC), 11-hydroxy-THC, and 11-nor-9-carboxy-THC in blood and plasma. *J Anal Toxicol* 1984;8:202–204.
50. McCurdy HH, Callahan LS, Williams RD: Studies on the stability and detection of cocaine, benzoylecgonine, and 11-nor-delta-9-tetrahydrocannabinol-9-carboxylic acid in whole blood using Abuscreen radioimmunoassay. *J Forensic Sci* 1989;34:858–870.
51. Romberg RW: The effect of sodium azide and other

preservatives on the stability of drugs in urine. *Syva Monitor* 1992;10:15.
52. Romberg RW, Past MR: Stability of THCCOOH and benzoylecgonine in frozen urine specimens. Abstract presented at the 43rd American Academy of Forensic Toxicology Annual Meeting, Anaheim, CA, February 1991.
53. Foltz RL: Analysis of cannabinoids in physiological specimens by gas chromatography/mass spectrometry. In: Baselt RC, ed: *Advances in analytical toxicology*, vol. 1. Foster City, CA: Biomedical Publications, 1984:125–157.
54. Cone EJ, Huestis MA: Urinary excretion of commonly abused drugs following unconventional means of administration. *Forensic Sci Rev* 1989;1: 122–139.
55. Cone EJ, Johnson RE: Contact highs and urinary cannabinoid excretion after passive exposure to marijuana smoke. *Clin Pharmacol Ther* 1986;40: 247–254.
56. Hingson RW, Scotch N, Mangione T, et al: Impact of legislation raising the legal drinking age in Massachusetts from 18 to 20. *Am J Public Health* 1983; 73:163–170.
57. Wechsler H, Rohman M, Kotch JB, et al: Alcohol and other drug use and automobile safety: a survey of Boston-area teen-agers. *J Sch Health* 1984;54: 201–203.
58. Valois RF: Student drug use and driving: a university sample. *Health Educ* 1986;17:38–42.
59. Sutton LR: The effects of alcohol, marihuana and their combination on driving ability. *J Stud Alcohol* 1983;44:438–445.
60. Soderstrom CA, Triffillis AL, Shankar BS, et al: Marijuana and alcohol use among 1023 trauma patients. *Arch Surg* 1988;123:733–737.
61. Zimmermann EG, Yeager EP, Soares JR, et al: Measurement of $\Delta 9$-tetrahydrocannabinol in whole blood samples from impaired motorists. *J Forensic Sci* 1983;28:957–962.
62. Poklis A, Maginn D, Barr JL: Drug findings in driving under the influence of drugs cases: a problem of illicit drug use. *Drug Alcohol Depend* 1987;20:57–62.
63. Lund AK, Preusser DF, Blomberg RD, et al: Drug use by tractor-trailer drivers. *J Forensic Sci* 1988;33: 648–661.
64. Alleyne BC, Stuart P, Copes R: Alcohol and other drug use in occupational fatalities. *J Occup Med* 1991;33:496–500.
65. Shannon HS, Hope L, Griffith L, et al: Fatal occupational accidents in Ontario, 1986–1989. *Am J Ind Med* 1993;23:253–264.
66. Moskowitz H: Attention tasks as skills performance measures of drug effects. *Br J Clin Pharmacol* 1984;18:51S–61S.
67. Manno JE, Kiplinger GF, Scholz N, et al: The influence of alcohol and marihuana on motor and mental performance. *Clin Pharmacol Ther* 1971;12:202–211.
68. Hansteen RW, Miller RD, Lonero L: Effects of cannabis and alcohol on automobile driving and psychomotor tracking. *Ann N Y Acad Sci* 1976;282: 240–256.
69. Bird KD, Boleyn T, Chesher GB, et al: Intercannabinoid and cannabinoid-ethanol interactions and their effects on human performance. *Psychopharmacology (Berl)* 1980;71:181–188.

70. Manno JE, Kiplinger GF, Haine SE, et al: Comparative effects of smoking marihuana or placebo on human motor and mental performance. *Clin Pharmacol Ther* 1970;11:808–815.
71. Kiplinger GF, Manno JE, Rodda BE, et al: Dose-response analysis of the effects of tetrahydrocannabinol in man. *Clin Pharm Ther* 1971;12:650–657.
72. Heishman SJ, Stitzer ML, Bigelow GE: Alcohol and marijuana: comparative dose effect profiles in humans. *Pharmacol Biochem Behav* 1989;31:649–655.
73. Evans MA, Martz R, Brown DJ, et al: Impairment of performance with low doses of marihuana. *Clin Pharmacol Ther* 1973;14:936–939.
74. Chesher GB, Bird KD, Stramarcos A, et al: A comparative study of the dose-response relationship of alcohol and cannabis on human skills performance. In: Harvey DJ, Paton W, Nahas GG, eds: *Marihuana '84*. Oxford: IRL Press Limited, 1985:621–627.
75. Reeve VC, Robertson WB, Grant J, et al: Hemolyzed blood and serum levels of $\Delta 9$-THC: Effects on the performance of road side sobriety tests. *J Forensic Sci* 1983;28:963–971.
76. Tinklenberg JR, Kopell BS, Melges FT, et al: Marijuana and alcohol: time production and memory functions. *Arch Gen Psychiatry* 1972;27:812–815.
77. Heishman SJ, Stitzer ML, Yingling JE: Effects of tetrahydrocannabinol content on marijuana smoking behavior, subjective reports, and performance. *Pharmacol Biochem Behav* 1989;34:173–179.
78. Dornbush RL, Fink M, Freedman AM: Marijuana, memory and perception. *Am J Psychiatry* 1971; 128:194–197.
79. Moskowitz H, Sharma S, McGlothlin W: Effect of marihuana upon peripheral vision as a function of the information processing demands in central vision. *Percept Mot Skills* 1972;35:875–882.
80. Braff DL, Silverton L, Saccuzzo DP, et al: Impaired speed of visual information processing in marijuana intoxication. *Am J Psychiatry* 1981;138:613–617.
81. Sharma S, Moskowitz H: Effect of marijuana on the visual autokinetic phenomenon. *Percept Mot Skills* 1972;35:891–894.
82. Kielholz P, Hobi V, Ladewig D, et al: An experimental investigation about the effect of cannabis on car driving behavior. *Pharmakopsychiatr Neuropsychopharmacol* 1973;6:91–103.
83. Barnett G, Licko V, Thompson T: Behavioral pharmacokinetics of marijuana. *Psychopharmacology (Berl)* 1985;85:51–56.
84. Varma VK, Malhotra AK, Dang R, et al: Cannabis and cognitive functions: A prospective study. *Drug Alcohol Depend* 1988;21:147–152.
85. Solowij N, Michie PT, Fox AM: Effects of long-term cannabis use on selective attention: an event-related potential study. *Pharmacol Biochem Behav* 1991;40:683–688.
86. Forney R, Martz R, Lemberger L, et al: The combined effect of marihuana and dextroamphetamine. *Ann N Y Acad Sci* 1976;281:162–170.
87. Foltin RW, Fischman MW, Pippen PA, et al: Behavioral effects of cocaine alone and in combination with ethanol or marijuana in humans. *Drug Alcohol Depend* 1993;32:93–106.
88. Attwood D, Williams R, McBurney L, et al: Cannabis, alcohol and driving: Effects on selected closed-course tasks. In: Goldberg L, ed: *Alcohol, drugs and*

traffic safety, vol. III, Stockholm: Almquist and Wiksell International, 1981:938–953.
89. Rafaelsen OJ, Bech P, Rafaelsen L: Simulated car driving influenced by cannabis and alcohol. *Pharmakopsychiatria* 1973;6:71–83.
90. Perez-Reyes M, Hicks RE, Blumberry J, et al: Interaction between marihuana and ethanol: effects of psychomotor performance. *Alcohol Clin Exp Res* 1988;1202:268–276.
91. Klonoff H: Marijuana and driving in real-life situations. *Science* 1974;186:317–324.
92. Smiley A, Moskowitz H, Ziedman K: *Effects of drugs on driving.* [Publication No. ADM85-1386.] Washington, DC: US Department of Health and Human Services, Government Printing Office, 1985.
93. Janowsky DS, Meacham MP, Blaine JD, et al: Marijuana effects on simulated flying ability. *Am J Psychiatry* 1976;133:384–388.
94. Yesavage JA, Leirer VO, Denari M, et al: Carryover effects of marijuana intoxication on aircraft pilot performance: a preliminary report. *Am J Psychiatry* 1985;142:1325–1329.
95. Leirer VO, Yesavage JA, Morrow DG: Marijuana, aging, and task difficulty effects on pilot performance. *Aviat Space Environ Med* 1989;60:1145–1152.
96. Moskowitz H: Marihuana and driving. *Accid Anal Prev* 1985;17:323–345.

SUGGESTED READING

1. Huestis MA, Cone EJ: Differentiating new marijuana use from residual drug excretion in occasional marijuana users. *J Anal Toxicol* 1998;22:445–454.
2. Huestis MA, Mitchell JM, Cone EJ: Urinary excretion profiles of 11-nor-9-carboxy-Δ9-tetrahydrocannabinol in humans after single smoked doses of marijuana. *J Anal Toxicol* 1996;20:441–452.
3. Johansson E, Ohlsson A, Lindgren JE, et al: Single-dose kinetics of deuterium-labelled cannabinol in man after intravenous administration and smoking. *Biomed Environ Mass Spectrom* 1987;14:495–499.
4. Reeve VC, Grant JD, Robertson W, et al: Plasma concentrations of Δ9-tetrahydrocannabinol and impaired motor function. *Drug Alcohol Depend* 1983;11:167–175.

SUGGESTED WEB SITE REFERENCES

1. International Cannabinoid Research Society: http://www.cannabinoidsociety.org.
2. Society of Forensic Toxicologists (to access its marijuana monograph): http://www.soft-tox.org.

SELF-ASSESSMENT QUESTIONS

1. Which of the following factors has or have contributed to the difficulty of development of sensitive analytical methods for measuring Δ^9-tetrahydrocannabinol (THC)?
   a. low drug concentration
   b. extensive drug metabolism
   c. high lipophilicity of the drug
   d. nonspecific adsorption of the drug
   e. all of the above

2. Which of the following statements is or are true of the smoking route of drug administration?
   a. It delivers a uniform, predictable drug dose.
   b. It has a bioavailability similar to that of the intravenous dose.
   c. It results in rapid delivery of drug to the brain.
   d. It allows individuals to titrate their drug dose.

3. Which of the following statements is or are true of marijuana smoking?
   a. The elimination half-life of THC is very short.
   b. 11-nor-9-Carboxy-Δ^9-tetrahydrocannabinol is the primary psychoactive component of marijuana
   c. Most of the THC is eliminated from the body in the urine.
   d. Significant amounts of drug (THC) are measured in the blood after a single puff of a marijuana cigarette.

4. What is the most commonly abused drug in the United States?
   a. heroin
   b. cocaine
   c. marijuana
   d. D-lysergic acid diethylamide (LSD)
   e. methamphetamine

5. Which of the following statements is true?
   a. The Δ^9 numbering system refers to the monoterpene numbering system.
   b. THC is highly soluble in water.
   c. Synthetic THC has been used to treat AIDS wasting disease.
   d. Approximately 75% of the THC present in a marijuana cigarette is destroyed by pyrolysis during smoking.

CHAPTER 6

Opioids

Jeri D. Ropero-Miller, Ph.D., and Bruce A. Goldberger, Ph.D., DABFT

LEARNING OBJECTIVES

After completing this chapter, the reader should be able to:

1. Differentiate between an opiate and an opioid.
2. Discuss the therapeutic indications, pharmacology, pharmacokinetics, and toxicology of opioid analgesics.
3. List the therapeutic and adverse side effects associated with these drugs.
4. Discuss the symptoms and treatment of withdrawal.
5. Discuss the analytical techniques for determining opioid exposure.

AN INTRODUCTION TO OPIOID ANALGESICS

Opioid analgesics are among the most effective and common medications for treatment of mild to severe pain. However, these drugs are often abused for their desirable central nervous system (CNS) effects, especially euphoria. Many opioids are highly addictive, leading to physical and psychological dependence.

The Drug Abuse Warning Network (DAWN) monitors national drug-abuse trends by surveying hospital emergency department records throughout the United States, and it then issues reports. From 1990 to 1996, the total number of drug-abuse mentions categorized as opioid analgesics increased 6.6% from 32,430 to 34,563. During this same period, trends in the medical use of Schedule II opioid analgesics (fentanyl, hydromorphone, meperidine, morphine, and oxycodone) used to treat severe pain increased; however, the abuse mentions of these opioid analgesics excluding morphine decreased compared with the total DAWN mentions for all drug categories (1).

The Substance Abuse and Mental Health Services Administration's 1996 National Household Survey on Drug Abuse reported an increasing trend in new heroin users from 1992 to 1995, with estimated "past month" heroin users increasing from 68,000 to 216,000 during this time (2).

National researchers participating in a biannual meeting of the Community Epidemiology Work Group sponsored by the National Institute on Drug Abuse reported an increase in heroin use among young populations. As indicated by nationwide emergency room mentions, heroin use among individuals age 18–25 years increased 51.4% from 1997 to 1999 (3).

BASIC CONCEPTS OF OPIATE AND OPIOID ANALGESICS

Definitions

Opiates and opioids have overlapping definitions throughout the literature (4–8). For this chapter, the following definitions will apply:

- "Opiates" are naturally occurring analgesic alkaloids derived from the opium poppy.
- "Opioids" are natural and semisynthetic alkaloid derivatives prepared from opium or synthetic surrogates whose actions mimic those of morphine. Therefore, opioids include opiates, synthetic opioids, and opiopeptins (endogenous neuropeptides).

General Description

This chapter describes common opioid analgesics including naturally occurring opiates such as morphine and codeine, as well as synthetic or semisynthetic agonists either defined by the Controlled Substance Abuse Act as having medicinal applications (Schedule II through V) or illicit opioids such as heroin that have no medicinal use (Schedule I). Table 6–1 lists the scheduling of opioid analgesics as of the 1998 revised Controlled Substance Abuse Act. Opioid antagonists, including naloxone, naltrexone, nalorphine, and levallorphan, will be referred to only in the treatment of opioid overdose. In addition, this chapter excludes all opioid antidiarrheals.

This chapter summarizes the pharmacology, pathophysiology, toxicology, and other facts of the common opioid analgesics. For discussion purposes, the opioids have been divided into two groups:

- Full agonist—opioids with affinity for all opioid receptors
- Mixed agonist-antagonist—opioids having agonistic pharmacological effects at one receptor and antagonistic effects at another receptor

GROUPS AND CHEMICAL STRUCTURE

Opioid analgesics are categorized into six groups according to chemical structure (see Figures 6–1 through 6–6):

- Phenanthrenes
- Morphinans
- Benzomorphans
- Cyclohexanols
- Phenylheptylamines
- Phenylpiperidines

Phenanthrenes (4,5-epoxymorphinans) are probably the best known and most characterized opioids. Phenanthrenes include both naturally occurring derivatives such as morphine and codeine, in addition to synthesized opioids such as heroin, hydromorphone, hydrocodone, and oxycodone. Opioids in this group possess the basic morphine structure, consisting of a γ-phenyl-N-methylpiperidine chemical "backbone" with various chemical modifications. These chemical modifications can include methylation or acetylation of C3 and C6 hydroxy groups, oxidation of the C6 hydroxy group to a ketone functionality (hydro-derivatives), saturation of the C7-C8 double bond (dihydro-derivatives), and hydroxylation at the C14 position (oxy-derivatives) (6).

The other five structural groups include drugs such as methadone and meperidine that have significant chemical modification (for example, the opening of the piperidine ring in the case of methadone) or are chemically unrelated to the phenanthrenes (such as propoxyphene), but produce similar analgesic effects due to their affinity for the opioid receptor and similar mechanism of action (6).

Subtle molecular differences can change the action of opioids including receptor affinity and lipid solubility (4). Examples include conversion of an agonist to an antagonist, an increase in lipid solubility (acetylation), and resistance to first-pass hepatic metabolism (methyl substitution) (4).

PHARMACOKINETIC VALUES

Opioids share common pathways for their pharmacokinetic properties: absorption, distribution, metabolism, and elimination. Table 6–2 summarizes many of the clinical pharmacokinetic values of opioids.

Absorption

Because opioids are well absorbed by most tissues, numerous routes of administration

TABLE 6–1. Opioid analgesics under the Schedule of Controlled Substances Act of the U.S. Code of Federal Regulations (revision 1998)

Schedule I	Schedule II	Schedule III	Schedule IV	Schedule V
Heroin	Alfentanil	Codeine	Butorphanol	Buprenorphine
Norlevorphanol	Alphaprodine	Dihydrocodeine	Dextropropoxyphene	Codeine
Normethadone	Bulk dextropropoxyphene	Ethylmorphine	Pentazocine	Dihydrocodeine
Opium derivatives	Codeine	Morphine		Ethylmorphine
Other nonmarketed synthetic narcotics	Dihydrocodeine	Nalorphine		Opium
	Ethylmorphine	Opium		
	Fentanyl			
	Hydrocodone			
	Hydromorphone			
	Levo-α-acetylmethadol (LAAM)			
	Levomethorphan			
	Levorphanol			
	Meperidine			
	Methadone			
	Morphine			
	Opium			
	Oxycodone			
	Oxymorphone			
	Thebaine			
	Sufentanil			

US Department of Justice Diversion Control Program. http://www.dea.gov

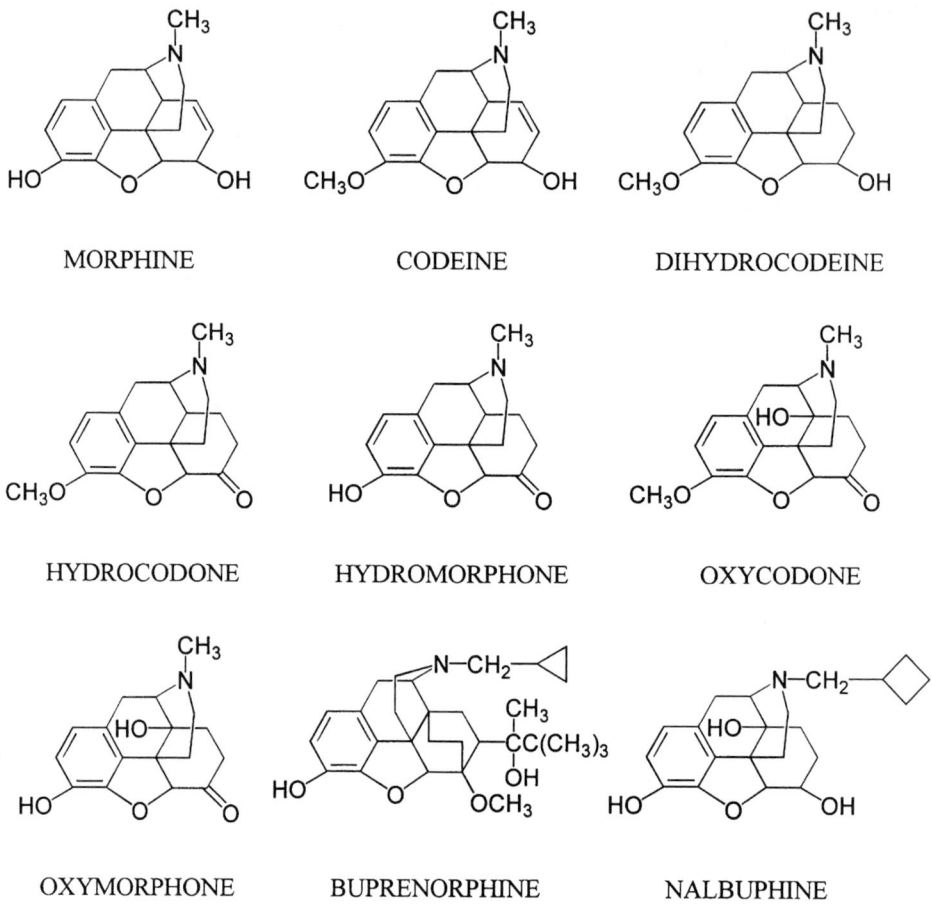

FIG. 6–1. Chemical structures of phenanthrenes.

are effective. Opioids are ingested orally, and the effective oral dose depends on the drug's bioavailability and the effect of first-pass metabolism. Parenteral administration of opioids by intramuscular, intravenous, or subcutaneous injection is common (4). Less-common routes of administration include transdermal absorption and epidural and intrathecal injections (9–12).

Distribution

The distribution of opioids depends on chemical and physiological factors in addi-

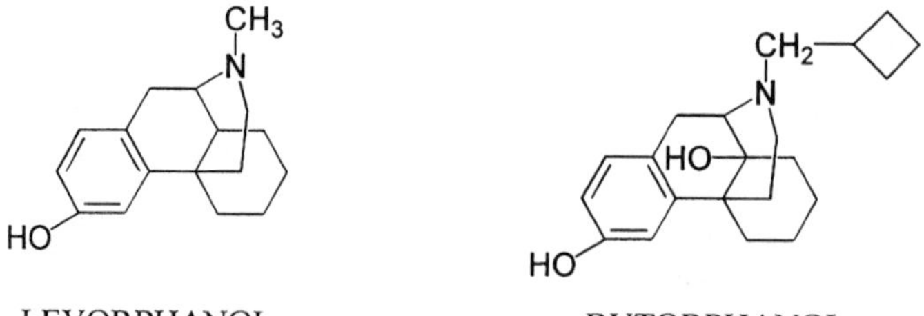

FIG. 6–2. Chemical structures of morphinans.

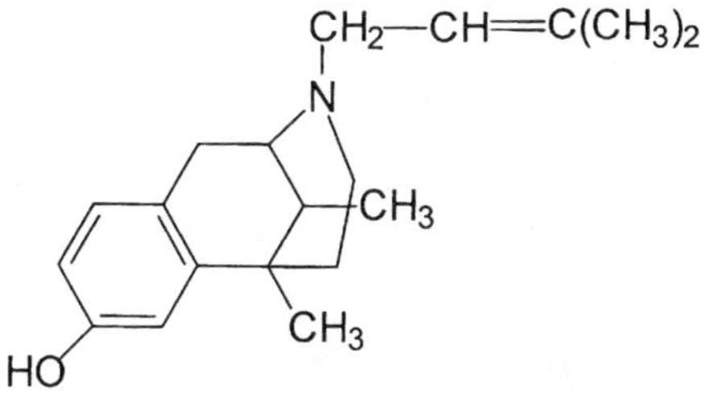

FIG. 6–3. Chemical structure of the benzomorphan pentazocine.

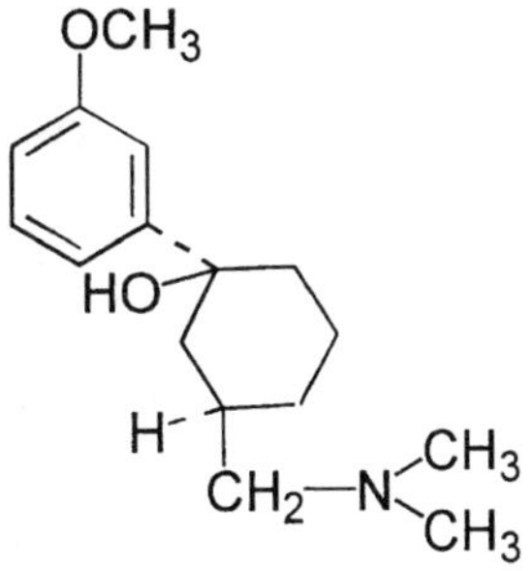

FIG. 6–4. Chemical structure of the cyclohexanol tramadol.

tion to specific properties of the drug. Some opioids including oxycodone, codeine, and buprenorphine demonstrate a volume of distribution consistent with the plasma compartment (that is, 3 L/70 kg), whereas other more lipophilic opioids such as heroin have a much higher volume of distribution (that is, 25 L/70 kg). Plasma protein binding greatly affects opioid distribution and varies largely among opioids, ranging from 20% to 95%. Opioids concentrate in the tissues of highly perfused organs such as the lungs, brain, kidney, liver, and spleen. Opioids further accumulate in lipid and skeletal muscle reservoirs and cross placental barriers to varying degrees (4).

Metabolism

Opioids are metabolized to a more water-soluble form in order to promote excretion. These metabolites are either pharmacologically active or inactive. Active metabolites may have analgesic effects that are stronger than the parent drug itself. Metabolic pathways include reduction, oxidation, N- and O-dealkylation, hydroxylation, and conjugation.

Hepatic biotransformation is divided into two major metabolic pathways, Phase I and Phase II reactions. Phase I reactions are nonsynthetic and oxidize or reduce a drug to its more polar form. Liver microsomes, known as the cytochrome P450 system, consist of enzymes located in the endoplasmic reticulum of hepatocytes. When an opioid binds to cytochrome P450, it is oxidized or reduced through an electron transport chain that uses the reduced form of nicotinamide-adenine dinucleotide phosphate (NADPH) as the proton carrier.

Phase II reactions, or synthetic reactions, increase the polarity of an opioid by enzymatically attaching, referred to as conjugation, a polar functional group such as a glucuronide or sulfate donor group.

Elimination

Most opioids and their metabolites are excreted in the urine, with only a small amount of glucuronide conjugates eliminated in the

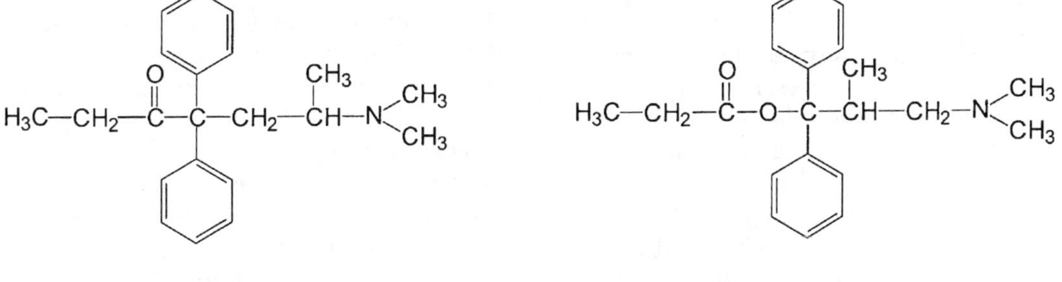

METHADONE

PROPOXYPHENE

FIG. 6–5. Chemical structures of phenylheptylamines.

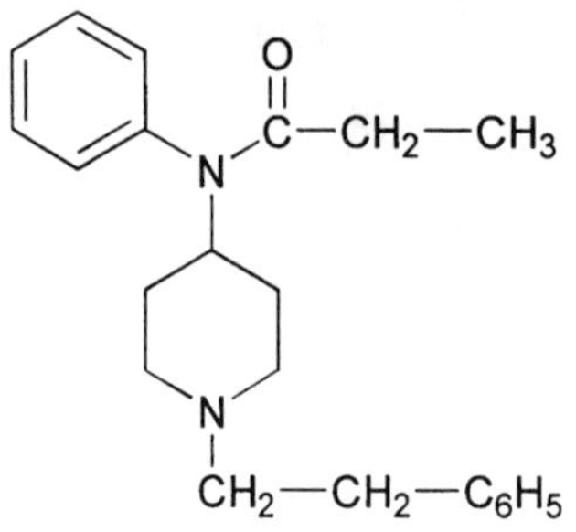

FENTANYL 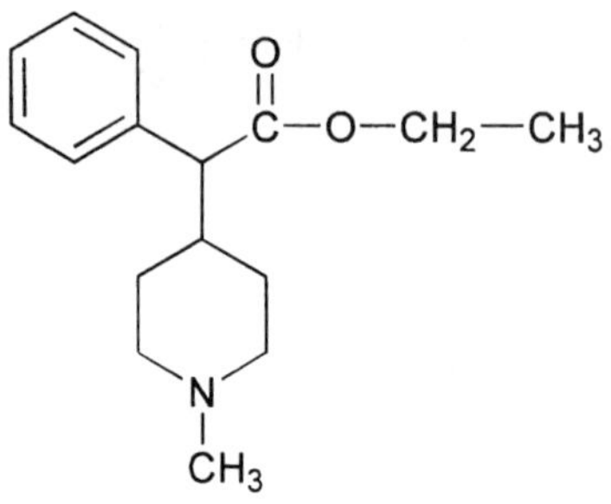   MEPERIDINE

FIG. 6–6. Chemical structures of phenylpiperidines.

feces or bile via enterohepatic circulation (5). Table 6–3 is a listing of opioids and their metabolites.

PHARMACODYNAMIC MECHANISM OF ACTION, TOLERANCE, AND DEPENDENCE

Based on their pharmacologic activity, opioids are classified as a:

- Full agonist
- Mixed agonist-antagonist
- Full antagonist

Opioid effects are mediated through specific receptors located at various sites in the CNS and peripheral organs. Opioid receptors are also located on several immune-response cell types such as neutrophils, lymphocytes, and monocytes that are commonly associated with endogenous opioid peptides produced during stress (6). Primary receptors include:

- µ-Receptors, responsible for euphoria, supraspinal analgesia, respiratory depression, miosis, reduced gastrointestinal motility, and physical tolerance and dependence
- κ-Receptors, mediating spinal analgesia, sedation, sleep, miosis, physical dependence, and limited respiratory depression
- δ-Receptors, mediating dysphoria, delusions, hallucinations, respiratory stimulation, and vasomotor stimulation

A less characterized receptor, σ, is purported to have effects similar to the δ-receptor. On a cellular level, opioid receptors exert their effects through changes in Ca^{2+} and K^+ flux associated with the cyclic adenosine monophosphate system of the nervous system (4).

Prolonged use of opioids leads to tolerance to the analgesic and CNS effects. Tolerance begins with the first dose, but is usually clinically insignificant until after ~2–3 weeks of chronic use. To minimize tolerance, opioids should be administered in small doses at frequent intervals. Cross-tolerance among opioids is a prevailing characteristic that must also be considered when administering these drugs (4).

Psychological and physical dependence makes opioid withdrawal and detoxification extremely painful and difficult. Typically, withdrawal signs from strong agonists appear within 6–8 hours after the last administration and peak at 36–72 hours. Psychological dependence produces strong cravings that can lead to pleas, demands, and manipulative behavior. Hallmarks of withdrawal reflecting the physical dependence include irritability, insomnia, anorexia, violent yawning and sneezing, gastrointestinal abnormalities, elevated heart rate, profuse sweating, and piloerection (7). Strong pains in the bones and muscles and uncontrollable muscle spasm are also consistent with withdrawal. The culmination of these symptoms after 36–72 hours is widely referred to as the abstinence or withdrawal syndrome. In most cases, these symp-

TABLE 6–2. Clinical pharmacology of opioids

Drug	Half-life (h)	V_d (L/kg)	Duration of analgesia (h)	Protein binding (%)	Therapeutic plasma concentration (ng/mL)	Toxic concentration (ng/mL)	Routes of administration	Addiction and abuse liability
Buprenorphine	2–4	2.5	4–8	96	0.3–0.7	ND	IM, IV, SC, E	Low
Butorphanol	2.9–8.4	5.0	3–4	83	1–2	ND	IM, IV, SC, N	Low
Codeine	1.9–3.9	3.5	3–4	7–25	10–100	>200	PO, SC, IM	Moderate
Dihydrocodeine	3.4–4.5	1.0–1.3	3–4	ND	60–84	800–1200	PO	Moderate
Fentanyl	3–12	3–8	1–1.5	79	1–3	>3	IM, IV, T, E	High
Heroin	0.08	25	3–5	20–35	NA	NA	PO, IM, IV, SC, N	High
Hydrocodone	3.4–8.8	3.3–4.7	3–5	25	2–24	>100	PO	Moderate
Hydromorphone	1.5–3.8	2.9	4–5	19	1–30	>100	PO, IM, IV, SC	High
Levorphanol	11–16	10–13	4–5	40	ND	>100	SC, PO	High
Meperidine	2–5	3.7–4.2	2–4	64	400–700	>1000	PO, IM, IV, SC	High
Methadone	15–30	4–5	4–6	87	100–400	>2000	PO, IM, IV, SC	High
Morphine	1.3–6.7	2–5	4–5	20–35	10–80	>200	PO, IM, IV, SC, T	High
Nalbuphine	1.9–7.7	2.4–7.3	3–6	ND	ND	ND	IM, IV, SC	Low
Opium	2–4	3–4	4–5	35	ND	ND	PO, PR, N	High
Oxycodone	4–6	1.8–3.7	3–4	45	10–100	>200	PO	Moderate
Oxymorphone	ND	2–3	3–4	ND	ND	ND	PR, IM, IV, SC	High
Pentazocine	2.1–3.5	4.4–7.8	3–4	61	50–200	>1000	PO, IM, IV	Low
Propoxyphene	8–24	12–26	4–5	78	100–400	>500	PO	Low
Tramadol	4.3–6.7	2.6–2.9	4–6	20	100–300	ND	PO, IM, IV, SC, E	Low

Vd, volume of distribution; ND, no data available; NA, not applicable. PO, oral; PR, rectal; IV, intravenous; IM, intramuscular; SC, subcutaneous; T, transdermal; E, epidural or intrathecal; N, smoking or intranasal. From Way WL, Way EL, Fields HL: Opioid analgesics and antagonists. In: Katzung BG, ed: *Basic and clinical pharmacology*, 6th ed. Norwalk, CT: Appleton and Lange, 1995:460–477; Baselt RC: *Disposition of toxic drugs and chemicals in man*, 5th ed. Foster City, CA: CTI, 2000; Ellenhorn MJ, Schonwald S, Ordog G, et al, eds: *Ellenhorn's medical toxicology: diagnosis and treatment of human poisoning*. Baltimore: Williams & Wilkins, 1997:405–447; Jaffe JH, Martin WR: Opioid analgesics and antagonists. In: Gilman AG, Rall TW, Nies AS, et al, eds: *The pharmacological basis of therapeutics*, 8th ed. New York: Pergamon Press, 1990; Burtis CA, Ashwood ER, eds: *Tietz textbook of clinical chemistry*, 2nd ed. Philadelphia: WB Saunders, 1994; Roscow CE: Buprenorphine: epidural and intrathecal use. In: Cowan A, Lewis JW, eds: *Buprenorphine: combating drug abuse with a unique opioid*. New York: Wiley-Liss, 1995:165–174; Lacy C, Armstrong LL, Lipsy RJ, et al: *Drug information handbook*. Hudson, OH: Lexi-Comp, 1993; Rowell FJ, Seymour RA, Rawlings MD: Pharmacokinetics of intravenous and oral dihydrocodeine and its acid metabolites. *Eur J Clin Pharmacol* 1983;25:419–424; Tietz NW: *Clinical guide to laboratory tests*, 3rd ed. Philadelphia: WB Saunders, 1995; Lee CR, McTavish D, Sorkin EM: Tramadol: a preliminary review of its pharmacodynamic and pharmacokinetic properties, and therapeutic potential in acute and chronic pain states. *Drugs* 1993;46:313–340; Knoben JE, Anderson PO, eds: *Handbook of clinical drug data*, 7th ed. Hamilton, IL: Drug Intelligence Publications, 1993; Uges DRA, ed: Therapeutic and toxic drug concentrations. *Bull Int Assoc Forensic Toxicol* 1996;26[Suppl]:1–34; *Physicians' desk reference*, 54th ed. Montvale, NJ: Medical Economics Data, 2000; Leiken JB, Paloucek FP, eds: *Poisoning & toxicology compendium*. Cleveland, OH:Lexi-Comp Inc., 1998.

TABLE 6-3. *Opioids and opioid metabolites*

Buprenorphine	Butorphanol	Codeine	Dihydrocodeine
Norbuprenorphine	Hydroxybutorphanol	Norcodeine	Dihydronorcodeine
Conjugated forms	Norbutorphanol	Morphine	Dihydromorphine
	Conjugated norbutorphanol	Conjugated codeine	Conjugated dihydrocodeine
		Conjugated norcodeine	Conjugated dihydronorcodeine
		Conjugated morphine	Conjugated dihydromorphine

Fentanyl	Heroin	Hydrocodone	Hydromorphone
Norfentanyl	Morphine	Norhydrocodone	Conjugated hydromorphone
Hydroxyfentanyl	6-Acetylmorphine	6-Hydrocodol	Conjugated 6-hydromorphol
Hydroxynorfentanyl	Conjugated morphine	6-Hydromorphol	
Despropionylfentanyl		Conjugated hydromorphone	
		Conjugated hydromorphol	

Levorphanol	Meperidine	Methadone	Morphine
Norlevorphanol	Normeperidine	EDDP	Normorphine
Conjugated levorphanol	Meperidinic acid	EMDP	Morphine-3-glucuronide
Conjugated norlevorphanol	Normeperidinic acid	Methadol	Morphine-6-glucuronide
	Conjugated meperidinic acid	Normethadol	Conjugated normorphine
	Conjugated normeperidinic acid	Conjugated forms	Morphine-3,6-diglucuronide
			Morphine-3-ethereal sulfate

Nalbuphine	Oxycodone	Oxymorphone	Pentazocine
Nornalbuphine	Noroxycodone	Norhydrocodone	*cis*-Hydroxypentazocine
Conjugated nalbuphine	Oxymorphone	6-β-Oxymorphol	*trans*-Hydroxypentazocine
	Conjugated oxycodone	Conjugated oxymorphone	*trans*-Carboxypentazocine
	Conjugated oxymorphone	Conjugated 6-β-oxymorphol	Conjugated forms
		Conjugated 6-α-oxymorphol	

Propoxyphene		Tramadol	
Norpropoxyphene		Nortramadol	O-Desmethylnortramadol and conjugated form
Dinorpropoxyphene		O-Desmethyltramadol and conjugated form	N-Desmethylnortramadol and conjugated form
Cyclic dinorpropoxyphene		N-Desmethyltramadol and conjugated form	O-Desmethyldinortramadol and conjugated form
Conjugated forms		Dinortramadol	N-Desmethyldinortramadol and conjugated form

EDDP, 2-ethylidene-1,5-dimethyl-3,3-diphenylpyrrolidine; EMDP, 2-ethyl-5-methyl-3,3-diphenylpyrroline. From Baselt RC: *Disposition of toxic drugs and chemicals in man*, 5th ed. Foster City, CA: CTI, 2000.

TABLE 6–4. *Chronology of opioid abstinence syndrome in humans*

Time	Symptoms
6–12 h	Lacrimation
	Yawning
	Rhinorrhea
	Perspiration
12–14 h	Irritability
	Piloerection (goose flesh)
	Restless sleep
	Weakness
	Mydriasis
	Tremor
	Anorexia
	Muscle twitching
36–72 h (peak of syndrome)	Increased irritability
	Increased heart rate
	Insomnia
	Hypertension
	Marked anorexia
	Hot and cold flashes
	Sneezing
	Alternating sweating and flushing
	Nausea and vomiting
	Piloerection
	Hyperthermia
	Hyperpnea
	Abdominal cramps
	Aching muscles
Syndrome duration	7–10 d

From Way WL, Way EL, Fields HL: Opioid analgesics and antagonists. In: Katzung BG, ed: *Basic and clinical pharmacology*, 6th ed. Norwalk, CT: Appleton and Lange, 1995:460–477; Ellenhorn MJ, Schonwald S, Ordog G, et al, eds: *Ellenhorn's medical toxicology: diagnosis and treatment of human poisoning.* Baltimore: Williams & Wilkins, 1997:405–447; Jaffe JH, Martin WR: Opioid analgesics and antagonists. In: Gilman AG, Rall TW, Nies AS, et al, eds: *The pharmacological basis of therapeutics*, 8th ed. New York: Pergamon Press, 1990.

toms follow a characteristic chronology, as shown in Table 6–4. The pharmacological activity of the opioid determines the severity of the withdrawal symptoms.

MEDICINAL USES

Therapeutic Application

Opioids have five primary therapeutic applications:

- Analgesia
- Anesthetic supplement
- Antidiarrheal
- Antitussive
- Detoxification in overdose of opioid agonists (antagonists)

Opioid analgesics are used to treat mild to severe pain, including pain accompanying myocardial infarction, pre- and postoperative periods, and cancer. They are also used to treat congestive heart failure (pulmonary edema) and anxiety. Opioids are effective analgesics because they have the ability to change pain perception by raising the threshold for pain. Subjective patient reports also indicate that they alter the patient's reaction to pain.

Opioids are commercially available as elixirs, pills, powders, suppositories, solutions, and transdermal patches (4,10,13). Illicit heroin is available in powdered form that is administered intranasally, intravenously, or smoked.

Precautions

Caution should be taken when administering opioids to special-risk patients who have reduced metabolic abilities or critical medical conditions that render them unable to tolerate stressful conditions. Opioid administration should be avoided or closely monitored in children, elderly people, pregnant women, and patients with any of the following conditions (10,13):

- Renal or hepatic dysfunction
- Debilitation
- Biliary abnormalities
- Hypothyroidism
- Urethral stricture
- Recent myocardial infarction
- Head trauma or intracranial pressure
- Premature pregnancy
- Pulmonary disease
- Bronchial asthma
- Acute alcoholism

Drug interactions also warrant precautionary measures when administering opioids. Some drugs enhance opioid actions, whereas others antagonize their effects. If drug interactions are not considered when determining an individual's therapeutic dose, an incorrect dose may be given and have serious consequences. Opioids have additive effects with the following drugs:

- CNS depressants
- General anesthetics
- Tranquilizers
- Sedative-hypnotics
- Tricyclic antidepressants
- Dextroamphetamine
- Monoamine oxidase (MAO) inhibitors

Respiratory depression may be greatly enhanced in patients receiving CNS depressants and opioids concomitantly. Coadministration of MAO inhibitors and opioids greatly increases the possibility of seizures, hyperpyrexic coma, and hypertension (4,10,13). Moreover, opioids may inhibit the metabolism and increase the serum concentrations of carbamazepine, phenobarbital, tricyclic antidepressants, and warfarin (13).

Because opioids are CNS depressants and have a sedative effect, they may impair an individual's mental and physical ability to drive a motor vehicle or operate machinery. While under the influence of opioids, patients should exercise caution in situations requiring full mental capacity (10).

TOXICOLOGY AND TREATMENT

Individuals using opioids often experience adverse effects in addition to the desired therapeutic effects (Table 6–5). Common complaints include nausea, vomiting, constipation, and mood swings. The most life-threatening adverse reaction, commonly observed after an overdose, is respiratory depression. Other risks associated with opioid toxicity include coma, hypothermia, seizures, and hypotension. Treatment of opioid overdose consists mainly of supportive and antidotal measures such as (6,13):

- Monitoring cardiovascular and respiratory status
- Ventilation to reestablish respiratory exchange

TABLE 6–5. *Therapeutic and adverse effects of opioids*

Central nervous system effects	
Nervous	Euphoria
	Analgesia
	Sedation
	Mental clouding and mood swings
Pulmonary	Respiratory depression
	Decreased responsiveness
Gastrointestinal	Nausea
	Vomiting
	Vertigo
Other	Cough suppression
	Miosis
	Truncal rigidity
	Flushing and warming of the skin
	Sweating and itching
Peripheral effects	
Cardiac	Bradycardia
	Orthostatic hypotension when system stressed
	Stroke
Gastrointestinal	Constipation
	Decreased motility
	Increased tone
	Decreased gastric secretions
	Biliary tract constriction of smooth muscle
Genitourinary	Decreased renal plasma flow
	Increased urethral and bladder tone
	Prolongation of labor
	Menstrual abnormalities
	Sexual dysfunction
Neuroendocrine	Increased antidiuretic hormone release

From Way WL, Way EL, Fields HL: Opioid analgesics and antagonists. In: Katzung BG, ed: *Basic and clinical pharmacology*, 6th ed. Norwalk, CT: Appleton and Lange, 1995:460–477; Ellenhorn MJ, Schonwald S, Ordog G, et al, eds: *Ellenhorn's medical toxicology: diagnosis and treatment of human poisoning.* Baltimore: Williams & Wilkins, 1997:405–447; Jaffe JH, Martin WR: Opioid analgesics and antagonists. In: Gilman AG, Rall TW, Nies AS, et al, eds: *The pharmacological basis of therapeutics*, 8th ed. New York: Pergamon Press, 1990.

TABLE 6–6. *Clinical symptoms and treatment of opioid overdose*

- Monitoring of cardiovascular and respiratory status
- Ventilation: reestablish respiratory exchange
- Oxygen: prevent anoxia
- Opioid antagonist (naloxone, naltrexone, nalorphine): counteract CNS and respiratory effects
- Intravenous fluids: counteract hypotension
- Therapeutic intervention
  - Doxapram: stimulate respiration
  - Vasopressors: treat hypotension
  - Anticonvulsants: treat seizures
- Other supportive measures
- Rewarming at <90 °F (<32.2 °C) (for hypothermia)
- Mechanical ventilation

From Way WL, Way EL, Fields HL: Opioid analgesics and antagonists. In: Katzung BG, ed: *Basic and clinical pharmacology*, 6th ed. Norwalk, CT: Appleton and Lange, 1995:460–477; Ellenhorn MJ, Schonwald S, Ordog G, et al, eds: *Ellenhorn's medical toxicology: diagnosis and treatment of human poisoning*. Baltimore: Williams & Wilkins, 1997:405–447; Jaffe JH, Martin WR: Opioid analgesics and antagonists. In: Gilman AG, Rall TW, Nies AS, et al, eds: *The pharmacological basis of therapeutics*, 8th ed. New York: Pergamon Press, 1990.

- Oxygen and anticonvulsants to combat seizure disorders
- Intravenous fluids and vasopressors to regain normal blood pressure
- Administration of an opioid antagonist (such as Naloxone) to counteract CNS and respiratory effects

Generally, hemodialysis and forced diuresis are ineffective in treating overdose patients. A more comprehensive description of overdose treatment is given in Table 6–6.

DETOXIFICATION

Opioid abusers normally require a treatment program for successful detoxification. Physicians and professional staff provide medical treatment, supervision, and counseling. Daily evaluations of the patient are essential. If the withdrawal symptoms are severe or prolonged, a methadone management protocol is followed to alleviate undesirable effects.

Methadone is a slower-acting opioid with a lower abuse liability that is frequently used in substitution pharmacotherapy. Patients are stabilized for 2–3 days before the methadone dosage is gradually decreased (6). However, some patients remain on methadone therapy for years. Buprenorphine and propoxyphene have also been investigated as alternatives in the treatment of opioid dependence.

Other treatments to alleviate withdrawal symptoms include fluids and electrolytes, antispasmodics (propantheline), sedative-hypnotics (phenobarbital), and antiadrenergics (clonidine) (6). For full recovery, an after-care program is highly advisable.

ANALYSIS OF OPIOIDS

Case 1

The patient was a 33-year-old woman admitted for delivery of a term, first-delivery baby. Prenatal visits and medical history were not significant, and prenatal ultrasound indicated healthy development of the fetus. Routine admission toxicology on a random urine at 1045 h indicated the following results:

Drug of Abuse Profile

- Emergency barbiturate screening test: none detected
- Cannabinoid screening: none detected
- Cocaine metabolite screening test: none detected

- Opiate-screening test: positive (300 ng/mL cutoff)

Test note: "Results are to be used for clinical evaluation only. Confirmation testing was not performed on the substance detected in this screen. A positive screening test is not necessarily indicative of drug presence."

The patient denied use of any medications. Despite concurrence by the patient's obstetrician-gynecologist, the well-baby nursery refused to admit the infant because of potential risk of withdrawal, and the infant was transferred to the neonatal intensive care unit for observation. The patient's physician called the laboratory for consultation. Review of the patient's opiate results indicated a weakly positive immunoassay result, just above the cutoff [+2 milliabsorbance units (mAU)/min]. The patient stated that she had consumed a single poppy-seed muffin for breakfast.

Because there was insufficient urine for confirmatory testing, a second random urine sample was collected at 1600 h. The results for this urine specimen were:

Drug of Abuse Profile

- Emergency barbiturate screening test: none detected
- Cannabinoid screening test: none detected
- Cocaine metabolite screening test: none detected
- Opiate-screening test: positive (300 ng/mL cutoff)

Test note: "Results are to be used for clinical evaluation only. Confirmation testing was not performed on the substance detected in this screen. A positive screening test is not necessarily indicative of drug presence."

The screening results indicated an even more positive opiate at +40 mAU/min. This urine specimen was sent out to a reference laboratory for confirmatory testing by gas chromatography–mass spectrometry (GC/MS).

The results of the confirmatory testing were:

- Morphine (total): 614 ng/mL (2149 nmol/L)
- Codeine: none detected
- 6-Acetylmorphine (6-AM): none detected (5 ng/mL detection limit)
- Morphine (free): 22 ng/mL (77 nmol/L) [5 ng/mL (17.5 nmol/L) detection limit]

Results were inconclusive for discrimination between street drug abuse, medication, and poppy-seed consumption. The baby was fine, and both patient and infant were discharged without incident.

Methods of Analysis

Opioid analgesics are analyzed by a variety of analytical techniques. Many specimen types require pretreatment steps including hydrolysis, precipitation, and/or extraction procedures to render them suitable for analysis. A general scheme illustrating opioid analysis is depicted in Figure 6–7.

Immunoassay

Immunoassay is the most commonly used screening technique for the identification of opioids in urine. Generally, urine specimens test positive by immunoassay as long as 3 days after opioid exposure (14). Often, these immunoassays can also be adapted for blood product analysis. Several "opiate" (phenanthrene-related compounds), methadone, fentanyl, and propoxyphene assays are commercially available, each one rapid and readily amenable for use with an automated chemistry analyzer. Commercial immunoassays include the:

- Enzyme-multiplied immunoassay technique (Emit®; Dade Behring, Deerfield, IL)
- Enzyme-linked immunosorbent assay (ELISA; Diagnostix, Mississauga, Ontario; Immunalysis Corp., San Dimas, CA; Neogen Corp., Lexington, KY; OraSure Technolgies, Inc., Bethlehem, PA)
- Fluorescence polarization immunoassay (FPIA; Abbott Diagnostics, Abbott Park, IL)

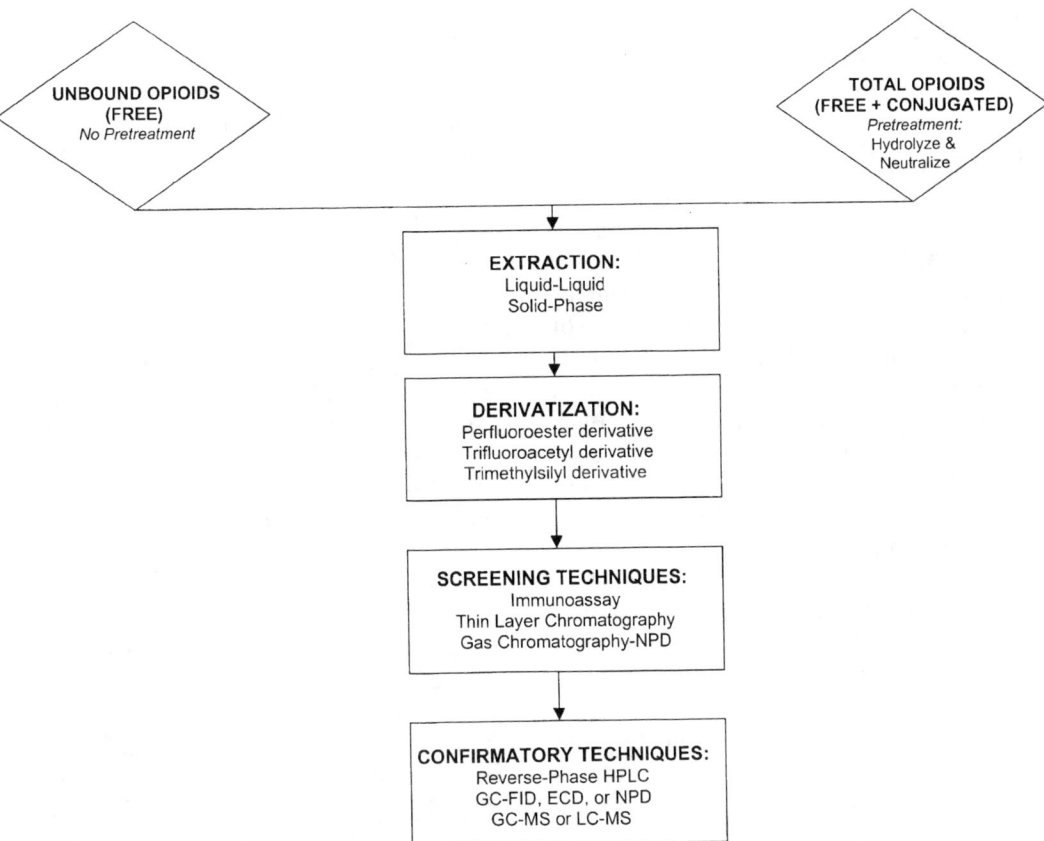

FIG. 6–7. Analysis of opioid analgesics. NPD, nitrogen-phosphorus detection; FID, flame-ionization detection; ECD electrochemical detection; LC/MS, liquid chromatography–mass spectrometry.

- Radioimmunoassay (RIA; Diagnostic Products Corporation, Los Angeles, CA)
- Kinetic interaction of microparticles in solution (KIMS®; Roche Diagnostic Systems)
- Cloned enzyme donor immunoassay (CEDIA®; Microgenics Corp., Fremont, CA)

Most "opiate" immunoassays use morphine as the calibrating analyte and are nonspecific, cross-reacting with morphine-like opioids and their metabolites. For example, hydromorphone and hydrocodone commonly produce positive immunoassay test results. Cross-reactivity varies with analyte concentration and reagent antibody specificity. For instance, the oxy-derivatives of morphine and codeine greatly deviate from the morphinan structure that reduces their potential cross-reactivity. The STC Technologies (Bethlehem, PA) ELISA screen, for example, is calibrated using morphine and its antibody will react with most phenanthrene-like compounds, but the assay does not cross-react significantly with oxymorphone or oxycodone. Furthermore, the methadone, fentanyl, and propoxyphene immunoassays are generally highly specific and do not cross-react with structurally unrelated compounds. Thus, immunoassays have to be evaluated for their cross-reactivity for the opioid of choice before using a particular assay. The analyst must know what the immunoassay results signify with respect to a presumptive-positive result.

Another interpretation issue for consideration by the analyst is potential interference from other drugs and exogenous analytes. Finally, the manufacturer's cutoff concentration for an immunoassay may not be suitable to the testing requirements of a particular laboratory.

Chromatography

Pretreatment and extraction are necessary for analysis by chromatographic techniques. Pretreatment of urine and blood specimens is required to obtain an accurate measure of the total opioids present including glucuronide conjugates. Pretreatment involves heated incubation for several hours with either β-glucuronidase and sulfatase enzymes or hydrochloric acid. Before extraction of the drug from the matrix, the hydrolysate must be neutralized (15).

Extraction of opioids from biological specimens commonly uses either liquid-liquid extraction (LLE) or solid-phase extraction (SPE). Most opioids are optimally extracted by LLE in the pH range of 8–10 from aqueous solution into an organic solvent. Organic solvent preparations include chloroform with an organic modifier such as isopropanol, *n*-butanol, or isoamyl alcohol.

SPE of opioids uses a sorbent made of copolymeric bonded-phase material with hydrophobic and cation-exchange properties. After column conditioning, the solid-phase sorbent sequentially binds and releases the drug analyte in a series of steps involving sample introduction, wash, and elution with elution solvent. A common SPE elution solvent is a mixture of methylene chloride/ isopropanol/ammonium hydroxide (78:20:2; vol./vol./vol.) (16,17).

Thin-Layer Chromatography

Thin-layer chromatography (TLC) is a rapid, simple, and inexpensive method widely used for the comprehensive screening of therapeutic and abused drugs in urine. For a 10-mL urine sample, the detection limits for most opioids by TLC are ~500–1000 ng/mL. Toxi-Lab® (Ansys, Inc., Irvine, CA) offers a commercial TLC system for the detection of many drugs including opioids with a stated approximate detection limit of 1000 ng/mL. A more sensitive and specific TLC system, the Toxi-Lab LTD™, detects codeine, dihydrocodeine, hydrocodone, hydromorphone, morphine, and 6-AM at a stated detection limit of 200 ng/mL.

Other Methods

High-performance liquid chromatography (HPLC) and gas chromatography (GC) are commonly used screening and confirmatory techniques. The method of choice for HPLC is reverse-phase chromatography coupled with coulometric electrochemical detection. Detection limits as low as 1–5 ng/mL can be achieved by this technique. Opioids are frequently measured by GC in combination with flame-ionization, nitrogen-phosphorus, electron-capture, or mass spectrometry detection. Many opioids require reaction with a derivatizing reagent to produce derivatives with adequate volatility and polarity. Trimethylsilyl-, perfluoroester-, and trifluoroacetyl- derivatives are the most common. GC/MS is the most widely accepted technique for forensic testing of opioids. The GC/MS detection limits are ~1–50 ng/mL (14).

All chromatographic assays should be investigated for potential interference due to structurally similar compounds. For example, interference with the measurement of morphine and codeine may occur due to the presence of opiate metabolites such as normorphine and norcodeine and synthetic opioids including hydromorphone and hydrocodone. Suitable levels of specificity are readily achievable with the formation of an appropriate analyte-derivative and optimized chromatographic conditions (16).

Interpretation

Blood

The presence of opioids and their metabolites in subjects suspected of drug impairment or overdose is frequently investigated because the results of drug and drug metabolite measurements are essential in developing diagnoses in these cases. The interpretation of toxicological findings in opioid-related deaths is complex for a number of reasons. Reports of opioid-related deaths indicate considerable variation in postmortem drug concentrations. Blood opioid concentrations in drug-related deaths overlap with those found in non-drug-related deaths due to homicides, suicides, and natural causes. Interpretation is further complicated by the subject's prior exposure to opioids and potential pharmacologic tolerance and the co-administration of other centrally acting drugs including ethanol and cocaine. Moreover, some opioids and their metabolites (for example, propoxyphene and norpropoxyphene) are cardiotoxic and require special evaluation of their safety if taken in combination with other medications or illicit drugs that are also cardiotoxic (for example, tricyclic antidepressants, mirtazapine, or cocaine). Postmortem redistribution of drug and drug metabolites must also be taken into consideration.

It is common practice to use therapeutic and toxic reference ranges for interpretation purposes. However, to use these values properly, one must know whether the reference range provided for the opioid is given as a "free" concentration or "total" concentration.

Depending on the drug detected, many laboratories measure free and total drug concentration (free drug plus conjugated metabolites) as a means of determining the mode of death. For example, rapid deaths are often characterized by a higher ratio of free morphine concentration to total morphine concentration.

Recent reports of patients administered large doses of morphine for treatment of severe pain associated with terminal disease have demonstrated free and total morphine concentrations in a range usually considered toxic. It is not uncommon for these patients to readily develop tolerance to the adverse CNS effects despite these high concentrations of drug.

Urine

Detectable concentrations of morphine and codeine in the urine usually occur as long as 48–72 hours after administration; however, route of administration and intraindividual differences in drug use, metabolism, and excretion patterns affect the detection time. Methadone and other opioids with longer elimination half-lives can be detected in the urine for 3–5 days after administration. Chronic administration of drugs will lead to an increase in detection times.

Morphine and codeine have been identified in the urine of individuals who have ingested food prepared with poppy seeds. Differentiating poppy seed ingestion from codeine, morphine and/or heroin use requires quantitative analysis. The following guidelines have been established to differentiate the source of morphine and codeine in urine:

- Total morphine concentration in urine >5000 ng/mL (>17,500 nmol/L) indicates of use of an opiate analgesic such as heroin, morphine, or codeine.
- Total morphine concentration in urine >1000 ng/mL (>3500 nmol/L), with no detectable amount of codeine, indicates morphine use. A morphine concentration >1000 ng/mL (>3500 nmol/L) and a morphine-codeine ratio <2 would exclude poppy seeds as the sole source for a positive opiate test result.
- Total codeine concentration in urine >300 ng/mL (>1002 nmol/L), with a morphine-to-codeine ratio of <2, is indicative of codeine use and rules out poppy seed ingestion.

Finally, unequivocal identification of heroin use is based upon the identification of 6-AM in the urine (18).

Alternative Matrices

Other biological tissues and fluids used for drugs-of-abuse testing include saliva, sweat, hair, nails, vitreous humor, cerebral spinal fluid, bile, meconium, amniotic fluid, breast milk, semen, sebum, stratum corneum, and organs such as kidney, brain, and liver (5,19–25).

Increasingly, alternative specimens are being used to enhance our interpretative capabilities, especially when evaluating difficult issues such as postmortem redistribution (liver drug concentration is a complementary specimen to resolve difficult cases); timing of drug exposure (hair can indicate recent vs. long-term use); and ease and reliability of point-of-care testing (detection of drugs in the breath and oral fluids).

In many situations, an analyst must exhibit caution when interpreting drug concentrations in these matrices due to the novelty of the matrix (comparison of oral fluids and breath concentrations to blood alcohol concentrations), lack of documented data (well-established toxic ranges for less commonly analyzed postmortem tissues), and unique issues (environmental contamination in hair) that are not associated with the more traditional matrices.

SPECIFIC AGENTS

Opium (Laudanum)

Opium (*Papaver somniferum*) is the original narcotic analgesic that all other opioids are derived from or patterned after for its pain-relieving effects. Derived from the Greek name "opion" for juice or sap, opium is comprised of more than 25 distinct alkaloids, including morphine and codeine, obtained from the milky exudates of the unripened opium poppy pod. The opium poppy is grown primarily in Southeast Asia, Mexico, and Southwest Asia. The weight percent composition of opium's major constituents is morphine (10%), noscapine (6%), papaverine (1%), codeine (0.5%), and thebaine (0.2%) (26).

Opium alkaloids have been used as drugs for hundreds of years; however, their medical use was not common until the isolation of morphine, codeine, and papaverine in the early 1800s. These discoveries, in conjunction with the invention of the hypodermic needle, led to the use of pure alkaloids rather than crude opium preparations such as tinctures. By World War II, opiate addiction became a problem, prompting the clinical field to introduce other opioids, like meperidine and methadone, which possess morphine-like analgesia but with less addiction potential (14,26). In 1909, the Opium Exclusion Act prohibited importation of opium into the United States (4).

Today, opium alkaloids (mixture) are still used in some geographical regions, especially Asia, to treat severe pain, under the trade names of Escopon®, Omnopon®, and Pantopon®. The products are supplied as 1 mL ampoules for intramuscular or subcutaneous injection. In addition to opium alkaloids, opium tinctures such as deodorized opium tincture (DTO, laudanum) and camphorated tincture of opium (paregoric), are used in the treatment of diarrhea or pain relief. Rectal suppositories of opium and belladonna (B&O Supprettes®) are also offered for the treatment of pediatric diarrhea (10,27).

Full Agonists

Codeine (Tylenol® with Codeine)

Codeine is a naturally occurring opioid agonist. Codeine is indicated for the relief of mild to moderate pain and is also an effective antitussive medication in low doses. Codeine is incorporated into a number of formulations, which include combination with acetaminophen or aspirin to increase its analgesic effect (5,10,13).

Codeine is administered orally or by intramuscular, intravenous, or subcutaneous injection in doses of 15–60 mg four to six times daily, not exceeding 360 mg in any 24-hour period. An oral dose is about two-thirds as effective as a parenteral dose (10,13).

First-pass metabolism significantly reduces the effectiveness of codeine. Codeine is metabolized by O-demethylation to morphine and N-demethylation to norcodeine. Codeine, morphine, and norcodeine are metabolized further, forming glucuronide and sulfate conjugates. Plasma morphine concentrations after codeine administration are about one-tenth the concentration of codeine. During the initial phase of elimination, concentrations of codeine in urine predominate, but over a 20- to 40-h period, morphine conjugates are the major products (5).

Dihydrocodeine (Synalgos®-DC)

Dihydrocodeine is a semisynthetic opioid analgesic, prepared by the hydrogenation of codeine, that has an affinity for the μ-receptors. Less efficacious than morphine, dihydrocodeine is indicated for mild to moderate pain relief. Dihydrocodeine capsules are administered orally, two to be taken every 4 hours as needed. Each capsule contains 16 mg of dihydrocodeine as the bitartrate salt, 30 mg of caffeine, and 356.4 mg of either acetaminophen or aspirin (10,13).

Dihydrocodeine is metabolized to nordihydrocodeine and dihydromorphine by N- and O-demethylation, respectively. Dihydrocodeine and its metabolites are further metabolized by conjugation (5,28).

Fentanyl (Sublimaze®)

Fentanyl, a synthetic opioid belonging to the phenylpiperidine family, was introduced in the United States in 1968 as an intravenous anesthetic and analgesic. Fentanyl is ~200 times as potent as morphine, but has comparable tolerance and physical dependence. Its rapid onset and short duration of action make it attractive for pre- and immediate postoperative medication and as a supplement for general anesthesia. Because of its availability and desirable effects, health-care professionals sometimes abuse fentanyl.

Fentanyl is available as an injectable solution. A typical adult dose is 50–100 μg administered intravenously or intramuscularly; children between 2 and 12 years are given a reduced dose of 2–3 μg/kg. Transdermal patches capable of providing continuous systemic delivery of fentanyl for 72 hours are also available. In this form, fentanyl is often administered at doses ranging from 25 to 100 μg/h for the management of chronic pain, for example, for a malignancy (10,13).

Fentanyl is metabolized primarily in the liver, producing the inactive metabolites despropionylfentanyl, norfentanyl, hydroxyfentanyl, and hydroxynorfentanyl (5).

There are several licit and illicit compounds that belong to the fentanyl family. We shall mention only the licit derivatives. Sufentanil, which is five to seven times as potent as fentanyl, is used as an analgesic and anesthetic in cardiac surgery (30- to 60-minute duration). Alfentanil, which has a very short duration of action (15 minutes) and is less potent than fentanyl, is used for minor surgeries. Lofentanil, which is used for prolonged respiratory depression in trauma patients, is 6000 times as potent as morphine. Carfentanil, which is 3200 times as potent as morphine, is used as an immobilizing agent for wild animals (6).

Heroin (Diacetylmorphine)

Heroin is a semisynthetic derivative, synthesized from morphine by the acetylation of the hydroxyl groups. Heroin is a widely abused opioid and is classified as a Schedule I substance by the United States Code of Federal Regulation. Although the medical use of heroin in the United States is prohibited, heroin is used in the United Kingdom for the management of severe chronic pain resulting from terminal illness (7).

Medicinal heroin is usually administered orally, or by intramuscular or subcutaneous injection, in doses of 5–10 mg every 4 hours. An oral dose is much less effective than a parenteral dose. Illicit heroin is typically administered by intravenous injection, smoking, or nasal insufflation.

Heroin is rapidly metabolized by deacetylation to 6-AM with a plasma half-life of approximately 5 minutes. 6-AM is further metabolized to morphine with an approximate plasma half-life of 6–25 minutes. The prolonged effect of heroin is due to the presence of free and conjugated morphine because heroin and 6-AM are rapidly metabolized. The analgesic effects of heroin are principally related to the combined effect of 6-AM and morphine (4–5).

The presence of 6-AM in urine is definitive evidence of heroin use. However, the detection time of 6-AM in urine is limited by its relatively short urinary half-life of 0.6 hours (5).

Hydrocodone (Vicodin®)

Hydrocodone is a semisynthetic analgesic and antitussive derived from codeine. Hydrocodone is six times as potent as codeine and is indicated for treatment of moderate to moderately severe pain. It is used postoperatively and in patients with pulmonary disease (5,10,13).

Hydrocodone is available as tablets (2.5–7.5 mg) that also contain acetaminophen (500–650 mg) or aspirin (500 mg) and as a cough syrup (1.7–5.0 mg/5 mL). A typical adult dose, in either form, is 5–10 mg every 4–6 hours, not exceeding 45 mg in any 24-hour period (10,13).

Hydrocodone is metabolized by O- and N-demethylation, forming hydromorphone and norhydrocodone, respectively, and by reduction of the 6-keto group, forming hydrocodol; hydrocodol is metabolized to hydromorphol. Hydromorphone and hydromorphol are further metabolized by conjugation (5).

Hydromorphone (Dilaudid®)

Hydromorphone, a hydrogenated ketone of morphine, is a semisynthetic opioid analgesic and antitussive. Hydromorphone has an addiction liability similar to morphine, although it is seven to 10 times as potent, and is indicated for treatment of moderate to severe pain due to cancer, surgery, trauma, biliary and renal colic, myocardial infarction, and burns (5,10,13).

Hydromorphone is available as the hydrochloride salt for oral, parenteral, and rectal administration. Tablets (2, 4, and 8 mg) may be taken orally every 4–6 hours in 2-mg doses.

For more severe pain, 4 mg may be prescribed. Parenterally, hydromorphone (1, 2, and 4 mg/mL) is normally administered in 1- to 2-mg doses every 4–6 hours, although a more concentrated injection is available (10 mg/mL) for patients already being prescribed opioid analgesics. Suppositories (3 mg) are given every 6–8 hours. Hydromorphone is also available as a cough syrup (1 mg/5 mL) with guaifenesin (100 mg/5 mL) and 5% alcohol (5,10,13).

Hydromorphone is metabolized to hydromorphol by reduction of the 6-keto group; both analytes are metabolized by conjugation (5).

Levorphanol (Levo-Dromoran®)

Levorphanol is a synthetic analgesic with properties and actions similar to those of morphine. Levorphanol, however, is four to five times as potent as morphine and has an analgesic effect that lasts longer. In addition to being used for moderate or severe pain relief in patients suffering from cancer, biliary and renal colic, trauma, and myocardial infarction, levorphanol is also a useful supplement to nitrous-oxide anesthesia. Used preoperatively, it allays apprehension and shortens recovery time (5,10,13).

Levorphanol is available as the tartrate salt in a solution (2 mg/mL) for subcutaneous injection or as a tablet (2 mg) for oral ingestion. The average adult dose of levorphanol is 2 mg every 6–8 hours (10,13).

Levorphanol is metabolized by N-demethylation to norlevorphanol; both analytes are metabolized further by conjugation (5).

Meperidine (Demerol®)

Meperidine is a synthetic analgesic, qualitatively similar to morphine, but only one-sixth to one-eighth as potent. The onset of action of meperidine is slightly faster than morphine, although the duration of action is shorter. Meperidine is indicated for the relief of moderate to severe pain, as preoperative medication, for support of anesthesia, and for obstetrical analgesia (6,10,13). The most prevalent meperidine addicts are health professionals, including physicians, nurses, and pharmacists (6).

Meperidine is administered to adults in 50- to 150-mg doses every 3–4 hours, as necessary, depending on its application. It is available as the hydrochloride salt in tablet form (50 and 100 mg), as a banana-flavored syrup (50 mg/5 mL) for oral use, and as a solution for parenteral use (25, 50, 75, and 100 mg/mL) (10,13).

Meperidine is metabolized to normeperidine by N-demethylation and to meperidinic acid and normeperidinic acid by hydrolysis of the ester functionalities of the parent compound and normeperidine, respectively. Meperidinic acid and normeperidinic acid are further metabolized by conjugation. Normeperidine accumulates in the plasma with chronic administration of meperidine. The metabolite possesses half the analgesic activity of meperidine, but is two to three times as toxic as the parent compound. It also appears to have a longer half-life (5).

Methadone (Dolophine®)

Methadone is a synthetic diphenylpropylamine opioid that has analgesic actions and potency similar to morphine when administered parenterally. Unlike morphine, however, tolerance and physical dependence develop slowly. Although methadone is indicated for severe pain relief, its primary application is the detoxification and/or maintenance treatment of narcotic addiction (10,13).

Methadone is available orally in tablets (5, 10, and 40 mg) and as a solution (10 mg/mL) for parenteral use. For pain relief, the average adult dose is 2.5–10 mg every 3–4 hours as necessary. For relief of withdrawal symptoms during detoxification, 15–20 mg/d is usually sufficient, although as much as 40 mg can be administered if symptoms are not suppressed. Maintenance of opiate dependence requires 20–120 mg/d. Maintenance patients (those receiving methadone for >21 days) may tolerate doses >200 mg/d, although doses as small as 50 mg have been known to be fatal to nontolerant adults (10,13,29).

Methadone is metabolized by mono- and di-N-demethylation, which, after subsequent spontaneous cyclization, leads to two pharmacologically inactive pyrrolidines. These compounds, in addition to methadone, are further metabolized by hydroxylation, resulting in the formation of *para*-hydroxypyrrolidines. Hydroxylation of methadone produces methadol, which is then metabolized to normethadol by N-demethylation (5).

Morphine (Roxanol™)

Morphine is a naturally occurring opioid agonist. Morphine is indicated for the relief of moderate to severe acute and chronic pain. Morphine is used to treat pain associated with acute myocardial infarction and relieve dyspnea associated with left ventricular failure and pulmonary edema. Morphine is also used as a preoperative medication as well as an alternative to skeletal muscle relaxants to suppress spontaneous respiration of patients on ventilators (5,10,13).

Morphine is administered orally, rectally, or by intramuscular, intravenous, or subcutaneous injection, typically in doses of 2.5–10 mg, until the desired effect is obtained. A controlled-release form is also available (MS Contin). Opiate-tolerant patients may require much larger doses. Morphine may also be administered by transdermal absorption and epidural and intrathecal infusion. An oral dose is about two-thirds as effective as a parenteral dose (10,13).

First-pass metabolism significantly reduces the effectiveness of morphine. Morphine is metabolized by glucuronidation and sulfation, N-demethylation, and *N*-oxide formation. Morphine-6-glucuronide is a pharmacologically active metabolite (5).

Oxycodone (Percocet®)

Oxycodone is a semisynthetic narcotic analgesic chemically derived from thebaine indicated for treatment of moderate to moderately severe pain. Analgesically, oxycodone is equipotent to morphine when administered subcutaneously (10,13).

Rarely is oxycodone given alone; combination formulations such as oxycodone with acetaminophen (325 mg), aspirin (325 mg), and/or caffeine are commercially available. The normal adult oral dose is 10–30 mg every 4 hours as indicated. A controlled-release form of oxycodone is also available (OxyContin®) (10,13).

Metabolism of oxycodone includes N- and O-demethylation, forming noroxycodone and oxymorphone, respectively. Glucuronide conjugation of oxycodone and oxymorphone leads to other polar metabolites (5).

Oxymorphone (Numorphan®)

Oxymorphone is a semisynthetic opioid derived from thebaine indicated for treatment of moderate to severe pain. Clinically effective in the treatment of severe pain, oxymorphone is commonly used in preoperative and obstetric analgesia and to treat anxiety and dyspnea associated with ventricular failure and pulmonary edema. The analgesic potency of oxymorphone is eight to 10 times that of morphine (13).

Parenteral injections or suppositories are the most common forms of administration. Adult doses depend on the route, with variations as follows: 1–1.5 mg every 4–6 hours (intramuscularly, subcutaneously), 0.5 mg every 4–6 hours (intravenously), and 5 mg every 4–6 hours (rectally). Low bioavailability resulting from first-pass metabolism makes oral administration ineffective. The therapeutic and toxic doses have not yet been determined (10,13).

Extensive metabolism of oxymorphone by reduction of the 6-keto group forms oxymorphol metabolites; both analytes are metabolized by conjugation (5).

Propoxyphene (Darvon®)

Propoxyphene, or dextropropoxyphene, has been available in the United States since 1957 for the treatment of mild to moderate pain. Propoxyphene is chemically similar to methadone; its dextrorotatory isomer is an analgesic, and the levorotatory isomer is an antitussive. Low abuse liability and mild withdrawal symptoms are two desirable characteristics of propoxyphene that have made it an alternative to methadone in detoxification and maintenance of narcotic dependence. However, propoxyphene and its metabolites are cardiotoxic, increasing the likelihood of overdose. Naloxone and nalorphine can alleviate the respiratory depression, but are somewhat ineffective for the treatment of cardiodysfunction (4).

Like oxycodone, propoxyphene is used in combination with non-opioid analgesics such as aspirin and acetaminophen to obtain an additive analgesic effect. Low efficacy makes propoxyphene unsuitable for treatment of severe pain. Propoxyphene is commercially available in tablets as the hydrochloride or napsylate salt. The common adult dosage is 65 mg, administered 4–6 times daily as needed, with a maximum daily dose of 600 mg (10,13).

Propoxyphene is metabolized to norpropoxyphene by N-demethylation. Further N-demethylation occurs, forming dinorpropoxyphene, which can subsequently dehydrate to form cyclic dinorpropoxyphene. Other polar metabolites are formed by ring hydroxylation, ester hydrolysis, and glucuronide conjugation. Norpropoxyphene is an active metabolite that sometimes contributes to both therapeutic and toxic effects more than the parent drug. Norpropoxyphene has a much longer half-life than propoxyphene (5).

Tramadol (Ultram®)

Tramadol is a synthetic analgesic, one-sixth to one-eighth as potent as morphine, indicated for moderate to moderately severe pain. The mode of action of tramadol is not completely understood, and at least two mechanisms, one opioid, appear applicable. There is less risk of respiratory depression with tramadol than with other opioid analgesics, an attractive feature. Tramadol is prepared as the hydrochloride salt and is available for oral use in tablets containing 50 mg or as an injectable solution for intramuscular and intravenous use. The recommended adult dose is 50–100 mg every 4–6 hours, not to exceed 400 mg/day (10,30,31).

Tramadol is extensively metabolized by N- and O-demethylation followed by glucuronide conjugation. The major metabolite of tramadol, mono-O-desmethyltramadol, is pharmacologically active and has stronger analgesic action than the parent drug (31).

Case 2

A 34-year-old nurse-anesthetist was found unconscious in a hospital bathroom. An empty 5-cc syringe was found nearby. The nurse was taken to the emergency department for treatment, and after the intravenous administration of glucose and naloxone, she regained consciousness. Physical examination was normal except for injection marks along the anticubital region of both arms. The nurse was uncooperative and would not discuss her substance-abuse problem.

Serum and urine specimens were obtained and sent to the clinical chemistry laboratory for drug analysis. The results of the drug screen, which included common therapeutic drugs and drugs of abuse, were negative. Analytical techniques used by the laboratory included immunoassay and TLC.

The nurse was transferred to the intensive care unit for monitoring. Further inquiry by risk management revealed missing ampoules of fentanyl in the outpatient surgical pharmacy. To complete the investigation, the serum and urine specimens were sent to a regional reference laboratory for fentanyl analysis. The results obtained by GC/MS were:

- Serum fentanyl: 4.5 ng/mL (13.4 nmol/L)
- Urine fentanyl: positive

Mixed Agonist-Antagonists

Buprenorphine (Buprenex®)

Buprenorphine is a semisynthetic opioid chemically derived from thebaine; it has a morphine-like structure. Although buprenorphine is a mixed agonist at the μ-receptor, its potency is 25–40 times that of morphine, a full agonist. Buprenorphine's actions persist longer than equal doses of morphine or meperidine due to its strong receptor affinity. Its high receptor affinity may also contribute to less-severe withdrawal symptoms and reduced effectiveness of naloxone reversal in the treatment of buprenorphine overdose. Moreover, buprenorphine is equipotent to naltrexone when used in the treatment of opioid overdose (5). Buprenorphine has been used in the treatment of cocaine and opioid dependence (32–34).

Buprenorphine is usually given by intramuscular or intravenous injection at a dosage of 1 mL (0.3 mg) every 6–8 hours. Epidural and intrathecal routes of administration have been used to treat acute and chronic pain of the legs and torso (5,6,10).

Buprenorphine is extensively metabolized by N-dealkylation to form an active metabolite, norbuprenorphine; both analytes are metabolized further to conjugated metabolites (5).

Butorphanol (Stadol®)

Butorphanol is a synthetic opioid analgesic used to treat moderate to severe pain since 1978. It has a benzomorphan structure, similar to levorphanol. Butorphanol use includes post- and preoperative anesthesia, balanced anesthesia, migraine headaches, obstetric labor, and musculoskeletal pain (10,13).

Butorphanol has agonist activity at the κ-receptor and antagonist activity at the μ-receptor. In addition to the usual CNS effects produced by opioid interaction with its receptor, butorphanol induces sedation and narcosis at 0.5 mg and 10–12 mg, respectively (10). The analgesic potency of butorphanol is four to eight times that of morphine (5,13).

Butorphanol is supplied as a tartrate salt in an injectable solution or a nasal spray. Parenteral administration at doses of 0.5–2 mg intravenously or 1–4 mg intramuscularly are given every 3–4 hours, as indicated. Transnasal delivery consists of 1–2 mg given every 3–4 hours (10,13).

Butorphanol is rapidly and well-absorbed

transnasally with a bioavailability of 60–70%, which is superior to its oral bioavailability of 5–17% (10). Metabolism of butorphanol yields a major inactive metabolite, hydroxybutorphanol, by hydroxylation of the cyclobutyl ring. A minor metabolite, norbutorphanol, is formed by N-dealkylation. Norbutorphanol is further metabolized by conjugation (5).

Butorphanol may precipitate withdrawal symptoms in opioid-dependent patients who have not been detoxified; consequently, it should be avoided in the treatment of these patients (13).

Nalbuphine (Nubain®)

Nalbuphine is a synthetic agonist-antagonist narcotic analgesic of the phenanthrene group, chemically related to oxymorphone and naloxone. Nalbuphine is commonly used to treat moderate to severe pain associated with pre- and postoperative analgesia, preanesthesia, balanced anesthesia, and childbirth. Nalbuphine is a strong κ-agonist and μ-antagonist that acts within 2–15 minutes after injection. Its analgesic duration lasts 3–6 hours with potency equivalent to morphine. Nalbuphine's antagonistic effects are one-fourth as potent as nalorphine's but 10 times as potent as pentazocine's. Nalbuphine has been effective in the treatment of respiratory depression occurring after postsurgical opioid administration given epidurally (10).

Nalbuphine is available as a hydrochloride salt solution that can be injected subcutaneously, intramuscularly, or intravenously. The normal adult dosage is 10 mg every 3–6 hours, as necessary. The single-dose amount should not exceed 20 mg, and the maximum daily dose is 160 mg (10,13).

Metabolism of nalbuphine includes N-dealkylation to form nornalbuphine and glucuronide conjugation of both analytes (5).

Pentazocine (Talwin®)

Pentazocine is a synthetic benzomorphan derivative chemically and structurally related to butorphanol. As an opioid antagonist, pentazocine is weakly active. Moreover, pentazocine weakly antagonizes morphine and meperidine, partially reversing the cardiovascular, respiratory, and behavioral depression induced by these drugs. One limitation to pentazocine treatment is its sedative effects. Pentazocine is indicated for the relief of moderate to severe pain and is commonly used for pre- and postoperative analgesia and as a supplement to surgical anesthesia (10,13).

Pentazocine is used with acetaminophen (650 mg) or aspirin (325 mg) to increase analgesic effectiveness and with naloxone (0.5 mg) to prevent misuse by intravenous injection. Pentazocine has narcotic analgesic properties similar to morphine and meperidine, with 30 mg of pentazocine being equipotent to 10 mg and 75–100 mg, respectively (10,13).

The adult parenteral dose is 30–60 mg given 6–8 times daily. The adult oral dose is slightly higher at 50–100 mg due to reduced bioavailability. The maximum daily dose is 360 mg (10,13).

The major chemical reaction during pentazocine metabolism is oxidation of either of the methyl groups on the dimethylallyl side chain to form cis- and trans-hydroxypentazocine. The trans-isomers can be further metabolized through carboxylation to form trans-carboxypentazocine. Glucuronide conjugation of the parent drug and metabolites also occurs (5).

Pentazocine has a relatively low potential for abuse and dependence; however, if combined with tripelennamine, it can be abused for its heroin-like effects. Unfortunately, tripelennamine potentiates pentazocine effects and lethality; thus, this combination is potentially deadly. Pentazocine precipitates withdrawal symptoms in opioid-dependent patients receiving opioids on a regular basis (6,13).

REFERENCES

1. Joranson DE, Ryan KM, Gilson AM, et al: Trends in medicinal use and abuse of opioid analgesics. *JAMA* 2000;283:1710–1714.

2. NIDA Infofax. http://www.nida.nih.gov/infofax/nationtrends.html. 1998.
3. Community Epidemiology Work Group. http://www.nida.nih.gov/CEWG/pubs.html. December 2000.
4. Way WL, Way EL, Fields HL: Opioid analgesics and antagonists. In: Katzung BG, ed: *Basic and clinical pharmacology*, 6th ed. Norwalk, CT: Appleton and Lange, 1995:460–477.
5. Baselt RC: *Disposition of toxic drugs and chemicals in man*, 5th ed. Foster City, CA: CTI, 2000.
6. Ellenhorn MJ, Schonwald S, Ordog G, et al, eds: *Ellenhorn's medical toxicology: diagnosis and treatment of human poisoning*. Baltimore: Williams & Wilkins, 1997:405–447.
7. Jaffe JH, Martin WR: Opioid analgesics and antagonists. In: Gilman AG, Rall TW, Nies AS, et al, eds: *The pharmacological basis of therapeutics*, 8th ed. New York: Pergamon Press, 1990.
8. Burtis CA, Ashwood ER, eds: *Tietz textbook of clinical chemistry*, 2nd ed. Philadelphia: WB Saunders, 1994.
9. Roscow CE: Buprenorphine: epidural and intrathecal use. In: Cowan A, Lewis JW, eds: *Buprenorphine: combating drug abuse with a unique opioid*. New York: Wiley-Liss, 1995:165–174.
10. *Physicians' desk reference*, 54th ed. Montvale, NJ: Medical Economics Data, 2000.
11. Bulow HH, Linnemann M, Berg H, et al: Respiratory changes during treatment of postoperative pain with high dose transdermal fentanyl. *Acta Anaesthesiol Scand* 1995;39:835–839.
12. Fiset P, Cohane C, Brown S, et al: Biopharmaceutics of a new transdermal fentanyl device. *Anesthesiology* 1995;83:459–469.
13. Lacy C, Armstrong LL, Lipsy RJ, et al: *Drug information handbook*. Hudson, OH: Lexi-Comp, 1993.
14. Goldberger BA: Opiates. Irving, TX: Abbott Laboratories, 1994. [Caplan YH, ed. Abused drugs monograph series.]
15. McCurdy HH: Enzymatic digestion of biological specimens for drug analysis. In: Liu RH, Goldberger BA, eds: *Handbook of workplace drug testing*. Washington, DC: AACC Press, 1995:45–66.
16. Foltz RL, Fentiman AF, Foltz RB: Morphine. In: Foltz RL, Fentiman AF, Foltz RB, eds: *GC/MS assays for abused drugs in body fluids*. [NIDA Research Monograph 32.] Rockville, MD: National Institute on Drug Abuse, 1980.
17. Cone EJ, Darwin WD: Rapid assay of cocaine, opiates and metabolites by gas chromatography-mass spectrometry. *J Chromatogr* 1992;580:43–61.
18. ElSohly MA, Jones AB: Origin of morphine and codeine in biological fluids. In: Liu RH, Goldberger BA, eds: *Handbook of workplace drug testing*. Washington DC: AACC Press, 1995:225–237.
19. Inoue T, Seta S, Goldberger BA: Analysis of drugs in unconventional samples. In: Liu RH, Goldberger BA, eds: *Handbook of workplace drug testing*. Washington DC: AACC Press, 1995:131–158.
20. Inoue T, Seta S: Analysis of drugs in unconventional samples. *Forensic Sci Rev* 1992;4:90–106.
21. Idowu OR, Caddy B: A review of the use of saliva in the forensic detection of drugs and other chemicals. *J Forensic Sci Soc* 1982;22:123–135.
22. Ropero-Miller JD, Goldberger BA, Cone EJ, et al: The disposition of cocaine and opiate analytes in hair and fingernails of humans following cocaine and codeine administration. *J Anal Toxicol* 2000; 24:546–554.
23. Levine B: Postmortem forensic toxicology. *Therapeutic Drug Monitoring and Toxicology In-Service Training and Continuing Education* 1994;15:7–13.
24. Prouty RW, Anderson WH: The forensic science implications of site and temporal influences on postmortem blood-drug concentrations. *J Forensic Sci* 1990;35:243–270.
25. Joseph RE Jr, Oyler JM, Wstadik AT, et al: Drug testing with alternative matrices I. Pharmacological effects and disposition of cocaine and codeine in plasma, sebum, and stratum corneum. *J Anal Toxicol* 1998;22:6–17.
26. Reisine T, Pasternak G: Opioid analgesics and antagonists. In: Gilman AG, Hardman JG, Limbird LE, eds: *Goodman and Gilman's the pharmacological basis of therapeutics*, 9th ed. New York: McGraw-Hill, 1996:521–553.
27. Leiken JB, Paloucek FP, eds: *Poisoning & toxicology compendium*. Cleveland, OH: Lexi-Comp Inc., 1998.
28. Rowell FJ, Seymour RA, Rawlings MD: Pharmacokinetics of intravenous and oral dihydrocodeine and its acid metabolites. *Eur J Clin Pharmacol* 1983;25:419–424.
29. Tietz NW: *Clinical guide to laboratory tests*, 3rd ed. Philadelphia: WB Saunders, 1995.
30. Lee CR, McTavish D, Sorkin EM: Tramadol: a preliminary review of its pharmacodynamic and pharmacokinetic properties, and therapeutic potential in acute and chronic pain states. *Drugs* 1993;46:313–340.
31. Stone JA: Tramadol. *Therapeutic Drug Monitoring and Toxicology In-Service Training and Continuing Education* 1995;16:221–223.
32. Mello NK, Mendelson JH: Buprenorphine suppresses heroin use by heroin addicts. *Science* 1980;207:657–659.
33. Johnson RE, Jaffe JH, Fudala PJ: A controlled trial of buprenorphine treatment for opioid dependence. *JAMA* 1992;267:2750–2755.
34. Mello NK, Kamien JB, Lukas SE, et al: Effects of intermittent buprenorphine administration on cocaine self-administration by rhesus monkeys. *J Pharmacol Exp Ther* 1993;264:530–541.

SELF-ASSESSMENT QUESTIONS

1. How long does the abstinence syndrome usually last?
   a. 8 hours
   b. 2–3 days
   c. 7–10 days
   d. 30 days
   e. 1 year

2. Which are among therapeutic and antidotal interventions for opioid overdose?
   a. anticonvulsants, naloxone, central nervous system (CNS) depressant
   b. anticonvulsants, respiratory stimulant, naloxone
   c. respiratory depressant, naloxone, vasopressor
   d. vasopressor, anticonvulsants, CNS depressant
   e. naloxone, vasopressor, sedative-hypnotic

3. Which opioid has a greater risk for overdose because of its cardiotoxicity?
   a. oxymorphone
   b. tramadol
   c. levorphanol
   d. nalbuphine
   e. propoxyphene

4. What opioid is most often used during patient detoxification and maintenance of strong opioid analgesics?
   a. nalbuphine
   b. methadone
   c. tramadol
   d. buprenorphine
   e. pentazocine

5. What is the common adverse effect associated with opioid analgesics?
   a. constipation
   b. mydriasis
   c. urinary incontinence
   d. tachycardia
   e. seizures

6. With which of the following drug or drug class(es) can opioids have additive effects?
   a. general anesthetics, cannabinoids, CNS depressants
   b. general anesthetics, tricyclic antidepressants, phenothiazines
   c. tricyclic antidepressants, sedative-hypnotics, cannabinoids
   d. CNS depressants, general anesthetics, tricyclic antidepressants
   e. tricyclic antidepressants, nalorphine, sedative-hypnotics

CHAPTER 7

Cocaine*

*Daniel Isenschmid, Ph.D., DABFT,
and Barry Levine, Ph.D., DABFT, DABCC*

LEARNING OBJECTIVES

After completing this chapter, the reader should be able to:

1. Describe the major pharmacologic and behavioral effects of cocaine.
2. Explain the mechanisms of action of cocaine.
3. Discuss the effects of route of administration on the pharmacokinetic and pharmacodynamic factors of cocaine.
4. Identify the major pathways of cocaine metabolism.
5. Identify the major cocaine product in common and unusual biological specimens.
6. Describe the different methods to identify and quantify cocaine and metabolites in different biological specimens.
7. List different mechanisms by which cocaine can produce its toxic or lethal effects.

GENERAL CHARACTERISTICS

History

Cocaine is an alkaloid that is found in the plant *Erythroxylon coca*, which grows principally in the northern South American Andes and to a lesser extent in India, Africa, and Java. The plant, which may reach 9 feet (2.7 m), favors higher elevations [up to 6000 feet (1829 m)] because, at lower elevations [<1500 feet (<457 m)], the alkaloid content is significantly diminished owing to more rapid growth. It takes about 2 years from the time of planting until the coca leaves may first be harvested, then depending on the altitude, the leaves may be harvested as many as three times a year. The leaves are dried and converted into a coca paste, which is eventually used to produce cocaine hydrochloride. The yields from 100 kg of coca leaves are about 1 kg and 800 g, respectively, for coca paste and cocaine hydrochloride.

In the mid-nineteenth century Carl Wohler, the chemist who synthesized urea, had coca leaves imported to Germany and presented them to his graduate student, Albert Niemann, to analyze. Niemann, as part of his doctoral dissertation, was the first to successfully isolate cocaine from the coca plant. From the 1860s through the turn of the century, cocaine appeared in various elixirs and tonics purported to have "magic" properties. Some of the more famous preparations included Vin Mariani, a mixture of wine and cocaine, and the original Coca-Cola, which at that time was a coca-based syrup supplemented with caffeine and marketed as a tonic and headache remedy. The Coca-Cola advertising phrase "the pause that refreshes" has often been attributed to the cocaine-con-

---

*Adapted from Isenschmid DS: Cocaine. In: Levine B, ed: *Principles of forensic toxicology*. Washington, DC: AACC Press, 1999:221–245.

taining cola but was not intended to refer to this early preparation. To this day, other by-products from the coca plant remain as some of the "secret" ingredients in the beverage.

Case Study

A 30-year-old black male was stopped by a police officer after driving erratically. As the driver exited his vehicle, he approached the police officer acting aggressively, combatively, and hyperactively. His speech was incoherent, and he threatened the initial officer and the back-up officers who arrived. He also displayed paranoia, because he was convinced that the officers were trying to kill him. When he took a swing at one of the officers, a struggle ensued. The driver also showed unexpected strength in that all the police officers were needed to subdue him. He was still acting this way as the officers transported him to the emergency department. As hospital personnel approached he suddenly stopped breathing. Resuscitation was unsuccessful, and the patient died. Postmortem toxicology indicated the following concentrations: blood cocaine 0.1 mg/L (0.33 µmol/L), blood benzoylecgonine 3.4 mg/L (11.7 µmol/L), urine cocaine 21 mg/L (69 µmol/L), and urine benzoylecgonine 200 mg/L (690 µmol/L).

Sources

Despite numerous early reports of cocaine toxicity including cardiac arrhythmias and fatalities, cocaine-containing products became ever more popular. In 1914, as cocaine abuse began to be viewed as a problem, the drug was labeled a narcotic (albeit incorrectly) under the Harrison Narcotic Act, and over-the-counter sales were discontinued. More recently, under the Controlled Substances Act, cocaine has been scheduled as a drug with some medicinal value but a high potential for abuse (Schedule II). It continues to have some medical use, almost exclusively limited to topical administration as a local anesthetic in ear, nose, and throat surgery (as the hydrochloride salt in 10–20% solutions) and in ophthalmological procedures (as a 1–4% solution).

Today, cocaine is one of the most commonly abused drugs. According to the American Association of Poison Control Centers Toxic Exposure Surveillance Systems, there were 4286 cocaine exposures reported throughout the country in 1999. Of these exposures, about 15% of the individuals were <20 years of age. Moreover, 87% of the cases required treatment in health-care facilities. It should be noted that due to limitations in data-collection methodology, these numbers probably represent an underestimation of total cocaine toxic exposures.

Cocaine sold on the street is in the form of the hydrochloride salt and crack. The salt form varies considerably in purity but today is usually at least 30% pure by the time it reaches the purchaser of gram-sized packets. The salt form is typically diluted ("cut") with agents such as mannitol, lactose, and sucrose to add bulk. In addition, readily available central nervous system (CNS) stimulants such as caffeine, phenylpropanolamine, and ephedrine and other local anesthetics such as lidocaine, procaine, and benzocaine are commonly used as diluents to simulate the actual drug. The cocaine powder supplied by dealers is often clumpy and first needs to be chopped. This is usually achieved using a mirror with a razor blade, after which the cocaine is arranged into thin lines about 30- to 60-mm long and 2-mm wide, resulting in an average dose of 25 mg, and then snorted through a straw or "tooter." Alternatively, cocaine may be snorted from a "coke spoon" or "bullet," a device in which a vial containing cocaine may be inverted over a closed chamber into which the cocaine falls and may then be rotated for convenient snorting. A single long fingernail may serve as a natural "coke spoon."

Crack is a free base form of cocaine that produces a characteristic crackling sound when smoked. It should not be confused with "free-basing," which is no longer commonly practiced. Free-basing was a process in which the user purified regular cocaine hydrochloride by mixing an aqueous solution of cocaine with baking soda or ammonia and adding ether, thereby extracting the free form of the drug into the organic solvent, which was then evaporated to dryness. The drug could then be smoked in a pipe (with the risk of igniting any remaining ether). Crack, on the other hand, is converted to the free base form before sale. It is prepared by adding baking soda to aqueous cocaine hydrochloride and heating it to remove the water. After heating, the mixture is cooled and filtered and the free base cocaine precipitates into small pellets or "rocks."

Chemistry

Cocaine (methylbenzoylecgonine) is an ester of benzoic acid and the amino alcohol methylecgonine, which contains a tropine moiety and is chemically, but not pharmacologically, related to atropine (Figure 7–1). The ecgonine portion of the molecule has four asymmetric carbon atoms and can exist as four racemates (eight optically active isomers). Cocaine is commercially synthesized from (–)-ecgonine in the presence of methanol and benzoic acid after hydrolysis of the ester alkaloids extracted from the plan material. In (–)-cocaine the benzoyl and methyl ester are located cis to the nitrogen bridge. (+)-Pseudococaine, with the methyl ester located trans to the nitrogen bridge, is also active.

Cocaine is structurally different from other local anesthetics by virtue of its tropine moiety, but possesses certain similarities. Like other local anesthetics, cocaine consists of a hydrophobic region and a hydrophilic region. The hydrophobic region contains a benzene ring, whereas the hydrophilic region consists of a tertiary amine. Cocaine is also similar to other local anesthetics due to its ester linkages, which allow the body to hydrolyze and deactivate the drug. However, the ester group is also susceptible to in vitro hydrolysis.

Effects

Cocaine is used medicinally as a topical local anesthetic; thus, its most important

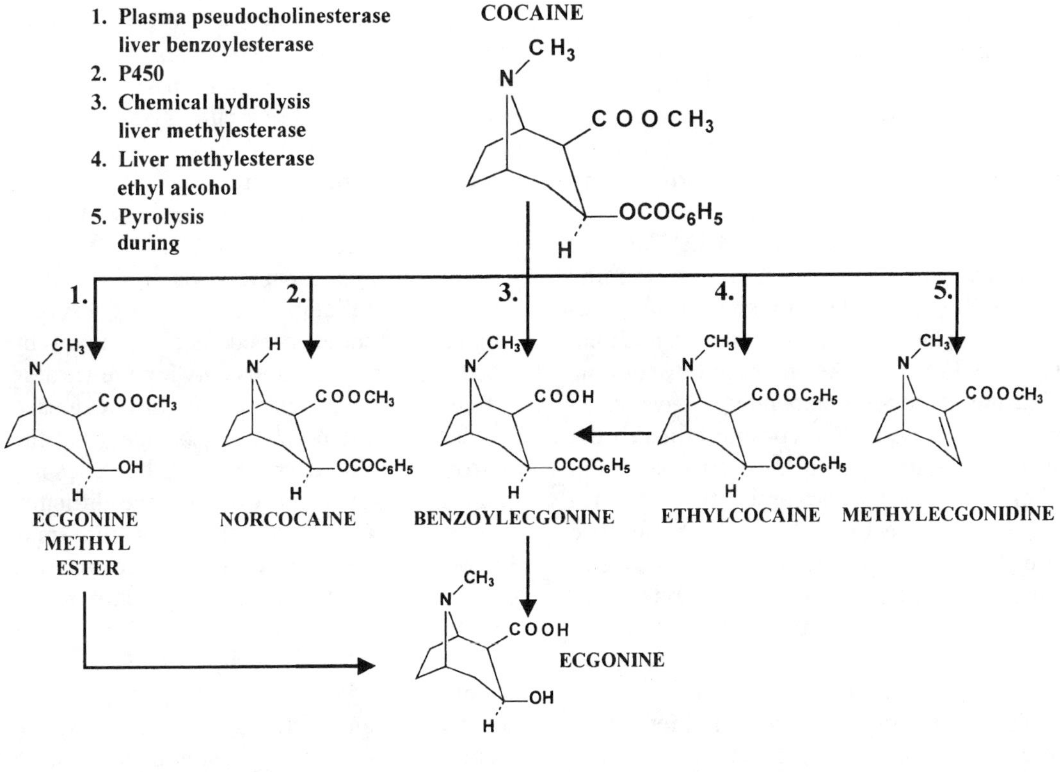

FIG. 7–1. Metabolism of cocaine.

mechanism of action clinically lies in the ability to block sodium channel conductance and thereby increase the threshold required to generate an action potential. However, unlike other local anesthetics, cocaine has the ability to block reuptake of the neurotransmitters norepinephrine (NE), dopamine (DA), and serotonin (5-HT). NE is responsible for the classic adrenergic effects seen with cocaine use including mydriasis, vasoconstriction, hypertension, tachycardia, and tachypnea. Although these effects may be most important toxicologically, the behavioral effects of cocaine appear to be mediated by its dopaminergic actions, which control the behavioral response to cocaine.

The desirable effects of cocaine include intense euphoria, psychic energy, heightened sexual excitement, and self-confidence (elevation of mood). Undesirable effects experienced by some users include paranoia, hallucinations, and dysphoria. The central stimulatory effects ("rush") are followed by depression ("crash"), and it is the positive reinforcement of the "rush" vs. the negative reinforcement of the "crash" that is the principal reason for the development of chronic cocaine abuse. After an acute dose of cocaine, brain concentrations of NE and DA are elevated briefly and are followed by a marked reduction to below-normal concentrations, correlating with the "rush" and "crash" experienced by the cocaine user. Cocaine prevents the reuptake of DA into the presynaptic dopaminergic neuron by binding to receptors on the DA transporter that is located on the dopaminergic nerve terminal (1). This DA reuptake, mediated by sodium, chloride, and energy-dependent active transport, is inhibited when cocaine binds to the sodium-binding site on the transporter and alters the chloride binding site, thus preventing the binding of both ions (2). Because translocation of DA across the membrane of the presynaptic neuron is inhibited, increased extracellular DA concentrations result in chronic stimulation of the DA receptor in the postsynaptic neuron.

The most common clinical manifestations after acute cocaine intoxication include profound CNS stimulation with psychosis and repeated grand-mal convulsions, ventricular arrhythmias, and respiratory dysfunction with Cheyne-Stokes breathing and ultimately respiratory paralysis. Other symptoms include mydriasis, hypertension leading to hypotension, and small muscle twitching. Cocaine's ability to cause increased muscular activity and vasoconstriction may produce extreme hyperthermia. The patient may also be in a coma. Acute myocardial infarctions have occurred after even a therapeutic dose of cocaine.

Symptoms of chronic cocaine use, other than psychiatric disturbances, include rhinitis (with possible nasal septum perforation), shortness of breath, cold sweats, tremors, violent protective behavior, distorted perception, tachycardia, tachypnea, dyspnea, and hyperkinetic behavior. The drug can injure cerebral arteries, and an acute hypertensive episode after a single dose in a chronic user can cause the vessels to rupture. Cocaine may induce seizures and angina and worsen pre-existing coronary heart disease to the point of a heart attack. Tolerance to the cardiovascular effects of cocaine do not occur and, during a cocaine "binge," the chronic cocaine user may self-administer more cocaine than the cardiovascular system can tolerate. Malnutrition, which can lower immune defenses and precipitate other diseases, and poor personal grooming may also be observed after chronic cocaine use.

The chronic cocaine user may become quite self-confident and egocentric. Psychological dependence becomes dominant by virtue of the drug's unusually high rewarding effects. The term "coked-out" has been used to describe the psychological changes in the chronic cocaine user. Thinking becomes impaired, and such individuals have difficulty concentrating and remembering; they may grow short-tempered and suspicious and may undergo a schizophrenic psychosis similar to that seen with amphetamine abuse. They may also become aggressive and experience attacks of paranoia or panic or auditory, visual, and tactile hallucinations. Symptoms may be severe enough to require hospitalization and antipsychotic pharmacotherapy.

PHARMACOKINETICS

Absorption

Cocaine may be administered intranasally (IN), by smoking (SM), intravenously (i.v.), and orally (PO). Cocaine is usually not administered PO because first-pass effects result in low bioavailability (about 20%) and reduced euphoric effects due to inefficient delivery to the brain (3). The i.v. route of administration, sometimes called "mainlining," is the only route that consistently produces 100% drug bioavailability. Bioavailability by both the IN and SM routes of administration has been shown to be quite variable. However, the convenience of these two routes of administration and the latter's rapid, intense onset of effects make them the most commonly used.

It has been suggested that IN bioavailability may be dose dependent with increased bioavailability at higher doses a function of the amount of drug available for absorption (4). Bioavailability estimates have ranged from as low as 25% (5) to as high as 94% (6). The IN route, though more efficient in delivering cocaine to the brain than the PO route of administration, results in delayed absorption due to the vasoconstrictive action of the drug and its propensity for being swallowed in various amounts during insufflation.

Smoked cocaine produces a rapid and intense high that is similar to the i.v. route, reflecting the efficiency of delivering the dose to the brain. Despite this, studies indicate that the average bioavailability by the SM route is only 57–70%, with considerable variations within studies (7,8). Interindividual variation in smoking technique, the temperature and nature of the smoked cocaine, and the construction of the pipe may all play a role in the erratic absorption of cocaine by the SM route. It has been postulated that cocaine may undergo pyrolysis in the pipe and that the first few "hits" provide greater bioavailability than later ones (8). Recovery of residual cocaine from pipes after smoking studies have demonstrated that approximately a quarter of the original dose of cocaine remains in the pipe (6).

Metabolism

Cocaine is metabolized primarily to benzoylecgonine and ecgonine methyl ester by different mechanisms (Figure 7–1). If cocaine is used with ethanol, ethylcocaine (cocaethylene), a metabolite produced by the transesterification of cocaine with ethanol, may be observed as well. These mechanisms of cocaine metabolism are not straightforward, however, and the interpretation of cocaine and metabolite concentrations in blood and other specimens is complicated by complex in vitro and in vivo metabolic reactions.

Stability studies support the long-held findings that cocaine is metabolized to ecgonine methyl ester via enzymatic hydrolysis by pseudocholinesterase and liver esterases and to benzoylecgonine via spontaneous hydrolysis at physiological and alkaline pH. Recent studies provide strong evidence that cocaine can be hydrolyzed to benzoylecgonine by liver carboxylesterases (9). Two types of esterases have been isolated and purified from nonspecific carboxylesterases in human liver. Liver methylesterase catalyzes the conversion of cocaine to benzoylecgonine and the transesterification of cocaine to ethylcocaine. In the absence of ethanol, this enzyme hydrolyzes cocaine exclusively to benzoylecgonine. In the presence of both cocaine and ethanol, transesterification of cocaine to ethylcocaine occurs about 3.5 times as fast as hydrolysis to benzoylecgonine (10).

A separate and distinct human liver esterase, benzoylesterase, was found to catalyze the conversion of cocaine to ecgonine methyl ester (9). When incubated with cocaine, this enzyme produced both ecgonine methyl ester and benzoylecgonine. However, the benzoylecgonine concentrations were equal to the buffer control, suggesting partial spontaneous hydrolysis. When benzoylesterase was incubated in the presence of cocaine and ethanol, no ethylcocaine was formed, and the ecgonine methyl ester concentrations produced were the same as without ethyl alcohol. Unlike methylesterase, cholinesterase inhibitors, such as sodium fluoride and physostigmine, can inhibit benzoylesterase.

Norcocaine, an N-demethyl metabolite of cocaine produced by liver cytochrome P-450, has received considerable study because of its conversion into a hepatotoxic metabolite. Norcocaine is metabolized to N-hydroxynorcocaine and then to norcocaine nitroxide. It was thought that the nitroxide was a free radical that led to hepatotoxicity, but it has since been demonstrated that a further oxidative product, norcocaine nitrosodium ion, is responsible (11). In humans, norcocaine is a minor metabolite and reports of hepatotoxicity attributed to cocaine use are rare. However, appreciable amounts of norcocaine have been quantified in plasma of subjects receiving chronic oral cocaine administration (12).

Both benzoylecgonine and ecgonine methyl ester are metabolized to ecgonine. Although the pathways have not been fully studied, enzymatic hydrolysis of the benzoyl ester and chemical hydrolysis of the methylester would be expected to varying degrees for these compounds. Ecgonine is difficult to extract from an aqueous matrix; hence, analysis of ecgonine has been limited. Originally thought to be a minor urinary metabolite, recent studies have reported significant quantities of ecgonine in urine and blood (13).

Anhydroecgonine methyl ester (methylecgonidine) has been identified as a unique cocaine metabolite after smoked cocaine administration. Although this metabolite has been reported to be produced in the injection port of a gas chromatograph, <1% conversion of cocaine to anhydroecgonine methyl ester occurs if the injection port of the gas chromatograph is maintained at 482 °F (250 °C) (14). Related metabolites, ecgonidine and norecgonidine methyl ester, have been identified in urine specimens (15).

Recently, ethylcocaine has received a great deal of attention. In addition to being formed by liver methyl esterase in simultaneous cocaine and ethanol users (9,10), it can also be formed by fatty acid ethyl synthase (16). Because it is structurally similar to cocaine, it was not surprising that this metabolite has similar physicochemical properties to cocaine and was found to enhance the euphoria associated with concurrent cocaine and alcohol use in animals. In vitro studies suggest that ethylcocaine may also be cardiotoxic (17,18). The appearance of ethylcocaine in the blood after simultaneous cocaine and ethanol use is delayed by 10–30 minutes (20). The average half-life of ethylcocaine based on various reports is about 120 minutes, slightly longer than average reports for cocaine (19,20). Metabolites of ethylcocaine include ecgonine ethyl ester, norethylcocaine, ethylecgonidine, and hydroxylated ethylcocaine. Many of these have been identified in postmortem blood and urine specimens. Anhydroecgonine ethyl ester has been identified in cocaine smokers also using ethanol. Ethylcocaine would also be expected to metabolize to benzoylecgonine by the same processes that convert cocaine to benzoylecgonine. Because liver methylesterase catalyzes the transesterification of cocaine to ethylcocaine 3.5 times as fast as it hydrolyzes cocaine to benzoylecgonine (10), enzymatic conversion of ethylcocaine to benzoylecgonine would not be expected until the cocaine and/or ethanol have been consumed. This may in part explain the longer half-life of ethylcocaine compared with cocaine.

Plasma Concentrations

Many pharmacokinetic studies have been performed with cocaine. Although considerable interindividual variation among subjects has been reported, several observations can be made. When the i.v. and SM routes administered bioequivalent doses of cocaine, similar absorption and elimination curves for cocaine were obtained (6,7,21). In studies in which two different doses of cocaine were given by the i.v., SM, or PO route, intrasubject data indicated that mean plasma cocaine concentrations were dose-related (22). When cocaine was administered IN, the dose and peak plasma concentration showed a poor correlation. In studies in which cocaine was administered IN or PO, peak plasma concentrations of cocaine were delayed between 30 and 60 minutes and an average of 60 minutes, respectively, after the last dose (4–7,12,22–25).

Peak plasma cocaine concentrations in most single-dose pharmacokinetic studies by the SM (up to 100 mg) (6,7,26), i.v. (up to 64 mg) (6,7,23,26), and IN (up to 100 mg) (4–7,22–24) routes generally averaged between 0.2 and 0.4 mg/L (0.66 and 1.3 µmol/L). Up to 1–2 mg/L (3.3–6.6 µmol/L) cocaine was measured in chronic multiple-dosing studies by the SM, i.v., and PO routes without adverse effects (12,27,28). In one study in which 100 mg and 200 mg of cocaine were administered i.v., mean peak plasma concentrations of 0.87 and 3.2 mg/L (2.9 and 10.5 µmol/L), respectively, were obtained with no toxicity reported (27).

Cocaine has been described to follow first-order elimination after i.v. and SM administration using both one- and two-compartment models. The average half-life for cocaine, based on the literature, for both routes of administration is about 60 minutes (4,7,8,26,27,29,30). Most investigators found the half-lives of i.v. and SM cocaine to be similar, although one found somewhat shorter half-lives after SM cocaine than for i.v. and IN cocaine administration (7). The similar elimination rates obtained by both the i.v. and SM route are consistent with the similar effects seen via the i.v. or SM route of administration. Generally, the half-life of cocaine does not appear to be dose-dependent, although one study did report dose-dependent kinetics after very high plasma cocaine concentrations were achieved after i.v. cocaine administration (27).

IN cocaine pharmacokinetics has generally been described using one- or two-compartment models with first-order input. In one study there was evidence for dose-dependent elimination (4). After IN cocaine administration, the average absorption half-life for cocaine was 12 minutes and the average elimination half-life 84 minutes based on published data (4,5,7,8,24,27). Recent pharmacokinetic studies using extended collection times and more sensitive analytical measurements suggest that the half-life of cocaine, especially by the SM and IN routes of administration, may be much longer than originally thought, averaging 4 to 5 hours, suggesting the possibility that small amounts of cocaine may be stored in lipid-soluble tissue, not unlike marijuana (6).

Benzoylecgonine appears in the plasma within 15–30 minutes after cocaine administration by the i.v., SM, and IN routes of administration (6,7,26). Based on the one pharmacokinetic study that has been extended for a sufficient period of time to allow the study of the pharmacokinetics of benzoylecgonine in plasma (6), the average half-lives of formation for i.v., SM, and IN routes of administration were 34, 29, and 112 minutes, respectively. The elimination half-lives were 347, 324, and 213 minutes, respectively. The rate of benzoylecgonine elimination was slow compared with its rate of formation, accounting for its longer half-life and accumulation in plasma while cocaine concentrations were decreasing. Increased plasma protein binding for benzoylecgonine compared with the negligible plasma protein binding of cocaine has been suggested to play a role in these findings that may also explain extended benzoylecgonine detection times seen in the urine of chronic high-dose cocaine users (31,32). More definitive investigations are required to clearly establish the mechanism(s) involved in the longer half-life in comparison to cocaine.

URINE CONCENTRATIONS

Cocaine and its metabolites are excreted into the urine almost exclusively by simple filtration. After a single dose of cocaine, 64–69% of a dose of cocaine was recovered in the urine within 3 days regardless of route of administration, 86% of this occurring within the first day (7). Some reports indicate that up to 90% of a dose may be recovered in a 24-hour urine specimen. A review of published data indicates that 1–9% of the drug is excreted unchanged depending on pH (the lower the urine pH, the greater amount in the urine), 26–54% as benzoylecgonine, 18–41% as ecgonine methyl ester, and 2–3% as ecgonine. Anhydroecgonine methyl ester and various demethylated and hydroxylated metabolites appear to be minor metabolites in urine. After continuous prolonged intravenous cocaine infusions the average urinary half-lives for cocaine, benzoylecgonine, and ecgonine methyl ester were 0.8, 4.5, and 3.1 hours, respectively, compared with 2.9, 4.8, and 4.9 hours, respectively, after a single smoked dose of cocaine (33).

Ecgonine, initially thought to be a minor urinary metabolite, has been shown to significantly exceed benzoylecgonine concentrations in the urine. One study showed that the concentration of ecgonine in the urine exceeded the benzoylecgonine concentration by a factor of five when ecgonine was present at or above 50 ng/mL (0.26 μmol/L) (34). In another study, three representative blood specimens collected from patients presenting to the emergency room with a cocaine overdose demonstrated that ecgonine concentrations were consistently higher than ecgonine methyl ester concentrations, sometimes as much as fivefold (35). However, pharmacokinetic studies measuring this metabolite have not yet been performed and its presence may be due to a result of hydrolysis of benzoylecgonine or ecgonine methyl ester.

OTHER SPECIMENS

Cocaine and metabolites may be detectable in hair longer than in urine. For this reason analysis of hair may be useful in determining a past history of cocaine use. Cocaine is present in higher concentrations than benzoylecgonine and ecgonine methyl ester in hair. Many important issues related to hair testing are still being studied including environmental contamination, washing techniques, sex or ethnic bias, quality-control procedures, proficiency testing, and the establishment of cutoff concentrations. Hair testing may be extremely useful in postmortem cases in which no other specimen is available and drug history is desired. In archeological studies, cocaine has even been found in the hair of ancient Incan skeletons.

Saliva (oral fluid) may be a useful specimen for the detection of recent cocaine use in clinical studies and workplace testing. Generally, a good correlation between saliva and plasma cocaine concentrations has been observed. Saliva cocaine concentrations have also been correlated with behavioral effects. In addition, saliva may be collected by direct observation without any invasive procedure. Appropriate specimen collection procedures are critical in minimizing contamination from the oral cavity, especially after SM and IN administration. The saliva-to-plasma cocaine ratio can also be affected by saliva pH and saliva flow rate after stimulation.

Sweat may be a useful, noninvasive specimen for monitoring drug use. Sweat may be collected with a collection patch for a predetermined period of time, allowing a continuous period of monitoring and/or accumulation of excreted drug. In sweat, cocaine is excreted primarily as the parent drug, which offers the added advantage of simple gas chromatographic analysis. Depending on the device used to collect sweat, any attempt at tampering with the collection device would be evident. Although wearing a sweat patch for monitoring cocaine use may provide for a wider detection window than urine does, correlation between accumulated sweat cocaine concentrations with impairment and time of use are not likely. Because trace amounts of cocaine can be detected in sweat after the i.v. administration of as little as 1 mg of cocaine, some concerns regarding passive exposure have been raised.

Meconium is the first stool specimen excreted by the neonate. Because meconium accumulates throughout gestation, it is a useful specimen with which to monitor fetal drug exposure from maternal drug use. Neonates appear to metabolize cocaine differently than adults do. In meconium, m-hydroxybenzoylecgonine is the metabolite most frequently present (36,37). Fortunately, this metabolite cross-reacts with many of the available immunoassays. However, it is important to include this analyte in confirmation procedures when analyzing meconium. In addition to parent cocaine, most of the metabolites of cocaine that have been reported in urine have also been reported in meconium, including anhydroecgonine methyl ester, which is suggestive of maternal crack use. Meconium analysis for the determination of cocaine in stillborn babies has also been shown to be useful. The presence of cocaine has been identified in meconium of a 17-week-old fetus, suggesting that fetal drug exposure can be determined early in gestation. Because meconium forms layers in the intestine as it is being deposited, it is not a homogenous specimen. As with other heterogeneous specimens, such as gastric contents, it is important that all available specimen be collected and thoroughly mixed before sampling. Caution should be given when interpreting cocaine and metabolite ratios if meconium is collected from a diaper because it may have been contaminated with urine.

ANALYSIS

Immunoassay

Immunoassays are commonly uses for screening purposes because they are readily amenable to large batch analysis, are relatively sensitive, and require little or no sample preparation. Because immunoassays are targeted to detect benzoylecgonine, they are particularly well suited for screening urine specimens. There are several types of immunoassays on the market. Depending on the product selected, immunoassays use the principle of fluorescence polarization immunoassay (FPIA), enzyme immunoassay (EIA), microparticle immunoassay (KIMS), radioimmunoassay (RIA), cloned enzyme donor immunoassay (CEDIA), or enzyme-linked immunosorbent assay (ELISA). In addition to their use for urine analysis, some of these assays have been successfully adapted to blood tissue analysis. Although all immunoassay techniques are targeted on benzoylecgonine, cross-reactivities vary considerably by manufacturer and analytical principle. Immunoassays that possess substantial cross-reactivity to cocaine and ethylcocaine, such as the Diagnostic Products Corporation (DPC, Los Angeles) RIA assay, are particularly useful for screening blood. The detection limit for cocaine and ethylcocaine, using the DPC RIA assay, is as low as 0.010 mg/L (0.033 µmol/L and 0.032 µmol/L, respectively), depending on lot. In addition, the assay can also readily detect benzoylecgonine at 0.100 mg/L (0.34 µmol/L). The FPIA assay shows considerable cross-reactivity to m-hydroxybenzoylecgonine, making it well suited to screening meconium. Although the cutoff concentration for benzoylecgonine in federal workplace drug testing is 0.300 mg/L, most immunoassays can detect lower concentrations reliably. In fact using 0.300 mg/L as the cutoff, 19% of cocaine-positive urines were missed in studies in pregnant women (38,39).

Chromatography

Before cocaine and its metabolites can be analyzed using chromatographic techniques, the drugs must be separated from the biological matrix. This may be accomplished either by liquid-liquid or solid-phase extraction procedures. The latter technique can be readily adapted to laboratory automation devices. There are several important issues that must be considered before selecting an extraction procedure. Cocaine and many of its metabolites are esters that are susceptible to hydrolysis in alkaline conditions and at elevated temperatures. In addition, plasma and liver esterases also contribute to hydrolysis. During sample preparation, it is critical that the amount of time the biological specimen remains in conditions that are unfavorable to cocaine stability be minimized; otherwise, esters may hydrolyze in vitro and complicate the interpretation of the analytical results. Consideration must also be given to the targeted analytes because there is a considerable range in polarity for cocaine and its metabolites.

Other than stability concerns, the extraction of cocaine and ethylcocaine is straightforward. These compounds may be readily extracted into *n*-butyl chloride at a pH of 8–9. A chloroform isopropanol mixture (9:1) is also commonly used. These conditions will also extract benzoylecgonine and ecgonine methyl ester, although not at optimal recoveries. Solid-phase extraction procedures, also used for these analytes, commonly use a protein precipitation step before a buffered supernatant is applied to the extraction column. Elution is typically accomplished with a mixture of methylene chloride, isopropanol, and ammonium hydroxide (78:20:2), although other elution solvents have been used as well, depending on the analytes targeted and the column used.

Thin-layer chromatography (TLC) is a simple technique that can be used to analyze for both cocaine and benzoylecgonine but is generally limited to the analysis of urine specimens due to a lack of sensitivity. In fact, TLC has higher detection limits than immunoassay, meaning that in some cases, an immunoassay screen followed by TLC confirmation may lead to some false-negative results. After extracting 5 mL of urine for basic drugs and spotting the concentrated extract on a silica TLC plate, one can expect a sensitivity of about 1–2 mg/L (3.4–6.9 μmol/L) for benzoylecgonine using the standard Davidow solvent with visualization by either Dragendorff's reagent and/or iodoplatinate. Sensitivity for benzoylecgonine may be increased up to 10-fold by an initial spraying with Dragendorff's reagent followed by an overspray with 20% sulfuric acid and exposure to iodine vapors. Sensitivity for cocaine by TLC is generally better than for benzoylecgonine using most basic drug extraction schemes. High-performance TLC results in considerably improved detection limits. As a qualitative tool, TLC may be used to confirm immunoassay results when positive forensic identification is not required. TLC is particularly useful for benzoylecgonine because it does not migrate up the plate very far, allowing it to be readily distinguished from many other basic drugs. Procedures for ecgonine methyl ester analysis by TLC have also been described with similar sensitivities. Ecgonine methyl ester can be extracted from urine using chloroform: isopropanol (9:1). After solvent evaporation and reconstitution with methylene chloride and methanol (1:1), separation is achieved on a standard silica plate using the Davidow procedure. Ninhydrin and iodoplatinate are used for visualization of ecgonine methyl ester, appearing as a deep blue spot with a R_f of 0.74.

Gas chromatography (GC) is the separation technique most frequently used for the analysis of cocaine and its metabolites. Cocaine, ethylcocaine, and their *N*-desmethyl metabolites can be readily assayed without derivatization using nitrogen-phosphorus detection and detection by mass spectrometry (MS) in both the electron-impact and positive chemical-ionization modes. Flame-ionization detection may also be used, but is not nearly as sensitive. Ecgonine methyl ester and related compounds can be detected without derivatization, but tend to tail on most analytical columns due to their free hydroxyl moiety.

Chromatography may be improved by derivatization to the *p*-nitrococaine or *p*-fluorococaine derivative using either *p*-nitrobenzoylchloride or *p*-fluorobenzoylchloride, respectively, in a modification of the Schotten-Baumann reaction. Benzoylecgonine and related compounds must be derivatized before analysis. Caution must be used not to derivatize metabolites to another targeted analyte. For example, some early procedures used derivatives that formed ethylcocaine from benzoylecgonine.

A variety of derivatization procedures have been used. Acylation procedures (such as pentafluoro) and silylation (such as trimethylsilyl) derivatize both benzoylecgonine and ecgonine methyl ester. Alkylation procedures (such as *n*-propyl) will derivatize *N*-desmethyl metabolites in addition to benzoylecgonine. Sequential derivatization allows for the simultaneous detection of multiple analytes. For example, ecgonine, ecgonine methyl ester, benzoylecgonine, and norcocaine can be derivatized with 1-propyliodide followed by *p*-nitrobenzoylchloride to yield *p*-nitro-*n*-propylcocaine, *p*-nitrococaine, *n*-propylcocaine, and *N*-propylcocaine, respectively. Electron-capture detectors are very sensitive and have been used for the analysis of benzoylecgonine after acylation. The major disadvantage to this detector is that cocaine must be reduced before acylation.

Detection by MS provides the highest specificity of all GC detectors and is nearly a requirement for the forensic confirmation of cocaine and metabolites today. A particular advantage with MS detection is that deuterated analogs of cocaine and its major metabolites are now readily available and, by virtue of having chemical characteristics nearly identical to their nondeuterated analogs, allow for excellent reproducibility and accurate analysis. If using other detectors, selection of internal standards should be carefully considered so that they undergo the same chemistry that the analyte undergoes.

Liquid chromatography (LC) is very useful because cocaine, ethylcocaine, and benzoylecgonine can be analyzed simultaneously without derivatization when using ultraviolet (UV) detection. Although UV detection has a sensitivity of at least 0.05 mg/L (0.17 µmol/L) for cocaine and benzoylecgonine, detection of other metabolites requires MS detection. In addition, UV detection is not as selective as other detection methods and does not allow the use of deuterated internal standards. The lack of selectivity can be a particular problem when trying to select an appropriate internal standard. For example, lidocaine may interfere with cocaine and meperidine may interfere with tetracaine, a commonly used internal standard. A further disadvantage of UV detection is that ecgonine methyl ester and related compounds as well as ecgonine do not possess a UV chromophore and would need derivatization for detection. On the other hand, LC coupled with MS may be the ideal separation and detection techniques for cocaine and its metabolites without need for derivatization. LC/MS is still not commonly used due to cost and maintenance issues. However, there are indications that this tool will become the next gold standard in forensic toxicology in the coming decade.

COMPLICATIONS AND CONTROVERSIES

Interpretation of Blood Concentrations

To attempt interpretation of cocaine and metabolite concentrations in blood, one must consider a large number of factors. As previously discussed, plasma cocaine concentrations in single-dose pharmacokinetic studies generally average between 0.2 and 0.4 mg/L (0.66 and 1.3 µmol/L). In some studies after chronic oral dosing and after repeated doses of smoked cocaine, plasma concentrations in excess of 1.0 mg/L (3.3 µmol/L) have been reported without adverse effects. In one instance a plasma concentration of 3.87 mg/L (12.7 µmol/L) was measured after i.v. cocaine administration without symptoms of toxicity. A search of the literature suggests that cocaine concentrations of less than 0.30 mg/L (0.99 µmol/L) are generally considered clinically therapeutic. However, therapeutic, toxic, and lethal cocaine concentrations clearly overlap, as observed by clinical and postmortem studies.

Tolerance and sensitization may play a significant role in the poor correlation observed. In a study of 130 patients presenting to an emergency room with acute cocaine toxicity, the mean plasma cocaine concentration was 0.34 mg/L (1.1 µmol/L) and ranged from 0.00 to 3.92 mg/L (0 to 12.9 µmol/L). The median cocaine concentration in these patients was only 0.07 mg/L (0.23 µmol/L). However, the mean and median benzoylecgonine concentrations in these patients were 1.57 mg/L and 1.06 mg/L (5.4 and 3.7 µmol/L), respectively. There was no correlation between the cocaine and metabolite concentrations in these patients and their clinical state or their outcome. However, the degree of symptoms of toxicity—most notably hyperthermia, heart rate, and psychosis—did provide a better predictor of a patient's outcome (13,40).

Interpretation of Urine Concentrations

There are few interpretative concerns for positive benzoylecgonine results once alternative medical explanations are ruled out, because drug concentrations cannot be correlated to impairment. Positive findings provide only an approximate window of time of use. Benzoylecgonine may be detected at the 300 ng/mL (1040 µmol/L) cut-off concentration for 2–4 days after cocaine use, depending on dose, frequency of use, urinary pH, and clearance (41,42). Positive immunoassay results in excess of 7 days have been reported after compulsive cocaine use with continuous positive results for up to 16 days (31,32). Quantitative benzoylecgonine concentrations corrected for creatinine in sequential urine specimens may be useful in substance abuse treatment programs, but single specimens collected in a forensic setting cannot be similarly interpreted.

In Vitro Stability

Cocaine contains two ester moieties, rendering it susceptible to hydrolysis in vitro and in vivo. In unpreserved blood, in vitro stability studies have shown that cocaine is hydrolyzed almost exclusively at the phenyl ester by plasma pseudocholinesterase to yield ecgonine methyl ester (43–47). The addition of sodium fluoride, while inhibiting enzymatic hydrolysis of cocaine to ecgonine methyl ester, does not prevent spontaneous chemical hydrolysis of cocaine to benzoylecgonine. The rate of hydrolysis of both esters has been shown to be temperature and pH dependent, with higher temperatures and pH increasing the rate of hydrolysis. The loss of cocaine in unpreserved blood can be dramatic. In blood fortified with cocaine at 2.0 mg/L (6.6 µmol/L), cocaine concentrations decreased to 0.64 mg/L (2.1 µmol/L) after storage at room temperature for 24 hours, with corresponding increases in ecgonine methyl ester concentrations. Even after the addition of 2% sodium fluoride, a 25% decrease in cocaine concentrations was observed at room and refrigerated temperatures within 5 and 80 days, respectively, with corresponding increases in benzoylecgonine concentrations. Evidence that the action of sodium fluoride may be reversible occurred late in the study with the appearance of some ecgonine methyl ester. Acidifying blood to pH 5 to inhibit chemical hydrolysis in conjunction with 2.0% sodium fluoride to inhibit enzymatic hydrolysis resulted in no cocaine loss over 200 days at refrigerated [39 °F (4 °C)] and frozen [5 °F (–15 °C)] temperatures and for at least 60 days at room temperature (48).

Benzoylecgonine and ecgonine methyl ester have ester moieties and are thus subject to temperature and pH-dependant hydrolysis as well. Benzoylecgonine is considerably more stable than ecgonine methyl ester in unpreserved blood (pH 7.4) at room temperature. A 50% loss of ecgonine methyl ester occurred in a 35-day period compared with a 25% loss for benzoylecgonine. Little loss of either compound was observed when the blood specimen was refrigerated for the same period of time (48). Ethylcocaine has been shown to be more stable than cocaine in blood stored at room temperature collected from rabbit studies. Similar observations were seen in liver, brain, and muscle collected from the same animals (49). Although the stability of ethylcocaine has not been studied in human blood, it would be expected that the same stability issues raised for cocaine would be observed for ethylcocaine.

Another factor related to the stability of cocaine that must be considered when interpreting cocaine and metabolite concentrations is the conditions to which the specimen was exposed during analysis. Primary specimens and aliquots should not be allowed to remain at room temperature for extended periods of time during the analytical process. Because cocaine is a basic drug, many procedures use basic conditions (pH 8–11) for the extraction of cocaine from the biological matrix into organic solvents. This process may be repeated during a back-extraction. The duration of exposure to these alkaline conditions, especially at room temperature, is critical and may contribute to artifactual hydrolysis of cocaine and its metabolites.

Impairment

There have been relatively few scientific studies on the effects of cocaine on driving performance. The euphoric effects of cocaine during acute intoxication may give a driver the feeling of increased mental and physical abilities. It has been suggested that this optimism may result in increased risk-taking behavior and perhaps increased probability of accidents (50), particularly in conjunction with ethanol use (51). In one study 62% of cocaine smokers have reported symptoms of suspiciousness, distrust, and paranoia (52). These effects have resulted in high-speed chases with police (53). Lapses of attention are also commonly reported ranging from ignoring changes in traffic signals to forgetting to replace gas caps at service stations.

Cocaine also appears to have a significant effect on vision. This may in part be related to cocaine's ability to produce mydriasis. After IN cocaine use, increased sensitivity to light, halos around bright objects, and difficulty focusing were reported by 43% of subjects (54). Over 34% of cocaine smokers reported blurred vision, often accompanied by glare recovery problems. Fifty percent of cocaine smokers and 18% of IN users reported hallucinations. "Snow lights," flashes or movements of light in the peripheral field of vision, was the most commonly reported hallucination. Reaction to "snow lights" included moving in their direction or trying to avoid or evade them. A statistically significant blue-yellow vision loss has been reported in subjects undergoing cocaine detoxification compared with control subjects (55). These findings may be related to decreased retinal neurotransmission as a result of dopamine depletion. Because there are normally high concentrations of dopamine in the retina, visual changes after chronic cocaine use are not unexpected.

By contrast, double-blind laboratory studies have demonstrated that low doses of cocaine may actually improve driving performance and counteract some of the performance decrements of ethanol (56–60). Similar observations have been suggested epidemiologically with other depressant drugs. In a study with more than 4000 subjects, an increased risk of injury was associated with use of psychoactive substances, but the risk was lower when cocaine was used with other depressant drugs (61). Although stimulants (amphetamines or cocaine) may acutely enhance performance of simple tasks, this may not remain the case as the complexity of the task increases (62). For example, cocaine-induced hyperexcitability has resulted in rapid steering or braking reactions due to sudden sounds such as horns or sirens (63). Because the effects of cocaine are brief, the "crash" which follows cocaine use may be a particularly dangerous time for driving.

Excited Delirium

A significant number of cases presenting to emergency departments has been associated with a cocaine-induced psychosis now commonly called cocaine-induced excited or agitated delirium (53). A classic presentation of this phenomenon is illustrated in the Case Study. This syndrome is characterized by severe hyperthermia [104–108 °F (40–42 °C)], extreme agitation and delirium, respiratory arrest, and sudden death. These individuals exhibit bizarre and violent behavior and extreme strength and are frequently seen running around—often naked—shouting, fighting, breaking things, and causing injury to themselves and/or others. In these cases, the stress from restraint may result in catecholamine surges on an already sensitized myocardium, resulting in arrhythmias. Excited delirium cases are frequently found in places where an individual may have attempted to cool down, such as the bathroom. Other evidence to attempt cooling such as wet towels and ice cube trays may also be present at the scene.

REFERENCES

1. Ritz MC, Lamb RJ, Goldberg SR, et al: Cocaine receptors on dopamine transporters are related to

self-administration of cocaine. *Science* 1987;237:1219–1233.
2. McElvain JS, Schenk JO: A multisubstrate mechanism of striatal dopamine uptake and its inhibition of cocaine. *Biochem Pharmacol* 1992;43:2189–2199.
3. Mayersoh M, Perrier D: Kinetics of pharmacologic response to cocaine. *Res Commun Chem Pathol Pharmacol* 1978;22:465–474.
4. Javaid JI, Musa MN, Fischman M, et al: Kinetic of cocaine in humans after intravenous and intranasal administration. *Biopharm Drug Dispos* 1983;4:9–18.
5. Wilkinson P, Van Dyke C, Jatlow P, et al: Intranasal and oral cocaine kinetics. *Clin Pharmacol Ther* 1980;27:386–394.
6. Cone E: Pharmacokinetics and pharmacodynamics of cocaine. *J Anal Tox* 1995;19:459–478.
7. Jeffcoat, AR, Perez-Reyes M, Hill JM, et al: Cocaine disposition in humans after intravenous injection, nasal insufflation (snorting), or smoking. *Drug Metab Dispos* 1989;17:153–159.
8. Cook CE, Jeffcoat R, Perez-Reyes M: Pharmacokinetic studies of cocaine and phencyclidine in man. In: Barnett G, Chiang CN, eds: *Pharmacokinetics and pharmacodynamics of psychoactive drugs.* Foster City, CA: Biomedical Publications, 1985:48–74.
9. Dean RA, Christian CD, Sample RHB, et al: Human liver cocaine esterases: ethanol-mediated formation of ethylcocaine. *FASEB J* 1991;5:2735–2739.
10. Brzezinski MR, Abraham TL, Stone CL, et al: Purification and characterization of a human liver cocaine carboxylesterase that catalyzes the production of benzoylecgonine and the formation of cocaethylene from alcohol and cocaine. *Biochem Pharmacol* 1994;48:1747–1755.
11. Kloss MA, Rosen G, Rauckman EJ: Cocaine-mediated hepatotoxicity. *Biochem Pharmacol* 1984;33:169–173.
12. Jufer RA, Walsh SL, Cone EJ: Cocaine and metabolite concentrations in plasma during repeated oral administration: development of a human laboratory model of chronic cocaine use. *J Anal Tox* 1998;22:435–444.
13. Blaho K, Logan BK, Winbery S, et al: Blood cocaine and metabolite concentrations, clinical findings, and outcome of patients presenting to an ED. *Am J Emerg Med* 2000;18:593–598.
14. Cone EJ, Hillsgrove M, Darwin WD: Simultaneous measurement of cocaine, cocaethylene, their metabolites, and "crack" pyrolysis products by gas chromatography-mass spectrometry. *Clin Chem* 1994;40:1299–1305.
15. Jenkins AJ, Goldberger BA: Identification of unique cocaine metabolites and smoking by-products in postmortem blood and urine specimens. *J Forensic Sci* 1997;42:824–827.
16. Heith A, Morse C, Tsujita T, et al: Fatty acid ethyl ester synthatase catalyzes the esterification of ethanol to cocaine. *Biochem Biophys Res Commun* 1995;208:549–554.
17. Xu Y, Crumb W, Clarkson C: Cocaethylene, a metabolite of cocaine and ethanol, is a potent blocker of cardiac sodium channels. *J Pharmacol Exp Ther* 1994;271:319–325.
18. Qiu Z, Morgan J: Differential effects of cocaine and cocaethylene on intracellular calcium and myocardial contraction in cardiac monocytes. *Br J Pharmacol* 1993;109:293–298.
19. McCance-Katz E, Price L, McDougle C, et al: Concurrent cocaine-ethanol ingestion in humans: pharmacology, physiology, behavior, and the role of cocaethylene. *Psychopharmacology (Berl)* 1993;111:39–46.
20. Perez-Reyes M, Jeffcoat A, Meyers A, et al: Comparison in humans of the potency and pharmacokinetics of intravenously injected cocaethylene and cocaine. *Psychopharmacology (Berl)* 1994;116:428–432.
21. Isenschmid DS, Fischman MW, Foltin RW, et al: Concentration of cocaine and metabolites in plasma of humans following intravenous administration and smoking of cocaine. *J Anal Toxicol* 1992;16:311–314.
22. Brogan WC, Lange RA, Glamann DB, et al: Recurrent coronary vasoconstriction caused by intranasal cocaine: possible role for metabolites. *Ann Intern Med* 1992;116:557–561.
23. Javaid JI, Fischman MW, Schuster H, et al: Cocaine plasma concentration: relation to physiological and subjective effects in humans. *Science* 1978;202:227–228.
24. Van Dyke C, Barash PG, Jatlow P, et al: Cocaine: plasma concentrations after intranasal application in man. *Science* 1976;191:859–861.
25. Van Dyke C, Ungerer J, Jatlow P, et al: Oral cocaine: plasma concentrations and central effects. *Science* 1978;200:211–213.
26. Isenschmid DS, Levine BS, Caplan YH: The role of ecgonine methyl ester in the interpretation of cocaine concentrations in postmortem blood. *J Anal Tox* 1992;16:319–324.
27. Barnett G, Hawks R, Resnick R: Cocaine pharmacokinetics in humans. *J Ethnopharmacol* 1981;3:353–366.
28. Isenschmid DS: *The role of ecgonine methyl ester in the interpretation of cocaine concentrations in post mortem blood.* Doctoral Dissertation, University of Maryland at Baltimore. 1991:100–112.
29. Chow MJ, Ambre JJ, Ruo TI, et al: Kinetics of cocaine distribution, elimination, and chronotropic effects. *Clin Pharmacol Ther* 1985;38:318–324.
30. Cone EJ, Kumor K, Thompson LK, et al: Correlation of saliva cocaine levels with plasma levels and with pharmacologic effects after intravenous cocaine administration in human subjects. *J Anal Toxicol* 1988;12:200–206.
31. Burke WM, Ravi NV, Dhopesh V, et al: Prolonged presence of metabolite in urine after compulsive cocaine use. *J Clin Psychiatry* 1990;51:145–148.
32. Weiss R, Gawin F: Protracted elimination of cocaine metabolites in long-term, high-dose cocaine abusers. *Am J Med* 1988;85:879–880.
33. Ambre J, Ruo T, Nelson J, et al: Urinary excretion of cocaine, benzoylecgonine, and ecgonine methyl ester in humans. *J Anal Toxicol* 1988;12:301–306.
34. Hornbeck CL, Barton KM, Czarny RJ: Urine concentrations of ecgonine from specimens with low benzoylecgonine levels using a new ecgonine assay. *J Anal Toxicol* 1995;19:133–138.
35. Smirnow D, Logan BK: Analysis of ecgonine and other cocaine biotransformation products in postmortem whole blood by protein precipitation-extractive alkylation and GC-MS. *J Anal Toxicol* 1996;20:463–467.

36. Steele BW, Bandastra ES, Wu NC, et al: m-Hydroxybenzoylecgonine: an important contributor to the immunoreactivity in assays for benzoylecgonine in meconium. *J Anal Toxicol* 1993;17: 348–352.
37. Lewis D, Moore C, Becker J, et al: Prevalence of m-hydroxybenzoylecgonine in meconium samples. In: Spiehler V, ed: *Proceedings of the 1994 Joint TIAFT/SOFT International Meeting.* Ann Arbor, MI: Omnipress, 1995:513.
38. Hicks JM, Morales A, Soldin S: Drugs of abuse in a pediatric outpatient population [Abstract]. *Clin Chem* 1990;36:1026.
39. Casanova OQ, Lombardero N, Behnke M, et al: Detection of cocaine exposure in the neonate. Analyses of urine, meconium and amniotic fluid from mothers and infants exposed to cocaine. *Arch Pathol Lab Med* 1994;118:988–993.
40. Logan BK: Considerations when trying to determine the role of cocaine toxicity in death. *Proceedings of the February 1998 California Association of Toxicologists Meeting* [a newsletter published by the California Association of Toxicologists] 1998: 19–24.
41. Ambre J: The urinary excretion of cocaine and metabolites in humans: a kinetic analysis of published data. *J Anal Toxicol* 1985;9:241–245.
42. Hamilton HE, Wallace JE, Shimek EL, et al: Cocaine and benzoylecgonine excretion in humans. *J Forensic Sci* 1977;22:697–707.
43. Inaba T, Stewart DJ, Kalow W: Metabolism of cocaine in man. *Clin Pharmacol Ther* 1978;23:547–552.
44. Jatlow PI: Cocaine: analysis, pharmacokinetics and metabolic disposition. *Yale J Biol Med* 1988;61:105–113.
45. Matsabura K, Maseda C, Fukui Y: Quantitation of cocaine, benzoylecgonine and ecgonine methyl ester by GC-CI-SIM after Extrelut extraction. *Forensic Sci Int* 1984;26:181–192.
46. Stewart DJ, Inaba T, Lucassen M, et al: Cocaine metabolism: cocaine and norcocaine hydrolysis by liver and serum esterases. *Clin Pharmacol Ther* 1979; 25:464–468.
47. Stewart DJ, Inaba T, Tang BK, et al: Hydrolysis of cocaine in human plasma by cholinesterase. *Life Sci* 1977;20:1557–1564.
48. Isenschmid DS, Levine BS, Caplan YH: A comprehensive study of the stability of cocaine and its metabolites. *J Anal Toxicol* 1989;13:250–256.
49. Moriya F, Hashimoto Y: Postmortem stability of cocaine and cocaethylene in blood and tissues of humans and rabbits. *J Forensic Sci* 1996;41:612–616.
50. Brookoff D, Cook CS, Williams C, et al: Testing reckless drivers for cocaine and marihuana. *N Engl J Med* 1994;331:518–522.
51. Byck R: The effects of cocaine on complex performance in humans. *Alcohol Drugs Driving* 1987;3:9–12.
52. Siegel RK: Cocaine smoking. *J Psychoactive Drugs* 1982;14:271–325.
53. Wetli CV, Fishbain DA: Cocaine-induced psychosis and sudden death in recreational cocaine users. *J Forensic Sci* 1985;30:873–880.
54. Siegel RK: Cocaine hallucinations. *Am J Psychiatry* 1978;135:309–314.
55. Desai P, Roy M, Brown S, et al: Impaired color vision in cocaine-withdrawn patients. *Arch Gen Psychiatry* 1997;54:691–694.
56. Higgins ST, Rush CR, Hughes JR, et al: Effects of cocaine and alcohol, alone and in combination, on human learning and performance. *J Exp Anal Behav* 1992;58:87–105.
57. Higgins ST, Rush CR, Bickel WK, et al: Acute behavioral and cardiac effects of cocaine and alcohol combinations in humans. *Psychopharmacology* 1993;111:285–294.
58. Foltin RW, Fischman MW, Pippen PA, et al: Behavioral effects of cocaine alone and in combination with ethanol and marihuana in humans. *Drug Alcohol Depend* 1993;32:93–106.
59. Foltin RW, Fischman MW: Ethanol and cocaine interactions in humans: cardiovascular consequences. *Pharmacol Biochem Behav* 1989;31:877–883.
60. Farré M, De La Torre R, Llorent M, et al: Alcohol and cocaine interactions in humans. *J Pharmacol Exp Ther* 1993;266:1364–1373.
61. Regidor E, Barrio G, de la Feunte L, et al: Non-fatal injuries and the use of psychoactive drugs among young adults in Spain. *Drug Alcohol Depend* 1996; 40:249–259.
62. Ellinwood EH, Nikaido AM: Stimulant induced impairment: a perspective across dose and duration of use. *Alcohol Drugs Driving* 1987;3:19–24.
63. Davis M: Cocaine: Excitatory effects on sensorimotor reactivity measured with acoustic startle. *Psychopharmacology (Berl)* 1985;86:31–36.

SELF-ASSESSMENT QUESTIONS

1. Matching
   ____ benzoylecgonine
   ____ cocaine
   ____ ecgonine methyl ester
   ____ ethylcocaine
   ____ methylecgonidine
   ____ *m*-hydroxybenzoylecgonine
   ____ norcocaine
   a. formation requires ethanol consumption
   b. in vitro blood cocaine hydrolysis product
   c. major cocaine product in hair
   d. major cocaine product in neonates
   e. major cocaine product in urine
   f. microsomal cocaine product
   g. cocaine pyrolysis product

2. List the two mechanisms of action of cocaine.

3. List three major effects of cocaine usage.

4. Which requires derivatization before analysis by gas chromatography–mass spectrometry?
   a. benzoylecgonine
   b. cocaine
   c. ecgonine methyl ester
   d. ethylcocaine

5. Describe the phenomenon of "rush" and "crash" as it applies to cocaine use.

6. Which of the following is a complication in the interpretation of blood cocaine concentrations?
   a. in vitro conversion to benzoylecgonine
   b. in vitro conversion to ecgonine methyl ester
   c. in vitro reaction with ethanol to form ethyl cocaine
   d. in vitro production of cocaine from benzoylecgonine

CHAPTER 8

Amphetamines*

Karla A. Moore, D.V.M., Ph.D., DABFT

LEARNING OBJECTIVES

After completing this chapter, the reader should be able to:

1. Describe the methods of methamphetamine synthesis.
2. Describe the major pharmacologic effects of amphetamines.
3. Describe the major pharmacokinetic characteristics of the amphetamines.
4. List the major methods in which the amphetamines can be analyzed in biological fluids, including the separation of enantiomers.
5. Explain how either an amphetamine screening test or a confirmation test can be positive without the tested person having used illicit methamphetamine.

INTRODUCTION

History

The compound "amphetamine" has come to represent a class of phenethylamine compounds that have varying degrees of sympathomimetic activity. "Sympathomimetic drugs" are those that mimic the actions of the endogenous neurotransmitters that stimulate the sympathetic nervous system. Edelano first synthesized amphetamine in

*Adapted from Moore KA: Amphetamines/ sympathomimetic amines. In: Levine B, ed: *Principles of forensic toxicology*. Washington, DC: AACC Press, 1999: 265–285.

1877. During the 1930s, Prinzmetal, Bloomberg, and others first used amphetamines clinically as a central nervous system (CNS) stimulant for the treatment of narcolepsy and depression, at which time its abuse potential first became evident. Since then, the ability of amphetamines to alleviate fatigue, improve performance of simple mental and physical tasks, elevate mood, increase confidence, and produce euphoria has led to their misuse and abuse.

Amphetamine use reached epidemic proportions during the late 1940s and early 1950s when these compounds were used by soldiers, factory workers, and prisoners of war in Japan during World War II. After World War II, a surplus on the Japanese market permitted sales without a prescription, with peak use occurring about 1954 (1). In the 1960s, methamphetamine abuse became a social problem in the United States. By 1970, 50% of legally manufactured amphetamine and related compounds were being sold illegally on the black market.

Increasing abuse of these drugs eventually led to their classification as a Schedule II controlled substance under the Controlled Substances Act (Public Law 91-513) of 1970. This classification limited the acquisition of these compounds through legitimate channels. This act stringently regulated the manufacture of these stimulants and forced manufacturers to decrease sales to retail pharmacies (2). Because methamphetamine is easily synthesized even in crude laboratories, it quickly became the "stimulant of choice," and a dramatic increase has occurred in the illicit production and use of methamphetamine hydrochloride over the past several years. Endemic areas for this increase in the United States include the Pacific Coast states and Hawaii. According to the American Association of Poison Control Centers Toxic Exposure Surveillance Systems, there were 16,684 amphetamine exposures reported throughout the country in 1999 (3). Of these exposures, approximately 69% of the individuals were <20 years of age. Furthermore, about one-half of these exposures required treatment in health-care facilities.

Synthesis

Because of its ease of manufacture and ready availability, methamphetamine has become the sympathomimetic amine of choice among stimulant abusers. Illicit methamphetamine is available as a water-soluble white crystalline powder (methamphetamine hydrochloride) known as "speed," "crank," "go," "crystal," or simply "meth." This compound is adulterated with a variety of substances such as sugars (usually added to give more volume to the final product) or cheaper stimulants (such as caffeine, phenylpropanolamine, or ephedrine or pseudoephedrine). Methamphetamine in this form can be used for intravenous injection, oral consumption, or, more rarely, nasal inhalation.

Methamphetamine is currently the most frequently encountered clandestinely produced controlled substance in the United States (4,5). Unlike *d*-amphetamine, *d*-methamphetamine is easy to synthesize. Laboratories producing methamphetamine continue to comprise the majority of all illicit drug synthesis laboratories seized by the Drug Enforcement Administration (DEA). California continues to lead the nation in the number of labs seized, with 1234 targeted in 1997. During the same year, DEA seized 1273 methamphetamine labs nationwide, up from 879 in 1996.

The most popular method of synthesis before the 1990s (>50% of the labs seized), probably because of its ease and economy, was reductive amination using phenyl-2-propanone (P2P), methylamine, aluminum foil, mercuric chloride catalyst, and alcohol. In the next most popular method (<10%), the product of an acetaldehyde-methylamine reaction was refluxed with benzylmagnesium chloride. The third most popular method (<10%) used the Leukart reaction, refluxing P2P with either methylamine and formic acid or *N*-methylformamide with hydrochloric acid. An important precursor in these syntheses is P2P, which, until 1981, was available commercially. Because of its importance in illicit synthetic methods, it is now listed as a Schedule II controlled substance. As a result of its control, some clandestine laboratories now synthesize only P2P.

During the late 1970s and early 1980s, the conversion of ephedrine to methamphetamine by reductive cleavage of the hydroxyl group either using thionyl chloride or hydriodic acid (HI) and red phosphorus was relatively uncommon; only 10 of the laboratories seized used this synthetic method. However, with the growing difficulty of obtaining precursors for the more popular reductive amination routes, and the increasing availability of (−)-ephedrine and (+)-pseudoephedrine both in this country and the Far East, methamphetamine synthesized from ephedrine has flooded the market.

From 1991 through 1993, the DEA seized >500 laboratories illicitly producing methamphetamine; the majority of these labs were on the West Coast and in Houston. The ephedrine reduction method was used in 81% of these labs, whereas the P2P method was used in only 16%. This discovery represents a direct reversal of trends noted in the 1980s. Forensic laboratories in New Mexico also reported that in the first 6 months of 1994 there was a "decided increase" in the number of cases of methamphetamine seizures, all using the ephedrine process. Additionally, the California Bureau of Narcotic Enforcement reports that the use of the precursor chemical ephedrine in California has rapidly increased with the availability of the chemical from Mexico. Word of the value of ephedrine as a precursor in methamphetamine manufacture has spread quickly in the Mexican drug community, causing it to become a "significant smuggling commodity," resulting in a significant price drop in the U.S. market.

In March 1994, the *Federal Register* documented the proposal by the DEA to make all regulated transactions of ephedrine, regardless of size, subject to the reporting and recordkeeping requirements of the Chemical Diversion and Trafficking Act of 1988 (CDTA). This proposal would subject all transactions involving bulk ephedrine and single-entity ephedrine drug products to the applicable provisions of the CDTA. The CDTA has reportedly already made the acquisition of commercially produced HI much more difficult. The illicit price of a gallon of HI in California has increased sharply to between $4000 and $7000. Lack of availability and this price increase has made violators seek alternative sources of HI and/or different processes of clandestinely manufacturing methamphetamine. It is likely that the tightening of ephedrine controls will soon shift the pendulum back to some of the older methods of methamphetamine synthesis.

Therapeutic Uses

Amphetamine (Dexedrine®, Ferndex®, Oxydess II®) and methamphetamine (Desoxyn®) are DEA Schedule II controlled substances. The substances in this schedule have a high abuse potential with severe psychic or physical dependence liability. Dosage forms include 5-, 10-, and 15-mg tablets and sustained-release capsules. Dexedrine® is also available as a 5 mg/mL elixir.

Clinical indications for these compounds include narcolepsy and attention deficit disorder. They can also act as appetite suppressants in the treatment of exogenous obesity, in which they function as a short-term adjunct to a regimen of weight reduction based on caloric restriction (6).

Effects

Methamphetamine causes CNS stimulation by displacing dopamine from nerve terminal storage vesicles. This release causes the hyperstimulation of dopaminergic receptor neurons in the synaptic cleft. Amphetamine and methamphetamine are substrates for the serotonin (5-HT), norepinephrine (NE) and dopamine (DA) transporters and lead to transmitter release by a process of transport-mediated exchange. Upon entry into the cytoplasm, the amphetamines further reduce accumulation of NE and DA in the synaptic vesicles. Catecholaminergic vesicles use an interior-acidic proton gradient for transmitter uptake. These drugs compete for protons with neurotransmitter already present in the granules. The resulting uncharged neurotransmitter then diffuses out of the granules down its concentration gradient. This mechanism causes a continuous release of neurotransmitter at low doses of stimulant, accounting for the locomotor stimulant and reinforcing effects of these compounds. Direct peripheral and organ stimulation at the various α- and β-adrenergic receptors also occurs, resulting in elevation of systolic and diastolic blood pressures and weak bronchodilator and respiratory stimulant action. At therapeutic doses, the heart rate may be reflexively slowed; large doses may produce cardiac arrhythmias.

Methamphetamine has greater CNS efficacy than *d*-amphetamine, most likely because of its greater ability to penetrate the CNS. The euphoric effects produced by methamphetamine, cocaine, and other designer sympathomimetic amines are difficult to distinguish clinically except for the much longer half-life of methamphetamine, sometimes as much as 10 times that of cocaine.

At low doses, methamphetamine-induced CNS stimulation manifests clinically as euphoria, increased alertness, intensified emotions, increased feeling of self-esteem and well-being, sensations of extreme physical and mental power, and allegedly increased sexuality. Most abusers begin by taking amphetamines orally, usually 150–250 mg/d. Doses in this range generally produce plasma concentrations of 0.1–0.2 mg/L (0.74–1.48 μmol/L). Progression to intravenous use is usually a result of the desire to intensify the euphoric feelings that often accompany use of these drugs.

In "toxic" doses [those producing plasma concentrations approaching 0.5 mg/L (3.70 μmol/L)], the amphetamines begin to produce unpleasant CNS symptoms such as anxiety, agitation, hallucinations, delirium, seizures, and death. Long-term, high-dose use of stimulants can both induce an acute psychotic state in previously healthy individuals and precipitate a psychotic episode in those with psychiatric illness. Hyperthermia may result from CNS-induced abnormalities, seizures, or muscular hyperactivity. Secondary rhabdomyolysis may also be seen.

Cardiovascular manifestations of toxicity begin to appear at plasma concentrations of 0.02 mg/L (0.14 µmol/L) and include hypertension, tachycardia, atrial and ventricular arrhythmias, and myocardial ischemia. Cerebrovascular accidents may also be seen, precipitated by elevated blood pressures or drug-induced vasospasms (1,7).

Evidence suggests that the neurochemical basis underlying the behavioral effects and increased motor activity associated with sympathomimetic amines involve dopaminergic systems. By enhancement of neurotransmitter release and blockade of reuptake, stimulants facilitate catecholaminergic neurotransmission. One of the defining characteristics of psychomotor stimulants (methamphetamine, amphetamine, cocaine, etc.) is their ability to elicit increases in spontaneous motor activity. At low doses, these drugs produce an alerting response characterized by increases in exploration and locomotion. As the dose increases, locomotor activity decreases and the behavioral patterns become more stereotyped (a continuous repetition of one or several types of behavior). The rewarding effects (those responsible for psychomotor stimulant abuse) are also thought to result from enhanced DA release in the limbic regions such as the nucleus accumbens (8).

CASE PRESENTATION

A 45-year-old white male works in a safety-sensitive position in the airline industry in California. As a condition of his employment, he is required to submit random urine specimens for screening for drugs of abuse. The urine specimens are to be tested according to the Department of Health and Human Services (HHS) guidelines. His initial urine specimen tests positive for methamphetamine and amphetamine at concentrations of 4700 and 1000 ng/mL (31.50 and 7.40 µmol/L), respectively. When the medical review officer investigates the positive result, he learns that the individual was recently diagnosed with Parkinson's disease and was prescribed selegiline (L-deprenyl) for the disease. Despite the apparent explanation for the positive result, the medical review officer requests that a chiral analysis for methamphetamine and amphetamine be performed. The chiral analysis indicates approximately equal concentrations of the d- and l-isomers of both compounds. Because selegiline is metabolized to only the l-isomers of amphetamine and methamphetamine, the medical review officer concludes that the l-methamphetamine and l-amphetamine resulted from the prescription use, but the presence of the d-isomers indicates abuse of methamphetamine. The employee is removed from his safety-sensitive position.

PHARMACOKINETICS

Amphetamine and methamphetamine share essentially the same pharmacokinetic profile. Methamphetamine is highly lipid-soluble and well absorbed orally, with a bioavailability of approximately 67% and a volume of distribution of 3.0–7.0 L/kg. A single oral methamphetamine dose of 0.125 mg/kg given to human male volunteers produced a peak plasma concentration of 0.020 mg/L (0.134 µmol/L) at 3.6 hours with an average elimination half-life of approximately 10 hours.

In another study that evaluated the effect of d-methamphetamine on circadian rhythms, 10 male volunteers were given 0.43 mg/kg d-methamphetamine during two test sessions, 1 week apart. Peak serum concentrations ranged from 0.06 to 0.3 mg/L (0.40 to 2.01 µmol/L) at 3.6 hours after administration during the "day" session and 0.03 to 0.08 mg/L (0.20 to 0.54 µmol/L) at 4.8 hours during the "night" session. After oral administration of 30 mg of amphetamine base to 8 adults, an average peak plasma concentration of 0.11 mg/L (0.81 µmol/L) was observed at 2.5 hours, decreasing to 0.08 mg/L (0.59 µmol/L) by 4.5 hours (9).

With intravenous use, doses are generally higher than those taken orally. During early use, 20- to 40-mg doses, three to four times a day, are sufficient. As tolerance develops, the dose and frequency of injections increase considerably. The drug is injected every 2–3 hours around the clock, during which the user never sleeps. Occasionally, these "runs" last as long as 10–12 days. After a run, users "fall out," that is, become so exhausted, disorganized, tense, or paranoid they cease using the drug and go to sleep. This sleep lasts 12–18 hours following a 3- to 4-day run. Upon awakening, the paranoid state is diminished, but lethargy may persist. At this point, a new run begins to terminate the lethargy. When use becomes well established, individual injections range from 100 to 300 mg. Long time abusers may use as much as 5000–15,000 mg/d (10).

Methamphetamine hydrochloride in the form called "ice" is abused by smoking. The term "ice" originated in the Far East and refers to the large crystals of methamphetamine hydrochloride produced through the ephedrine-reduction method. The popularity of ice is due to the immediate clinical effects of euphoria resulting from the drug's rapid absorption from the lungs. In one study, the bioavailability of an average inhaled dose of 21.8 mg was approximately 90%. Smoked doses of approximately 20 mg of methamphetamine hydrochloride produced peak plasma concentrations of 0.04 mg/L (0.27 µmol/L) in less than an hour. Although methamphetamine concentrations after smoking peak rapidly, they remain high for a considerable period of time, declining with a half-life of 11–12 hours. The volume of distribution was 3.24 L/kg for the smoked dose and 3.73 L/kg for the intravenous dose. Because of the more rapid and intense drug effect, patients describe a high distinct from that produced by snorting or ingesting the drug (11,12).

Methamphetamine is excreted in the urine largely as unchanged parent drug, under normal conditions, up to 45% of a dose in a 24-hour period. Approximately 7% of an administered dose undergoes N-demethylation to amphetamine (Figure 8–1). Amphetamines are also metabolized in the liver via aromatic hydroxylation. Accumulated hydroxylated metabolites have been implicated in the development of amphetamine psychosis. Elimination of the sympathomimetic amines is highly dependent on urinary pH. Urinary acidification to a pH <5.6 decreases the plasma half-life from 11–12 hours to 7–8 hours. Alkalinization increases the half-life to 18–34 hours. For every one-unit increase in urinary pH, there is an average 7-hour increase in plasma half-life. In acidic urine, up to 76% of a dose can be found as unchanged drug, whereas in alkaline urine <2% of a given dose will be detected as unchanged drug and <0.1% as amphetamine. Approximately 15% of a dose is excreted as p-hydroxymethamphetamine. After the ingestion of 10 mg of methamphetamine, urine concentrations of methamphetamine can range from 0.5 to 4.0 mg/L (3.35 to 26.8 µmol/L) during the first 24 hours. Long-time abusers, who may be using as much as 5000 to 15,000 mg/d, can have urinary methamphetamine concentrations of >20 mg/L (134 µmol/L) and amphetamine concentrations >10 mg/L (73.96 µmol/L) (9,13).

ANALYSIS

Thin-Layer Chromatography

There are two popular thin-layer chromatography (TLC) systems in use today for detection of a wide range of drugs and their metabolites. The Toxi-Lab® (Irvine, CA) AB System is a complete TLC broad-spectrum drug-detection system. It contains all the necessary equipment to extract, concentrate, and detect a broad spectrum of >300 drugs. Identification is accomplished by dipping the TOXI-GRAM® through a series of reagents and viewing under ultraviolet light. The sympathomimetic amines [ephedrine (EPH), pseudoephedrine (PE), phenylpropanolamine (PPA), amphetamine, phentermine, and methamphetamine] have similar color characteristics and/or position on Toxi-Lab® A (R_f 0.2–0.3). However, a difference can be observed between the over-the-counter sym-

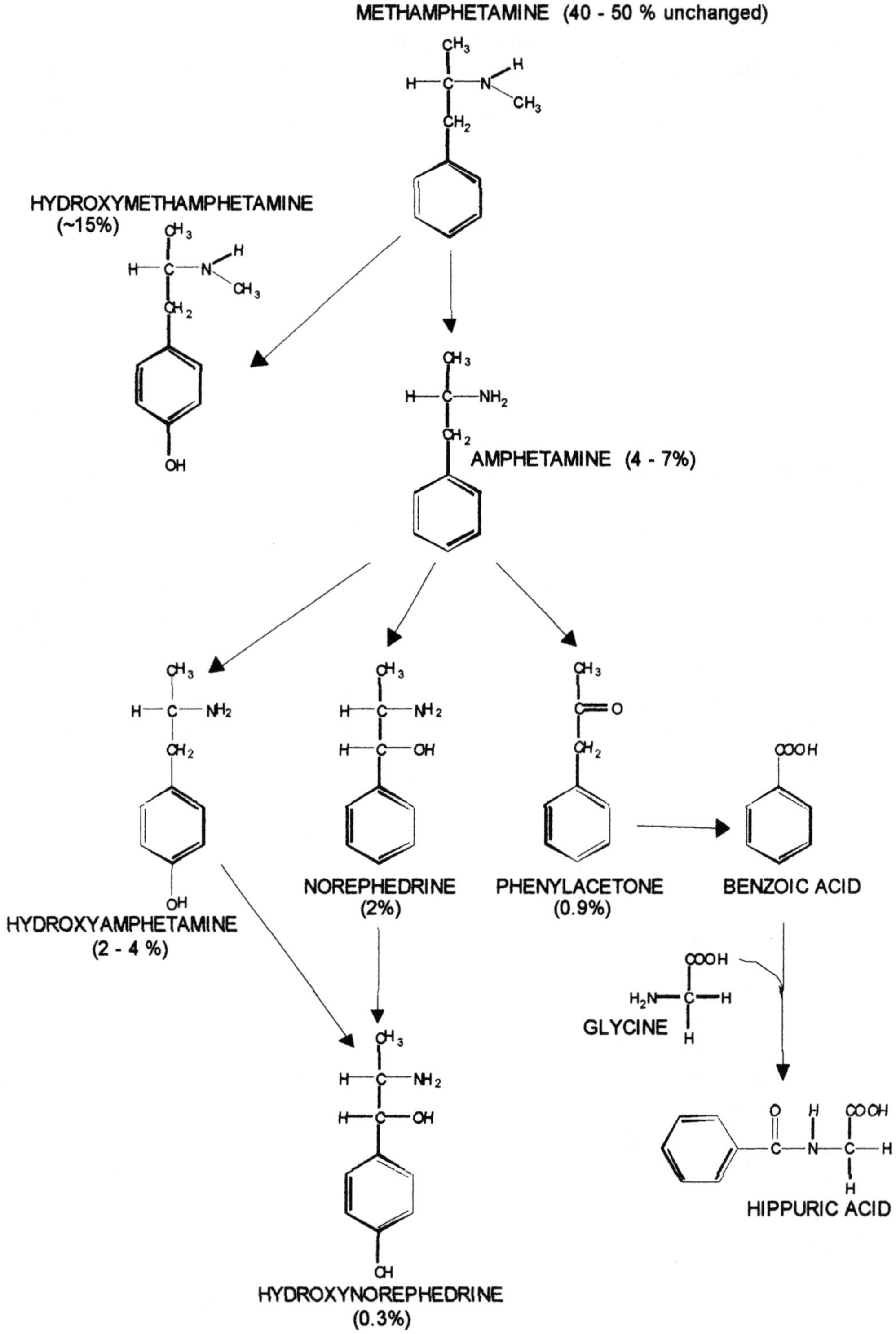

FIG. 8–1. Metabolism of methamphetamine.

pathomimetic amines (EPH, PE, and PPA) and the illicit or prescription compounds (phentermine, amphetamine, and methamphetamine) at "Stage I" (sulfuric acid dip); the over-the-counter drugs form green centers, and the illicit or prescription sympathomimetic amines form brown or gold centers.

The Toxi-Lab® system also has two special procedures for the detection and differentiation of sympathomimetic amines. One procedure involves developing the TOXI-GRAM® in acetone and ammonium hydroxide (NH_4OH). EPH-PE, PPA, amphetamine, phentermine, and methamphetamine will have R_f values of 0.9, 0.85, 0.7, 0.4, and 0.25, respectively. EPH and PPA can be further separated by adding 0.2 mL of acetaldehyde to 100 mL of TOXI-LAB A developing solvent and then combining 3.0 mL of this solution with 15 mL of concentrated NH_4OH. This solvent system is then used to develop the TOXI-GRAM®. R_f values for EPH-PE, PPA, amphetamine, phentermine, and methamphetamine are 0.83, 0.65, 0.55, 0.38, and 0.23, respectively. Sensitivity for most of the drugs in this class using the Toxi-Lab® system is 0.5 mg/L.

The Drug-Skreen® II System from Eppendorf-Brinkmann (Westbury, NY) is another TLC system available for the detection of a wide range of drugs and their metabolites. In this system, extraction of the drugs is accomplished at a single pH (9.5), and drug separation is by TLC using silica gel precoated plastic sheets. Identification is accomplished with a sequential spraying technique. Sympathomimetic amines are visualized under ultraviolet light (254 nm) after 0.1% ninhydrin application to the plate. Amines will generally appear as pink spots with a R_f of approximately 0.39. PPA appears as a red spot and elutes just before amphetamine (R_f 0.27).

Immunoassay

Since the introduction of the EMIT-dau® assays in the mid-1970s, immunoassays based upon the competitive binding of labeled drugs or sites for specific drug antibodies have proven successful for rapid urine screening for amphetamines and other drugs of abuse.

Syva Co. (now part of Behring Diagnostics, San Jose, CA) developed EMIT®, which uses an enzyme-multiplied immunoassay technology. The EMIT-dau® amphetamine class assay (EC) contains polyclonal antibodies and has a cutoff calibrator of 300 ng/mL (2.22 µmol/L) d-amphetamine. The assay readily cross-reacts with ephedrine, pseudoephedrine, phentermine, and other phenylisopropylamines. The EC assay is indicated when low detection limits for amphetamines and/or high cross-reactivity to other structurally related stimulants is desired. Thus, EC is used in athletic drug testing and clinical toxicology laboratories and when screening patients in substance-abuse rehabilitation programs.

The EMIT-dau® monoclonal amphetamine-methamphetamine assay (EM) contains monoclonal antibodies, has a cutoff calibrator of 1000 ng/mL (6.70 µmol/L) d-methamphetamine, and is configured for low-volume random-access analyzers. The assay displays much less cross-reactivity with phenylisopropylamine derivatives than the EC assay. However, the assay readily detects low concentrations of designer drugs methylenedioxyamphetamine (MDA) and methylenedioxymethamphetamine (MDMA). EM is indicated when HHS guidelines of a 1000 ng/mL cutoff calibrator are required, and high selectivity for only amphetamine, methamphetamine, MDA, or MDMA is desired.

The EMIT®-II amphetamine-methamphetamine assay (EII) contains monoclonal antibodies with a cutoff calibrator of 1000 ng/mL (6.70 µmol/L) d-methamphetamine, but, unlike the EM assay, requires 1000 ng/mL (7.40 µmol/L) d-amphetamine to produce a positive response. The EII assay has low cross-reactivity to phenylisopropylamine derivatives and is configured for high-volume, high-speed analyzers when HHS guidelines are required and high selectivity for amphetamine or methamphetamine is desired. In an effort to further confront the issue of cross-reactivity of these assays with EPH, PE, and PPA, Syva Co. also supplies an optional amphetamine confirmation kit. The kit contains sodium hydroxide and sodium periodate. which are added to the specimen before testing. These reagents eliminate cross-reacting compounds with a hydroxyl group in the α-position through oxidative cleavage of the α- and β-carbon bonds.

Abbott Laboratories (North Chicago, IL) developed the TDx/FLx/AxSYM immunoassays that use fluorescence polarization immunoassay (FPIA) technology. Currently, the TDx/FLx/AxSYM Amphetamine/Methamphetamine II Assay is widely used by clinical laboratories for the detection of amphetamine and/or methamphetamine in urine obtained from emergency departments and substance abuse rehabilitation programs. These FPIA systems contain monoclonal antibodies, have a cutoff calibrator of 300 ng/mL (2.22 µmol/L) d-amphetamine, and are configured for low-volume random-access analyzers such as Abbott TDx and FLx analyzers and the high-volume, high-speed AxSYM analyzer. The assay displays low cross-reactivity with EPH, PE, and other phenylisopropylamine derivatives.

Radioimmunoassays (RIA) are available to screen for amphetamines and related compounds. For example, Diagnostic Products Corporation (DPC, Los Angeles) has a double-antibody amphetamine assay and a coated-tube methamphetamine assay. RIA assays typically use d-amphetamine as the calibrator for amphetamine assays and d-methamphetamine as the calibrator for methamphetamine assays.

Roche Diagnostics Systems, Inc. (Somerville, NJ), introduced an immunoassay system that uses the theory of kinetic interaction of microparticles in solution (KIMS). The Roche ONLINE® KIMS assay for amphetamines contains both amphetamine and methamphetamine monoclonal antibodies. d-Amphetamine is used as the kit calibrator. With a 1000 ng/mL (7.40 µmol/L) cutoff, this assay will give a positive result for urine samples containing amphetamine at or above the cutoff and urine samples containing 200 ng/mL (1.48 µmol/L) amphetamine and 500 ng/mL (3.35 µmol/L) methamphetamine.

Samples containing methamphetamine in the absence of amphetamine will be negative at a 1000 ng/mL cutoff. With a 500 ng/mL cutoff, the assay will report as "positive" samples containing either 500 ng/mL amphetamine (3.7 µmol/L) or methamphetamine (3.35 µmol/L). At a 500 ng/mL cutoff, there is approximately 30% cross-reactivity with MDMA and MDA and 47% cross-reactivity with dl-methamphetamine. Using the 1000 ng/mL (7.4 µmol/L) calibrator, only MDA shows any significant cross-reactivity (30%).

The advent of immunoassays has facilitated the rapid screening of forensic specimens for abused drugs in both urine drug testing and postmortem laboratories. Unfortunately, urine specimens are not available for all postmortem cases. This lack has led to the modification of commercially available immunoassays for use on alternate specimens such as blood, bile, and liver. For example, many laboratories currently use RIA to screen blood and tissue specimens for abused drugs. Although this method continues to provide reliable analytical data, the problems and costs related to handling and disposing of radioactive material persists. EMIT® has been in use for many years and has been specifically applied to blood specimens. More recently, cloned enzyme donor immunoassay (CEDIA®) and enzyme-linked immunosorbent assay (ELISA) have become other commercially available enzyme immunoassay (EIA) methodologies applicable to specimens other than urine. Other strengths of EIA methods include sensitivity and ease of automation. Additionally, because of the adaptability of microtitration plate technology to the EIA methodology, ELISA has proved adequate for analyzing microliter quantities of sample (14).

The microtitration plate technology used to detect drugs of abuse is a competitive, solid-phase, heterogeneous, enzyme immunoassay. Competitive assays are generally used for small molecules and are typical for other technologies such as CEDIA®, EMIT®, and RIA. In the ELISA methodology, the microplate technology uses antibodies immobilized onto the surface of a microtitration plate. Free drug and drugs conjugated to an enzyme (horseradish peroxidase) compete for binding to the antibody. After a short incubation, the plate is washed to remove all unbound enzyme conjugate and sample debris. A substrate solution is added that produces a colored product in the presence of bound enzyme conjugate. The amount of color produced is measured in absorbance (450 nm) via a spectrophotometer and measurements taken are inversely proportional to the amount of free drug that was present in the original sample. This technique is robust and reasonably insensitive to normal temperature variation.

STC Technologies, Inc. (Bethlehem, PA), manufactures both an "Amphetamine-Specific Micro-Plate ELISA" and a "Methamphetamine-Specific Micro-Plate ELISA" intended for the qualitative determination of amphetamine and methamphetamine in serum. These assays can be modified for the analysis of blood and tissues and, with the use of urine calibrators, can be used for urine screening as well. The cross-reactivity of these kits with various other structurally related compounds is listed in Table 8–1.

CEDIA® is a trademark method originally developed by Roche Diagnostics/Boehringer Mannheim Corporation (Somerville, NY) but currently manufactured and distributed by Microgenics Corporation (Fremont, CA). This methodology is designed to be used primarily on the Hitachi 717 high-speed, high-volume autoanalyzers. CEDIA® drugs-of-abuse assays are enzyme immunoassays that use recombinant DNA technology. They are based on the enzyme β-galactosidase (from *Escherichia coli*), which has been genetically engineered into two inactive fragments: the Enzyme Donor (ED) and the Enzyme Acceptor (EA). ED and EA spontaneously reassociate to form a fully active enzyme that, in the assay format, cleaves chlorophenolred-β-galactoside (CPRG) to chlorophenolred (CPR) and galactose. CPRG does not absorb significant energy at $\lambda = 570$ nm, whereas this is the λ_{max} for CPR. Therefore, production of CPR is easily measured. A secondary wavelength (660 nm) for CPR can also be used to correct for minor changes in sample absorbance. The drug in the specimen competes with drug conjugated to the ED fragment of β-galactosidase for antibody binding sites. If drug is present in the specimen, it will bind to antibody, leaving the inactive enzyme fragments free to form active enzyme. If drug is not present in the specimen, antibody will bind to drug conjugated to the ED fragment, inhibiting the reassociation of the inactive β-galactosidase fragments, and no active enzyme will be formed. The amount of enzyme formed and resultant absorbance change are proportional to the amount of drug present in the specimen. Cross-reactivity of the CEDIA® DAU Amphetamines assay [1000 ng/mL (7.4 µmol/L) cutoff] with various drugs in this class is listed in Table 8–1.

Gas Chromatography–Mass Spectrometry

In 1988, HHS set stringent guidelines for laboratories to follow for accurate and precise analysis of five classes of drugs of abuse (amphetamines, cannabinoids, cocaine, opiates, and phencyclidine). These guidelines required gas chromatography–mass spectrometric (GC/MS) confirmation of these drugs of abuse as well as GC/MS quantifications (15). Since then, GC/MS has become the gold standard method for the identification of sympathomimetic amines, especially in light of the ongoing problem of interference from noncontrolled over-the-counter phenylisopropylamines. In order to avoid labs' reporting false-positive amphetamine-methamphetamine results, HHS also instructed their certified labs in December 1990 that methamphetamine may be reported positive only if the specimen also contains 200 ng/mL amphetamine (16). All Department of Defense and HHS-certified laboratories now use GC/MS selected-ion monitoring for quantification of these compounds.

Because of their extreme volatility at the high temperatures encountered in GC/MS systems, compounds in this class require derivatization before analysis. Although

TABLE 8–1. Cross-reactivity (%) of various phenethylamines in different immunoassays

Compound	EMIT monoclonal	EMIT Emit II	TDx®	DPC® Coat-a-Count (meth)	DPC® double antibody (amph)	Roche On-Line® (KIMS)	STC microplate ELISA (meth)	STC microplate ELISA (amph)	CEDIA® DAU (amph)
d-Amphetamine	250.0	100.0	100.0	0.3	100.0	100.0	1.3	100.0	101.0
d-Methamphetamine	100.0	100.0	100.0	100.0	0.0	0.5	100.0	<1	100.0
Phenylpropanolamine	1.3	0.4	<0.1	0.1	0.5	0.7	<0.1	<0.1	0.3
l-Ephedrine	2.0	0.7	<0.1	0.9	0.05	<0.2	1.4	–	0.4
Pseudephedrine	1.0	0.3	<0.03	0.3	0.05	<0.2	1.8	<0.1	0.6
Phentermine	–	50.0	35.0	0.04	0.17	<0.1	–	<0.1	1.9
MDA	–	33.0	151.0	<0.1	138.0	32.0	2.8	>100.0	1.9
MDMA	–	17.0	99.0	>200.0	0	0.2	1513	<1	69.0

meth, methamphetamine; amph, amphetamine.

TABLE 8–2. GC/MS ion characteristics for HFBA derivatives of sympathomimetic amines

Compound	Retention time (min)[a]	Ions monitored	Comments
a-Phenethylamine	4.6	104, 118, 99, 149/150	
dl-Amphetamine	4.7	240, 118, 91	
Phentermine	4.8	254, 91	No 210 ion
Phenylpropanolamine	5.4	240, 160, 330, 275	
dl-Methamphetamine	5.5	254, 118, 210	
Ephedrine	5.7	254, 210, 344	No 118 or 91 ions
Pseudephedrine	6.0	254, 210, 344	No 118 or 91 ions
Phenylephrine	6.5	240, 169	No 91 ion
MDA	7.1	135, 162, 240, 169	No 91 ion
MDMA	7.7	162, 254, 210, 162	No 118 or 91 ions
Benzphetamine	7.9	148, 91, 65, 149	Not derivatized

[a]HP-1 capillary column (12 m × 0.2 mm (internal diameter) × 0.33 μm). Initial oven temperature = 100 °C; injection port = 250 °C; ramped from an initial temperature of 100 °C (held for 2 minutes) to 280 °C at 20 °C/min.

there are many derivatives available for GC/MS use, those most commonly used with primary and secondary amines include heptafluorobutyric anhydride (HFBA), pentafluoropropionic anhydride (PFPA), trifluoroacetic anhydride, and 4-carbethoxyhexfluorobutyryl chloride. HFBA and PFPA are currently the most commonly used derivatizing reagents in forensic drug-testing labs for this class of compounds. Table 8–2 provides GC/MS ion characteristics for HFBA derivatives of amphetamine, methamphetamine, and other amines.

Phentermine also presents some unique analytical challenges. In particular, it demonstrates significant cross-reactivity with most amphetamine or methamphetamine immunoassays, even at concentrations as low as 2.5 mg/L [16.75 μmol/L; urinary concentrations of phentermine after therapeutic doses range from 5 to 25 mg/L (33.50 to 167.53 μmol/L)]. Underivatized phentermine exhibits chromatographic behavior similar to amphetamine in many GC systems and an electron ionization spectra that is indistinguishable from methamphetamine. However, the trifluoroacetyl or propionic acid derivatives of phentermine, amphetamine, and methamphetamine are readily resolved and identified. Alternately, chemical ionization (CI) GC/MS readily distinguishes phentermine from methamphetamine because phentermine yields a distinct 133 m/z ion not found in the CI spectrum of methamphetamine. However CI technology is not currently readily available to many forensic laboratories.

Liquid Chromatography

Liquid chromatographic methods have been described for monitoring therapeutic dosing of these compounds. However, these methods are not commonly used for forensic applications.

COMPLICATIONS AND CONTROVERSIES

Enantiomer Differences

Amphetamine and methamphetamine contain a chiral carbon and, therefore, exist as enantiomers. In many cases, pharmaceutical companies in the development of drugs have exploited the presence of the chiral carbon and its ability to bind with chiral receptors. d-Methamphetamine and d-amphetamine are used therapeutically for narcolepsy, attention deficit disorder, and weight loss. l-Methamphetamine has one-tenth the CNS stimulation effect of the d-isomer but has greater peripheral vasoconstrictive properties and thus has been used in over-the-counter nasal inhalers. Even though the immunoassays currently available for use by federally certified civilian laboratories and the Department of Defense urine drug-testing facilities are relatively stereospecific, occasional false-positives have been reported from individuals taking over-the-counter cold medications.

The goals of forensic and urine drug-testing laboratories should include preventing such false-positive reports and detecting all true-positive cases of illicit drug abuse. Many screening and confirmation techniques were failing one or both of these goals. EMIT®, TDx®, and Toxilab® have the required sensitivity but are susceptible to false-positives from nasal inhaler use. Standard GC/MS procedures are unable to distinguish between enantiomers. It became clear that an additional step was necessary before reporting positive methamphetamine results.

Chromatographic identification of d- and l-isomers depends on the enantiomers reacting with optically active substrates to form diastereomers. This can be accomplished in one of two ways:

- A chiral, optically active stationary phase can be used to form transient diastereomers with the optically active drug causing different partition coefficients.
- A chiral derivatizing reagent may be used to achieve separation on an achiral stationary phase.

Chiral stationary phases are available for both HPLC and GC systems. However, chiral derivatizing reagents have gained popularity because they allow chiral analysis on the same instrument used for other routine analyses. The main disadvantage of using a chiral derivatizing reagent is the difficulty of ensuring the optical purity of the derivatizing reagent; four possible isomers can result from the reaction of an asymmetric sample with an asymmetric reagent rather than the two desired. The four derivative isomers consist of two diastereomeric pairs of chromatographically unresolvable enantiomers. Experiments must also be done to demonstrate both that racemization is not occurring during the derivatizing process and that stereoselective formation of one pair of diastereomers is not occurring.

Amines have been converted to diastereomeric amides with several chiral acid chloride reagents for GC analysis, such as O-methylmandelyl chloride, α-phenylbutyryl chloride or anhydride, and N-trifluoroacetyl-L-prolyl chloride (L-TPC). L-TPC is the most popular of this type of reagents. First used as a chiral derivatizing reagent for the separation of amino acids, it has since become popular for both the on-column and pre-injection derivatization of amphetamine and methamphetamine. There have been some reports of racemization with the use of L-TPC. Amide derivatives can also be prepared with acyclimidazole reagents and with chiral acids in the presence of N,N'-carbonyldiimidazole. Diastereomeric sulfonamide derivatives have been created using (+)-camphor-10-sulfonyl chloride and menthyl carbamates, such as menthyl chlorformate. These reagents have been used for sympathomimetic amines and amino acid analysis (17).

Cross-Reactivity of Immunoassays

In the previous section, a discussion of the different types of immunoassays for amphetamine and methamphetamine was included. One complication encountered to a greater extent with immunoassays for amphetamine and methamphetamine than with other immunoassays is the cross-reactivity to other sympathomimetic amines such as EPH, PE, PPA, phentermine, MDA, and MDMA. Table 8-1 summarizes the different cross-reactivities of the various sympathomimetic amines with the different immunoassays. The ability to detect substances other than amphetamine and methamphetamine is an advantage in clinical laboratories that perform emergency toxicology testing. The goal of this testing is to include or exclude as many substances as possible in a short time. Therefore, a negative screening result that can rule out a large number of drugs within a class is preferable to a very specific immunoassay that would eliminate only a few substances.

Therapeutic Drugs Converted to Amphetamine or Methamphetamine

There are several drugs that are metabolized to amphetamine and methamphetamine. This group includes benzphetamine, clobenzorex, selegiline, famprofazone, fenethylline, and fenproporex. There are several ways to document the use of these drugs as an explanation for a positive urinalysis result. Examination of medical history is a straightforward approach. Alternatively, it may be possible to identify parent drug in the urine specimens. Unfortunately, doing so is not always possible given the short half-lives of many of these drugs. Instead, the identification of a metabolite unique to these drugs may provide analytical verification that these drugs were ingested.

Enantiomer analysis can be used to document selegiline use. Prescription selegiline is the *l*-enantiomer; it is converted in the body exclusively to *l*-methamphetamine and *l*-amphetamine. Therefore, any findings of the *d*-isomers of these compounds would be inconsistent with selegiline use. Conversely, fenethylline is a racemic for the amphetamine portion of the molecule and one would expect to find equal amounts of the *d*- and *l*-isomers in the urine (18).

REFERENCES

1. Ellenhorn MJ, Barceloux DG: *Medical toxicology—diagnosis and treatment of human poisoning.* New York: Elsevier Science Publishing Co., 1988:625–641.
2. Morgan JP, Kagan DV: Street amphetamine quality and the controlled substances act of 1970. *J Psychedelic Drugs* 1978;10(4):303–317.
3. Litovitz TL, Klein-Schwartz W, White S, et al: 1999 annual report of the American Association of Poison Control Centers Toxic Exposure Surveillance System. *Am J Emerg Med* 2000:18(5):517–574.
4. Frank RS: The clandestine drug laboratory situation in the United States. *J Forensic Sci* 1983;28:18–31.
5. Allen AC, Kiser WO: Methamphetamine from ephedrine: I. Chloroephedrines and aziridines. *J Forensic Sci* 1987;32:953–962.
6. Burnham TH, Short RM, eds: CNS stimulants: amphetamine. In: Olin BR, ed: *Drug facts and comparisons 2000.* St. Louis, MO: Facts and Comparisons, A Wolters Kluwer Company, 2000:770–773.
7. Hoffman BB, Lefkowitz RJ: Catecholamines and sympathomimetic drugs. In Gilman AG, Goodman LS, Rall TW, et al, eds: Goodman and Gilman's pharmacological basis of therapeutics, 8th ed. New York: Pergamon Press, 1990:187–220.
8. Samanin R, Garattini S: Neurochemical mechanism of action of anorectic drugs. *Pharmacol Toxicol* 1993;73:63–68.
9. Baselt RC, Cravey RH, eds: Amphetamine. In: Baselt RC, ed: *Disposition of toxic drugs and chemicals in man*, 5th ed. Foster City, CA: Chemical Toxicology Institute, 2000:49–51.
10. Lee CYS, Heffez LB, Mohammadi H: Crystal methamphetamine abuse: a concern to oral and maxillofacial surgeons. *J Oral Maxillofac Surg* 1992;50:1052–1054.
11. Cook CE: Pyrolytic characteristics, pharmacokinetics, and bioavailability of smoked heroin, cocaine, phencyclidine and methamphetamine. *NIDA Res Monogr* 1991;115:6–23.
12. Cook CE, Jeffcoat AR, Hill JM, et al: Pharmacokinetics of methamphetamine self-administered to human subjects by smoking S-$(+)$-methamphetamine hydrochloride. *Drug Metab Dispos* 1993;21:717–723.
13. Caldwell J: The metabolism of amphetamines in mammals. *Drug Metab Rev* 1976;5(2):219–280.
14. Moore KA, Werner C, Zannelli RM, et al: Screening postmortem blood and tissues for nine classes of drugs of abuse using automated microplate immunoassay. *Forensic Sci Intl* 1999;106:93–102.
15. Department of Health and Human Services; Alcohol, Drug Abuse and Mental Health Administration: Mandatory Guidelines for Federal Workplace Drug Testing Programs; Final Guidelines: Notice. *Fed Regist* 1988;53(69; April 11):11978–11989.
16. Department of Health and Human Services; Alcohol, Drug Abuse and Mental Health Administration: Mandatory Guidelines for Federal Workplace Drug Testing Programs; Final Guidelines: Notice. *Fed Regist* 1994;59(110; June 9):29908–29993.
17. Cody J: Issues pertaining to monitoring the abuse of amphetamines in workplace drug testing. *Forensic Sci Rev* 1994;6:81–96.
18. Cody J: Metabolic precursors to amphetamine and methamphetamine. *Forensic Sci Rev* 1993;5:109–127.

SELF-ASSESSMENT QUESTIONS

1. What are the two primary precursor compounds for methamphetamine synthesis?

2. Describe the structural modification on the sympathomimetic amine structure that accounts for hallucinogenic activity.

3. List two general analytical approaches in the separation of amphetamine enantiomers.

4. List two reasons a subject might test positive in the urine assay for methamphetamine without having abused illicit methamphetamine.

5. Which technique is not ordinarily used in amphetamine and methamphetamine analysis in biological specimens?
   a. gas chromatography–mass spectrometry
   b. immunoassay
   c. liquid chromatography
   d. thin-layer chromatography

6. What is the major urinary product of methamphetamine intake?
   a. amphetamine
   b. hydroxyamphetamine
   c. hydroxymethamphetamine
   d. methamphetamine

7. Explain the pharmacologic differences between d-methamphetamine and l-methamphetamine.

CHAPTER 9

γ-Hydroxybutyrate

Daniel J. Cobaugh, Pharm.D., DABAT, and Sarah Kerrigan, Ph.D.

LEARNING OBJECTIVES

After completing this chapter, the reader should be able to:

1. *Describe the history of abuse of γ-hydroxybutyrate (GHB)–related substances in the United States.*
2. *Describe the common clinical manifestations of GHB toxicity.*
3. *Explain the approach to medical management of these patients in the emergency department.*
4. *Describe common methods of toxicological analysis of GHB.*
6. *Describe the limitations of GHB interpretation in antemortem and postmortem samples.*

INTRODUCTION

γ-Hydroxybutyrate (GHB) and its metabolic precursors, γ-butyrolactone (GBL) and 1,4-butanediol (BD), recently emerged as new recreational euphoriants. By the late 1990s, GHB gained notoriety in the popular press and was popularized as the "date rape drug" by the media.

Despite its only recent visibility and incorporation into contemporary drug culture, the effects of GHB-related substances have been studied for more than five decades.

In 1947, the central nervous system (CNS) depressant effects of GBL were documented. In 1960, GHB was synthesized as an orally active analog of γ-aminobutyric acid (GABA) (1). GHB is an endogenous neuromodulator found in mammalian tissue. It satisfies many of the criteria for consideration as a neurotransmitter, having specific receptor sites, endogenous synthesis, and heterogeneous distribution in the CNS (2). Its ability to produce sleep and reversible coma aroused interest in the drug as a surgical anesthetic. However, the lack of analgesic effect and seizure-like activity prevented widespread use of the drug.

Although GHB (sodium oxybate) is not currently an approved drug in the United States, it does have accepted medical uses elsewhere for anesthesia, resuscitation, sleep disorders, and treatment of substance dependence. Use of the drug within the United States is limited to clinical trials approved by the Food and Drug Administration (FDA) for the treatment of certain disorders such as narcolepsy. GHB has been shown to decrease daytime sleepiness and episodes of cataplexy, sleep paralysis, and hypnagogic, or dream-like hallucinations, in narcoleptic patients (3). In October of 2000, Orphan Medical Inc. submitted a new drug application to the FDA for GHB under the trade name Xyrem® for the therapeutic management of cataplexy. GHB has also been studied in the prevention of hypoxic injury (4–5).

Beginning in the late 1980s, GHB was sold in health-food stores as a dietary supplement that enhanced bodybuilding, hastened weight loss, and induced sleep. In the 1990s, it became popular as a recreational drug due to its intoxicating and euphoric effect. In

1990, the San Francisco Poison Center reported several cases of GHB exposure that resulted in coma and seizures (6). In that same year, FDA began an intense investigation of GHB distribution after numerous cases of GHB-related illness were reported (7). Between 1995 and 1998, the American Association of Poison Control Centers reported one to two deaths per year from GHB and its analogs. In 1999 alone, 10 GHB deaths were reported by poison centers (8). In eight of these cases, GHB was the substance involved. One case involved GBL, and one case involved BD. Since 1992, the Drug Abuse Warning Network (DAWN) has reported over 5000 GHB-related emergency department (ED) visits. A recent DAWN report on "club drugs" indicated a significant rise in GHB-related episodes between 1994 and 1999. The most dramatic increase occurred between 1997 and 1999, during which time the number of GHB-related visits increased from 762 to 2973 (Figure 9–1).

In November 1990, after widespread reports of poisonings and adverse reactions, the FDA designated GHB an investigational new drug and banned its distribution outside of approved clinical trials. In 1999, the FDA issued warnings about health hazards associated with the GHB-related products GBL and BD and urged the companies responsible for manufacturing and distributing these products to voluntarily recall the items. Gradually, individual states across the country designated GHB a scheduled drug. Federal action came in April 2000, when salts and isomers of GHB were placed into Schedule I of the federal Controlled Substances Act. As a result, manufacturing or selling GHB is punishable by as much as 20 years imprisonment. Schedule I substances are defined as those with high potential for abuse, no current medically accepted use in the United States, and a lack of accepted safety for use under medical supervision. If the FDA approves Xyrem® for the treatment of cataplexy, GHB will likely revert to a Schedule III substance with Schedule I consequences for illegal use.

Analogous laws in some states designate GBL as an illicit intoxicant, but both GBL and BD remain legal alternatives to GHB in parts of the country. Although BD is not a controlled substance, the FDA declared it to be a Class I Health Hazard, having potentially life-threatening risk. In 2000, an amendment to the federal Controlled Substances Act made GBL a List I chemical (9). Consequently, any person who imports, exports, or distributes GBL must register with the Drug Enforcement Administration and provide necessary documentation. However, the emergence of new metabolic precursors and analogs compound the regulatory, legis-

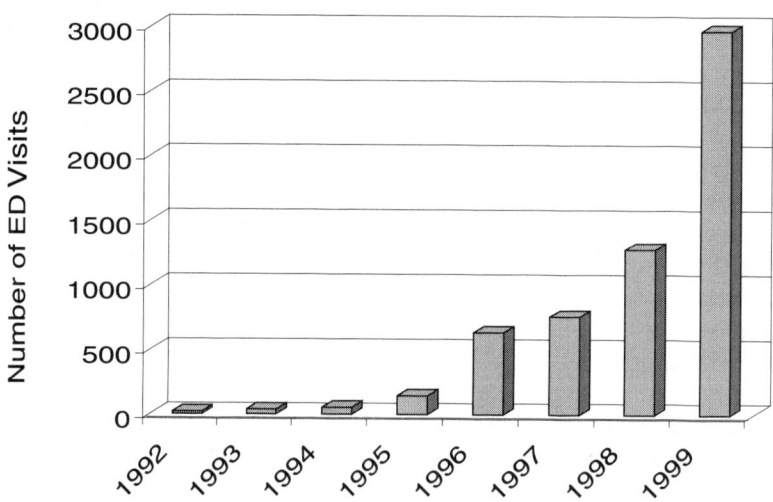

FIG. 9–1. Number of emergency department visits involving GHB. Source: Drug Abuse Warning Network.

lative, and toxicological issues surrounding GHB, posing a considerable ongoing challenge.

Regulation of the sale and distribution of GHB has been hampered by the prompt appearance and ready availability of metabolic precursors and analogs (10). Numerous sites on the Internet describe the clandestine synthesis and pharmacological effects of GHB alternatives, such as 4-hydroxyvalerate (also called 4-methyl-GHB) and GHB-aldehyde. Emergence of these supposedly legal alternatives is a direct consequence of GHB prohibition. These alternatives are typically homologous molecules (such as hydroxyvaleric acid and hydroxycaproic acid), substituted alcohols, and lactones. Because of the widespread industrial use of many of these relatively noncomplex chemicals, regulation at the state or federal level becomes extremely challenging.

CASE REPORT

A 22-year-old man was at a rave with a group of friends. At 0100 h, he ingested approximately three "capfuls" of GHB along with an alcoholic beverage. Forty-five minutes after the ingestion, his friends found him unconscious in his car. Drying vomitus was noted around his mouth and in the interior of the car. The paramedics were called, and on arrival they found the patient to be minimally responsive and shaking uncontrollably. The cardiac monitor showed sinus bradycardia with a rate of 56 beats/min, and his respiratory rate was 8 breaths/min. Intravenous access was obtained, and his respirations were assisted with a bag-valve mask. The paramedics administered naloxone and dextrose without response.

In the emergency department (ED), the patient could not be aroused and his vital signs were heart rate 48 beats/min, blood pressure 100/60 mmHg, respiratory rate 6 breaths/min, and temperature 98.6 °F (37 °C). The electrocardiogram showed sinus bradycardia. His pupils were dilated and sluggishly reactive. An arterial blood gas obtained shortly after arrival in the ED showed the following: pH 7.28, pCO$_2$ 54 mmHg, pO$_2$ 58 mmHg, and HCO$_3^-$ 20 mEq/L (20 mmol/L). The patient was intubated, and activated charcoal 50 g was administered via a nasogastric tube. An arterial blood gas obtained after intubation showed correction of his acidosis. Toxicology screening revealed a blood ethanol concentration of 180 mg/dL (30 mmol/L) and GHB in the urine. A chest X-ray obtained after intubation showed bilateral infiltrates.

The patient was transferred to the intensive care unit. Six hours after presentation, he began to awaken, became agitated, and extubated himself. Subsequent chest X-rays showed improvement. He was observed for an additional 24 hours in the intensive care unit and then discharged.

OCCURRENCE AND USES

GHB is encountered as an odorless, colorless liquid or hygroscopic powder (often as the sodium or potassium salt). It is frequently distributed or sold in water bottles, mouthwash bottles, and other common household vessels. Dyes or flavorings may be added to mask the sometimes salty or soapy taste of GHB or the solvent-like aroma of GBL and BD, which are encountered as liquids. Doses are typically administered in "capfuls" of liquid that contain largely unknown concentrations of drug. An illicit dose of 1 teaspoon (5 mL) of powdered GHB (2.5 g in a 70-kg adult, or 35 mg/kg) is common (11).

γ-Hydroxybutyric acid behaves as both an acid and an alcohol. Hydroxy acids with a sufficient number of carbon atoms may undergo intramolecular esterification. Both γ and δ hydroxy acids can lose water spontaneously to form cyclic esters known as lactones (Figure 9–2). Treatment with alkali rapidly cleaves the ring to produce an open-chain hydroxy acid. This property is widely utilized during the clandestine manufacture of GHB. Despite the fact that GBL is efficiently converted to GHB in vivo, clandestine synthesis is common, and recipes are widely available on the Internet. GBL is converted into GHB

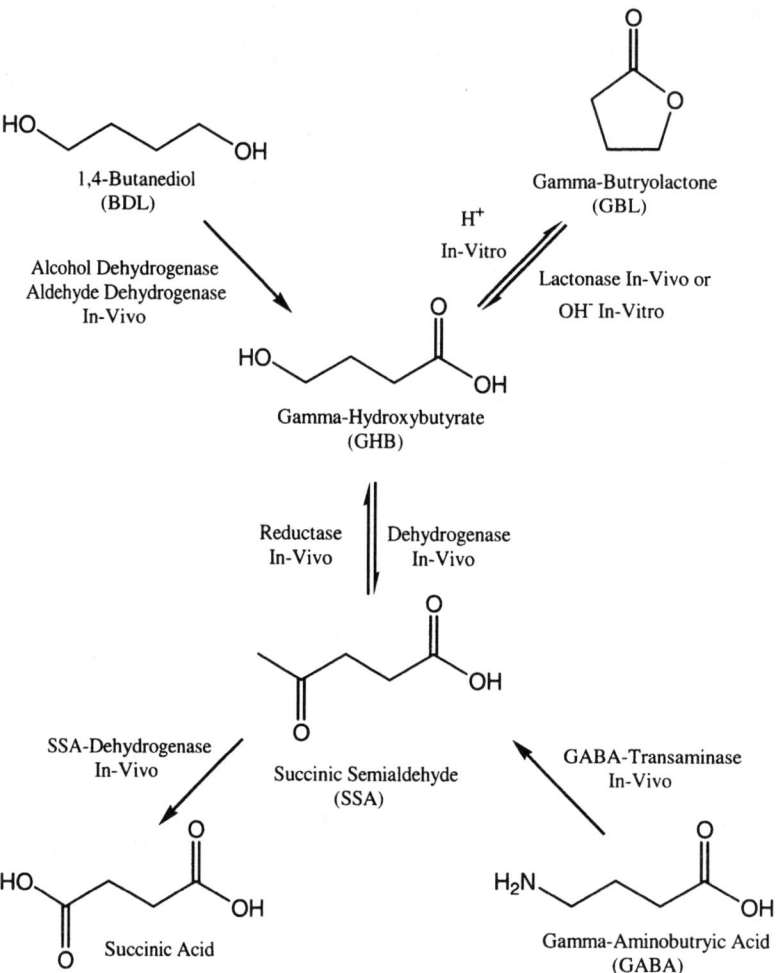

FIG. 9–2. In vitro and in vivo transformations of GHB and structurally related analogs. From Doherty JD, Hattox SE, Snead OC, et al: Identification of endogenous gamma-hydroxybutyrate in human and bovine brain and its regional distribution in human, guinea pig and rhesus monkey brain. *J Pharmacol Exp Ther* 1978;207:130–139.

in the presence of alkali, such as sodium or potassium hydroxide. Conversion rates are typically 70–80%, and the reaction is usually complete in less than an hour. Care must be taken to neutralize the pH using dilute acid before consumption.

GBL is a clear oily liquid that is available in hardware stores and through chemical supply houses. It is a component of many industrial cleaners, paint removers, wood cleaners, textile aids, and drilling oils. It is a solvent for polymers such as cellulose acetate, methyl methacrylate, and polystyrene and can be found in nail polish remover and glue removers. Intoxication after ingestion of GBL in "acetone-free" nail polish remover has been reported (12). Lactones are widely used as flavors and aromas in food, drinks, and cosmetics, and GBL has been detected in certain wines (13). BD is an industrial chemical and solvent used universally in the manufacture of organic chemicals. It is also encountered in pine needle oil spray (14).

Much of the difficulty associated with identification of GHB analogs and precursors in dietary supplements is directly attributable to poor or deceptive labeling. There is no standardization or regulation for listing chemical ingredients in dietary supplements. The National Institute of Science and Technology lists more than 30 chemical names for GBL alone (Table 9–1), making it particularly easy for a manufacturer to confound an unwitting consumer.

Both GBL and BD are metabolic precursors of GHB (Figure 9–2). GBL is converted to GHB by lactonases in the blood, and BD is metabolized to GHB by alcohol and acetaldehyde dehydrogenase (15,16). Because both of these substances are converted to GHB in vivo, the clinical manifestations and toxicology of these prodrugs are identical to those of GHB.

DEMOGRAPHICS OF GHB USE

Popular drug culture classifies GHB as an entheogen, meaning a substance that generates spirituality within (17). This term is more usually reserved for natural psychomimetics such as peyote, psilocybin, and traditional hallucinogens used for shamanic inebriation. Nevertheless, widespread popularity and use of the drug transects the stereotypical socioeconomic and demographic boundaries that are often associated with illicit drug use.

It is claimed that GHB and its metabolic precursors produce a wide range of beneficial effects including increased libido, restful sleep, mood enhancement, weight loss, and muscular development. The perceived effects of the drugs are reflected in the common street names, such as "Easy Lay," "good hormones at bedtime," and "weight belt cleaner," among others (Table 9–1). GHB has been particularly popular among people active in bodybuilding and the rave scene. It was illicitly marketed in gymnasiums and fitness centers as an alternative to anabolic steroids and as a replacement for L-tryptophan. Studies have indicated that GHB mediates the release of growth hormone, possibly by some cholinergic mechanism. Although GHB-induced release of growth hormone has been demonstrated in humans (18,19), there is no experimental data that link GHB to increased muscle mass or fat catabolism.

Drug surveillance organizations have classified GHB as a "club drug." These drugs, which include lysergic acid diethylamide (LSD), ketamine, methylenedioxymethamphetamine (MDMA, Ecstasy), and flunitrazepam (Rohypnol®), are popular at dance clubs and raves. GHB-related substances are less expensive alternatives to MDMA, which may cost as much as $40 per dose. One dose or "capful" of GHB typically costs between $5 and $10. The low cost and deceptive marketing that promotes GHB-related substances as low-risk or natural raise concern that these compounds will emerge as new gateway drugs.

PHARMACOLOGIC MECHANISM

A combination of both GABAergic and dopaminergic systems appear to be involved in GHB's mechanism of action (20). GHB has been reported to both stimulate and inhibit dopamine release (21). It behaves as a GABA-B agonist and produces a biphasic dopamine response, inhibiting release of dopamine at low doses and promoting release at higher doses. The effect of GHB on the cerebral dopaminergic system is not yet completely understood. GHB has distinct neurophysiological and pharmacological actions that may be attributed to the activation of specific GHB receptors that have yet to be elucidated. GHB has also been shown to cause the formation of an opiate-like substance in the striatum and to inhibit norepinephrine release from the hypothalamus. Anesthesia is thought to be due to a general suppressant action on the entire cerebrospinal axis, and muscular relaxation to be due to direct effects on the spinal cord (22).

PHARMACOKINETICS AND TOXICOKINETICS

Once ingested, GBL and BD are rapidly metabolized to GHB (Figure 9–2). GBL is converted to GHB in vivo by a rapidly acting

TABLE 9–1. *Chemical names and common street names of GHB analogs*

GHB
  Chemical names
    4-Hydroxybutyrate
    Butanoic acid, 4-hydroxy-
    γ-Hydroxybutyrate
    Sodium oxybate
  Common names
    Cherry Menth
    Degreaser plus
    Lye
    Easy Lay
    Everclear
    Fantasy
    G
    G-Caps
    G-Riffick
    Gamma-G
    Gamma-OH
    GBH
    Geebers
    Georgia Home Boy
    Gib
    Great Hormones at Bedtime
    Liquid E
    Liquid Ecstasy
    Liquid X
    Natural Sleep-500
    Nature's Quaalude
    Organic Quaalude
    Salty water
    Somatomax
    Scoop
    Soap
GBL
  Chemical names
    Butyrolactone
    2(3H)-Furanone, dihydro-
    γ-Hydroxybutyric acid cyclic ester
    γ-Hydroxybutyric lactone
    γ-Hydroxybutyrolactone
    Butanoic acid, 4-hydroxy-, γ-lactone
    Butyric acid lactone
    Butyryl lactone
    Dihydro-2(3H)-furanone
    Tetrahydro-2-furanone
    1,4-Butanolide
    4-Butanolide
    4-Butyrolactone
    4-Deoxytetronic acid
    4-Hydroxybutanoic acid lactone
    4-Hydroxybutyric acid lactone
    Dehydro-2(3H)-furanone
    1,4-Butyrolactone
    α-Butyrolactone
    Butyric acid, 4-hydroxy-, γ-lactone
    Dihydro-2-furanone
    1,2-Butanolide
    4-Hydroxybutanoic acid, γ-lactone
    γ-Lactone
    2-Oxolanone
    1-Oxacyclopentan-2-one
    2-Oxotetrahydrofuran
    γ-Butanolactone
  Common names
    Blue Nitro
    Firewater
    Furanone Extreme
    Gamma-G
    GBL
    GH-Release
    Insom-X
    Invigorate
    Jolt
    Liquid Libido
    Notro Vitality
    Regenerize
    RenewTrient
    Remforce
    Revivarant
    Revivarant-G
BD
  Chemical names
    1,4-Butanediol
    1,4-Butylene glycol
    1,4-Dihydroxybutane
    1,4-Tetramethylene glycol
    1,4-BD
    Butane-1,4-diol
    Butanediol
    Diol 14B
    Sucol B
    Tetramethylene glycol
    Tetramethylene 1,4-diol
  Common names
    Biocopia PM
    Borametz
    BVM
    Cherry FX Bomb
    Enliven
    FX
    Inner G
    Lemon FX Drop
    NRG3
    Orange FXPro G
    Pine Needle Extract
    Promusol
    Rest-eze
    Revitalize-Plus
    Serenity
    Somato-Pro
    Thunder Nectar
    Weight Belt Cleaner

lactonase found in blood and liver. Animal studies indicate the half-life of the conversion of GBL to GHB in plasma to be approximately 1 minute (15). BD is converted to GHB by alcohol dehydrogenase and aldehyde dehydrogenase via an intermediate, 4-hydroxybutanal. Because of the rapid and extensive conversion of both BD and GBL, the toxicological profile of these substances is analogous to that of GHB.

GHB is a low-potency depressant drug that is rapidly absorbed from the gastrointestinal tract. Within about 15 minutes of administration, the user may experience drowsiness, dizziness, euphoria, nausea, visual disturbances, or unconsciousness. The duration of action is typically 3 hours, and peak serum concentrations are observed 0.5–2 hours after oral ingestion. GHB readily crosses the blood-brain barrier and the placenta (23). GHB is not bound to plasma proteins and is moderately distributed to tissue, following a two-compartment distribution model. In the central compartment, the volume of distribution is estimated at 0.4 L/kg and the half-life is 0.3–1 hour. Nonlinear elimination kinetics observed in some individuals indicate that oral absorption and elimination of GHB is a capacity-limited process (24). Higher doses of drug appear to increase absorption and elimination half-lives and the time to peak plasma concentration. Less than 5% of the dose is eliminated unchanged in the urine, and approximate detection times in blood and urine are 6–8 and 12 hours, respectively (25,26). GBL appears to be more rapidly and completely absorbed than GHB. Co-ingestion of BD and ethanol has been shown to increase tissue concentrations of GHB, suggesting that there might be in vivo competition between the two for alcohol dehydrogenase (27). Radioisotope studies in animals indicated that GHB undergoes oxidative metabolism and is eliminated as carbon dioxide, through respiration.

CLINICAL PRESENTATION

The primary effects of GHB-related substances result from their ability to cause dose-related CNS depression (28). GHB doses of 10 mg/kg have resulted in mild CNS effects including amnesia and hypotonia. Drowsiness, dizziness, and euphoria have been observed in the 20–30 mg/kg range, and profound CNS depression and respiratory depression have occurred with doses above 50 mg/kg (29,30). A unique characteristic of GHB is spontaneous recovery from coma, which may last only a few hours. Users frequently experience euphoria with loss of inhibitions and sedation. In contrast, combative or aggressive behavior has been observed in some individuals who have taken GHB. Emergence delirium, characterized by myoclonic jerking motions, confusion, agitation, and combative behavior, is usually transient (30 minutes), but may intensify with stimulation.

Concurrent use of other sedative or CNS depressant drugs, such as ethanol, opiates, benzodiazepines, or neuroleptics, may potentiate the effects of GHB or its metabolic precursors (31). Multiple drug use is common, and GHB is often used with other intoxicants, particularly alcohol. This practice complicates the identification of GHB intoxication in a clinical setting. In one study, of 88 patients who received hospital treatment in San Francisco for GHB intoxication, 39% of cases involved co-ingestion of ethanol and 28% involved another drug, most commonly amphetamine, Ecstasy, cocaine, or heroin (32). GHB has also been used to ameliorate the unwanted side effects of chronic methamphetamine use (33).

In Chin's study of GHB-intoxicated patients, a Glasgow Coma Scale (GCS) score of 3 was observed in 28% of the patients and 32% had scores between 4 and 8. Interestingly, 19% of the patients in this series presented with a GCS of 14 or 15. Often, patients require intubation in the field or on presentation to the ED and are completely awake within 4 to 6 hours. In the Chin series, the mean duration of intubation was 179 minutes. This short length of intubation is consistent with findings in another case series reported by Li et al. in which the length of intubation ranged from 2 to 6 hours (34).

Bradycardia, defined as a heart rate of 55

TABLE 9–2. *Common clinical effects of GHB*

System	Clinical effect
HEENT	Nystagmus
Cardiovascular	Bradycardia
	Orthostatic hypotension
	Hypertension
Respiratory	Respiratory depression
	Cheyne-Stokes respirations
Neurologic	Amnesia
	Hypotonia
	Somnolence
	Drowsiness
	Dizziness
	Euphoria
	Headache
	Confusion
	Ataxia
	Withdrawal syndrome
	Seizure-like activity
	Coma
Gastrointestinal	Nausea
	Vomiting
	Increased salivation
	Esophageal burns
Genitourinary	Incontinence
	Hematuria
Acid-base	Acidosis
Fluid and electrolyte	Hypokalemia
	Hypernatremia
Dermatologic	Profuse sweating
Musculoskeletal	Hypotonia
Endocrine	Mild hyperglycemia

HEENT, head, eyes, ears, nose, and throat.
From Centers for Disease Control and Prevention: Multi-state outbreak of poisonings associated with illicit use of gamma-hydroxybutyrate. *MMWR Morb Mortal Wkly Rep* 1990;39:861–863.

beats/min or less, was observed in 36% of the patients in Chin's series. One patient had a rate of 24 beats/min. Bradycardia was often associated with deterioration in mental status. Patients with bradycardia had a median initial GCS score of 4, compared with a median initial score of 9.5 for patients without bradycardia. Eleven percent of the patients were initially hypotensive—defined as a systolic blood pressure of less than 90 mmHg. In all of these cases of hypotension, co-ingestion of ethanol, another drug, or both was involved. Hypotension was accompanied by bradycardia in 6 out of 10 patients.

A CO_2 measurement of 45 mmHg or greater was observed in 70% of the patients, and pH measurements ranged from 7.24 to 7.34. Hypothermia and emesis were observed in approximately one-third of the cases. Vomiting is often associated with emergence from coma, but it can be seen at any time after ingestion of GHB. Chin noted that 85% of the cases of vomiting occurred in patients with a GCS of 8 or less.

The first cases of exposure to the GHB precursor GBL were reported in Europe. Five pediatric cases of ingestion of GBL-containing glue removers were presented at the European Association of Poisons Control Centers' 13th Congress in 1988. All five children experienced mental status changes, and three cases of bradycardia and respiratory depression were reported. Andersen and Netterstrom reported coma, bradycardia, and respiratory depression in two patients after ingestion of nail polish remover that contained GBL and ethanol (35). Rambourg-Schepens and Buffet reported similar clinical manifestations in an adult and a child who ingested a nail polish remover containing GBL (12).

In February 1999, the Centers for Disease Control and Prevention reported 34 exposures to GBL in Minnesota, New Mexico, and Texas (36). In the 14 New Mexico cases, the patients ranged in age from 14 to 36 years old. Nine were male, and five were female. Products implicated included Firewater, Blue Nitro, and Renewtrient. The mean dose of GBL was 3 ounces (85 mL). The clinical effects seen in these patients—mental status changes, respiratory depression, bradycardia, and vomiting—are consistent with the toxic effects of GHB.

Esophageal burns are an interesting clinical effect associated with GHB. Dyer and Reed reported a patient who ingested a liquid that was thought to be GHB (37). Soon after ingestion she complained of burning of her mouth, tongue, and esophagus. The patient was seen in the ED and was referred for an endoscopy that revealed a swollen epiglottis, esophageal erythema, and ulceration of the esophagus. This patient was hospitalized for 17 days and required multiple esophageal dilations over a 10-week period.

Along with acute toxicity, a GHB withdrawal syndrome has been described. This withdrawal syndrome consists of anxiety, agitation, tremors, insomnia, disorientation, paranoia, auditory and visual hallucinations, tachycardia, elevated blood pressure, and ocular-motor changes.

In a review of the literature describing GHB withdrawal, Craig et al. described eight case reports of GHB withdrawal (38). In two of the eight, GHB was the only substance involved. The first patient, a 23-year-old woman, was reported at the time of presentation to have ingested 1.5 caps of GHB liquid every 3 hours for 1 year. On presentation she was experiencing anxiety, tremors, hallucinations, paranoia, agitation, delirium, mild hypertension, and tachycardia. She was treated with propranolol, benzodiazepines, and phenothiazines. Her withdrawal period lasted for 9 days. The second patient was a 30-year-old male with a history of ingestion of 5–6 ounces (34–41 g) of GHB per day. He had a 2.5-year history of GHB use. This patient presented with anxiety, tremors, diaphoresis, hallucinations, paranoia, nystagmus, impaired memory, hypertension, and tachycardia. He was treated with benzodiazepines and thiamine and his withdrawal subsided after 8 days.

TREATMENT

The differential diagnosis of the poisoned patient with mental status changes can be complex. Examples of implicated substances include drugs of abuse, sedative-hypnotics, antidepressants, toxic alcohols, and carbon monoxide. GHB-related substances now need to be considered as part of the differential.

As always, the ABCs—airway, breathing, and circulation—are the first priority. Intravenous access should be established, and these patients should be placed on a cardiac monitor as well as a pulse oximeter. They should be closely observed for changes in neurologic, respiratory, and cardiovascular status. Given the incidence of vomiting accompanied by CNS depression, aspiration precautions should be taken. Due to the potential for CNS depression, intubation may be necessary, particularly if co-ingestion of other CNS depressant substances has occurred.

Antagonists, such as naloxone and flumazenil, have had limited, if any, effect in reversing the sedating effects of these substances. Early reports from the anesthesiology literature discuss physostigmine and neostigmine as potential antidotes. Recently, anecdotal cases reports describing successful use of physostigmine to reverse GHB were published (39). Physostigmine should be used with great caution, especially in polysubstance exposures. It is contraindicated if there is a history of cyclic antidepressant exposure. If it is used, it should be administered slowly over 5 minutes to avoid precipitation of seizures.

There is limited, if any, role for gastric decontamination after ingestion of GHB, GBL, or BD. Ipecac syrup–induced emesis is contraindicated due to the CNS depression associated with these substances. Use of ipecac syrup in a patient with a compromised airway could result in aspiration of vomitus and a pneumonitis. Because absorption of GHB and GHB analogs is rapid, activated charcoal and gastric lavage would be of limited benefit. Position statements on gut decontamination developed by the American Academy of Clinical Toxicology and the European Association of Poisons Centres and Clinical Toxicologists suggest that gastric lavage should be instituted within 60 minutes of ingestion, and activated charcoal is most effective when administered within 60 minutes of ingestion (40,41). In recent polysubstance ingestions, activated charcoal and gastric lavage should be considered.

Seizures should be treated with benzodiazepines such as diazepam or lorazepam. The potential need to treat seizures argues further against the use of the benzodiazepine antagonist flumazenil. Use of flumazenil could block the benzodiazepine receptors and inhibit therapeutic anticonvulsant effects. Symptomatic bradycardia can be treated with atropine.

Hypotension can be treated with fluids and vasopressors. Nonsedating antiemetics, such as ondansetron, could be used to treat vomiting. Centrally acting antiemetics with CNS depressant effects, such as prochlorperazine, should be avoided because they can blur the clinical picture.

In symptomatic patients, electrolyte and blood glucose values should be measured. In patients with respiratory depression, an arterial blood gas should be obtained. A urine toxicology screen may be helpful in determining the presence of other substances. Reliable screening tests for GHB are not yet widely available. Techniques such as gas chromatography–mass spectrometry (GC/MS) have been described, but because these tests are rarely available within the time frame of the intoxication, they may be of limited benefit in clinical decisionmaking.

Some authors suggest that patients can be discharged if they are asymptomatic after 6 hours of ED observation. If hospital admission is required, subsequent treatment of GHB toxicity is primarily a continuation of the supportive care that was initiated in the prehospital setting and the ED. Given the relatively rapid reversal of mental status changes, these patients should be monitored closely so that they do not risk self-extubation. The clinical course can be complicated in polysubstance exposures, if the patient has aspirated and developed a pneumonitis, or if the patient has experienced significant hypoxia. Under these circumstances, more advanced care will be required.

TOXICOLOGICAL ANALYSIS

Due to the efficient transformation of both BD and GBL in vivo, toxicological analyses are commonly targeted toward GHB. In antemortem samples, metabolic precursors like BD and GBL are frequently undetected. However, in postmortem samples after GBL or BD overdose, precursors may be detectable. GBL is not detected in blood or urine after administration of GHB, indicating that lactonization of GHB does not occur in vivo (43).

In vitro conversion of GHB to GBL has been readily utilized for toxicological analyses. GBL is more amenable to conventional methods of extraction and gas chromatography than GHB, which is considerably more polar and less volatile. Acidification of the specimen results in lactonization, after which a solvent (such as methylene chloride or chloroform) can be used to extract the GBL (43). These processes typically result in conversion and recovery of about 70% of the GHB as GBL. However, care must be taken during evaporation of the lactone to prevent sample loss. Despite its low molecular mass (86 Da), GBL has been analyzed directly using GC/MS. A disadvantage of this indirect approach to GHB determination is that the characteristic ions have low mass-to-charge ratios and therefore limited specificity. In addition, a separate extract that does not undergo acidification should also be analyzed to ensure that intact GBL is not present in the sample.

GHB can be analyzed directly using either liquid-liquid or solid-phase extraction techniques (44). Due to the small size and polar moieties, extracts tend to be nonspecific, containing many other endogenous carboxylates and polar molecules. Derivatization is necessary before chromatographic analysis. Silylation of the hydroxyl and carboxylate moieties using common derivatizing agents is widely used. Analysis of GHB without in vitro conversion to the lactone may be advantageous from a methodological and legislative perspective. Derivatization of GHB increases thermal stability, reduces volatility, and increases the molecular weight, enhancing confidence in mass spectral identification.

Care must be taken to prevent inadvertent lactonization of GHB to GBL by exposure to acid or heat. Interference from other structurally similar endogenous substances should also be considered, particularly β-hydroxybutyric acid, which is excreted in large quantities in the urine of people with diabetes and during ketoacidosis. Urea is another common interfering agent. Silylation of both urea and GHB can produce derivatives with similar chromatographic and mass spectral characteristics. These and other interferences can be overcome using chemical ionization GC/MS (45).

Illicit doses of GHB are reported to be approximately 35 mg/kg (11), although dose varies considerably among individuals, depending on experience with the drug, tolerance, and intended use. Oral doses of 25 and 50 mg/kg produced average peak plasma concentrations of 55 and 90 mg/L (0.53 and 0.87 mmol/L), respectively (46). Narcoleptic patients receiving oral doses of 50 mg/kg produced peak plasma concentrations in the range of 48–125 mg/L (0.46–1.20 mmol/L), and 25 mg/kg produced dizziness or drowsiness with an average peak plasma concentration of 80 mg/L (0.77 mmol/L) at 0.5 h (3,24). Blood GHB concentrations less than 50 mg/L (0.48 mmol/L) have been associated with euphoria, uninhibited behavior, lightheadedness, and arousal. GHB concentrations in excess of 50 mg/L (0.48 mmol/L) have been observed in patients experiencing sleepiness, slurred speech, and loss of consciousness, and seizure-like activity, coma, and death have been observed in patients with GHB concentrations in excess of 250 mg/L (2.40 mmol/L). Although tolerance and interindividual differences make it impossible to predict the effect of the drug from the concentration, a dose-dependent relationship does exist. Blood and urine GHB concentrations reported in the scientific literature are summarized in Table 9–3.

DRUG-FACILITATED SEXUAL ASSAULT

The CNS depressant effects of GHB—rapid onset, memory loss, and fast metabolism—make the drug particularly effective as an agent to induce chemical submission, most notably drug-facilitated sexual assault. Liquid GHB, which is readily concealed in water bottles or similar nonsuspicious vessels, can immobilize a person or render them comatose within a relatively short time. It is claimed that GHB increases libido. However, it is more likely that the perception of enhanced sexual arousal and performance is the result of decreased social inhibitions that are associated with the drug.

A recent study concluded that although nearly two-thirds of urine specimens obtained from sexual assault victims contained alcohol or drugs, less than 3% of these were attributable to GHB or Rohypnol® (47). Another study of 1179 alleged cases of sexual assault indicated that GHB was present in approximately 4% of urine samples collected from the victims in 49 states. The highest rate of GHB use was in California (8%), compared with 0–6% for all other states (48). However, GHB-facilitated sexual assault may be underreported: Screening tests for GHB are limited, and many laboratories do not routinely test for the drug (49). GHB has a very short half-life and may be undetectable in urine by 12 hours. Delays in collection of biological evidence and unwillingness of the victim to report the crime compound the problem.

ENDOGENOUS GHB

Concentrations of endogenous GHB in the human brain are extremely low compared with what is needed to produce a pharmacologic response, reported to be at least 100-fold higher (50). Micromolar concentrations are found in some areas of the mammalian brain, with specific GHB-binding sites concentrated in the hippocampus and cortical areas (51). GHB is also present in non-neuronal tissues. It has been detected in the kidney at concentrations 10-fold those found in brain. Heart and skeletal muscle was also found to contain GHB at concentrations fivefold those of neuronal tissue (52). It undergoes metabolic conversion to succinic acid semialdehyde (SSA) and succinate before entering the tricarboxylic acid cycle. GHB is both derived from and catabolized to SSA by different enzymes (Figure 9–2), and there is evidence to suggest that mammalian neuronal tissue can reduce SSA to GHB as well as convert GABA to GHB.

TABLE 9-3. Blood and urine concentrations of GHB

Blood (mg/L)	Urine (mg/L)	Observations and situation
33	714	Driving under the influence, confusion, disorientation, nystagmus[a]
221	2200	Nonfatal GHB overdose[a]
130	1600	Confused, ataxic, vomiting[a]
—	1975	Driving under the influence, ataxic, confused, nystagmus, asleep[b]
—	1086	Driving under the influence, nystagmus, unconscious[c]
33	—	Driving under the influence[d]
73	—	Driving under the influence[d]
125 (serum) (BAC 0.13)	—	Nausea, dizziness, coma[e]
47	308	Drug-facilitated sexual assault, memory loss[f]
157	—	Driving under the influence, asleep behind the wheel[f]
>260	—	Heavy sleep[g]
156–260	—	Moderate sleep[g]
52–156	—	Light sleep[g]
<52	—	Wakefulness[g]
280	6171	Fatality: postmortem samples[h]
648 (heart), 330 (peripheral)	—	Fatality: postmortem samples[i]
1473 (heart), 761 (femoral)	407	Fatality: postmortem samples[j]

BAC, blood alcohol concentration.

[a] Couper F, Logan BK: Determination of gamma-hydroxybutyrate (GHB) in biological specimens by gas chromatography—mass spectrometry. *J Anal Toxicol* 2000;24:1–7.

[b] Stephens BG, Baselt RC: Driving under the influence of GHB? *J Anal Toxicol* 1994;18;357–358.

[c] Fromhold S, Busby C: GHB. A brief review with a few case histories. *DRE* 1998;10:2–4.

[d] Pan YM, Wall WH, Solomons E: GHB in Georgia (1996–1998). Presented at the American Academy of Forensic Sciences Annual Meeting, Orlando, FL, 1999.

[e] Louagie HK, Verstraete AG, Soete CJD, et al: A sudden awakening from a near coma after combined intake of gamma-hydroxybutyric acid (GHB) and ethanol. *J Toxicol Clin Toxicol* 1997;35:591–594.

[f] LeBeau M, Montgomery M, Miller M, et al: Analysis of biofluids for gamma-hydroxybutyrate (GHB) and gamma-butyrolactone (GBL) by headspace GC-FID and GC/MS. *J Anal Toxicol* 2000;24:421–428.

[g] Helrich M, McAslan T, Skolnik S, et al: Correlation of blood levels of 4-hydroxybutyrate with state of consciousness. *Anesthesiology* 1964;25:771–775.

[h] Kraner J, Plassard J, McCoy D, et al: A death from ingestion of 1,4-butanediol, a GHB precursor. Presented at the Society of Forensic Toxicologists Annual Meeting, poster #39, Milwaukee, WI, 2000.

[i] Dixon MM, Kalasinsky KS, Kish SJ, et al: GHB overdose case: examination of blood, brain and hair. Presented at the Society of Forensic Toxicologists Annual Meeting, poster #38, Milwaukee, WI, 2000.

[j] Anderson D, Muto J, Andrews J: Case report: postmortem tissue distribution of gamma-hydroxybutyrate (GHB) and gamma-butyrolactone (GBL) in a single fatality. Presented at the Society of Forensic Toxicologists Annual Meeting, poster #37, Milwaukee, WI, 2000.

Arbitrary cutoff concentrations are frequently used when reporting GHB. There are no mandated guidelines, so institutions that analyze for GHB are responsible for establishing their own thresholds. In antemortem blood and urine samples, cutoffs of 5 or 10 mg/L (0.05 or 0.10 mmol/L) are widely used, whereas postmortem cutoffs in excess of 50 mg/L (0.48 mmol/L) have been recommended due to artifactual GHB production. In a selection of approximately 100 non-GHB-related deaths, average heart and femoral blood was found to contain on average 12 mg/L (0.12 mmol/L) [range, 2–36 mg/L (0.02–0.35 mmol/L)] and 11 mg/L (0.11 mmol/L) [range, 2–48 mg/L (0.02–0.46 mmol/L)] GHB, respectively. In deaths attributed to GHB, postmortem blood concentrations were 98–596 mg/L (0.94–5.93 mmol/L) (53). Postmortem urine is less susceptible to artifactual GHB production, containing on average 5 mg/L (0.05 mmol/L) [range, 0–14 mg/L (0–0.13 mmol/L)] GHB. As such, urine is the preferred specimen for GHB-related death investigation.

Endogenous concentrations of GHB in antemortem blood and urine are typically <10 mg/L (0.10 mmol/L). However, storage conditions and preservatives play an important role in the artifactual production of GHB in biological specimens. Absence of sodium fluoride and storage at room temperature may artifactually increase the GHB concentration. Postmortem blood stored for 40 days at 77 °F (25 °C) in the absence of sodium fluoride showed elevated GHB concentration [9–433 mg/L (0.09–4.16 mmol/L)] compared with preserved blood stored at 39 °F (4 °C) [5–77 mg/L (0.05–0.74 mmol/L)] (54). Yellow-top Vacutainers™ and blood-collection tubes that contain trisodium citrate, citric acid and dextrose have been shown to artifactually increase the GHB concentration in antemortem blood samples (55).

Some authors have attributed postmortem increases in GHB and GABA to the decreased activity of the Krebs cycle and enzymatic changes that occur after death. Artifactual GHB may be produced from GABA and SSA during anoxic conditions (67). Oxidation of GHB to SSA is the rate-limiting step in the catabolic pathway of GHB (Figure 9–2) (56). Conversion of GHB to SSA proceeds at approximately 1/1000 of the rate at which SSA is converted to succinate by SSA dehydrogenase. Factors that regulate either of the enzymes involved, particularly GHB dehydrogenase, can influence tissue concentrations of endogenous GHB as well as the physiologic response from exogenously administered GHB (57). It is expected that inhibition of GHB dehydrogenase in vivo will increase endogenous concentrations of GHB. This increase has been demonstrated using drugs that are known to inhibit this enzyme, such as barbiturates, diphenylhydantoin, valproic acid, and salicylates (58). Interestingly, it has also been shown that GHB dehydrogenase is inhibited by endogenous substances including ketone bodies, α-ketoglutarate, branched ketoacids derived from amino acid degradation, and degradation products of phenylalanine (59). This is perhaps the most compelling explanation for the postmortem and artifactual production of GHB. However, in the presence of certain organic acidemias, antemortem blood and urine samples can also accumulate extremely high concentrations of endogenous GHB.

SUCCINIC SEMIALDEHYDE DEHYDROGENASE DEFICIENCY

A rare metabolic anomaly was identified in 1981 in which a deficiency in SSA dehydrogenase caused an accumulation of GHB and SSA. This disorder caused severe psychomotor retardation, ataxia, and convulsions (60). There have been several reports of this rare disorder, known as 4-hydroxybutyric aciduria. In response to the defect, physiological fluids accumulate large quantities of GHB that produces numerous neuropharmacological effects. Difficulty associated with speech, psychomotor retardation, hypertonia, and hyperkinesis have been reported (61,62). In one study, the following abnormalities were observed in 23 patients: motor delay, including fine motor skills, 78%; language delay, 78%; hypotonia, 74%; mental delay, 74%; seizures, 48%; decreased or absent reflexes, 39%; ataxia, 30%; behavioral problems, 30%; hyperkinesis, 30%; and electroencephalographic abnormalities, 26% (63). The disease is believed to be the result of genetic defects in the SSA gene (64). Urinary concentrations of GHB can reach 1000-fold those of normal subjects (65,66). Clinical improvements in some individuals have been observed after treatment with vigabatrin (66).

CONCLUSIONS

GHB and GHB analog toxicity present new challenges to the health-care professional. Although the clinical course of GHB toxicity can be relatively mild, there is a risk for significant morbidity and mortality. This risk is enhanced in polysubstance exposures. Early supportive care is the mainstay of treatment for patients exposed to these substances. Efforts to prevent the use of GHB-related substances, through education and regulatory actions, must be a major component of our overall approach to these new toxic hazards.

GHB and its analogs and metabolic precursors pose a formidable challenge to clinicians, toxicologists, and law enforcement and legislative bodies. Interpretation of toxicological findings is often complicated by the presence of endogenous GHB and delay in specimen collection as well as storage and preservation issues. The popularity of the drug, its widespread appeal, and the ready availability of unregulated alternatives to GHB ensure its continued use as a recreational intoxicant for some time to come.

REFERENCES

1. Laborit H: Sodium 4-hydroxybutyrate. *Int J Neuropharmacol* 1964;3:433–452.
2. Beardsley PM, Balster RL, Harris LS: Evaluation of the discriminative stimulus and reinforcing effects of gammahydroxybutyrate (GHB). *Psychopharmacology* 1996;127:315–322.
3. Scharf M, Lai AA, Branigan B, et al: Pharmacokinetics of gammahydroxybutyrate (GHB) in narcoleptic patients. *Sleep* 1998;21;507–514.
4. Kolin A, Brezina A, Mamelak M, et al: Cardioprotective action of sodium gamma-hydroxybutyrate against isoproterenol induced myocardial damage. *Int J Exp Pathol* 1993;74:275–281.
5. Kolin A, Brezina A, Mamelak M: Cardioprotective effects of sodium gamma-hydroxybutyrate (GHB) on brain induced myocardial injury. *In Vivo* 1991;5:429–431.
6. Dyer JE, Galbo MJ, Andrews KM: 1,4-Butanediol "pine needle oil": overdose mimics toxic profile of GHB [Abstract]. *J Toxicol Clin Toxicol* 1997;35:554.
7. Dyer JE: Gamma-hydroxybutyrate: a health-food product producing coma and seizure like activity. *Am J Emerg Med* 1991;9:321–324.
8. Centers for Disease Control and Prevention: Multi-state outbreak of poisonings associated with illicit use of gamma-hydroxybutyrate. *MMWR Morb Mortal Wkly Rep* 1990;39:861–863.
9. Drug Enforcement Administration, Justice Department: Placement of gamma-butyrolactone in List I of the Controlled Substances Act (21 U.S.C. 802(34)). Final rule (2000). *Fed Regist* 2000;65:21645–21647.
10. Fowkes SW: The emergence of GHB alternatives. *Smart Life News* 1998;6(9):1,10–12.
11. Baselt R: Gamma-hydroxybutyrate. In: Baselt R: *Disposition of toxic drugs and chemicals in man.* Foster City, CA: Chemical Toxicology Institute, 2000:386–388.
12. Rambourg-Schepens MO, Buffet M: Gamma-butyrolactone poisoning and its similarities to gamma-hydroxybutyric acid: two case reports. *Vet Hum Toxicol* 1997;39:234–235.
13. McCabe E, Layne EC, Sayler DF, et al: Synergy of ethanol and a natural soporific—gamma hydroxy-butyrate. *Science* 1971;171:404–406.
14. Litovitz TL, Klein-Schwartz W, White S, et al: 1999 Annual Report of the American Association of Poison Control Centers Toxic Exposure Surveillance System. *Am J Emerg Med* 2000;18:517–574.
15. Roth RH, Giarman NJ: Gamma-butyrolactone and gamma-hydroxybutyric acid I: distribution and metabolism. *Biochem Pharmacol* 1966;15:1333–1348.
16. Irwin RD: NTP summary report on the metabolism, disposition, and toxicity of 1,4-butanediol. [Toxicity Report Series No. 54; NIH Publ. 96-3932.] Bethesda, MD: National Toxicology Program, 1996.
17. Erowid: http://www.Erowid.org, 2001.
18. Takahara J, Yunoki S, Yakushiji W, et al: Stimulatory effects of gamma-hydroxybutyric acid on growth hormone and prolactin release in humans. *J Clin Endocrinol Metab* 1977;44:1014–1018.
19. Volpi R, Chiodera P, Caffarra P, et al: Muscarinic cholinergic mediation of the GH response to gamma-hydroxybutyric acid: neuroendocrine evidence in normal and parkinsonian subjects. *Psychoneuroendocrinology* 2000;25:179–185.
20. Cash CD: What is the role of the gamma-hydroxybutyrate receptor? *Med Hypotheses* 1996;47:455–459.
21. Feigenbaum JJ, Howard SG: Does gamma-hydroxybutyrate inhibit or stimulate central DA release? *Int J Neurosci* 1996;88:1–2.
22. Doherty JD, Stout RW, Roth RH: Metabolism of (1-14C) gamma hydroxybutyric acid by rat brain after intraventricular injection. *Biochem Pharmacol* 1975;24:469–474.
23. Vickers MD: Gamma-hydroxybutyric acid. *Int Anesthesiol Clin* 1969;7:75–89.
24. Palatini P, Tedeschi L, Frison G, et al: Dose-dependent absorption and elimination of gamma-hydroxybutyric acid in healthy volunteers. *Eur J Clin Pharmacol* 1993;45:353–356.
25. Hoes MJ, Guelen PJM: Gamma-hydroxybutyrate as hypnotic. *L'Encephale* 1980;6:93–99.
26. Ferrara SD, Tedeschi L, Frison, et al: Effect of moderate or severe liver dysfunction on the pharmacokinetics of gamma-hydroxybutyric acid. *Eur J Clin Pharmacol* 1996;50:305–310.
27. Poldrugo F, Barker S, Basa M, et al: Ethanol potentiates the toxic effects of 1,4-butanediol. *Alcohol Clin Exp Res* 1985;9:493–497.
28. POISINDEX® Editorial Staff: Gamma hydroxybutyrate. In: Smith R, Toll Hill L, Hurlbut KM, eds: *POISINDEX® System.* Englewood, CO: MICROMEDEX Inc. (Edition expires 5/31/2000.)

29. Mamelak M: Gammahydroxybutyrate: an endogenous regulator of energy metabolism. *Neurosci Biobehav Rev* 1989;13:187–198.
30. Suner S, Szlatenyi CS, Wang RY: Pediatric gamma hydroxybutyrate intoxication. *Acad Emerg Med* 1997;4:1041–1045.
31. Poldrugo F, Snead OC, Barker S: Chronic alcohol administration produces an increase in liver 1,4-butanediol concentration. *Alcohol Alcohol* 1985;20:251–253.
32. Chin RL, Sporer KA, Cullison B, et al: Clinical course of γ-hydroxybutyrate overdose. *Ann Emerg Med* 1998;31:716–722.
33. Galloway GP, Frederick SL Jr, Staggers FE Jr, et al: Gamma-hydroxybutyrate: an emerging drug of abuse that causes physical dependence. *Addiction* 1997;92:89–96.
34. Li J, Stokes SA, Woeckener A: A tale of novel intoxication: seven cases of γ-hydroxybutyric acid overdose. *Ann Emerg Med* 1998;31:723–728.
35. Andersen MB, Netterstrom B: Loss of consciousness after ingestion of nail varnish remover [Abstract]. *Ugeskr Laeger* 1992;154:3064.
36. Centers for Disease Control and Prevention: Adverse effects associated with ingestion of gamma-butyrolactone—Minnesota, New Mexico, and Texas, 1998–1999. *MMWR Morb Mortal Wkly Rep* 1999;48:137–140.
37. Dyer JE, Reed JH: Alkali burns from illicit manufacture of GHB [Abstract]. *J Toxicol Clin Toxicol* 1997;35:553.
38. Craig K, Gormez HF, McManus JL, et al: Severe gamma-hydroxybutyrate withdrawal: a case report and literature review. *J Emerg Med* 2000;18:65–70.
39. Yates SW, Viera AJ: Physostigmine in the treatment of gamma-hydroxybutyric acid overdose. *Mayo Clin Proc* 2000;75:401–402.
40. American Academy of Clinical Toxicology and European Association of Poisons Centres and Clinical Toxicologists: Position statement: gastric lavage. *J Toxicol Clin Toxicol* 1997;35:711–719.
41. American Academy of Clinical Toxicology and European Association of Poisons Centres and Clinical Toxicologists: Position statement: single-dose activated charcoal. *J Toxicol Clin Toxicol* 1997;35:721–741.
42. Ferrara SD, Zotti S, Tedeschi L, et al: Pharmacokinetics of gamma-hydroxybutyric acid in alcohol dependent patients after single and repeated oral doses. *Br J Clin Pharmacol* 1992;34:231–235.
43. Doherty JD, Hattox SE, Snead OC, et al: Identification of endogenous gamma-hydroxybutyrate in human and bovine brain and its regional distribution in human, guinea pig and rhesus monkey brain. *J Pharmacol Exp Ther* 1978;207:130–139.
44. McCusker R, Paget-Wilkes H, Chronister, CW, et al: Analysis of gamma-hydroxybutyrate (GHB) in urine by gas chromatography-mass spectrometry. *J Anal Toxicol* 1999;23:301–305.
45. Kerrigan S, Doyle R, Phillips W: Determination of gamma-hydroxybutyrate in blood and urine using chemical ionization gas chromatography/mass spectrometry. Presented at the American Academy of Forensic Sciences Annual Meeting, Seattle, WA, 2001.
46. Ferrara SD, Tedeschi L, Frison G, et al: Therapeutic gamma-hydroxybutyric acid monitoring in plasma and urine by gas chromatography-mass spectrometry. *J Pharm Biomed Anal* 1993;11:483–487.
47. Slaughter L: Involvement of drugs in sexual assault. *J Reprod Med* 2000;45:425–430.
48. ElSohly M, Salamone SJ: Prevalence of drugs used in cases of alleged sexual assault. *J Anal Toxicol* 1999;23:141–146.
49. Badcock NR, Zotti R: Rapid screening test for gamma-hydroxybutyric acid (GHB, Fantasy) in urine [Letter]. *Ther Drug Monit* 1999;21:376.
50. Cash C: Gamma-hydroxybutyrate: an overview of the pros and cons for it being a neurotransmitter and/or a useful therapeutic agent. *Neurosci Behav Rev* 1994;18:291–304.
51. Castelli M, Mocci I, Langlois X, et al: Quantitative autoradiographic distribution of gamma-hydroxybutyric acid binding sites in human and monkey brain. *Brain Res Mol Brain Res* 2000;78:91–99.
52. Nelson T, Kaufman E, Kline J, et al: The extraneural distribution of gamma-hydroxybutyrate. *J Neurochem* 1981;37:1345–1348.
53. Anderson DT, Kuwahara T: Endogenous gamma-hydroxybutyrate (GHB) levels in postmortem specimens. Presented at the California Association of Toxicologists Quarterly Meeting, Las Vegas, NV, November 1997.
54. Stephens BG, Coleman DE, Baselt RC: In vitro stability of endogenous gamma-hydroxybutyrate in postmortem blood [Commentary]. *J Forensic Sci* 1999;44:231–231.
55. LeBeau M, Montgomery M, Jufer R, et al: Elevated GHB in citrate-buffered blood. *J Anal Toxicol* 2000;24:383–384.
56. Roth RH: Formation and regional distribution of gamma-hydroxybutyric acid in mammalian brain. *Biochem Pharmacol* 1970;19:3013–3019.
57. Kaufman EE, Nelson T: An overview of gamma-hydroxybutyrate catabolism: the role of the cytosolic NADP(+)-dependent oxidoreductase EC 1.1.1.19 and of a mitochondrial hydroxy-acid-oxoacid transhydrogenase in the initial, rate-limiting step in this pathway. *Neurochem Res* 1991;16:965–974.
58. Kaufman E, Nelson T: Evidence for the participation of a cytosolic NADP+ dependent oxidoreductase in the catabolism of gamma-hydroxybutyrate in vivo. *J Neurochem* 1987;48:1935–1941.
59. Kaufman E, Nelson T: Regulation and properties of an NADP+ oxidoreductase which functions as gamma-hydroxybutyrate dehydrogenase. *J Neurochem* 1983;40:1639–1646.
60. Jakobs C, Bojash M, Monch E, et al: Urinary excretion of gamma-hydroxybutyric acid in a patient with neurological abnormalities. The probability of a new inborn error of metabolism. *Clin Chim Acta* 1981;111:169–178.
61. Gibson KM, Hoffmann G, Nyhan WL, et al: 4-Hydroxybutyric aciduria in a patient without ataxia or convulsions. *Eur J Pediatr* 1988;147:529–531.
62. Onkenhout W, Maaswinkel-Mooij PD, Poorthuis BJ: 4-Hydroxybutyric aciduria: further clinical het-

erogeneity in a new case. *Eur J Pediatr* 1989;149:194–196.
63. Gibson KM, Christensen E, Jakobs C, et al: The clinical phenotype of succinic semialdehyde dehydrogenase deficiency (4-hydroxybutyric aciduria): case reports of 23 new patients. *Pediatrics* 1997;99:567–574.
64. Chambliss KL, Hinson DD, Trettel F, et al: Two exon-skipping mutations as the molecular basis of succinic semialdehyde dehydrogenase deficiency (4-hydroxybutyric aciduria). *Am J Hum Genet* 1998;63:399–408.
65. Rahbeeni Z, Ozand PT, Rashed M, et al: 4-Hydroxybutyric aciduria. *Brain Dev* 1994;16[Suppl]:64–71.
66. Al-Essa MA, Bakheet SM, Patay ZJ, et al: Clinical, fluorine-18 labeled 2-fluoro-2-deoxyglucose positron emission tomography (FDG PET), MRI of the brain and biochemical observations in a patient with 4-hydroxybutyric aciduria; a progressive neurometabolic disease. *Brain Dev* 2000;22:127–131.

SELF-ASSESSMENT QUESTIONS

GHB, γ-hydroxybutyrate

GBL, γ-butyrolactone

BD, 1,4-butanediol

1. Which of the following statements about GHB and GHB analogs is correct?
   a. GHB is a synthetic substance that was synthesized from human γ-aminobutyric acid (GABA).
   b. GHB was used as a sedative-hypnotic agent in the United States in the 1940s and 1950s.
   c. GBL was used experimentally as an anesthetic agent in Europe in the 1940s and 1950s.
   d. GBL has been used in the formulation of nail polish removers and artificial nail removers.
   e. In the 1980s, BD was sold in health-food stores for its anabolic effects.

2. Which of the following statements about the pharmacology of GHB is correct?
   a. GHB has neuromodulatory effects on both dopaminergic and GABAergic systems.
   b. GHB does not have an effect on human GABA receptors.
   c. GHB has limited pharmacologic effect on the cerebrospinal axis.
   d. GHB has not been shown to affect norepinephrine release.
   e. GHB has no effect on dopamine regulation in the human brain.

3. Which of the following statements about the therapeutic use of GHB is correct?
   a. Orphan Medical Inc. is investigating the use of GHB for the treatment of ethanol withdrawal.
   b. GHB is currently being investigated in the United States as an agent to reverse opiate and opioid withdrawal.
   c. The use of GHB as an anesthetic agent was limited by its lack of analgesic effect.
   d. GHB has been shown to be ineffective in the treatment of narcolepsy.
   e. GHB has anabolic effects due to its ability to increase growth hormone.

4. Which of the following statements about GHB is correct?
   a. A drug-facilitated sexual assault is more likely to involve GHB than ethanol.
   b. Instructions for the synthesis of GHB can be found on the Internet.
   c. GHB is classified as a Schedule II substance under the U.S. Food, Drug, and Narcotic Act.
   d. GHB is most often abused by mixing it with marijuana and smoking.
   e. GHB toxicity cases seen in U.S. emergency departments have not increased since 1992.

5. Which of the following statements about the pharmacokinetics of GHB is correct?
   a. The effects of GHB are delayed and occur 4–6 hours after exposure.
   b. GHB has a long half-life that ranges from approximately 12 to 14 hours.
   c. GHB is not known to cross the blood-brain barrier or into the placenta.
   d. BD exerts its toxic effects without being metabolized to GHB.
   e. GBL is more rapidly absorbed from the gastrointestinal tract than GHB.

6. Which of the following statements about the toxicologic effects of GHB is correct?
   a. Co-ingestion of central nervous system depressants may enhance GHB toxicity.
   b. Coma from GHB toxicity usually persists for 24–48 hours after exposure.
   c. GHB-induced central nervous system depression is not dose-related.
   d. Status epilepticus is observed in most patients with GHB toxicity.
   e. GBL and BD appear to induce less central nervous system toxicity than GHB.

7. Which of the following statements about the toxicologic effects of GHB is correct?
   a. Emergence delirium and combative behavior may last for several hours.
   b. Common adverse effects include bradycardia, hypothermia, and emesis.
   c. Acidosis occurs due to the endogenous conversion of GHB to other organic acids.
   d. Hypotension has not been observed in GHB toxicity.
   e. A GHB withdrawal syndrome has not been described in the literature.

8. Which of the following statements about the medical management of GHB exposure is correct?
   a. GHB-poisoned patients often require prolonged intubation and ventilatory support.
   b. Ipecac syrup–induced emesis can safely be used in the treatment of GHB toxicity.
   c. Activated charcoal is ineffective in adsorbing GHB-related substances in the gastrointestinal tract.
   d. The anticholinergic effects of GHB support use of gastric lavage several hours after exposure.
   e. Atropine has been shown to effectively reverse GHB-related bradycardia.

9. Which of the following statements about the medical management of GHB exposure is correct?
   a. Flumazenil has been shown to effectively reverse GHB toxicity.
   b. Naloxone has been shown to effectively reverse GHB toxicity.
   c. Anecdotal case reports describe the effective use of physostigmine for GHB toxicity.
   d. Patients with GHB toxicity should be observed for at least 24 hours after exposure.
   e. GHB-induced seizures do not respond to benzodiazepines or barbiturates.

10. Which of the following statements about the GHB analysis is correct?
    a. Artifactual elevation of GHB concentration can occur postmortem.
    b. Quantitative GHB concentrations should be obtained to guide medical management of GHB-poisoned patients.
    c. On-site screening tests for GHB are readily available.
    d. Any detectable concentration of GHB in blood or urine is indicative of exogenous GHB exposure.
    e. GBL may be present after administration of GHB.

CHAPTER 10

Point-of-Care Testing for Drugs of Abuse*

Alan H.B. Wu, Ph.D., DABCC

LEARNING OBJECTIVES

After completing this chapter, the reader should be able to:

1. Cite the incidence of drug use among emergency department patients and in the general population.
2. Describe the rationale for point-of-care testing (POCT) for drugs of abuse testing in various therapeutic management approaches.
3. List the advantages of POCT for workplace drug testing.
4. Describe the current approaches used for POCT.
5. Describe the differences in antibody specificities for POCT vs. drug testing in a central lab.
6. Describe the issues of at-home drug testing for testing urine and for detecting drug residues.

RATIONALE FOR POINT-OF-CARE DRUG TESTING IN URINE

The Emergency Department

The major advantage of testing urine for drugs of abuse at the point of care is reduction in turn-around times for obtaining results. For drug testing in the emergency department (ED), detection of specific drugs contributing to a patient's symptoms may be important in the emergent therapeutic management of that patient. Certain therapeutic measures are specifically designed for particular toxidromes, such as naloxone for opioid. Table 10–1 lists this and other therapeutic measures that can be taken by ED physicians in cases of drug overdoses. If the results of a drug screen can be made routinely available within the first few minutes after patient presentation, such as at the point of care, a decision to administer the appropriate therapeutic measure can be immediately made.

On the other hand, if results of the screen take longer than 30 minutes, there may be a delay in administering the countermeasures, or such measures might be given without knowledge of a positive drug screen result. A negative drug screen result in the ED might also have merit for ruling out use of the commonly abused drugs as a cause for the presenting symptoms. This may lead to use of other perhaps more invasive or expensive diagnostic studies, such as computerized tomography scanning of the head.

The incidence of finding a positive result in a urine drug screen in a patient who presents with acute symptoms is regularly tabulated by the Drug Abuse Warning Network and the National Hospital Ambulatory Medical Care Survey (1,2). In 1993, the incidence

---

*Parts of this chapter were adapted from Wu AHB: Point-of-care drug testing. In: Kost GD, ed: *Principles and practice of point-of-care testing*. Philadelphia: Lippincott Williams & Wilkins, 2001.

TABLE 10–1. Acute therapeutic management of drug overdoses

Drug	Drug management
Opiates	Opioid antagonist (naloxone)
Cocaine and PCP	Seizure and agitation control (benzodiazepines)
Benzodiazepines	Flumazenil
Acetaminophen	N-Acetylcysteine
Salicylates	Peritoneal or hemodialysis
Tricyclic antidepressants	Physostigmine
Marijuana	None (acute clinical problems)

of drug-related ED visits was 2.7% for alcohol and 1.1% for all other drugs. It is important to note that finding a positive urine drug screen may be incidental to the reason for the ED presentation. A more revealing statistics might be the correlation of ED drug incidence with that of the general population. Using data from 1993, Table 10–2 illustrates that opiates (heroin in particular) and cocaine are associated with a disproportionate increase in ED admissions, suggesting that these drugs are more detrimental to human health than other drugs, such as tetrahydrocannabinol (THC), that have a high incidence of use within the general population, but are infrequently found in an ED population. Differences in drug-testing strategies may also contribute to these figures.

Workplace Drug Testing

Workplace drug testing for federal employees is regulated by the Substance Abuse and Mental Health Services Administration (SAMHSA) (3). The current regulation calls for screening of urine with an immunoassay, followed by confirmation analysis of the same specimen by gas chromatography/mass spectrometry (GC-MS).

The advantages of point-of-care testing (POCT) for drugs in the workplace are different. Testing on-site enables results to be obtained while the individual is still within the collection area. The donor could, in theory, witness the testing process as a check on the chain-of-custody. In certain drug-testing situations that require people to have a negative urine result before resumption of duties, such as after a road accident, subjects who have a negative urine drug test result can immediately return to work. Samples that produce a positive result will require that the sample be sent to a laboratory that can perform the necessary confirmation analysis (for example, by GC-MS). Due to subject confidentiality, it may be the most ethical

TABLE 10–2. Incidence of positive drug results for emergency visits and the general US population[a]

Drug	ED incidence[b]	General population[c]	ED drug index[d]
Alcohol	2.7%	49.6%	5
Cocaine	0.14%	0.6%	23
Heroin and morphine	0.090%	0.1%	90
THC	0.032%	4.3%	0.7
Benzodiazepines	0.58%	0.3%	19
Analgesics	0.081%	0.7%	12
PCP	0.007%	0.2%	3
LSD	0.004%	0.8%[d]	5

LSD, lysergic acid diethylamide.
[a] Data given in percent of the adolescent and adult US population ($\geq$12 y).
[b] Drug Abuse Warning Network, 1993.
[c] National Institute on Drug Abuse Household Survey on Drug Abuse, 1993. Reported usage over the previous month (indicating current user).
[d] Ratio of ED incidence of drug use to incidence in the general population $\times$ 100.

course for an outside individual to conduct the testing rather than a fellow employee.

Although the current federal guidelines for workplace drug testing do not permit POCT, there are experimental programs that are being investigated. For example, the US Post Office has been experimenting with on-site testing for their employees (4). Using the Roche TesTcup (Branchburg, NJ), there are approximately 1000 collection sites currently in use.

POINT-OF-CARE TESTING APPROACHES

Spot Tests

The original approach for POCT was the spot test. A spot test is performed without using laboratory instrumentation and makes use of a visual or colorimetric endpoint. Table 10–3 lists two classes of spot tests that have been developed for clinical toxicology purposes. A comprehensive list of toxicologic spot tests is available from the 1986 (2nd edition) *Tietz Textbook of Clinical Chemistry* (5). Simple spot tests require no additional sample preparation procedures and can be performed at the bedside. A few drops of an appropriate reagent are added to a urine sample. The production of a characteristic color suggests the presence of a particular drug or drug class. Complex spot tests may require additional steps, such as extraction, or color development such as heating, before end-point detection. These tests are not useful at the point of care but can be used by trained technologists in a central laboratory.

Although the advantages of simple spot tests are rapid turn-around times and low costs, these tests do suffer from low analytical sensitivity and nonspecificity for the target drug. In addition, spot tests have not been waived by the Clinical Laboratory Improvement Amendments; therefore, use of these tests at the point of care requires documentation for quality control and proficiency. There are no workplace drug-testing applications for spot tests.

Urine-Based Immunoassay POCT Devices

Commercial development of POCT devices for urine drugs of abuse testing has undergone a series of improvements over the past 5–7 years, since the introduction of the first single-analyte POCT assay, OnTrak (Roche Diagnostics, Somerville, NJ). OnTrak uses latex agglutination, with the absence of agglutination being indicative of a positive result. Multianalyte devices were soon developed, enabling simultaneous analysis of 2–8 drugs within a single panel. Many manufacturers have produced a five-drug panel targeting the classes that are tested under federal guidelines ("SAMHSA-5": amphetamines, cocaine, marijuana or THC, opiates, and phencyclidine). Most of these POCT devices use a negative-indicating reaction, that is, the absence of a band at the test zone indicates a positive result (Figure 10–1).

TABLE 10–3. *Spot tests for drugs-of-abuse testing*

Drug(s)	Reagent(s) and procedure
Simple spot tests (bedside test)	
Acetone	Sodium nitroprusside (dipstick test)
Imipramine, desipramine	Acid dichromate (Forrest reagent)
Paraquat	Sodium bicarbonate and sodium dithionite
Phenothiazines	Ferric chloride, perchloric, nitric acid
Salicylates	Ferric nitrite (Trinder's reagent)
Complex spot tests (lab test)	
Acetaminophen	Cold acid, sodium nitrite, and α-naphthol reagent
Chloral hydrate	Ether extract, alkaline pyridine, and boil
Ethchlorvynol	Acidify, chloroform extraction, evaporation, reaction with $HgNO_3$
Heavy metals	Acidify and boil sample with copper wire (Reinsch test)
Volatiles	Conway microdiffusion dish

A.

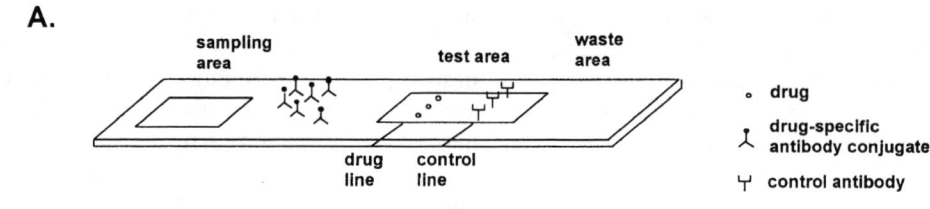

Diagram of device regions

B.

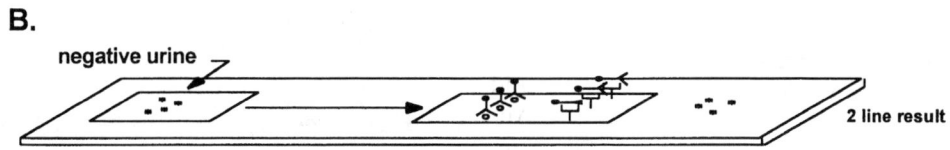

Addition of drug-free urine

C.

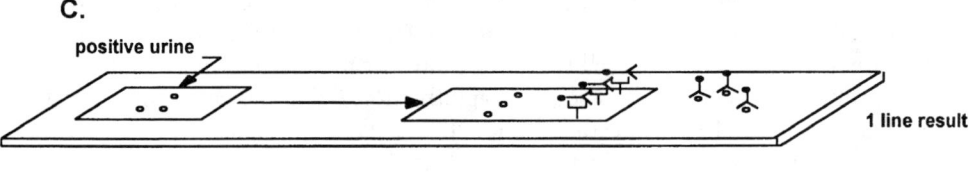

Addition of a positive urine sample

D.

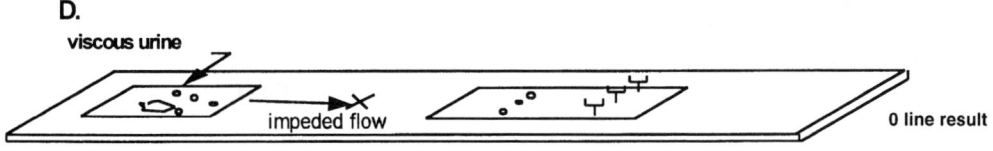

Interference by a viscous adulterant

FIG. 10–1. Schematic representation of a negative-indicating POCT device for urine drug testing. *(A)* Device regions. *(B)* Addition of a urine sample containing no drugs produces a two-line result, as the colored drug-specific antibody conjugate binds to the drug line. *(C)* Addition of a urine sample containing the drug at a concentration above the threshold results in binding of the drug to conjugated antibody, instead of the immobilized drug area, and only one line is produced. *(D)* The possible interference of POCT by a viscous adulterant. From Wu AHB: Point-of-care drug testing. In: Kost GD, ed: *Principles and practice of point-of-care testing*. Philadelphia: Lippincott Williams & Wilkins, 2001.

The Biosite Triage DOA device (San Diego, CA) is the only positive-indicating POCT device (Figure 10–2) and consists of a panel of seven or eight drugs (with tricyclic antidepressants) (6). Triage requires incubation of sample with antibodies, a transfer, and a wash step. Single- and multiple-unit urine "dipstick-like" systems have been developed that do not require the addition of any reagents or wash steps. These devices do not require a fixed aliquot of sample to be pipetted onto the unit.

The simplest POCT device to date is one in which the testing device has been combined with the urine collection cup itself. Rapid Drug Screen (American Bio Medica, Ancramdale, NY) makes use of a urine cup whose lid can be replaced with a slotted cap. Once in place, the POCT device can be directly inserted into urine. TesTcup (Roche) has the testing panel built into the front of the device. The collection cup is tilted and a volume of urine enters the testing area. These combination cup–testing devices obviate the need to aliquot or even open the urine specimens (in the case of TesTcup), which can reduce the aliquot chain-of-custody documentation needed for workplace drug testing.

Antibody Sensitivity and Specificity Comparisons

As in any immunoassay, the sensitivity and specificity of the drug test for a specific drug, metabolite, or drug analog depends on the antibodies chosen for the assays. Tests on a point-of-care platform are no exception. Using data obtained from manufacturers' package inserts, Table 10–4 compares the specificity for drug analogs and metabolites for Enzyme Multiplied Immunoassay Technique (EMIT, Dade Behring, San Jose, CA), used in the central laboratory, vs. Triage DOA and First Check (Irving, CA) point-of-care tests.

Considerable differences are observed. For amphetamines, the EMIT assay was more sensitive to the *l*-methamphetamine isomer, which is available as an over-the-counter medication (for example, Vicks inhaler). For cocaine, although the principal metabolite, benzoylecgonine, is usually targeted for drugs-of-abuse testing, the First Check assay also shows a high degree of cross-reactivity with free cocaine, which may be useful if urine is collected immediately after cocaine use (that is, before there is substantial accumulation of the metabolite). For phencyclidine (PCP), only the Triage assay is sensitive to thienylcyclohexylpiperidine, a PCP analog. The specificity of these assays is similar for opiate metabolites and synthetic analogs and for the family of barbiturates, with the exception of phenobarbital, which has high cross-reactivity in the Triage system.

Considerable differences are observed for the benzodiazepines. In particular, the Triage system was designed to be sensitive to glucuronide metabolites of the benzodiazepines rather than the parent compound. Because these metabolites are the major forms of the drug that are excreted in urine, the sensitivity of Triage was shown to be superior to assays that use antibodies directed against the parent compound, such as Abbott (Abbott Park, IL) ADx (97.5% vs. 75.6%, respectively) (7).

Comparative Studies against Central Laboratory Testing

There have been only a few published studies on assay accuracy of POCT devices. In one study, Triage DOA was compared with GC-MS on 606 positive and 325 negative samples. The study found sensitivities and specificities ranging from 93% to 100% and 95% to 100%, respectively (8). Slightly lower specificities were obtained in another study conducted on a pediatric population (87–91% for barbiturates, THC, cocaine, and opiates) (9). For amphetamines and PCP, the specificity was low at 53% and 40%, and was due to the presence of sympathomimetic amines and diphenhydramine metabolite, respectively, substances known to cross-react with the antibodies used. Table 10–5 lists the sensitivity and specificity of Triage, OnTrak, and EZ-Screen (Environmental Diagnostics, Burlington, NC) (10). Similar results for THC, cocaine, and opiates were also obtained for Frontline (Roche) (11). These studies

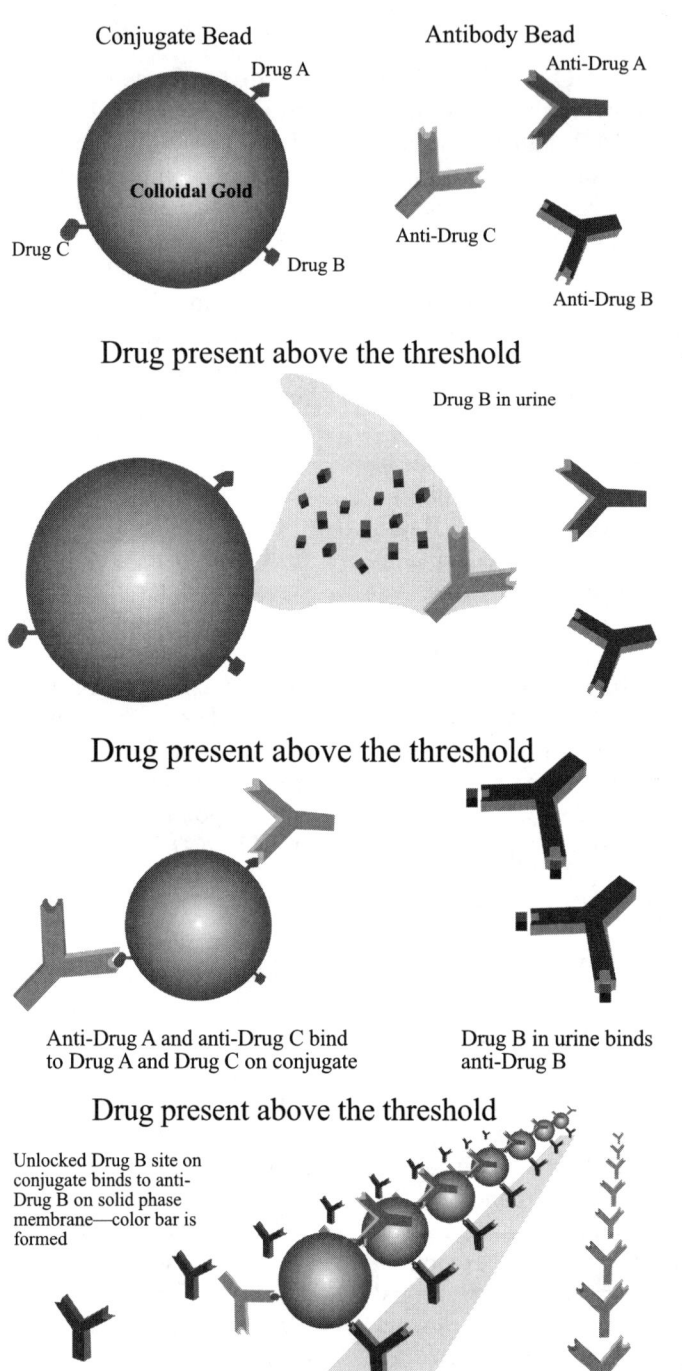

FIG. 10–2. Simplified schematic diagram of the Biosite Triage Device. This process is repeated for a total of eight drugs. Step 1: Identification of the colloidal gold bead conjugated to target drugs and specific antisera to these drugs. Step 2: Addition of urine containing drug B (*squares*) at above-threshold concentrations. The sample is negative for drugs A (*triangle*) and C (*circle*). Step 3: Antibodies to drug B bind to the bead. Step 4: Unbound drug conjugated to the bead are captured by the solid phase to produce a line. No line is produced for drugs A and C, which were absent in the original sample. Reproduced with permission from Preston Publications, Division of Preston Industries, Inc. Wu AH, Wong SS, Johnson KG, et al: Evaluation of the Triage system for emergency drugs of abuse testing in urine. *J Anal Toxicol* 1993;17:241–245.

TABLE 10–4. *Percent cross-reactivities for analog drugs and metabolites for EMIT, Triage, and First Check*

Drug	EMIT	Triage	First Check
Amphetamines			
l-Methamphetamine	50	10	NA
l-Amphetamine	17	3	NA
3,4-Methylenedioxyamphetamine	33	33	50
3,4-Methylenedioxymethamphetamine	17	33	NA
Cocaine			
Free cocaine	<1	10	60
Ecgonine methylester	<1	NA	NA
Opiates			
Morphine-3-glucuronide	37	61	100
Codeine	125	100	120
6-Monoacetylmorphine	60	100	NA
Oxycodone	7	1	NA
Hydromorphone	60	75	43
PCP			
Thienylcyclohexylpiperidine	2	100	1
Phencyclohexamine	<1	5	NA
THC			
Δ^8-THC	NA	17	<1
11-OH-Δ^9-THC	90	10	NA
8-OH-Δ^9-THC	110	NA	NA
8,11-diOH-Δ^9-THC	123	<1	50
Barbiturates			
Amobarbital	100	100	15
Barbital	13	33	15
Butabarbital	40	100	60
Butalbital	100	100	150
Pentobarbital	100	100	30
Phenobarbital	13	66	6
Benzodiazepines			
Alprazolam	166	100	6000
Chlordiazepoxide	25	6	30
Clonazepam	80	1	<1
Diazepam	182	67	300
Lorazepam	27	66	6
Nitrazepam	100	60	30
Temazepam	105	100	600

NA, data not available.

show that POCT is an acceptable alternative to automated drug-of-abuse testing in a central laboratory.

The studies described above used urine samples submitted for routine drug testing. In most cases, the concentrations of the drugs were either completely negative or at concentrations that exceeded the cutoff by a substantial amount. It is likely in these studies that there were only a few urine samples that were at or very near the assay's cutoff concentration. In order to determine the absolute accuracy of POCT devices, Taylor et al. evaluated five POCT devices using quality-control materials established at 25% below and 25% above the stated threshold (12). In contrast to the good correlations shown in Table 10–5, Table 10–6 shows that the absolute accuracy of these devices varied from assay to assay. For some drugs, the cutoff was set inappropriately too low, whereas for other classes, the cutoff was set inappropriately too high. It is possible that variability in the interpretation of the visual endpoints may contribute to the inaccuracy of these devices. The "steepness" of the signal vs. drug

TABLE 10–5. Sensitivity and specificity of POCT and EMIT for drugs-of-abuse testing

Drug class	EMIT		Triage		OnTrak		EZ-Screen	
	sens.	spec.	sens.	spec.	sens.	spec.	sens.	spec.
Amphetamines	61	99	44	97	47	96	55	84
Barbiturates	67	99	67	99	73	99	62	94
Benzodiazepines	89	87	88	90	79	92	NA	NA
Cocaine	94	99	95	99	94	96	73	89
Opiates	93	97	84	99	95	94	95	89
THC	90	98	57	98	74	98	84	95

sens., sensitivity (%); spec., specificity (%). Adapted from Ferrara SD, Tedeschi L, Frison G, et al. Drugs-of-abuse testing in urine; statistical approach and experimental comparison of immunochemical and chromatographic techniques. *J Anal Toxicol* 1994;18:278–291.

concentration curve is also a factor in assay accuracy. These data suggest that there may be some compromise for POCT relative to the central laboratory.

Oral-Fluids–Based POCT

A major problem with the way urine drug testing programs are administered today is adulteration (see below). As a result, there is substantial interest in using alternative samples for drug testing. One sample that has received significant interest is oral fluids. An important advantage of saliva analysis is that the collection can be witnessed, and the specimen is in full view of the collector at all times. Therefore, it is difficult for the donor to adulterate the sample after collection, or during collection, as it is unlikely that an individual will put an adulterant into the mouth. The disadvantage is that it is difficult to obtain large volumes of saliva that is useful for GC-MS confirmation and retesting (if requested). Moreover, the concentration of metabolites in saliva is lower than in urine; thus, screening and confirmation assays must have higher assay sensitivity (13).

Two manufacturers have recently announced POCT devices for saliva. Avitar (Canton, MA) has a device for opiates, marijuana, or cocaine, and STC (Bethlehem, PA), in cooperation with LabOne Inc., has released a device for the SAMHSA-5 group. Special collection devices have been developed to facilitate collection of sufficient volumes of oral fluid.

At-Home Drug Testing

POCT devices for drugs are now available through the World Wide Web for consumer

TABLE 10–6. Summary of the evaluation of five POCT devices for drugs of abuse[a]

Drug class	Observation
Amphetamines	Poor sensitivity to methamphetamine for all assays (25% above cutoff consistently negative)
Cocaine	Inappropriately low cutoff for all assays (25% below cut off consistently positive)
Carboxy-THC	Inappropriately low cutoff for all assays except TesTcup (which had 60% false-negatives at 25% above cutoff concentrations)
Opiates	Mostly accurate except for PharmScreen and Status DS (which had 100% and 60% false-positives at 25% below cutoff concentration, respectively)
PCP	Mostly accurate except for Status DS (which had 60% false-negative at 25% above cutoff concentrations)

[a]Assays were PharmScreen (PharmHem Labs, Menlo Park, CA), TesTcup (Roche), Status DS (Orion Diagnostics, Somerset, NJ), Rapid Drug Screen (American Bio Medica Corp.), and Accusign DOA 2 (Princeton BioMedi-Tech, Princeton, NJ).

drug testing. Many individuals use these to prescreen their own urine samples before a scheduled urine drug test (for example, a pre-employment drug test). Individuals who are positive might consider urine adulteration. One way to eliminate the effectiveness of this practice is to perform more random screening. Parents may benefit from these devices by checking urine samples from their children.

The Food and Drug Administration has established an advisory panel to examine the issues of at-home drug testing (14). Current panel recommendations are that all presumptive-positive results by immunoassay be confirmed by GC-MS. Pricing for at-home testing devices might include the expenses for mailing the sample to a certified drug laboratory and performing the confirmation test. Some members of the panel suggested that the urine drug screen panel itself might be set up such that a positive result is reported as "indeterminate," without identification as to the specific presumptively positive drug class. In this way, the tested individual might not be prematurely ostracized before having the results of the confirmation analysis.

On-site testing for drug residues may be the next level of drug detection. American Bio Medica is marketing Drug Detector, a swipe pad designed to sample surfaces for the presence of powdered drug residues (15). An aerosol is applied to the pad that would turn a color in the presence of marijuana, cocaine, heroin, and methamphetamine. The objective of this testing is to confront adolescents whom adults suspect of possible drug use with direct evidence. Such results are not legally binding because the source of the drug cannot be attributed to one individual. For example, one study demonstrated a high prevalence of US currency is contaminated with cocaine (79% of samples tested had >0.1 mg cocaine) (16). It is likely that only a few bills are heavily contaminated by drug dealers, and that most of the other bills are contaminated by counting and sorting machines used at financial institutions, not by widespread illicit drug dealing (17).

SAMPLE ADULTERATION IN WORKPLACE DRUG TESTING

Due to the major consequences an employee might face in failing a urine drug test, such as employment dismissal or suspension, urine "adulteration" has become an increasingly popular practice. In vivo adulteration refers to the intentional ingestion of a drug or substance, such as a diuretic, for the purpose of accelerating the clearance of the drug or dilution of the drug concentration in urine with an excess consumption of water (18). In vitro adulteration refers to the addition of a foreign substance to invalidate results of urine drug testing. Initial adulterants were household items such as soap, detergents, and bleach that were available at urine-collection stations. These adulterants alter the pH, ionic strength, or surface tension such that they interfere with immunoassay detection methods. More sophisticated adulterants are now sold over the Internet. Many of these compounds are designed to destroy the intended targets, making them unavailable for detection by either immunoassay screening or GC-MS confirmation methods (19). Due to the increasing proliferation of these products, drug-testing laboratories have begun to examine urine for evidence of adulteration. In addition to testing for the target drug classes, samples are also tested for creatinine, specific gravity, pH, glutaraldehyde, nitrite, and chromate. Presence of these analytes at concentrations that exceed physiologic or pathologic conditions can be an indicator that the urine sample was adulterated.

With regard to immunoassay screening, the effect of adulterants is generally independent of the platform used for the testing (that is, independent of whether it is POCT or standard methods used in a central laboratory). This platform independence is particularly true for adulterants that oxidize drugs to other compounds that are unrecognized by the antibody used. However, if on-site drug testing is conducted by individuals who do not normally handle large volumes of urine samples, the chance of detecting an adulterated sample (for example, by physical appearance) is diminished. In this situation, it is useful for those conducting on-site testing to also perform on-site adulteration testing. Urine dipsticks for the major adulterant indicators are available (for example, Chimera Chemical Co., Asheville, NC).

For a more in-depth discussion of adulteration, see chapter 11, "Urine Adulteration before Testing for Drugs of Abuse."

Case Study

An individual seeking a new job is asked by the potential employer to submit a urine sample for a drug test. This company is not required to conduct testing under federal guidelines because these employees do not work in public "safety-sensitive" positions. Therefore, on-site drug testing is performed by the employer's health clinic. The individual is instructed to remove his overcoat and to privately donate a urine sample. After several minutes, the individual emerges from the toilet and gives the urine cup to the collector, who notes a temperature of 92 °F (33.3 °C). Although this is below expected body temperature, it is within the acceptable limits of 90–100 °F (32.2–37.8 °C). The collector takes an aliquot of the urine and tests the sample using a POCT device for the SAMHSA-5 classes of drugs. Because the marijuana test line becomes positive, the urine sample is sealed in the presence of the donor, with donor initials on the seal, and sent to the toxicology laboratory for confirmation analysis. The lab receives the specimen 2 days later and performs the GC-MS analysis. The sample produces a negative result for 9-carboxy-THC metabolite and for the deuterated internal THC standard used for the analysis. The analyst confirms to the certifying scientist that the internal standard was added to the aliquot of urine before extraction. The certifying scientist suspects that this urine was adulterated, and tests the sample for nitrite. A nitrite concentration of 1040 mg/mL confirms the presence of the adulterant nitrite (20). Based on this finding, the individual's application for employment was denied by the firm.

REFERENCES

1. Substance Abuse and Mental Health Administration: Preliminary estimates from the Drug Abuse Warning Network. Advance Report no. 8. Rockville, MD: US Department of Health and Human Services, 1994.
2. Substance Abuse and Mental Health Services Administration: National Household Survey on Drug Abuse: Population estimates 1994. Rockville, MD: US Department of Health and Human Services, 1995.
3. Department of Health and Human Services: Mandatory guidelines for federal workplace drug testing programs; final guidelines notice. *Fed Regist* 1988; 53:11969–11989.
4. Postal Service to trial onsite testing of all new employees. *Drug Det Report* 1998;8:65–66.
5. Blank RV, Decker WJ: Analysis of toxic substances. In: Tietz NW: *Fundamentals of clinical chemistry*, 2nd ed. Philadelphia: Saunders, 1986: 1679–1683.
6. Buechler KF, Moi S, Noar B: Simultaneous detection of seven drugs of abuse by the Triage panel for drugs of abuse. *Clin Chem* 1992;37:1678–1684.
7. Koch TR, Raglin RL, Scheree K, et al: Improved screening for benzodiazepine metabolites in urine using the Triage Panel for drugs of abuse. *J Anal Toxicol* 1994;18:168–172.
8. Wu AHB, Wong SS, Johnson KG, et al: Evaluation of the Triage system for emergency drugs-of-abuse testing in urine. *J Anal Toxicol* 1993;17:241–245.
9. Valentine JL, Komoroski EM: Use of a visual panel detection method for drugs of abuse: clinical and laboratory experience with children and adolescents. *J Pediatr* 1995;126:135–140.
10. Ferrara SD, Tedeschi L, Frison G, et al: Drugs-of-abuse testing in urine: statistical approach and experimental comparison of immunochemical and chromatographic techniques. *J Anal Toxicol* 1994; 18:278–291.
11. Wennig R, Moeller MR, Haguenoer JM, et al: Development and evaluation of immunochromatographic rapid tests for screening of cannabinoids, cocaine, and opiates in urine. *J Anal Toxicol* 1998; 22:148–155.
12. Taylor EH, Oertli EH, Wolfgang JW, et al: Accuracy of five on-site immunoassay drugs-of-abuse testing devices. *J Anal Toxicol* 1999;23:119–124.
13. Cone EJ, Hillsgrove M, Darwin WD: Simultaneous measurement of cocaine, cocaethylene, their metabolites, and "crack" pyrolysis products by gas chromatography-mass spectrometry. *Clin Chem* 1994;40: 1299–1305.
14. Confirmation process urged for on-site home drug tests. *Drug Det Report* 1997;7(13):3.
15. Onsite test detects drug residue without specimens. *Drug Det Report* 2000;10(2):3.
16. Oyler J, Darwin WD, Cone EJ: Cocaine contamination of United States paper currency. *J Anal Toxicol* 1996;20:213–216.
17. Jourdan TH, Wang D: Cocaine contamination of U.S. currency. [Abstract B29]. American Academy of Forensic Sciences Annual Meeting, New York, 1997.
18. Wu AHB: Integrity of urine specimens submitted

for toxicologic analysis: adulteration, mechanisms of action, and laboratory detection. *Foren Sci Rev* 1998;10:47–65.
19. Wu AHB, Sexton K, Cassella G, et al: Adulteration of urine by "Urine Luck." *Clin Chem* 1999;45:1051–1057.
20. ElSohly MA, Feng S, Kopycki WJ, et al: A procedure to overcome interferences caused by the adulterant "Klear" in the GC-MS analysis of 11-nor-Δ^9-THC-9COOH. *J Anal Toxicol* 1997;21:240–242.

SELF-ASSESSMENT QUESTIONS

1. Which drugs are the most significant from a toxicologic viewpoint and warrant emergency testing?

2. What are the advantages of point-of-care testing (POCT) for the emergency department? For workplace drug testing?

3. What is the current regulation concerning POCT in the workplace?

4. What are the differences between antibody specificities for POCT vs. central lab assay?

5. Describe the differences between a positive- and a negative-indicating POCT device.

CHAPTER 11

Urine Adulteration before Testing for Drugs of Abuse*

Alan H.B. Wu, Ph.D., DABCC

LEARNING OBJECTIVES

After completing this chapter, the reader should be able to:

1. *Define and describe the various forms of urine adulteration that are currently being practiced in conjunction with drugs of abuse testing.*
2. *Describe the responsibilities of the collector and drug-testing laboratory with regards to detecting and reporting adulterated samples.*
3. *Describe the mechanism of the various forms of urine adulterants.*
4. *Describe how adulterants work on immunoassays and gas chromatography–mass spectrometry assays.*
5. *Describe the strategies available for detecting in vitro adulteration.*

DEFINITION OF ADULTERATION

Webster's New Collegiate Dictionary defines the term adulterate as "to corrupt, debase, or make impure by addition of a foreign or inferior substance." In the context of urine drug testing, the term adulteration refers to the intentional or unintentional addition of a foreign substance to urine that invalidates or obscures urine drug analysis. For the purposes of this chapter, this term is broadened to include any action, willful or inadvertent, that makes a urine drug test result an inaccurate reflection of drug use. This includes the substitution of urine and the ingestion of drugs, substances, or fluids designed to alter the natural biological metabolism or clearance of drugs from the body. It also includes the inadvertent ingestion of food or beverages that contain drugs or similar substances that can produce a positive result. Although in most cases, the intent is to make the drug test falsely negative, there have been situations in which adulteration by pharmaceutical tablets was designed to make the result falsely positive (1).

PREVALENCE OF ADULTERATION PRACTICES

The extent to which submitted samples may have been adulterated has been examined in several studies. One investigator in the United Kingdom estimated the frequency of possible adulterated or diluted specimens submitted to a toxicology laboratory by measuring urine pH and creatinine (2). Using a normal pH range of 5.0–8.0 and a creatinine cutoff of <20 mg/dL, 10 of 50 (20%) samples were outside the pH range, and 40 of 50 (80%) were below the creatinine cutoff.

These high estimates, however, are not supported by data from laboratories certified

---

*Updated and modified from Wu AHB: Integrity of urine specimens for toxicologic analysis—adulteration, mechanisms of action, and laboratory detection. *Foren Sci Rev* 1998;10:47–65.

by the Substance Abuse and Mental Health Services Administration (SAMHSA). In a study conducted at Med Tox Laboratories (St. Paul, MN) involving 7700 specimens, only 4.6% of urine samples had creatinine concentrations <20 mg/dL (3). Nevertheless, this figure is considerably higher than the 0.5% of truck drivers who have low creatinine concentrations. The difference may better reflect the true percentage of tested individuals who attempt to adulterate their urine. Because urine collection from truck drivers is performed at roadside or at weigh stations, they do not have the opportunity to consume large amounts of fluids to intentionally dilute urine.

The frequency of adulteration among samples submitted for clinical toxicology and medical compliance program testing has not been established. The data from workplace drug-testing practices are not necessarily applicable. On the one hand, the frequency of positive results for drugs of abuse is higher for clinical toxicology; therefore, one might expect a greater need to adulterate. Conversely, the consequences of a positive urine drug test result to the donor are likely to be less because testing is on clinical grounds and not for denial or retention of employment. Also, patients who are admitted to the emergency department in an obtunded or comatose state will be catheterized and have no opportunity to adulterate their urine sample.

SAMHSA AND DEPARTMENT OF TRANSPORTATION REGULATIONS AND EFFECT ON ADULTERATION

Mandatory guidelines for workplace urine drug testing in federal agencies are administered by SAMHSA (4) and those for regulated transportation industries by the Department of Transportation (DOT) (5). Federal workplace drug testing has recommendations for testing of urine for creatinine, specific gravity, pH, and nitrites. Although workplace testing procedures are not applicable to clinical toxicology, SAMHSA and DOT have defined cutpoints that indicate the presence of adulterated, substituted, or diluted specimens. These concentrations are summarized in Table 11–1. Under these guidelines, the term "substitution" refers to the replacement of the donor's urine with a non-urine liquid that may not contain creatinine or may have specific gravity and pH that are outside the limits of normal human urine. Unfortunately, new regulations are necessarily slow to develop and cannot keep pace with the new developments in the area of commercial adulterant practices. Several states in the United States have enacted legislation making it illegal to sell urine adulterants. Until more states pass such laws or there is a federal mandate, the practice of in vitro adulteration will continue and likely grow.

ADULTERATION PRACTICES

Substitution

For this section, the term "substitution" is broadened from the SAMHSA definition to include the practice whereby a urine specimen from a drug-abusing donor to be tested is switched by the donor with urine from a drug-free individual. Urine substituted in this manner will likely pass SAMHSA regulations because it will contain the necessary creatinine, pH, and specific gravity values. Pre-assayed urine that has been certified to be drug free can be purchased through suppliers who make this product available to drug-testing laboratories in preparation of calibrators and negative controls, or through the World Wide Web for the sole purpose of substitution. (See Table 11–2.)

Two procedures are currently in use for detection of substituted urine: monitoring urine temperature immediately after collection, and careful witnessing of the collection of urine itself. Unfortunately, acceptable temperatures can be achieved if the substituted urine is stored in the axilla, in the vaginal cavity, or next to the scrotum just prior to donation (6). Effective same-gender

TABLE 11–1. SAMHSA definitions for urine adulteration[a]

Parameter	Limits	SAMHSA Interpretation
Creatinine, specific gravity	≤5 mg/dL and ≤1.001	Substituted
Creatinine, specific gravity	≤5 mg/dL and ≥1.020	Substituted
Creatinine, specific gravity	<20 mg/dL and <1.003	Diluted
Creatinine, specific gravity	≤5 mg/dL and 1.003 <SG <1.019	Specimen unsuitable for testing
pH	≤3 or ≥9	Adulterated, pH too high or low
Nitrite	≥500 mg/L	Adulterated, nitrite too high

[a]Program document numbers 35 and 37, Research Triangle Institute, Research Triangle Park, NC, 1998.

witnessed collection requires close observation of urination, an unpleasant duty for most individuals. If the tested individual is not closely monitored during a void, substitution can still occur; the donor can conceal a pouch of drug-free urine and release its contents directly to the urine cup. Self-catheterization is the most extreme manner for urine substitution. In this situation, an individual first completely voids, then inserts a catheter through the urethra, and adds a small amount of drug-free urine directly back into the bladder (7). The urine sample might be someone else's urine, or urine from the donor collected during a time when the individual had abstained from drug use. The hazards of this system aside, individuals who go to this extreme will likely escape detection unless there is enough suspicion to warrant DNA analysis.

TABLE 11–2. Forms of adulterations for urine drug testing

Form	Description
Substitution	Replacement of a freshly voided urine sample with a drug-free sample
In vivo	Ingestion of water, substance, or drug to hasten excretion or inhibit excretion
In vitro	Addition of a foreign substance directly to a voided sample to invalidate test results
Inadvertent	Ingestion or exposure to food or preparation that results in a positive urine drug test result in an individual who is not abusing a drug

In Vivo Adulterants

Dilutional

In vivo adulteration is the intentional ingestion of fluids, substances, and/or drugs designed to dilute urine or to hasten or increase the metabolism and/or excretion of drugs in the body. Water is an effective in vivo dilutional adulterant. Patients who suffer from psychogenic polydipsia routinely consume large volumes of water, which can dilute urine of electrolytes by up to 10-fold. Under this condition, drugs present at or near the cutoff concentrations will be diluted to produce negative results. Drugs such as phencyclidine (PCP) and tetrahydrocannabinol (THC) are not excreted in very high concentrations into urine and can be diluted by excessive fluid intake. Other drugs such as opiates, amphetamines, and cocaine produce drug concentrations that can exceed 10 times the cutoff concentrations, particularly when urine is donated soon after the drug is used. For these individuals, dilutional adulterants might not be effective in producing the desired negative result, because it is not possible to consume enough water to reduce urine drug concentrations to that extent.

Diuretics are drugs designed to enhance the clearance of water and sodium. Diuretics are used for the treatment of heart failure, hypertension, hepatic ascites, pulmonary edema, and renal edema. There are many types of natural and synthetic diuretics. Some of these have been used to adulterate urine. It should be noted that some diuretics give urine an unusual color indicating the presence of an unnatural condition. Xanthine compounds are diuretics found in popular beverages and include caffeine (coffee and tea), theophylline (tea), and theobromine (cocoa). Xanthines increase blood flow to the kidney and may produce diuretic action through an increase in glomerular filtration pressure. The effect of xanthine diuretics is minimal compared with other diuretic drugs.

The ingestion of tea made from golden seal root was thought to be effective in producing an in vivo adulterant; however, the alkaloid present in this tea interferes only with thin-layer chromatographic methods, and not immunoassays or gas chromatography–mass spectrometry (GC/MS) analysis (8). It should be noted that although the intent of tea is to dilute urine, if the tea is made from coca-leaf (illegal in the United States), positive results for cocaine metabolite can occur (9).

Loop diuretics are among the most effective agents for inducing large quantities of water loss and sodium excretion. Drugs such as furosemide act on the ascending limb of the loop of Henle and inhibit chloride reabsorption, while maintaining plasma potassium concentrations. Although these drugs may be effective in diluting urine drug concentrations before a test, they can have significant adverse effects such as a rapid reduction of blood volume, orthostatic hypotension, headache, dizziness, nausea, vomiting, and diarrhea.

Very high plasma concentrations of osmotically active substances will also induce water loss. Commercial osmotic agents are available for the purpose of in vivo adulteration. Test Free contains 64 g of glucose and heterogeneous carbohydrates and is an osmotic diuretic absorbable by the gastrointestinal system. The individual is instructed to consume large volumes of water and to empty his/her bladder two or three times before submitting a urine specimen for drug testing.

Acidic or Basic Food, Drugs, or Liquids

Food, drugs, or liquids that are highly acidic (such as vinegar) or basic (sodium bicarbonate) do not in themselves produce sufficient changes in urine pH to interfere with immunoassay screening procedures. However, acidification or alkalinization can affect the metabolism and rate of clearance for drugs. Alkalinization of urine containing weakly acidic drugs will ionize the drug, resulting in increased water solubility and urinary clearance:

$$HA \text{ (weak acid)} \xrightarrow{\text{base}} H^+ + A^-$$

Acidification of urine containing weakly basic drugs will also enhance clearance in the reverse manner:

$$RNH_2 \text{ (weak base)} \xrightarrow{\text{acid}} RNH_3^+$$

For the basic drug amphetamine, as much as 74% of the parent compound is excreted in urine that is slightly acidic. In alkaline urine, only 1% is excreted as the parent compound, with a higher proportion of the various hydroxylated and conjugated metabolites excreted (10). In a similar way, acidification has been reported to enhance the excretion of the weakly basic drug PCP (11).

Miscellaneous Drugs and Substances

There are many other drugs and substances that can cause negative interferences with drug assays and could be considered candidates for an in vivo adulterant by a drug abuser. Salicylates have been shown to negatively interfere with enzyme-multiplied immuno technique (EMIT®; Behring Diagnostics Inc, San Jose, CA) assays (12). However, under therapeutic salicylate concentrations, the false-negative effects of salicyluric acid are minimal. Ibuprofen can also produce a false-negative result for THC by GC/MS if there is an inadequate amount of derivatization reagent used (13). Because most drug-testing laboratories are aware of this problem, ingestion of ibuprofen would not be an effective adulterant strategy. Other drugs and substances that interfere at the level of immunoassay detection include fluorescein used in retinal angiography (14) and the antibiotic metronidazole (15). Although these compounds do not result in falsely negative screening results for urine containing the targeted drugs, they can invalidate the test such that a re-collection may be required. Thus, the drug-positive individual may escape detection on that particular instance and might temporarily abstain from use before a re-collection. Other in vivo substances have been suggested as potential adulterants to produce falsely negative results, but have proven to be ineffective. These include high concentrations of vitamin C (ascorbic acid) (16) and herbal teas (8).

Inadvertent Adulterants

Consumption of certain foods and drinks can lead to the inadvertent true-positive on a drug screen and a false-positive indication of drug abuse. The finding of positive urine results for opiates by individuals eating bakery products containing poppy seeds before a drug test is well known and described (17,18). Depending on the source, poppy seeds contain varying concentrations of codeine and morphine ranging from 0 to 78.8 and 0 to 964 µg/g, respectively (19). Therefore, poppy seed consumption will produce a true-positive opiate result in urine, and a false-positive indication of drug abuse by the donor.

The purposeful or inadvertent ingestion of Cannabis seed oil will produce a positive result for marijuana much in the same manner as poppy seeds do for opiates (20). These seeds are available in hemp shops, in health food stores, and from oil manufacturers. In one study, twice-daily ingestion of cold-pressed hemp seed oil resulted in 9-carboxy-THC metabolite concentrations ranging from 41 to 68 ng/mL by GC/MS (21). Urine was positive even 2 days after the oil was discontinued. In a similar study, hemp seed snack bars and cookies made from hemp seed flour and butter were given to volunteers to determine their potential for producing positive urine results. Although some urines were screened positive, none had 9-carboxy-THC concentrations that exceeded the GC/MS cutoff concentration of 15 ng/mL.

Food produced from farm animals is naturally rich in creatinine. This byproduct of muscle metabolism in the animal is identical to that produced in humans. Therefore, the consumption of meat temporarily increases serum and urine creatinine concentrations. Eating meat or adding creatinine to collected urine may obscure the laboratory's attempts to identify diluted urine specimens if creatinine is the only dilutional marker used.

There are several examples whereby prescription drugs can produce falsely positive immunoassay results for targeted drugs of abuse. Well-known examples include sympathomimetic amines in the amphetamine assay, hydrocodone and hydromorphone in the opiate assay, and cyclobenzaprine in the tricyclic antidepressant assay. These drugs have similar chemical structures to the targeted drug and cross-reactivities are due to the inability of the antibody to recognize the subtle differences. There are several examples of cross-reactivities occurring for a drug (or metabolite) that is structurally unrelated to the target drug (Figure 11–1) (22,23). These interferences can be assay specific, that is, they do not uniformly affect all commercial immunoassays. These drugs do not produce falsely positive results by GC/MS analysis.

In Vitro Adulterants

Household Items

Household cleaners and over-the-counter pharmaceutical products were once popular in vitro adulterants because they were readily available in bathroom closets, pockets, and purses. For individuals who were "spot" screened, that is, they were instructed to immediately report to a urine-collection station, there was little time to prepare for a drug test. Studies were conducted on the major commercial immunoassays for drugs of abuse (24). In some cases, there are discrepancies in the data reported in the literature, because

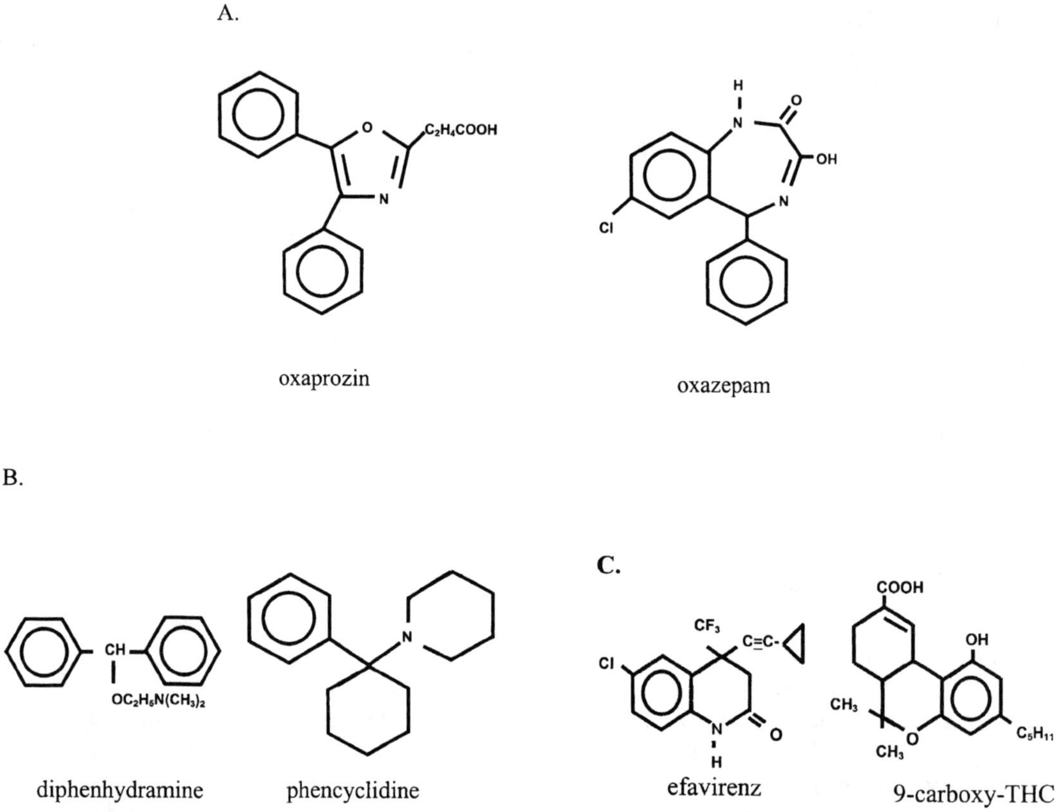

FIG. 11–1. False immunoassay results for (*A*) oxaprozin in the EMIT benzodiazepine assay, (*B*) diphenhydramine in the original Abbott TDx PCP assay, and (*C*) efavirenz in the CEDIA THC assay.

the studies were performed at different concentrations and under different analysis conditions. However general conclusions can be made, as summarized in Table 11–3. The reader is referred to the original literature or reviews (7,24) for a discussion of specific household items.

Commercial Adulterants

The paranoia surrounding drug testing has spurned a cottage industry of companies that are formulating and marketing in vivo and in vitro adulterants. These products are widely available through the World Wide Web. UrinAid was among the first product to be sold for this purpose. The active ingredient in UrinAid has been identified to be glutaraldehyde (25). At concentrations of ≥0.75%, glutaraldehyde interferes with most laboratory-based immunoassays by producing negative absorbance rates that are below the rates expected for drug-free urine (26). Mary Jane's Super Clean 13 contains alkylephoxysulfonate, a component found in Joy® dishwashing detergent (7). As shown in Table 11–4, Joy® can produce false-negative results for some drug classes, and false-positive results for others.

An adulterant containing nitrites, Klear, contain 1 g of potassium nitrite. Samples containing THC will screen positive by immunoassay, but will confirm negative when GC/MS is performed. This result occurs because before confirmation analysis, nitrite oxidizes 9-carboxy-THC under the acidic extraction conditions used to other compounds (27). Nitrite will also interfere with the analysis of deuterated internal standards for 9-carboxy-

TABLE 11-3. Effect of various adulterants on immunoassays for drugs of abuse[a]

Adulterant	Amphetamines	Barbiturates	Benzodiazepines	Cocaine	Opiates	PCP	THC
Ammonia							+R
Ascorbate							−R
Bicarbonate	−R	−R			−E	−F	
Bleach	−all	−C	−E, −C	−C	−all	−all	−E, −F, −C
Detergent	−C	+F, −C	−C	+F, −C	−C	−C	
Drano®	−E, −C	−E, −C	−E, −C	−all	−E, −C	−C	−R, −E, −C
Golden seal	−C		−F				−R, −E, −C
Hand soap	+F	+F	−E				−R, −E
Joy®	+F	+F	+R, −E, +F		−E		+R, −E, +F
Lime	−R				−R		−R
Peroxide			−E, +F				
Phosphate				−R, −F	−F	−R	
Salt	−E	−E	−E	−E	−E	−E	−E
Vanish®	−R				−R		−R
Vinegar							−E, −F
Visine®			−E				−R, −E, −C

R, RIA; E, EMIT; F, FPIA; C, CEDIA.
[a]Key: Adulterant reduces (−) or increases (+) response to drug listed by methodology.

THC analysis. Certifying scientists who observe such interference can then suspect the presence of this adulterant in a particular urine sample. Laboratories not using deuterated THC analogs as internal standards may not know that the target drug has been destroyed.

Urine Luck contains pyridinium chlorochromate (28). Like nitrite, the chromate in these products also oxidizes 9-carboxy-THC. This reaction can occur at neutral pH; therefore, adulteration begins from the moment urine is collected. Thus, immunoassay screening tests will be negative for THC, and the laboratory may not know of its presence. Chromates have a yellow color that is more intense than typical urine.

The latest generation of commercial urine

TABLE 11-4. Commercial in vitro adulterants[a]

Product	Class	Active ingredient
Mary Jane's Super Clean 13	Surfactant	Alkylephoxysulfonate
UrineAid	Fixative	Glutaraldehyde
Amber-13	Acid	1.7 mol/L HCl
THC-Free	Acid	2.1 mol/L HCl
Clear Choice	Fixative	Glutaraldehyde
Klear	Oxidant	Potassium nitrite
Whizzies	Oxidant	Sodium nitrite
Randy's Klear	Oxidant	Nitrite
Urine Luck	Oxidant	Pyridinium chlorochromate
LL418	Oxidant	Pyridinium chlorochromate
Sweet Pee's Spoiler	Oxidant	Pyridinium chlorochromate
Randy's Klear II	Oxidant	Pyridinium chlorochromate
Urine Luck II	Oxidant	Chromate
Stealth	Oxidant	Peroxidase
Formula 2000	Unknown	Unknown
Randomizer	Unknown	Chloride
Urine Klear	Unknown	Unknown
Crystal Klear	Unknown	Unknown

[a]Davis KH. Adulterant update. DOT/HHS Laboratory Director's Meeting, December 1999.

TABLE 11-5. *Causes of unusual urine appearances and colors*[a]

Appearance	Pathologic	Nonpathologic
White, cloudy	Chyle, pus, bacteria	Phosphates, urates, cellular elements, carbonates, mucus
Smoky or turbid	Erythrocytes	
Yellow to orange	Direct bilirubin, urobilin	Normal color Dietary: carrots, food color, rhubarb, senna Drugs: riboflavin, serotonin, sulfasalazine, Pyridium
Pink to red	Heme pigments, porphyrins, erythrocytes	Dietary: beets, cascara, food coloring Drugs: methyldopa, phenacetin, phenolphthalein
Purple	Porphyrins	
Brown	Bilirubin, myoglobin	Drugs: levodopa, quinine, resorcinol Dietary: rhubarb metabolites
Black	Melanin, homogentisic acid	
Blue-green	Oxidized bilirubin complex	Drugs: amitriptyline, methylene blue, thymol, vitamin B
Gray		Drugs: flurazolidone, nitrofurantoin, nitrofurazone

[a]Adapted from Graff L: *A handbook of routine urinalysis.* Philadelphia: Lippincott, 1982.

adulterants is Stealth. The active ingredient is thought to be a peroxidase, which also oxidizes THC immediately after urine collection. Stealth is more difficult to detect than nitrites or chromates, however, because the enzyme itself breaks down and becomes inactive after a few hours. Thus, if GC/MS analysis is performed, typically 24–48 hours after collection, the inactivated adulterant will not be able to oxidize the deuterated internal standard. Table 11–5 summarizes the commercial adulterants, class of action, and active ingredient that are or have been available for sale.

Mechanisms of Action for Adulterants

Although all drug-screening immunoassays make use of specific antibodies directed against drug groups, their detection schemes differ from one to another. Therefore the extent to which an adulterant interferes with the assay depends on the specific assay used. An adulterant that interferes in one assay will not necessarily work in another. An individual whose urine is being tested cannot dictate which immunoassay will be used. Moreover, if a second laboratory is used for a retest, the immunoassay screening step is unnecessary.

Most adulterants do not change the drugs present in the sample, but rather, they alter the means by which they are detected. Therefore, most adulterants do not affect the GC/MS analysis of the drug and it may be possible to confirm the presence of a drug even if the immunoassay is negative or unsuitable for testing.

Photometric Interferences

Homogeneous immunoassays such as EMIT and cloned enzyme-donor immunoassay (CEDIA, Microgenics Inc., Fremont, CA) use enzyme activity for the determination of drug concentrations. Adulterants that absorb at this wavelength or inhibit the absorption of the reduced form of nicotinamide-adenine dinucleotide phosphate (NADH) will cause an interference. For example, the salicylate metabolite, salicyluric acid, interferes with the EMIT assay by reducing the apparent molar absorptivity of NADH at 340 nm (29). Because salicyluric acid does not absorb at 376 nm, no interference is observed when this wavelength is used. Metronidazole interferes with EMIT assays because it contributes to high absorptivity background at 340 nm, thereby exceeding the absorbance limits of the spectrophotometers used for these test systems (15). Neither salicyluric nor metronidazole produce interferences with CEDIA, because this assay measures the activity of β-galactosidase activity, whose product is monitored at 570 nm; or with radioimmunoassay (RIA), which does not use photometric detection; or with fluorescence polarization immunoassay (FPIA), which measures fluorescence emission between 525 and 550 nm. On the other hand, the use of fluorescein as a dye for radiologic imaging produces high background fluorescence that interferes with all FPIA assays (14). In general, RIA procedures are most immune from adulterants because they are removed by the bound antibody separation step before detection.

Alterations in pH

Most enzymes have a narrow pH range by which the activity is optimized. Therefore, enzyme immunoassays are highly subject to the effects of adulterants that alter the pH of urine specimen (such as vinegar, lemon juice, bicarbonate, and alkaline detergents). The degree of pH change depends on the buffering capacity of the urine sample and on the amount and pH of the adulterant used. The pH also affects molar absorptivity and relative fluorescence intensities. Although the effect of varying pH on RIA is not expected to be as great, extreme pH changes can produce falsely negative as well as falsely positive results (30).

Antibody-Antigen Interactions

All immunoassays rely on the specific interaction between antibodies from the assay and the antigens (drugs) from the specimens. This delicate balance can be disrupted by changes in pH, ionic strength, viscosity, and surface tension. For example, table salt at high concentrations can disrupt this balance in EMIT assays by changing the ionic strength (31). Soap and detergents prevent antibody binding, notably for THC assays (32). Not all assays for a given manufacturer are affected to the same degree. It is likely that the degree of avidity between the antibody used for the specific assay and its target antigen determines how much the assay is interfered with by the presence of these adulterants.

Miscellaneous Interactions

In vitro adulterants have the potential to interfere with immunoassays in unpredictable ways. For example, one ingredient in Visine® is benzalkonium chloride (33). This substance has been shown to promote the sequestration of THC to micelle bodies, making them unavailable for binding to THC-specific antibodies. The in vitro addition of protein A from *Staphylococcus aureus* will bind major subclasses of immunoglobulin G antibodies and will also interfere with immunoassays.

Interferences in Point-of-Care Drug-Testing Systems

All of the previously published data were conducted on these laboratory-based immunoassays. However, point-of-care (POC) drug-testing devices are now widely available (34). Most of these assays make use of immunochromatography, whereby antibodies are immobilized onto a solid support in a strip to capture and detect drugs as they pass through in the aqueous (mobile) phase. To date, there have been no studies conducted on the performance of these devices in the presence of in vitro adulterants. Because the design of these assays is significantly different from the solution-based laboratory assays, it is possible that the adulterants discussed in the previous section will not behave in the same manner. For example, adulterants that retard or impede flow to the targeted areas will produce interferences in POC devices that will not be observed on instrument-based immunoassays (see Figure 11–2). POC devices have built-in procedural positive controls that enable the analyst to see whether the sample has migrated to an

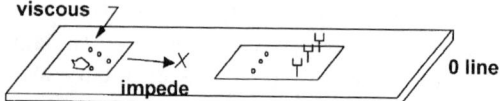

FIG. 11–2. Mechanism for interference of a viscous adulterant using POCT devices. See Chapter 10 for a description of the normal operation of the POCT device.

area where the antibodies are immobilized. On the other hand, the POC devices are designed to be used by individuals with considerably less experience and training than licensed technologists employed in certified toxicology laboratories. Therefore detection of potential adulterants and troubleshooting of problematic specimens will be greatly limited in the on-site testing environment. For adulterants that act by interfering with antibody-antigen reactions, it is likely that POC testing (POCT) will be affected the same as laboratory-based immunoassays.

Gas Chromatography–Mass Spectrometry Analysis

Although there are a lot of studies conducted on the effect of household adulterants on immunoassays, there are few studies that examine the effects of these substances on GC/MS analysis. There are some in vitro adulterants that can interfere with both immunoassay screening and GC/MS confirmation analysis. In one study using household bleach, an inverse relationship was demonstrated between the amount of bleach added to urine specimens and the concentration of 9-carboxy-THC recovered (35). The addition of bleach oxidized the opiates to compounds that were no longer detectable by GC/MS. Detergent and Drano® also interfere with GC/MS analysis of opiates, THC, and cocaine, whereas there is no effect for Visine® (36). Vinegar produced a reduction in the GC/MS quantification, but the drug was recoverable. It is likely that vinegar disrupts the pH necessary for optimal extraction of the drug, rather than converting it to a new compound. In a recent study, denture-cleaning tablets have been shown to produce false-negative results for amphetamine, benzoylecgonine, and 9-carboxy-THC (37). Although GC/MS confirmation was not performed, it was postulated that the sodium perborate present in these tablets may have oxidized these targeted drugs to new compounds.

LABORATORY DETECTION OF ADULTERANTS

The collectors and accessioners at a testing laboratory have the responsibility of detecting urine that may have been substituted or adulterated. This can be accomplished with a series of physical or chemical tests or with DNA analysis.

Physical Appearance

Urine can also present as a cloudy or milky mixture as a result of sediment formation. Sediment contains crystals, casts, and other cellular elements (38). Acidic or basic adulterants may cause the precipitation of crystals from the urine. Other adulterants such as hand soap may cause urine to have a cloudy or turbid appearance. However, sedimentation naturally occurs when urine stands for a few hours, particularly when cold. Therefore, no conclusions concerning adulteration of these specimens can be made when sediments are present. On the other hand, urine that forms bubbles when shaken or refracts light to give it a rainbow appearance is unnatural and suggests adulteration with detergents.

The color of urine is normally clear to yellow depending on its concentration. Clear urine is the most dilute. The first morning urine void, which is not usually collected for drug testing, is the most concentrated. Urine that is an unusual color might be the result of either in vivo or in vitro adulteration of the specimen by the donor. For example, brown urine can be caused by the use of the diuretic golden seal tea. Blue urine might originate from the dye used in the toilet water of the collection station. Unfortunately, most of the unusual colors observed in urine can also be caused by various dietary, physiologic, and pathologic conditions. Table 11–5 summarizes some of the causes of unusual urine colors. Because of these factors, a urine specimen of an unusual color does not necessarily indicate an adulteration.

Urine can have a wide range of odors as the result of both normal physiology and pathologic conditions. Urine from individuals with a urinary tract infection may contain urea-splitting microorganisms that will produce urine with an ammoniacal odor. Fruity odors are the result of ketones, which are released in starvation and diabetic ketoacidosis. Certain congenital diseases such as phenylketonuria, maple syrup urine disease, or isovaleric acidemia produce unusual odors. Specimens from patients with inborn errors of metabolism, however, are very unlikely to be submitted for routine drug analysis. The odor of many commercial disinfectants and cleaners is rather distinctive and can be used to denote adulteration. The odor of bleach is unmistakable. Likewise, a lemon scent to urine is likely due to detergents or cleaners added after collection. Urine odor is becoming less important as there is shift in adulteration practices away from household items and towards commercial adulterants, which are odorless.

Tests for Urine Dilution

Creatinine is a natural degradation product of creatine metabolism that originates from skeletal muscles. In the absence of renal disease, serum creatinine concentrations are related to muscle mass, and values are fairly stable among individuals. Slightly higher values are observed in individuals with a high-meat diet, as it contains a high concentration of animal creatinine. The normal range for 24-hour urinary creatinine is 14–26 mg $\cdot$ kg^{-1} $\cdot$ d^{-1} for males and 11–20 mg $\cdot$ kg^{-1} $\cdot$ d^{-1} for females. Because urine collected for drug analysis is not performed over 24 hours, and the sex of the donor is often not known, these clinical limits cannot be used. Instead, SAMHSA has established a normal limit of 20 mg/dL for federal drug testing.

An interesting study was conducted by Luceri et al. involving urine samples submitted to an addiction treatment center that were suspicious because of below-normal creatinine concentration (<45 mg/dL) (39). When 100 samples were concentrated three- to sixfold by evaporation, 27% of these samples were positive for one drug or another upon retest. The authors consider these 27 samples to have been originally adulterated by dilution and suggest that concentrating urine by evaporation may be a means to reduce the number of false-negative results. This approach may be acceptable for drug testing in controlled settings such as drug rehabilitation or prison parole programs, but is not currently permitted for workplace drug testing.

The specific gravity is also the density of a liquid and is standardized against distilled water (1.000). The specific gravity is an indicator of the concentration of dissolved materials in the urine and is related to the number and weight of the particles in solution. There are two common methods for measuring specific gravity. Refractometry measures the refractive index of urine, that is, the ratio of the velocity of light in air to that in urine (38). The dipstick method is based on the pKa change of certain pretreated polyelectrolytes in relation to ionic strength, a surrogate for specific gravity. The higher the ionic strength, the higher the [H$^+$] and lowering of pH, which is what is actually measured on the dipstick pad. The dipstick specific gravity is not as accurate as the refractometer, because highly buffered alkaline urine will produce falsely low results. Thus, inaccurate readings can occur with adulterants that change the pH of urine.

Specific gravity values >1.030 can be pathologically caused by dehydration, proteinuria, glycosuria, adrenal insufficiency, liver disease, and congestive heart failure and after use of radio-opaque dyes. High values might also be caused by addition of adulterants such as sodium chloride. Low specific gravity occurs in diabetes insipidus, glomerulonephritis, pyelonephritis, and other kidney diseases. Diuretic use such as caffeine, alcohol, and Lasix (furosemide) will result in low specific gravity. Some individuals consume diuretics or excessive amounts of water just before a urine drug test in an effort to dilute their urine to below cutoff concentrations. Urine that is significantly <1.000 indicates substantial adulteration of the specimen by an organic solvent with a lower specific gravity than that of water (such as methanol and ethanol).

pH

The pH of urine from a normal individual is usually between 6.0 and 7.0, although values between 4.5 and 8.0 can be observed as the result of disease. The pH of urine can be measured with a pH meter or colorimetrically. Persistently acidic urine can be caused by respiratory and metabolic acidosis, uremia, and severe diarrhea. It can also be caused by starvation and diets high in ascorbic acid and citric fruits. Persistently alkaline urine can be caused by respiratory and metabolic alkalosis, urinary tract infections, and diets that are high in vegetables, milk, and other dairy products, and can result after medications such as sodium bicarbonate and acetazolamide. In patients with a renal tubular acidosis, the urine pH is alkaline and the blood is acidic.

The adulteration of urine by foreign substances can produce significant changes in urine pH. The addition of lemon juice and vinegar lowers pH, whereas the addition of basic substances such as bleach, detergents, and soap increases pH.

The pH of urine also determines the suitability of the specimen for immunoassay testing. Most enzyme immunoassays (for example, EMIT) require the pH to be within a range of 5–8. When a specimen is determined to be outside of these limits, the manufacturers of immunoassays recommend that the pH be adjusted back to these limits with the addition of either acid or base. However, pH adjustment introduces foreign substances and diluents to the original sample or aliquot and is therefore not recommended for forensic testing. SAMHSA does not recommend pH adjustments of urine before analysis.

Glutaraldehyde

Laboratory procedures for the analysis of glutaraldehyde are available. A commercial glutaraldehyde assay is available through Chimera Research and Chemicals, Inc. (Seminole, FL). The assay makes use of the Tollen's reagent. In the classical reaction, aldehydes react with the reagent to reduce silver, producing the corresponding acid and elemental silver (40). In the Chimera kit, a modification is made such that a colorimetric endpoint is produced.

Nitrite

Nitrites can be detected using colorimetric assays adapted to automated chemistry analyzers. In one commercial assay (Chimera), nitrites react with 0.2% sulfanilamide and 0.65% naphthylene-ethylene-diamine to form a diazonium salt, which couples with an indicator to yield a colored product. The nitrite detection limit is 75 mg/L, well below that concentration intended for adulteration, and well above that expected for nitrites produced by urinary bacterial infections. In a study conducted by Urry et al., urine specimens positive by the urinalysis strip test had nitrite concentrations of <15 mg/L. Specimens positive by culture for nitrate-reducing microorganism were <36 mg/L, and specimens from patients on medications that might metabolize to nitrite were <6 mg/L (41). In contrast, the SAMHSA cutoff for samples adulterated with nitrites is 500 mg/L. The nitrite test on a urine dipstick can also be used for routine urinalysis. This dipstick is designed for urine nitrite concentrations that are much lower than what is adulterated by commercial urine adulterants.

Oxidizing Adulterants

Dipstick indicators for urine adulterants that act by oxidation such as chromate and peroxidase are currently being developed. These tests act on the oxidizing capability of the tested adulterants and will only be useful while the oxidants themselves are reactive.

GENETIC MARKER ANALYSIS FOR IDENTITY

Once the substitution of urine has escaped detection by the collectors, the only manner by which toxicology laboratories can determine whether the urine sample is from the donor in question is to perform genetic testing. Such testing is also important when claims of sample mix-ups are made by donors who produce positive results for drugs of abuse.

Medical review officers and attorneys must be aware of the options available to them. Currently, the best approach for urine identity testing is DNA analysis. DNA can be extracted from squamous and transitional epithelial cells excreted into urine from the lining of the urinary tract (41). Urine from females has a higher excretion rate than males, due to vaginal shedding. Results of urine typing can be compared against a fresh blood sample where the DNA is extracted from leukocytes collected in lavender tubes containing ethylenediaminetetraacetic acid (EDTA). Extracted DNA is amplified by polymerase chain reaction (PCR). A commercial dot-blot assay is available and widely used for forensic purpose (PM+DQA1, Perkin Elmer, Foster City, CA). This assay requires very little DNA (<1 ng), and short DNA sequences (<240 bases) can be tested. The application of this PCR dot-blot assay has been evaluated on forensic urine samples (42). The ability to type urine was not affected by the presence of biological substances such as albumin, glucose, and *Escherichia coli* or in vitro adulterants such as bleach and saponin (43). DNA is extracted from the pellet of a centrifuged urine sample, thereby permitting the supernatant to be retested for the drugs of abuse without recovery losses.

REFERENCES

1. Johnson CA, Cary P: Intentional adulteration of urine specimens for drugs of abuse testing to produce false positive results. *J Anal Toxicol* 1990; 14:195–196.
2. George S, Braithwaite RA: An investigation into the extent of possible dilution of specimens received for urinary drugs of abuse screening. *Addiction* 1995;90:967–970.
3. More people drinking fluids to avoid testing positive. *Drug Detect Report* 1993;3(9):5.
4. Department of Health and Human Services: Mandatory Guidelines for Federal Workplace Drug Testing Programs; final guidelines notice. *Fed Regist* 1994;59:29908–29931.
5. Department of Transportation: Procedure for Transportation Workplace Drug Testing Programs; final rule. *Fed Regist* 2000;65:79461–79479.
6. Person N, Ehrenkrantz JR: Fake urine samples for drug analysis: hot, but not hot enough. *JAMA* 1988;259:841.
7. Cody JT: Adulteration of urine specimens. In: Liu RH, Goldberger BA, eds: *Handbook of workplace drug testing*. Washington DC: AACC Press, 1995: 181–207.
8. Morgan JP: Affidavit. In: Zeese KB, ed: *Drug testing legal manual*. New York: Clark Boardman, 1987.
9. ElSohly MA, Stanford DF, ElSohly HN: Coca tea and urinalysis for cocaine metabolites. *J Anal Toxicol* 1986;10:256.
10. Baselt RC, Cravey RH: *Disposition of toxic drugs and chemicals in man*, 4th ed. Foster City, CA: Chemical Toxicology Institute, 1995:44–46.
11. Done AK, Aronow R, Miceli JN, et al: Pharmacokinetic observations in the treatment of phencyclidine poisoning. In: Rumack BH, Temple AR, eds: *Management of the poisoned patient*. Princeton, NJ: Science Press, 1997.
12. Wagener RE, Linder MW, Valdes R: Decreased signal in Emit assays of drugs of abuse in urine after ingestion of aspirin: potential for false-negative results. *Clin Chem* 1994;40:608–612.
13. Brunk SD: False negative GC/MS assay for carboxy THC due to ibuprofen interference. *J Anal Toxicol* 1988;12:290–291.
14. Inloes R, Clark D, Drobnies A: Interference of fluorescein, used in retinal angiography, with certain clinical laboratory tests. *Clin Chem* 1987;33: 2126–2127.
15. Tamayo CL, Tena T: High concentration of metronidazole in urine invalidates EMIT results. *J Anal Toxicol* 1991;15:159.
16. Schwarz RH, Bogema S: Ingestion of megadoses of ascorbic acid will not produce 'clean' urine from marijuana smokers. Arch Path Lab Med 1988; 112:769.
17. Hayes LW, Krasselt WG, Mueggler PA: Concentrations of morphine and codeine in serum and urine after ingestion of poppy seeds. *Clin Chem* 1987;33: 806–808.
18. Bowie LJ, Kirkpatrick PB: Simultaneous determination of monoacetylmorphine, morphine, codeine, and other opiates by GC/MS. *J Anal Toxicol* 1989; 13:326–329.
19. ElSohly MA, Jones AB: Origin of morphine and codeine in biological fluids. In: Liu RH, Goldberger BA, eds: *Handbook of workplace drug testing*. Washington DC: AACC Press, 1995:225–237.
20. Struempler RE, Nelson G, Urry FM: A positive cannabinoids workplace drug test following the ingestion of commercially available hemp seed oil. *J Anal Toxicol* 1997;21:283–285.
21. Fortner N, Fogerson R, Lindman D, et al: Mari-

juana-positive urine test results from consumption of hemp seeds in food products. *J Anal Toxicol* 1997;21:476–481.

22. Camara PD, Audette L, Velletri K, et al: False-positive immunoassay results for urine benzodiazepine in patients receiving oxaprozin (Daypro). *Clin Chem* 1995;41:115–116.

23. Levine BS, Smith ML: Effects of diphenhydramine on immunoassays of phencyclidine in urine. *Clin Chem* 1990;36:1258.

24. O'Connor E, Ostheimer D, Wu AHB: Limitations of forensic urine drug testing by methodology and adulteration. *Therapeutic Drug Monitoring and Toxicology In-Service Training and Continuing Education* 1993;14:277–288.

25. Sansom HL, Freser MD, Botelho C, et al: Detection of urine specimens adulterated with UrinAid. 23rd annual meeting of the Society of Forensic Toxicologists, October 1993, Phoenix, AZ. Abstract 39.

26. George S, Braithwaite RA: The effect of glutaraldehyde adulteration of urine specimens on Syva EMIT II drugs-of-abuse assays. *J Anal Toxicol* 1996;20:195–196.

27. Tsai JSC, ElSohly MA, Tsai SF, et al: Investigation of nitrite adulteration on the immunoassay and GC-MS analysis of cannabinoids in urine specimens. *J Anal Toxicol* 2000;24:708–714.

28. Wu AHB, Sexton K, Cassella G, et al: Adulteration of urine by "Urine Luck." *Clin Chem* 1999;45:1051–1057.

29. Linder MW, Valdes R: Mechanism and elimination of aspirin-induced interference in Emit II d.a.u. assays. *Clin Chem* 1994;40:1512–1516.

30. Cody JT, Schwarzhoff RH: Impact of adulterants on RIA analysis of urine for drugs of abuse. *J Anal Toxicol* 1989;13:277–284.

31. Kim HJ, Cerceo C: Interference by NaCl with the EMIT method of analysis for drugs of abuse. *Clin Chem* 1976;22:1935–1936.

32. Duc TV: EMIT tests for drugs of abuse: interference by liquid soap preparations. *Clin Chem* 1985;31:658–659.

33. Pearson SD, Ash KO, Urry FM: Mechanism of false-negative urine cannabinoid immunoassay screens by Visine eyedrops. *Clin Chem* 1989;35:636–638.

34. Wu AHB, Forte E, Casella G, et al: The CEDIA assays for screening drugs of abuse in urine and the affect of adulterants. *J Forensic Sci* 1995;40:614–618.

35. Baiker C, Serrano L, Lindner B: Hypochlorite adulteration of urine causing decreased concentration of Δ^9-THC-COOH by GC/MS. *J Anal Toxicol* 1994;18:101–103.

36. Cassella G: In vivo and in vitro interferences in drug testing. MS Thesis, University of Connecticut, Storrs, CT, 1998.

37. Stolk LM, Scheijen JL: Urine adulteration with denture-cleaning tablets. *J Anal Toxicol* 1997;21:403.

38. Graff L: *A handbook of routine urinalysis*. Philadelphia: Lippincott, 1982:11.

39. Luceri F, Godi F, Messeri G: Reducing false-negative tests in urinary drugs-of-abuse screening. *J Anal Toxicol* 1997;21:244–245.

40. Shiner RL, Fuson RC: *The systematic identification of organic compounds. A laboratory manual.* New York: Wiley, 1948:145.

41. Urry FM, Komaromy-Hiller G, Staley B, et al: Nitrite adulteration of workplace urine drug testing specimens. I. Sources and associated concentrations of nitrite in urine and distinction between natural sources and adulteration. *J Anal Toxicol* 1998;22:89–95.

42. Tsongalis G, Anamani DE, Wu AHB: DNA fingerprinting for identification of urine specimen donors by polymerase chain reaction amplification typing of the HLA DQA locus. *J Forensic Sci* 1996;41:1031–1034.

43. Linfert DR, Wu AHB, Tsongalis G: The effect of pathologic substances and adulterants on the DNA typing of urine by the PM1 + DQα assay kit. *J Forensic Sci* 1998;43:1041–1045.

SELF-ASSESSMENT QUESTIONS

1. Which immunoassay is least affected by in vitro adulteration?
   a. enzyme-multiplied immuno technique
   b. fluorescence polarization immunoassay
   c. radioimmunoassay
   d. cloned enzyme-donor immunoassay
   e. all about the same

2. According to the Substance Abuse and Mental Health Services Administration, what values of specific gravity and/or creatinine are consistent with a substituted sample?
   a. creatinine <20 mg/dL only
   b. specific gravity <1.003
   c. creatinine ≤5 mg/dL and specific gravity ≤1.001
   d. creatinine ≤5 mg/dL or specific gravity ≤1.001
   e. creatinine ≤5 mg/dL or specific gravity ≥1.020

3. Which form of in vitro adulterant is characterized by low tetrahydrocannabinol (THC) concentrations on immunoassay screening and low or negative THC and deuterated THC internal standards by gas chromatography–mass spectroscopy (GC/MS)?
   a. acid/base
   b. oxidants
   c. surfactants
   d. fixative
   e. none of the above

4. Why were household cleaners the first in vitro adulterants to be used?
   a. They were inexpensive.
   b. They were available at a moment's notice.
   c. They were effective for all drugs.
   d. Their presence could not be detected by the laboratory.
   e. They were especially effective on point-of-care testing devices.

5. Which is true for in vivo adulterants?
   a. They can lead to serious health effects.
   b. They are easier for the laboratory to detect.
   c. They can act by interfering with immunoassays.
   d. They can act by interfering with GC/MS assays.
   e. They are not available on the World Wide Web.

6. Which of the following is not an example of an inadvertent adulteration that can lead to a true-positive urine result for a drug of abuse?
   a. consumption of poppy seed products
   b. secondhand smoke of a marijuana cigarette
   c. consumption of hemp seed oil
   d. dermal exposure of benzocaine

CHAPTER 12

Volatile Alcohols: Ethanol, Methanol, and Isopropanol

Thomas G. Rosano, Ph.D., DABCC, DABFT

LEARNING OBJECTIVES

After completing this chapter, the reader should be able to:

1. List the sources and chemical properties of the volatile alcohols.
2. Identify the clinical features of acute intoxication with volatile alcohols and of chronic intoxication with ethanol.
3. Describe the toxicokinetics of volatile alcohols and apply principles to estimations of dose and blood concentrations.
4. Associate the laboratory findings with ethanol, methanol, and isopropanol intoxication.
5. Describe the laboratory methods for measurement of the volatile alcohols.
6. Specify the laboratory's role in the clinical management of volatile alcohol poisoning.

OVERVIEW

Ethanol, methanol, and isopropanol are volatile alcohols that have similar chemical properties and share a capacity to cause human intoxication and even death. Approximately 4% of the 1.3 million nonpharmaceutical exposures reported in 1999 to the American Association of Poison Control Centers (AAPCC) were due to the volatile alcohols. Alcohols rank sixth as a cause of fatal poisonings; only analgesics, antidepressants, stimulants, cardiovascular drugs, and sedatives and hypnotics cause more deaths (1). The incidence of ethanol intoxication predominates, with >50% of the adult population of the United States reported to be ethanol users. Even with a current trend toward decreasing per capita consumption, ethanol in combination with other drugs is involved in 35% of drug-related emergency department (ED) visits and in 37% of drug-related deaths (2,3). The economic effect of ethanol use in the United States alone was $184.6 billion for 1998 (4).

The clinical toxicology laboratory plays an important role in assessment and treatment of patients intoxicated with volatile alcohols. Severe intoxication with ethanol, methanol, or isopropanol can lead to respiratory depression and convulsions, but the presentation of intoxication differs in clinical symptoms, metabolic disturbance, lethal dose, and other laboratory findings. Table 12–1 outlines and compares clinical and toxicologic essentials of acute poisoning with volatile alcohols. Chronic intoxication with ethanol is an additional sociomedical problem that affects a significant portion of the population. A wide range of laboratory tests are used in detection, monitoring, and management of acute volatile alcohol poisoning as well as in chronic intoxication with ethanol. An interpretation of routine and toxicology laboratory data is based on information and principles presented in this case-oriented chapter.

TABLE 12–1. Comparison of clinical and toxicologic parameters of volatile alcohols

	Ethanol	Methanol	Isopropanol
Odor	+	−	+ (acetone)
Central nervous system depression	+	+	+
Convulsion	+	+	+
Lethal dose (approximate)	5–6 g/kg (3 g/kg peds)	1–5 g/kg	3–4 g/kg
Threshold of lethal blood concentration (approximate)	350–500 mg/dL	150–200 mg/dL	128–200 mg/dL
Time to peak blood concentration without a meal	15–90 min	Similar to ethanol	Similar to ethanol
Half-life in blood ($T_{1/2}$)	2–14 h (zero order)	2–24 h (zero order)	3–6.5 h (first-order)
Volume of distribution (V_d)	0.6–0.7 L/kg	0.6–0.7 L/kg	0.6–0.7 L/kg
Metabolic acidosis	None to mild lactic acid or ketoacidosis	Delayed, severe acidosis from formic acid	None to mild lactic acid acidosis
Metabolite	Acetaldehyde	Formic acid	Acetone (also CNS depressant)
Anion gap	±	+	±
Osmolal gap	+	+	+
Osmolal equivalent of 1 mg/dL alcohol	0.23 mOsmol/kg	0.34 mOsmol/kg	0.17 mOsmol/kg

+, present; −, absent; peds, pediatric patients; CNS, central nervous system.

ETHANOL INTOXICATION

Case Presentations

Case 1: Altered Mental Status in a Chronic Alcoholic

A 65-year-old homeless man was transported to the ED after being found on the roadside in an incoherent state with signs of vomiting. Medical history was significant for alcohol abuse and a recent ED visit for intoxication and facial lacerations. On physical exam the patient was a thin, unkempt white male for whom vital signs were measured, including blood pressure (158/62 mmHg), pulse (100 beats/min), respiration (28 breaths/min) and temperature (98 °F; 36.7 °C).

His skin showed erythema of the palm and face, without evidence of angiomata or gynecomastia. The sclera of the eyes were clear, with evidence of bilateral orbital abrasions. The neurological exam revealed an arousable, lethargic patient, disoriented to person, place, and time. The patient also displayed inappropriate behavior and was incontinent. The abdominal examination was benign.

At the bedside, a fingerstick glucose concentration was 121 mg/dL (6.7 mmol/L), and the blood alcohol estimation by breath testing was 56 mg/dL (12 mmol/L). Hematology tests showed an elevated white blood cell count of $10.3 \times 10^3/\mu L$ ($10.3 \times 10^9/L$), a mean corpuscular volume of 76 µm³ (96.6 fL), and a mean corpuscular hemoglobin of 33.5 pg, with a normal hematocrit of 45.0% (0.45) and platelet count (228/mm³).

Blood chemistries revealed normal sodium 138 mEq/L (138 mmol/L), potassium 3.8 mEq/L (3.8 mmol/L), chloride 99 mEq/L (99 mmol/L), total carbon dioxide 25 mEq/L (25 mmol/L), anion gap 14 mEq/L (14 mmol/L), blood urea nitrogen (BUN) 17 mg/dL (6.1 mmol/L), creatinine 0.8 mg/dL (71 µmol/L), total bilirubin 1.4 mg/dL (24 µmol/L), lipase 54 mIU/mL (54 U/L), and aspartate aminotransferase 21 U/L, in addition to a serum ethanol concentration of 69 mg/dL (15 mmol/L) and a mild elevation in osmolality [310 mOsm/kg (310 mmol/L)] and amylase (190 U/L). Screening for other volatile alcohols was negative.

After the initial evaluation, a computer tomography of the head was performed and showed right frontotemporal contusion with right chronic, subdural hemorrhage, including an acute component. The patient was admitted with a diagnosis of closed head injury and received initial treatment with Dilantin (phenytoin) [300 mg by mouth (P.O.) each day], thiamine (200 mg P.O.), folate (1 mg P.O.), and multivitamins (P.O. each day).

Although the laboratory information in this case did not lead to the diagnosis, laboratory testing was essential in ruling out potential causes of the patient's altered mental state.

SELF-ASSESSMENT QUESTIONS SET 1

The following self-assessment questions may assist in the analysis of case 1:

1. Does the serum ethanol concentration account for the degree of altered mental status?

2. What is the relation between serum and breath ethanol measurements, and do the measurements agree in this case?

3. Is the osmolal gap consistent with the serum ethanol concentration?

4. What is the metabolic role of thiamine, and why is it indicated in this case?

Case 2: Mouthwash Poisoning in a Child

In this published case report (5), a 33-month-old female child was found in a stuporous state by her parents, and an empty 16-ounce (473-mL) mouthwash bottle (18.5% ethanol vol./vol.) was found nearby with approximately 11 ounces (325 mL) of mouthwash unaccounted for by the parents. The child received intravenous 5% dextrose en route to the hospital, and on admission, the 13.1-kg child was comatose and responsive only to deep pain. Vital signs included blood pressure (88/50 mmHg), pulse (125 beats/min), and respiration (28 breaths/min).

Laboratory findings 2 hours after the ingestion included sodium 139 mEq/L (139 mmol/L), potassium 3.6 mEq/L (3.6 mmol/L), chloride 106 mEq/L 106 mmol/L), total carbon dioxide 9 mEq/L (9 mmol/L), BUN 24 mg/dL (8.6 mmol/L), glucose 97 mg/dL (5.4 mmol/L), serum ethanol 310 mg/dL (67 mmol/L), arterial blood gas pH 7.18, pCO_2 25 mmHg, and pO_2 102 mmHg.

The child was treated with gastric lavage and administration of intravenous fluids with bicarbonate supplement. Eight hours after ingestion the serum ethanol concentration was 130 mg/dL (28 mmol/L), and after 18 hours the child was responding normally with normal blood gases and electrolytes.

SELF-ASSESSMENT QUESTIONS SET 2

The following self-assessment questions may assist in the analysis of case 2:

1. Estimate the oral dose of ethanol in 11 ounces (325 mL) of a mouthwash with 18.5% vol./vol. ethanol.

2. How does the amount of ethanol in 11 ounces (325 mL) of a mouthwash containing 18.5% ethanol compare with that in an alcoholic beverage?

3. What is the dose of ethanol per kilogram of body weight in this case, and how does this dose relate to the range of fatal dose reported in the literature?

4. What is the estimated amount (grams) of ethanol in the body of a 13.1-kg female child with a V_d of 0.70 and a blood ethanol concentration of 310 mg/dL (3.10 g/L)?

Sources of Ethanol

Alcoholic beverages including beers, wines, and distillate spirits contain ethanol in the concentration range of 0.5% to 95%. Beers are produced by fermenting cereal grains including barley, rice, and wheat. The content of ethanol in commercially produced beers is variably labeled by manufacturers, and a recent study by Logan and co-workers of >400 beer and malt beverages showed significant variability in alcohol concentration (6). Ales averaged 5.51% (vol./vol.) ethanol but ranged from 2.9% to 12.7%. Lagers averaged 5.32%, with progressively increasing mean concentrations in light (4.13%), ice (6.07%), and malt liquors (7.23%). Malt beverages averaged 4.26% content of ethanol, and nonalcoholic beers containing <0.5% ethanol are also available commercially.

Wines are commonly produced from fermented grapes and range in ethanol content from 7% to 14%, unless fortified by the addition of ethanol distillates. Distilled alcoholic beverages have ethanol content that exceeds the natural fermented concentration and include brandy with 40–43% ethanol and liquors with 40–95%. Distilled beverages that are sweetened or flavored are called liqueurs. Although consumption of alcoholic beverages is common, intoxication with nonbeverage products containing ethanol may also occur with clinical consequences as presented above in the second case study. The ethanol content of representative products are shown in Table 12–2.

Epidemiology and Health-Related Consequences of Ethanol Use

Surveillance data on alcohol use, abuse, and health-related consequences are available in the *Tenth Special Report to the United States Congress on Alcohol and Health* (4). Figure 12–1 shows the per capita consumption trends in the United States from 1977 to 1997. Ethanol usage has been on a downward trend for all beverage types since 1982, with the greatest decrease in the use of distilled products. Survey data also show a decrease in the rate of heavy drinking and an increase in abstention rates. Although the rate of ethanol use has decreased, results from the 1999 National Household Survey

TABLE 12–2. *Ethanol content of ethanol in some nonbeverage products*

Mouthwashes	10–27% vol./vol.
Aftershaves	15–80%
Rubbing alcohols	70–90%
Paint strippers	25%
Perfumes and colognes	25–95%
Dishwashing detergents	1–10%
Denatured alcohols	90–99.9%
Glass cleaners	10%
Elixirs and cough preparations	2–25%
Hair tonics	25–60%
Solid can fuels	>60%
Extracts	40–90%

on Drug Abuse estimate there are 105 million alcohol users over the age of 12, with about 45 million users engaged in binge drinking (2). The prevalence of abuse and dependence is age dependent, as shown in Figure 12–2, and is greatest among individuals between 18 and 29 years of age.

Ethanol-related morbidity and mortality result primarily from trauma and chronic disease. Ethanol-associated injuries include motor vehicle accidents, drowning, air crashes, occupational injury, criminal violence, and suicide. Traffic fatalities are a leading cause of death for people up to 35 years of age, and in 1998, 30% of traffic fatalities involved a driver or nonoccupant with a blood ethanol concentration of ≥0.10 g/dL (2.2 mmol/L) (7). Figure 12–3 shows the consistent contribution of ethanol use to fatal accidents from 1977 to the most recent report for 1998. Ethanol is also involved in approximately 50% of fire fatalities, and use of ethanol is also a major risk factor in fatal drowning (8). Ethanol involvement is also common in homicide, suicide, and a substantial portion of domestic violence (9). In violent crimes, 24% of the 11.1 million crimes committed annually involve an offender who had been drinking alcohol before the offense (10). In addition to traumatic injury, ethanol use is also causally linked to diseases that include hepatitis; cirrhosis; cancer of the mouth, larynx, and esophagus; cardiomyopathy; hypertension; arrhythmia; Korsakoff's syndrome; and fetal alcohol syndrome (4,11).

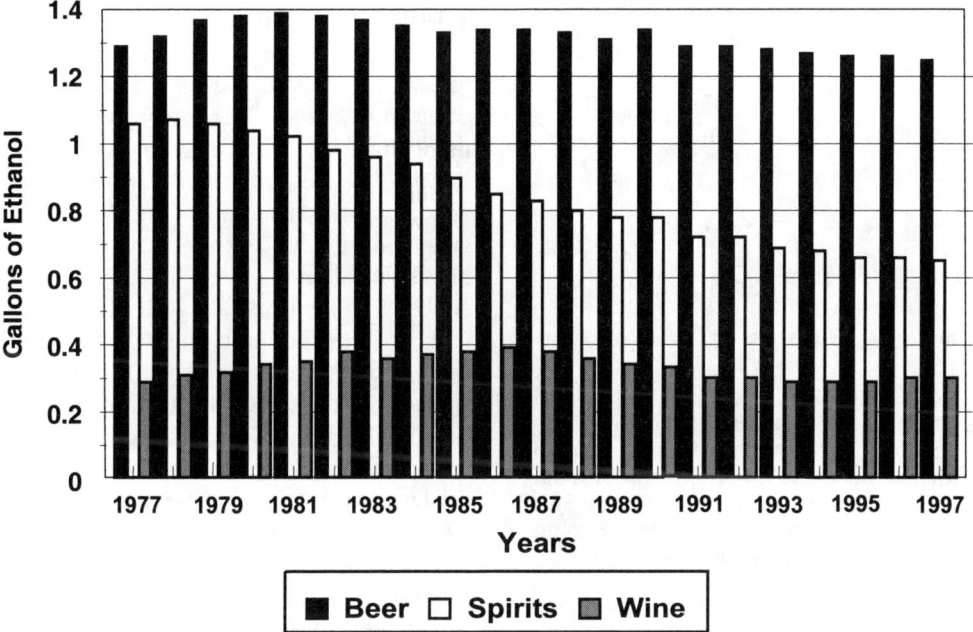

FIG. 12–1. United States per capita consumption trends in ethanol from 1977 to 1997. Data from Surveillance Report #51, National Institute on Alcohol Abuse and Alcoholism, Division of Biometry and Epidemiology, U.S. Department of Health and Human Services, Public Health Service, National Institutes of Health, December 1999.

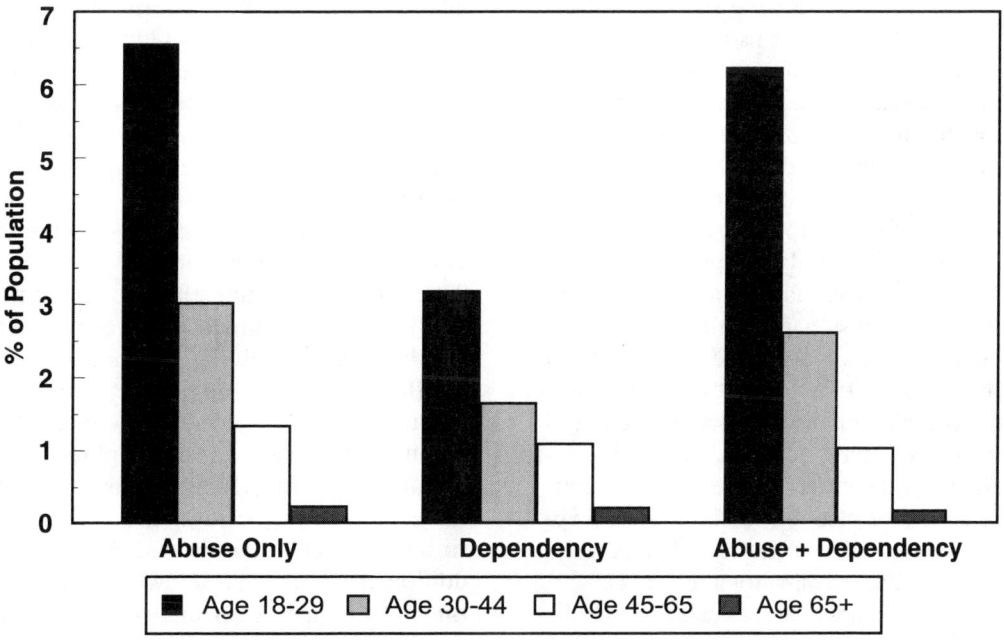

FIG. 12–2. Prevalence of ethanol abuse and dependence. Adapted from epidemiology data in the Ninth Special Report to the U.S. Congress on Alcohol and Health from the Secretary of Health and Human Services, U.S. Department of Health and Human Services, Public Health Service, National Institutes of Health, National Institute on Alcohol Abuse and Alcoholism, June 1997.

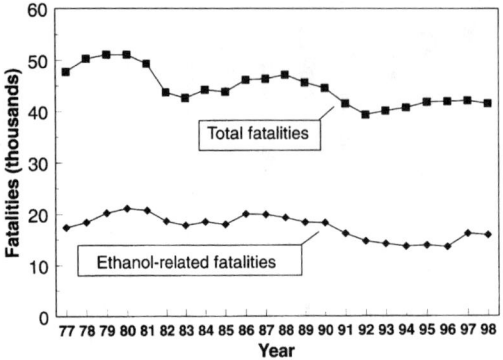

FIG. 12–3. Contribution of ethanol use to fatal accidents from 1977 to 1998. Adapted from epidemiology data in the Ninth Special Report to United States Congress on Alcohol and Health from the Secretary of Health and Human Services, U.S. Department of Health and Human Services, Public Health Service, National Institute on Alcohol Abuse and Alcoholism, June 1997 and Annual Reports of the National Traffic Highway Safety Administration, Fatality Analysis Reporting System.

Ethanol Toxicodynamics in Acute Intoxication

Alcohol produces a wide spectrum of effects on the central nervous system (CNS) by interaction with specific receptors on neuronal membranes. Alcohol is a CNS depressant with effects proportional to the ethanol concentration in blood. The acute intoxication effect on the CNS is characterized by confusion, ataxia, impaired judgment, and emotional lability and may progress to stupor, coma, and eventually death. The altered brain function results primarily from effects on neuronal receptors, producing reduced excitatory activity of the neurotransmitter glutamate and enhanced inhibitory action of the γ-aminobutyric acid (GABA) and glycine neurotransmitters (12,13). The potentiating effects of alcohol on serotonin receptors may also mediate the rewarding effects that lead to chronic alcohol abuse (14).

Ethanol tolerance does occur and may be due to both an increased rate of metabolism (metabolic tolerance) and a resistance to ethanol effects at the cellular level (functional tolerance). For example, during a drinking episode, a more pronounced CNS effect occurs during the rise in blood ethanol concentration due to the acute functional tolerance (Mellamby effect). Chronic heavy drinkers may have less performance impairment than abstainers or occasional drinkers because of metabolic and functional tolerance. Deterioration in driving skills that result from alcohol's CNS depressant effects has been demonstrated at blood ethanol concentrations of ≤ 0.05 g/dL (1.1 mmol/L), with progressive impairment at higher concentrations. Effects on reaction time, visual tracking, mental concentration, attention time, information processing, perception, and psychomotor function may be involved in alcohol-related driving impairment. Table 12–3 shows the acute effects of ethanol on nonalcoholics with progressive increase in blood alcohol concentration.

The most life-threatening effects of severe ethanol intoxication are respiratory and cardiovascular. Respiratory depression may lead to pulmonary failure in severe intoxication, and asphyxiation from aspiration of vomitus may also occur. Prominent depression of cardiovascular function occurs with high doses of ethanol, and atrial fibrillation as well as atrial-ventricular block may occur from acute overdose. Moderate doses of ethanol cause a minor increase in pulse rate and blood pressure, as well as an increase in high-density lipoprotein (HDL) cholesterol concentrations in blood.

Ethanol has a wide spectrum of acute effects on other organ systems. With external application to the skin, ethanol is an astringent (irritant) and rubefacient (coolant, bactericidal), whereas internal use induces vasodilation with redness and flushing of the face and other areas. Acute intoxication also impairs eye movement causing diplopia, nystagmus, and altered vision that may contribute to impaired performance in critical visual tasks such as motor vehicle operation. In the gastrointestinal (GI) system consumption of beverages with ethanol concentrations >20% stimulates gastric secretions, whereas concentrations >40% may causes direct irritation of the GI tract. These effects are influenced by food intake, drinking patterns, and tolerance level. Other organ sys-

TABLE 12–3. *Acute intoxication effects of ethanol*

Blood ethanol concentration (g/dL)	Acute intoxication effects
0.02–0.05	Most individuals experience mild euphoria but may have no obvious signs. Diminished fine motor function occurs.
0.05–0.15	Euphoria with reduction in judgment, motor function, and reaction time
0.15–0.30	Obvious signs of intoxication including impaired balance, speech, reaction time, emotional stability, vision, and comprehension
>0.25	Marked loss in motor function and impaired consciousness, associated with death in uncomplicated adult cases
>0.40	Respiratory depression and concentrations at which most fatalities occur

To convert g/dL to mmol/L, multiply the value by 21.71.

Note: It is illegal to operate a motor vehicle in all states with a presumptive blood ethanol concentration of >0.10 g/dL. The presumptive concentration is 0.08 g/dL in at least 18 states, including Alabama, California, Florida, Hawaii, Idaho, Illinois, Kansas, Maine, New Hampshire, New Mexico, New York, North Carolina, Oregon, Texas, Utah, Vermont, Virginia, and Washington.

tem effects of acute ingestion include skeletal myopathies with rhabdomyolysis and nonpancreatic hyperamylasemia of salivary origin (15). Ethanol is also a diuretic that inhibits renal antidiuretic hormone secretion, especially during the rise in blood ethanol concentration.

Acute ethanol intoxication can also cause metabolic effects. A reversible fatty infiltration of the liver in acute overdose is due to overloading hepatic metabolic function. The hypoglycemia that may result from acute ethanol ingestion may lead to seizures and coma, especially in children, and dietary and metabolic factors that lead to a low liver glycogen content may predispose children as well as adults to an ethanol–induced hypoglycemic episode. The metabolic acidosis may occur (16) as demonstrated in case study 2, in which the child presented to the ED with a severe acidosis [pH <7.2 and bicarbonate <10 mEq/L (10 mmol/L)]. The lactic acid acidosis may be due to depletion of nicotinamide-adenine dinucleotide from ethanol metabolism and inhibition of hepatic gluconeogenesis.

Ethanol Toxicodynamics in Chronic Intoxication

The heart and GI system are primary targets of chronic ethanol use, and long-term heavy drinking may lead to cardiomyopathy with ventricular enlargement and dysfunction. Congestive heart failure and sudden death may result from heavy ethanol use over time. Recent studies point to toxic effects of alcohol on myocytes from reactive oxygen species and detrimental changes in surface receptors (17,18). Coagulopathies may also result from hepatic and nutritional disorders. It has been shown, however, that deaths from cardiovascular disease may be reduced by 30–40% for men and women who consume one or more drinks daily compared with nondrinkers, but the potential benefit vs. risk is greatest with moderate ethanol use (11).

In the GI system, chronic alcohol abuse is associated with cancer of the mouth, esophagus, pharynx, larynx, and liver. Ethanol can also cause hepatitis, acute and chronic pancreatitis, and cirrhosis. Cirrhosis may lead to portal hypertension, which results in esophageal varices, ascites, and splenomegaly. The incidence of cirrhosis over time does correlate with the per capita use of ethanol, as shown in Figure 12–4. Alcoholic use is also linked to increased risk of acetaminophen liver toxicity, due possible to alcohol-induced cytochrome P450 2E1 in Kupffer cells (19).

Neurologic status may also be affected by chronic heavy use of alcohol. In long-term alcoholics there is a continuum of brain damage from moderate deficits to severe psychosis (20). Thiamine deficiency, which occurs in

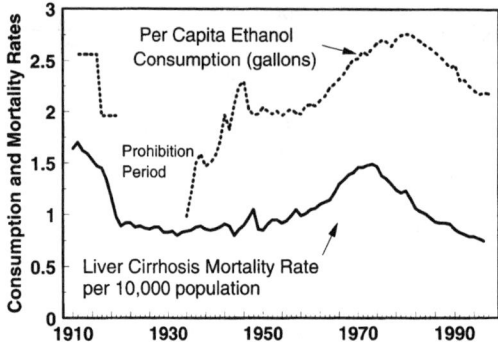

FIG. 12–4. Correlation of cirrhosis with per capita use of ethanol. Data from Surveillance Reports Number 51 and 52, National Institute on Alcohol Abuse and Alcoholism, Division of Biometry and Epidemiology, U.S. Department of Health and Human Services, Public Health Service, National Institutes of Health, December 1999.

poorly nourished alcoholics, is the cause of Wernicke's encephalopathy. Korsakoff's psychosis may also develop and is characterized by antiretrograde amnesia with the inability to retain new information. Alcoholic polyneuropathy may result from the combination of nutritional deficiency and direct ethanol toxicity in chronic heavy ethanol users. Other major CNS effects of ethanol include cerebral degeneration, dependency, and withdrawal with delirium, tremors, and seizures.

Alcoholism is also associated with endocrine and hematologic manifestations. The most common endocrine manifestations of alcoholism include male hypogonadism, gynecomastia, decreased libido, impotence, palmar erythema, and spider angiomata. Hypoglycemia and alcoholic ketoacidosis are important metabolic abnormalities in the acute care setting. Anemia is common and may be due to folate and B_{12} deficiencies (megaloblastic) or iron deficiency. Thrombocytopenia, vitamin K deficiency, and coagulopathies also may result from chronic ethanol abuse.

Heavy maternal alcohol consumption during pregnancy may lead to fetal alcohol syndrome, which is characterized by congenital facial abnormalities, prenatal growth retardation, and damage to the developing CNS resulting in neurodevelopmental abnormalities. Alcohol exerts its effect on the fetus at multiple sites including direct effects on nerve, bone, and cartilage development and indirect effects on placental function and blood flow.

Ethanol Toxicokinetics

Absorption of ethanol in the GI tract occurs by passive diffusion across a concentration gradient that depends on contact time, alcohol concentration, blood flow to the region, and surface area. Limited absorption of ethanol occurs in the mouth and esophagus, with up to 25% absorption in the stomach and >75% absorption in the small intestine in the fasting individual. Delayed absorption may occur with food intake, fatty foods, nausea, shock, anticholinergic agents, fear, panic, stimulants, gastric fibrosis, emotional disturbances, and GI irritation from high ethanol concentrations (>40%). Absorption is most rapid at ethanol concentrations of 10–30% and is increased by factors that increase GI motility, blood flow, and gastric emptying. Figure 12–5 summarizes a study of the time required to reach peak blood ethanol concentration after rapid consumption of ethanol on an empty stomach.

The V_d of ethanol in individuals with the same body weight is affected by gender (male

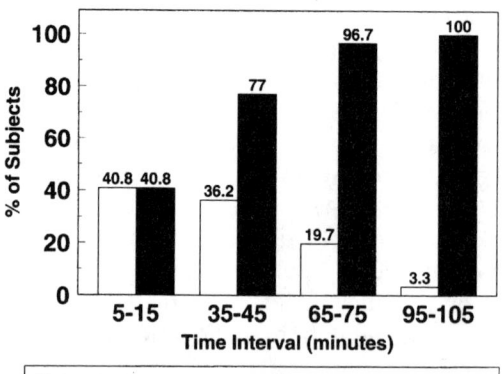

FIG. 12–5. A study of time required to reach peak blood ethanol concentration after rapid consumption. Adapted with permission from Jones AW, Jonsson KA, Neri A: Peak blood-ethanol concentration and the time of its occurrence after rapid drinking on an empty stomach. *J Forensic Sci* 1991;36:376–385.

TABLE 12–4. *Distribution of ethanol in body tissue and fluid compared with blood*

Specimen	Multistudy average	Range of averages	Number of studies
Urine	1.29	1.01–1.44	15
Serum or plasma	1.16	1.12–1.18	4
Vitreous humor	1.14	0.99–1.34	8
Saliva	1.13	1.10–1.20	4
Cerebrospinal fluid	1.08	0.92–1.18	4
Skeletal muscle	0.90	0.89–0.91	2
Brain	0.84	0.62–1.24	9
Kidney	0.66	—	1
Liver	0.60	0.56–0.63	3

Adapted from Garriott JC, ed. *Medicolegal aspects of alcohol*, 3rd ed. Tuscon, AZ: Lawyers & Judges Publishing Co.,1996.

> female), age (decreases V_d), and adiposity (decreases V_d). V_d averages 0.62–0.79 L/kg for males and 0.55–0.66 L/kg for females (21). The V_d (L/kg) is ~6% higher than Widmark's r value due to the density of blood (~1.06). Table 12–4 summarizes the average ratio of ethanol in body tissue and fluids to whole blood. The difference between serum ethanol concentration and that in whole blood is a function of the relative water content of the specimens. The ethanol dose as well as the concentration in body fluids and tissue can be estimated by the application of several toxicokinetic formulas, as shown in this chapter's Appendix.

The clearance rate of ethanol from blood ranges from 10 to 23 mg/dL (2.2 to 5.0 mmol/L) per hour and occurs under zero-order kinetics except at very low blood concentrations [<0.03 g/dL [0.65 mmol/L)] and possibly at very toxic concentrations. Alcohol dehydrogenase (ADH) is the rate-limiting step in the major pathway for metabolism of ethanol under normal conditions, as shown in Figure 12–6. Class I ADH is the main hepatic enzyme involved in oxidation of ethanol to acetaldehyde, and the enzyme operates under zero-order kinetics (independent of substrate) at blood ethanol concentrations >20–30 mg/dL (4.3–6.5 mmol/L).

Genetic polymorphisms of class I ADH may explain the interindividual variation in the rate of ethanol metabolism. Class IV ADH is expressed in human stomach and has much higher k_m value for ethanol than does class I ADH. The role of gastric ADH in first-pass metabolism and sex-related differences in the rate of ethanol metabolism remains unresolved. Oxidation of acetaldehyde to the nontoxic metabolite acetate is catalyzed by aldehyde dehydrogenase (ALDH). ALDH2 in mitochondria exhibits polymorphism. An inactive allelic form of ALDH2 is expressed in ~50% of Chinese and Japanese and may result in acetaldehyde accumulation and unpleasant effects (flushing, nausea) associated with ethanol use.

The microsomal ethanol-oxidizing system (MEOS) is a minor pathway of ethanol metabolism under normal conditions, but heavy use of ethanol induces the P-450 gene (CYP2E1) responsible for metabolism

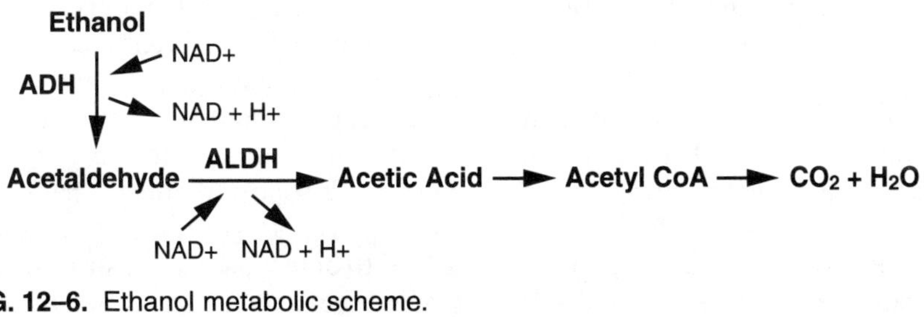

Ethanol → (ADH, NAD+ → NAD + H+) → Acetaldehyde → (ALDH, NAD+ → NAD + H+) → Acetic Acid → Acetyl CoA → $CO_2 + H_2O$

FIG. 12–6. Ethanol metabolic scheme.

through the MEOS. Nontolerant adults metabolize ~7–10 g of ethanol per hour, resulting in a decrease in blood ethanol of 10–23 mg/dL (2.2–5.0 mmol/L) per hour. With induction of MEOS, clearance of blood ethanol may increase to >30 mg/dL (6.5 mmol/L) per hour. CYP2E1 is polymorphic, and a rare allele (c2) has higher base-line activity and may contribute to liver damage through increased generation of free radicals from ethanol metabolism (22).

More than 90% of ethanol is eliminated by metabolism, and the remainder is eliminated unchanged along with water in urine, breath, saliva, and perspiration. Although the urine to blood ethanol concentration ratio averages 1.3, there is considerable intra- and interindividual variation in the ratio due to factors including diuresis and stage of ethanol disposition in the body. Urine ethanol concentrations therefore may not correlate with the blood ethanol concentration at a specific point in time. Breath and blood ethanol concentrations do correlate, and the relationship is described by Henry's law, which states that a dissolved volatile (ethanol in blood) when in contact with a closed air space (alveolar air) forms an equilibrium, which results in a constant ratio of blood to alveolar air ethanol at a given temperature. At an oral temperature of 93.2 °F (34 °C), the blood to breath ratio of ethanol is ~2100, and breath alcohol determination is widely used in law enforcement, emergency medicine, and government-regulated testing under US Department of Transportation regulations.

Methods of Ethanol Analysis

Specimens for ethanol measurement include blood, serum, plasma, urine, saliva, breath, vitreous humor, cerebral spinal fluid, and tissue. The primary specimens differ for clinical (serum, plasma), law enforcement (breath, whole blood), and postmortem (whole blood, vitreous humor, urine, tissue) applications. Blood tubes should be filled and kept stoppered during processing to prevent volatilization of ethanol. Blood collected under sterile conditions without preservatives should be analyzed within 4 hours. For storage up to 2 days at 41 °F (5 °C), addition of potassium oxalate monohydrate (5.0 mg/mL of blood) plus sodium fluoride (1.5 mg/mL of blood) is recommended, and storage may be extended indefinitely if samples are maintained at temperatures of ≤–4 °F (≤–20 °C). For up to 2 days of transport in a nonrefrigerated condition, a higher concentration of sodium fluoride (10 mg/mL of blood) is recommended (23). Do not use alcohol swabs or pads to prepare venipuncture sites for blood collection for either clinical or forensic purposes.

There is a wide range of methods for ethanol analysis, varying in specificity and cost. Microdiffusion methods that rely on the volatility of ethanol and oxidation by a chromic acid reagent are relatively time consuming, nonspecific, and primarily of historical value. The osmolal gap, determined by the difference between the measured and calculated osmolality, is readily determined in the clinical laboratory but is nonspecific because other low-molecular-weight chemicals may produce an osmolal gap.

Gas chromatography (direct injection and headspace analysis) and enzymatic methods are the primary methods currently used for ethanol measurement in fluid samples. Gas chromatography allows simultaneous determination of ethanol and other volatiles including methanol, isopropanol, and acetone (24). The breath alcohol analysis is used as a noninvasive technique of determining the alcohol content of blood. Direct breath alcohol analysis is performed with either infrared spectrometry or electrochemical oxidation (fuel cell) detection techniques. Analysis of expired air for ethanol has been used for >50 years, and instruments used in the United States are calibrated in units of grams of ethanol per 210 L of breath, corresponding to grams of ethanol per 100 mL of blood. For law enforcement or US Department of Transportation monitoring purposes, breath alcohol is reported only to the first two decimal places. Passive breath sensors may be used to detect alcohol in air from the vicinity of a driver's mouth or from the interior of an

TABLE 12–5. Methods of alcohol measurement used at the Albany Medical Center

Technique (sample)	Method essentials	Application
GC/FID direct injection (serum, plasma)	1:10 sample dilution Glass column 5% carbowax Column temp 90 °C Internal standard n-propanol	Clinical detection and quantification of methanol, ethanol, isopropanol, and acetone
Enzymatic Abbott Axsym (serum)	Radiative energy attenuation ADH/diaphorase coupled	Clinical quantification of ethanol
GC/FID head space (whole blood, urine, vitreous humor, tissue)	1:10 sample dilution Dual capillary column Column temp 40 °C Internal standard n-propanol	Clinical and forensic detection and quantification of methanol, ethanol, isopropanol, and acetone
Fuel cell electrochemical (breath)	Electrochemical oxidation of ethanol to acetic acid, producing electrons and detectable current flow	Evidential breath ethanol testing by certified breath alcohol technicians for regulated testing under DOT standards

GC/FID, gas chromatography with flame ionization detection; DOT, Department of Transportation; ADH, alcohol dehydrogenase.

automobile. Fluid and breath analysis of ethanol has been reviewed (21). Table 12–5 lists the methods currently used in a university-based clinical and forensic toxicology laboratory where clinical, medical examiner, and US Department of Transportation testing for ethanol is performed.

Units of measurement for ethanol vary in clinical and forensic testing protocols. Ethanol concentration in fluid specimens is commonly reported in clinical toxicology as mg/dL, in forensic toxicology as g/dL or gram percent (wt./vol.), and in the literature as mmol/L. For conversion of units, 100 mg/dL is equivalent to 0.10 g/dL, 0.10% (wt./vol.), and 21.7 mmol/L.

Other Laboratory Measurements

Other useful laboratory determinations in the evaluation and management of ethanol intoxication include glucose (hypoglycemia), electrolytes (hypokalemia, anion gap), blood gases (acidosis), magnesium (hypomagnesemia), creatine kinase (rhabdomyolysis), BUN, creatinine (renal function), alanine aminotransferase, aspartate aminotransferase, bilirubin (liver function), osmolality (osmolar gap), amylase (pancreatitis), qualitative urinary ketones (ketosis), coagulation studies (liver function, coagulation disorders), and hemogram (anemia, leukopenia, thrombocytopenia). As an indicator of recent alcohol use, the ratio of 5-hydroxytryptophol to 5-hydroxyindole-3-acetic acid in urine provides a specific and more sensitive method of detecting recent alcohol consumption than does ethanol (25). Biological markers to screen for alcoholism including carbohydrate-deficient transferrin, γ-glutamyltransferase, and mean corpuscular volume (26,27). More recently, studies have shown that platelet adenylyl cyclase activity discriminates between subjects with and without alcohol dependence in a population of subjects who had not consumed significant quantities of ethanol recently (28).

METHANOL INTOXICATION

Case Presentations

Case 3: Methanol Poisoning and Treatment

In this published case report from the Albany Medical Center (29), a 45-year-old 80-kg white male with a history of chronic ethanol abuse, but with no other significant medical history, was hospitalized after a reported ingestion of two-thirds of a gallon (~2.5 L) of windshield-wiper solvent (30% methanol) 20 hours before admission. On

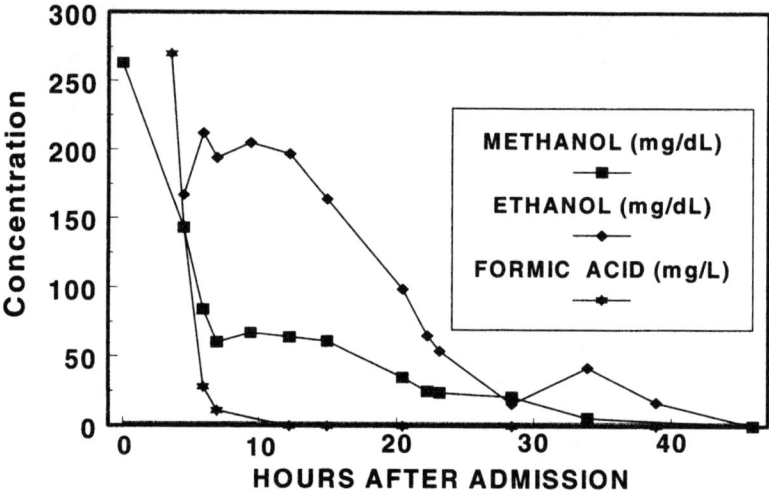

FIG. 12–7. Serial monitoring of methanol, ethanol, and formic acid in a case of methanol poisoning. Reprinted with permission from Burns AB, Bailie GR, Eisele G, et al: Use of pharmacokinetics to determine the duration of dialysis in management of methanol poisoning. *Am J Emerg Med* 1998:16;538–540.

admission, the patient was unconscious and severely hypothermic (92 °F; 33.3 °C), with fixed and dilated (3-mm) pupils. Mild papilledema was noted on ophthalmic exam.

Admission laboratory results included sodium 154 mEq/L (154 mmol/L), potassium 4.4 mEq/L (4.4 mmol/L), chloride 106 mEq/L (106 mmol/L), total carbon dioxide 6 mEq/L (6 mmol/L), anion gap 46 mEq/L (46 mmol/L), osmolality 465 mOsm/L (465 mmol/L), BUN 14 mg/dL (5.0 mmol/L), creatinine 1.6 mg/dL (141 µmol/L), glucose 225 mg/dL (12.5 mmol/L), and creatine kinase 4662 U/L. Arterial blood gases on 50% O_2 revealed a pH of 6.76, pCO_2 38 mmHg, and pO_2 180 mmHg. An initial alcohol screen was positive for methanol at a serum concentration of 260 mg/dL (82 mmol/L). Ethanol, isopropanol, and ethylene glycol were not detected. The initial formate concentration, measured after initial treatment, was 27 mg/dL (5.9 mmol/L).

Initial treatment included a 100-g intravenous ethanol loading dose followed by 25 g/h intravenous infusion and 650 mEq of intravenous sodium bicarbonate over a 4-hour period. High-flux hemodialysis was started 2 hours after admission for an initial duration of 5.5 hours, during which time the ethanol infusion was increased to 50 g/h. Two subsequent 4-hour dialyses were performed to attain a methanol concentration of <50 mg/dL (16 mmol/L). Serial monitoring of methanol, ethanol, and formic acid for the patient is shown in Figure 12–7.

The clinical toxicology laboratory played a vital role throughout this entire treatment course. Serial methanol and ethanol measurements were performed to evaluate the adequacy and time frame of treatment. Measurement of serum formate concentrations is also valuable in cases of methanol poisoning, but methods are not readily available in the emergency setting. In this case, serum formate concentration was measured retrospectively.

SELF-ASSESSMENT QUESTIONS SET 3

The following self-assessment questions may assist in the analysis of case 3:

1. What is the osmolal gap, and is the gap consistent with the concentration of methanol at the time of admission?

2. What is the relative clearance rate of methanol and formic acid during hemodialysis?

3. What is the role of ethanol treatment?

4. What is the clearance rate of methanol with and without hemodialysis?

5. What is the severity and cause of acidosis in the case?

Methanol Intoxication Statistics and Sources

In 1999 the AAPCC reported 976 exposures to methanol, including 6 fatalities (1). Methanol is used primarily as a solvent for industrial purposes and may be found in a wide range of consumer products, including gasline antifreezes (99–100% vol./vol.), windshield-washer fluids (17–99%), windshield de-icers (4–89%), duplicating fluids (60–90%), paint removers (3–50%), model airplane fuels (43–77%), carburetor fluids (1–38%), ethanol denaturants (2–5%), solid can fuels (<4%), and glass cleaners (1–40%).

Methanol Toxicodynamics and Toxicokinetics

Acute methanol exposure by ingestion or inhalation may cause severe intoxication. Initial clinical findings may include nausea, abdominal pain, neurologic signs of lethargy, confusion, and an ethanol-like hangover. After a latent period, severe untreated cases may progress to anion gap metabolic acidosis, coma, seizure, and ultimately respiratory and circulatory failure. Visual signs include reduced visual acuity and pronounced whiteness in the visual field (snow field effect), with funduscopic signs of peripapillary edema and hyperemia. Metabolic accumulation of formic acid is a major cause of methanol toxicity. In addition to causing a metabolic acidosis, formic acid inhibits cytochrome oxidase and leads to histotoxic hypoxia. The toxic optic neuropathy, especially, is produced by formic acid. Lactic acid accumulation is sometimes found in methanol-poisoned patients, but is due to methanol-induced hypotension and is not a direct metabolic effect.

Methanol is rapidly absorbed by dermal, respiratory, or GI routes. A peak blood concentration of methanol is reached within 30–60 minutes after ingestion, and methanol is rapidly distributed in the body water with a V_d similar to ethanol (~0.6–0.7 L/kg). Approximately 75–85% of methanol is metabolized (Figure 12–8), with 10–20% excretion by the lungs and 3–5% excreted by the kidneys. Methanol clearance from the blood follows zero-order kinetics with a half-life of 2–24 hours, and the half-life can be extend beyond 30 hours with co-ingestion or infusion of ethanol (30). Formaldehyde is a toxic metabolite but has a half-life of only a few minutes or less. Formic acid elimination follows zero-order kinetics with a half-life up to 2–24 hours, resulting in significant clinical toxicity from formate accumulation. Elimination enhancement by dialysis significantly reduces the half-life of methanol (<3.5 hours) and especially formate (45–165 minutes), whereas therapeutic administration of ethanol competes for the ADH catalytic site and limits formate production.

Laboratory Measurements

The gas chromatographic techniques for ethanol may also be used to detect and quan-

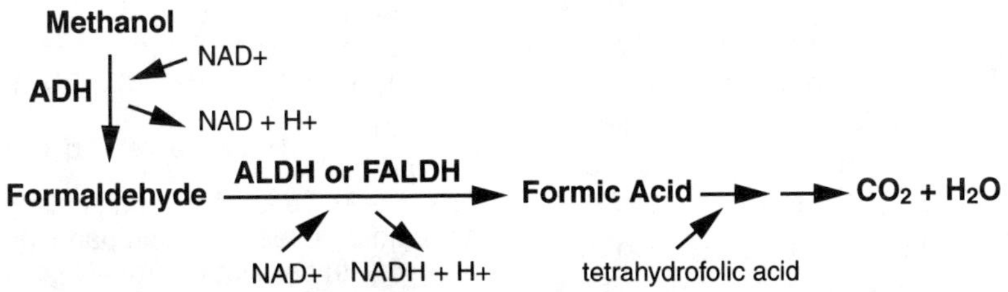

$$\text{Methanol} \xrightarrow[\text{ADH}]{\text{NAD+} \to \text{NAD + H+}} \text{Formaldehyde} \xrightarrow[\text{NAD+} \to \text{NADH + H+}]{\text{ALDH or FALDH}} \text{Formic Acid} \xrightarrow{\text{tetrahydrofolic acid}} CO_2 + H_2O$$

FIG. 12–8. Methanol metabolic scheme.

tify methanol in plasma, serum, or whole blood. Concomitant measurement of methanol and ethanol by gas chromatography is used routinely in the diagnosis and treatment of methanol overdose. Low serum or plasma concentration of methanol [<5 mg/dL (<0.156 mmol/L)] may result from ethanol abuse (inhibition of endogenous methanol metabolism) or from consumption of denatured ethanol or congeners in alcoholic beverages. Methanol concentrations >25 mg/dL (>0.78 mmol/L) are generally considered toxic, but the degree of toxicity is better related to the formate concentration in blood. Analysis of formic acid can be performed by gas chromatography (31), but is not generally available in most clinical toxicology laboratories.

Osmolality should be measured by a freezing-point depression method. In determining osmolal gap, it should be understood that there is a wide variation in the "normal" osmolal gap and the formulas used in calculating osmolality. Caution should be used in the interpretation of these calculations (32,33). Although a normal osmolal gap [for example, <20 mOsm/kg (<20 mmol/L)] cannot be used to rule out intoxication, an increased gap does suggest the presence of a toxic alcohol or alcohols.

ISOPROPANOL INTOXICATION

Case Presentations

Case 4: Isopropanol Intoxication

In this published case report (34), a 46-year-old black woman was brought to the emergency room in Grade V coma without a medical history. The patient had a fruity breath, no signs of trauma, rectal temperature of 95 °F (35 °C), pulse rate of 100 beats/min, 24 respiration/min, and a blood pressure of 100/80 mmHg. Neurological exam revealed equal, slowly reactive 2-mm diameter pupils, flaccid extremities, no response to pain, symmetrically diminished deep tendon reflexes, and flexed toes. The remainder of the physical exam was unremarkable, and the patient was transferred to the intensive care unit.

Admission laboratory values included arterial blood pH of 7.35, pCO$_2$ 41 mmHg, O$_2$ 99 mmHg (room air), white blood cell count 11.8 × 10^3/μL (11.8 × 10^9/L), hematocrit 41% (0.41), sodium 140 mEq/L (140 mmol/L), potassium 3.7 mEq/L (3.7 mmol/L), chloride 106 mEq/L (106 mmol/L), total carbon dioxide 24 mEq/L (24 mmol/L), anion gap 10 mEq/L (10 mmol/L), BUN 80 mg/dL (28.6 mmol/L), creatinine 10 mg/dL 884 μmol/L), glucose 141 mg/dL (7.8 mmol/L), and negative serum ketones. The admission ethanol, isopropanol, and acetone concentrations in serum were 13 (2.8), 200 (33), and 12 (2.1) mg/dL (mmol/L), respectively. Serial isopropanol and acetone concentrations are shown graphically in Figure 12–9. Nicotine and diphenhydramine were the only other drugs identified by the serum and urine drug screens in this case. A chest roentgenogram revealed atelectasis at the

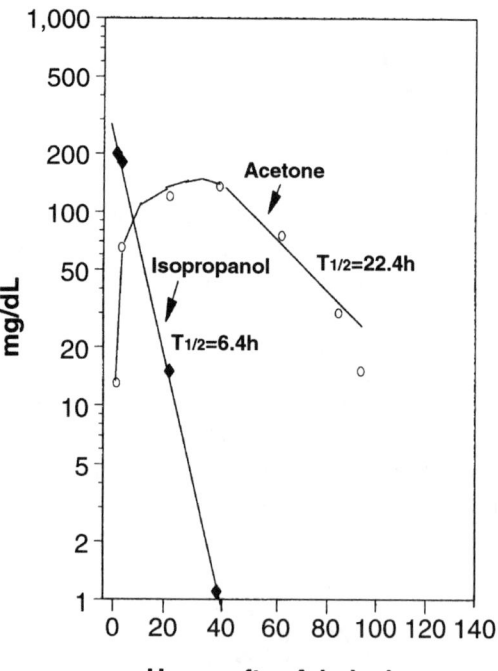

FIG. 12–9. Serial monitoring of isopropanol and acetone in a case of isopropanol poisoning. Reprinted with permission from Natowicz M, Donahue J, Gorman L, et al: Pharmacokinetic analysis of a case of isopropanol intoxication. *Clin Chem* 1985;31:326–328.

lung bases, and both an electrocardiogram and a computer tomography of the head were normal.

The patient received supportive management. On the day after admission, she gradually awakened and showed spontaneous extremity movement but did not obey commands or respond to deep pain. Two days after admission, the patient was alert and responsive and was identified as a chronic schizophrenic who had missed her last dose of antipsychotic medication.

SELF-ASSESSMENT QUESTIONS SET 4

The following self-assessment questions may assist in the analysis of case 4:

1. How would the level of CNS depression in this case compare with ethanol intoxication at a similar blood ethanol concentration?

2. How do the elimination kinetics of isopropanol and acetone compare?

3. What effect would co-ingestion of ethanol have on isopropanol elimination and serum acetone concentrations?

Isopropanol Intoxication Statistics and Sources

Approximately 19,000 isopropanol exposures, including 1 fatality, were reported to the AAPCC in 1999 (1). Isopropanol is used in some rubefacients (rubbing alcohol 70–90% vol./vol.) and may also be present in significant amounts in some ethanol denaturants (5%), de-icers (70–80%), glass cleaners (1–14%), cements (5–20%), paint strippers (2–11%), and several skin and hair products.

Isopropanol Toxicodynamics and Toxicokinetics

The effects of acute intoxication with isopropanols are primarily GI and CNS. Abdominal pain, vomiting, gastritis, and hematemesis may occur. Isopropanol produces a greater CNS depression than equivalent ethanol concentrations in blood, and acetone also contributes to the central depression. Ingestion or inhalation may result in inebriation, headache, lethargy with rapid development of CNS depression, and prolonged coma in severe poisonings. Hypotension and hypothermia may also occur. Ketosis without acidosis is common.

Isopropanol is rapidly absorbed from respiratory or GI routes. A peak blood concentration of isopropanol is reached within 30–60 minutes after ingestion, and isopropanol is rapidly distributed in the body water with a V_d similar to that of ethanol (0.6–0.7 L/kg). Isopropanol elimination follows first-order kinetics with a reported half-life ranging from 3 to 6.4 hours (34,35). Approximately 80% of absorbed isopropanol is metabolized, with excretion of the remainder by the kidneys and to a small extent in saliva and intestines. The route of isopropanol metabolism is shown in Figure 12–10. Acetone, the major metabolite, is excreted primarily by the kidneys but also expired through the lungs. Acetonemia may also be produced in starvation ketosis, and endogenous formation of isopropanol with serum concentrations up to 30 mg/dL have been observed in acetonemic patients with diabetes mellitus (36).

After isopropanol intoxication, the acetone concentration in blood depends on the time after ingestion and the co-administration of ethanol (37). In the case study, the

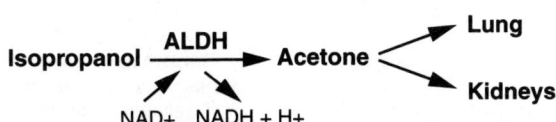

FIG. 12–10. Isopropanol metabolic scheme.

serum concentrations of isopropanol and acetone were consistent with a recent and significant isopropanol ingestion, with an isopropanol concentration of 200 mg/dL (33 mmol/L) and an acetone concentration of 12 mg/dL (2.1 mmol/L). Acetonemia without acidosis is common in isopropanol intoxication, as was demonstrated in the case study. Acidosis may, however, occur not as a result of the production of an acid metabolite but due to the hypotension and hypoperfusion of tissue that can lead to a lactic acid acidosis.

Laboratory Measurements

The gas chromatographic techniques cited for ethanol and methanol may also be used to detect and quantify isopropanol and acetone.

REFERENCES

1. Litovitz TL, Klein-Schwartz W, White S, et al: 1999 annual report of the American Association of Poison Control Centers Toxic Exposure Surveillance System. *Am J Emerg Med* 2000;18:517–574.
2. National Household Survey on Drug Abuse: SAMHSA Series H-3 (http://www.samhsa.gov/OAS/household99.htm).
3. Drug Abuse Warning Network: *Detailed emergency department (ED) tables 1999 and annual medical examiner data 1998* (http://www.samhsa.gov/OAS/).
4. National Institute on Alcohol Abuse and Alcoholism: *Tenth special report to United States Congress on alcohol and health from the Secretary of Health and Human Services.* Washington, DC: U.S. Department of Health and Human Services, Public Health Service, National Institutes of Health, 2000 (http://silk.nih.gov/silk/niaaa/publication/10report).
5. Weller-Fahy ER, Berger LR, Troutman WG: Mouthwash: a source of acute ethanol intoxication. *Pediatrics* 1980;66:302–305.
6. Logan BK, Cave GA, Distefano S: Alcohol content of beer and malt beverages: forensic considerations. *J Forensic Sci* 1999;44:1292–1295.
7. National Highway Traffic Safety Administration: *Traffic safety facts 1998: alcohol.* Washington, DC: U.S. Department of Transportation, National Center for Statistics and Analysis, 1998.
8. Howland J, Hingson R, Levensen S, et al: Alcohol use and aquatic activity: Massachusetts. *JAMA* 1993;264:19–20.
9. Murdoch D, Pihl RO, Ross D: Alcohol and crimes of violence: present issues. *Int J Addict* 1990;25:1065–1081.
10. Greendeld LA: *Alcohol and crime: an analysis of national data on the prevalence of alcohol involvement in crime.* Report prepared for Assistant Attorney General's National Symposium on Alcohol Abuse and Crime. Washington, DC: U.S. Department of Justice, 1998.
11. Thun MJ, Peto R, Lopez AD, et al: Alcohol consumption and mortality among middle-aged and elderly U.S. adults. *N Engl J Med* 1997;337:1705–1714.
12. Diamond I, Gordon AS: Cellular and molecular neuroscience of alcoholism. *Physiol Rev* 1997;77:1–20.
13. Korpi ER: Role of $GABA_A$ receptors in the actions of alcohol and in alcoholism: recent advances. *Alcohol Alcohol* 1994;29:115–129.
14. Lovinger DM, Zhou Q: Alcohols potentiate ion current mediated by recombinant $5-HT_3RA$ receptor expression in a mammalian cell line. *Neuropharmacology* 1994;33:1567–1572.
15. Block RS, Weaver DW, Bowman DL, et al: Acute alcohol intoxication: significance of the amylase level. *Ann Emerg Med* 1983;12:294–296.
16. Leung AK: Ethyl alcohol ingestion in children: a 15-year review. *Clin Pediatr* 1986;25:617–619.
17. Husain K, Somani SM: Response of cardiac antioxidant system to alcohol and exercise training in the rat. *Alcohol* 1997;14:301–307.
18. Strasser RH, Nuchter I, Rauch B, et al: Changes in cardiac signal transduction systems in chronic ethanol treatment preceding the development of cardiomyopathy. *Herz* 1996;21:232–240.
19. Koivisto T, Mishin VM, Mak KM, et al: Induction of cytochrome P4502E1 by ethanol in rat Kupffer cells. *Alcohol Clin Exp Res* 1996;20:207–212.
20. Butterworth RF: Pathophysiology of alcoholic brain damage: synergistic effects of ethanol, thiamine deficiency and alcoholic liver disease. *Metab Brain Dis* 1995;10:1–8.
21. Garriott JC, ed: *Medicolegal aspects of alcohol*, 3rd ed. Tucson, AZ: T Lawyers & Judges Publishing Company, Inc., 1996.
22. Tsutsumi M, Takada A, Wang JS: Genetic polymorphism of cytochrome P4502E1 related to development of alcohol liver disease. *Gastroenterology* 1994;107:1430–1435.
23. Dubowski KM: *Blood alcohol testing in the clinical laboratory. NCCLS Approved Guideline* T/DM6A. Vol. 17, No. 14. Wayne, PA: NCCLS, 1997.
24. Tagliaro F, Lubli G: Chromatographic methods for blood alcohol determination. *J Chromatogr* 1992;580:161–190.
25. Helander A, Beck O, Jones AW: Laboratory testing for recent alcohol consumption: comparison of ethanol, methanol, and 5-hydroxytryptohol. *Clin Chem* 1996;42:618–624.
26. Salaspuro M: Biological state markers of alcohol abuse. *Alcohol Health Res World* 1994;18:131–135.
27. Sibler H: Carbohydrate-deficient transferrin in serum: a new marker of potential harmful alcohol consumption reviewed. *Clin Chem* 1991;37:2029–2037.
28. Menninger JA, Baron AE, Conigrave KM, et al: Platelet adenylyl cyclase activity as a trait marker of alcohol dependence. WHO/ISBRA Collaborative Study Investigators. International Society for Biomedical Research on Alcoholism. *Alcohol Clin Exp Res* 2000;24:810–821.
29. Burns AB, Bailie GR, Eisele G, et al: Use of pharmacokinetics to determine the duration of dialysis in management of methanol poisoning. *Am J Emerg Med* 1998;16:538–540.

30. Palatnick W, Redman LW, Sitar DS, et al: Methanol half-life during ethanol administration: implications for management of methanol poisoning. *Ann Emerg Med* 1995;26:202–207.
31. Abolin C, McRae JD, Tozer TN, et al: Gas chromatographic head-space assay of formic acid as methyl formate in biological fluids: potential application to methanol poisoning. *Biochem Med* 1980;23:209–212.
32. Osterloh JD, Kelly TJ, Khayam-Bashi H, et al: Discrepancies in osmolal gaps and calculated alcohol concentration. *Arch Pathol Lab Med* 1996;120:637–641.
33. Demedts P, Theunis L, Wauters A, et al: Excess serum osmolality gap after ingestion of methanol: a methodology-associated phenomenon? *Clin Chem* 1994;40:1587–1590.
34. Natowicz M, Donahue J, Gorman L, et al: Pharmacokinetic analysis of a case of isopropanol intoxication. *Clin Chem* 1985;31:326–328.
35. Daniel DR, McAnalley BH, Garriott JC: Isopropyl alcohol metabolism after acute intoxication in humans. *J Anal Toxicol* 1981;5:110–112.
36. Bailey DN: Detection of isopropanol in acetonemic patients not exposed to isopropanol. *Clin Toxicol* 1990;28:459–466.
37. Kelner M, Bailey DN: Isopropanol ingestion: interpretation of blood concentrations and clinical findings. *J Toxicol Clin Toxicol* 1983;20:497–507.

APPENDIX

Toxicokinetic Formulae

1. Calculate blood ethanol concentration at time zero:

   $$C_p \text{ (g/L)} = [D \text{ (g)}]/[V_d \text{ (L/kg)} \times W \text{ (kg)}]$$

   where
   C_p = blood concentration at time zero
   D = dose of ethanol
   V_d = volume of distribution
   W = body weight

2. Calculate the amount of ethanol in the body:

   $$E_t \text{ (g)} = C_t \text{ (g/L)} \times V_d \text{ (L/kg)} \times W \text{ (kg)}$$

   where
   E_t = ethanol in body at time t
   C_t = blood ethanol concentration at time t
   V_d = volume of distribution
   W = body weight

3. Calculate the amount of ethanol in a drink:

   $$D \text{ (g)} = F \times \text{Vol (mL)} \times 0.8 \text{ (g/mL)}$$

   where
   D = grams of ethanol in the drink
   F = fraction of ethanol (% vol./vol.)
   Vol = volume in milliliters (1 ounce = ~30 mL)
   0.8 = approximate specific gravity of ethanol
   (actual specific gravity = 0.789 at 20 °C).

CHAPTER 12 ADDENDUM

Point-of-Care-Testing for Alcohol

Tai C. Kwong, Ph.D., DABCC

The use of point-of-care-testing (POCT) for alcohol (mostly breath alcohol, and to a more limited extent saliva alcohol) is gaining in popularity. Many clinical services such as emergency rooms, occupational health departments, drug-treatment centers, and psychiatry wards now routinely perform alcohol POCT. The primary reason for choosing POCT for alcohol over blood-alcohol testing provided by the central laboratory is one common to all POCTs—the advantage of rapid turn-around time. In outpatient clinics, having access to alcohol results in a few minutes can expedite management and disposition decisions. Performing an alcohol test in the presence of the patient and the instantaneous display of test result is a powerful tool in confronting a patient about his or her self-reported drinking status. An added advantage is that alcohol POCTs do not require venipuncture, thus eliminating the need for highly trained phlebotomists. Moreover, the simplicity of a breath- or saliva-alcohol test device means that more operators can be trained.

An alcohol POCT should have assay performance appropriate for its intended clinical use. Laboratorians should consult their clinical colleagues to determine the appropriate assay performance requirements. For example, in a drug-treatment clinic in which an alcohol test is used to check a patient's self-reported alcohol use, a positive or negative result may suffice. In some emergency departments, where the breath-alcohol test may be used to ensure legal sobriety before release of the patient, greater demand for test accuracy and precision will be expected. The use of the alcohol test for such a purpose has obvious legal implications even though testing is conducted in a clinical setting. The fine points of breath-alcohol test accuracy and its relationship to blood alcohol, hotly debated among forensic toxicologists, now become more relevant and should be taken into consideration. Decision on disposition of patients who have been drinking should be based on clinical criteria, not breath- or blood-alcohol concentrations.

BREATH-ALCOHOL TESTING

Henry's law is the basic principle upon which the breath-alcohol test is based (1). It states that in a closed system under constant temperature, the concentration of a gas in air above the liquid is proportional to the concentration of the gas dissolved in the liquid. The distribution of the gas in the gas and liquid phases will be determined by the partition ratio. In the breath-alcohol test, the assumption is that an equilibrium state exists between the alcohol in the blood perfusing the lung and that in the alveolar (breath). If the partition ratio for alcohol is known, blood-alcohol concentration can be determined indirectly by measuring alcohol concentration in breath. For forensic purposes, a partition ratio of 2100 has been adopted, although val-

ues have been reported to range from 830 to 9000 (2,3).

The breath-alcohol test has limitations due to deviations from Henry's law (1,4). The human body is not a closed system with constant temperature as required if Henry's law is to hold. Consequently, the concentration of alcohol in the alveolar is not necessarily the same as that in exhaled breath. Temperature changes affect the concentration of alcohol in the vapor phase; body temperature has a diurnal variation and different people have different normal temperatures, although 99.5 °F (37.5 °C) is often cited as the average (5,6). In addition, there is a negative temperature gradient of expired air from the initial internal body temperature of about 99.5 or 100.4 °F (37.5 or 38 °C) to 94.3 °F (34.6 °C) of the last part of exhaled breath (1).

The breath test assumes that the alcohol concentration in exhaled breath reaching the measuring device is the same as that in the lungs. But this does not always hold true because there is continuous exchange of alcohol in breath with fluid during passage through the airways on exhalation. The commonly assumed relationship of breath-alcohol concentration with breath volume has breath-alcohol concentration reaching a plateau at the end-expiratory breath. In reality, breath-alcohol concentration does not reach the constant concentration level (plateau) that is assumed to be the same as alveolar alcohol concentration. Instead, it continues to increase with a gentle slope. The conclusion is that there is re-equilibrium of alcohol in breath and the capillaries lining the airways (2).

Alcohol is more soluble in water than in lipids and other substances. Therefore, the hematocrit will affect breath-alcohol results because the partition ratio changes with the hematocrit. Blood with a higher hematocrit has a lower water content and a higher alcohol concentration in the aqueous component, giving a higher alcohol concentration in the vapor phase. Therefore, for a given blood-alcohol concentration, an individual with a higher hematocrit will have a higher breath-alcohol concentration than another individual with a lower hematocrit (4). Because a fixed partition ratio of 2100 is used by testing devices to calculate blood-alcohol concentration, a higher hematocrit will result in a higher blood-alcohol readout on the device. Normal variation in hematocrit of test subjects can produce error in the range of 10–14% (4). Therefore, a range of breath-alcohol results is associated with a given blood-alcohol concentration due to variation in hematocrit.

In addition to the physiological variables, the breath-alcohol test is also vulnerable to sampling errors. An insufficient collection will yield a breath sample that is not end expiratory breath. Sample contamination by residual alcohol in the mouth after recent consumption of alcoholic beverage or regurgitation of alcohol from the stomach can grossly distort breath test results because the actual quantity of alcohol measured in breath is very small. A 15-minute pretest deprivation period and careful subject preparation as part of the testing procedure can eliminate contamination problems and is the standard of practice in forensic breath testing (7). In many busy clinical areas, the 15-minute waiting period may not be feasible. If this important safeguard is bypassed, users must exercise caution in interpreting breath test results.

To obviate the problem of a breath-alcohol concentration that can be associated with a range of blood-alcohol concentrations, some forensic toxicologists have advocated dissociating breath alcohol from blood alcohol in forensic testing, as well as redefining legislation and driving impairment in terms of measured breath alcohol, not blood-alcohol concentration. They proposed reporting breath test result directly as breath alcohol in grams of alcohol per 210 liters of breath without any reference to the blood-alcohol concentration of the individual (1). It is not in the purview of this talk to assess the pros and cons of reporting breath-alcohol results in forensic testing. It is important to point out, though, that physicians and nurses are familiar with clinical laboratory results in concentration units per 100 mL of blood (or serum), not those based on 210 L of breath.

Breath-Alcohol Devices

Most of the breath-alcohol instruments currently in use utilize designs based on either electrochemical oxidation (fuel cell) or infrared spectrophotometry. The former are small, portable instruments preferred by users in clinical settings, whereas the latter instruments, particularly those meeting the requirements of evidentiary testing, are used more commonly in law-enforcement situations.

Electrochemical Oxidation (Fuel Cell)

In such a device, breath flows past an electrode to which a voltage is applied. Ethanol at the electrode surface is oxidized to acetic acid (via acetaldehyde) and the net movement of electrons (current) is proportional to ethanol concentration (1). These devices are subject to interference; methanol, isopropanol, n-propanol, and acetaldehyde are potential interfering substances, but acetone does not respond measurably.

These devices are small, portable, and easy to use, requiring minimal operator judgement or manipulation. Analysis is rapid, with a result available in 2 minutes. After repeated positive results, the recovery time to achieve base-line reading gets progressively longer due to delay in the slow oxidation of the accumulated acetic acid.

Infrared Spectrometry

Instruments based on infrared spectrometry are the most common instruments in use in enforcement of traffic laws (1,8). Different chemical bonds and functional groups absorb infrared radiation at characteristic wavelengths. The resulting spectrum can be used to identify and quantify the unknown compound. In breath analyzers, only two or three wavelengths are used, and these wavelengths are not unique to ethanol; many organic compounds have significant absorption at these wavelengths. The potential for interference by these other chemicals, however, is greatly reduced by the limited number of volatile compounds encountered in human breath. The major potential interference comes from acetone, which can accumulate to high concentrations in poorly controlled diabetic patients. In the design of the popular Intoxilyzer (CMI, Inc., Owensboro, KY) instrument, selectivity for ethanol is enhanced by the used of two wavelengths, 3.48 and 3.99 μm, corresponding to the vibrations of the methyl group. The former wavelength measures ethanol concentration in breath. The latter wavelength is for the detection of acetone, the presence of which will distort the relative absorption of ethanol at 3.48 and 3.39 μm; ethanol concentration is adjusted before the final reading is displayed.

There have been reported cases of false-positive results due to volatile compounds with methyl group or similar groups such as methylene and methine; various petroleum-based fuels (kerosene); cough drop vapor (menthol); tetrahydrofuran; and toluene (8). Therefore, a critical part of the interpretation of a positive breath-alcohol test is to assess whether patients have had prior exposure to potentially interfering substances.

Quality Assurance for Breath-Alcohol Test

POCT for alcohol in a clinical setting must satisfy the same quality-assurance and quality-control requirements mandated for other POCTs. Frequently, non-laboratory personnel, such as physicians, nurses, drug treatment counselors, or clerical staff, perform the clinical breath-alcohol tests. These professionals may have inadequate training and experience in test performance and may also lack appreciation for the importance of quality assurance for test reliability. In many locales, breath-alcohol tests, similar to other "off-site" clinical laboratory tests, may fall under licensure of the clinical laboratories, and clinical laboratory professionals may find themselves responsible for the breath-alcohol service.

Clinical laboratory professionals should communicate frequently with users of breath-alcohol tests and take into consideration their usage patterns, expectations, and recommendations in designing a quality-assurance program. Due to the long history and extensive use of forensic breath-alcohol test in traffic law enforcement, forensic scientists have in their possession well-established testing processes, government-approved instruments and personnel-training programs, and well-defined quality-assurance programs (7). Users of clinical breath-alcohol test can benefit from the collective experience of the forensic community in developing their own quality-assurance measures that can best meet the demands of the health-care environment.

A quality-assurance program for breath-alcohol test should include the following components: selection of appropriate breath-alcohol devices, proper system validation and deployment, protocols to monitor and evaluate the overall quality of the total testing process (preanalytic, analytic, and postanalytic), the effectiveness of policies and procedures, and operator-training and competency-assessment programs.

Selection of Breath-Alcohol Devices

It is strongly recommended that laboratories consider using only those devices listed in the National Highway Traffic Safety Administration (NHTSA) Conforming Product List (9). Selection should be based on features that best meet the requirements of the clinical services, and accuracy and precision should meet or exceed performance required for intended clinical use. These devices should be able to measure breath-alcohol concentration within a clinically relevant analytical range and should be able to prevent a false-positive result due to acetone up to 0.02% weight per volume (wt./vol.). It is recommended that breath-alcohol test results be displayed in concentration units of grams of alcohol per 1000 milliliters of blood and not as grams of alcohol per 210 liters of breath. In clinical settings, physicians and nurses are accustomed to interpreting medical laboratory test results in concentration units per 100 milliliters of blood.

Analytical System Validation and Deployment

The clinical laboratory has the responsibility for validating that device performance meets or exceeds specifications before release for clinical use at near-patient sites. NHTSA-approved breath-alcohol simulator and certified alcohol solutions or certified dry gas alcohol standards should be used for validation (10). Accuracy and precision studies should be performed at clinically relevant alcohol concentrations. Device specificity for breath alcohol (ethanol) should be challenged with aqueous solutions of volatiles (acetone, methanol, isopropanol) at concentrations likely to be encountered clinically. Claims for calibration stability must be verified by use of suitable quality-control (accuracy check) materials each day the device is used, preferably under routine operating conditions. Vapor delivered from a wet breath-alcohol simulator or a certified dry gas standard is suitable quality control material. The alcohol concentration indicated on the dry gas cylinder label must be corrected to account for altitude before use. This correction is accomplished by selecting a correction factor that corresponds to altitude from a table supplied by the manufacturer.

Quality Control Program

The quality-control program must monitor the performance of the device(s) and the operators who may use them. Each device must be checked for accuracy each day of use with a dry gas standard and an air blank. The air-blank challenge is frequently a design feature of the device, but a breath sample from an alcohol-free operator may be used to verify device performance in the analysis of an alcohol-free sample. The dry gas standard should have a certified ethanol concentration that is adjusted for atmospheric pressure. The recovery of alcohol must be within the accuracy tolerance established by the manufacturer. Unfortunately, no external proficiency survey program exists to assess accuracy and interlaboratory performance.

Essential Testing Procedures

Performance of a breath test should follow a series of procedural steps, each of which is critical to the reliability of test results.

- Use of test device under manufacturer-recommended environmental conditions
- Use of a properly calibrated device
- Verification that the blank and alcohol accuracy checks (quality control) recoveries are within specifications
- Use of an air check or blank breath test immediately before the patient test
- Confirmation of patient identification
- Observation of the patient to ascertain that residual alcohol and foreign objects are cleared from the mouth
- Instruction of patient on proper delivery of a deep-lung sample
- Documentation of test date, time, device, quality-control results, patient identification, and test results
- Prompt and accurate reporting of results

Administrative and Educational Components

A comprehensive, up-to-date, standard operating procedure manual for the breath-alcohol testing must be available and must include the laboratory's policies and complete instructions for procedures. The procedure manual must be available to, and used by, personnel in the laboratory and at the testing site. The training and competency evaluation of staff at the point of care is a challenge because the personnel are not laboratory staff and the testing site is not within the laboratory's organizational structure. Most breath-alcohol device manufacturers have NHTSA-approved training-program materials. These are valuable resources that should be supplemented and/or modified to conform to the clinical environment. The competency of approved operators should be evaluated periodically.

SALIVA-ALCOHOL TESTING

The use of saliva for alcohol measurement has considerable appeal relative to other biological samples. Its advantages are the noninvasive nature of sample collection, ready availability of saliva in sufficient quantity for analysis, and the ease of test performance.

Alcohol is a low-molecular-weight compound that does not bind to plasma proteins, is water-soluble, and is distributed to all body fluids in proportion to the water content of the body fluid. Theoretically, if whole blood and saliva water contents are 850 and 994 g/L water, respectively, the saliva-blood alcohol ratio should be 1.17. However, the experimentally determined value is 1.07. This discrepancy can amount to approximately 9% difference in blood-alcohol concentration calculated from saliva-alcohol concentration (11),

A commercially available device for measuring alcohol in saliva (STC Diagnostics, Bethlehem, PA) has been evaluated for clinical use; it is also approved by NHTSA as a screening device for alcohol in bodily fluid (the Q.E.D.A150 device only) (12). This device is based on enzymatic oxidation of alcohol by alcohol dehydrogenase. Saliva-alcohol concentration is read as a color bar against a concentration scale printed on the device. Saliva-alcohol results agree well with alcohol concentrations in venous blood (13,14) and end-expired breath (13,15) of healthy volunteers dosed with moderate amounts of alcohol.

In a clinical study conducted with patients admitted into a detoxification program, blood-alcohol concentration and breath-alcohol concentrations agreed very well ($r = 0.97$, $n = 52$) (16). Saliva-alcohol results, however, did not correlate with venous blood-alcohol concentrations as well, particularly those at high concentrations ($r = 0.75$, $n = 36$). The investigators attributed the problem to the difficulty of saliva collection from highly intoxicated patients. There were numerous failures in obtaining specimens, hence the smaller number of study saliva specimens compared with blood (36 vs. 52). Moreover, several of the samples failed the built-in sample volume-control mechanism designed to indicate sufficient amount of sample has been applied for proper functioning of the device. When sufficient volume of saliva was applied to the device, saliva-alcohol results agreed better with those of blood. Thus, the use of saliva alcohol is problematic with highly intoxicated patients. Saliva-alcohol testing can have a role in other clinical settings in which patients are less intoxicated or the intent of testing is to assess a patient's drinking history or the degree of intoxication without the need to strive for the accuracy required for forensic analysis.

REFERENCES

1. Dubowski KM: The technology of breath-alcohol analysis. [DHHS Publication No. (ADM) 92-1728.] Washington, DC: U.S. Department of Health and Human Services, National Institute on Alcohol Abuse and Alcoholism, 1992.
2. Hlastala MP: The alcohol breath test: a review. *J Appl Physiol* 1998;84:401–408.
3. Alobaidi TAA, Hill DW, Payne JP: Significance of variations in blood: breath partition coefficient of alcohol. *Br Med J* 1976;2:1479–1481.
4. Labianca DA: The chemical basis of the breathalyzer. *J Chem Educ* 1990;67:259–261.
5. Fox GR, Hayward JS: Effect of hypothermia on breath-alcohol analysis. *J Forensic Sci* 1987;32:320–325.
6. Fox GR, Hayward JS: Effect of hyperthermia on breath-alcohol analysis. *J Forensic Sci* 1989;34:836–841.
7. Dubowski KM: Quality assurance in breath-alcohol analysis. *J Anal Toxicol* 1994;18:306–311.
8. Labianca DA: Breath alcohol analysis via infrared spectrophotometry. In: Fitzgerald EF, ed: *Intoxication test evidence*, 2nd ed. Deerfield, IL: Clark, Boardman, Callagher, 1995:31-1–31-17.
9. National Highway Traffic Safety Administration: Highway safety programs; model specifications for devices to measure breath alcohol. *Fed Regist* 1999;64:30097–30100.
10. National Highway Traffic Safety Administration: Highway safety programs; model specifications for calibrating units for breath alcohol testers; con-

forming products list of calibrating units. *Fed Regist* 1997;62:43146–43125.
11. Jones AW: Inter- and intra-individual variation in the saliva/blood alcohol ratio during ethanol metabolism in man. *Clin Chem* 1979;25:1394–1398.
12. National Highway Traffic Safety Administration: Highway safety programs; conforming product list of screening devices to measure alcohol in bodily fluids. *Fed Regist* 1995;60:442214–442215.
13. Jones AW: Measuring ethanol in saliva with the QED enzymatic test device: Comparison of results with blood- and breath-alcohol concentrations. *J Anal Toxicol* 1995;19:169–74.
14. Christopher TA, Zeccardi JA: Evaluation of the QED saliva alcohol test: a new, rapid, accurate device for measuring ethanol in salvia. *Ann Emerg Med* 1992,21:1135–1137.
15. Bates ME, Martin CS: Immediate, quantitative estimation of blood alcohol concentration from saliva. *J Studies Alcohol* 1997:58:531–538.
16. Bendtsen, P, Hultberg J, Carlsson M, et al: Monitoring ethanol exposure in a child setting by analysis of blood, breath, saliva, and urine. *Alcohol Clin Exp Res* 1999;23:1446–1451.

CHAPTER 13

Glycols

Bryan A. Wolf, M.D., Ph.D., Anne F. Eder, M.D., Ph.D., and Leslie M. Shaw, Ph.D., DABCC

LEARNING OBJECTIVES

After completing this chapter, the reader should be able to:

1. Describe the epidemiology of ethylene glycol intoxication.
2. List the typical clinical features of ethylene glycol intoxication.
3. List the laboratory findings associated with ethylene glycol poisoning.
4. Describe the methods used to measure ethylene glycol concentrations.
5. Describe the role of the laboratory in the management of patients poisoned with ethylene glycol.

INTRODUCTION

Ethylene glycol, a member of the family of glycols, is an important toxicological problem in medical practice (1–3). Although the incidence of ethylene glycol poisoning is not very common, it can present various analytical challenges for the toxicology laboratory. Glycols are alcohols containing two hydroxyl groups (diols) and include ethylene glycol, diethylene glycol, propylene glycol, and glycerol (Table 13–1). Diethylene glycol has been implicated in various episodes of mass poisonings. Propylene glycol, a commonly used diluent in pharmaceutical products, can be associated with toxicity. Ethylene glycol intoxication is often compared with methanol poisoning because it produces similar signs and symptoms, and in the emergency room, the treatment strategies for both intoxications are similar (2).

The first report of fatal ethylene glycol intoxication was described in 1930. However, it was not until the Elixir Sulfanilamide tragedy of 1937 that a clear appreciation of the toxicity of glycols emerged. Elixir Sulfanilamide was a liquid preparation of sulfanilamide containing 72% diethylene glycol. During a 4-week period, 353 patients ingested this elixir, and 105 died. The fatal dose of diethylene glycol was estimated at 38 g in children and 71 g in adults. This massive episode of poisoning led to the implementation of the 1938 Food, Drug and Cosmetic Act, which required proof of safety before the marketing of new drugs.

In 1999, ethylene glycol was responsible for at least 6281 poisonings and 25 fatalities in the United States, while other glycols contributed an additional 3424 poisonings with 1 reported fatality (4). Nearly 15% of the individuals poisoned with ethylene glycol were children ≤6 years old.

There are three common scenarios of intoxication:

- Alcoholics who drink it as substitute for alcohol
- Suicide attempts
- Young children who drink it out of curiosity

Diethylene glycol poisoning is not common in the United States. However, there

TABLE 13–1. Major metabolites and structures of glycols and related compounds

Compound	Structure	Toxicity	Metabolites
Ethylene glycol	CH_2-CH_2 $\;\;\|\;\;\;\;\;\;\;\;\|$ $OH\;\;\;\;OH$	Neurological, cardiopulmonary, renal—oxaluria	Glycoaldehyde Glycolate Glyoxylate Oxalate Lactate
Diethylene glycol	$CH_2-CH_2-O-CH_2-CH_2$ $\;\;\|\;\|$ $OH\;OH$	Neurological, cardiopulmonary, renal—no oxaluria	2-Hydroxyethyoxyacetate
Propylene glycol	$CH_2-CH-CH_3$ $\;\;\|\;\;\;\;\;\;\;\|$ $OH\;\;\;OH$	Safe for use in pharmaceuticals; hyperosmolality, lactic acidosis	Lactate Pyruvate
Glycerol	$CH_2-CH-CH_2$ $\;\;\|\;\;\;\;\|\;\;\;\;\|$ $OH\;\;OH\;\;OH$	Safe for use in pharmaceuticals; hyperglycemia, glycosuria	Pyruvate Glucose

have been incidents of mass poisoning worldwide. These include reports of 51 deaths in Bangladesh and 47 deaths in Nigeria among children who ingested a paracetamol elixir containing diethylene glycol. Other episodes include ingestion of a tainted sedative by children in South Africa (7 deaths), ingestion of contaminated glycerol by adults in India (14 deaths), ingestion of tainted wine in the Netherlands, and the topical application of a 1% silver sulfadiazine burn preparation containing diethylene glycol in Spain (5 deaths) (5–6).

Case 1

A 58-year-old man presented to the emergency room after a suicide attempt. He reported ingesting about "20 ounces [56 mL] of half-strength antifreeze," cutting his wrists with a dull knife, and falling down a flight of stairs. One empty and one half-full 1-gallon [4.5-L] containers of Prestone antifreeze, reportedly half-strength, were found near the patient.

In the emergency room, the patient was confused but otherwise neurologically intact. Physical examination revealed stable vital signs and superficial lacerations of both wrists. Laboratory values on admission are given in Table 13–2, and include a metabolic acidosis with increased anion gap and osmolal gap. A diagnosis of ethylene glycol poisoning was made on the basis of history, physical examination, and confirmatory laboratory examination. The initial toxicology screen revealed ethylene glycol, caffeine, and nicotine. The serum ethylene glycol concentration on admission, 7910 mg/L (127 mmol/L), was determined by an enzymatic reaction that uses glycerol dehydrogenase from *Enterobacter aerogenes* on the Hitachi (San Jose, CA) 704 automated analyzer. Treatment included charcoal, intravenous ethanol infusion (1 g/kg loading dose, 100 mg/kg per hour maintenance dose), elective intubation for airway protection, and emergency hemodialysis. The serum ethanol concentration was maintained between 77 and 194 mg/dL (17 and 42 mmol/L). The serum ethylene glycol concentration progressively decreased from 7910 mg/L (127 mmol/L) to 150 mg/L (2 mmol/L) after 28 hours of therapy (including 16 hours of hemodialysis). Serum creatinine concentrations were within the reference range, and arterial pH was never <7.21 throughout the hospital stay. The patient recovered completely within 3–5 days without discernible neurological, renal, or cardiac sequelae (7).

TABLE 13–2. Laboratory findings on presentation

Test	Reference range	Case 1	Case 2
Serum ethylene glycol, mg/L (mmol/L)	—	7910 (127)	None detected
Arterial blood gases			
pH	7.35–7.45	7.23	7.29
pCO_2, mmHg	35–45	16	21
HCO_3^-, mEq/L and mmol/L	22–26	7	11
Anion gap, mEq/L and mmol/L	12–16	26	20
Lactate, mg/dL (mmol/L)	6.3–18.9 (0.7–2.1)	Not reported	9.9 (1.1)
Serum osmolality, mOsm/kg H_2O and mmol/kg H_2O			
Measured	270–290	434	None detected
Calculated		264	None detected
Gap	<10	170	None detected
Creatinine, mg/dL (μmol/L)	0.5–1.7 (44–150)	0.9 (80)	2.5 (221)
Urea nitrogen, mg/dL (μmol/L)	8.1–24.9 (2.9–8.9)	5.0 (1.8)	19.0 (6.8)
Urinalysis		Rare calcium oxalate crystals	Hematuria, rare, amorphous crystals

Reprinted with permission from Eder AF, McGrath CM, Dowdy YG, et al: Ethylene glycol poisoning: toxicokinetic and analytic factors affecting laboratory diagnosis. *Clin Chem* 1998;44:168–177.

Case 2

An 18-year-old man with a history of asthma presented to the emergency room complaining of nausea, vomiting, diffuse abdominal pain, and malaise over the previous 2 days. His vital signs were blood pressure 150/100 mmHg, pulse 84, and rectal temperature 37.9 °C (100.7 °F). He was lethargic but arousable and oriented to person, time, and place. Medications included theophylline (Theo-Dur) and epinephrine (Primatene Mist; 5.5 g/L). Laboratory values on admission are given in Table 13–2. Results of the urine and serum toxicology screens were negative. The patient was admitted for metabolic acidosis and renal failure. Renal function deteriorated, and serum creatinine peaked at 11.9 mg/dL (1052 μmol/L) on the fifth hospital day. Ultrasound revealed enlarged edematous echogenic kidneys, compatible with acute renal failure. Results of an arteriogram for renal vasculitis and a skin biopsy for Henoch-Schönlein purpura were negative.

Renal biopsy revealed acute tubular necrosis and deposition of calcium oxalate crystals, supporting a diagnosis of ethylene glycol poisoning. The patient denied ethylene glycol or antifreeze ingestion and suicide attempts. Ethylene glycol was not detected in stored serum samples from admission or taken on the fifth hospital day, as measured by gas chromatography (GC). Treatment with hemodialysis for 3 weeks resulted in complete recovery of renal function. Twenty-four months after this admission, the patient demonstrated no clinical or laboratory evidence of renal dysfunction [serum creatinine, 1.1 mg/dL (97 μmol/L); blood urea nitrogen, 12 mg/dL (4.3 mmol/L)] (7).

Discussion of Case Studies

As shown by Case 1, obtaining an accurate history is often the most valuable diagnostic tool. In Case 1, ethylene glycol intoxication was recognized immediately because the patient could clearly relate the details of his suicide attempt. In contrast, in Case 2, the diagnosis was delayed because the patient did not report ethylene glycol intoxication. Although family members suspected that he had attempted suicide, he repeatedly denied intentional ethylene glycol intoxication, and there was no evidence of intentional or accidental poisoning. The diagnosis of ethylene glycol intoxication was made only after a renal biopsy was performed to investigate the cause of acute renal failure.

Denial of ethylene glycol ingestion occurs because the patient is concealing intentional poisoning or is unaware of consumption. For example, in the United States, a child undergoing an extensive medical evaluation for a possible inherited genetic disease had in fact been poisoned with ethylene glycol by a caregiver (8). In Germany, a man unknowingly drank contaminated water from a radiator for 6 weeks and consumed enough ethylene glycol to result in intoxication and renal failure (9). In other cases, a patient cannot relate his or her history because of profound neurological depression. If the patient is comatose, the differential diagnosis should include diabetic ketoacidosis, alcoholic ketoacidosis, renal failure, and ingestion of methanol and other toxic compounds.

GENERAL CHARACTERISTICS

Ethylene glycol (1,2-ethanediol) is a colorless, odorless, relatively nonvolatile liquid. It has a boiling point of 197 °C. It is commonly found in automotive products such as antifreeze, windshield de-icers, and coolants; detergents; paints; polishes; cosmetics; and juices and wines as an improperly added preservative or sweetener.

There are several other glycols, and their characteristics are listed below.

Diethylene Glycol

Diethylene glycol can cause significant clinical toxicity including nausea, vomiting, metabolic acidosis, renal failure, hepatomegaly, pulmonary edema, hypertension, coma, and seizures. It has been implicated in cases in which diethylene glycol was used during pharmaceutical manufacture or distribution instead of the more expensive but less toxic diluent propylene glycol (5,6,10–13).

Propylene Glycol

Propylene glycol is considerably less toxic than ethylene glycol and is used in pharmaceuticals. Propylene glycol has been associated with central nervous system (CNS), renal, hematological, and cardiac toxicity, mainly in low-birth-weight infants (intravenous multivitamin preparations, silver sulfadiazine ointment), in patients with impaired renal function (nitroglycerin preparations), or in patients who receive rapid infusion of an intravenous medication containing propylene glycol (14). Polyethylene glycol has not been associated with clinical toxicity.

Other Glycols

Other potentially toxic glycols include diethylene dioxide, ethylene glycol esters (ethylene glycol diformate and ethylene glycol monoacetate), and ethylene glycol monoalkyl ethers (methyl cellusolve, ethylene glycol methyl ether).

PHARMACOKINETICS AND TOXICOKINETICS

The clinical features of ethylene glycol poisoning are summarized in Table 13-3. Classically, there are three stages. In the initial stage, neurological symptoms of CNS depression occur 30 minutes to 12 hours after ingestion. The second stage occurs 12–24 hours after ingestion and includes cardiopulmonary manifestations. Finally, the third stage occurs 24–72 hours after ingestion and consists of manifestations of acute renal failure. The severity of each stage and the progression from one stage to the next depend on the amount of ethylene glycol ingested as well as the timing of therapy. In some cases the stages may overlap, one stage may predominate, or one or more of the stages may not be clinically apparent. In other cases, symptoms may occur several days after ingestion (15).

Ethylene glycol itself causes only mild inebriation, resembling ethanol intoxication (2). The severe clinical features of ethylene glycol intoxication are due not to the parent compound but to toxic metabolites. The lag time for development of clinical signs of toxicity reflects the time needed for the toxic

TABLE 13–3. Clinical stages of ethylene glycol intoxication

Stage	Onset after ingestion (hours)	Manifestations
1	0.5–12	Neurological
		Patient appears inebriated but lacks alcoholic fetor
		Nausea, vomiting, hematemesis
		Nystagmus, papilledema
		Depressed reflexes
		Convulsions
		Coma
2	12–24	Cardiopulmonary
		Tachypnea
		Tachycardia
		Hypertension
		Congestive heart failure
		Pulmonary edema
3	24–72	Renal
		Flank pain and costovertebral angle tenderness
		Acute tubular necrosis

Data from Eder AF, Wolf BA: Glycol poisoning. *Therapeutic Drug Monitoring and Toxicology* 1996;17:175–187; and Brown CG, Trumbull D, Klein Schwartz W, et al: Ethylene glycol poisoning [Clinical conference]. *Ann Emerg Med* 1983;12:501–506.

metabolites to accumulate. Thus, in Case 1, the only clinical manifestation of ethylene glycol intoxication was mild CNS depression because the patient was evaluated within hours of ingestion. In contrast, in Case 2, there was substantial renal failure caused by the metabolites of ethylene glycol, and the poisoning was not detected for several days (7).

The toxicokinetic parameters for ethylene glycol are shown in Table 13–4. Ethylene glycol is rapidly absorbed from the gastrointestinal tract, and symptoms of poisoning can occur as early as 30 minutes after ingestion. Ethylene glycol is metabolized in the liver by alcohol dehydrogenase according to the pathway shown in Figure 13–1. The severity of ethylene glycol intoxication depends on the dose ingested and the formation of toxic metabolites (16) that include glycoaldehyde, glycolate, glyoxylate, oxalate, and lactate. The elimination half-life of ethylene glycol is increased at least fivefold in the presence of ethanol because both compounds compete for the active site of alcohol dehydrogenase. The enzyme has a much greater affinity for ethanol than for ethylene glycol or methanol, and concentration of ethanol as low as 50 mg/dL (11 mmol/L) will saturate the enzyme (17). To ensure competitive inhibition of alcohol dehydrogenase activity, the serum concentration of ethanol should be maintained

TABLE 13–4. Toxicokinetic parameters for ethylene glycol

Lethal dose[a] (mL/kg)	1.4–1.6
Lethal serum concentration (g/L)	0.3–4.3
Volume of distribution (L/kg)	0.5–0.8
Half-life ($t_{1/2}$) (hours)	3–8.6
$t_{1/2}$ during ethanol infusion (hours)	17–18
$t_{1/2}$ during hemodialysis (hours)	2.5–3.5
Mean renal clearance[b] (mL/min)	0.75–27.5
Mean hemodialysis clearance (mL/min)	156–210

[a] Although the lethal dose is considered to be 100 mL for a 70-kg adult, survival has been reported after ingestions from 0.27 to 2 L. Death has been reported after ingestion of 30 mL.

[b] Lower value obtained in a patient with renal failure; upper value obtained in a patient with normal renal function.

Reprinted with permission from Eder AF, McGrath CM, Dowdy YG, et al: Ethylene glycol poisoning: toxicokinetic and analytic factors affecting laboratory diagnosis. *Clin Chem* 1998;44:168–177.

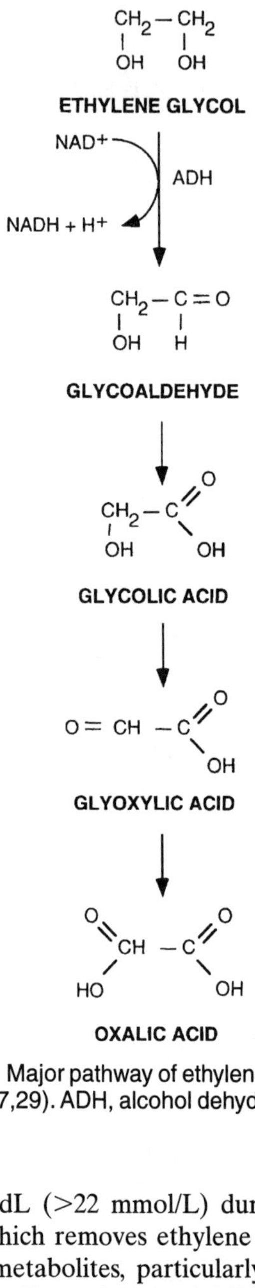

FIG. 13–1. Major pathway of ethylene glycol metabolism (17,29). ADH, alcohol dehydrogenase.

>100 mg/dL (>22 mmol/L) during hemodialysis, which removes ethylene glycol and its toxic metabolites, particularly glycolate and oxalate. Fomepizole, an inhibitor of alcohol dehydrogenase, can also be used to inhibit the formation of toxic metabolites (18,19).

Toxicokinetic evaluation of ethylene glycol elimination during hemodialysis and ethanol infusion was performed in Case 1 (Figure 13–2). Assuming a peak concentration of 7910 mg/L, the volume of distribution was calculated to be 0.5 L/kg. Using a noncompartmental pharmacokinetic model with extravascular administration of the historical dose, the elimination half-life was determined to be 3.79 hours. The clearance during hemodialysis was 113 mL/min (7).

MANAGEMENT—THE ROLE OF THE LABORATORY

The major features of the laboratory diagnosis of ethylene glycol poisoning are shown in Table 13–5 and include the presence of a metabolic acidosis with anion gap, elevated osmolality with osmolal gap, the detection of calcium oxalate crystalluria, and the detection of serum ethylene glycol.

Anion and Osmolal Gaps

The anion gap refers to the calculated difference between the sum of the measured cations and the sum of the measured anions. Most laboratories routinely measure only sodium and potassium, which account for ~95% of the extracellular cations, and chloride and bicarbonate, which account for ~85% of the extracellular anions. The anion gap can be calculated by the following equation (20):

$$\text{Anion gap} = (\text{sodium}) - (\text{chloride} + \text{bicarbonate})$$

Because the sum of the measured cations does not equal the sum of the measured anions in healthy individuals, the normal anion gap is generally considered to be 7–15 mEq/L (7–15 mmol/L). Electroneutrality dictates that anions must be present to balance the cationic charges, but these anions, such as sulfate, phosphate, and serum proteins, are not routinely measured. When plasma pH is altered by the production of organic acids, the bicarbonate is consumed, and the additional unmeasured anions augment the anion gap. There are a finite number of disease states that result in metabolic acidosis accompanied by an increased anion gap (Figure 13–3). In ethylene glycol poisoning, glycolic acid accounts for as much as 96% of the anionic gap in patients poisoned with ethylene glycol

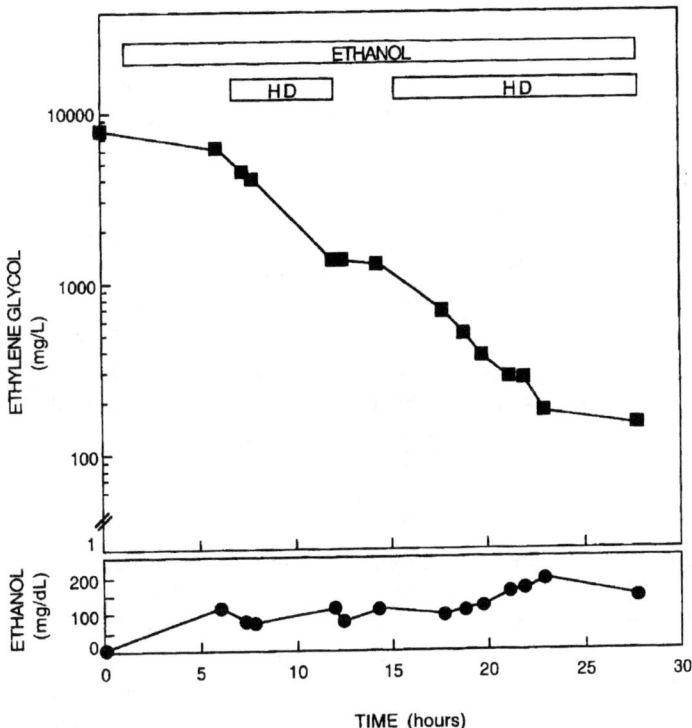

FIG. 13–2. Toxicokinetic evaluation of ethylene glycol elimination. Serum ethylene glycol (squares) and serum ethanol (circles) concentrations during treatment of the patient described in Case 1. The duration of therapy with intravenous ethanol and the two periods of hemodialysis (HD) are indicated by the bars above the graphs. To convert to SI units (mmol/L), multiply ethylene glycol values in mg/L by 0.0161 and ethanol values in mg/dL by 0.217. Reprinted with permission from Eder AF, McGrath CM, Dowdy YG, et al: Ethylene glycol poisoning: toxicokinetic and analytic factors affecting laboratory diagnosis. *Clin Chem* 1998;44:168–177.

(21). Lactic acid also appreciably accumulates (16,21).

The osmolal gap, like the anion gap, reflects the presence of unmeasured molecules in the plasma. Osmolality, like other colligative properties, is based on the number of solute particles in a given weight of solvent. In healthy individuals, serum osmolality is

TABLE 13–5. *Laboratory features of ethylene glycol intoxication*

Feature	Laboratory finding
Severe anion gap metabolic acidosis	pH <7.20
	CO_2 <8 mEq/L (<8 mmol/L)
	Anion gap >20 mEq/L (>20 mmol/L)
Osmolality	Osmolal gap >10 mOsm/kg H_2O (>10 mmol/kg H_2O)
Urinalysis	Calcium oxalate crystalluria
	Low specific gravity
	Proteinuria
	Microscopic hematuria
Leukocytosis	White blood count $10 \times 10^3/\mu L - 40 \times 10^3/\mu L$
Hypocalcemia	Calcium <8.0 mg/dL (<2 mmol/L)

Reprinted with permission from Eder AF, Wolf BA: Glycol poisoning. *Therapeutic Drug Monitoring and Toxicology* 1996;17:175–187.

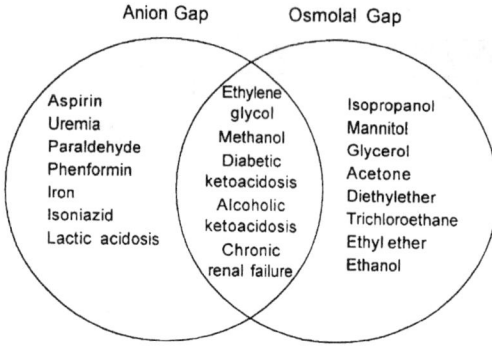

FIG. 13–3. Causes of elevated anion gaps and osmolal gaps. Reprinted with permission from Eder AF, Wolf BA: Glycol poisoning. *Therapeutic Drug Monitoring and Toxicology* 1996;17:175–187.

determined by the concentration of sodium, urea nitrogen, and glucose and is approximated by the following formula (22):

Calculated plasma osmolality
(mOsm/kg H_2O) = [1.86 × sodium (mEq/L)] + [urea nitrogen (mg/dL)/2.8] + [glucose (mg/dL)/18] + 9

Osmolality, measured by vapor-point elevation or freezing-point depression, is 282–300 mOsm/kg H_2O (282–300 mmol/kg H_2O) in a healthy individual. The difference between measured and calculated osmolality, the osmolal gap, is due to the presence of other solutes such as calcium, lipids, and proteins, which are not considered in the above formula. Significant elevation of the osmolal gap, generally considered to be >10 or 15 mOsm/kg H_2O (>10 or 15 mmol/kg H_2O), suggests the presence of osmotically active particles other than sodium, urea nitrogen, and glucose.

Low-molecular-weight substances that achieve appreciable plasma concentration to affect the osmolality and osmolal gap include ethylene glycol, methanol, ethanol, acetone, and other toxins (22,23). In a series of five patients poisoned with ethylene glycol, the osmolal gap ranged from 21 to 49 mOsm/kg H_2O (21 to 49 mmol/kg H_2O), whereas plasma ethylene glycol concentrations ranged from 25 to 255 mg/dL (4 to 41 mmol/L) (24).

The contribution to the osmolal gap due to the presence of ethylene glycol or ethanol can be calculated: Each 100 mg/dL (16 mmol/L) increment in ethylene glycol concentration contributes ~16 mOsm/kg H_2O (~16 mmol/kg H_2O), and each 100 mg/dL (22 mmol/L) of ethanol contributes 22 mOsm/kg H_2O (22 mmol/kg H_2O) to the osmolal gap. The corrected osmolal gap, or the residual osmolality, can be used to follow patients who have simultaneously ingested ethylene glycol and ethanol or to follow patients receiving ethanol infusion for ethylene glycol poisoning. In clinical cases, osmolality should be measured by freezing-point depression because vapor-point elevation underestimates volatile solutes such as ethanol, methanol, and other alcohols (25). An elevated osmolal gap may also reflect pseudohypernatremia due to elevated serum proteins or lipids.

In an appropriate clinical setting, the presence of metabolic acidosis with an increased anion and osmolal gap is highly suggestive of ethylene glycol poisoning or methanol poisoning. Neither osmolal gaps nor anion gaps, however, are universally present in cases of ethylene glycol poisoning, and their absence cannot be used to rule out toxic alcohol or glycol ingestion.

The amount of ethylene glycol consumed and the time course of both ingestion and medical intervention will affect the development of the anion and osmolal gaps. In two series of ethylene glycol–intoxicated individuals, metabolic acidosis was initially apparent in 50% and 86% of cases (26,27). An osmolal gap may not be apparent late in the course of poisoning, whereas an anion gap may not be evident early in the clinical course (28). These observations reflect the compounds that are the source of the gaps. As ethylene glycol is metabolized, its contribution to the osmolal gap diminishes. Because the major metabolites of ethylene glycol are charged particles, they are counterbalanced by sodium and are taken into consideration in the formula to calculate the serum osmolality (23). Although nonionic metabolites may contribute to the osmolal gap, the concentration of ethylene glycol accounts for most of the osmolal gap in cases of ethylene glycol poisoning (28). Conversely, the generation of unmeasured acidic metabolites during the course of the disease augments the anion gap.

In addition to the timing considerations, an anion gap may not be present for additional reasons. Simultaneous ingestion of ethanol impairs metabolism of ethylene glycol and delays the appearance of an anion gap. Simultaneous ingestion of bromide masks the anion gap because bromide is not distinguished from chloride in some assays (29). Similarly, an anion gap was not seen in a patient who had ingested both ethylene glycol and lithium carbonate, possibly because of the additional bicarbonate load (30).

The simultaneous presence of both an anion gap metabolic acidosis and an osmolal gap in a comatose patient is not specific for ethylene glycol or methanol intoxication (Figure 13-3). Anion and osmolal gaps may be present in other clinical settings such as diabetic ketoacidosis, alcoholic ketoacidosis, chronic renal failure, multiple organ failure, and critical illness (7,31).

Calcium Oxalate Crystalluria

A distinctive laboratory finding in cases of ethylene glycol poisoning is calcium oxalate crystalluria (7). The other uncommon clinical setting characterized by calcium oxalate nephrolithiasis is primary hyperoxaluria, an inherited metabolic disorder associated with early onset renal failure and death. Calcium oxalate crystals can assume different forms: calcium oxalate monohydrate (whewellite) crystals are needle shaped (spindle or prism shaped), and calcium oxalate dihydrate (weddellite) crystals are envelope shaped (dipyramidal, octahedral). Other forms of calcium oxalate include dumbbell, ovoid, and elliptical crystals. The most common type in cases of ethylene glycol poisoning is likely calcium oxalate monohydrate (32).

Detection of the monohydrate form of calcium oxalate crystals in urine provides supportive evidence for the diagnosis of ethylene glycol poisoning. Monohydrate calcium oxalate crystals are the thermodynamically stable form under normal physiological conditions, whereas dihydrate calcium oxalate crystals form only at high concentrations of calcium and oxalate. The dihydrate form is metastable and will undergo transformation to the monohydrate form. Because urinalysis is rapid and easy, repetitive urine microscopy is very useful in the differential diagnosis of an anion gap metabolic acidosis of unknown origin. Other frequently reported findings on urinalysis in cases of ethylene glycol poisoning include low specific gravity, proteinuria, and microscopic hematuria.

Ethylene Glycol

The diagnosis of ethylene glycol intoxication is confirmed by quantification of serum ethylene glycol concentrations. The availability of rapid, specific enzymatic tests facilitates screening for ethylene glycol poisoning in the emergency room (33,34). The University of Pennsylvania and other institutions currently include ethylene glycol in the comprehensive drug screen for overdose patients (35).

The serum osmolal gap has been used to approximate serum ethylene glycol concentration during therapy. Specific methods to measure ethylene glycol in serum or urine include colorimetric, fluorometric, chromatographic, and enzymatic assays. The analytical method of choice for ethylene glycol is GC, based on flame-ionization detection of ethylene glycol itself or of a derivative (36-38). Commonly, ethylene glycol is analyzed as the boronic ester derivatives using packed or capillary columns. Typically, underivatized ethylene glycol is difficult to analyze because of poor chromatographic behavior and low sensitivity for flame-ionized detectors. However, a method for direct injection of ethylene glycol on a wide-bore capillary column has been described (37,38). Advantages of this method include elimination of the derivatization step, resolution of diethylene glycol and other glycols as well as other polar drugs, and extended analytical lifetime of the column. Finally, a GC technique has been developed that allows for simultaneous determination of ethylene glycol and other diols.

An appropriate internal standard, such as propylene glycol, 1,3-propanediol, 2,3-butanediol, or 1,2-butanediol, must be used with GC. Recently, the use of propylene glycol as an internal standard has been discouraged because propylene glycol in some intravenous drug preparations may result in underestimation of the concentration of ethylene glycol in serum, and propylene glycol itself may be responsible for clinical toxicity (39–41).

Although GC is considered the gold standard for detecting ethylene glycol, there have been notorious cases of misdiagnosis. In one case, propionic acid was mistakenly identified as ethylene glycol, and a mother was falsely accused of poisoning her infant son, who, in fact, had an inherited metabolic disease, methylmalonic acidemia (42). If ethylene glycol is detected in a sample, confirmation by mass spectrometry or another method is often recommended because identification based on retention time alone has led to misidentification not only of propionic acid, but also of propylene glycol and 2,3-butanediol as ethylene glycol (42,43). Moreover, in sera from ketoacidotic diabetic patients, methanol-like products are generated by the oxidation-reduction derivatization procedure, which may be interpreted as evidence of ethylene glycol poisoning (44). The use of verifier samples (containing ethylene glycol, propylene glycol, and an internal standard) should be encouraged to minimize some of these problems.

Because the toxicity of ethylene glycol is due to its oxidation to organic acids, several methods have been developed to measure these metabolites. Glycolic acid, the major metabolite of ethylene glycol, has been measured by high-performance liquid chromatography (HPLC) and GC as well as by a colorimetric method and isotachophoresis (45). The colorimetric method uses sulfuric acid and chromotropic acid and is relatively specific for glycolic acid. In contrast, the four major acidic metabolites of ethylene glycol can be measured simultaneously by isotachophoresis (46).

The first enzymatic assay described for ethylene glycol measurement used yeast alcohol dehydrogenase (47). This enzyme oxidizes ethanol, other alcohols, and ethylene glycol, and the production of the reduced form of nicotinamide-adenine dinucleotide phosphate (NADH) in this reaction is monitored spectrophotometrically at 340 nm. The major disadvantage is the significant interference observed with other alcohols including methanol, ethanol, and isopropanol. Although these volatile alcohols can be eliminated from the specimen by heating for 30 minutes at 100 °C, incomplete removal and the additional pretreatment required detract from the clinical utility of this assay.

A more specific screening procedure for detecting ethylene glycol evolved from an observation that ethylene glycol interfered with a commercial assay for triglyceride determination (48). The DuPont (Wilmington, DE) aca triglyceride assay for triglycerides uses lipase and glycerol dehydrogenase (GDH) in the following reactions, in which NAD^+ stands for the oxidized form of nicotinamide-adenine dinucleotide:

$$Triglycerides \xrightarrow{lipase} glycerol + free\ fatty\ acids$$

$$Glycerol + NAD^+ \xrightarrow{GDH} dihydroxyacetone + NADH$$

The change in absorbance at 340 nm, due to the formation of NADH, is proportional to the total amount of glycerol, and thus of triglycerides, in the sample. Glycols interfere with triglyceride determination because they, too, are oxidized by GDH. This interference has been exploited to develop a convenient and rapid screening method for detecting ethylene glycol in serum. This method involves pretreating the serum sample with lipase, glycerol kinase, and other cofactors to effectively enzymatically modify triglycerides and glycerol, but not ethylene glycol. After pretreatment of sera, only ethylene glycol, if present, interacts with GDH. The ethylene glycol concentration is then estimated by the DuPont aca triglyceride method.

Another rapid method for measuring ethylene glycol in biological specimens also exploits the positive interference of ethylene glycol in the triglyceride assay on the DuPont aca discrete analyzer (49,50). This technique measures triglycerides by two different enzymatic assays, the DuPont aca triglyceride assay, which measures both ethylene glycol and triglycerides, and a Boehringer Mannheim method that measures only triglycerides. The Boehringer Mannheim method on the Hitachi 717 (Boehringer Mannheim, Montreal, Canada) is an enzymatic end-point method that involves lipase, glycerol kinase, glycerol peroxidase, and peroxidase. Ethylene glycol does not interfere with the Boehringer Mannheim method. The ethylene glycol concentration is calculated from the difference between these two assays, or the "triglyceride gap."

The enzyme used in the DuPont aca triglyceride assay, GDH purified from *Enterobacter aerogenes*, has been used in a specific, straightforward assay for ethylene glycol. This enzyme specifically oxidizes ethylene glycol producing NADH, which is measured spectrophotometrically.

$$\text{Ethylene glycol} + \text{NAD}^+ \xrightarrow{\text{glycerol dehydrogenase}} \text{glycoaldehyde} + \text{NADH}$$

No cross-reaction is observed with ethanol, methanol, *n*-propanol, isopropanol, acetaldehyde, lactate, glyoxal, glycolic acid, glyoxylic acid, or oxalic acid. Interfering substances included glycoaldehyde and glycerol. Both compounds may compete with ethylene glycol for the active site of the enzyme. In cases of ethylene glycol poisoning, glycoaldehyde is one of the metabolites; however, it has a short half-life and would not interfere with the assay. The interference from glycerol likely has no practical relevance and may be disregarded based on expected serum glycerol concentrations [<4.6 mg/dL (<0.5 mmol/L)]. Interestingly, GDH from *Aspergillus niger* or *Cellulomonas* species lacks the required specificity for ethylene glycol assay. The kinetic enzymatic assay using glycerol dehydrogenase from *Enterobacter aerogenes* allowed analysis of serum ethylene glycol concentrations to be completed in 5 minutes.

This manual method to measure serum ethylene glycol was semiautomated on the Monarch 2000 centrifugal analyzer (Instrumentation Laboratory, Lexington, MA), and automated on the multiparametric analyzer Cobas Mira (Roche Diagnostic Systems, Branchburg, NJ), allowing rapid quantification of ethylene glycol. These methods are suitable for emergency screening (33,34). Moderately hemolyzed, severely lipemic, or grossly icteric plasma specimens were evaluated; however, none demonstrated interference in the ethylene glycol assay. In addition, no significant interference from ethanol, methanol, isopropanol, acetone, lactate, or glycerol was detected. This evaluation, however, involved supplementing normal sera with these compounds. Sera from critically ill patients or patients with lactic acidosis, diabetic ketoacidosis, or kidney failure were not evaluated. Abnormal production of NADH may occur in some disease states and will interfere with the assay, resulting in falsely positive values for ethylene glycol. In patients with increased serum lactate and lactate dehydrogenase, extraneous production of NADH from the oxidation of lactate to pyruvate catalyzed by lactate dehydrogenase may interfere with the assay, resulting in falsely positive values for ethylene glycol (35).

REFERENCES

1. Jacobsen D, McMartin KE: Methanol and ethylene glycol poisonings. Mechanism of toxicity, clinical course, diagnosis and treatment. *Med Toxicol* 1986;1:309–334.
2. Turk J, Morrell L: Ethylene glycol intoxication [Clinical conference]. *Arch Intern Med* 1986;146:1601–1603.
3. Parry MF, Wallach R: Ethylene glycol poisoning. *Am J Med* 1974;57:143–150.
4. Litovitz TL, Klein-Schwartz W, White S, et al: 1999 annual report of the American Association of Poison Control Centers Toxic Exposure Surveillance System. *Am J Emerg Med* 2000;18:517–574.
5. Hanif M, Mobarak MR, Ronan A, et al: Fatal renal failure caused by diethylene glycol in paracetamol elixir: the Bangladesh epidemic. *BMJ* 1995;311:88–91.
6. Okuonghae HO, Ighogboja IS, Lawson JO, et al:

Diethylene glycol poisoning in Nigerian children. *Ann Trop Paediatr* 1992;12:235–238.
7. Eder AF, McGrath CM, Dowdy YG, et al: Ethylene glycol poisoning: toxicokinetic and analytic factors affecting laboratory diagnosis. *Clin Chem* 1998; 44:168–177.
8. Woolf AD, Wynshaw-Boris A, Rinaldo P, et al: Intentional infantile ethylene glycol poisoning presenting as an inherited metabolic disorder. *J Pediatr* 1992;120:421–424.
9. Kaiser W, Steinmauer HG, Biesenbach G, et al: Chronic ethylene glycol poisoning [German]. *Dtsch Med Wochenschr* 1993;118:622–626.
10. Gerin M, Patrice S, Begin D, et al: A study of ethylene glycol exposure and kidney function of aircraft de-icing workers. *Int Arch Occup Environ Health* 1997;69:255–265.
11. Scalzo AJ: Diethylene glycol toxicity revisited: the 1996 Haitian epidemic. *J Toxicol Clin Toxicol* 1996; 34:513–516.
12. Wax PM: Elixirs, diluents, and the passage of the 1938 Federal Food, Drug and Cosmetic Act. *Ann Intern Med* 1995;122:456–461.
13. Wax PM: It's happening again—another diethylene glycol mass poisoning. *J Toxicol Clin Toxicol* 1996; 34:517–520.
14. Yorgin PD, Theodorou AA, Al-Uzri A, et al: Propylene glycol-induced proximal renal tubular cell injury [Review]. *Am J Kidney Dis* 1997;30:134–139.
15. Underwood F, Bennett W:. Ethylene glycol intoxication. Prevention of renal failure by aggressive management. *JAMA* 1973;226:1453–1454.
16. Gabow PA, Clay K, Sullivan JB, et al: Organic acids in ethylene glycol intoxication. *Ann Intern Med* 1986;105:16–20.
17. Peterson CD, Collins AJ, Himes JM, et al: Ethylene glycol poisoning: pharmacokinetics during therapy with ethanol and hemodialysis. *N Engl J Med* 1981; 304:21–23.
18. Jacobsen D: New treatment for ethylene glycol poisoning. *N Engl J Med* 1999;340:879–881.
19. Brent J, McMartin K, Phillips S, et al: Fomepizole for the treatment of ethylene glycol poisoning. *N Engl J Med* 1999;340:832–838.
20. Dorwart WV, Chalmers L: Comparison of methods for calculating serum osmolality from chemical concentrations, and the prognostic value of such calculations. *Clin Chem* 1975;21:190–194.
21. Jacobsen D, Ovrebo S, Ostborg J, et al: Glycolate causes the acidosis in ethylene glycol poisoning and is effectively removed by hemodialysis. *Acta Med Scand* 1984;216:409–416.
22. Glasser L, Sternglanz PD, Combie J, et al: Serum osmolality and its applicability to drug overdose. *Am J Clin Pathol* 1973;60:695–699.
23. Hoffman RS, Smilkstein MJ, Howland MA, et al: Osmol gaps revisited: normal values and limitations. *J Toxicol Clin Toxicol* 1993;31:81–93.
24. Jacobsen D, Bredesen JE, Eide I, et al: Anion and osmolal gaps in the diagnosis of methanol and ethylene glycol poisoning. *Acta Med Scand* 1982;212: 17–20.
25. Eisen TF, Lacouture PG, Woolf A: Serum osmolality in alcohol ingestions: differences in availability among laboratories of teaching hospital, nonteaching hospital, and commercial facilities. *Am J Emerg Med* 1989;7:256–259.
26. Karlson-Stiber C, Persson H: Ethylene glycol poisoning: experiences from an epidemic in Sweden. *J Toxicol Clin Toxicol* 1992;30:565–574.
27. Moriarty RW, McDonald RHJ: The spectrum of ethylene glycol poisoning. *Clin Toxicol* 1974;7:583–596.
28. Jacobsen D, Hewlett TP, Webb R, et al: Ethylene glycol intoxication: evaluation of kinetics and crystalluria. *Am J Med* 1988;84:145–152.
29. Heckerling PS: Ethylene glycol poisoning with a normal anion gap due to occult bromide intoxication. *Ann Emerg Med* 1987;16:1384–1386.
30. Leon M, Graeber C: Absence of high anion gap metabolic acidosis in severe ethylene glycol poisoning: a potential effect of simultaneous lithium carbonate ingestion [Review]. *Am J Kidney Dis* 1994;23:313–316.
31. Eder AF, Wolf BA: Glycol poisoning. *Therapeutic Drug Monitoring and Toxicology* 1996;17:175–187.
32. Jacobsen D, Akesson I, Shefter E: Urinary calcium oxalate monohydrate crystals in ethylene glycol poisoning. *Scand J Clin Lab Invest* 1982;42:231–234.
33. Mahly M, Lardet G, Vallon JJ: Automated Cobas Mira kinetic enzymatic assay for ethylene glycol applied to emergency situations. *J Anal Toxicol* 1994;18:269–271.
34. Standefer J, Blackwell W: Enzymatic method for measuring ethylene glycol with a centrifugal analyzer. *Clin Chem* 1991;37:1734–1736.
35. Eder AF, Dowdy YG, Gardiner JA, et al: Serum lactate and lactate dehydrogenase in high concentrations interfere in enzymatic assay of ethylene glycol. *Clin Chem* 1996;42:1489–1491.
36. Smith NB: Determination of serum ethylene glycol by capillary gas chromatography. *Clin Chim Acta* 1984;144:269–272.
37. Edinboro LE, Nanco CR, Soghioan DM, et al: Determination of ethylene glycol in serum utilizing direct injection on a wide-bore capillary column. *Ther Drug Monit* 1993;15:220–223.
38. Livesey JF, Perkins SL, Tokessy NE, et al: Simultaneous determination of alcohols and ethylene glycol in serum by packed- or capillary-column gas chromatography. *Clin Chem* 1995;41:300–305.
39. LeGatt DF, Tisdell RH: Ethylene glycol quantification: avoid propylene glycol as an internal standard [Letter]. *Clin Chem* 1990;36:1860–1861.
40. Apple FS, Googins MK, Resen D: Propylene glycol interference in gas-chromatographic assay of ethylene glycol [letter]. *Clin Chem* 1993;39:167.
41. Smith NB: Internal standards in gas-chromatographic analyses for ethylene glycol in serum [Letter]. *Clin Chem* 1993;39:2020.
42. Shoemaker JD, Lynch RE, Hoffmann JW, et al: Misidentification of propionic acid as ethylene glycol in a patient with methylmalonic acidemia. *J Pediatr* 1992;120:417–421.
43. Jones AW, Nilsson L, Gladh SA, et al: 2,3-Butanediol in plasma from an alcoholic mistakenly identified as ethylene glycol by gas-chromatographic analysis. *Clin Chem* 1991;37:1453–1455.
44. Bjellerup P, Kallner A, Kollind M: GLC determination of serum-ethylene glycol, interferences in ketotic patients [Letter]. *J Toxicol Clin Toxicol* 1994; 32:85–87.
45. Fraser AD, MacNeil W: Colorimetric and gas chromatographic procedures for glycolic acid in serum: the major toxic metabolite of ethylene glycol. *J Toxicol Clin Toxicol* 1993;31:397–405.

46. Ovrebo S, Jacobsen D, Sejersted OM: Determination of ionic metabolites from ethylene glycol in human blood by isotachophoresis. *J Chromatogr* 1987;416:111–117.
47. Eckfeldt JH, Light RT: Kinetic ethylene glycol assay with use of yeast alcohol dehydrogenase. *Clin Chem* 1980;26:1278–1280.
48. Ryder KW, Glick MR, Jackson SA: Emergency screening for ethylene glycol in serum. *Clin Chem* 1986;32:1574–1577.
49. Blandford DE, Desjardins PR: A rapid method for measurement of ethylene glycol. *Clin Biochem* 1994;27:25–30.
50. Hansson P, Masson P: Simple enzymatic screening assay for ethylene glycol (ethane-1,2-diol) in serum. *Clin Chim Acta* 1989;182:95–101.

SELF-ASSESSMENT QUESTIONS

1. What are common sources of ethylene glycol?
   a. antifreeze
   b. paints
   c. detergents
   d. polishes
   e. all of the above

2. To what is the severity of ethylene glycol intoxication mainly due?
   a. accumulation of ethylene glycol in the central nervous system
   b. generation of toxic metabolites
   c. accumulation of ethylene glycol in the kidney
   d. inhibition of alcohol dehydrogenase
   e. all of the above

3. Which of the following glycols is not usually toxic?
   a. ethylene glycol
   b. diethylene glycol
   c. polyethylene glycol
   d. propylene glycol
   e. ethylene glycol monoalkyl ethers

4. What are important laboratory diagnostic features of ethylene glycol poisoning?
   a. history of ethylene glycol ingestion
   b. elevated osmolality with osmolal gap
   c. elevated serum creatinine
   d. metabolic alkalosis
   e. all of the above

5. Which of the following laboratory findings is the most suggestive of ethylene glycol poisoning in a comatose patient?
   a. metabolic acidosis with osmolal gap
   b. unexplained osmolal gap
   c. crystals in the urine
   d. arterial pH <7.20
   e. all of the above

6. Which is true for the gas chromatographic analysis of ethylene glycol?
   a. It is the gold standard.
   b. It can be confused with propionic acid.
   c. It can be confused with propylene glycol.
   d. It should be confirmed by another method such as mass spectrometry.
   e. all of the above

7. What does the enzymatic assay for analysis of ethylene glycol measure?
   a. glycoaldehyde and glycerol
   b. ethanol
   c. methanol
   d. lactate
   e. all of the above

CHAPTER 14

Psychotropic Agents: The Benzodiazepines

Albert D. Fraser, Ph.D., FCACB, DABCC, DABFT

LEARNING OBJECTIVES

After completing this chapter, the reader should be able to:s

1. List the major clinical indications for the benzodiazepines.
2. Describe current prescribing patterns for and use of benzodiazepines in North America.
3. List the major risk factors for benzodiazepine dependence.
4. List some advantages and limitations of immunoassay screening for benzodiazepine metabolites.
5. Understand the laboratory issues relevant to flunitrazepam use and detection.

INTRODUCTION

Benzodiazepines were developed by the Hoffmann–La Roche Company in Switzerland and first marketed in 1961 with the licensing of chlordiazepoxide (Librium®). Today, benzodiazepines are widely prescribed around the world as anxiolytic agents, sedative-hypnotics, muscle relaxants, and anticonvulsant agents. When introduced, benzodiazepines offered several advantages over earlier drugs used for the treatment of anxiety and sleeping disorders. By the early 1980s, benzodiazepines such as diazepam (Valium®) were the most highly prescribed central nervous system (CNS)–active drugs in the world. The popularity of benzodiazepines based on the total number of prescriptions dispensed (1) in 1999 is shown in Table 14–1. The chemical structure of several widely prescribed benzodiazepines is found in Figure 14–1. In the past 15 years, psychiatrists and governmental regulatory authorities worldwide have carefully scrutinized benzodiazepine usage. This scrutiny was spurred by numerous reports of psychological and physiologic addiction, misuse and abuse, and adverse effects associated with long-term use and/or withdrawal of benzodiazepines (2–4).

All benzodiazepines exert to a varying degree some pharmacological activity as hypnotic agents, anxiolytic activity, anticonvulsant action, muscle-relaxant activity, and amnesic activity. There are large differences, however, in potency among the various benzodiazepines due to differences in binding affinity to benzodiazepine receptor subtypes.

CASE PRESENTATION

A 15-year-old female, who had been at a school dance, was brought to the emergency department of a children's hospital in a semiconscious state. The emergency room physician noted a strong smell of alcohol on her breath and ordered a urine drug screen. No street drugs or drug paraphernalia were found in her purse. The patient's parents were contacted and questioned about their

TABLE 14–1. Benzodiazepines in the top 200 drugs prescribed in the United States (1999)

Drug	Manufacturer	Rank
Alprazolam	Greenstone	45
Alprazolam	Geneva	72
Lorazepam	Mylan	78
Clonazepam	Teva Pharm	86
Diazepam	Mylan	127
Temazepam	Mylan	159
Lorazepam	ESI Lederle	161
Alprazolam	Purepac	178
Lorazepam	Purepac	199

daughter's medical history and street drug use. The hospital laboratory analyzed a urine specimen for eight drugs-of-abuse classes by immunoassay and for alcohols by gas chromatography (GC) and performed serum assays for acetaminophen and salicylates. The qualitative ethyl alcohol screen was strongly positive and the only immunoassay test that screened positive was the benzodiazepine assay.

The attending physician requested quantitative analysis of ethyl alcohol in serum and spoke to the parents about possible sources of benzodiazepine prescription drugs. A grandmother who visited recently had rheumatoid arthritis and was having trouble sleeping. She had prescriptions for lorazepam and a new drug for her arthritis but the mother did not remember the name of the second drug.

The serum ethyl alcohol concentration was 125 mg/dL (27.13 mmol/L), and the tests for acetaminophen and salicylates were negative. A urine sample was analyzed for benzodiazepine confirmation on an automated high-performance liquid chromatography (HPLC) system with diode-array detection. The technologist was unable to confirm any common benzodiazepine or benzodiazepine metabolites. The mother was again questioned about the availability of other drugs and remembered that the grandmother's prescription medication for arthritis was called Daypro® (oxaprozin).

The patient was kept for observation overnight and returned home the following morning with her parents.

CLINICAL USES OF BENZODIAZEPINES

The prevalence of insomnia is common in the general public (particularly in the elderly and in women). In the United Kingdom

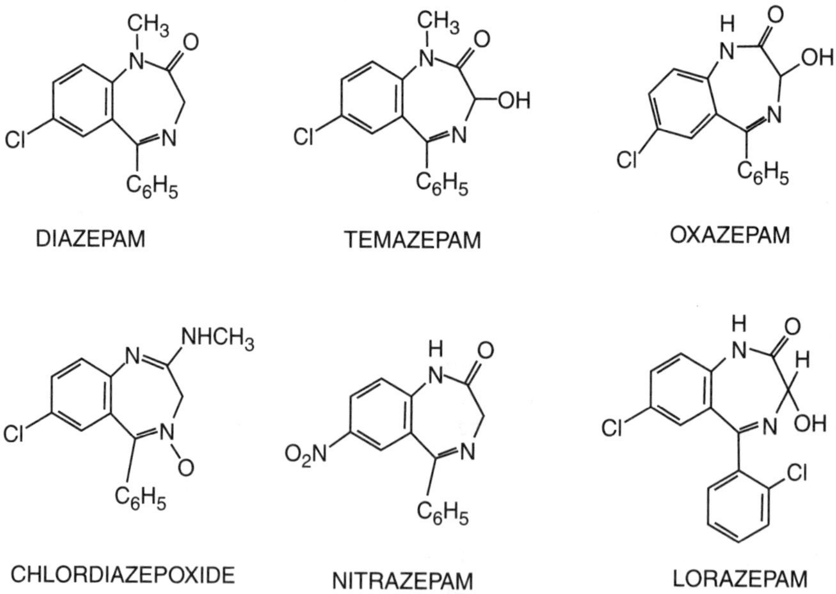

FIG. 14–1. Chemical structure of various benzodiazepines.

~40% of the elderly population complain of disturbed sleep, and British women >65 years of age consumed ~40% of all benzodiazepine hypnotics supplied by family physicians in the early 1980s (5). The major limitations of benzodiazepines as hypnotic agents include the rapid development of tolerance to their hypnotic effects and rebound insomnia (when treatment stopped) that commonly occurs upon withdrawal of the drug. The elderly are more susceptible to CNS depression due to benzodiazepine hypnotics (such as triazolam) and also metabolize benzodiazepines slower than younger individuals.

On the advice of the Committee on Safety of Medicines, the product licenses for triazolam were suspended in 1991 in the United Kingdom (6–7). The reasons for removing triazolam from the British market included the assertion that the manufacturer failed to report the 30% incidence of adverse side effects in an early protocol. Secondly, since the initial release of triazolam in the United Kingdom, adverse effects including rebound insomnia and dose-related adverse effects such as amnesia, delirium, behavioral disturbances, affective disorders, and psychosis had been attributed to triazolam. Despite several appeals of this decision by the manufacturer (Upjohn Pharmaceutical Co., Kalamazoo, MI) and subsequent review by the UK regulatory authorities, revocation of the license to market Halcion® (triazolam) was declared permanent in June 1993.

At the same time, the US Food and Drug Administration (FDA) reviewed the clinical use of triazolam and decided not to remove triazolam from the American market (8). The FDA required revised labeling for triazolam emphasizing that this drug is indicated for short-term (7–10 days) treatment of insomnia, and it recommended a lower daily dose in the elderly (0.125 mg). Despite the adverse publicity about triazolam use in the elderly, benzodiazepines remain important drugs for the treatment of insomnia.

Benzodiazepines are potent anxiolytic agents and are effective in both anxious patients and patients undergoing stress. Major advantages of benzodiazepines in the treatment of anxiety include rapid onset of action and less toxicity than alternative therapeutic agents. Tolerance to the anxiolytic effects of benzodiazepines develops much more slowly than tolerance to their hypnotic effects. Limitations of benzodiazepines include the fact that long-term users of standard therapeutic doses may show cognitive deficits and benzodiazepines may cause or aggravate depression.

Benzodiazepines such as diazepam are the preferred agents for the treatment of status epilepticus and for control of seizures due to a drug overdose. Clonazepam and clobazam (a 1,5-benzodiazepine) are available in several countries as anticonvulsant agents. Limitations of long-term use of these benzodiazepines for seizure control include the development of tolerance in many patients, sedation, and psychomotor impairment.

The most prevalent symptoms of benzodiazepine overdoses in children were reviewed by Wiley and Wiley (9). In a series of 46 children hospitalized for benzodiazepine ingestion, ataxia was observed in 87% of cases, lethargy in 57%, altered mental status in 70%, and respiratory depression in 9%. The most common benzodiazepines reported in patients with a reliable history of benzodiazepine ingestion were lorazepam, clonazepam, and alprazolam. In the 13 patients with a history of lorazepam ingestion in this study, only 1 of 13 urine immunoassay tests screened positive for the benzodiazepines compared with 4 of 4 patients with known alprazolam ingestion and 2 of 2 patients with a history of temazepam and triazolam overdose. The clinical observations in the case presented earlier were consistent with a benzodiazepine overdose even though the confirmation method was negative for benzodiazepines such as lorazepam.

CLINICAL TOXICITY AND THE BENZODIAZEPINES

Commonly reported adverse effects of benzodiazepines include physiological dependence; impairment of recall, learning, and performance; and drug-drug interaction with other drugs such as ethyl alcohol. Deliberate self-poisoning with drugs is a common method of choice for individuals attempting suicide. Psychotropic drugs (especially benzodiazepines and barbiturates) are often used in self-poisoning.

A study in Sweden (10) investigated the prevalence of drug overdose in all suicide attempts in a 3-year period in the greater Lund-Orup area. Benzodiazepines were the most commonly used drugs in attempted and completed suicides. In this group, 51% took an overdose of benzodiazepines. The most popular benzodiazepines were diazepam, flunitrazepam, and oxazepam. Toxicological screening after drug-overdose fatalities showed that benzodiazepines were most often detected (55% of cases), followed by analgesics and antidepressants. Flunitrazepam was found in 39% of fatalities and was considered the sole cause of death in certain cases. The authors concluded that physicians should be more cautious in prescribing psychotropic agents and analgesics. The risk of suicide attempts after benzodiazepine use was also studied in subjects recently prescribed a benzodiazepine and/or an antidepressant drug compared with unexposed control subjects. Individuals considered at greatest risk of attempting suicide were males <40 years of age who were prescribed a benzodiazepine and were not taking any antidepressant medication.

FATALITIES DUE TO BENZODIAZEPINE POISONING

Benzodiazepines have a wide margin of safety based on experimental studies in animals. In the rat, for example, the lethal dose of alprazolam (LD_{50}) is 331–2171 mg/kg. The effective therapeutic dose for alprazolam (ED_{50}) is 0.5 mg/kg, which results in a therapeutic ratio (LD_{50}/ED_{50}) for alprazolam of 662–4342. Based on this ratio, one would anticipate that death solely due to a benzodiazepine overdose would be very rare. Clinical experience over three decades has demonstrated that benzodiazepines are very safe drugs due to this wide (LD_{50}/ED_{50}) ratio. There have been, however, published reports of fatalities due to nitrazepam, flunitrazepam, triazolam, alprazolam, and flurazepam use in which the major causative agent leading to death was considered the benzodiazepine taken as an overdose (11). In some of these cases, ethyl alcohol was also involved.

Fatal poisonings attributed to benzodiazepines in the United Kingdom were reported (12). The authors investigated the number of deaths attributed directly to benzodiazepines from the Office of Population Census and Surveys and from the Registrar General for Scotland from 1980 to 1989. The number of deaths was calculated per million prescriptions for each benzodiazepine, as were the estimated number of deaths per million patients and the estimated number of deaths per 1000 kg of diazepam-equivalents consumed. The authors stated that benzodiazepines resulted in 5.9 deaths per million prescriptions. In this 10-year period, 1576 fatalities were attributed to benzodiazepines in the United Kingdom. The major hypnotic agents implicated were flurazepam and temazepam, and the major anxiolytic drugs considered leading to death after an overdose were prazepam and alprazolam. These authors concluded that benzodiazepines were much less toxic than barbiturates. Temazepam toxicity was considered much greater than that of other benzodiazepines based on this study (temazepam was implicated in 491 out of 921 benzodiazepine hypnotic deaths). These studies indicated that the benzodiazepines were safe drugs compared with several alternative medications. Also, the number of benzodiazepine-related fatalities are very low relative to the widespread use of these drugs. In a very small percentage of cases, however, benzodiazepines can cause serious and life-threatening toxicity after an overdose of only a benzodiazepine.

IMMUNOASSAY SCREENING OF BENZODIAZEPINES

Immunoassays are the most common method for screening biological fluids for benzodiazepine use in the emergency set-

TABLE 14–2. *Benzodiazepines with a long half-life (20–40 h) and long-acting metabolites*

Drug	Metabolite	Metabolite half-life (h)
Diazepam	Nordiazepam	36–200
Flurazepam	Desalkylflurazepam	40–250
Chlordiazepoxide	Nordiazepam	36–200
Prazepam	Nordiazepam	36–200

ting. A brief summary of benzodiazepine metabolism and elimination half-life is found in Table 14–2. The metabolism of diazepam to nordiazepam and to oxazepam is found in Figure 14–2. Manufacturers of diagnostic kits for screening of benzodiazepines and metabolites in urine currently use a cut-off concentration of either 200 or 300 ng/mL. Is there a pharmacokinetic basis for the selection of these urine cutoff values? Unfortunately, there is no pharmacokinetic basis for the selection and application of these cutoff values. The commercial cutoff values used today are intended to detect medium to short half-life benzodiazepines with or without any active metabolites (Tables 14–3 and 14–4). These cutoff values were not based on urine analysis after single or multiple drug doses in human volunteers. In addition, no attempt was made to account for urine dilution when these cutoff values were introduced. These cutoff values are triggered by highly varying concentrations of benzodiazepines or metabolites, depending on their immunological cross-reactivity. More specifically, one should be aware of the limitations of using a 200 or 300 ng/mL cutoff value for benzodiazepines today for the following reasons:

1. These cut-off values were established many years ago when most benzodiazepines were prescribed in doses of 5 to 20 mg/d. For example, a cutoff value of 300 ng/mL may be appropriate when testing for diazepam (300 ng/mL = 1050 nmol/L), temazepam (998 nmol/L), or oxazepam (1046 nmol/L) use, but this cut-off value is too high for more potent benzodiazepines given in much lower doses (such as alprazolam, triazolam, and flunitrazepam). It has been shown that opti-

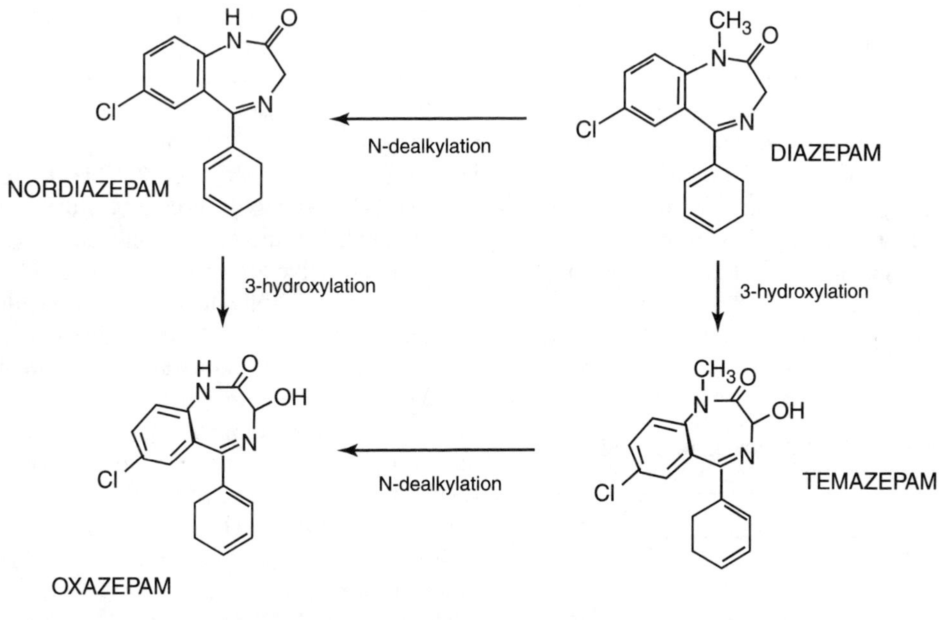

FIG. 14–2. Benzodiazepine metabolic pathway: diazepam to oxazepam.

TABLE 14–3. *Benzodiazepines with a medium to short half-life (10–20 h) and active metabolites*

Drug	Metabolite
Bromazepam	3-Hydroxybromazepam
Estazolam	4-Hydroxyestazolam
Flunitrazepam	7-Aminoflunitrazepam
Alprazolam	α-Hydroxyalprazolam

TABLE 14–5. *Benzodiazepines with an ultrashort half-life and active metabolites*

Drug	Metabolite
Midazolam	α-Hydroxymidazolam
	4-Hydroxymidazolam
	α-Hydroxymidazolam glucuronide
Triazolam	α-Hydroxytriazolam
	4-Hydroxytriazolam

mal detection of alprazolam and triazolam use requires use of a 100 ng/mL (324 nmol/L and 291 nmol/L, respectively) urine screening cutoff value (13–14). Examples of potent benzodiazepines with very short elimination half-lives and active metabolites are found in Table 14–5.

2. The antibodies included in most commercial immunoassays for benzodiazepines are directed toward the free or nonconjugated form of the benzodiazepine metabolite. The nonconjugated form of these drugs is how the drug is consumed orally, but not how the kidney excretes the drug after absorption and metabolism. Most benzodiazepines are primarily excreted as glucuronide conjugates (15). Examples include α-OH-alprazolam glucuronide (after alprazolam consumption) and lorazepam glucuronide after lorazepam use. Because of poor cross-reactivity toward the conjugated metabolites, many benzodiazepine metabolites will not be detected unless glucuronide hydrolysis is performed before screening. Many investigators (16–19) have advocated enzymatic hydrolysis of benzodiazepine conjugates (with β-glucuronidase) before immunoassay urine screening. One of the major diagnostic suppliers now markets a urine benzodiazepine kit that incorporates β-glucuronidase.

In the case presented earlier, the comatose patient had access to lorazepam. The clinical chemist investigated whether the immunoassay used for benzodiazepine screening in his laboratory would detect lorazepam. He found that most commercially available benzodiazepine immunoassays (with a screening cutoff of 200 or 300 ng/mL) were unable to detect lorazepam use by urine testing (15). Lorazepam detection by immunoassay, however, could be improved by, first, using a 100 ng/mL screening cutoff and, second, incorporating an enzymatic hydrolysis step with β-glucuronidase in the screening procedure before urine immunoassay testing for benzodiazepines and metabolites.

Another concern with any immunoassay screening test for drugs of abuse in urine is the possibility of false-positive results due to undesirable cross-reactivity toward another drug or substance that is not a benzodiazepine. In the case presentation, the clinical chemist found that the arthritis medication prescribed for the grandmother (Daypro®) would cause a false-positive immunoassay screening result (20–22). Daypro® is a popular medication that is ranked #150 in the top 200 list of most highly prescribed drugs in 1999 (1). Unfortunately, no one has developed an effective pretreatment method to eliminate this unwanted cross-reactant (probably a metabolite of oxaprozin) in urine benzodiazepine immunoassays. Therefore, in any clinical situation in which a false-positive benzodiazepine screening assay is unacceptable, all presumptive-positive immunoassay results must be confirmed by a specific benzodiazepine confirmation method such as HPLC or GC–mass spectrometry (GC/MS) before reporting the result.

Divanon et al. (23) studied the frequency of actual benzodiazepine overdoses identi-

TABLE 14–4. *Benzodiazepines with a medium to short half-life and no active metabolites*

Drug	Half-life (h)
Lorazepam	10–20
Nitrazepam	15–35
Oxazepam	4–15
Temazepam	8–14

fied by the enzyme-multiplied immuno technique (EMIT®) serum benzodiazepine assay and how the immunoassay results related to the pharmacological nature of individual benzodiazepines consumed. In 588 serum analyses performed in their laboratory, there were 285 presumptive-positive EMIT screening results and no false-positive results (all serum specimens were analyzed by a more sensitive GC confirmation procedure). There were, however, 20 false-negative EMIT results [alprazolam (n = 6), bromazepam (n = 10), clobazam (n = 1), and flunitrazepam (n = 3)]. Using a 200 ng/mL (697 nmol/L) oxazepam cutoff calibrator, Divanon et al. found that bromazepam and lorazepam serum standards were only positive above 400 and 1000 ng/mL (1265 nmol/L and 3113 nmol/L), respectively. These authors concluded that poor cross-reactivity and/or low therapeutic-to-toxic serum values may lead to false-negative screening for bromazepam, flunitrazepam, alprazolam, clonazepam, and lorazepam (Table 14–6). From another perspective, some benzodiazepines may have excellent immunoreactivity (oxazepam, nordiazepam, and temazepam) in the assay and also have higher therapeutic serum values. For these three drugs, it is possible to obtain misleading screen-positive results when the serum concentration is at therapeutic concentrations. These authors described the limitations of the EMIT® serum benzodiazepine assay and strongly recommended that one should not calculate semiquantitative values for serum benzodiazepines by immunoassay and attempt to relate these results to the severity of clinical toxicity.

Flunitrazepam (Rohypnol®) is a benzodiazepine marketed legally in many countries. It is not an approved drug in Canada or the United States. Rohypnol® (roofies) have often been called a date-rape drug and its use was reported in many clinical and forensic toxicology cases (particularly in the states of Florida and Texas in the mid 1990s).

A recent study (24) evaluated five immunoassay kits for detection of flunitrazepam metabolites in male volunteers after single oral doses of 1–3 mg of flunitrazepam. The

TABLE 14–6. *Benzodiazepines that give false-negative screen results in the EMIT serum assay due to poor cross-reactivity and/or low toxic serum concentrations*

Bromazepam
Fluritrazepam
Alprazolam
Clonazepam
Lorazepam

immunoassays evaluated included the EMIT d.a.u. and EMIT II (both from Behring Diagnostics, San Jose, CA), Abbott (Abbott Park, IL) fluorescence polarization immunoassay (FPIA), Cozart Bioscience Ltd. (Abington, Oxford, UK) Auto-Lyte, and Roche Diagnostic Systems (Branchburg, NJ) On-Line systems. After a single 1-mg dose, none of the immunoassays were positive. The maximum 7-aminoflunitrazepam concentration (after a 1-mg dose) was 15 ng/mL (52.9 nmol/L) 14 hours postdose (GC/MS) in one volunteer. After a 2-mg dose, the EMIT II assay was positive at 10 hours postdose and the Roche On-Line assay was positive at 10 and 19 hours postdose. At 19 hours postdose (2 mg), the maximum 7-aminoflunitrazepam value was 33 ng/mL (116 nmol/L) by GC/MS. After a single 3-mg dose, the EMIT II assay screened positive at 8, 14, and 18 hours postdose, whereas the Roche On-Line screened positive at 8, 14, 18, 33, and 40 hours postdose. The maximum 7-aminoflunitrazepam concentration after the 3-mg dose was 48 ng/mL (169 nmol/L) by GC/MS at 14 hours postdose.

IDENTIFICATION AND CONFIRMATION OF BENZODIAZEPINES

The most popular confirmation methods for benzodiazepine or benzodiazepine metabolites are HPLC and GC/MS. Many clinical laboratories use HPLC with ultraviolet (UV) detection to confirm the presence of benzodiazepines and metabolites in serum and urine. HPLC offers the advantage that commercial systems that can be automated are available for drug analysis. Another advantage of HPLC analysis is that biological specimen extracts do not require chemical derivatization before analysis. Liquid chromatographic (LC) methods are acceptable for benzodiazepine confirmation as long as the user is aware of their analytical limitations. Some of the more potent and rapidly metabolizing benzodiazepines (such as triazolam, flunitrazepam, and estazolam) may not be detected by HPLC. High therapeutic and toxic concentrations of these potent short-acting benzodiazepines are often below the detection limit of HPLC methods with UV detection. Benzodiazepines are generally well separated by liquid chromatographic methods and do not require chemical derivatization before analysis.

When available, the preferred method for benzodiazepine identification and confirmation is GC/MS or LC/MS. One can detect many benzodiazepines at much lower concentrations by GC/MS, especially when the GC/MS is operated in the negative chemical-ionization mode. GC/MS, however, does require chemical derivatization before analysis of most benzodiazepine metabolites.

Identification of specific benzodiazepines and metabolites is often essential in clinical and forensic toxicology. Therefore, it is recommended that benzodiazepine glucuronide conjugates be hydrolyzed by enzymatic hydrolysis, not by acid treatment. Acid hydrolysis leads to formation of benzophenones that can be used to confirm that a benzodiazepine was present. Identification of a specific benzodiazepine in urine, however, requires enzymatic hydrolysis with β-glucuronidase before GC/MS analysis. A comprehensive GC/MS method for benzodiazepine confirmation was recently published by Raymon et al. (25). Their method included the following benzodiazepine calibrators at concentrations from 50 to 1000 ng/mL: 7-aminoflunitrazepam, 7-aminoclonazepam, alprazolam, α-OH-alprazolam, desalkylflurazepam, hydroxyethylflurazepam, diazepam, nordiazepam, oxazepam, lorazepam, temazepam, triazolam, and α-OH-triazolam. Urine specimens were hydrolyzed enzymatically with β-glucuronidase at 37 °C for 120 minutes before derivatization with N-methyl-N-(tert-butyldimethylsilyl)-trifluoroacetamide derivative and automated analysis by GC/MS in the selected ion monitoring mode. A list of ions selected for benzodiazepine monitoring is found in Table 14–7. This method provides excellent separation of benzodiazepine peaks and good sensitivity required for low-dose benzodiazepines such as flunitrazepam and triazolam.

ElSohly et al. (26) described a GC/MS method for flunitrazepam and 7-aminoflunitrazepam in whole blood and plasma after 2-mg doses of flunitrazepam in volunteers. They reported that flunitrazepam [with a limit of detection (LOD) of 5 ng/mL (16.5 nmol/L)] was detected in whole blood but not in plasma. The major metabolite (7-aminoflunitrazepam) was detected in both whole blood and plasma up to 12 hours after a 2-mg dose in volunteers. Peak flunitrazepam ranged from 3.5 to 6.7 ng/mL (11.2 to 21.4 nmol/L) in whole blood and <LOD in plasma. Peak 7-aminoflunitrazepam ranged from 2.8 to 3.2 ng/mL (9.9 to 11.2 nmol/L) in whole blood and 10.3 to 12.2 ng/mL (36.3 to 42.9 nmol/L) in plasma. Using a heptafluorobutyrate derivatization procedure of the benzophenones, quantitative ions were 513, 274, and 453 for 7-aminoflunitrazepam, flunitrazepam, and 7-aminonorflunitrazepam, respectively. The maximum retention time was 8.5 minutes for the flunitrazepam benzophenone.

GC/MS methods are popular methods for benzodiazepine confirmation in clinical toxicology laboratories today. GC/MS does require derivatization of benzodiazepines to improve spectral definition and to reduce breakdown of the drug on the analytical column. The most common derivatives for benzodiazepines and metabolites are trimethylsilyl and t-butyldimethylsilyl derivatives. Most studies of benzodiazepines by GC/MS used electron impact ionization with selected ion monitoring. Although chemical ionization methods are not as widely available, negative chemical ionization offers much greater sensitivity and specificity (27) for many benzodiazepines (Table 14-8) than electron impact or positive chemical ionization modes.

If one has access to a liquid chromato-

TABLE 14–7. Selected ion monitoring for benzodiazepines

Drug	Relative RT	Quantitative ion	Qualifier ion(s)
Diazepam	0.778	283	256
Desalkylflurazepam-TBDMS	0.793	345	347
Nordiazepam-TBDMS	0.801	327	329
7-Aminoflurazepam	0.937	255	254, 282
Oxazepam d_5-diTBDMS	1.000	462	464, 518
Oxazepam-diTBDMS	1.002	457	459, 513
Temazepam-TBDMS	1.075	357	283, 359
7-Aminoclonazepam-TBDMS	1.106	342	344, 399
Lorazepam-TBDMS	1.157	491	493, 513
2-Hydroxyethylflurazepam-TBDMS	1.199	389	391
Alprazolam	1.271	279	308, 273
Triazolam	1.395	313	342, 315
α-OH-Alprazolam-TBDMS	1.639	381	382, 383
α-OH-Triazolam-TBDMS	1.797	415	417

RT, retention time; TBDMS, *tert*-butyldimethylsilyl derivative; diTBDMS, di-*tert*-butyldimethylsilyl derivative.

graph mass spectrometer (LC-MS) or tandem MS (MS-MS), these systems offer the best chromatographic separation of benzodiazepines and metabolites (without chemical derivatization). Also, LC-MS and MS-MS can be used directly without enzymatic hydrolysis of benzodiazepine glucuronide conjugates when analyzing urine specimens (28). Detection limits for 10 benzodiazepines (29) ranged from 10 to 200 pg on-column by MS-MS. One can anticipate more LC-MS and MS-MS methods for benzodiazepines in biological fluids as these instruments drop in price and become more available in clinical toxicology laboratories. An LC-MS or MS-MS (30) benzodiazepine confirmation method (one that included analysis of lorazepam and/or lorazepam glucuronide) would be the most definitive method to confirm and/or rule out the presence of lorazepam in the case presented earlier.

CONCLUSIONS

The benzodiazepines have provided major therapeutic advances for a variety of clinical disorders since being introduced almost 40 years ago. Unfortunately, until the early 1980s, there was limited awareness of the potential long-term adverse effects of benzodiazepine use, especially in the elderly. Use of benzodiazepines in the elderly population remains controversial after the removal of triazolam in the United Kingdom several years ago. In many clinical situations, overuse and abuse of these drugs are areas of concern. The benzodiazepines have also become popular with the drug-abusing populations in modern society. The licensing, clinical use, and misuse of benzodiazepines varies significantly among various countries of the world (31). Commercial immunoassays for benzodiazepines and metabolites have provided many advantages to laboratories performing urine drug testing. The number of approved benzodiazepines with variable chemical structure and metabolic pathways continues to present a challenge to diagnostic companies to develop antibodies that cross-react with most if not all benzodiaze-

TABLE 14–8. Comparative signal-to-noise ratios of commonly encountered benzodiazepines based on ionization modes

Drug	NCI	PCI	EI
Oxazepam-diTMS	1719	NS	NS
Lorazepam-TMS	547	NS	NS
Nordiazepam-TMS	115	25	22
α-OH-Triazolam-TMS	2686	25	53

NCI, negative ion chemical ionization; PCI, positive ion chemical ionization; EI, electron impact ionization; TMS, trimethylsilyl derivative; diTMS, di-trimethylsilyl derivative; NS, not specified.

pines in clinical use. Commercial immunoassay products that have a screening cutoff of 200 or 300 ng/mL and do not incorporate enzymatic hydrolysis in their kits are unable to detect many benzodiazepines in common use today.

In most situations, presumptive-positive immunoassays for benzodiazepines should be confirmed by another analytical method. The recommended method for benzodiazepine identification and confirmation is GC/MS or LC/MS. HPLC is an acceptable confirmation method in most cases (especially with diode-array detection) in which definitive characterization of the drug is not required. Many HPLC methods lack the analytical sensitivity required to detect the more potent benzodiazepines. LC/MS and MS-MS systems are very sensitive and specific methods for quantification and/or confirmation of benzodiazepines in biological specimens. Unfortunately, these systems are not widely available in many clinical toxicology laboratories.

REFERENCES

1. RxList. The top 200 prescriptions for 1999 by number of US prescriptions dispensed. Web site: http://www.rxlist.com/top200.htm.
2. Woods JH, Winger G: Current benzodiazepine issues. *Psychopharmacology (Berl)* 1995;118:107–115.
3. Griffiths RR: Commentary on review by Woods and Winger: Benzodiazepines: long term use among patients is a concern and abuse among polydrug abusers is not trivial. *Psychopharmacology (Berl)* 1995; 118:116–117.
4. Salzman C: The benzodiazepine controversy: therapeutic effects versus dependence, withdrawal, and toxicity. *Harv Rev Psychiatry* 1997;4:279–282.
5. Taylor D: Current usage of benzodiazepines in Britain. In: Freeman H, Rue Y, eds: *The benzodiazepines in current clinical practice.* London: Royal Society of Medicine Services: 1987:13–18.
6. Brahams D: Triazolam licensing in UK. *Lancet* 1993;341:1587.
7. Dyer C: New report criticises Upjohn over Halcion. *BMJ* 1994;308:1321–1322.
8. New Halcion labeling. *FDA Med Bull* 1992;22:7–8.
9. Wiley CC, Wiley JF: Pediatric benzodiazepine ingestion resulting in hospitalization. *J Toxicol Clin Toxicol* 1998;36:227–231.
10. Ekedahl AM: Medicine self-poisoning and the sources of the drugs in Lund, Sweden. *Acta Psychiatr Scand* 1994;89:255–261.
11. Drummer OH, Ranson D: Sudden death and benzodiazepines. *Am J Forensic Med Pathol* 1996;17: 336–342.
12. Serfaty M, Masterton G: Fatal poisonings attributed to benzodiazepines in Britain in the 1980's. *Br J Psychiatry* 1993;163:386–393.
13. Fraser AD, Meatherall R: Comparative evaluation of five immunoassays for the analysis of alprazolam and triazolam metabolites in urine: effect of lowering the screening and GC-MS cut-off values. *J Anal Toxicol* 1996;20:217–223.
14. Fraser AD, Meatherall RC: Improved cross-reactivity to α-OH-triazolam in the BMC CEDIA dau urine benzodiazepine assay. *Ther Drug Monit* 1998; 20:331–334.
15. Meatherall R, Fraser AD: Comparison of four immunoassays for the detection of lorazepam in urine. *Ther Drug Monit* 1998;20:673–675.
16. Beck O, LaFolie P, Hjemdahl P, et al: Detection of benzodiazepine intake in therapeutic doses by immunoanalysis of urine: two techniques evaluated and modified for improved performance. *Clin Chem* 1992;38:271–275.
17. Beck O, Lin Z, Brodin K, et al: The On-Line screening technique for urinary benzodiazepines: comparison with EMIT, FPIA, and GC-MS. *J Anal Toxicol* 1997;21: 554–557.
18. Meatherall RC: Benzodiazepine screening using EMIT II and TDx: urine hydrolysis pretreatment required. *J Anal Toxicol* 1994;18:385–390.
19. Meatherall RC: Optimal enzymatic hydrolysis of urinary benzodiazepine conjugates. *J Anal Toxicol* 1994;18:382–324.
20. Fraser AD, Howell P: Oxaprozin cross-reactivity in three commercial immunoassays for benzodiazepines in urine. *J Anal Toxicol* 1998;22:50–54.
21. Camara PD, Audette L, Velletri P, et al: False positive immunoassay for urine benzodiazepine in patients receiving oxaprozin (Daypro'). *Clin Chem* 1995;45:115–116.
22. Matuch-Hite T, Jones P Jr, Moriarity J: Interference of oxaprozin with benzodiazepines via enzyme immunoassay technique. *J Anal Toxicol* 1995;19:130.
23. Divanon F, Debruyne D, Moulin M: Benzodiazepines: toxic serum concentrations in positive enzyme immunoassay responses. *J Anal Toxicol* 1999; 22:559–566.
24. Barrett AM, Walshe K, Kavanagh PV, et al: A comparison of five commercial immunoassays for the detection of flunitrazepam and other benzodiazepines in urine. *Addict Biol* 1999;4:81–87.
25. Raymon LP, Steele BW, Walls HC: Benzodiazepines in Miami-Dade County, Florida driving under the influence (DUI) cases (1995–1998) with emphasis on Rohypnol®: GC-MS confirmation, patterns of use, psychomotor impairment, and results of Florida legislation. *J Anal Toxicol* 1999;23:490–498.
26. ElSohly MA, Feng S, Salamone, SJ, et al: GC-MS determination of flunitrazepam and its major metabolite in whole blood and plasma. *J Anal Toxicol* 1999;23:486–489.
27. Fitzgerald RL: Analytical toxicology of the benzodiazepines. *Therapeutic Drug Monitoring and Toxicology* 1995;16:169–186.
28. Drummer OH: Methods for the measurement of benzodiazepines in biological samples. *J Chromatogr B* 1998;713:201–225.
29. Verweij AMA, Hordijik ML, Lipman PJL: Liquid chromatographic-thermospray tandem mass spec-

trometric quantitative analysis of some drugs with hypnotic, sedative and tranquilizing properties in whole blood. *J Chromatogr B* 1996;686:27–34.
30. Weinmann W, Lehmann N, Muller C, et al: Identification of lorazepam and sildenafil as examples for the application of LC/ionspray MS and MS/MS with mass spectra library searching in forensic toxicology. *Forensic Sci Int* 2000;113;339–344.
31. Fraser AD: Use and abuse of the benzodiazepines. *Ther Drug Monit* 1998; 20:481–489.

SELF-ASSESSMENT QUESTIONS

1. What are the primary clinical indications for prescribing a benzodiazepine?

2. What are the most significant limitations of long-term benzodiazepine use?

3. List four benzodiazepines in the "Top 200 Drugs" prescribed in the United States last year.

4. Describe four risk factors for benzodiazepine dependence.

5. What are the major analytical considerations when screening and confirming flunitrazepam in biological fluids?

6. List the advantages and limitations of immunoassays and chromatographic methods for benzodiazepine analysis in biological fluids.

CHAPTER 15

Antidepressant Drugs

Paul J. Orsulak, Ph.D., Pei-Ke Liu, M.D., and Linda C. Akers, M.S.

LEARNING OBJECTIVES

After completing this chapter, the reader should be able to:

1. *Summarize recommended therapeutic ranges for antidepressant drugs.*
2. *Explain the relationships between serum concentrations, therapeutic response, and toxicity.*
3. *Describe drug interactions and other factors that alter serum concentrations of antidepressant drugs.*
4. *Understand clinical factors that affect response to antidepressants.*

INTRODUCTION

Tricyclic antidepressant medications were originally introduced in the 1950s. They include amitriptyline, doxepin, nortriptyline, imipramine, desipramine, protriptyline, trimipramine, and clomipramine. Maprotiline and amoxapine are tetracyclic antidepressants that were introduced in the early 1980s, as were the triazolopyridine compound trazodone and the triazolobenzodiazepine alprazolam, which is marketed as an anxiolytic drug. Fluoxetine, a chemically distinct antidepressant, was introduced in 1988. Bupropion, another antidepressant with unique pharmacologic properties, was introduced in 1989. More recently, additional drugs have become available for treatment of depression (sertraline, paroxetine, fluoxamine, and others). Antidepressant medications have also been used to treat childhood enuresis, anorexia nervosa, bulimia, migraine headaches, some forms of chronic pain, childhood depressions, panic disorders, and attention-deficit disorders, although these applications are less well established (1).

Tricyclic antidepressants, although very effective for treating depression, are also associated with a very high rate of mortality. Antidepressants are second only to analgesics in the number of fatalities and represent the fifth largest category of toxic substances involved in emergency episodes. Analgesics and sedative-hypnotics, as well as cleaning substances and animal and insect bites, cause more exposures in adults. In 1999 as in previous years, antidepressants were the second largest category causing fatalities in the United States, after analgesics (2). Non-tricyclic antidepressants accounted for 80% of toxic exposures, but the tricyclic antidepressants were the primary substance implicated in 75% of the antidepressant fatalities.

This chapter updates and discusses current guidelines to form the basis for using serum (or plasma) concentration measurements of antidepressants. It also reviews salient pharmacologic and clinical issues and data that must be taken in account to properly use therapeutic drug monitoring measurements for optimal clinical benefit.

Approximately 60–70% of depressed patients treated with a single antidepressant medication respond well to treatment. The high percentage who fail or who cannot tolerate available compounds has prompted the continuing search for new and more specific agents (3). Effective antidepressant therapy has also been enhanced by refinements in the diagnosis of depression (4,5), by identification of subgroups of patients who may be more or less responsive to antidepressant treatment alone (6–8), and by identification of therapeutic serum concentrations for at least some of the antidepressants more than a decade ago (9–12).

It has been suggested that improved response may occur if antidepressant serum concentration measurements are used to ensure that adequate medication is used (11,13) and the duration of treatment is extended (14).

Most studies of serum or plasma concentrations over the past two decades have focused on correlations between plasma or serum concentrations of antidepressants and their therapeutic effects (9,10,15,16). The monitoring of antidepressant serum concentrations as an adjunct to the clinical management of depressed patients has been developing over recent years. Wider acceptance of these measurements has been hampered, however, by this earlier focus on only the relationship between serum (or plasma) concentrations and antidepressant response alone. Alteration of this focus to a broader one that includes issues of toxicology, drug interactions, generic drug substitutions, and a broader knowledge of pharmacokinetic issues has taken this area far away from the original, more narrowly focused discussions (10,12). The broader attention to relationship between serum (or plasma) concentrations and clinical effects other than just antidepressant effects has triggered development of a series of guidelines for use of serum concentrations more like those used in other fields of medicine.

Table 15–1 summarizes the recommended therapeutic ranges for each antidepressant. They do not require the unequivocal understanding of an exacting numerical relation between blood concentration and antidepressant response alone.

Blood measurements can provide unequivocally useful information rapidly when

TABLE 15–1. Target therapeutic ranges for the antidepressant drugs available in the United States

Drug	Concentration range (µg/L)	Concentration range (nmol/L)
Amitriptyline plus nortriptyline	80–250	288–901
Amoxapine plus 8-hydroxyamoxapine	200–600	637–1912
Clomipramine	70–200	222–635
Desmethylclomipramine	150–300	499–997
Clozapine	100–700	306–2142
Desipramine	125–300	469–1126
Doxepin plus desmethyldoxepin	150–250	537–895
Fluoxetine plus norfluoxetine	200–700	647–2263
Imipramine plus desipramine	150–250	535–892
Maprotiline	200–600	721–2163
Nortriptyline	50–150	190–570
Paroxetine	20–200	61–607
Protriptyline	70–260	266–988
Sertraline	30–200	98–653
Desmethylsertraline	No established range	No established range
Trazodone	800–1600	2152–4303
Trimipramine	150–250	510–849
Venlafaxine	20–150	72–541
Desmethylvenlafaxine	100–600	380–2278

Concentrations >450 µg/L for tricyclic antidepressants (amitriptyline, nortriptyline, imipramine, desipramine, protriptyline, doxepin, and trimipramine) should be considered potentially toxic. For all others, concentrations greater than twice the therapeutic range should be considered potentially toxic. All such concentrations should be evaluated carefully.

a patient fails to show adequate clinical response or no response at all. Often the justification has been made that a drug has failed, when in fact it is the trial that has failed. Given the availability of reliable blood concentrations measurements, it is unsatisfactory to conclude that a patient has failed without documenting that the drug used has at least reached the bloodstream in adequate concentrations. The presence or absence of side effects can also be a valuable indication of trial adequacy, but again, serum concentration can easily disclose whether a patient is particularly susceptible to anticholinergic side effects such that an adequate trial cannot be achieved.

More recently, understanding of some of the relationships between serum concentrations and toxicity has evolved (17,18). These relationships now form the basis for wider application of therapeutic monitoring not only of the older tricyclic drugs such as nortriptyline and imipramine, but for all antidepressants. Measurement of the tricyclic and other antidepressants in serum provides an accurate and useful indicator of optimal dosage and can enable the clinician to correct easily for differences in metabolism, alterations in serum concentrations as a result of drug interactions, and failure to achieve adequate serum concentrations due to noncompliance (9,10,17,19,20). Bourin et al. (21) also reviewed the issue of blood concentration monitoring and concluded that even though exact relationships are not defined for amitriptyline and clomipramine, blood concentrations are useful in cases of treatment failure, the elderly, and overdosage or toxicity.

Case 1: Amitriptyline Dose Response

Ms. L, a 66-year-old white woman, was evaluated after a 3-month depressive episode manifested by anergia, anhedonia, psychic retardation, decreased productivity, mild diurnal variation, and weight loss. Her depression had proved resistant to treatment, both with amitriptyline and imipramine in doses as high as 350 mg/d in successive trials. The patient noted that on high doses of the medication she "felt worse" but did not exhibit even minimal anticholinergic side effects.

The patient was placed on a combination of therapy including 75 mg/d of amitriptyline and 12 mg/d of perphenazine. After 3 weeks, she experienced distinct improvement, although some symptoms persisted. The patient continued this program of treatment, with supportive psychotherapy, for nearly 1 year without complete remission. Evaluation of the patient's medication program was initiated. The amitriptyline was increased to 100 mg/d, after which serum concentrations were 212 ng/mL [amitriptyline 120 ng/mL (433 nmol/L) and nortriptyline 92 ng/mL (350 nmol/L)]. Even though the serum concentrations were within the therapeutic range, the patient's mild but persistent depression indicated that an increase in dosage was required. Amitriptyline was increased from 100 to 125 mg/d.

Upon increase in medication dosage, the patient did not improve but rather became more depressed. No side effects were reported, but the concentration of amitriptyline plus nortriptyline was 373 ng/mL.

The antidepressant dose was reduced over a period of 10 days to 75 mg/d. This reduction in antidepressant dosage resulted in a striking clinical improvement. Upon reexamination 3 months later, the patient remained improved even though her medication had been reduced further to 50 mg/d of amitriptyline and 8 mg/d of perphenazine. Total serum antidepressant concentration at this time was 170 ng/mL.

Several factors in this case point to the value of monitoring the serum concentrations of amitriptyline even though the exact therapeutic range has not been firmly established. Individual differences in drug metabolism due either to genetic factors or to drug interactions can lead to poor correlation between administered dose and serum concentrations of the drug. In this case, a 25% increase in the dosage of amitriptyline from 100 to 125 mg/d resulted in a nearly twofold increase in serum concentration, from 212 to 373 ng/mL (from 765 to 1346 nmol/L). During the episode outlined here, the small increase in dosage and blood concentration actually resulted in decreased clinical effect.

In general, serum measurements of these drugs can be used to determine patient compliance, assess the clinical situation of a patient who has failed to respond, and determine the causes of exaggerated responses. Therapeutic monitoring is also useful in patients who are at risk for significant side effects and for whom polypharmacy is required, and to maximize clinical response over time. Monitoring antidepressants can aid the physician in determining a proper course of action when drug toxicity is confused with the underlying disorder or when large interindividual differences in drug metabolism, such as those seen in elderly patients, lead to poor correlation between administered dose and serum concentration. Serum concentration measurements are also useful when the pharmacokinetic consequences of polypharmacy and drug interactions are unknown.

Case 2: Response to Nortriptyline

Mr. A. was a 66-year-old white man with a major depressive disorder. He was 5 foot 9 inches (1.74 m) tall and weighed 138 pounds (62.6 kg). Physical, laboratory, and neurological examinations were all within normal limits. There was no history of either drug abuse or extensive treatment with medications for prior conditions.

Treatment with nortriptyline was begun at 50 mg/d. His serum concentration on day 5 was 25 ng/mL (95 nmol/L), and the dose was increased to 75 mg/d administered in a single dose at bedtime. A plasma concentration on day 15 was 43 ng/mL (163 nmol/L). Nortriptyline dosage was increased to 125 mg/d at bedtime, but by day 26, serum concentration of nortriptyline had risen to only 62 ng/mL (236 nmol/L). The patient had experienced no remission of his depressive symptoms and was exhibiting no side effects to the antidepressant medication.

Perphenazine (8 mg/d) was then added to this patient's drug regimen, but 5 days later the nortriptyline serum concentration was only 70 ng/mL (266 nmol/L). Over the next week, perphenazine dosage was increased to 48 mg/d and nortriptyline was continued at 125 mg/d, but by day 46 no increase in nortriptyline serum concentration [63 ng/mL (239 nmol/L)] was observed, and the patient continued to show no clinical improvement or side effects.

Nortriptyline was discontinued and desipramine was begun. The dosage of desipramine was rapidly increased to 200 mg/d administered in a single dose at bedtime. Serum concentrations of desipramine drawn at ~0800 on several mornings were completely negative for desipramine, nortriptyline, and all other medications. Because the patient was an inpatient, compliance was relatively easy to monitor, and ward staff members were confident that the patient was taking the medication as prescribed. On days 78 and 79, an additional attempt to document compliance was conducted. Two hundred milligrams of desipramine were administered orally at 0800, and a sample was drawn 6 hours later. On both days, serum concentrations of desipramine were ~20 ng/mL (~75 nmol/L). The guidelines for therapeutic monitoring of tricyclic antidepressants suggest that serum concentrations are useful when large individual differences in absorption, distribution, or metabolism exist. Serum concentrations are also useful when a patient is nonresponsive or when poor compliance is suspected.

The patient in this case failed to respond to two trials of antidepressant drugs, at doses that were reasonable for the patient's clinical condition, age, weight, and physical condition. However, after the trial with perphenazine in conjunction with nortriptyline, we suspected that the patient was showing unusual absorption characteristics because perphenazine usually increases serum concentrations of nortriptyline by blocking conversion of nortriptyline to 10-hydroxynortriptyline. Unusually poor absorption was also suspected when the concentrations of desipramine drawn 12 hours after relatively large (200-mg) single doses yielded undetectable concentrations of drug. Although poor compliance was suspected, it was ruled out by observing the patient ingest the drug. A single-dose pharmacokinetic study, determining peak time, peak serum concentration, and initial half-life—all as expected—confirmed this unusual case of poor antidepressant absorption.

This case illustrates the utility of therapeutic monitoring of antidepressants when treating nonresponsive patients. Assuming that a correct diagnosis has been made and that tricyclic antidepressant therapy is in order, it is essential to verify that a patient is receiving an adequate trial of a particular antidepressant before switching medications, altering the dose, or performing other pharmacologic actions. In this case, serum concentrations of nortriptyline and desipramine were instrumental in identifying an unusual circumstance that probably prevented this patient from achieving serum concentrations high enough to result in clinical improvement.

THERAPEUTIC DRUG CONCENTRATION MONITORING

Measurement of serum concentration of antidepressants can aid the practitioner in determining a proper course of action whenever response to treatment is not as expected. This rationale goes further than previous indications for monitoring only drugs with well-defined relationships between antidepressant response and serum or plasma concentrations (10). The growing understanding of the pharmacology of antidepressant drugs, the relationship between serum concentrations and drug interactions, and the relationships between serum concentrations and antidepressant effects, side effects, and toxic effects that has evolved over the past 5 years of experience now forms the basis of current guidelines for therapeutic monitoring of the tricyclic, tetracyclic, and other antidepressants (17–19).

For therapeutic monitoring to be effective, several conditions must exist (see Table 15–2):

- There must be some relationship between serum (or blood) concentration and clinical effect (17).
- There has to be an appropriate rationale for therapeutic monitoring of drug concentrations (for example, a danger of

TABLE 15–2. Indications for monitoring blood concentrations of antidepressants

A therapeutic range has been established
An optimal therapeutic range or a minimal therapeutic concentration exists
Response is not as expected
- Significant side effects are observed
- Expected side effects do not occur
- Toxicity due to elevated blood concentrations may have occurred
Response is uneven or unstable
- Patient may be noncompliant
- Brand or generic substitution has occurred
- Unusual drug interactions may be present
Unusual or altered metabolism is known or suspected
- Age effects—children, adolescents, elderly
- Race effects
- Concurrent medications may induce or inhibit metabolism
- Medical illness causes altered metabolism
- Lifestyle has changed (weight loss or gain, smoking)

From Depression Guideline Panel: *Depression in primary care*: vol. 1, *Detection and diagnosis clinical practice guideline*, Number 5. [AHCPR Publication No. 93-0550.] Rockville, MD: U.S. Department of Health and Human Services, Public Health Service, Agency for Health Care Policy and Research, 1993; and Depression Guideline Panel: *Depression in primary care*: vol. 2, *Treatment of major depression clinical practice guideline*, Number 5. [AHCPR Publication No. 93-0551.] Rockville, MD: U.S. Department of Health and Human Services, Public Health Service, Agency for Health Care Policy and Research, 1993.

toxicity without clear clinical signs and symptoms).
- There must exist practical methods for measuring the drug that are both analytically and pharmacologically valid. (A valid analytic method is one that offers the precision, accuracy, and sensitivity necessary to provide useful information; pharmacologic validity is achieved when the method measures the drugs and/or cross-reactivities that compromise the data provided. At present, such methods are readily available for all antidepressants.)

The therapeutic ranges for antidepressant response for some antidepressants such as nortriptyline, imipramine, and desipramine are well defined, whereas those for others are less well understood (10,11,15,16). However, the expected serum concentrations (22) that occur in most patients treated successfully are now reasonably well established for most of the antidepressants (see Table 15–1). These values allow us to use serum concentration measurements effectively for most antidepressants to aid the management of patients.

Serum concentration measurements are clearly useful and indicated for a drug such as nortriptyline, which has a defined "therapeutic window," that is, a range of concentrations [50–150 µg/L (190–570 nmol/L)] in which optimal clinical response is expected and in which the drug concentrations both below and above this range are associated with less favorable clinical response (10,12,23). Georgotas et al. (24) also suggested that lower plasma concentrations of nortriptyline, particularly during the early weeks of treatment, may delay treatment response.

When taken together, all of the studies of imipramine indicate that more favorable clinical response occurs when plasma concentrations of imipramine plus its metabolite desipramine are in the range of 200–250 µg/L. Higher concentrations generally produce increased side effects but no further clinical response, although much higher concentrations occasionally are seen in patients with favorable responses (10,12,23).

Later studies by Brosen et al. (25) examined steady-state concentrations of imipramine and its metabolites in 17 hospitalized depressed patients and found that 11 of 12 endogenously depressed patients responded to imipramine when plasma concentrations were optimized but that response was poor in nonendogenous depressed patients. Brosen et al. (25,26) also documented in a study of depressed patients treated with imipramine that kinetics of this drug are both dose dependent and highly variable across patients. Doses ranging from 50 to 400 mg/d were required to optimize plasma concentrations. There was also a disproportionate rise in desipramine concentrations and to a lesser extent imipramine concentrations as dose increased. Similar observations were made by Ereshefsky et al. (27) in a report on 347 patients treated to steady state with imipramine.

The therapeutic range for desipramine (when used alone) is quite wide [125–300 µg/L (469–1125 nmol/L)]. Studies (28–30) have indicated that a minimum threshold of at least 115–125 µg/L (431–469 nmol/L) is necessary in both adult and geriatric patients. Perry et al. (12) found a similar range of 108–158 µg/L (405–593 nmol/L) based on their review of selected studies to 1987.

The therapeutic range for amitriptyline plus its metabolite nortriptyline of ~80–250 µg/L is generally accepted, although controlled studies often yielded conflicting results or failed to find a relation between blood concentration and antidepressant response (10,12). Baumann et al. (31,32) also failed to find a correlation between plasma concentrations of amitriptyline and nortriptyline and clinical response even when hydroxylated metabolites were taken into account in this group of 16 endogenously depressed inpatients. These authors also found no correlation between either total or non–protein-bound amitriptyline and nortriptyline and clinical response in 30 endogenous depressed patients treated with 150 mg/d of amitriptyline for 21 days. However, the reliance solely on the relation between blood concentrations and antidepressant response as measured by a change on selected scales (such as the Hamilton Rating Scale) may be too narrow. The concentrations required to produce toxicity are similar to those found for other antidepressants (>450 µg/L), and issues of compliance and drug concentrations can still be assessed using plasma or serum concentrations of amitriptyline.

Therapeutic ranges, given in Table 15-1, for other antidepressants are based on limited data, but experience over the past 5 years indicates that these values do still serve as reasonable clinical guidelines when a patient's response is not as expected by the clinician.

Even in the absence of well-defined therapeutic ranges, monitoring can be useful and is indicated in many situations. Serum concentration measurements can be useful in a patient who has failed to respond to the drug prescribed in a reasonable period of time and in determining the causes of exaggerated or unusual responses to antidepressants. It is inappropriate to conclude that a patient is a nonresponder unless it can be demonstrated that the steady-state serum concentration was within the expected range for a reasonable period of time. Similarities observed between depressive and toxic symptoms in patients with deteriorating clinical status (18,33,34) often make it difficult for the clinician to correctly assess a patient without quantitative blood concentration information.

For example, several studies (35,36) have now demonstrated that central nervous system (CNS) toxicity (also referred to as toxic delirium) secondary to treatment with tricyclics including amitriptyline, desipramine, and imipramine occurs most frequently at serum concentrations (of parent and metabolite) >450 µg/L. This finding was also reported in children (18). In a further review of this phenomenon, Preskorn and Jerkovich (37) concluded that risk factors for developing CNS toxicity secondary to tricyclic antidepressants were, in order of importance, plasma antidepressant concentration, age (with the elderly at greater risk), and sex (with females at greater risk). From a review of 976 patients treated with tricyclic antidepressants, 58 (6%) developed tricyclic reduced CNS toxicity, which is consistent with other literature, but plasma concentrations >450 µg/L increased the risk >10-fold compared with concentrations <450 µg/L. Further, they pointed out that this syndrome, which includes affective, psychotic, and cognitive symptoms, can precede the toxic delirium by up to 2 weeks and can lead to erroneous clinical decisions.

Tamayo et al. (38) analyzed the incidence of potentially toxic serum concentrations (>400 µg/L) in a group of 196 monitored patients on standard doses of imipramine, amitriptyline, nortriptyline, maprotiline, and clomipramine. Antidepressant concentrations ranged from 403 to 1776 µg/L. Early clinical symptoms, however, suggested that serum concentrations were excessive in only 23% of these cases. The most significant risk factors for developing high plasma concentrations in this group were concurrent use of neuroleptics, advanced age, and daily doses >2.5 mg/kg. Taken together, these studies point to an unequivocal need to avoid such blood concentrations when adequate antidepressant response can be sustained at lower doses and serum concentrations. Notable also is the finding that cardiotoxicity, as evidenced by electrocardiogram (EKG) changes, can also occur at tricyclic concentrations >450 µg/L (19).

Serum concentrations are also useful to verify that a patient has received an adequate trial of antidepressant before a decision is made to change medication or alter dosage. Serum concentration measurements are particularly indicated when alterations in medications result in doses that exceed the recommend limits because they provide objective evidence to substantiate the need and safety of an unusual course of treatment.

Serum concentrations can also be used to assess compliance—an important issue both when initiating therapy and during the course of long-term maintenance therapy. Studies (39) have shown that as many as 50% of the patients treated with antidepressants fail to take the prescribed medication at some point during the course of treatment. Perel (40) found that within-individual differences in plasma concentrations measured repeatedly by over time were related to poor compliance, and he found that patients with large intraindividual variation in plasma concentrations tended to have recurrence of depressive episodes, even when they appeared to be taking medications reliably.

Therapeutic monitoring can identify patients with slow or rapid metabolism and is useful for patients in whom there is a clear need to keep dosage and concentration at a minimum effective concentration. These patients include children, pregnant women, nursing mothers, and patients with hepatic, inflammatory, or renal disease who may exhibit seriously altered metabolism or protein binding. Therapeutic monitoring of antidepressants can also aid the physician in determining a proper course of action when large interindividual differences in drug metabolism lead to a poor correlation between administered dose and serum concentration (27,33). Sparteine oxidation phenotype has also been used (26) to characterize the ability of individuals to metabolize imipramine, and such studies confirm the wide interindividual differences that can occur. Clearance rates for imipramine can be twice as high in rapid metabolizers as in slow metabolizers, and clearance of desipramine can vary more than eightfold in the same individuals. Slow desipramine metabolizers can have half-lives as long as 97 hours compared with 17 hours in rapid metabolizers, who hydroxylate both imipramine and desipramine more quickly. Such large differences can easily lead to toxic accumulation even at modest doses.

Further need to use plasma concentration measurement comes from studies suggesting that nonlinear kinetics do occur in patients treated with antidepressants (25,41). In the latter study, doses of imipramine of 50–350 mg/d were required in rapid metabolizers, whereas doses as low as 20 or 25 mg/d were sufficient in slow metabolizers to achieve therapeutic serum concentrations. Occurrence of nonlinear kinetics that yields unpredictable relationships between dose and blood concentration was particularly pronounced in patients with diabetic neuropathy symptoms. Musa (42) found nonlinear kinetics of trimipramine similar to those found for other tricyclics within the therapeutic range for this drug as well and concluded that blood concentrations are useful in determining the presence of this effect.

Similar effects are likely for all of the antidepressants because of their common metabolism via the P450 system. Studies by Ereshefsky et al. (27) have confirmed the occurrence of wide (20-fold) interpatient variability in steady-state plasma concentrations for both imipramine and doxepin at fixed doses. They further showed a decline in clearance of doxepin as a function of steady-state concentration such that at upper dosage ranges for doxepin, metabolic capacity may be exceeded, yielding greater than expected increases in blood concentrations upon increments in dosage. These studies also identified greater than threefold difference in clearance of doxepin in patients receiving concurrent inducers or inhibitors of hepatic metabolism.

Serum or plasma concentration measurements may be particularly useful when treating children or adolescents. Geller et al. (43–45) noted very large differences in pharmacokinetic measurements in this group, with up to a fivefold variation in half-life in postpubertal subjects. They also noted that prepubertal subjects had shorter half-lives than adults and thus would require more frequent dosing. Preskorn et al. (46) examined steady-state concentrations of imipramine and its metabolites in hospitalized children and found that whereas imipramine varied only 12-fold on similar doses, desipramine concentrations varied up to 72-fold. These authors also found that intraindividual differences were linearly correlated with dose so that plasma concentrations could be used rationally to ensure an adequate but not toxic trial of medication.

More recent attempts to model the kinetics of imipramine and its metabolites in adolescents (47) found significantly increased residence times with increasing age. In adolescent patients, elimination time as measured by half-lives nearly doubled from age 12 to age 18, pointing again to the need to carefully assess dosage requirements in this age group, particularly when daily doses as high as 5 mg/kg might be required. Such doses would produce toxicity in adults. That none of these variables can be predicted in an individual makes serum concentration measurements of antidepressants the only rational means of ensuring an appropriate and safe trial.

Significant differences in plasma concentrations of antidepressants may result when generic preparations of antidepressants are substituted for brand-name drugs (48,49). Monitoring patients who exhibit variable or changing clinical response that otherwise cannot be explained may eliminate problems that result simply because different generic preparations are being dispensed with each prescription. Even altering dose schedules from "once at bedtime" to divided (four times a day) can result in unpredictable blood concentrations (27) in certain patients.

Therapeutic monitoring of tricyclic antidepressant serum concentrations can be useful in managing the treatment of elderly patients. Geriatric patients can attain extremely high plasma concentrations of tricyclic antidepressants when treated with only modest doses, whereas conventional doses can lead to toxic concentrations in such patients (34,50). More recently, Katz et al. (51) studied the pharmacokinetics of nortriptyline in 22 elderly patients with an average age of 84 in an institutional setting and described as medically frail. They found no clinically significant group differences in kinetics between these subjects and younger individuals, and a typical average relationship plasma dose and plasma concentrations [an average dose of 80 mg/d yielded an average blood concentration of 100 µg/L (380 nmol/L)]. However, individual dose requirements varied 20-fold, again pointing to the need to measure plasma concentrations to establish dosage for each individual patient.

Reports on the cardiac side effects of tricyclic antidepressants (19,52–54) have focused attention on the need to avoid toxic plasma concentrations of these drugs in patients with cardiac disease, and there is general agreement that plasma concentration measurements can be useful in determining the optimal dose to be given to such patients (19). As mentioned earlier, significant EKG changes are associated with plasma concentrations >450 µg/L.

Race can also play a significant role in determining the steady-state plasma tricyclic antidepressant concentrations in patients taking these drugs. Early studies by Ziegler and Biggs (55) showed that black patients achieved higher steady-state plasma concentrations of nortriptyline than did white patients receiving similar doses. Japanese patients may require as little as half of the dose usually given to whites (56) to attain the serum blood concentration. Other Asian populations (57) and Hispanics (58) may also respond to less drug (and thus to lower blood concentrations) than whites due to hypersensitive receptors. Time required to attain peak blood concentration may also be shorter in Asian patients (57), leading to different or unusual side-effect profiles for these patients.

The presence of other medications and drug-drug interactions may also significantly alter steady-state plasma concentrations of tricyclic antidepressants (27,59,60). Serum concentration measurements may therefore be important when the pharmacokinetic consequences of polypharmacy and drug interactions are unknown. Whenever a patient is being treated with multiple medications, nonlinear dose-blood concentrations relationships often result (27), yielding unpredictable and potentially dangerous serum concentrations. For example, reports that appeared soon after the introduction of fluoxetine in 1988 (61–65) indicated that coadministration of this drug caused elevations in previously stable tricyclic antidepressant blood concentrations, causing toxic reactions, presumably through inhibition of oxidative metabolism. Subsequent case reports reviewed by Ciraulo and Shader (66) have substantiated this phenomenon.

Hydrocortisone, neuroleptics, and methylphenidate are known to inhibit the metabolism of tricyclic antidepressants by the liver, thus producing higher plasma concentrations (38,67,68), whereas barbiturates, some anticonvulsants, chloral hydrate, and glutethimide stimulate liver microsomal activity, thus lowering tricyclic antidepressant concentrations in plasma (69). Benzodiazepines appear to have no effect on the rate of tricyclic antidepressant metabolism (70,71). Alprazolam, in contrast, reduces clearance of at least imipramine, resulting in a 25% increase in serum concentrations of imipramine and desipramine (72).

Cimetidine, but not ranitidine, coadministered with doxepin or desipramine inhibits the metabolism of these drugs. For doxepin, cimetidine causes an increase in elimination half-life from 19 to 26 hours in healthy volunteers with concurrent increases in plasma concentrations (73–75). Cimetidine, but not ranitidine, also increases the half-life of imipramine from 13 hours to 23 hours in all of 12 volunteers pretreated with this drug through impairment of demethylation (76,77).

Quinidine also reduces clearance of orally administered imipramine by 35% and desipramine by 85%, but unlike the cimetidine effect, this blockade of metabolism appears to occur at the 2-hydroxylation step (78). Slow metabolism of clomipramine has also been reported with concurrent administration of allopurinol, an inhibitor of hepatic drug metabolism (79). Like other tricyclics, clomipramine is also subject to nonlinear kinetics, resulting in longer half-lives than expected in some chronically treated patients (80). Metabolic abnormalities such as Behçet's syndrome or other malabsorption syndromes can also cause marked reduction in absorption and thus in plasma concentrations of amitriptyline (and presumably other lipophilic antidepressants) (81).

Alcoholics have significantly greater clearance of imipramine and desipramine than do healthy individuals, with significantly decreased half-lives, according to one study (82). Half-lives for imipramine were decreased from 19.6 hours to 10.9 hours, and half-lives for desipramine were decreased from 22.5 hours to 16.5 hours. Free drug concentrations were also decreased in alcoholics. Taken together, these observations point to the need for higher doses of imipramine and desipramine in detoxified alcoholics and to the need for periodic plasma concentration measurements to assess the status of such patients being treated with tricyclic antidepressants.

Weight loss may also cause increased plasma concentrations (83). Cigarette smoking lowers steady-state plasma concentrations of tricyclic antidepressants, presumably by induction of liver-metabolizing enzymes (84,85).

Finally, therapeutic monitoring contributes to minimizing the cost of treatment and care (11). Noncompliance or inappropriately low concentrations of drugs will delay the onset of clinical response, increasing the length of stay for hospitalized patients. Inappropriately high concentrations of antidepressant that are seen in many patients treated with routine doses (86) and nonlinear kinetics due to drug interactions (87) may delay response or cause toxicity, leading to additional care.

DISCUSSION

Even though further studies of the relationship between serum concentrations of antidepressants and their therapeutic efficacy are required, it is clear that these determinations can help the clinician to arrive more quickly at the optimal dosage of some antidepressant drugs and that therapeutic monitoring of serum concentrations can eliminate some of the uncertainty associated with the use of these drugs in clinical practice. The traditional practice of accepting failure to respond to an antidepressant as sufficient reason to change medication or course of treatment might fall by the wayside if every treatment failure was monitored and ascertained only after an adequate trial had been used.

Recent advances in the development of methods for analyzing antidepressants have resulted in dramatic increases in the availability of these determinations. The application of therapeutic drug monitoring in the treatment of patients with depressive disorders may begin to enhance clinical care by ensuring that a therapeutic antidepressant serum concentration has been obtained; by reducing the risk of treating patients with toxic dosages; and by eliminating some of the uncertainty associated with the use of these drugs in clinical practice. It is clear that serum concentration determinations can aid the clinician in arriving more quickly at the optimal dosage of some of these drugs for many patients. Despite our understanding of the multiple side effects of the antidepressant drugs (see Table 15–3) and drug interactions, patients whose clinical condition suggests that they would respond to these drugs often fail to do so. In this population, monitoring of antidepressant concentrations can become an important therapeutic tool. Failure of a patient to respond, poor compliance, clinician or patient limitation of doses resulting from fear of cardiovascular or other side effects, or drug interactions can all be corrected through appropriate pharmacological management. Further studies of antidepressant pharmacokinetics and pharmacodynamics to identify the most significant correlations with clinical effects will also increase the utility of therapeutic drug monitoring in depressed patients.

REFERENCES

1. Orsulak PJ, Waller D: Antidepressant drugs: additional clinical uses. *J Fam Pract* 1989;28:209–216.
2. Litovitz TL, Klein-Schwartz W, Caravati EM, et al: 1998 Annual Report of the American Association of Poison Control Centers Toxic Exposure Surveillance System. *Am J Emerg Med* 2000;18:517–574.
3. Ostrow D: The new generation antidepressants: promising innovations or disappointments? *J Clin Psychiatry* 1985;46:2–30.
4. Nelson JC, Charney DS: Primary affective disorder criteria and the endogenous-reactive distinction. *Arch Gen Psychiatry* 1980;37:787–793.
5. Nelson JC, Charney DS, Quinlan DM: Evaluation of the DSM-III criteria for melancholia. *Arch Gen Psychiatry* 1981;38:555–559.
6. Glassman AH, Shostak M, Kanton SJ: Imipramine and delusional depressions. *Am J Psychiatry* 1979;136:462–463.
7. Charney DS, Nelson JC: Delusional and non-educational unipolar depression: further evidence for distinct subtypes. *Am J Psychiatry* 1981;138:328–333.
8. Spiker DG, Weiss JC, Dealy RS, et al: The pharmacological treatment of delusional depression. *Am J Psychiatry* 1985;142:430–436.
9. Amsterdam J, Brunswick D, Mendels J: The clinical application of tricyclic antidepressant pharmacokinetics and plasma levels. *Am J Psychiatry* 1980;137:653–662.
10. Glassman AH, Schildkraut JJ, Orsulak PJ, et al: Tricyclic antidepressant blood level measurements and clinical outcome: an APA task force report. *Am J Psychiatry* 1985;142:155–162.
11. Preskorn SH: Tricyclic antidepressant plasma level monitoring: an improvement over the dose-response approach. *J Clin Psychiatry* 1986;47[1 Suppl]:24–30.
12. Perry PJ, Pfohl BM, Holstad SG: The relationship between antidepressant response and tricyclic antidepressant plasma concentrations. A retrospective analysis of the literature using logistic regression analysis. *Clin Pharmacokinet* 1987;13:381–392.
13. Quitkin FM: The importance of dosage in prescribing antidepressants. *Br J Psychiatry* 1985;147:593–597.
14. Quitkin FM, Rabkin JG, Ross D, et al: Duration of antidepressant drug treatment. *Arch Gen Psychiatry* 1984;41:238–245.
15. Scoggins BA, Maguire KP, Norman TR, et al: Measurement of tricyclic antidepressants. Part II. Applications of methodology. *Clin Chem* 1980;26:805–815.
16. Van Brunt N: The clinical utility of tricyclic antidepressant blood levels: a review of the literature. *Ther Drug Monit* 1983;5:1–10.
17. Preskorn SH, Dorey RC, Jerkovich GS: Therapeutic drug monitoring of tricyclic antidepressants. *Clin Chem* 1988;34:822–828.
18. Preskorn SH, Weller E, Jerkovich G, et al: Depression in children: concentration-dependent CNS

TABLE 15–3. *Symptoms of tricyclic antidepressant overdose*

Anticholinergic effects
- Severe mouth dryness
- Dilated pupils
- Hyperpyrexia
- Urinary retention
- Intestinal stasis

Central nervous system effects
- Excitement, restlessness
- Myclonus, hyperreflexia
- Seizures
- CNS depression
- Drowsiness
- Areflexia
- Hallucinations
- Respiratory depression
- Hypothermia
- Coma

Cardiovascular effects
- Sinus tachycardia
- Interventricular conduction defects
- Ventricular dysrhythmias
- Ventricular tachycardia
- Ventricular fibrillation
- Arteriovenous conduction defects
- Bradycardia
- Cardiac arrest

toxicity of tricyclic antidepressants. *Psychopharmacol Bull* 1988;24:140–142.
19. Preskorn SH, Kent TA: Mechanisms and interventions in tricyclic antidepressant overdoses. In: Stancer HC, Garfinkel PE, Rakoff V, eds: *Guidelines for the use of psychotropic drugs.* New York: SP Medical and Scientific Books. 1984:63–75.
20. Voris JC, Morin C, Kiel JS: Monitoring outpatient's plasma antidepressant-drug concentrations as a measure of compliance. *Am J Hosp Pharm* 1983; 40:119–129.
21. Bourin MS, Kergueris MF, Lapierre YD: Therapeutic monitoring of treatment with antidepressants. *Psychiatr J Univ Ott* 1989;14:460–462.
22. Orsulak PJ: Therapeutic monitoring of antidepressant drugs: current methodology and applications. *J Clin Psychiatry* 1986;47:39–50.
23. Balant-Gorgia AE, Balant LP, Garrone G: High blood concentrations of imipramine or clomipramine and therapeutic failure: a case report study using drug monitoring data. *Ther Drug Monit* 1989; 11:415–420.
24. Georgotas A, McCue RE, Cooper TB, et al: Factors affecting the delay of antidepressant effect in responders to nortriptyline and phenelzine. *Psychiatry Res* 1989;28:1–9.
25. Brosen K, Gram LF, Klysner R, et al: Steady-state levels of imipramine and its metabolites: significance of dose-dependent kinetics. *Eur J Clin Pharmacol* 1986;30:43–49.
26. Brosen K, Otton SV, Gram LF: Imipramine demethylation and hydroxylation: impact of the sparteine oxidation phenotype. *Clin Pharmacol Ther* 1986;40:543–549.
27. Ereshefsky L, Tran-Johnson T, Davis CM, et al: Pharmacokinetic factors affecting antidepressant drug clearance and clinical effect: evaluation of doxepin and imipramine—new data and review. *Clin Chem* 1988;34:863–880.
28. Nelson JC, Jatlow PI, Quinlan DM: Subjective complaints during desipramine treatment. Relative importance of plasma drug concentrations and the severity of depression. *Arch Gen Psychiatry* 1984; 41:55–59.
29. Nelson JC, Mazure C, Quinlan DM, et al: Drug-responsive symptoms in melancholia. *Arch Gen Psychiatry* 1984;41:663–668.
30. Nelson JC, Jatlow PI, Mazure C: Desipramine plasma levels and response in elderly melancholic patients. *J Clin Psychopharmacol* 1985;5:217–220.
31. Baumann P, Jonzier-Perey M, Koeb L, et al: Amitriptyline pharmacokinetics and clinical response: II. Metabolic polymorphism assessed by hydroxylation of debrisoquine and mephenytoin. *Int Clin Psychopharmacol* 1986;1:102–112.
32. Baumann P, Jonzier-Perey M, Koeb L, et al: Amitriptyline pharmacokinetics and clinical response: I. Free and total plasma amitriptyline and nortriptyline. *Int Clin Psychopharmacol* 1986;1:89–101.
33. Appelbaum PS, Russell GV, Orsulak PJ, et al: Clinical utility of tricyclic antidepressant blood levels: a case report. *Am J Psychiatry* 1979;136:339–341.
34. Preskorn SH, Simpson S: Tricyclic-antidepressant-induced delirium and plasma drug concentration. *Am J Psychiatry* 1982;139:822–823.
35. Preskorn SH, Biggs JT: Use of tricyclic antidepressant blood levels. *New Engl J Med* 1978;298:166.
36. Preskorn SH, Weller EB, Weller RA, et al: Plasma levels of imipramine and adverse effects in children. *Am J Psychiatry* 1983;140:1332–1335.
37. Preskorn SH, Jerkovich GS: Central nervous system toxicity of tricyclic antidepressants: phenomenology, course, risk factors, and role of therapeutic drug monitoring. *J Clin Psychopharmacol* 1990; 10:88–93.
38. Tamayo M, Fernandez de Gatta MM, Gutierrez JR, et al: High levels of tricyclic antidepressants in conventional therapy: determinant factors. *Int J Clin Pharmacol Ther Toxicol* 1988;26:495–499.
39. Johnson D: Study of the use of antidepressant medication in general practice. *Br J Psychiatry* 1974; 125:186–212.
40. Perel J: Compliance during tricyclic antidepressant therapy: pharmacokinetic and analytical issues. *Clin Chem* 1988;34:881–887.
41. Sindrup SH, Brosen K, Gram LF: Nonlinear kinetics of imipramine in low and medium plasma level ranges. *Ther Drug Monit* 1990;12:445–449.
42. Musa MN: Nonlinear kinetics of trimipramine in depressed patients. *J Clin Pharmacol* 1986;29:746–747.
43. Geller B, Cooper TB, Chestnut EC, et al: Preliminary data on the relationship between nortriptyline plasma level and response in depressed children. *Am J Psychiatry* 1986;143:1283–1286.
44. Geller B, Cooper TB, McCombs HG, et al: Double-blind, placebo-controlled study of nortriptyline in depressed children using a "fixed plasma level" design. *Psychopharmacol Bull* 1989;25:101–108.
45. Geller B, Cooper TB, Graham DL, et al: Double-blind placebo-controlled study of nortriptyline in depressed adolescents using a "fixed plasma level" design. *Psychopharmacol Bull* 1990;26:85–90.
46. Preskorn SH, Bupp SJ, Weller EB, et al: Plasma levels of imipramine and metabolites in 68 hospitalized children. *J Am Acad Child Adolesc Psychiatry* 1989;28:373–375.
47. Dell RB, Hein K, Ramakrishnan R, et al: Model for the kinetics of imipramine and its metabolites in adolescents. *Ther Drug Monit* 1990;12:450–459.
48. Ostroff R, Dochester J: Tricyclics, bioequivalency and clinical response. *Am J Psychiatry* 1978;135: 1560–1561.
49. Rosenbaum JF, Falk WE, Gastfriend DR, et al: Acute distress after switch from Norpramin to generic desipramine. *Am J Psychiatry* 1989;146:122.
50. Nies A, Robinson DS, Friedman MJ, et al: Relationship between age and tricyclic antidepressant plasma levels. *Am J Psychiatry* 1977;134:790–793.
51. Katz IR, Simpson GM, Jethanandani V, et al: Steady state pharmacokinetics of nortriptyline in the frail elderly. *Neuropsychopharmacology* 1989;2: 229–236.
52. Bigger JT, Kantor SJ, Glassman AH, et al: Cardiovascular effects of tricyclic antidepressant drugs: In: Lipton MA, DiMascio A, Killam KF, eds: *Psychopharmacology: a generation of progress.* New York: Raven Press, 1978:1033–1046.
53. Glassman AH, Bigger JT, Giardina EV, et al: Clinical characteristics of imipramine-induced orthostatic hypotension. *Lancet* 1979;1:468–472.
54. Kantor SJ, Glassman AH, Bigger JT, et al: The cardiac effects of therapeutic plasma concentrations of imipramine. *Am J Psychiatry* 1978;135:534–538.

55. Ziegler VE, Biggs JT: Tricyclic plasma levels. Effect of age, race, sex, and smoking. *JAMA* 1977;238:2167–2169.
56. Kishimoto A, Hollister L: Nortriptyline kinetics in Japanese and Americans. *J Clin Psychopharmacol* 1984;4:171–172.
57. Pi EH, Simpson GH, Cooper TB: Pharmacokinetics of desipramine in Caucasian and Asian volunteers. *Am J Psychiatry* 1986;143:1174–1176.
58. Gaviria M, Gil AA, Javaid JI: Nortriptyline kinetics in Hispanic and Anglo subjects. *J Clin Psychopharmacol* 1986;6:227–231.
59. Hollister LE: Interaction of psychotherapeutic drugs with other drugs and disease states. In: Lipton MA, DiMascio A, Killam KF, eds: *Psychopharmacology: a generation of progress*. New York: Raven Press, 1978:987–992.
60. Perel JM, Stiller RL, Glassman AH: Studies on plasma level/effect relationships in imipramine therapy. *Commun Psychopharmacol* 1978;2:429–439.
61. Aranow AB, Hudson JI, Pope HG, Jr, et al: Elevated antidepressant plasma levels after addition of fluoxetine. *Am J Psychiatry* 1989;146:911–913.
62. Bell IR, Cole JO: Fluoxetine induces elevation of desipramine level and exacerbation of geriatric nonpsychotic depression [Letter]. *J Clin Psychopharmacol* 1988;8:447–448.
63. Dista Products Company: Prozac (fluoxetine) Package insert update. Dista Marketing letter No. 33, 1988. [Dista Products Company, a division of Eli Lilly and Company, Lilly Corporate Center, Indianapolis, IN 46285]
64. Rudorfer MV, Potter WZ: Combined fluoxetine and tricyclic antidepressant. *Am J Psychiatry* 1989;146:562–564.
65. Vaughan DA: Interaction of fluoxetine with tricyclic antidepressants. *Am J Psychiatry* 1988;145:1478.
66. Ciraulo DA, Shader RI: Fluoxetine drug-drug interactions: I. Antidepressants and antipsychotics. *J Clin Psychopharmacol* 1990;10:48–50.
67. Gram LF, Over KF: Drug interaction: inhibitory effect of neuroleptics on metabolism of tricyclic antidepressants in man. *Br Med J* 1972;1:463–465.
68. Nelson JC, Jatlow PI: Neuroleptic effect on desipramine steady-state plasma concentrations. *Am J Psychiatry* 1980;137:1232–1234.
69. Alexanderson B, Evans DAP, Sjoqvist F: Steady-state plasma levels of nortriptyline in twins: influence of genetic factors and drug therapy. *Br Med J* 1969;4:764–768.
70. Otani K, Nordin C, Bertilsson L: No interaction of diazepam on amitriptyline disposition in depressed patients. *Ther Drug Monit* 1987;9:120–122.
71. Silverman G, Braithwaite RA: Benzodiazepines and tricyclic antidepressant plasma levels. *Br Med J* 1973;3:18–20.
72. Ereshefsky L, Antal EJ, Wells BG, et al: Multi-center evaluation of the kinetic and clinical interaction of alprazolam and imipramine [Abstract]. *Clin Pharmacol Ther* 1986;39:178.
73. Abernethy DR, Todd EL: Doxepin-cimetidine interaction: increased doxepin bioavailability during cimetidine treatment. *J Clin Psychopharmacol* 1986;6:8–12.
74. Steiner E, Spina E: Differences in the inhibitory effect of cimetidine on desipramine metabolism between rapid and slow debrisoquin hydroxylators. *Clin Pharmacol Ther* 1987;42:278–282.
75. Sutherland DL, Remillard AJ, Haight KR, et al: The influence of cimetidine versus ranitidine on doxepin pharmacokinetics. *Eur J Clin Pharmacol* 1987;32:159–164.
76. Spina E, Koike Y: Differential effects of cimetidine and ranitidine on imipramine demethylation and desmethylimipramine hydroxylation by human liver microsomes. *Eur J Clin Pharmacol* 1986;30:239–242.
77. Wells BG, Pieper JA, Self TH, et al: The effect of ranitidine and cimetidine on imipramine disposition. *Eur J Clin Pharmacol* 1986;31:285–290.
78. Brosen K, Gram LF: Quinidine inhibits the 2-hydroxylation of imipramine and desipramine but not the demethylation of imipramine. *Eur J Clin Pharmacol* 1989;37:155–160.
79. Balant-Gorgia AE, Balanti L, Zysset T: High plasma concentrations of desmethylclomipramine after chronic administration of clomipramine to a poor metabolizer. *Eur J Clin Pharmacol* 1987;32:101–102.
80. Kuss HJ, Jungkunz G: Nonlinear pharmacokinetics of chlorimipramine after infusion and oral administration in patients. *Prog Neuropsychopharmacol Biol Psychiatry* 1986;10:739–748.
81. Chaleby K, El-Yazigi A, Atiyeh M: Decreased drug absorption in a patient with Behcet's syndrome. *Clin Chem* 1987;33:1679–1681.
82. Ciraulo DA, Barnhill JG, Jaffe JH: Clinical pharmacokinetics of imipramine and desipramine in alcoholics and normal volunteers. *Clin Pharmacol Ther* 1988;43:509–518.
83. Jobson K, Burnett G, Linniola M: Weight loss and a concomitant change in plasma tricyclic levels. *Am J Psychiatry* 1978;135:237–238.
84. Linniola M, George L, Guthrie S, et al: Effect of alcohol consumption and cigarette smoking on antidepressant levels of depressed patients. *Am J Psychiatry* 1981;138:841–842.
85. Perry PJ, Browne JL, Prince RA, et al: Effects of smoking on nortriptyline plasma concentrations in depressed patients. *Ther Drug Monit* 1986;8:279–284.
86. Garvey MJ, Tuason VB, Johnson RA, et al: Elevated plasma tricyclic levels with therapeutic doses of imipramine. *Am J Psychiatry* 1984;141:853–856.
87. Dugas JE, Bishop DS: Nonlinear desipramine pharmacokinetics: a case study. *J Clin Psychopharmacol* 1985;5:43–45.

SELF-ASSESSMENT QUESTIONS

1. Central nervous system toxicity of tricyclic antidepressants should be considered when serum concentrations exceed what value?
   a. 200 ng/mL
   b. 300 ng/mL
   c. 400 ng/mL
   d. 450 ng/mL
   e. 500 ng/mL

2. What can serum concentrations of antidepressant drugs be used to do?
   a. assess compliance
   b. monitor drug interaction
   c. identify nonlinear kinetics
   d. ensure safe use of antidepressant drugs
   e. all of the above

3. True or false: Accurate therapeutic ranges have been developed for all antidepressants currently on the market in the United States.
   a. true
   b. false

4. True or false: Therapeutic ranges of tricyclic and other antidepressants can be used to accurately predict clinical outcome.
   a. true
   b. false

5. Which of the following drugs generally increase serum concentrations of antidepressants?
   a. alcohol
   b. neuroleptics
   c. fluoxetine
   d. methylphenidate
   e. all of the above

6. Which of the following drugs tend to reduce tricyclic antidepressant plasma concentrations?
   a. barbiturates
   b. carbamazepine
   c. chlorohydrate
   d. none of the above
   e. all of the above

7. What are the most significant risk factors for developing high plasma concentrations?
   a. age
   b. concurrent use of neuroleptics
   c. concurrent use of anticonvulsants
   d. a and b
   e. a and c

CHAPTER 16

Agents for the Treatment of Bipolar Disorder

Paul J. Orsulak, Ph.D., M.B.A.

LEARNING OBJECTIVES

After completing this chapter, the reader should be able to:

1. Discuss the uses of anticonvulsant drugs such as carbamazepine and valproic acid in the treatment of bipolar disorder.
2. Describe the rationale for serum concentration monitoring of anticonvulsants in the treatment of bipolar disorder.
3. Describe drug interactions and side effects associated with anticonvulsant treatment of bipolar patients.

INTRODUCTION

Bipolar disorder afflicts an estimated 0.4–1.2% of the general U.S. population. Unlike unipolar depression, bipolar disorder afflicts men and women almost equally (1). The onset typically occurs early in the third decade, but can also start late in life. Manic episodes, which can last from days to months, usually begin suddenly and subside fairly abruptly, typically switching into a prolonged depression lasting months. The course of the illness between episodes is not well characterized.

Bipolar I disorder is defined as a history of one or more manic episodes and one or more major depressive episodes with "normal" euthymic intervening intervals. Episodes in bipolar I disorder are subclassified as either mixed, manic, or depressed, depending on the clinical features of the current episodes or, if the disorder is in partial or full remission, the most recent episode. In bipolar II disorder, only hypomania occurs, although the depressions may be just as significant as in bipolar I disorder. Rapid cycling of at least four discrete affective episodes per year occurs in both bipolar I and bipolar II disorders. There are no manic or hypomanic episodes in unipolar disorder. Unipolar depression may have recurrent episodes of depression.

Genetic research in bipolar disorder suggests that there are several modes of inheritance. One group of researchers has reported on a subgroup of early-onset bipolar patients (mean age 25) who have an X-linked genetic defect. Another group has found evidence of an association with an autosomal dominant gene on the short arm of chromosome 11.

An apparent increased prevalence of bipolar disorder in the higher socioeconomic classes is controversial and may represent overdiagnosis of schizophrenic disorders in lower socioeconomic classes.

For proper diagnosis, a manic episode needs to be distinguished from other psychopathological events that can mimic mania, such as a schizophrenic episode. An important distinction is that the manic patient, unlike the schizophrenic patient, is free from hallucinations and delusions, except those that have occurred in the context of the manic episode and have not lasted for >2 weeks beyond the episode. During an episode of severe mania or depression, however, patients may exhibit symptomatology that is indistinguishable from schizophrenia. A thorough history obtained from reliable friends or family is crucial to making valid diagnoses.

Mania cannot be discretely discriminated from certain periods either enjoyed during normal reactive pleasure or associated with psychotic phenomena from any cause. Most patients who are severely manic with psychotic features cannot be reliably discriminated from patients with psychosis from other causes.

During the passage from normalcy (euthymia) to severe mania, patients usually pass through a hypomanic stage that is similar to normal reactive happiness. Typically, bipolar patients next show classic manic symptoms with either euphoria-grandiosity or paranoia-hostility predominating, finally reaching an undifferentiated psychotic state. During recovery, the patient generally passes down through the same stages.

The ideal pharmacologic agent for the treatment of bipolar disorder would possess acute and prophylactic antimanic as well as antidepressant properties. Historically, lithium and neuroleptics have been the mainstays for treating patients with bipolar disorder. Efficacy of lithium has been fairly well documented, but the high incidence of side effects may interfere with patient compliance.

For those who cannot tolerate or do not respond to these traditional medications, the choice of other treatment alternatives has been limited until recently (2,3). Currently, carbamazepine and valproic acid are considered effective alternatives for treating bipolar illness (4,5), with valproic acid (Depakote®) having Food and Drug Administration (FDA) approval for this purpose (6–9). These agents have been used for many years in the treatment of seizure disorders, and their toxicology is well understood. Although anticonvulsants are listed among agents responsible for fatalities, the number is relatively small, with the entire group accounting for 14 fatalities in 1999 (10).

LITHIUM

Lithium has been used mostly for treatment of acute mania and long-term prophylaxis of recurrent mania and depressive episodes. There are >100 studies involving 3000 patients with acute mania in 20 countries indicating that lithium is effective in 60% of patients. Although there are numerous reports of lithium enhancing the antidepressant effect of tricyclic antidepressants (TCAs) and monoamine oxidase inhibitors (MAOIs), it has a more modest antidepressant effect when used alone in most patients.

For treating and preventing mixed affective states and schizoaffective disorder, lithium is usually inadequate, and adjunctive neuroleptic therapy is often required, both acutely and for prophylaxis.

Clinical experience indicates that as many as 40% of patients in the acute manic phase of bipolar disorder (DSM-III-R) fail to respond adequately to treatment with lithium. Additionally, lithium alone may be even less effective in correcting manic behavior in certain patients: the agitated or psychotic patient, the patient with mixed mania and depression, and the patient with rapid cycling. The extent of this lithium-resistant patient population highlights the need for alternative treatment options. The toxic side effects that often occur with lithium therapy preclude continued treatment for many patients, thus further necessitating the need for alternative treatment options.

Lithium has a high incidence of side effects. However, the frequency of such symptoms varies as a function of the patient's affective state; some side effects are greater during euthymia and others during affective illness. Side effects include tremor, nausea, polyuria, acne, and hypothyroidism. Patient concern and compliance will consequently vary during different stages of the illness. Adverse events resulting from lithium do occur but most are readily treated. Fatalities, however, are rare. In 1999, only six fatalities were reported involving lithium, with four being reported as intentional suicide and two reported as therapeutic error (10).

Although neuroleptics are widely used in

the acute treatment of bipolar disorder, their utility is limited by potentially significant side effects (akathesia, dystonias, akinesia, dyskinesias, orthostasis, anticholinergic effects, and sedation). Most often, neuroleptic drugs are discontinued after the acute episode because therapy with these agents requires close monitoring.

Clonazepam is a benzodiazepine anticonvulsant that may be used for selected seizure types, as an antipanic agent, and to rapidly reduce activated psychotic symptoms in patients able to take oral medication. Initial dosage ranges from 0.25 to 2 mg every 2–4 hours, with lower doses for elderly or organically impaired patients. Clinical correlation is necessary to determine the daily dose. Concomitant use of antimanic or antipsychotic treatments should be begun, and all are compatible with clonazepam. Once behavioral control is obtained, the dose is reduced using patient sedation as a marker. Clonazepam may also be used for outpatients for acute intervention at the first signs of suspected mania. Withdrawal effects are rare because of its relatively long half-life (~40 hours).

The three most common side effects to clonazepam are ataxia, drowsiness, and paradoxical behavior changes, including disinhibition. The first two tend to be dose related and subside with chronic administration. Less frequently reported adverse reactions include treatment-emergent depression and sexual dysfunction.

TCAs clearly can exacerbate the "switch" process in bipolar patients and also induce rapid cycling. MAOIs are less likely to do so. There are numerous case reports of some patients becoming responders as blood concentrations of lithium are increased [that is, from 0.6 to 1.2 mEq/L (0.6 to 1.2 mmol/L)]. This is commonly observed for acute manic episodes and less so for prophylaxis. Some literature suggests that "subclinical" hypothyroidism may predispose bipolar patients to have a labile mood and/or incomplete response to lithium. The best indicator of this is an elevation of serum thyroid-stimulating hormone (TSH) (using an ultrasensitive assay), which may be at the upper range of normal limits.

Lithium also causes a benign increase in the leukocyte count (WBC) of ~25% and commonly has antithyroid effects, usually detected by an increase in TSH. Long-term lithium therapy is in rare cases associated with a decrease in renal function (reduced glomerular filtration rate), most sensitively measured in the blood by serum creatinine concentration.

Case 1: A Case of Lithium Poisoning

A 36-year-old woman was rushed to the emergency department in a stuporous state. When she was aroused, her speech was slurred, but she was able to complain of blurred vision. The patient was unable to provide the remainder of her history. Her family and a clinic psychiatrist who had last seen her several months before admission were helpful. The woman had a history of emotional problems for which she had been seen by numerous psychiatrists over the past few years. Because she had not continued with one psychiatrist long enough, no definite diagnosis was made. The patient's purse contained bottles of phenothiazines, a TCA, lithium carbonate, and several analgesic mixtures.

The patient was sweating and was poorly nourished. Her blood pressure was 140/85 mmHg, her pulse was 110 beats/min, respirations were 18/min, and her temperature was 99.1 °F (37.3 °C).

There were no visible signs of head trauma. An external exam of the eye was normal, and extraocular movements were normal, but bilateral horizontal nystagmus was present. The pupils were equal and reactive to light. The cranial nerves tested were within normal limits. The patient's heart rate was regular, with no murmurs, thrills, heaves, or gallop sounds.

Neurologic exam revealed that the patient was agitated on stimulation. She had faint tremor and fasciculations in both upper extremities. There was no motor weakness, but the deep tendon reflexes were exaggerated. Twice during the exam, clonus of both lower extremities was present, and choreoathetoid movements were observed.

At this point, blood samples were drawn and a urine specimen collected. The patient was given 50 mL of 50% dextrose in water and 100 mg of thiamine intravenously. Her clinical condition did not change.

The initial laboratory data obtained revealed normal values for glucose, creatinine, electrolytes, calcium, thyroid function tests, liver function tests, complete blood count (CBC), arterial blood gases, and carboxyhemoglobin.

The electrocardiogram (ECG) showed a regular sinus rhythm at 110 beats/min with a normal QRS complex, nonspecific ST segment, and T-wave changes in addition to U-waves. The chest radiograph was normal. A lumbar puncture was performed. All cerebrospinal fluid determinations (glucose, total protein, microscopic examination) were normal.

The differential diagnosis in this patient is complex given the multitude of symptoms and signs. Psychosis could explain some of the clinical findings but is a diagnosis of exclusion that would be considered only after an organic cause is excluded. The following should be high on the list for differential diagnosis: sedative-hypnotic or alcohol withdrawal, endocrine dysfunction such as thyrotoxicosis or hypoglycemia, cerebrovascular accidents, and stimulant abuse. The presence of the abnormal muscle movements are more consistent with a toxic manifestation of exposure to organophosphates, cholinergic substances, xanthines, ergot alkaloids, barium salts, fluorides, heavy metals, lithium, strychnine, quaternary ammonium compounds, nicotine, succinylcholine, scorpion and black widow spider venom, and/or snake bites.

Given this patient's known psychiatric history and possession of several psychopharmacologic medications, the most likely cause is an overdose of a combination of cyclic antidepressants, neuroleptic agents, and lithium.

Emergency drug analyses revealed a serum lithium concentration of 3.6 mEq/L (3.6 mmol/L), which is much greater than the therapeutic range of 0.6–1.2 mEq/L (0.6–1.2 mmol/L). Lithium intoxication is often difficult to establish on purely clinical grounds because the onset of symptoms is often insidious and the neuropsychiatric manifestations are nonspecific. The anticholinergic side effects of cyclic antidepressants were not present in this patient, nor were the extrapyramidal effects of phenothiazines.

This patient was treated with activated charcoal for absorption of organic drug present in the gastrointestinal tract before the toxicologic analysis data were available on the assumption that a multiple-drug ingestion was involved. Once the serum lithium concentration value was reported, hemodialysis was instituted for enhanced removal of lithium. The patient's altered mental status and neuromuscular abnormalities disappeared within the next 11–12 days. Although lithium clearly appeared to be the major contributor to this toxic event, other drugs could not be ruled out. Often an emergency drug screen, which focuses on drugs of abuse, does not look for or detect psychoactive agents such as antidepressants or neuroleptic drugs. The use of activated charcoal was appropriate course of action.

ANTICONVULSANTS

As early as 1966, the amide form of valproic acid as an acute and prophylactic agent for patients with affective disorders was used. Similarly, in 1973, the use of carbamazepine was reported. Today, both valproic acid and carbamazepine have emerged as useful agents in bipolar illness. These anticonvulsants appear to be selectively effective for specific psychiatric disorders. Anticonvulsants are also associated with fewer side effects than the traditional antimanic agents lithium and neuroleptics, and they have minimal effect on cognitive function.

Current research and clinical experience suggest that the response rate to anticonvulsant therapy for bipolar disorder is most promising in certain patients. These potentially treatment-responsive patients frequently present with organic affective disorders, tend to be refractory to lithium therapy, exhibit rapid cycling, and/or may be noncompliant with lithium therapy because of associated side effects.

All three major specific antimanic therapies (lithium, valproic acid, and carbamazepine) are administered by increasing initial doses in a stair-step fashion, using clinical response in conjunction with blood concentrations to achieve optimal dosage.

Japanese investigators first suggested the use of carbamazepine in treating affective disorders in the early 1970s. Favorable preliminary results prompted several investigators in the United States to study carbamazepine in psychiatric disorders.

A double-blind study of carbamazepine vs. chlorpromazine in patients with acute mania was performed. Ballenger (11) was one of the first to report results in two double-blind, placebo-controlled trials of carbamazepine in affective disorders. These studies showed a 63–70% moderate to marked improvement in patients with acute mania.

Numerous open and several controlled studies have been published in the psychiatric literature investigating the use of carbamazepine for major affective disorders. Despite the variability in patient selection and study design, marked to moderate improvement in manic and depressive symptoms has been reported in ~60% of the patients treated. The antimanic response appears to be greater than the antidepressant response.

Even though carbamazepine is a TCA, like amitriptyline and imipramine, it has a slow and often incomplete antidepressant effect. Clinical opinion and case studies suggest that it is more effective in atypical bipolar treatment than lithium, both acutely and prophylactically. The response of the symptoms of mania to carbamazepine does not depend on, and is not a function of, the presence of neurologic findings. Response, or lack thereof, to one anticonvulsant is not predictive of response to another, so each should be tried and evaluated independently.

The positive effects of carbamazepine appear to be greater in nonclassic mania, rapid cycling, dysphoric-mixed states, and cases without genetic factors. In the early 1980s, Brennan et al. (12) reported that four out of five, or 80%, of patients responded to valproic acid, and Bowden et al. (9) then reported that six out of eight, or 75%, in another cohort showed marked response to valproic acid. In a more recent study, 36 patients, 17 of whom received valproic acid, were studied in the largest double-blind, placebo-controlled trial of an antimanic agent to date. Nine of 17 patients showed a marked response to valproic acid. In these three blind, controlled studies, 10 out of 14, or 71%, of acutely manic patients showed a marked response (complete remission of symptoms) to valproic acid.

Several other investigators have reported on the use of valproic acid in bipolar illness, with moderate to marked response in >50% of the treated patients. Examination of the current literature reveals 12 studies reporting on the use of valproic acid in bipolar or schizoaffective disorder. In those studies, 297 patients received the drug on an open basis and 31 under blind and placebo-controlled conditions. At least one-third of the patients responded inadequately to or could not tolerate conventional treatments. Patients experiencing significant improvement with valproic acid therapy were 185 of 328, or 56%, and some patients were successfully maintained on valproic acid for periods ranging from several months to many years.

Essential information required for successful drug therapy with anticonvulsants includes pharmacokinetics, drug interactions, and nontherapeutic effects of administered drugs.

The metabolic breakdown of carbamazepine (98%) occurs in the hepatic microsomal oxidative enzyme system (P450). Clinical complications from this system include lowered concentrations of coadministered drugs as well as carbamazepine itself (before autoinduction). In addition, with chronic administration, the half-life of carbamazepine is decreased, and more frequent dosing becomes necessary.

Valproic acid is metabolized primarily via β-, ω-, and ω_1-oxidative pathways in the liver, and a small amount is excreted in the urine as glucuronide. This route of biotransformation generally results in increased concentrations of coadministered drugs.

Positive outcomes are most likely to be achieved when the patient's individual characteristics are considered along with the pharmacologic profile of the therapeutic agent. Thus, all three primary antimanic drug therapies require laboratory monitoring. The incidence and severity of laboratory abnormalities are both low.

Carbamazepine decreases the WBC by ~15%; in some patients it depresses the WBC more in a dose-related manner. In very rare instances (<1:100,000), it produces bone marrow suppression that can progress to aplastic anemia. When this occurs, it usually is in the first 6 months of treatment.

Liver function test (LFT) results [serum glutamic-oxaloacetic transaminase (SGOT), serum glutamate-pyruvate transaminase (SGPT), or alkaline phosphatase] are often elevated in patients on carbamazepine or valproic acid therapy. These elevations are considered benign and do not necessitate a change in treatment unless they exceed twice the upper limit of normal. In cases with such excessive concentrations, the clinician should retest and determine whether the medication should be reduced in dosage or discontinued, or whether a consultation should be obtained (see Table 16–1).

When two or more drugs are administered concomitantly, certain drug interactions can occur that may require dosage adjustments to maintain therapeutic serum concentrations. Enzyme-inducing drugs may increase hepatic metabolism, resulting in lower plasma concentrations of coadministered drugs. On the other hand, drugs that inhibit enzyme induction may result in increased concentrations of coadministered drugs.

An example of the former would be a patient receiving haloperidol and carbamazepine. Carbamazepine, a potent enzyme inducer, could cause a lowering of the haloperidol concentration. An example of the latter would be a patient receiving erythromycin and carbamazepine. The hepatic enzyme inhibition caused by erythromycin can result in increased carbamazepine concentrations.

Carbamazepine absorption is rapid, but with marked variation from patient to patient. Biotransformation via induction of the P450 system is the main route of metabolism; only 2% is excreted unchanged in the urine.

Autoinduction is the term used to describe the phenomenon whereby carbamazepine accelerates its own biotransformation. When therapy is started at a constant dose, the serum drug concentration initially increases but then declines as autoinduction occurs (via the P450 system). The half-life of carbamazepine drops from ~35 hours (range 18–65) after a single dose to ~10–25 hours after 2–4 weeks of oral administration. As a result, the dose may have to be increased after 2–4 weeks to maintain clinical response. Because of wide variations in plasma concentrations and degree of autoinduction, treatment must be individualized.

Induction of the hepatic P450 enzyme system by carbamazepine also results in the lowering of plasma concentrations of other concomitantly administered drugs. For example, carbamazepine has been reported to decrease haloperidol concentrations by 40–60%, as well as to decrease the effectiveness of oral contraceptives. Therefore, clinicians should be informed about these drug interactions.

Carbamazepine is expected to reduce the WBC and platelet count by ~15%, resulting in a new baseline count. If the WBC falls below threshold [$2 \times 10^3/\mu L$ to $3 \times 10^3/\mu L$ ($2-3 \times 10^9/L$) is commonly used], the carbamazepine dosage can be reduced or discontinued temporarily and the WBC followed closely. The WBC should increase quickly. Dosage can then be increased if necessary. Should the WBC decrease to $2 \times 10^3/\mu L$ ($2 \times 10^9/L$), again, carbamazepine therapy should be stopped until the WBC

TABLE 16–1. *Rationale for monitoring antimanic drugs*

- Therapeutic confirmation
- Suspected toxicity or overdose
- Absence of therapeutic response
- Drug failure or poor dose-response correlation
- Noncompliance
- Monitoring of active metabolites
- Suspected drug interactions

starts to increase. It is important to note that a suppression of WBC and platelet count is not a precursor of the rare aplastic anemia phenomenon.

The most common dose-related side effects for patients taking carbamazepine are neurologic and sensory. For patients taking valproic acid, however, the most common dose-related side effect is gastrointestinal. The use of the enteric-coated formulation (Depakote®) can minimize these side effects. Dose-related side effects can often be managed clinically by careful titration of dosing in response to the transient side effects. Periodic blood tests are required for all three major antimanic medications, but the frequency varies depending on the drug.

Monitoring blood concentrations of valproic acid is useful during the initial dose titration. CBC and LFTs are usually obtained to detect any early (albeit benign) changes at monthly intervals for several months, then less frequently.

Carbamazepine concentrations are useful in initial dose titrations. Because carbamazepine induces its own metabolism (thereby reducing its own blood concentration) between the second and fourth weeks, blood concentrations should be obtained weekly for the first 2 months, then less frequently. In clinical settings this is commonly done every 3–6 months. Because the depression in the WBC is most evident after the second week of therapy, monitoring the CBC at 2-week intervals for 2–3 months is reasonable, and then at 3- to 4-month intervals. If an acute infectious syndrome develops, a CBC should be obtained immediately to rule out an aplastic process.

Idiosyncratic reactions occur with both carbamazepine and valproic acid, and fortunately these are rare occurrences. These rare events are best predicted by clinical signs and symptoms rather than by specific laboratory testing. The clinical signs of bone marrow suppression include infection, fever, pallor, weakness, and/or petechiae. Hepatotoxicity may occur with either carbamazepine or valproic acid. The clinical signs include loss of appetite, nausea, vomiting, periorbital edema, abdominal pain, easy bruising, and/or malaise. When patients present with suspected idiosyncratic reactions, the offending agent must be discontinued immediately as supportive measures are begun.

Retrospective analyses have shown that the risk of fatal hepatotoxicity with valproic acid is very low, particularly among patients receiving valproic acid monotherapy. The primary risk of fatal hepatic dysfunction is for patients under the age of 2 who receive valproic acid as part of a multidrug regimen (for example, multiple anticonvulsants). There have been no reports of fatal hepatotoxicity among patients over the age of 10 receiving valproic acid monotherapy. A more recent analysis of the reports of fatal hepatic dysfunction has revealed that the incidence of fatal hepatic dysfunction has actually fallen despite an increase in the number of individuals receiving valproic acid. Benign elevations of SGOT, SGPT, and alkaline phosphatase occur with carbamazepine therapy and do not necessitate discontinuance unless concentrations exceed double the upper limit of normal. Most patients can tolerate 400–600 mg/d initially, but some require very slow titration or tolerate only low doses. Serum concentration is a better indicator of potential toxicity. It is not closely correlated with response.

Recent data suggest that the rate of fetal abnormalities in epileptic women on carbamazepine therapy is slightly elevated. A rash develops in ~10% of patients on carbamazepine therapy. The rash may require discontinuation of the medication in psychiatric patients if it persists after switching to another formulation of carbamazepine.

If the WBC drops to $3 \times 10^3/\mu L$ ($3 \times 10^9/L$), the dosage of carbamazepine is usually decreased. If the count drops to $2 \times 10^3/\mu L$ ($2 \times 10^9/L$), carbamazepine therapy commonly is suspended until the count returns to higher concentrations.

Because carbamazepine is a TCA related to amitriptyline and imipramine, one can expect some similar side effects.

Monitoring blood for hematologic and hepatic function may be carried out according to many paradigms. After stabilization of dose, some clinicians may monitor as often as monthly, although idiosyncratic reactions are usually sudden in onset. Periodic monitoring of long-term patients two to four times a year is common practice.

Transient, benign elevations of LFTs are common, and the dosage is usually maintained unless such elevations reach twice the upper limit of normal. The medication is then discontinued. In nonepileptics, abrupt discontinuation is not associated with withdrawal seizures, although a brief taper over a few days may obviate mild withdrawal symptoms seen with other TCAs.

To initiate valproic acid therapy, begin with 250–500 mg twice a day and then titrate the dose to achieve a serum concentration of 50–120 µg/mL (347–832 µmol/L).

First-trimester exposure to valproic acid has specifically been linked to neural tube defects such as spina bifida. About 1% of babies born to women who receive valproic acid early in pregnancy are born with such defects. α-Fetoprotein analysis and ultrasound before the 20th week of gestation can be used to detect an open neural tube in the developing fetus when therapeutic intervention is an option, should tests prove positive.

To detect potentially serious but rare hepatic dysfunction in patients receiving valproic acid, LFTs are performed before treatment and regularly during treatment. It is also essential to observe any clinical signs or symptoms of liver abnormalities, because these may be better indicators of true hepatic dysfunction than laboratory measurements (Table 16–2).

One may observe elevations of hepatic function tests that are usually benign and transient. If there are no clinical signs of illness, one usually then follows LFTs as indicated clinically (usually every 1–2 weeks) until they stabilize or revert to normal values. If test abnormalities reach twice the upper limit of normal, then discontinuation of medication is indicated. Changes in CBC are rare, although these may include thrombocytopenia.

The very rare occurrence of hepatic failure has been reported in young children who were usually on multiple anticonvulsants. Periodic blood tests are not a safeguard from idiosyncratic reactions. Clinical signs of serious illness or liver dysfunction should alert the clinician to pursue evaluation.

Valproic acid plasma concentrations are generally obtained as a guideline to avoid high concentrations associated with an increased frequency of side effects and to document the concentrations at which individual patients respond.

Blood tests are not absolutely indicated on a weekly basis initially, but serve to document tolerance to treatment. Once a stable dose has been achieved, less frequent monitoring is appropriate.

The range of dosages required to reach therapeutic concentrations in patients receiving valproic acid is 750–1500 mg/d; most patients tend to cluster between 750 and 1250 mg/d. A clinical response to valproic acid is generally observed with serum concentrations between 50 and 100 µg/mL (347 and 693

TABLE 16–2. *Laboratory measures during treatment of mania*

	Week									
	1/2	1	1-1/2	2	3	4	5	6	7	8
Lithium										
Concentration	x	x	x	x	x		x			x
Carbamazepine										
Concentration	x	x	x	x	x	x				x
CBC	x			x		x		x		x
LFTs						x				x
Valproic acid										
Concentration	x	x	x	x						x
CBC						x				x
LFTs						x				x

μmol/L), although others suggested an upper limit of 120 μg/mL (832 μmol/L) for patients with bipolar disorder. It is important to note that these ranges are based on clinical observations of response. Because of the interindividual variation in patients' response to valproic acid (slow vs. rapid metabolizers), dosing as well as monitoring should be tailored to the clinical response of the patient.

REFERENCES

1. Goodwin FK, Jamison KR: *Manic depressive illness.* New York: Oxford University Press, 1990.
2. Prien RF, Gelenberg AJ: Alternatives to lithium for preventive treatment of bipolar disorder. *Am J Psychiatry* 1989;146(7):840–848.
3. Sachs GS: Adjuncts and alternatives to lithium therapy for bipolar affective disorder. *J Clin Psychiatry* 1989;50[Suppl]:31–39.
4. Emrich HM, Okuma T, Muller AA, eds: *Anticonvulsants in affective disorders.* Amsterdam: Excerpta Medica, 1984.
5. Kravitz HM, Fawcett J: Carbamazepine in the treatment of affective disorders. *Med Sci Res* 1987; 15:1–8.
6. McElroy SL, Keck PE, Pope HG, et al: Sodium valproate: its use in primary psychiatric disorders. *J Clin Psychopharmacol* 1987;7(1):16–24.
7. McElroy SL, Keck PE, Pope HG, et al: Valproate in the treatment of rapid-cycling bipolar disorder. *J Clin Psychopharmacol* 1988;8:275–279.
8. Calabrese JR, Delucchi GA: Spectrum of efficacy of valproate in 55 patients with rapid-cycling bipolar disorder. *Am J Psychiatry* 1990;147:431–434.
9. Bowden CL, Janicak PG, Orsulak P, et al: Relationship of serum valproate concentration to response in mania. *Am J Psychiatry* 1996;153:765–767.
10. Litovitz TL, Klein-Schwartz W, White S, et al: 1999 Annual Report of the American Association of Poison Control Centers Toxic Exposure Surveillance System. *Am J Emerg Med* 2000;517–574.
11. Ballenger JC: The use of anticonvulsants in manic depressive illness. *J Clin Psychiatry* 1988;49[Suppl]: 21–25.
12. Brennan MJW, Sandyk R, Borsook D: Use of sodium valproate in the management of affective disorders: basic and clinical aspects. In: Emrich HM, Okuma T, Muller AA, eds: *Anticonvulsants in affective disorders.* Amsterdam: Excerpta Medica, 1984: 56–65.

SELF-ASSESSMENT QUESTIONS

1. Bipolar disorder afflicts what percent of the U.S. general population?
   a. 1%
   b. 2%
   c. 5%
   d. 10%
   e. unknown

2. True or false: Genetic research in bipolar disease suggests that there are several modes of inheritance and that bipolar disease is a genetically related disorder.
   a. true
   b. false

3. What has been the most widely used treatment for bipolar disorder until recently?
   a. lithium
   b. lithium plus neuroleptics
   c. neuroleptics
   d. lithium plus antidepressants
   e. anticonvulsants

4. Which are the most widely used anticonvulsants for treatment of bipolar disease?
   a. valproic acid
   b. carbamazepine
   c. phenytoin
   d. a and b
   e. all of the above

5. What is the therapeutic range for valproic acid when treating bipolar disease?
   a. 50–75 µg/mL
   b. 50–100 µg/mL
   c. 50–120 µg/mL
   d. 50–150 µg/mL
   e. 50–200 µg/mL

CHAPTER 17

Antiepileptic Drugs

Steven J. Soldin, Ph.D., FACB, FCACB, Andrew Volosov, Ph.D., and Offie Porat Soldin, Ph.D., M.B.A.

LEARNING OBJECTIVES

After completing this chapter, the reader should be able to:

1. Describe the disposition of antiepileptic drugs in patients with epilepsy.
2. Describe the incidence of drug overdoses in the United States for these agents.
3. Determine when to recommend measurement of "free" drug concentrations in plasma.
4. Discuss drug-drug interactions and the importance of metabolites.

INTRODUCTION

Epilepsy has been defined as electrical storms in the brain. Approximately 1 in 11 people will experience a seizure and some 3% of the population will have recurrent, unprovoked seizures. There are two categories of seizures:

- Those arising in one cerebral hemisphere (called partial or focal seizures)
- Those involving both hemispheres (called generalized seizures)

Drug therapy using anticonvulsants is the mainstay of epilepsy treatment. The therapeutic goal is to reduce seizure frequency, preferably so that the patient becomes seizure-free, with minimal side effects. The selection of the antiepileptic drug is usually based on the type of seizures, patient's sex and age, concurrent medical conditions, and concomitant medications. Even with optimized treatment, only around 75% of patients achieve desirable seizure control, that is, one-fourth of all epileptic patients have therapy-resistant (refractory or intractable) forms of epilepsy and consequently suffer from uncontrolled seizures (1).

The first breakthrough in antiepileptic therapy occurred in 1912, when the anticonvulsant properties of phenobarbital were discovered. Although highly sedative, this drug found a wide antiepileptic use. Nowadays, phenobarbital is still in clinical use for partial and generalized tonic-clonic seizures, in spite of severe adverse effects.

The biggest landmark in the treatment of epilepsy was the discovery of phenytoin. It was the first time that a new anticonvulsant was discovered using a newly developed animal model for screening potential antiepileptics (2). Phenytoin was introduced in 1938 and is still the most widely used anticonvulsant drug, probably because of its nonsedative properties. It is commonly used to treat general motor (tonic-clonic, grand mal) and focal seizures and is less effective in the treatment of complex partial seizures. Anticonvulsants accounted for 14 deaths in the United States in 1999, with 1 due to phenytoin. There were 4045 cases of phenytoin exposure, with most falling within the <6-year and >19-year age groups (3).

Carbamazepine was approved for use in the United States in 1974 and is structurally related to the tricyclic antidepressant drugs. It is effective as an anticonvulsant in the treatment of partial complex and generalized tonic-clonic seizures. Although adults receive daily dosages of 7–20 mg/kg, in prepubertal children daily dosage requirements can be higher (10–30 mg/kg) due to the greater activity of the hepatic microsomal enzyme system in the latter. The dose is usually increased over the first few weeks because of autoinduction of its own metabolism. There were no deaths (6132 incidences of known exposure) in the United States in 1999. Most of these exposures were in the <6-year and >19-year age groups.

While using valproic acid (VPA) to solubilize potential antiepileptic agents, Meunier found that VPA possessed antiepileptic activity itself (4). VPA was approved as an antiepileptic drug in 1978. It is commonly used to treat absence seizures as well as generalized and partial seizures. It has also been approved for such nonseizure conditions as bipolar manic depressive episodes and as a prophylactic agent for migraine headaches (4,5). There were 8 deaths (8743 cases of known exposure) due to VPA in the United States in 1999 (3). Most of these were in the <6-year and >19-year age groups.

CASE STUDIES

Case 1

A 13-year-old girl who was receiving phenytoin therapy for her epilepsy drug presented in the emergency room with seizures. A stat phenytoin concentration was 43 mg/L (170 µmol/L). A review of the history showed a phenytoin concentration 6 months ago of 15 mg/L (59 µmol/L). Her seizures were controlled in the emergency department with a benzodiazepine and her dose was held until her concentration was 15 mg/L (59 µmol/L). A small reduction in dose from 225 mg/d to 200 mg/d was found to provide this patient with a therapeutic concentration.

This case illustrates that phenytoin can induce seizures at concentrations >40 mg/L (>158 µmol/L) (6). It also illustrates the saturation kinetic phenomenon of phenytoin and the slowing down of phenytoin metabolism when a patient goes through puberty. This necessitates close monitoring of the child's drug concentrations and appropriate downward adjustment of the drug dose when necessary.

Case 2

This patient was a 10-year-old boy with end-stage renal disease. He was receiving phenytoin to control his seizures. On his most recent hospitalization, he manifested signs and symptoms of phenytoin toxicity, which included nystagmus, ataxia, and confusion. His serum phenytoin was found to be 15 mg/L (59 µmol/L). However, his "free" phenytoin was found to be 3.5 mg/L (14 µmol/L). His phenytoin dose was reduced to give a free phenytoin in the therapeutic range of 1–2 mg/L (4–8 µmol/L).

Phenytoin is usually 90% protein bound. However, in renal failure, byproducts build up in the blood and displace phenytoin from its protein-binding sites. In these cases it is necessary to make dosage adjustments to bring the free phenytoin concentration into the therapeutic range [1–2 mg/L (4–8 µmol/L)].

Case 3

This patient was an 18-year-old woman with known epilepsy. Her movements were uncoordinated even though she was on a multiple-drug regimen including valproate and lamotrigine. Her steady-state plasma trough concentrations are given in Table 17–1.

Although her serum concentrations of phenytoin, carbamazepine, lamotrigine, and valproate were all within the therapeutic range for those drugs, the ratio of carbamazepine epoxide to carbamazepine (5.5 to 7.8, or 70.5%) was high, and her total carbamazepine plus carbamazepine epoxide concentration was 13.3 mg/L (also high), no doubt accounting for her clinical signs and symptoms of toxicity.

TABLE 17–1. Steady-state trough concentrations for case 3, an 18-year-old woman with epilepsy

Drug	Steady-state plasma trough concentration [mg/L (µmol/L)]	Therapeutic range [mg/L (µmol/L)]
Phenytoin	20 (79)	10–20 (40–79)
Carbamazepine	7.8 (33)	4–12 (17–51)
Carbamazepine epoxide	5.5[a]	—
Lamotrigine	3.3 (13)	1–12 (4–47)
Valproate	60 (416)	50–100 (347–693)

[a]SI value not given for metabolite.

Case 4

A 10-year-old girl presented with absence seizures. Before VPA therapy was initiated, her baseline liver function tests [alanine aminotransferase (ALT), alkaline phosphatase, bilirubin total and direct] were within the reference range. Two months after initiating VPA therapy, both liver ALT and alkaline aminotransferase were >97.5th percentile. Her VPA steady-state trough concentration was within the therapeutic range [75 mg/L (520 µmol/L)]. Her VPA dose was decreased.

VPA can often cause an elevation of liver function tests. Reducing the VPA dose usually results in liver function tests returning to normal. The VPA dose can then be gradually increased to achieve optimal serum concentrations.

MECHANISM OF ACTION

Phenytoin

Phenytoin's mechanism of action is not clear. However, phenytoin stabilizes membranes in the brain (and thereby suppresses seizures) and in the heart (and hence suppresses arrhythmias). It has been suggested that phenytoin suppresses seizures by blocking post-tetanic potentiation by influencing synaptic transmission. The mechanisms postulated for this effect include alteration of (7–9):

- Ion fluxes associated with depolarization
- Repolarization
- Membrane stability
- Calcium uptake in presynaptic terminals
- Na^+K^+-ATP–dependent ionic membrane pump

Carbamazepine

The mechanism of action of both carbamazepine and carbamazepine epoxide is through inhibition of the voltage-dependent fast sodium channel. Carbamazepine is a tricyclic compound that is effective against psychomotor and grand mal seizures and also relieves symptoms of trigeminal neuralgia (7–9).

Valproic Acid

VPA was originally thought to act by increasing concentrations of the neuroinhibitor γ-aminobutyric acid (GABA) and it has been postulated that VPA potentiates postsynaptic GABA response. It has also been suggested that VPA exerts a direct effect on the cellular membrane. Possibly VPA functions through a combination of all of these mechanisms (7–9).

PHENYTOIN DISPOSITION

The usual daily dose of phenytoin prescribed to adults is 4–6 mg/kg. Prepubertal children require somewhat higher daily doses (5–10 mg/kg) to achieve the same steady-state concentrations due to the greater activity of the hepatic microsomal enzyme system during this age range. Phenytoin undergoes saturation kinetics, that is, each individual has a threshold serum and plasma concentration beyond which the enzymes involved in its metabolism become "saturated." Any further slight increase in dose can cause the serum concentration to increase out of proportion to the dose administered. When saturation occurs, the metabolism of phenytoin changes from a first-order (drug concentration–dependent) process to a zero-order (drug concentration–independent) process (6,10).

Phenytoin protects against seizures at serum and plasma concentrations of 10–20 mg/L (40–79 µmol/L), although higher concentrations may be needed in some cases (10). It can precipitate seizures at concentrations >40 mg/L (>158 µmol/L) (6). Because the relationship between serum or plasma concentration and clinical efficacy and toxicity is good, therapeutic drug monitoring and optimizing the dosage regimen to achieve a therapeutic concentration and avert toxicity is an important adjunct to therapy.

PHARMACOKINETICS OF PHENYTOIN

The pharmacokinetics of phenytoin are shown in Table 17–2.

Toxic side effects include nystagmus, dysarthria, diplopia, ataxia, and exacerbation of seizures. Also note that chronic use can lead to hirsutism and gum hypertrophy. Taking phenytoin during pregnancy is contraindicated and can lead to the fetal hydantoin syndrome. Folate and vitamin D deficiency may necessitate vitamin supplementations and if uncorrected leads to megaloblastic anemia. High concentrations lead to drowsiness, confusion, and coma (6,9,10).

Absorption

Absorption rate and bioavailability depend on the formulation used. Approximately 90% of an oral dose is absorbed. Peak concentrations are achieved in 2–8 hours. For treatment of status epileptics, phenytoin is usually administered via the intravenous (i.v.) route (see "Fosphenytoin" section).

Metabolism

The major biotransformation pathway consists of metabolism to arene oxide via the cytochrome oxidase system enzyme arene oxidase (Figure 17–1). Arene oxide is spontaneously converted to 5-p-hydroxyphenyl-5-phenylhydantoin (HPPH). This pathway accounts for 60–80% of phenytoin elimination. Saturation of this enzyme occurs at low concentrations of phenytoin, a substrate whose Michaelis-Menten constant is low, resulting in saturation of the system, usually within the therapeutic range of phenytoin. Once saturation has occurred, any further slight increase in dose results in a disproportionate increase in serum or plasma concentration with concomitant toxic side effects. The activity of this hepatic microsomal system is very age dependent, being low at 0–3 months of age, ap-

TABLE 17–2. *Pharmacokinetics of phenytoin*

Protein binding (%)	~90, temperature dependent
Daily maintenance dose (mg/kg)	Adults, 4–6
	Prepubertal children, 5–10
Therapeutic range [mg/L (µmol/L)]	10–20 (40–79)
Toxic plasma concentrations [mg/L (µmol/L)]	20–30 (79–119): nystagmus
	30–40 (119–158): ataxia
	>40 (>158): lethargy, seizures
Chronic use	Gum hypertrophy, hirsutism
Saturation kinetics	Yes
Half-life (h)	Adults, ~24
	Prepubertal children, ~12–24
Apparent volume of distribution (L/kg)	0.6–0.98 (adults, children)

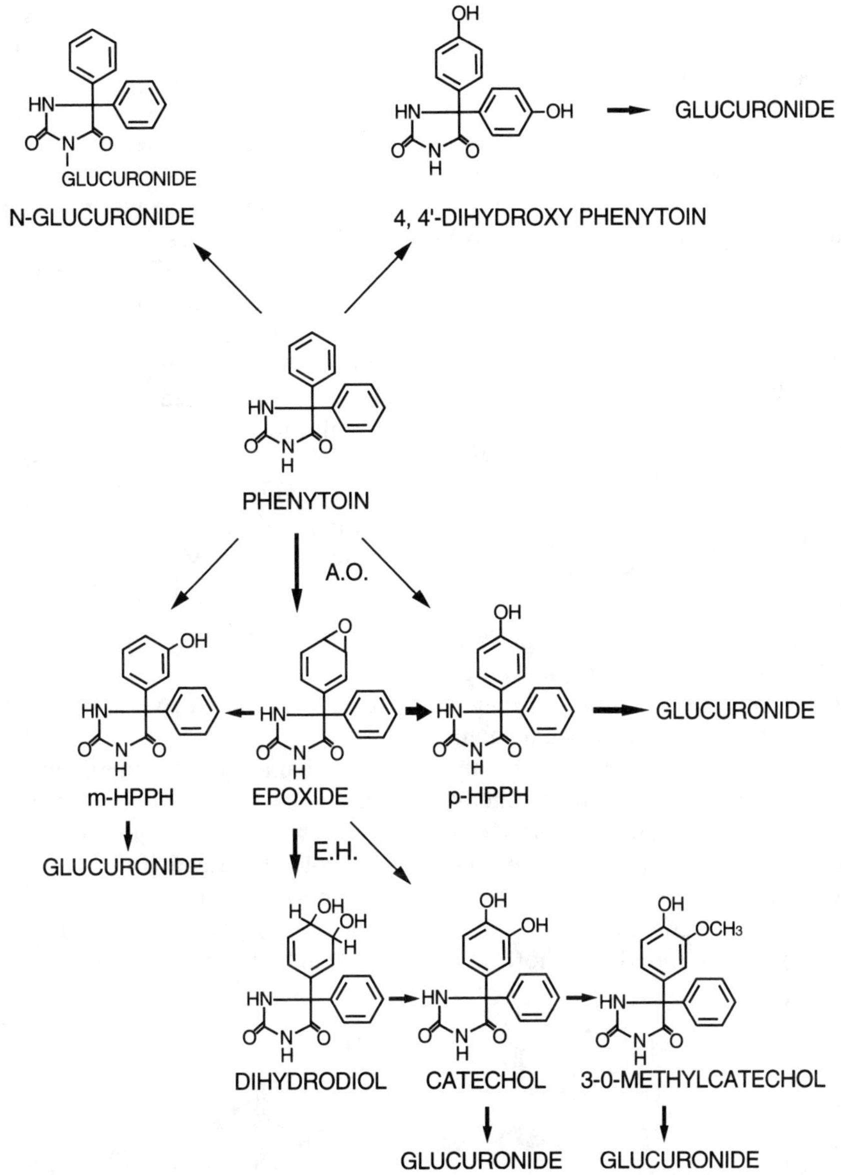

FIG. 17–1. Phenytoin metabolism.

proximately double that of the adult from 4 months to puberty, and declining to adult values after puberty. It follows that patients approaching puberty need to be followed closely with frequent monitoring of serum or plasma concentrations and appropriate downward adjustment of dose where required. Less than 5% of the dose is excreted unchanged in the urine, with 60–70% being excreted as HPPH conjugated with glucuronic acid. HPPH is converted by the enzyme epoxide hydrolase to the dihydrodiol. The diol accounts for 7–11% of phenytoin metabolites recovered from urine (6,9,10).

Drug Interactions

Drug interactions with phenytoin are fairly complex. In general, drugs that enhance the activity of the hepatic microsomal enzyme system will speed up clearance and decrease half-life of phenytoin and vice versa (for example, phenobarbital, and carbamazepine).

Several drugs inhibit the metabolism of phenytoin, thereby causing increased concentrations of phenytoin and potential toxicity. These drugs include disulfiram, isoniazid, sulthiame, bishydroxycoumarin (dicumarol), and phenyramidol.

FOSPHENYTOIN

Fosphenytoin (3-phosphoryloxymethyl-5, 5-phenytoin) is a prodrug of phenytoin. It is rapidly converted to phenytoin, formaldehyde, and phosphate by phosphatases in the liver, erythrocytes, and other tissues after both i.v. and intramuscular (IM) administration. The mean half-life of conversion is 15 minutes. Status epilepticus, a neurologic emergency, occurs ~60,000 times a year in the United States. Emergency departments require a means to rapidly administer and reach therapeutic concentrations of phenytoin without trauma and pain to vessels from i.v. or muscles from IM use of phenytoin. Fosphenytoin holds great promise to avoid these unpleasant side effects.

Fosphenytoin is indicated for short-term parenteral administration. It can be used for the control of generalized convulsive status epilepticus, for the prevention of seizures during neurosurgery, and as a short-term substitute for oral phenytoin. A 375-mg dose of fosphenytoin is equivalent to a 250-mg dose of sodium phenytoin. Fosphenytoin is associated with less pain and phlebitis than phenytoin, and, unlike phenytoin, the drug can be given safely by IM injection. Therapeutic drug monitoring is simplified if specimens are drawn after all the fosphenytoin has been converted to phenytoin. Using the 5 half-life concept and i.v. conversion half-life of 15 minutes and IM conversion half-life of 40 minutes, one would wait a minimum of 75 minutes after i.v. administration and 200 minutes after IM injection before drawing the specimen (11,12).

CARBAMAZEPINE DISPOSITION

Carbamazepine is metabolized by carbamazepine epoxidase to carbamazepine-10, 11-epoxide, a metabolite that has significant anticonvulsant activity and is approximately equipotent to the parent drug (Figure 17–2). The 10,11-epoxide is then further metabolized to transcarbamazepine diol, which is in-

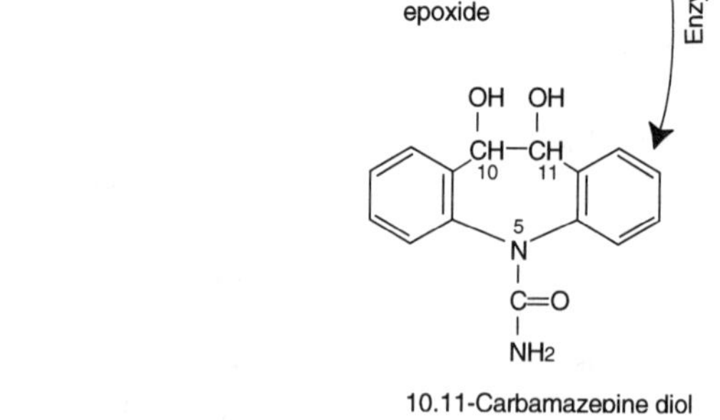

FIG. 17–2. Carbamazepine metabolism.

active and is excreted both free and in conjugated form.

Normally, with monotherapy the carbamazepine epoxide steady-state trough concentrations are ~20–25% of the parent drug. However, when polytherapy is used, and especially when carbamazepine is administered with valproate and/or lamotrigine, the epoxide concentrations increase significantly to ~50% of the parent drug.

The pharmacokinetics of carbamazepine are shown in Table 17–3.

Toxicity

At concentrations >12 mg/L (>51 μmol/L) toxic effects include nystagmus, blurred vision, unsteadiness, drowsiness, dizziness, nausea, vomiting, and headache. It is possible that some patients with a therapeutic concentration of carbamazepine may present with clinical toxicity due to the metabolite concentration of the 10,11-epoxide (13–15). It is also possible that some patients with subtherapeutic concentrations of carbamazepine may have adequate seizure control due to the presence of the active metabolite carbamazepine-10,11-epoxide.

Drug Interactions

The most common drug interactions involve coadministration of other anticonvulsant drugs such as phenobarbital and phenytoin, which increase hepatic microsomal enzyme activity and decrease the half-life of carbamazepine.

VALPROIC ACID DISPOSITION

VPA is extensively metabolized by the liver; 3–7% of a dose is excreted in urine as unchanged drug. VPA is metabolized by conjugation (glucuronidation), β-oxidation, and α-hydroxylation to >10 metabolites. One of these, 4-en-valproic acid, has been associated with hepatotoxicity (16). The main metabolite, 2-n-propyl-3-oxo pentanoic acid, possesses anticonvulsant activity comparable to VPA (4,5). Therapeutic range is 50–150 mg/L (346–1040 μmol/L) (4,5). An overview of the literature reveals that concentrations >100 mg/L (>693 μmol/L) are frequently needed in refractory cases, especially in partial seizures and atonic or "atypical absence seizures." In the psychiatric literature, a therapeutic range of 50–125 mg/L (346–866 μmol/L) is recommended for control of the mania of bipolar disease (4,17).

The pharmacokinetics of VPA are shown in Table 17–4.

Toxicities of VPA

The common toxicities include nausea, weight gain (44%), vomiting, tremor (10%), dyspepsia, and transient hair loss (12%). Uncommon toxicities include idiosyncratic, non–dosage-related hepatic failure, which

TABLE 17–3. *Pharmacokinetics of carbamazepine*

Absorption (%)	70–80 of the oral dose is absorbed
Protein binding (%)	65–80 bound
Daily maintenance dose (mg/kg)	15–20, children
	7–15, adults
Therapeutic range [mg/L (μmol/L)]	4–12 (17–51)
Toxic plasma concentrations [mg/L (μmol/L)]	>12 (>51)
Half-life (h)	8–28, neonates
	5–30, children and adults
Half-life of carbamazepine-10,11-epoxide (h)	5–6
Time to peak concentration (h)	3
Apparent volume of distribution (L/kg)	0.8–1.9

TABLE 17–4. *Pharmacokinetics of VPA*

Daily dose (mg/kg)	Adults 10–45
	Children 10–60
Dosing interval	Usually given in 1 to 4 doses/d
Time to peak concentration (h)	Syrup 1/2–1
	Tablets 1/2–2
	Enteric-coated tablets 3–8
Elimination half-life (h)	Neonates 15–60
	Children 5–15
	Adults 6–17
Route of elimination	95% hepatic metabolism
Protein binding (%)	~90 at 0–80 mg/L
	<90 at higher concentrations
Therapeutic range [mg/L (µmol/L)]	50–150 (346–693)
Clearance (mL · h^{-1} · kg^{-1})	Adults, 5–10
	Children, 5-30
	Higher clearance at higher serum concentrations and in children
Volume of distribution (L/kg)	0.1–0.5
Bioavailability (oral and rectal) (%)	90–100

occurs in 1 in 600 children <2 years old on polytherapy. For VPA given as monotherapy to patients >10 years old, only 1 incident of hepatic failure was reported between 1978 and 1993. Other uncommon toxicities are thrombocytopenia (concentration related), acute hemorrhagic pancreatitis (rare), teratogenicity (open neural tube defects in 1–2% of fetuses) (18), and hyperammonemia (16).

Drug Interactions

Patients receiving valproate and/or lamotrigine may also have higher concentrations of the active metabolite carbamazepine-10,11-epoxide. VPA inhibits metabolism of phenobarbital. VPA also inhibits the breakdown of carbamazepine-10,11-epoxide. Because this metabolite is active a patient may exhibit signs of carbamazepine toxicity without a change in concentration of carbamazepine when VPA is added to the drug regimen (13–15). VPA is displaced from its binding sites by salicylate.

FREE DRUG MEASUREMENTS

Phenytoin

Numerous drugs (VPA, salicylates, thiazides, and endogenous compounds in patients with renal failure) displace phenytoin from its binding site on albumin.

In the presence of these compounds, the free fraction of phenytoin increases, placing the patient at increased risk for toxicity, and symptoms of toxicity become apparent. Free phenytoin concentrations can be measured by using either equilibrium dialysis or, more commonly, devices with molecular cutoff filters such as the Amicon (Millipore, Bedford, MA) Centrifree micropartition system or the Worthington Diagnostics (Freehold, NJ) "ultrafree" system. Temperature affects the degree of binding of phenytoin to albumin, with drug binding decreasing as temperature increases. Control of temperature is therefore important. An 18 °F (10 °C) drop in temperature, for example, from 98.6 °F to 80.6 °F (37 °C to 27 °C), can result in an ~25% decrease in the "free fraction" (19,20). Because the patient is usually at 98.6 °F (37 °C), it is recommended to use this temperature when assessing free phenytoin concentrations. The free concentrations of phenytoin have been shown to be significantly increased in patients with human immunodeficiency virus (21).

It should be noted that the temperature effect on percent free concentrations varies from drug to drug. For example, it is minimal for carbamazepine.

Carbamazepine

Carbamazepine is bound to plasma proteins (65–80% bound), whereas the epoxide binding is less (50–60% bound). Because the free fraction is pharmacologically active and because the epoxide has a greater percentage in the free form and is equipotent to the parent drug, the epoxide presumably contributes significantly to the anticonvulsant effect.

Valproic Acid

VPA is 90–95% protein bound, predominantly to albumin at concentrations up to 80–100 mg/L (554–693 μmol/L). At higher concentrations the free fraction increases, and measurement of the free fraction is sometimes recommended (20–22). Diseases that produce an alteration in protein binding (liver disease, chronic renal failure, burns, and protein-losing enteropathy) may also require monitoring of free drug concentrations.

METHODS OF ANALYSIS

Phenytoin

The three commonly used methods of analysis for phenytoin include immunoassays, high-performance liquid chromatography (HPLC), and gas-liquid chromatography (GC). HPLC and GC methods have the advantage of separating phenytoin from its major metabolite, HPPH. Some of the immunoassays cross-react with the metabolite. Nevertheless, in surveys such as the College of American Pathologists' proficiency-testing programs, by far the most laboratories are using immunoassays to monitor phenytoin concentrations in patients.

Carbamazepine

Most laboratories use immunoassay-based methods to measure carbamazepine concentrations. HPLC is also used and is the method of choice for measurement of the active metabolite carbamazepine-10,11-epoxide (23). In a recent publication Shen et al. (15) show that most immunoassay systems cross-react with this metabolite, the degree of cross-reactivity varying from one system to the next. This could be of considerable clinical concern, especially in patients receiving VPA or lamotrigine because the ratio of active metabolite to parent drug is much higher in this group of patients.

Valproic Acid

Immunoassays and GC are the most commonly used procedures for the measurement of VPA.

The great majority of laboratories use an immunoassay procedure. In a recent College of American Pathologists' proficiency-testing program, only 5 of 3348 laboratories used GC. The analytical range of the method used should at least cover the whole therapeutic range [50–150 mg/L (346–1040 μmol/L)].

NEW ANTIEPILEPTIC DRUGS

Epilepsy is a chronic disease, and to date no cure has been discovered. Antiepileptic treatment concentrates on dealing with the symptoms, that is, controlling seizures, rather than treating the cause of the disease. In most cases, epileptic patients have to take drugs from the moment of the first seizure and take them for the rest of their lives. Hence, it is especially important that antiepileptic drugs be well tolerated, without significant adverse effects. Unfortunately, this is not the case with the major existing antiepileptics. This fact represents the incentive for continuing the search for better-tolerated and less-toxic new drugs.

Oxcarbazepine

Oxcarbazepine, the 10-keto-analog of carbamazepine, is a new antiepileptic drug that has been approved for the treatment of partial onset seizures and generalized tonic-clonic seizures (24). It was introduced in Europe in the early 1990s and was approved by the Food and Drug Administration in late 1999. Compared to carbamazepine, oxcarbazepine shows a reduced allergenic potential, possibly improved central nervous system tolerability, linear and less variable pharmacokinetics, little or no susceptibility to interactions mediated by inhibition of cytochrome P450 (CYP) isoenzymes, and a much reduced propensity to cause liver enzyme induction (25,26).

Mechanism of Action

The pharmacological activity of oxcarbazepine is primarily exerted through its 10-monohydroxy metabolite (MHD). The precise mechanism by which oxcarbazepine and MHD exert their antiepileptic effect is unknown; however, in vitro studies indicate that both compounds block voltage-sensitive sodium channels, resulting in stabilization of hyperexcited neural membranes and inhibition of repetitive neuronal firing. This mechanism of action is thought to be similar to that of carbamazepine (27).

Pharmacokinetics and Metabolism

The scientific incentive for development of oxcarbazepine as a new antiepileptic drug was based on its theoretically advantageous metabolic pathway compared with the metabolic pathway of its structurally related compound carbamazepine. Unlike carbamazepine, which is metabolized by CYP3A4 and CYP2C8 to a stable carbamazepine-10,11-epoxide, oxcarbazepine undergoes rapid presystemic metabolic 10-keto-reduction mediated by cytosol arylketone reductase to the active MHD, which is then partly conjugated with glucuronic acid and partly converted to carbamazepine-10,11-*trans*-dihydrodiol before excretion in urine (28). The metabolic pathway of oxcarbazepine is considered to be nonoxidative, noninducible, and much less prone to drug interactions than the metabolic pathway of carbamazepine. In humans, the efficacy of the keto-reductive metabolic pathway is much higher than in other species and the blood concentration of the parent drug is negligible compared with that of MHD (29).

After oral administration, oxcarbazepine is completely absorbed and extensively metabolized to its pharmacologically active metabolite MHD. In fact, oxcarbazepine undergoes a presystemic (first-pass) metabolic conversion to MHD and effectively serves as a prodrug of MHD. MHD represents the active anticonvulsant entity and is responsible for the antiepileptic activity after administration of oxcarbazepine. In a mass balance study in humans, only 2% of total radioactivity in plasma was due to unchanged oxcarbazepine, with approximately 70% present as MHD, the remainder attributable to minor metabolites.

Pharmacokinetic parameters of oxcarbazepine and MHD are summarized in Table 17–5.

Methods of Measurement

Plasma concentrations of oxcarbazepine and MHD are usually measured by HPLC. Because MHD is a chiral molecule and consequently has two stereoisomers, chiral HPLC assays find increasingly wider use for separate quantification of MHD enantiomers. Although somewhat time-consuming and costly, these assays may provide more rigorous and detailed information about pharmacokinetic behavior of oxcarbazepine and its metabolites (30,31). However, this issue is beyond the scope of this chapter.

Drug Interactions

Although to a significantly lesser extent than carbamazepine, oxcarbazepine can inhibit CYP2C19 and CYP3A4/5, with potentially important effects on plasma concentrations of other drugs. However, in vitro studies in human liver microsomes demonstrated that oxcarbazepine and MHD have little or no capacity to function as inhibitors for other most common CYP enzymes (32).

TABLE 17-5. *Pharmacokinetics of oxcarbazepine*

	Oxcarbazepine	MHD
Absorption (%)	100	
Protein binding (%)		40
Dose (g/d)	≤1200	
Half-life (h)	<2	8–12
Time to peak concentration (h)		~4.5
Apparent volume of distribution (L)		49 (adults)

Clinically significant interactions with oral contraceptives were detected during treatment with oxcarbazepine. The mean AUC values of two hormonal components, ethinylestradiol and levonorgestrel, were decreased by 48–52% and 32–52%, respectively. Therefore, concurrent use of oxcarbazepine with hormonal contraceptives may render these contraceptives less effective.

Toxicity

Oxcarbazepine is generally considered to be less toxic and better tolerated than carbamazepine. However, treatment with oxcarbazepine can cause adverse effects. The most pronounced and clinically significant adverse effect that may develop during oxcarbazepine use is hyponatremia [sodium serum concentrations <125 mEq/L (<125 mmol/L)]. In controlled epilepsy studies, approximately 25% of oxcarbazepine-treated patients had sodium concentrations <125 mEq/L (<125 mmol/L) at some point during treatment, compared with none of the patients on placebo or active control (carbamazepine, phenytoin, VPA). Most patients who developed hyponatremia were asymptomatic, but cases of symptomatic hyponatremia were reported during postmarketing use. Sodium concentrations returned to normal within a few days after discontinuation of oxcarbazepine, without additional treatment.

Other adverse effects associated with administration of oxcarbazepine include dizziness, diplopia, fatigue, nausea, ataxia, and abdominal pain. These effects are similar to those usually associated with carbamazepine treatment.

FELBAMATE

Felbamate is one of a new generation of antiepileptic drugs. It is structurally unique, with a broad spectrum of efficacy. Launched in 1993 in the United States, it is approved for adjunctive or monotherapy in adults with partial seizures with or without secondary generalization. It was effective in patients refractory to other antiepileptic drugs and was particularly beneficial as add-on therapy in children suffering from Lennox-Gastaut syndrome, being the first drug shown to be effective in treating this condition in controlled trials. Lennox-Gastaut syndrome is a childhood disorder with multiple seizure types, slow spike-wave electroencephalograms, mental retardation, and resistance to standard therapy with antiepileptic drugs (33). Although felbamate has considerable efficacy in treating many types of seizures, it is limited by side effects and idiosyncratic reactions and has been relegated to the role of a second-line medication to be used only when other agents have failed and seizures are severe enough to risk serious side effects.

Mechanism of Action

Felbamate, 2-phenyl-1,3-propanediol dicarbamate, has a unique dual mechanism of action. It enhances the GABA system while inhibiting excitatory amino acid responses. The mechanism by which felbamate exerts its anticonvulsant activity is unknown. In animal studies designed to detect anticonvulsant activity, felbamate has properties in common with other marketed anticonvulsants. Its unique clinical profile is thought to be due to interaction with N-methyl-D-aspartate (NMDA) receptors, resulting in decreased excitatory amino acid neurotransmission (34). Receptor-binding studies in vitro indicate that felbamate has weak inhibitory effects on GABA-receptor binding and benzodiazepine-receptor binding and is devoid of activity at the MK-801 receptor binding site of the NMDA receptor-ionophore complex.

Changes of the electroencephalogram paroxysmal pattern during felbamate therapy in Lennox-Gastaut syndrome suggest a selective effect of felbamate on the ictal atypical spike and wave pattern. The differential effect of felbamate on ictal patterns may be a reflection of a different action on the excitatory and inhibitory systems (35). Felbamate was effective in inducing a functional recovery from the damage caused by ischemia in rat neocortical slices. These results suggest that a noticeable neuroprotection can be obtained during glucose and O_2 deprivation by preventive therapeutic use of felbamate (36).

Pharmacokinetics and Metabolism

Felbamate is absorbed well orally with bioavailability greater than 90%. Absorption is not affected by food (37). It is metabolized by the hepatic CYP system and has a half-life of 20–23 hours, unaltered after multiple doses (38). About 40–50% of the absorbed dose appears unchanged in urine, and an additional 30–40% is present as unidentified metabolites and conjugates. About 15% is present as parahydroxyfelbamate, 2-hydroxyfelbamate, and felbamate monocarbamate, which do not have significant anticonvulsant activity. Felbamate C_{max} and AUC are proportionate to dose after single and multiple doses over a range of 100- to 800-mg single doses and 1200- to 3600-mg daily doses. C_{min} (trough) blood concentrations are also dose proportional. Multiple daily doses of 1200, 2400, and 3600 mg gave C_{min} values of 30 ± 5, 55 ± 8, and 83 ± 21 µg/mL (126 ± 21, 231 ± 34, and 348 ± 88 µmol/L). Felbamate gave dose-proportional steady-state peak plasma concentrations in children ages 4–12 over a range of 15, 30, and 45 mg/kg per day with peak concentrations of 17, 32, and 49 µg/mL (71, 32, and 206 µmol/L) (39). The pharmacokinetics of felbamate are shown in Table 17–6.

Drug Interactions

Felbamate may be better tolerated than the standard agents because of more favorable pharmacokinetic characteristics and lack of interactions with drugs other than antiepileptics. Felbamate affects the steady-state concentrations of other antiepileptic drugs that depend on hepatic metabolism. Felbamate inhibits the metabolism of phenytoin, phenobarbital, VPA, and carbamazepine epoxide, resulting in increases in their concentrations (Table 17–7). Therefore, dos-

TABLE 17–6. Felbamate pharmacokinetics

Daily dose	Adults (>14 y) 1200–3600 mg
	Children (2–14 y) 15–45 mg/kg
Dosing interval	Usually given in 3 to 4 doses/d
Route of elimination	Hepatic metabolism
	40–50% excreted unchanged in urine
Bioavailability (%)	>90
Protein binding (%)	22–25; mostly to albumin
	Dependent on albumin concentration
Half-life (h)	20–23
Apparent volume of distribution (mL/kg)	756 ± 82 after a 1200-mg dose
Therapeutic range [mg/L (µmol/L)]	20–60 (84–252)
Clearance (mL · h^{-1} · kg^{-1})	Single 1200-mg dose: 26 ± 3
	Multiple daily doses of 3600 mg: 30 ± 8

TABLE 17–7. *Effects of felbamate on standard anticonvulsant drug concentrations*

Standard anticonvulsant	Effect of felbamate
Phenytoin	Increase 25%
Valproic acid	Increase 25%
Carbamazepine	Decrease 30%
Carbamazepine epoxide	Increase 55%
Phenobarbital	Increase 25%

age of any of these drugs must be lowered appropriately when given concurrently with felbamate (40). The addition of phenytoin or carbamazepine reduces felbamate concentrations by 40%. Because felbamate is only 25% protein bound, minimal binding effects occur with protein-bound drugs (41).

Toxicity

The most common side effects include anorexia, weight loss, vomiting, insomnia, nausea, headache, dizziness, and somnolence (39). The clinical effects of overdose included epigastric distress and tachycardia. Felbamate is considered a category C medication during pregnancy. Because it is found in breast milk, breastfeeding is not recommended for patients taking felbamate. Two very serious potential toxic effects of felbamate are aplastic anemia and hepatic failure (39,42,43). The risk of aplastic anemia in patients taking felbamate is 100 times that in the general population, but a recent identification of a reactive metabolite atropaldehyde and human leukocyte antigen studies suggest that high-risk patients can be identified. Because of the serious potential toxicity, felbamate should be reserved for the rare treatment of patients with epilepsy that is difficult to control. Pediatric use is approved only for adjunctive therapy in children with Lennox-Gastaut syndrome.

Methods of Measurement

HPLC is the method of choice for measurement of felbamate in plasma and urine (44), although GC can also be used (45). Because felbamate is usually used as an add-on treatment, these methods have the advantage of separating felbamate from other co-administered antiepileptics, as well as metabolites.

Treatment

Felbamate therapy should be initiated slowly and doses increased in increments of 300–400 mg. Target maintenance dose is 1800–3600 mg/d, given in two or three doses a day when used as an add-on drug and in three or four doses a day when used as monotherapy. Because felbamate may cause aplastic anemia or fulminant hepatic failure, it should be used only in selected patients unresponsive to other measures, accompanied by regular tests (every 2–4 weeks) of hepatic function and serial blood counts.

CONTROVERSIES

- We generally measure total drug concentrations in the clinical laboratory. Should we be measuring free phenytoin concentrations routinely?
- Should we be measuring "free" carbamazepine concentrations?
- Should we be monitoring carbamazepine epoxide concentrations?
- Should we be measuring free VPA concentrations routinely?

REFERENCES

1. Elwes RDC, Johnson AL, Shorvon SD, et al: The prognosis for seizure control in newly diagnosed epilepsy. *N Engl J Med* 1984;311:944–947.
2. Merritt HH, Putnam TJ: A new series of anticonvulsant drugs tested by experiments on animals. *Arch Neurol Psychiatry* 1938;39:1003–1015.
3. Litovitz LT, Klein-Schwartz W, White S, et al: 1999 Annual Report of the American Association of Poison Control Centers Toxic Exposure Surveillance System. *Am J Emerg Med* 2000;18:517–574.
4. Steele BW: Valproic acid [Continuing Education Program]. Northfield, IL: Therapeutic Drug Monitoring/Endocrinology Resource Committee, College of American Pathologists, November 1998.
5. Schumacher GE, ed: *Therapeutic drug monitoring.* Norwalk, CT: Appleton and Lange, 1995.
6. Soldin SJ, Kwong TC: Therapeutic drug monitoring and clinical toxicology in a pediatric hospital. In: Soldin SJ, Rifai N, Hicks JM, eds: *Biochemical basis*

of pediatric disease, 3rd ed. Washington, DC: AACC Press, 1998:533–569.
7. Fariello R, Smith MC: Valproate: mechanism of action. In: Levy R, Mattson R, Meldrum B, et al, eds: *Antiepileptic drugs*, 3rd ed. New York: Raven Press, 1989:567–575.
8. Levy RH, Mattson RH, Meldrum B: *Antiepileptic drugs*, 4th ed. New York: Raven Press, 1995.
9. Pippenger CE, Perry JK, Kutt H, eds: *Antiepileptic drugs: quantitative analysis and interpretation.* New York: Raven Press, 1978.
10. Levine M, Chang T: Therapeutic drug monitoring of phenytoin. Rationale and current status [Review]. *Clin Pharmacokinet* 1990;19:341–358.
11. Dasgupta A, Handy BC, Datta P: Mathematical models to calculate fosphenytoin concentrations in the presence of phenytoin using phenytoin immunoassays and alkaline phosphatase. *Am J Clin Pathol* 2000;113:87–92.
12. Witte DL: Fosphenytoin [Continuing Education Program]. Northfield, IL: Therapeutic Drug Monitoring/Endocrinology Resource Committee, College of American Pathologists, May 1998.
13. Al-Qudah AA, Hwang PA, Giesbrecht E, et al: Contribution of carbamazepine-10, 11-epoxide to neurotoxicity in epileptic children on polytherapy. *Jordan Med J* 1991;25:171–177.
14. Potter JM, Donnely A: Carbamazepine-10,11-epoxide in therapeutic drug monitoring. *Ther Drug Monit* 1998;20:652–657.
15. Shen S, Elin RJ, Soldin SJ: Characterization of cross-reactivity by carbamazepine 10,11-epoxide with carbamazepine assays. *Clin Biochem* 2001;34:157–158.
16. Kondo T, Ishida M, Kaneko S, et al: Is 2-propyl-4-pentenoic acid, a hepatotoxic metabolite of valproate, responsible for valproate-induced hyper ammonemia? *Epilepsia* 1992;33:550–554.
17. McElroy SL, Keck PE Jr, Pope HG Jr, et al: Valproate in the treatment of bipolar disorder: literature review and clinical guidelines. *J Clin Psychopharmacol* 1992;12:36S–41S.
18. Koren G, Kennedy D: Safe use of valproic acid during pregnancy. *Can Fam Physician* 1999;45:1451–1453.
19. Ratnaraj N, Goldberg VD, Hjelm M: Temperature effects on the estimation of free levels of phenytoin, carbamazepine and phenobarbitone. *Ther Drug Monit* 1990;12:465–472.
20. Soldin SJ: Free drug measurements, when and why? An overview. *Arch Pathol Lab Med* 1999;123:822–833.
21. Dasgupta A, McLemore JL: Elevated free phenytoin and free valproic acid concentrations in serum of patients infected with human immunodeficiency virus. *Ther Drug Monit* 1998;20:63–67.
22. Gugler R, von Unruh GE: Clinical pharmacokinetics of valproic acid. *Clin Pharmacokinet* 1980;5:67–83.
23. Soldin SJ, Hill JG: High performance liquid chromatographic analysis of anticonvulsant drugs using radial compression column. *Clin Biochem* 1980:13:99–101.
24. Dam M, Ostergaard LH: Oxcarbazepine. In: Levy RH, Mattson RH, Meldrum BS, eds: *Antiepileptic drugs*, 4th ed. New York: Raven Press; 1995:997–1010.
25. Schachter SC: Oxcarbazepine. In: Eadie MJ, Vajda FJE, eds: *Antiepileptic drugs. Pharmacology and therapeutics. Handbook of experimental pharmacology.* Berlin: Springer-Verlag, 1999:319–330.
26. Lloyd P, Flesch G, Dieterle W: Clinical pharmacology and pharmacokinetics of oxcarbazepine. *Epilepsia* 1994;35[Suppl 3]:10–13.
27. McLean MJ, Schmutz M, Wamil AW, et al: Oxcarbazepine: mechanisms of action. *Epilepsia* 1994;35[Suppl 3]:5–9.
28. Schutz H, Feldmann KF, Faigle JW, et al: The metabolism of 14C-oxcarbazepine in man. *Xenobiotica* 1986;16:769–778.
29. Feldmann KF, Brechbuhler S, Faigle JW, et al: Pharmacokinetics and metabolism of GP 47 680, a compound related to carbamazepine, in animals and in man. In: Meinardi H, Rowan AJ, eds: *Advances in epileptology.* Amsterdam: Swets and Zeitlinger BV, 1978:290–294.
30. Flesch G, Francotte E, Hell F, et al: Determination of the R-(−) and S-(+) enantiomers of the monohydroxylated metabolite of oxcarbazepine in human plasma by enantioselective high-performance liquid chromatography. *J Chromatogr* 1992;581:147–151.
31. Volosov A, Xiaodong S, Perucca E, et al: Enantioselective pharmacokinetics of 10-hydroxycarbazepine after oral administration of oxcarbazepine to healthy Chinese subjects. *Clin Pharmacol Ther* 1999;66:547–553.
32. Tartara A, Galimberti CA, Manni R, et al: The pharmacokinetics of oxcarbazepine and its active metabolite 10-hydroxy-carbazepine in healthy subjects and in epileptic patients treated with phenobarbitone or valproic acid. *Brit J Clin Pharmacol* 1993;36:366–368.
33. Felbamate Study Group in Lennox-Gastaut Syndrome: Efficacy of felbamate in childhood epileptic encephalopathy (Lennox-Gastaut syndrome). *N Engl J Med* 1993;328:29–33.
34. Kleckner NW, Glazewski JC, Chen CC, et al: Subtype-selective antagonism of N-methyl-D-aspartate receptors by felbamate: insights into the mechanism of action. *J Pharmacol Exp Ther* 1999;289:886–894.
35. Marciani MG, Spanedda F, Placidi F, et al: Changes of the EEG paroxysmal pattern during felbamate therapy in Lennox-Gastaut syndrome: a case report. *Int J Neurosci* 1998;95(3–4):247–253.
36. Siniscalchi A, Zona C, Guatteo E, et al: An electrophysiological analysis of the protective effects of felbamate, lamotrigine, and lidocaine on the functional recovery from in vitro ischemia in rat neocortical slices. *Synapse* 1998;30(4):371–379.
37. Palmer KJ, McTavish D: Felbamate. A review of its pharmacodynamic and pharmacokinetic properties, and therapeutic efficacy in epilepsy. *Drugs* 1993;45:1041–1065.
38. Faught E, Sachdeo RC, Remler MP, et al: Felbamate monotherapy for partial-onset seizures: an active-control trial. *Neurology* 1993;43:688–692.
39. Felbamate. Package insert. Cranbury, NJ: Wallace Laboratories, 2001.
40. Sachdeo R, Kramer LD, Rosenberg A, et al: Felbamate monotherapy: controlled trial in patients with partial onset seizures. *Ann Neurol* 1992;32:386–392.
41. Jensen PK: Felbamate in the treatment of refractory partial-onset seizures. *Epilepsia* 1993;34[Suppl 7]:S25–S29.

42. Clinical update: Felbatol (felbamate). Cranbury, NJ: Wallace Laboratories, 1995.
43. Ben-Menachem E: Expanding antiepileptic drug options: clinical efficacy of new therapeutic agents. *Epilepsia* 1996;37[Suppl 2]:S4–S7.
44. Behnke CE, Reddy MN: Determination of felbamate concentration in pediatric samples by high-performance liquid chromatography. *Ther Drug Monit* 1997;19:301–306.
45. Gur P, Poklis A, Saady J, et al: Chromatographic procedures for the determination of felbamate in serum. *J Anal Toxicol* 1995;19:499–503.

SELF-ASSESSMENT QUESTIONS

1. Phenytoin is bound to proteins. Which of the binding percentages most accurately reflects the situation for phenytoin at 98.6 °F (37 °C)?
   a. 20%
   b. 40%
   c. 60%
   d. 70%
   e. 90%

2. Which is true for phenytoin?
   a. The concentration of phenytoin in plasma is proportional to the phenytoin dose over the concentration range 0–60 mg/L (0–238 µmol/L).
   b. Phenytoin undergoes saturation kinetics.
   c. Phenytoin is weakly protein bound.
   d. The metabolism of phenytoin is independent of patient age.

3. Which of the following most accurately reflects the plasma protein binding of carbamazepine?
   a. 10–20%
   b. 30–40%
   c. 40–50%
   d. 65–80%
   e. 80–95%

4. What happens to the half-life of carbamazepine when phenobarbital is added to the patient's drug regimen?
   a. usually unaffected
   b. usually shortened
   c. usually increased

5. Prepubertal children have a much higher clearance of valproic acid than adults do.
   a. true
   b. false

6. Valproic acid is strongly bound to plasma proteins. At 0–80 mg/L (0–554 µmol/L), what is the approximate binding?
   a. 80%
   b. 70%
   c. 90–95%
   d. 100%

7. What happens to the ratio of carbamazepine epoxide to carbamazepine when valproate and/or lamotrigine are added to the drug regimen?
   a. unchanged
   b. increased
   c. decreased

CHAPTER 18

Digoxin and Other Cardiac Glycosides

Arti N. Shah, M.S., M.D., Leslie M. Shaw, Ph.D., DABCC, and Sandy Shah, D.O.

LEARNING OBJECTIVES

After completing this chapter, the reader should be able to:

1. *Describe the incidence of poisoning cases due to cardiac glycoside medicines and cardiac glycoside–containing plants.*
2. *Characterize the pharmacokinetics of digoxin and digitoxin.*
3. *Discuss the factors that contribute to difficulties in the measurement of digoxin.*
4. *Describe the toxicity of digoxin in the typical acute pediatric poisoning case and chronic overdosing in the elderly patient with congestive heart failure.*
5. *Discuss three controversies and questions about digoxin.*

INTRODUCTION

Two major classes of cardiac glycosides have been identified: those produced by glands in the skin of poisonous toads (bufadienolides) and those contained in plant tissue (cardenolides). More than 400 naturally occurring cardioactive glycosides have been identified. Exposure to plants containing glycosides can occur through multiple mechanisms including ingesting sap, berries, leaves, blossoms, seeds, or teas brewed from plant parts; drinking contaminated water; eating food prepared with or stirred by poisonous plant parts; and even inhaling smoke from burning plants. Common plant sources of cardiac glycosides are illustrated in Table 18-1.

Cardiac glycosides have been in use for centuries as therapeutic agents. The cardiac glycosides of medicinal importance are derived from *Digitalis lanata* (digoxin, digitoxin, deslanoside) and *Digitalis purpurea* (digitoxin, digitalis). They are often called "digitalis" because of their derivation from the digitalis (foxglove) plant.

Accounts of the benefits of glycosides can be found in the Roman literature, and the use of an extract from the oleander tree is described in Indian Ayurvedic medicine centuries before the birth of Christ (1). The purple foxglove was noted in a Welsh treatise on medicinal plants (*Meddygon myddmai*, 12th and 13th century A.D.) and was described in great detail in the famous herbal compiled by Leonard (*Historia stirpium*, Basel, 1542). Foxglove was an official drug of the London Pharmacopoeia in 1650 (2). The ability of digitalis glycosides to ameliorate the symptoms of heart failure was first recognized in the 18th century by William Withering, an English physician and botanist, who obtained his clue regarding the medicinal uses and properties from a Shropshire midwife (3). After 10 years study of the foxglove, Withering published in 1785 the first reliable clinical account of the therapeutic and toxic effects of digitalis leaves in the historically significant monograph titled *An Account of the Foxglove and Some of its Medicinal Uses, with Practical Remarks on*

TABLE 18–1. *Common plant sources of cardiac glycosides*

Common name	Scientific name	Poisonous part
Purple foxglove	*Digitalis purpurea*	Leaf
Grecian foxglove	*Digitalis lanata*	Leaf
—	*Strophanthus* species	Seed
Lily-of-the-valley	*Convallaria majalis*	All parts
Oleander	*Nerium oleander*	All parts
Yellow oleander	*Thevetia neriifolia*	All parts
Squill or sea onion	*Urginea maritima*	Bulb
Dogbane (wild ipecac)	*Apocynum cannabinum*	Root, sap
Member of a squill family	*Hispidus* species	Seeds

Dropsy and Other Diseases. He recommended that

> it be continued until it acts either on the kidneys, the stomach, the pulse, or the bowel . . . let it be stopped upon the first appearance of any of these effects and I will maintain that the patient will not suffer from its exhibition, nor the practitioner be disappointed in any reasonable effects (4).

He further gave a most accurate account of digitalis intoxications:

> The foxglove, when given in very large and repeated doses, occasions sickness, vomiting, purging, giddiness, confused vision, objects appearing green and yellow; increase secretions of the urine, with frequent motions to part with it; slow pulse, even as slow as 35 in a minute, cold sweats, convulsions, syncope, death (4).

Despite its widespread use in clinical practice for the last two centuries, controversy continues concerning the use of digoxin in patients with heart failure in sinus rhythm. The widespread use of digoxin and the low therapeutic index (the therapeutic index is also referred to as the toxic:therapeutic ratio, for which a small increase in the dose beyond the therapeutic dose may result in toxicity) contributed to the high incidence of toxicity and mortality observed before the use of therapeutic drug monitoring. Monitoring serum digoxin concentrations combined with other clinical data aids the physician in adjusting digoxin dosage to achieve optimal therapeutic effects while minimizing the risk of toxic effects.

Digoxin is the prototype cardiac glycoside that is extensively used in Western medicine for the treatment of congestive heart failure and atrial fibrillation (AF). In 1999 digoxin ranked as the 15th most commonly prescribed drug in North America and it remains the number 1 drug for which serum concentrations are monitored (5).

In 1999, of the 2,201,156 exposures to toxic substances reported by the American Association of Poison Control Centers (AAPCC), there were 2810 exposures due to cardiac glycoside medicines (accounting for 6.4% of all cardiovascular drug exposures) and 2498 exposures due to cardiac glycoside–containing plants (accounting for 2.19% of all plant exposures) (6). Digoxin represents the most commonly prescribed cardiac glycoside in the United States, and among the plant exposures oleander is estimated to represent approximately 25% of cases. The AAPCC reported 20 deaths from the 2810 exposures to digoxin and other cardiac glycoside medicines, whereas there were no reported cases of death from the 2498 exposures to cardiac glycoside–containing plants. Since its inception in 1983, AAPCC has reported only one case of death from cardiac glycoside–containing plants. Mortality is rare from plant exposure, but case reports documenting fatalities from oleander, foxglove, squill, and other related plants have been published. Morbidity associated with exposure to cardiac glycoside–containing plants tends to be less severe than that associated with digoxin and other cardiac glycoside medicines. Table 18–2 summarizes the demographic profile of exposure cases, and Figure 18–1 demonstrates temporal trends in exposures to and deaths

TABLE 18–2. *Demographic profile of cardiac glycoside exposure cases*

	Cardiac glycoside medicines	Plants containing cardiac glycoside
Number of exposures	2810	2498
Total cases[a]	43,856	113,864
Age (y)		
<6	812 (28.89%)	1784 (71.42%)
6–19	128 (4.56%)	365 (14.61%)
>19	1799 (64.02%)	318 (12.73%)
Reason		
Unintentional	2247 (79.96%)	2389 (95.63%)
Intentional	285 (10.14%)	84 (3.36%)
Other	1 (0.03%)	3 (0.12%)
Adverse reaction	234 (8.33%)	19 (0.76%)
Outcome		
None	964 (34.31%)	981 (30.27%)
Minor	191 (6.80%)	134 (18.64%)
Moderate	408 (14.52%)	28 (1.12%)
Major	153 (5.44%)	3 (0.12%)
Death	20 (0.71%)	0 (0.00%)

[a]Total cases in either the cardiovascular drug or plant categories.
Modified from Litovitz TL, Klein-Schwartz W, Cobaugh DJ, et al: 1999 annual report of the American Association of Poison Control Centers Toxic Exposure Surveillance System. *Am J Emerg Med* 2000;18:517–574.

from cardiac glycoside overdose as reported by participating centers of the AAPCC.

The accidental ingestion of cardiac glycoside–containing plants, especially among children, is a common occurrence in certain parts of the world, in particular, the Mediterranean Basin, Australia, and Southeast Asia (7,8).

Nerium oleander exposure represents approximately 25% of cardiac glycoside plant ingestions. sAll parts of this plant have been shown to contain several cardiac glycosides that mimic digoxin in their pharmacologic effects and toxicologic manifestations. As is the case for digoxin overdoses, in severe cases of

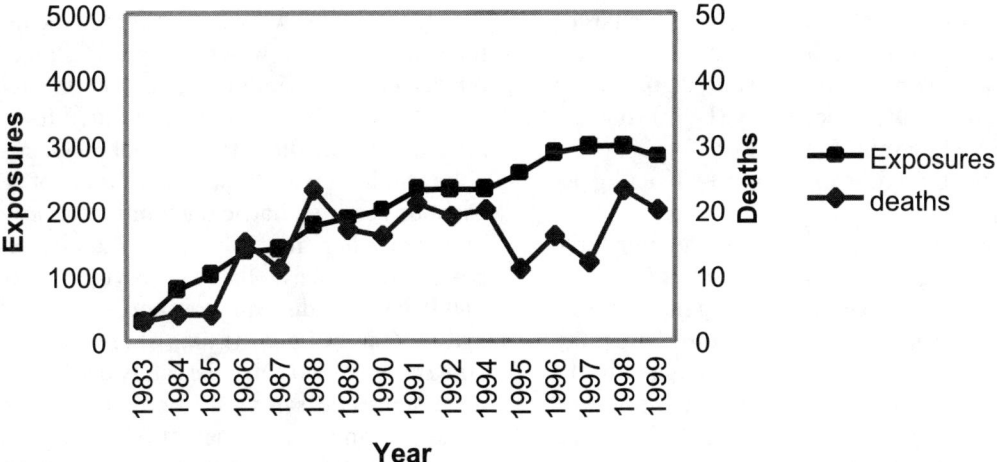

FIG. 18–1. Temporal trends in exposures to and deaths from cardiac glycoside overdose as reported by participating centers of the AAPCC.

cardiac glycoside plant ingestions, administration of digoxin-specific Fab antibody fragments was shown to be an effective treatment (9). As discussed in the section "Challenges in Measuring Digoxin," the plant glycosides oleandrin and oleandrigenin cross-react with several digoxin immunoassays, thereby providing a means for their detection in poisoning cases.

There are also reports of toxicity from toad venom poisoning resembling digoxin toxicity. The most toxic components of this venom are steroids (bufadienolides) similar in structure to digoxin. Dried toad venom is used in China as a traditional medicine known as Chan Su and is a major component of Kyushin, a popular traditional medication used in other Asian countries (10,11). The venoms of the Colorado River toad (*Bufo alvarius*) and that of the cane toad (*Bufo marinus*) are similar to that of the Chinese toad (12). In the United States, Brubacher and associates reported the first effective use of digoxin Fab fragments in the treatment of patients poisoned by toad venom (12).

CLINICAL PRESENTATIONS

Case 1: Acute Intoxication

Patient AB was a 6-week-old white female (2.3 kg in weight) with trisomy 21 and complete common atrioventricular (AV) canal and duodenal atresia who was discharged from the hospital 4 days before readmission. She was taking furosemide and digoxin (0.25 mL) via G-tube twice a day (b.i.d.) (digoxin liquid 50 µg/mL; therefore, 0.25 mL = 12.5 µg, for a total dose of 25 µg/d, or 10 µg/kg per day).

The parents inadvertently administered 2.5 mL (125 µg b.i.d.) for 2–3 days (48 µg/kg per dose). The patient received her last dose of digoxin approximately 12 hours before admission to the hospital. She was seen in the cardiology clinic with some vomiting and electrocardiogram (ECG) notable for bradycardia with a heart rate = 70, some junctional escape beats with a PR interval of 0.12–0.14 seconds, and a QRS interval of 0.06 seconds.

The serum digoxin concentration in a serum sample drawn approximately 16 hours after the last dose was 14.3 µg/L (18.3 nmol/L). Electrolytes were within normal limits including K^+.

In light of the amount of digoxin taken (48 µg/kg per dose × 4–6 doses), the symptomatology of vomiting, ECG changes of bradycardia, and junctional escape beats together with the elevated serum digoxin value of 14.3 µg/L (18.3 nmol/L), treatment with digoxin-specific Fab fragments was instituted. The patient received one-third of a vial of digoxin-specific Fab fragments, which resulted in a heart rate of 150, a PR interval of 0.12 seconds, and a QRS interval of 0.06 seconds. The dose of Fab was calculated based on steady-state serum digoxin concentrations. The formula used was:

$$\text{Dose (in number of 40-mg vials)} = (\text{serum digoxin concentration in µg/L} \times \text{weight in kg}) \div 100$$

Case 2: Acute Intoxication

A 2-year-old boy ingested some of his grandparent's digoxin. He was asymptomatic at presentation to the emergency department 1 hour later and had a normal ECG. The initial serum digoxin concentration was 1.9 µg/L (2.4 nmol/L). He remained asymptomatic. A repeat ECG about 6 hours later remained normal and the digoxin concentration measured at that time was not higher than the initial value but was lower, namely, 1.6 µg/L (2.4 nmol/L). He was admitted and monitored over the next 24 hours and continued to remain asymptomatic with a normal ECG.

Lewander et al. found in a study of 41 cases of acute pediatric digoxin ingestion in previously healthy children (not taking digoxin themselves) that an asymptomatic child, with a digoxin concentration <2.0 µg/L (2.6 nmol/L) at the time of presentation and a normal ECG, may be safely discharged if no signs and symptoms or ECG changes occur after a 6-hour observation (13). These patients are at very low risk for subsequent symptoms or ECG changes. Further, even among those children with higher concentrations at the time of presentation, ranging up to 11.6 μg/L (14.9 nmol/L), only mild, transient ECG disturbances occurred; with resting bradycardia and first- or second-degree AV block, all children did well.

Still, there are rare case reports of children with massive acute digoxin ingestions who develop malignant ventricular dysrhythmias. Although the isolated cases of severe digoxin poisoning in the pediatric population require use of digoxin-specific Fab antibody fragments, the findings in this study do not support routine use of digoxin antidote in acute pediatric digoxin poisoning. Based on their results, the authors concluded that in this pediatric population without underlying cardiac disease (13):

- Most acute digoxin intoxications were not severe.
- Signs and symptoms on presentation predict a digoxin concentration greater than 2 ng/mL (2.6 nmol/L).
- A correlation between serum potassium and digoxin concentrations was not observed.
- Non–life-threatening bradycardia and conduction disturbances were the only abnormal ECG changes noted.
- A serum digoxin concentration greater than 2 ng/mL (2.6 nmol/L) in the absence of signs or symptoms or ECG abnormalities soon after ingestion does not accurately predict the potential for toxicity or serum digoxin concentration later in the course.

Case 3: Chronic "Therapeutic" Intoxication

Patient CD was an 87-year-old white female with a history of congestive heart failure, myocardial infarction, and type 2 diabetes who presented with a 3-day history of nausea, poor appetite, and "not feeling well." The patient stated that she vomited the dry toast she forced herself to eat in the morning. She did not report any chest or abdominal pain. She denied fevers, chills, night sweats, and gastrointestinal (GI) and genitourinary symptoms. The patient was taking warfarin, glyburide, furosemide, digoxin (0.25 mg once a day), and potassium.

On physical examination, the patient was pleasant but appeared in mild distress. Her vital signs were temperature = 98.5 °F (36.0 °C), blood pressure = 130/85 mmHg, heart rate = 52 beats/min, and respiratory rate = 20. The examination of the cardiovascular system revealed irregularly irregular rhythm without murmurs, rubs, or gallops. Her abdomen was soft and nondistended with mild periumbilical tenderness without any rebound or guarding. Her abdominal examination was negative for masses or hepatosplenomegaly. She reported no flank tenderness. On neurological examination, the patient was alert and awake with difficulty in hearing. The remainder of the neurological examination was unremarkable with a motor score of 5/5 in all extremities, intact light touch sensation, and normal speech pattern and content.

Her ECG showed AF with a rate of 48 beats/min and nonspecific lateral T wave changes. The laboratory studies revealed Na^+ = 143 mEq/L (143 mmol/L), K^+ = 5.6 mEq/L (5.6 mmol/L), Cl = 100 mEq/L (100 mmol/L), HCO_3 = 27 mEq/L (27 mmol/L), blood urea nitrogen (BUN) = 51 mg/dL (18.2 mmol/L), creatinine = 1.7 mg/dL (150 μmol/L), glucose =147 mg/dL (8.2 mmol/L), and serum digoxin = 7.7 μg/L (9.9 nmol/L) approximately 7 hours after the last dose.

The patient was treated with five vials of digoxin-specific antibody fragments. Sixty minutes later her heart rate was 68 and she had significant improvement in her nausea. Three days later, serum predose digoxin concentration was 1.6 μg/L (2.0 nmol/L) and Mrs. CD was discharged on digoxin 0.125 mg once a day.

GENERAL CHARACTERISTICS OF DIGOXIN

Digitalis and related cardiac glycosides that exert their potent inotropic and electrophysiologic effects on the heart are found naturally in several plants and in the venom and skin of certain toads. Table 18–3 illus-

TABLE 18–3. Sources of cardiac glycosides of clinical importance

Plant source	Precursor glycoside	Split off by enzymatic and mild alkaline hydrolysis	Glycoside	Split off by acid hydrolysis	Aglycone, or genin
Digitalis					
D. purpurea (leaf)	Purpurea-glycoside A (desacetyldigilanid A)	Glucose	Digitoxin	Digitoxose	Digitoxigenin
D. lanata (leaf)	Lanatoside A (digilanid A)	Glucose + acetic acid	Digitoxin	Digitoxose	Digitoxigenin
	Lanatoside B (digilanid B)	Glucose + acetic acid	Digitoxin	Digitoxose	Digitoxigenin
	Lanatoside C (digilanid C, Cedilanid	Glucose + acetic acid	Digitoxin	Digitoxose	Digitoxigenin
Strophantus					
S. kombe (seed)	K-strophanthin-β	Glucose	Cymarin	Cymarose	Strophanthidin
S. gratus (seed)	—	—	Ouabain (G-strophanthin)	Rhamnose	Ouabagenin (G-strophanthin)

Reproduced with permission from Smith TW, Antman EM, Friedman PL, et al: Digitalis glycosides: mechanisms and manifestations of toxicity (Part I). *Prog Cardiovasc Dis* 1984;26(5):413–458.

Structure of digoxin

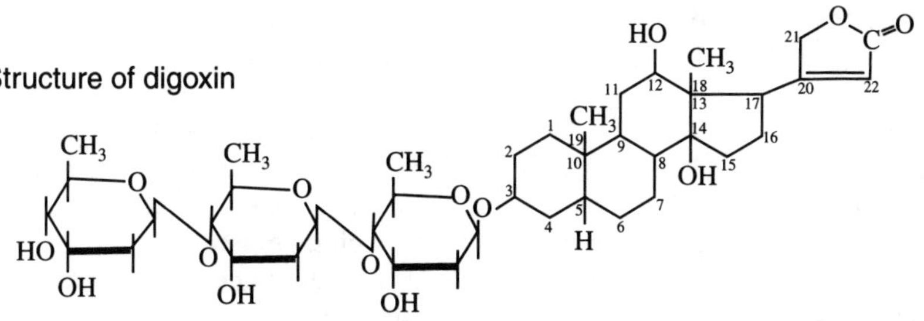

Structure of key cardiac glycosides

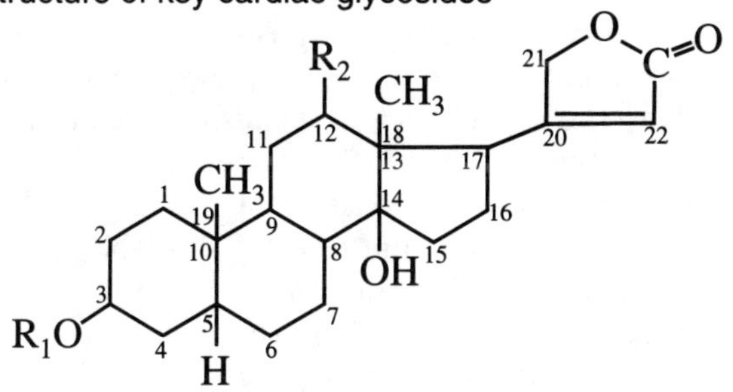

Compound	R_2	R_1 sugars at 3 position
Digoxin	OH	Tridigitoxose
Digitoxin	H	Tridigitoxose
Deslanoside	OH	Tridigitoxose-glucose
Lanatoside C	OH	Digitoxose-acetyldigitoxose-glucose

FIG. 18–2. Structure of digoxin (*top*) and key cardiac glycosides (*bottom*). Ouabain has only one sugar attached to C3, namely, 6-deoxy-α-mannose. The C19 position has $-CH_2OH$ instead of $-CH_3$, whereas an $-OH$ group is on C1, C3, and C11. Reproduced with permission from Soldin SJ: Digoxin—issues and controversies. *Clin Chem* 1986;32:5–12.

trates sources of cardiac glycosides of clinical importance.

Leaves of the plant *Digitalis lanata* contain precursors termed lanatosides A, B, and C. After mild alkaline hydrolysis (removal of an acetyl group) and enzymatic hydrolysis (removal of glucose), these precursors yield digitoxin and digoxin. Removal of glucose from lanatoside A and C yields acetyldigitoxin and acetyldigoxin, respectively. Removal of an acetyl group from lanatoside C produces desacetyl lanatoside C. The leaves of *Digitalis purpurea* contain precursors termed desacetyldigilanid A. These precursors lack the acetyl group found in precursors from *Digitalis lanata*. Enzymatic hydrolysis of desacetyldigilanid A leads to production of digitoxin.

Digoxin is described chemically as (3β, 5β,12β)-3-[(O-2,6-dideoxy-β-D-ribo-hexopyranosyl-(1→4)-O-2,6-dideoxy-β-D-ribo-hexopyranosyl-(1→4)-2,6-dideoxy-β-D-ribo-hexopyranosyl)oxy]-12,14-dihydroxycard-20(22)-enolide. Its molecular weight is 780.95, and its structural formula is presented in Figure 18–2.

Cardiac glycosides consist of a characteristic steroid nucleus and a lactone (known as an aglycone or genin part of the molecule) coupled with one to four sugar moieties. An α,β-unsaturated five-membered, occasionally six-membered, lactone ring, is positioned at C17 of the steroid (cyclopentanoperhydrophenanthrene) nucleus. At C3 and C4, a β-oriented hydroxyl substitution is usually present. The sugar moieties are usually attached at the C3 position, and they influence the activity of the cardiac glycosides by affecting solubility, absorption, half-life, metabolism, distribution, and toxicity. If the sugar moieties are removed, the aglycone still retains digitalis activity, but the compound's binding affinity, relative potency, and duration of action are reduced. In therapeutically useful glycosides such as digoxin there is an –OH group at C12, and the steroid moiety is completely saturated. The aglycone moiety contains the pharmacologic activity, whereas the sugar molecules enhance water solubility and cell penetrability. Thus, the sugar portion of the structure influences glycoside potency and the dose-response relationship.

Digoxin exists as odorless white crystals that melt with decomposition above 446 °F (230 °C). The drug is practically insoluble in water and in ether; slightly soluble in diluted (50%) alcohol and in chloroform; and freely soluble in pyridine. Digoxin is commercially available as tablets, capsules, oral elixir, and an injection formulation.

CLINICAL PHARMACOLOGY

Mechanism of Action

The precise mechanism of action of digoxin is not fully understood. The pioneering studies of Skou and Repke addressed the potential relationship between sodium and calcium with respect to the cardiac Na^+K^+-ATPase (14). They indicated that the activity of Na^+K^+-ATPase in excitable tissues is probably regulated by Na^+-Ca^{2+} antagonism. It has been suggested that digitalis increases intracellular calcium indirectly by inhibition of sodium-potassium transport, and that the digitalis effect is not due to a direct inhibition of Na^+K^+-ATPase. It is now generally accepted that an increase in intracellular Ca^{2+} results from the Na^+-Ca^{2+} exchange system and at therapeutic concentrations digitalis glycosides increase the availability of Ca^{2+} to myocardial contractile proteins (actin, myosin) after contraction. Although the contractile proteins and the troponin-tropomyosin system are directly involved in muscular contraction, it is not clear how digoxin supplements their action. Digoxin does not directly influence these proteins or the cellular mechanism that provides energy for contraction, nor does it directly influence contraction in skeletal muscle.

The sequence of events suggested then and since by several investigators is that cardiac glycoside binding to the inhibitory site on sarcolemmal Na^+K^+-ATPase inhibits the outward transport of Na^+, leading to increased $[Na^+]_i$ through the Na^+-Ca^{2+} exchange mechanism, which leads to enhanced Ca^{2+} influx or reduced Ca^{2+} efflux or both. The consequent rise in $[Ca^{2+}]_i$ increases the amount of Ca^{2+} available to the contractile elements and thus to a positive inotropic effect (15). The proposed mechanisms are shown in Figure 18–3 and Figure 18–4.

The most important pharmacologic effects of digitalis on cardiac muscle is to shift its force-velocity relationship upward (positive inotropic effect) via the sequence of events mentioned above. This action is demonstrated in patients with nonfailing as well as failing hearts. In the absence of heart failure, increased force of contraction does not increase cardiac output. It remains unchanged or slightly decreased. However, in patients with a failing heart, when the force of contraction is increased, the stroke volume and the cardiac output will be increased and the elevated left ventricular end-diastolic pressure and volume will be decreased, leading to a reduction in pulmonary and systemic venous pressure. Also, systolic emptying is more complete, and diastolic heart size is reduced.

In patients with congestive heart failure, an increased cardiac output will decrease sympathetic tone, thereby reducing the heart rate and causing diuresis in the edematous patient. Myocardial oxygen consumption is increased after digitalis administration because of its ability to augment myocardial contraction. In patients without heart failure, there is no change in the physical size of the heart, but wall tension is increased, leading to enhanced energy and oxygen requirements. In patients with heart failure, increased

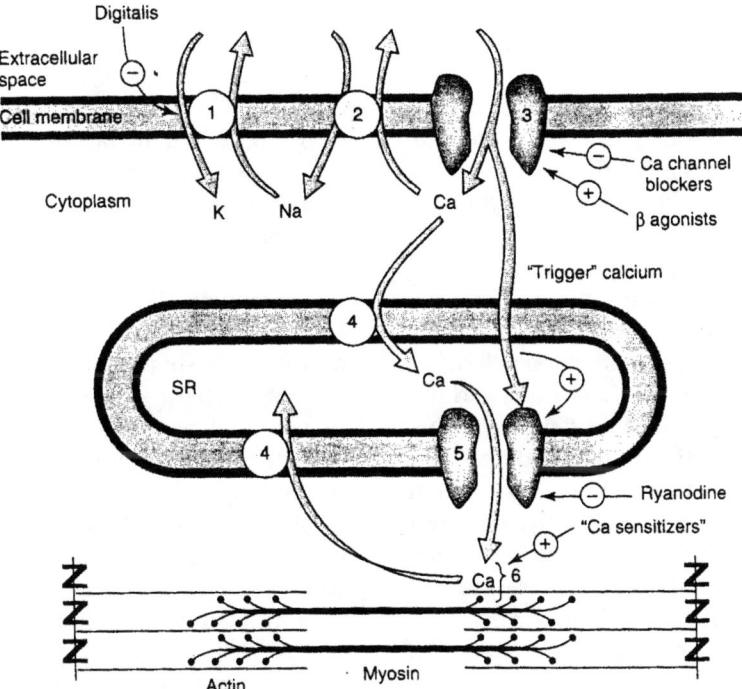

FIG. 18–3. Cellular mechanism. Schematic diagram of a cardiac sarcomere with the cellular components involved in contraction. (1) Na+K+-ATPase; (2) Na+C2+ exchanger; (3) Voltage-gated calcium channel; (4) Calcium pump in the wall of the sarcoplasmic reticulum (SR); (5) Calcium release channel in the SR; (6) Site of calcium interaction with troponin-tropomyosin system. Reproduced with permission from Katzung BG, ed: *Basic and clinical pharmacology*, 6th ed. Norwalk, CT: Appleton & Lange, 1995.

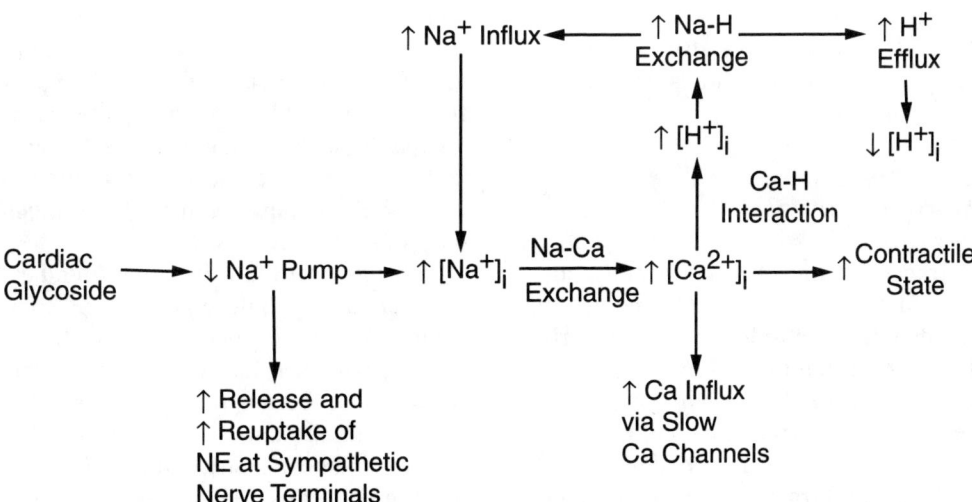

FIG. 18–4. Mechanisms of modulation of myocardial function by cardiac glycosides. Reproduced with permission from Smith TW: Digitalis. Mechanism of action and clinical use. *N Engl J Med* 1988;318:358–365.

stroke volume caused by digitalis use results in a decrease in size and wall tension of the myocardium. The end result is reduced oxygen consumption. Cardiac glycosides do not decrease coronary blood flow, and in patients with known heart failure the restoration of efficient heart action may improve coronary circulation.

In addition to its inotropic effects, digoxin also exerts a significant influence on the electrical properties of the heart. It increases the slope of the phase 4 depolarization, reduces the action potential duration, and decreases the maximal diastolic depolarization. The decrease in action potential duration is probably the result of increased potassium conductance, which is caused by increased intracellular calcium. Shortening of the action potential contributes to the shortening of atrial and ventricular refractoriness.

Cardiac glycosides decrease conduction velocity through the AV node and prolong the effective refractory period of the AV node by increasing vagal activity, by a direct effect on the AV node, and by a sympatholytic effect. In patients with supraventricular tachyarrhythmias such as atrial flutter or AF, digoxin decreases the number of atrial depolarizations that reach the ventricle, leading to a decrease in ventricular rate. Cardiac glycosides shorten the effective refractory period of the atria and increase conduction velocity by a reflex increase in vagal tone and by a direct effect on the atria. Table 18–4 lists major actions of digitalis on cardiac electrical functions.

Pharmacokinetics

Differences in the pharmacokinetics and pharmacodynamic properties among the various digitalis glycosides are due to variations in water or lipid solubility and polarity caused by both the diversity in formulation of the product and the chemical structure (Table 18–5). Digoxin is absorbed within 2–6 hours from the GI tract after an oral dose because it is poorly soluble in water. Absorption of digoxin from Lanoxin tablets is 60–80% complete compared with an identical intravenous (i.v.) dose or Lanoxicaps. When Lanoxin tablets are taken after meals, the rate of absorption is slowed, but the total amount of digoxin absorbed is usually unchanged. When taken with meals high in bran fiber, the amount absorbed from an oral dose may be reduced.

Approximately 10% of the general population harbors the enteric bacterium *Eubacterium lentum*, which can convert oral digoxin into inactive metabolites (for example, dihydrodigoxin). This can lead to decreased bioavailability, with patients requiring a higher than normal maintenance dose. Treatment of such patients with antibiotics can potentially result in a sudden increase in bioavailability and digitalis toxicity. This interaction is significantly reduced if digoxin is given as digoxin solution (Lanoxicaps) in capsules. In the latter formulation digoxin is more rapidly absorbed in the upper GI tract.

After drug administration, a 6- to 8-hour distribution phase is observed. This is followed by a much more gradual serum concentration decline, which depends on digoxin elimination from the body (16). Figure 18–5 illustrates the relationship between tissue and plasma digoxin concentrations after its i.v. administration.

Once absorbed into blood, the drug is widely distributed to tissues with the highest concentration achieved in the kidney, heart, diaphragm, and liver. The lowest concentrations are found in the plasma and brain (Figure 18–6) (17). Although digoxin concentration appears to be low in skeletal muscles compared with kidney and heart, total digoxin content is greatest in this tissue because skeletal muscle constitutes as much as 40% of lean body mass (18).

In the myocardium, digoxin is found in the sarcolemma-T tubule system bound to a receptor. Because of its extensive distribution in body tissues, it has a large apparent volume of distribution (average of 6.3 L/kg). Only small amounts of digoxin are distributed into adipose tissue, and therefore serum digoxin concentrations are not significantly altered by large changes in fat tissue weight. Its distribution space therefore correlates with lean or ideal body weight.

TABLE 18–4. Major actions of digitalis on cardiac electrical functions

Variable	Atrial muscle	A-V node	Purkinje system, ventricles
Effective refractory period	↓ PANS	↑ PANS	↓ Direct
Conduction velocity	↑ PANS	↓ PANS	Negligible
Automaticity	↑ Direct	↑ Direct	↑ Direct
ECG before arrhythmias	Negligible	↑ PR interval	↓ QT interval, T wave inversion, ST segment depression
ECG during arrhythmias	Atrial tachycardia, atrial fibrillation	A-V nodal tachycardia, A-V blockade	Premature ventricular contractions, bigeminy, ventricular tachycardia, ventricular fibrillation

PANS, parasympathetic actions; Direct, direct membrane actions; ↓, decreased; ↑, increased.
Reproduced with permission from Katzung BG, ed: *Basic and clinical pharmacology*, 6th ed. Norwalk, CT: Appleton & Lange, 1995.

TABLE 18–5. *Pharmacokinetic parameters and serum concentrations of digitalis glycosides*

	Digoxin (by mouth)	Digoxin (intravenous)	Digitalis (by mouth)
Onset (min)	30–120	5–30	60–240
Peak (h)	2–6	1–5	8–12
Half-life	30–40 h[a]	30–40 h[a]	5–9 d
GI absorption (%)	60–100	60–100	>90
Plasma protein binding (%)	20–40	20–40	>90
Percent metabolized	<20	<20	>80
Major route of elimination	Renal	Renal	Hepatic 32%, renal (metabolites)
Therapeutic serum concentrations (ng/mL)	0.8–2.0[b]	0.8–2.0[b]	9–25
Toxic serum concentrations (ng/mL)	>2.5[c]	>2.5[c]	>35
Volume of distribution (L/kg)	6–7[d]	6–7[d]	0.6

To convert digoxin concentrations in ng/mL to nmol/L, multiply by 1.281.

[a]This range is for patients with normal renal function. The half-life increases as renal function deteriorates.

[b]This range of concentrations represents that which is most commonly used in the United States.

[c]The risk for toxicity increases as trough concentrations rise above 1.4 ng/mL.

[d]The average volume of distribution is decreased below this range in patients with chronic renal failure, defined as creatinine clearance <10 mL/min. In the latter the average volume of distribution is 4.8 L/kg.

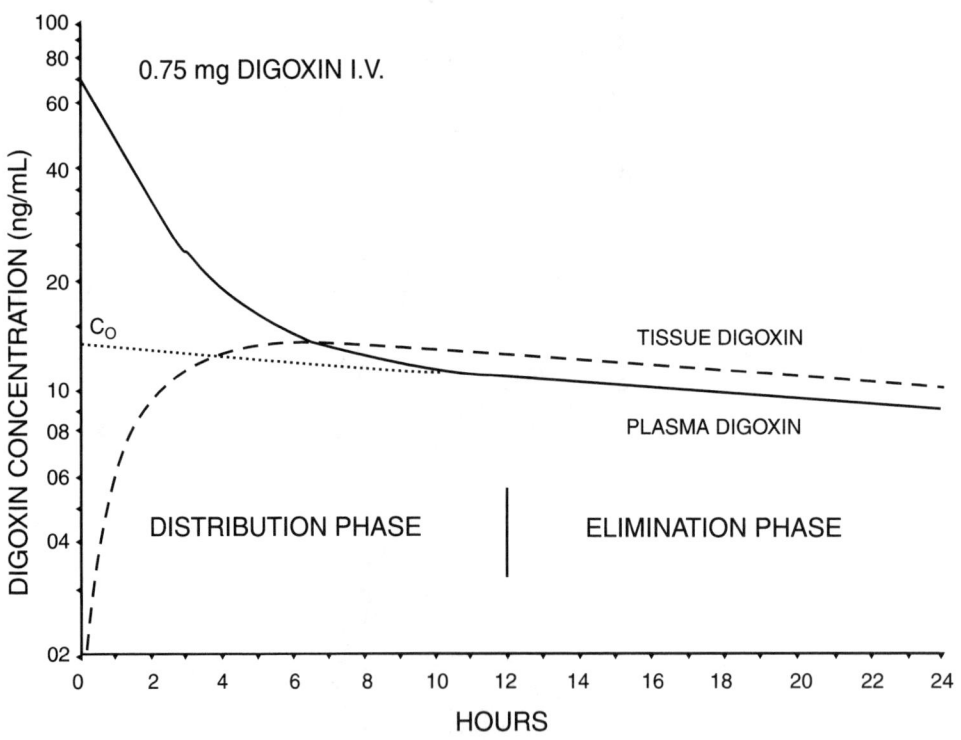

FIG. 18–5. Relationship between tissue and plasma digoxin concentrations after administration of an intravenous (I.V.) dose of the drug. Reproduced with permission from Soldin SJ: Digoxin—issues and controversies. *Clin Chem* 1986;32:5–12.

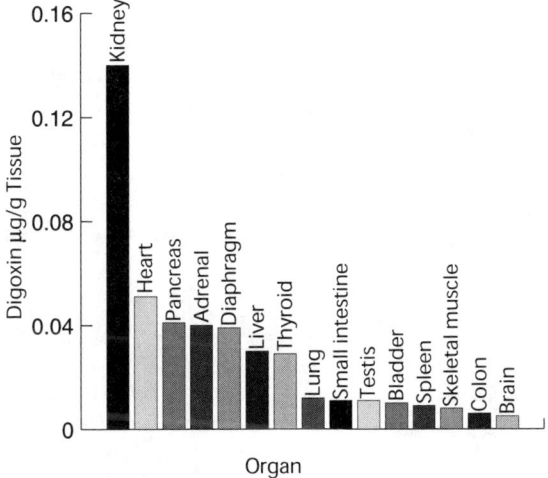

FIG. 18–6. Tissue concentrations of digoxin in a patient who received 1.0 mg of tritium labeled digoxin 5.5 hours before death. Reproduced with permission from Doherty JE: Clinical use of digitalis glycosides: an update. *Cardiology* 1985;72:225–254.

Depending on their polarity and lipid solubility, cardiac glycosides undergo varying degrees of hepatic metabolism, enterohepatic circulation, and renal filtration and reabsorption (See Figure 18–7 and Figure 18–8). Elimination of digoxin follows first-order kinetics (the quantity of digoxin eliminated at any time is proportional to the total body content). Digoxin, which is more polar than digitoxin, undergoes less enterohepatic circulation and is not metabolized extensively before excretion as digoxin. After i.v. administration of digoxin, 50–70% of it is excreted unchanged by the kidneys, largely as parent drug and active metabolites, which include digoxigenin, bisdigitoxoside, digoxigenin mono-digitoxoside, and dihydrodigoxin. Renal excretion of digoxin is proportional to glomerular filtration rate and largely independent of urine flow. In patients with renal impairment, significant accumulation of digoxin may occur, leading to toxic concentrations of drug. Cardioactivity of digoxin metabolites and their activities relative to digoxin are summarized in Table 18–6.

Less polar glycosides such as digitoxin undergo more enterohepatic circulation and are metabolized extensively before excretion. Digitoxin is inactivated by hepatic degradation, 50–80% to inactive metabolites, which are excreted by the kidneys, and ~8% is converted to digoxin.

These drugs have long half-lives (digoxin 30–40 hours in patients with normal renal function vs. digitoxin 118–216 hours). Thus, for digoxin, without the use of i.v. or oral loading doses, near–steady-state concentrations and fully developed clinical effects are achieved in 4 half-lives, or about 1 week after instituting maintenance therapy in patients with normal or close to normal renal function. For digitoxin, this time period will be correspondingly greater. After discontinuation of therapy, several days are required for complete dissipation of effects (6–8 days for digoxin and 3–5 weeks for digitoxin). Digoxin is not effectively removed by hemodialysis, exchange transfusion, or cardiopulmonary bypass because most of the drug is found in tissues and not in blood. Because of digitoxin's greater than 90% plasma protein binding, it is not effectively removed by peritoneal or hemodialysis.

CLINICAL USE

Cardiac glycosides are used mainly in the prophylactic management and treatment of patients with heart failure, to control the ventricular rate in patients with atrial flutter

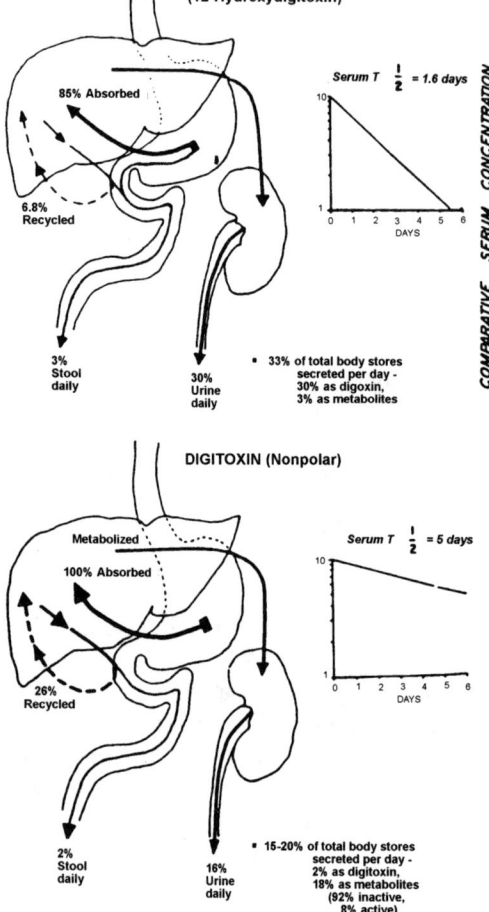

FIG. 18–7. Digoxin and digitoxin pharmacokinetics, showing average values for absorption, excretion, enterohepatic circulation, and half-life. From Doherty JE: Digitalis glycosides: pharmacokinetics and their clinical implications. *Ann Intern Med* 1973;79:229–238.

or AF, and in treatment and prevention of recurrent paroxysmal atrial tachycardia. Digoxin is the most commonly used cardiac glycoside, primarily because it can be administered by various routes, it has an intermediate duration of action, and the pharmacokinetics of digoxin in patients with or without renal disease have been extensively studied.

Although digoxin therapy has been used for centuries in patients with heart failure and atrial fibrillation, its use has been plagued with controversy (19). There is renewed interest in the use of digoxin due to its effectiveness in improving symptoms in patients with chronic heart failure. In 1994, more than 1.5 million patients were receiving digoxin for heart failure.

Congestive Heart Failure

The inotropic actions of digoxin increase cardiac output, which results in diuresis and generalized relief of the symptoms of right heart failure caused by systemic venous congestion (for example, peripheral edema) and the symptoms of left-sided heart failure caused by pulmonary congestion (for example, dyspnea, orthopnea, and cardiac asthma). Digoxin is more effective in low output failure secondary to hypertension, coronary artery or atherosclerotic heart disease, primary myocardial disease, nonobstructive cardiomyopathies, and valvular heart disease. Digoxin is less effective in high-output heart failure caused by arteriovenous fistula, bronchopulmonary insufficiency, infection, anemia, and hyperthyroidism.

The Prospective Randomized Study of Ventricular Failure and the Efficacy of Digoxin (PROVED) study analyzed 88 patients with New York Heart Association (NYHA) class II or III heart failure with mean left ventricular ejection fraction (LVEF) of 28% who were receiving digoxin and diuretics. These patients were randomly assigned to either withdrawal from or continuation of digoxin. Patients who continued to receive digoxin had a significantly higher LVEF than those who were withdrawn from digoxin. The withdrawal of digoxin was associated with a significant worsening of maximal exercise capacity and an increased rate of treatment failure (20).

The Randomized Assessment of (the Effect of) Digoxin on Inhibitors of the Angiotensin-Converting Enzyme (RADIANCE) Study analyzed 178 patients with mild to moderate chronic stable congestive heart failure (CHF) (NYHA class II or III; LVEF <30%) who were receiving long-term therapy with diuretics, digoxin, and angiotensin-converting enzyme (ACE) inhibitors. Pa-

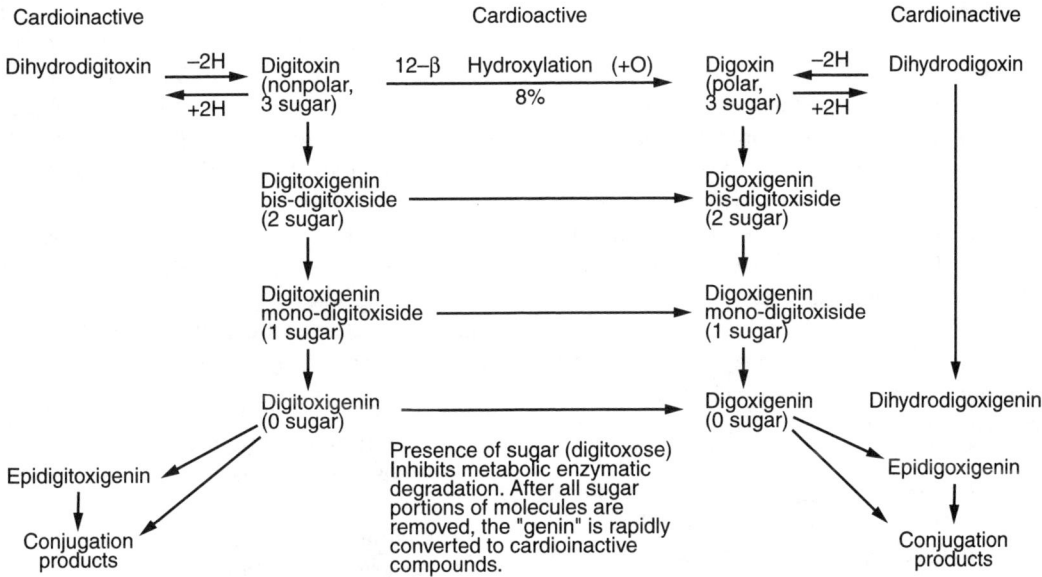

FIG. 18–8. Metabolic pathways of digoxin and digitoxin. (Note: Digoxin is a part of the metabolic pathway of digitoxin). Doherty JE: Digitalis glycosides: pharmacokinetics and their clinical implications. *Ann Intern Med* 1973:79:229–238.

tients were randomly assigned to withdrawal from or continuation of digoxin. Discontinuation of digoxin significantly increased the rate of worsening CHF sufficient to require withdrawal from the study. In addition, various measures of functional capacity or cardiac function (NYHA class, LVEF, weight change) worsened with withdrawal of digoxin (21).

The Digitalis Investigation Group (DIG) trial (22) evaluated the effect of digoxin on mortality and hospitalization in patients with moderate to severe chronic CHF and sinus rhythm. It showed that digoxin had no effect on overall mortality in patients receiving diuretics and ACE inhibitors; however, it did reduce the number of hospitalizations and the combined outcome of death or hospitalization attributable to worsening heart failure. In clinical practice, according to the DIG trial, digoxin therapy is likely to affect the frequency of hospitalization, but not survival.

The Randomized Aldactone Evaluation Study (RALES) (23) investigators evaluated 1663 patients who had severe heart failure (NYHA class III or IV) and an LVEF of no more than 35% and who were being treated with an ACE inhibitor, a loop diuretic, and in most cases digoxin. The use of spironolactone in addition to digoxin improved mortality compared with treatment with digoxin alone. They reported that the blockade of aldosterone receptors by spironolactone, in addition to standard therapy, substantially reduces the risk of both morbidity and death among patients with severe heart failure. This reduction in risk of death was due to a significant decrease in the risk of both death from progressive heart failure and sudden cardiac death from cardiac causes.

TABLE 18–6. *Percentage cardioactivity of digoxin metabolites*

Metabolite	Activity relative to digoxin (%)
Dihydrodigoxigenin	2
Dihydrodigoxin	2–6
Digoxigenin	4–21
Digoxigenin mono-digitoxide	66
Digoxigenin bis-digitoxide	77

Reproduced with permission from Soldin SJ: Digoxin—issues and controversies. *Clin Chem* 1986;32:5–12.

Atrial Fibrillation

Digoxin is the oldest therapy for treatment of AF and it is the most frequently prescribed drug for ventricular rate control in the patient with this arrhythmia. AF encompasses a variety of discrete clinical syndromes, including paroxysmal, chronic, acute, and postoperative (24). Despite its wide use in the treatment of AF, few studies exist regarding its efficacy in AF.

In paroxysmal AF, therapy is aimed at preventing paroxysms from occurring and at controlling ventricular rate during paroxysms. To date, there are no data that suggest that digoxin is helpful in achieving these treatment goals. In 1990, Rawles et al. (25) examined 139 episodes of AF occurring during Holter monitoring in 72 patients with paroxysmal AF. Of these 72 patients, 31 patients were taking digoxin and 41 patients were not. In these two groups, digoxin did not reduce frequency of paroxysms; also, ventricular rate during paroxysms was not reduced. Instead, an association was found with more episodes of fibrillation lasting 30 minutes or longer. The observation that digoxin has no effect on controlling ventricular rate in paroxysmal AF yet controls heart rate in chronic AF can be explained by the high sympathetic tone occurring at the onset of a paroxysm, a situation in which digoxin is minimally effective because the catecholamines release overcomes the vagotonic effects of the drug (26). Digoxin decreases the ventricular rate primarily by enhancing the vagal activity on the AV node (27). Because the resting autonomic tone of the heart is predominantly vagal, digoxin further enhances this tone and effectively controls the ventricular rate at rest. However, under conditions of exercise or stress, the dominance of vagal tone is removed and digoxin becomes less effective.

Therapeutic Drug Monitoring

Digoxin exerts its therapeutic effects at low serum concentrations in a very narrow range [0.8–2.0 ng/mL (1.2–2.6 nmol/L)]. Its use is complicated by the need to maintain therapeutic concentrations while keeping the risk of toxicity low. Therefore, there is a clinical need to measure the circulating concentrations of digoxin to establish and maintain a therapeutic dose and to avoid digoxin toxicity. Much controversy exists as to appropriate indications for digoxin therapeutic drug monitoring. The following are recommended indications for cost-effective use of digoxin serum concentration monitoring (28):

- Validation of initial dose of therapy
- Validation of patient compliance
- Assessment of the effect of a change in renal function on digoxin concentrations
- Validation of initial doses when interacting drugs are administered
- Assessment of effects of altered hemodynamics on drug distribution and elimination
- Assessment of nonresponse
- Prevention as well as diagnosis of toxicity

DETERMINATION OF SERUM DIGOXIN CONCENTRATION

Various analytical methods have been developed for measuring digoxin in the pharmaceutical industry. These methods include bioassay, Na^+K^+-ATPase receptor assay, colorimetry, fluorometry, gas-liquid chromatography, high-performance liquid chromatography (HPLC), polarography, and enzymatic methods. However, these methods lack the sensitivity and specificity needed to measure the low therapeutic concentrations of digoxin in serum. In 1969, Smith and associates reported the use of radioimmunoassay for digoxin and showed a correlation between clinical toxicity and serum digoxin concentrations (24).

Currently, immunoassays remain the most widely used method for measurement of digoxin in serum and biological fluids. For routine measurement of digoxin concentrations, several commercially available immunoassays are widely used including radioimmunoassay, fluorescence polarization, chemiluminescence, and enzyme immunoassays. There are also more sophisticated methods available using HPLC with digoxin detection by immunoassay and HPLC–mass spectrometry. Despite these advances in the measurement of serum digoxin concentration, the accurate measurement of digoxin in routine clinical practice remains difficult and challenging.

Dissatisfaction with the available immunoassays for digoxin led to a re-evaluation of digoxin receptor assays using human heart ATPase (29). The Na^+K^+-ATPase receptor assay involving human heart tissue was more sensitive and specific for measuring digoxin activity than the earlier version, which initially involved dog kidney preparations (30). The specificity of the receptor assay using human heart ATPase was significantly better than the immunoassays at the time it was developed, with minimal interference from digoxin-like immunoreactive factors (DLIFs) (28) and cross-reactivity of metabolites being proportional to their pharmacological activity relative to digoxin (28). Nevertheless, the receptor assay is tedious, requiring human heart ATPase, and is unsuitable for use in the routine clinical laboratory (31). A soluble or cloned receptor assay may be one direction in which further work could proceed to provide a receptor-based digoxin assay (30).

Challenges in Measuring Digoxin

Factors that contribute to the difficulties in the measurement of digoxin are the following (32):

- Low concentrations
- Steroid-like nucleus
- Endogenous DLIFs
- Digoxin metabolites
- Mixtures of digoxin and DLIFs
- Other cardenolide-like compounds
- Presence of antidote (Fab)

The low concentrations of digoxin in blood and its low molecular weight of 780 place constraints on its measurement by immunoassays (30). Compared with many other therapeutically monitored drugs that are administered in the mg/kg quantities, the daily dosage of digoxin is in the low μg/kg range (33). The need for these low doses is based on the relatively low concentrations of such inhibitors needed for Na^+K^+-ATPase inhibition (24).

Digoxin's structure is very similar to that of many exogenous and endogenous steroid molecules. Most immunoassays routinely used to measure digoxin have reasonably low cross-reactivities (<1%) with the known physiologic hormones at their normal concentrations in blood (34). However, the major cross-reactivity problems seem to be with the newly discovered endogenous counterparts to digoxin (24).

In 1985, Valdes introduced the term DLIF to describe endogenous substances that cross-react with antidigoxin antibodies (35). These substances are able to competitively displace labeled digoxin from antibody, causing false-positive results and affecting the accuracy of digoxin monitoring tests. DLIFs have been reported in neonates, cord blood, pregnant women, renal failure patients, patients with moderate to severe liver failure, and individuals who have exercised strenuously for a long period of time. The following is a list summarizing the physiological and pathological conditions associated with elevations of DLIFs (30).

- Renal failure
- Liver disease or failure
- Critically ill without renal or liver failure
- Diabetes-related stress
- Being a newborn infant
- Pregnancy
- Preterm labor
- Pregnancy-induced hypertension
- Hypertension
- Strenuous exercise
- Volume expansion
- Myocardial infarction
- Cor pulmonale (enlargement of right ventricle)
- Mucocutaneous lymph node syndrome

DLIF and its related congeners (digoxin

analogs) have been shown to occur naturally and are found in human serum. Because the fraction of DLIF not bound to protein increases in the conditions listed above, their detection is more prevalent by immunoassays.

The available digoxin immunoassays show different susceptibilities to DLIFs; in fact, with one solid-phase radioimmunoassay, both heparinized blood and urine samples of healthy volunteers were reported to contain DLIFs (36). New-generation immunoassays have reduced the interference from DLIF (37). Strategies to reduce DLIF activity include changing the incubation temperature and time for the immunoassay kits and measuring the unbound fraction of digoxin (38). In the latter case, preparations of plasma ultrafiltrate selectively decrease the contribution of DLIF because of significantly tighter binding of plasma proteins compared with the much weaker binding of digoxin.

In 1994 Miller et al. reported variability in the cross-reactivities of the digoxin metabolites in digoxin immunoassays (39). They compared cross-reactivities of digoxin metabolites in various commercially available digoxin immunoassays with their reported individual biological potencies. This comparison demonstrated that:

- The bioactivity of digoxin metabolites decreases with the loss of sugars.
- The biological activity of digoxigenin is approximately 10% of digoxin's.
- The chemically reduced metabolite of digoxin, dihydrodigoxin, has negligible bioactivity.
- Some antibodies to digoxin mimic the response of the receptor.

Polar metabolites of digoxin, which account for approximately one-third of metabolic products, can also interfere in digoxin immunoassays. Gault et al. presented evidence that 3β-digoxigenin is metabolized via 3-keto-digoxigenin to 3α-(epi)digoxigenin, which is conjugated, in part, to 3-epi-glucuronide and 3-epi-sulfate. They postulated that the sequence of hydrolysis, oxidation, and conjugation leading to polar end-metabolites is a major route of biotransformation of digoxin (40). The biological activities of these metabolites have not been defined (41). These metabolites are responsible for substantial analytical variation among several digoxin assays. Immunoreactivity due to these metabolites leads to inaccurate measurements and overestimation of bioactive digoxin in serum (32).

Several commercial digoxin immunoassays required pretreatment of serum for the precipitation of protein. Such treatment can release the majority of DLIF bound to protein, creating more interference. These effects vary from one serum sample to the next. However, as noted above, several new-generation immunoassays have reduced DLIF interference. These immunoassays have eliminated the pretreatment step, thus reducing the contribution of treatment-induced release of DLIF to the digoxin measurement.

The cross-reactivities of a suspected interferent are elevated by examining its response in an assay and comparing results with those of the standards. The interference caused by DLIFs in digoxin immunoassays should be assessed by its addition to serum containing known concentrations of digoxin (or vice versa) and subsequent measurements of the recovered immunoreactivity (42). Published studies suggest bidirectional interference (positive or negative) depending on the assay configuration tested when both types of compounds are present at varying concentrations 43,44).

Other cardenolide-like compounds structurally similar to digoxin can cross-react in the digoxin immunoassays and thus interfere with the accurate measurement of digoxin in serum. This interference can be taken advantage of in detecting poisoning caused by their ingestion. A recent study demonstrates that the extent of cross-reaction is highly variable, not only among assays but also within any one assay for the two structures, oleandrin and oleandrigenin (45).

Specimens from patients treated with digoxin Fab fragments give misleading values for digoxin concentrations by most immunoassays tested to date (46). The Fab fragments tend to interfere with various steps in the immunoassay, leading to inaccurate results. Some investigators suggest that during treatment with Digibind (digoxin immune Fab), it is helpful to measure the concentration of unbound digoxin in the patient's serum. Although no definitive evidence has yet been presented, the rationale here is that removal of all bioactive digoxin is not clinically efficacious, and the ability to titrate the unbound digoxin is proposed as the optimal approach to monitoring this pharmacologically active fraction (28).

Interpretation of Serum Digoxin Concentrations

It is essential that the physician use the serum digoxin concentration as a potentially valuable tool, but not depend on the assay alone for clinical decisionmaking. To provide a useful scheme for the interpretation of serum digoxin concentration, Cauffield et al. recommend consideration of the following 10 questions (47).

- How should digoxin be dosed?
- When are serum digoxin determinations indicated?
- What is the therapeutic range of serum digoxin concentration?
- How soon after the patient's last digoxin dose was the digoxin concentration obtained?
- What is the patient's clinical status?
- How long has the patient received the current dosage of digoxin?
- Have changes occurred in the patient's renal function?
- Is the patient receiving medications that interact with digoxin?
- Is the patient taking medications that interfere with the assay used to detect digoxin?
- Does the patient have serum electrolyte abnormalities?

An algorithm for interpreting serum digoxin concentrations is presented in Figure 18–9.

Sample Timing

An important characteristic of digoxin pharmacokinetics is the longer time it takes to reach distributional equilibrium compared with other therapeutically monitored drugs. It takes at least 6–8 hours for digoxin in serum to reach distributional equilibrium with the drug in tissues. After distributional equilibrium is reached the concentration in serum parallels that in tissues. Many studies have documented that up to 52% of digoxin determinations are inappropriate because samples were obtained within the first 6 hours after the last dose of the drug (48). The generally accepted guideline for appropriate sampling is to obtain serum samples at least 8 hours after the last digoxin dose but preferably at 12 hours or later. One practical strategy that works for both inpatients and outpatients is to include on the prescription label a recommendation to take the daily digoxin dose at 1700 h. Doing so will ensure that postdistributional samples are always obtained when digoxin monitoring is required (47).

In suspected digoxin poisoning cases a sample at the time of hospitalization is always appropriate. Interpretation of the digoxin concentration in this setting requires taking into account the possibility, especially in cases of acute ingestion, that the drug is still in the absorption and distribution phase. Continuous clinical monitoring and follow-up concentrations would be required as part of the care of such emergency patients.

DIGOXIN TOXICITY

The therapeutic-to-toxic ratio of digoxin is very narrow, and digoxin toxicity remains a potential problem. Any circumstances that increase the body concentration of digoxin or modify cardiac sensitivity to digoxin can lead to digitoxin toxicity. Drug interactions may increase digoxin concentrations, leading to toxicity. Some pathological conditions

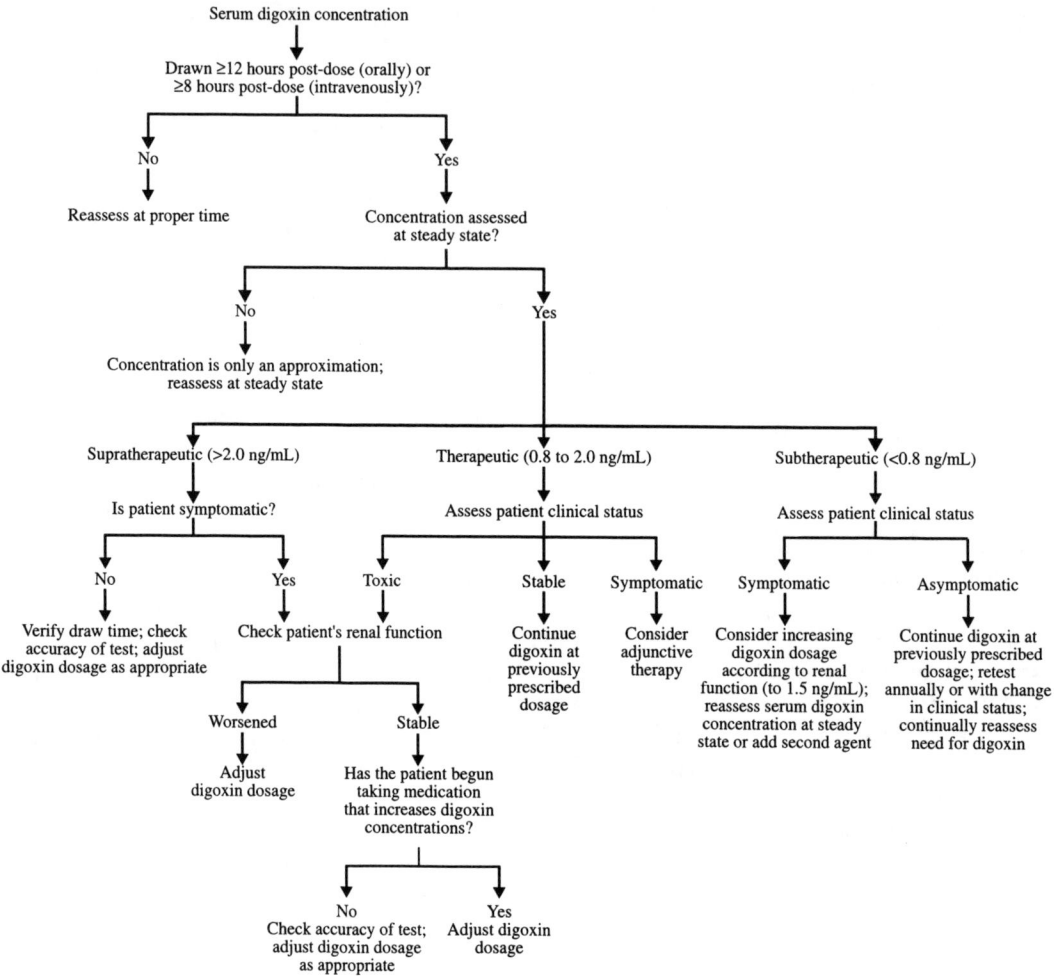

FIG. 18–9. An algorithm for interpreting digoxin concentrations. Modified from Cauffield JS, Gums JG, Grauer K. The serum digoxin concentration: ten questions to ask. *Am Fam Physician* 1997;56(2):495–503,509–510.

such as myocardial ischemia or infarction can increase myocardial sensitivity to digitalis, predisposing to toxicity. Various metabolic disorders (for example, acidosis and alkalosis) and electrolyte abnormalities (for example, hypokalemia) can alter the sensitivity of the myocardium to a given concentration of digoxin. Table 18–7 contains factors predisposing patients to digitalis intoxication.

There are many reported drug interactions with digoxin, and the list continues to expand. The quinidine-digoxin interaction is a well-known one in clinical practice. Quinidine is often added to digoxin in the treatment of AF to convert patients to normal sinus rhythm. This concomitant administration of quinidine generally causes serum digoxin concentrations to increase up to two- to threefold. This increase probably is due to decreased renal and extrarenal clearance. Recent studies have provided evidence that the *p*-glycoprotein multidrug transporter is responsible for the export of digoxin from the renal tubule to the tubular lumen and that this process is inhibited by drugs such as quinidine and verapamil (49–51). Thus, inhibition of digoxin export by the *p*-glycoprotein multidrug transporter by certain concomitant medications is now emerging as the most important contributing

TABLE 18–7. *Factors that predispose patients to digoxin intoxication*

Patient-related factors	Electrolyte abnormalities	Drugs
Old age	Hypokalemia	Diuretics
Severe heart disease	Hypernatremia	Steroids
Myocardial ischemia	Hypercalcemia	Reserpine
Myocardial infarction	Hypomagnesemia	Catecholamines
Myocarditis	Alkalosis	Quinidine
Recent cardiac surgery		Verapamil
Cor pulmonale		Amiodarone
Renal failure		Cyclosporine
Hemodialysis		
Hypothyroidism		
Anorexia		
Amyloidosis		

mechanism for those digoxin drug-drug interactions that result in increased digoxin concentrations. The verapamil-digoxin interaction can increase serum digoxin concentrations by 60–75%, and amiodarone can also increase serum digoxin concentrations by 70–100% in adults and up to 70–800% in children (52).

Bile acid sequestrates (such as cholestyramine and colestipol) and some broad-spectrum antibiotics, antacids, and kaolin-pectin mixture (such as Kaopectate) may decrease absorption of digoxin. Table 18–8 lists drugs that can alter digoxin concentrations.

Even though the incidence of severe adverse reactions has declined with improved understanding of pharmacology, a trend toward using lower dosages, and the availability of methods to measure digoxin concentrations, the adverse or toxic effects of cardiac glycosides are significantly more common than with other drugs with similar widespread use. Digoxin toxicity is uncommon with serum concentrations below 1.4 ng/mL (1.8 nmol/L) and is present in approximately 70% of patients with steady-state, postabsorption, postdistribution concentrations above 3.0 ng/mL (3.8 nmol/L) (53,54). Digitalis causes a wide range of adverse effects, which can be divided into those of cardiac origin and those associated with systems other than the heart (extracardiac). Table 18–9 shows clinical manifestations of digoxin toxicity.

When symptoms develop in a patient with cardiovascular disease who is taking digoxin and is presumed to be intoxicated, it is not easy to differentiate between symptoms caused by an underlying clinical condition. Although the various symptoms associated with digitalis toxicity may vary among the available purified glycosides and their incidence is difficult to determine, we have been able to approximate their relative frequency as a result of an error in the formulation of the drug that took place in February 1969. A pharmaceutical firm in the Netherlands prepared tablets that contained 0.20 mg of digitoxin and 0.05 mg of digoxin instead of the intended 0.25 mg of digoxin tablets without digitoxin. Lely and Van Enter studied 179 patients who took these tablets and suffered from digitalis intoxication in the town of Veenendaal, Netherlands (55). Extreme fatigue and serious eye conditions were observed in 95% of these patients. Twelve patients experienced transient psychosis. Other prominent symptoms included weakness, nausea, diarrhea, vomiting, and abnormal dreams. Table 18–10 reports prevalence of signs and symptoms associated with digoxin intoxication observed by Lely and Van Enter over a 10-week study period (54).

Piergies and Worwaf reported a concurrent audit of 92 patients with plasma or serum digoxin concentrations of 3.0 ng/mL (3.8 nmol/L) or more (56). Evidence of digoxin toxicity was present in 44 of these patients, and premature blood sampling accounted for the high concentrations in 30 nontoxic patients. Another 14 patients tolerated high digoxin concentrations without apparent ad-

TABLE 18–8. *Drugs that alter digoxin concentrations*

Drugs that may increase digitalis serum concentrations
- Alprazolam
- Aminoglycosides (oral)[a]
- Amiodarone
- Anticholinergics
- Benzodiazepines
- Bepridil
- Captopril
- Cyclosporine
- Diltiazem
- Diphenoxylate
- Erythromycin
- Esmolol
- Felodipine
- Flecainide
- Hydroxychloroquine
- Ibuprofen
- Indomethacin
- Itraconazole
- Nifedipine
- Omeprazole
- Propafenone
- Propantheline
- Quinidine
- Quinine
- Tetracycline
- Tolbutamide
- Verapamil

Drugs that may decrease digitalis serum concentrations
- Aminoglutethimide
- Aminoglycosides (oral)
- Aminosalicylic acid
- Antacids
- Antihistamines
- Antineoplastics
- Barbiturates
- Cholestyramine
- Colestipol
- Hydantoins
- Oral hypoglycemic agents
- Kaolin-pectin
- Metoclopramide
- Neomycin
- Penicillamine
- Rifampin
- Sucralfate
- Sulfasalazine

[a]Occurs in <10% of patients.

verse effects. Digoxin toxicity observed in 44 patients was manifested primarily by GI symptoms (23%) and ECG rhythm and conduction abnormalities (43%). Both ECG abnormalities and GI symptoms were observed in 27% of these patients. Central nervous system symptoms were encountered less frequently (7%).

Intoxication can occur as a result of an acute single overdose, accidental ingestion in children, or chronic accidental overmedication in patients with heart disease. There are two common clinical presentations of digitalis toxicity: acute toxicity and chronic toxicity. Acute toxicity occurs mainly in the younger population due to unintentional overdose or accidental ingestion. After an acute overdose, patients frequently develop nausea and vomiting, bradycardia, hyperkalemia, and/or AV block. In chronic toxicity, patients are older, with an underlying heart illness. These patients are often hypokalemic and hypomagnesemic owing to concurrent diuretic therapy. They more commonly present with ventricular arrhythmias (for example, ectopy, bidirectional ventricular tachycardia, or ventricular fibrillation). Clinical cases were presented earlier under "Clinical Presentations" that are representative of acute and chronic digoxin intoxication. Table 18–11 summarizes the differences between the two usual presentations of digitalis intoxication.

In digitalis intoxication, the physical examination of the patient is usually unremarkable with the exception of vital signs, particularly the pulse. In 1976, Wellens (57) listed the following four signs that are suggestive of digitalis intoxication.

- Appearance of a slow heart rate in a patient with a previously normal or fast heart rate.
- Appearance of a fast heart rate in a patient with a previously normal heart rate.
- Appearance of a regular rhythm in a patient with a previously irregular rhythm.
- Appearance of a regularly irregular rhythm.

Therapeutic Measures for Digoxin Poisoning

Initial treatment of acute digoxin poisoning is comparable to that of any toxicologic emergency. Therapy includes providing gen-

TABLE 18–9. *Clinical manifestations of digoxin intoxication*

Cardiac manifestations
- Sino-atrial • Bradycardia, tachycardia, conduction disturbances, multifocal pacing, sinus arrest
- Atrial • Ectopic pacemaker activity, tachycardia, paroxysmal tachycardia, flutter, fibrillation, arrest
- A-V conduction • 1°, 2°, 3° block, ectopic pacemaker activity
- Ventricular • Ectopic pacemaker activity, premature ventricular contraction, tachycardia, fibrillation, asystole, cardiac arrest

Extracardiac manifestations
- General • Drowsiness, dizziness, headache, anorexia, nausea, vertigo, syncope, respiratory depression, seizures
- Mood and perception • Confusion, depression, disorientation, euphoria, delirium, hallucinations
- Visual • Distorted color vision, blurred vision, scotoma

1°, first degree; 2°, second degree; 3°, third degree.

TABLE 18–10. *Prevalence of signs and symptoms of acute and chronic digoxin toxicity*

Manifestation	Prevalence (%)
Fatigue	95
Visual symptoms	95
Weakness	82
Nausea	81
Anorexia	80
Psychic complaints	65
Abdominal pain	65
Dizziness	59
Abnormal dreams	54
Headache	45
Diarrhea	41
Vomiting	40

Modified from Lely A, van Enter C: Large-scale digitoxin intoxication. *Br Med J* 1970;3:737–740.

eral supportive care, preventing further GI absorption, increasing excretion, administering digoxin-specific antibodies, and treating specific complications such as dysrhythmias and electrolyte abnormalities. Table 18–12 summarizes treatment for digoxin intoxication.

The current recommendations for severe digoxin toxicity call for the use of Digibind®, which is lyophilized sheep anti-digoxin Fab antibody fragments. Indications for the use and dosing of Digibind® are listed in Table 18–13 and Table 18–14, respectively. Massive doses of digoxin poisons the membrane bound Na^+K^+-ATPase system in the myocardium and in all other body tissues, leading to a disturbance of electrolyte gradients. The abnormal membrane function results in

TABLE 18–11. *Comparison of two classic presentations of digoxin intoxication*

	Acute intoxication	Chronic intoxication
Age	Young	Old
Cardiovascular	No underlying heart disease	Underlying heart disease
	Paroxysmal atrial tachycardia with block	Premature ventricular contractions more common
	Bradycardia more common	
Gastrointestinal	Nausea	Anorexia
	Vomiting	
Central nervous system	Seizures	Drowsiness
		Lethargy
		Disorientation
Electrolyte abnormalities	Hyperkalemia	Hypokalemia
Digoxin concentrations	Higher	Can be within therapeutic range

TABLE 18–12. Summary of treatment for digitalis intoxication

General therapy
- Discontinue digitalis
- Monitor cardiac rhythm
- Determine electrolytes and serum digoxin concentrations
- Observe for hemodynamic compromise
- Multiple-dose activated charcoal
- Steroid-binding resins for digitoxin and for patients with renal insufficiency

Management of dysrhythmias
- Fab-digoxin therapy
- Supraventricular and ventricular dysrhythmias
- Hyperkalemia [serum K^+ >5.0 mEq/L (5.0 mmol/L)]
- Hypotension unresponsive to fluids
- Potassium and magnesium replacement indicated in the presence of hypokalemia and contraindicated in the presence of preexisting hyperkalemia, bradycardia, and atrioventricular block
- Phenytoin or lidocaine for ventricular dysrhythmias in the presence of atrioventricular block (when Digibind is unavailable)
- Atropine for sinoatrial and atrioventricular conduction abnormalities
- Pacemaker (external preferred) for persistent bradycardia and atrioventricular block, if no response to immunotherapy or above management
- Cardioversion with lowest effective energy for immediate therapy of unstable ventricular tachycardia or defibrillation for ventricular tachycardia

Management of hypotension or low-output state
- Fab-digoxin therapy
- Vasopressors to maintain arterial pressure
- Control-associated dysrhythmias

Modified from Karkal SS, Ordog GJ, Wasserberg J. Dealing rapidly and effectively with a complex cardiac toxidrome. *Emerg Med Rep* 1991;12:29–44.

hyperkalemia. The absence of the normal Na^+K^+ gradient reduces the resting membrane potential and can lead to the failure of the conduction system to operate as a pacemaker, ultimately resulting in a complete loss of any cardiac electrical activity.

The use of purified digoxin-specific Fab fragments in human poisoning was first reported in 1976 by Smith et al. (58). Initial case reports demonstrating the success of Fab fragments in patients with severe digoxin (59) and digitoxin intoxication (60) and cardiac arrest have been followed by multicenter trials (61,62) and a postmarketing surveillance study (63).

Fab fragments are cleaved from sheep antibodies (raised against a digoxin-protein conjugate) by papain and separated from the more immunogenic Fc fragments by affinity chromatography (64). Because Fab fragments have a mass of approximately 50 kDa they tend to remain in the extracellular milieu. Initially, Fab fragments bind to the circulating digoxin and then they are distributed extravascularly, where these fragments bind to free digoxin in the interstitial space. This creates a concentration gradient promoting the removal of digoxin from its cellular binding site (58). The pharmacologically inactive digoxin-Fab complex is formed, which is then eliminated by renal and nonrenal pathways (47).

Treating massive digoxin overdose is a

TABLE 18–13. Indications for administration of digoxin-specific antibody fragments

Severe ventricular dysrhythmia
Progressive bradydysrhythmia unresponsive to atropine
Potassium concentration ≥5 mEq/L (≥5 mmol/L) in setting of suspected digoxin toxicity
Rapidly progressive cardiac or gastrointestinal symptoms or a rising potassium concentration
Desire to confirm the diagnosis

TABLE 18-14. *Calculation of equimolar dose of Digoxin-specific Fab fragments*

Dose based on steady-state serum digoxin concentrations

$$\text{Dose (in number of 40-mg vials)} = \frac{(\text{Serum digoxin concentration in ng/mL})(\text{Weight in kg})}{100}$$

(For Digitoxin):

$$\text{Dose (in number of 40-mg vials)} = \frac{(\text{Serum digitoxin concentration in ng/mL})(\text{Weight in kg})}{1000}$$

Dose based on acute ingestion of known amount

$$\text{Number of 40-mg vials} = \frac{\text{Total digitalis load in mg}}{0.5 \text{ mg}}$$

Dose based on acute ingestion of unknown amount

For acute ingestion give 20 vials (800 mg) for adults and children (monitor for fluid overload in children).
Alternatively, 400 mg can be administered followed by an additional 400 mg if clinically indicated.

Dosage for toxicity during chronic therapy

For adults, 6 vials (228 mg) usually is adequate to reverse most cases of toxicity.
In infants and children (≤20 kg) a single vial is usually sufficient.

Physician's desk reference, 52nd ed. Montvale, NJ: Medical Economics Company, 1998.

---

challenge. Hyperkalemia is the earliest potentially lethal presentation of digoxin poisoning. Hemodialysis, the treatment of choice for refractory hyperkalemia, may temporarily attenuate the excess serum potassium concentration but does not appreciably reduce digoxin concentration because of its large volume of distribution. The digoxin-specific polyclonal antibody fragments (Fab) neutralize digoxin toxicity by reversing the Na^+K^+-ATPase receptor binding of digoxin and augmenting its excretion.

CONTROVERSIES AND QUESTIONS

The following are three current controversies and issues regarding digoxin.

- Its clinical utility in treatment of heart failure in patients with sinus rhythm has been controversial. Recent prospective randomized controlled trials (PROVED, RADIANCE, DIG trial) showed that treatment with digoxin leads to improvement of the signs and symptoms of CHF, increased exercise tolerance, improvement in LVEF, less need for medication, and reduction in the rate of hospitalization. These trials have reconfirmed and further defined the value of digoxin in contemporary treatment of CHF.
- There is a continuing evolution of the development of methods for measurement of digoxin and the pharmacologically active metabolites. There is a need to define the most reliable and accurate measurement for digoxin and concentration-effect relationship in CHF patients.
- The need for reevaluation of the therapeutic range for digoxin is currently being debated. Although there has been a generally accepted therapeutic range for digoxin in clinical practice, reevaluating this range is an important issue due to two developments. The first is that today's most frequently used immunoassays are less subject to interferences, particularly DLIFs, than the earlier commonly used procedures. Therefore, the measured concentration values are generally lower than was previously the case due to the improved accuracy. The second reason is that digoxin is often given with an ACE inhibitor and a diuretic. Establishing an effective range is challeng-

ing because of many intraindividual physiological and analytical variables.

REFERENCES

1. Poole-Wilson PA, Robinson K: Digoxin—a redundant drug in congestive cardiac failure. *Cardivasc Drugs Ther* 1989;2:733–741.
2. Kennedy RH, Seifen E: Cardiac toxicology of digitalis. In: Baskin SL, ed: *Principles of cardiac toxicity.* Boca Raton, FL: CRC Press, 1991:217–274.
3. Tanz RD: Cardiac glycosides and other drugs used in the treatment of congestive heart failure. In: Craig CR, Stitzel RE, eds: *Modern pharmacology*, 3rd ed. Boston: Little, Brown, 1990:307–321.
4. Goodman LS: Cardiovascular and hematologic agents—digitalis. In: Haddad LM, Shannon MW, Winchester JF, eds: *Clinical management of poisoning and drug overdose*, 3rd ed. Philadelphia: WB Saunders, 1998:1001–1020.
5. Top 200 brand-name drugs by prescription in 1999. *Drug Topics* 2000;144(5):64–70.
6. Litovitz TL, Klein-Schwartz W, White S, et al: 1999 annual report of the American Association of Poison Control Centers Toxic Exposure Surveillance System. *Am J Emerg Med* 2000;18(5):517–574.
7. Shaw D, Pearn J: Oleander poisoning. *Med J Aust* 1979;2:267–269.
8. Ansford AJ, Morris H: Fatal oleander poisoning. *Med J Aust* 1981;1:360–361.
9. Sadafi R, Levy I, Amitai Y, et al: Beneficial effect of digoxin-specific Fab antibody fragments in oleander intoxication. *Arch Intern Med* 1995;155:2121–2125.
10. Chen KK, Kovarikova A: Pharmacology and toxicology of toad venom. *J Pharm Sci* 1967;56:1535–1541.
11. Huang KC: Chan Su. In: Huang KC, ed: *The pharmacology of Chinese herbs.* Boca Raton, FL: CRC Press, 1993:114–117.
12. Brubacher JR, Ravikumar PR, Bania T, et al: Treatment of toad venom poisoning with digoxin specific Fab fragments. *Chest* 1996;110:1282–1288.
13. Lewander WJ, Gaudreault P, Einhorn A, et al: Acute pediatric digoxin ingestion. A ten-year experience. *Am J Dis Child* 1986;140:770–773.
14. Repke K: Metabolism of cardiac glycosides. First International Pharmacology Meeting, Stockholm, 1961. In: Wilbrant W, ed: *New aspects of cardiac glycosides*, vol. 3. Oxford, UK: Pergamon Press, 1963:46.
15. Smith TW, Antman EM, Friedman PL, et al: Digitalis glycosides: Mechanisms and manifestations of toxicity, Part II. *Prog Cardiovasc Dis* 1984;26:495–540.
16. Soldin SJ: Digoxin—issues and controversies. *Clin Chem* 1986;32:5–12.
17. Doherty JE: Clinical use of digitalis glycosides: an update. *Cardiology* 1985;72:225–254.
18. Doherty JE, Perkins WH, Flanigan WJ: The distribution and concentration of tritiated digoxin in human tissues. *Ann Intern Med* 1967;66:116–124.
19. Withering W: An account of foxglove and some of its medical uses with practical remarks on dropsy and other diseases. In: Willins F, Keyes T, eds: *Classics of cardiology.* New York: Henry Schuman, 1941:231–252.
20. Uretesky BF, Young JB, Shahidi FE, et al, on behalf of PROVED Investigative Group: Randomized study assessing the effect of digoxin withdrawal in patients with mild to moderate chronic congestive heart failure: results of PROVED trial. *J Am Coll Cardiol* 1993,22:955–962.
21. Packer M, Gheorghiade M, Young JB, et al, for the RADIANCE study: Withdrawal of digoxin from patients with chronic heart failure treated with angiotensin converting inhibitors. *N Engl J Med* 1993, 329:1–7.
22. The Digitalis Investigative Group: The effect of digoxin on mortality and morbidity in patients with heart failure. *N Engl J Med* 1997;336:525–533.
23. Pitt B, Zannad F, Remme WJ, et al: The effect of spironolactone on morbidity and mortality in patients with severe heart failure. *N Engl J Med* 1999; 341:709–717.
24. Smith TW, Butler VP, Haber E: Determination of therapeutic and toxic digoxin concentrations by radioimmunoassay. *N Engl J Med* 1969;281: 1212–1216.
25. Rawles JM, Metcalfe MJ, Jennings K: Time of occurrence, duration, and ventricular rate of paroxysmal atrial fibrillation: the effect of digoxin. *Br Heart J* 1990;63:225–227.
26. Falk RH, Leavitt JI: Digoxin for atrial fibrillation: a drug whose time has gone? *Ann Intern Med* 1991; 114:573–575.
27. Pritchett EL: Management of atrial fibrillation. *N Engl J Med* 1992;326:1264–1271.
28. Lewis RP: Clinical use of serum digoxin concentrations. *Am J Cardiol* 1992;69:97G–107G.
29. Bednarczyk B, Soldin SJ, Gasinska I, et al: Improved receptor assay for measuring digoxin activity. *Clin Chem* 1988;34:393–397.
30. Manchester L, Giesbrecht E, Soldin SJ: Receptor radioligand system for measuring digoxin activity. *Ther Drug Monit* 1987;9:61–66.
31. Soldin SJ: Receptor assays in the clinical laboratory. *Clin Biochem* 1996;29:439–444.
32. Jortani SA, Valdes R Jr: Digoxin and its related endogenous factors. *Crit Rev Clin Lab Sci* 1997; 34:225–274.
33. Kelly RA, Smith TW: Pharmacological treatment of heart failure. In: Hardman JG, Limbird LE, Molinoff PB, et al, eds: *Goodman and Gilman's pharmacological basis of therapeutics*, 9th ed. New York: McMillan, 1996:809–838.
34. Lau BWC, Valdes R Jr: Criteria for identifying endogenous compounds as digoxin-like immunoreactive factors in humans. *Clin Chim Acta* 1988; 175:67–78.
35. Valdes R Jr: Endogenous digoxin-like immunoreactive factors: impact on digoxin measurements and potential physiological implications. *Clin Chem* 1985;31:1526–1532.
36. Balzan S, Clerico A, del Chicca MG, et al: Digoxin-like immunoreactivity in normal human plasma and urine, as detected by a solid-phase radioimmunoassay. *Clin Chem* 1984;30:450–451.
37. Azzazy HME, Duh SH, Maturen A, et al: Multicenter study of Abbott AxSYM® Digoxin II assay and comparison with 6 methods for susceptibility

to digoxin-like immunoreactive factors. *Clin Chem* 1997;43: 1635–1640.
38. Yannakou L, Diamandis EP, Souvatzoglou A: Effects of incubation time and temperature on the interference of DLIFs in digoxin immunoassays. *Ther Drug Monit* 1987;9:461–466.
39. Miller JJ, Straub RW, Valdes R Jr: Digoxin immunoassay with cross-reactivity of digoxin metabolites proportional to their biological activity. *Clin Chem* 1994;40:1898–1903.
40. Gault MH, Kalra J, Longerich L, et al: Digoxigenin biotransformation. *Clin Pharmacol Ther* 1982;31: 695–704.
41. Gault HM, Longerich LL, Loo JCK, et al: Digoxin biotransformation. *Clin Pharmacol Ther* 1984;35: 74–81.
42. Miller JJ, Hess PP, Valdes R Jr: The effect of digoxin on digoxin-like immunoreactive factor (DLIF) interference varies among digoxin assays [Abstract]. *J Clin Immunoassay* 1992;15:63.
43. Metheke ML, Valdes R Jr: A physician's office-based digoxin assay (Seralyzer) evaluated for interference by endogenous digoxin-like immunoreactive factors. *Ann Clin Lab Sci* 1989;19:168–174.
44. Jortani SA, Miller JJ, Helm RA, et al: Unexpected suppression of digoxin values caused by DLIF [Abstract]. *Clin Chem* 1996;42:S124.
45. Jortani SA, Helm RA, Valdes R Jr: Na+, K+-ATPase inhibition of Na+, K+-ATPase by oleandrin and oleandrigenin and their detection by digoxin immunoassays. *Clin Chem* 1996;42:1654–1658.
46. Rainey PM: Effects of digoxin immune Fab (ovine) on digoxin immunoassays. *Am J Clin Pathol* 1989; 92:779–786.
47. Cauffield JS, Gums JG, Grauer K: The serum digoxin concentration: ten questions to ask. *Am Fam Physician* 1997;56(2):495–503,509–510.
48. Bernard DW, Bowman RL, Grimm FA, et al: Nighttime dosing assures postdistribution sampling for therapeutic drug moniroting of digoxin. *Clin Chem* 1996;42:45–49.
49. Hori R, Okamura N, Aiba T, et al: Role of p-glycoprotein in renal tubular secretion of digoxin in the isolated perfused rat kidney. *J Pharmacol Exp Ther* 1993;266:1620–1625.
50. De Lannoy IA, Koren G, Klein J, et al: Cyclosporin and quinidine inhibition of renal digoxin excretion: evidence for luminal secretion of digoxin. *Am J Physiol* 1992;263(4 Pt 2):F613–F622.
51. Okamure N, Hirai M, Tanigawara Y, et al: Digoxin-cyclosporin A interaction: modulation of the multidrug transporter P-glycoprotein in the kidney. *J Pharmacol Exp Ther* 1993;266:1614–1619.
52. McEvoy GK, ed: *AHFS drug information 2001.* Cardiac drugs: Cardiac glycosides [monograph]. Bethesda, MD: American Society of Health-System Pharmacists, 2001:1500–1510.
53. Smith TW, Haber E: Digoxin intoxication: the relationship of clinical presentation to serum digoxin concentration. *J Clin Invest* 1970;49:2377–2386.
54. Aronson JK, Graham-Smith DG, Wigley FM. Monitoring digoxin therapy. The use of plasma digoxin therapy. The use of plasma digoxin concentration measurements in the diagnosis of digoxin toxicity, *Q J Med* 1978;186:111–122.
55. Lely A, van Enter C: Large-scale digitoxin intoxication. *Br Med J* 1970;3:737–740.
56. Piergies AA, Worwag EM, Atkinson AJ: A concurrent audit of high digoxin plasma levels. *Clin Pharmacol Ther* 1994;55:353–358.
57. Wellens HJJ: The electrocardiogram in digitalis intoxication. In: Yu PN, Goodman IF, eds: *Progress in cardiology.* Philadelphia: Lea & Febinger, 1976: 271–290.
58. Smith TW, Haber E, Yeatman L, et al: Reversal of advanced digoxin intoxication with Fab fragments of digoxin-specific antibodies. *N Engl J Med* 1976; 294:443–456.
59. Murphy DJ, Bremmer WF, Haber E, et al: Massive digoxin poisoning treated with Fab fragments of digoxin-specific antibodies. *Pediatrics* 1982;70(3): 472–473.
60. Ligman H, Vincent JL, Hallemans R: Treatment of severe digitoxin intoxication by digoxin-specific Fab antibody fragments. *Acta Cardiol* 1984;39:301–305.
61. Antman EM, Wenger TL, Butler VP Jr, et al: Treatment of 150 cases of life threatening digitalis intoxication with digoxin-specific Fab antibody fragments: final report of a multicenter study. *Circulation* 1990;81:1744–1752.
62. Smolarz A, Roesch S, Lenz E, et al: Digoxin specific antibody (Fab) fragments in 34 cases of severe digitalis intoxication. *Clin Toxicol* 1985;23:327–340.
63. Hickey AR, Wenger TL, Carpenter VP, et al: Digoxin immune Fab therapy in the management of digitalis intoxication; safety and efficacy results of an observational surveillance study. *J Am Coll Cardiol* 1991;17:247–260.
64. Curd J, Smith TW, Jaton JC, et al: The isolation of digoxin-specific antibody and its use in reversing the effects of digoxin. *Proc Natl Acad Sci U S A* 1971; 68:2401–2406.

290 / CHAPTER 18

SELF-ASSESSMENT QUESTIONS

1. Which of the following characterizes the incidence of cardiac glycoside poisoning in the United States in 1998?
   a. The incidence of cardiac glycoside medicine poisoning is the greatest.
   b. The incidence of plant cardiac glycoside poisoning is the greatest.
   c. The incidence of poisoning cases involving cardiac glycoside medicines is about the same as that for plant cardiac glycosides.
   d. The number of deaths caused by plant cardiac glycosides is greater than that caused by cardiac glycoside medicines.
   e. There were no deaths caused by either the cardiac glycoside medications or the plant cardiac glycosides.

2. In which are cardiac glycosides not found?
   a. lily-of-the-valley
   b. oleander
   c. *Digitalis lanata*
   d. cactus
   e. *Digitalis purpurea*

3. Which of the following pairs of pharmacokinetic parameters for digoxin and digitoxin are not correct?

	Digoxin	Digitoxin
a. half-life (h)	30–40	5–9
b. plasma protein binding (%)	20–40	>90
c. volume of distribution (L/kg)	6–7	0.6
d. % metabolized	<20%	>80%
e. route of administration	by mouth, intravenous	by mouth only

4. Which tissue contains the highest total amount of digoxin present in patients receiving this medication?
   a. heart
   b. kidney
   c. liver
   d. skeletal muscle
   e. small intestine

5. Which is not a toxic effect of digoxin?
   a. bradycardia
   b. mydriasis
   c. atrioventricular-conduction block
   d. premature ventricular contractions
   e. distorted color vision

6. Which factor does not contribute to the difficulties in digoxin measurement?
   a. mixtures of digoxin and digoxin-like immunoreactive factors
   b. low concentrations
   c. presence of the antidote (Fab)
   d. terpene-like nucleus
   e. other cardenolide-like compounds

7. Which is not an important factor that needs to be taken into account in order to effectively interpret digoxin concentration?
   a. Have changes occurred in the patient's renal function?
   b. Is the patient taking medications that interact with digoxin?
   c. Does the patient have abnormal electrolytes?
   d. How soon after the patient's last dose of digoxin was the digoxin concentration obtained?
   e. the rapid distribution kinetics of digoxin

8. Briefly describe three areas of controversy involving digoxin.

CHAPTER 19

Calcium Channel Blockers: An Overview

Sandy Shah, D.O., Arti N. Shah, M.S., M.D., and Leslie M. Shaw, Ph.D., DABCC

LEARNING OBJECTIVES

After completing this chapter, the reader should be able to:

1. Describe the incidence of poisoning cases due to calcium channel blockers (CCBs).
2. List five classes of CCBs and list currently approved CCBs for clinical use in each class.
3. Describe the role of calcium in mechanism of actions of CCBs.
4. Describe major features of CCB pharmacokinetics.
5. Discuss the analytical approaches for detection of the major classes of CCBs and measurement of their concentration in serum.
6. Describe the pathophysiology of CCB toxicity.
7. Describe the major manifestations of CCB toxicity and its treatment options.
8. Describe the lessons learned from recent clinical trials about the use of CCBs in patients with cardiovascular diseases.

INTRODUCTION

In the 31 years since Albrecht Fleckenstein introduced the concept of "calcium antagonism," we have witnessed exponential advances in our knowledge of the pharmacology and clinical applications of calcium antagonists. Initially, the focus was on the mechanisms of action of these agents in cardiac and vascular smooth muscle and on their use in the management of cardiovascular disorders. Subsequently, research expanded into the effects of calcium antagonists on other organ systems, exploring a wider range of potential therapeutic applications. This chapter will provide a historical overview and discuss mechanisms of action, clinical pharmacokinetics, laboratory analysis, cardiovascular effects, clinical indications, toxicity, and current controversies.

The idea of calcium channel inhibition began with Lindner in 1960, who noted that prenylamine, a newly developed coronary vasodilator synthesized by the Hoechst Company in Frankfurt, Germany, depressed cardiac performance in canine heart-lung preparations (1). In 1962, Hass and Hartfelder reported that verapamil, a putative coronary vasodilator, possessed negative chronotropic and inotropic effects that were not observed with other vasodilatory agents such as nitroglycerine (2).

A symposium held on the Island of Capri in the Mediterranean Sea at the end of October 1967 was attended by cardiologists who were invited to discuss the possible mechanism of action of prenylamine. It is here where Winifred Nayler and Albrecht Fleckenstein and his colleagues introduced a calcium concept of prenylamine action (3,4).

One year before the Capri symposium, Fleckenstein demonstrated that prenylamine was able to block calcium-dependent excitation-contraction coupling in heart muscle (5). In Capri, Nayler developed a similar concept, which postulated that prenylamine inhibits the calcium permeation across cellular membranes (3).

In 1969, a new class of drugs was identified as calcium antagonists (6). These drugs restricted in active heart muscle calcium-dependent adenosine triphosphate utilization, contractile energy expenditure, and oxygen requirement of the beating heart without impairing sodium-dependent action-potential parameters. Verapamil (Iproveratril) and prenylamine were members of this new drug family first reported in 1964.

In 1968, Ferdinand Dengel, the chief chemist in Knoll Company in Ludwigshafen, West Germany, synthesized gallopamil. It was the 600th verapamil derivative tested in Dengel's laboratory and therefore called D-600. Gallopamil was the second drug after verapamil to meet the strict experimental criteria established by Fleckenstein for characterizing the action mechanism of highly specific calcium antagonists (7–9).

In 1969, Fleckenstein was given two compounds named Bay a1040 and Bay a7168 by Professor Kronberg, a leading pharmacologist at the Bayer Company. Both of these compounds were strong vasodilators and had significant negative inotropic effects on the myocardium. These compounds appeared to have a mechanism of action similar to that of verapamil and D-600. The chemical structures of these two compounds were kept secret for more than 3 years, and later they were named nifedipine (Adalat) and niludipine. The first reports concerning the fundamental Ca^{2+} antagonistic properties of nifedipine (Bay a1040) (10,11) were reported by Fleckenstein in the early 1970s and niludipine (Bay a7168) (12) in 1979. Nifedipine was the first of the dihydropyridines to be used clinically.

Nicardipine was first synthesized by Yamanouchi Pharmaceutical Co., Tokyo, Japan, in the early 1970s and introduced as a cerebral vasodilator in 1976 (13). Nicardipine has been marketed in Japan since 1981 as Perdipine (YC-93) and elsewhere as Vasonase, Cardene, Nicodel, Nerdipine, and Dacarel (RS-69216). In the United States this drug is approved for use in angina pectoris and systemic hypertension.

In 1966 in Japan, the Tanabe Seiyaku Company began work on the synthesis of 1,5-benzothiazepine derivatives as central nervous system antidepressants (14). In 1968, the company discovered and began development of a novel series of 1,5-benzothiazepines that were found to have potent coronary vasodilatory properties (15,16). The D-cis isomer of a group of DL-cis-3-acetoxy-2-(4-methoxy phenyl) derivatives was finally selected, named CRD-401, and further developed. This drug, known as diltiazem, was approved in Japan for the treatment of angina in 1973 and of hypertension in 1982. This was the same year in which it (Cardizem, Marion Laboratories) was first approved in the United States for the treatment of angina after its licensing by Marion Laboratories in 1976. Later, this drug was approved for hypertension, atrial fibrillation or flutter, and paroxysmal supraventricular arrhythmia.

Dozens of additional calcium antagonists (calcium channel blockers, CCBs) were tested in the1960s and early 1970s but most were never approved for clinical use. Scientists at the Centre Européan de Recherché Mauvernay in Riom, France, synthesized bepridil, an agent with additional properties beyond calcium channel blockade, in 1972. Bepridil has been shown to be effective for angina but because of concerns regarding development of life-threatening cardiac arrhythmias, it is currently indicated as a second-line agent for patients with angina who are not optimally controlled on or who are intolerant of standard therapy.

There were three main factors limiting the efficacy of the available CCBs:

- Side effects such as headache, leg edema, and constipation that occur before maximal effect of drug is observed
- Insufficient plasma concentration because of pharmacokinetic problems such as high first-pass metabolism or a half-life that is too short

- Cardiac reflexes that oppose the therapeutic effects of the drug

Hoffmann–La Roche Ltd. developed a program to discover a new CCB without these problems. This led to the discovery of mibefradil (Posicor®) in 1986, which is the first representative of a new class of CCBs and the first T-channel blocker. The main properties of this drug are listed below:

- Chemically, mibefradil, which is a single enantiomer, belongs to a new chemical class (tetralene derivatives).
- Mibefradil blocks both L- and T-voltage–operated calcium channels.
- Mibefradil has no negative inotropic properties at biologically relevant plasma concentrations, even on failing hearts.
- In vivo, mibefradil does not produce reflex tachycardia, but rather decreases heart rate slightly.
- During long-term therapy in humans, mibefradil has an optimal pharmacokinetic profile with a half-life of 10–15 hours and nearly complete absorption.

Mibefradil was approved for the treatment of chronic stable angina and hypertension on June 16, 1997. More than 25 drugs have been found to be potentially dangerous if used with Posicor®. Due to the complexity of the prescribing information needed to address these interactions, Roche Laboratories withdrew the drug from the US market on June 8, 1998. It is important to point out that the withdrawal was not a result of the T-channel–blocking effect, but rather due to the way in which mibefradil is metabolized (by cytochrome P4503A4–catalyzed oxidation, as well as by the 2D6 isoenzyme) (17). Before the release of mibefradil for general clinical use, it was recognized that concomitant administration with terfenadine, cisapride, astemizole, and cyclosporine could affect their plasma concentrations (17).

There are five distinct classes of CCBs that have been evaluated for clinical use (Table 19–1):

- Phenylalkylamines (such as verapamil)
- Benzothiazepines (such as diltiazem)
- Dihydropyridines (such as nifedipine)
- Diarylaminopropylethers (such as bepridil)
- Tetralol derivatives (such as mibefradil)

Although more than 35 CCBs are available for clinical use in various parts of the world, in the United States eleven of these agents are approved for clinical use. These include verapamil, diltiazem, amlodopine, felodopine, isradipine, nifedipine, nicardapine, nimodipine, nisoldipine, bepridil, and mibefradil. Each of these agents has distinct structures, pharmacological properties, and actions (18). The chemical structures of these agents differ substantially and are illustrated in Table 19–1.

In 1999, of the 2,201,156 exposures to toxic substances reported to the American Association of Poison Control Centers (AAPCC), there were 8844 exposures due to CCBs (accounting for 20.16% of all cardiovascular drug exposures) (19). AAPCC reported 61 deaths from these 8844 exposures. It is not possible to know which CCB was most used in the exposure and death cases. Table 19–2 summarizes the demographic profile of CCBs' exposure cases, and Figure 19–1 demonstrates temporal trends in exposures to and deaths from CCB overdose as reported by participating centers of the AAPCC. A 1998 survey found that Norvasc (amlodipine), Cardizem CD (diltiazem), and Procardia XL (nifedipine) were among the top 25 prescription drugs dispensed by pharmacies (20). Table 19–3 lists the CCBs found in the top 200 brand name drug list.

MECHANISM OF ACTION

Role of Calcium

Calcium is the most abundant electrolyte in the human body. The average adult contains more than half a kilogram of calcium, and most of this is located in bone (99%) (21). In the soft tissues, calcium is 10,000 times as concentrated in the extracellular fluids (22). This preference for the extracellular seems odd because calcium seems most involved with smooth muscle

TABLE 19–1. *Chemical classifications and chemical structures of calcium channel antagonists currently approved for therapeutic use in the United States*

Class	Name	Structure	Trade name
Phenylalkylamines	Verapamil	$C_{27}H_{38}N_2O_4$	Calan, Covera-HS, Isoptin, Verelan
Benzothiazepines	Diltiazem	$C_{22}H_{26}N_2O_4S$	Cardizem, Dilacor, Tiazac
Dihydropyridines	Amlodipine	$C_{20}H_{25}ClN_2O_5$	Norvasc
	Felodipine	$C_{18}H_{19}Cl_2NO_4$	Plendil
	Isradipine	$C_{19}H_{21}N_3O_5$	Dynacirc
	Nicardipine	$C_{26}H_{29}N_3O_6 \cdot HCl$	Cardene
	Nifedipine	$C_{17}H_{18}N_2O_6$	Adalat, Procardia
	Nimodipine	$C_{21}H_{26}N_2O_7$	Nimotop
	Nisoldipine	$C_{20}H_{24}N_2O_6$	Sular

TABLE 19–1. *Chemical classifications and chemical structures of calcium channel antagonists currently approved for therapeutic use in the United States (continued)*

Class	Name	Structure	Trade name
Diarylaminopropylethers	Bepridil	$C_{24}H_{34}N_2O \cdot HCl \cdot H_2O$	Vascor
Tetralene derivatives	Mibefradil[a]	$C_{29}H_{38}FN_3O_3 \cdot 2HCl$	Posicor[a]

[a] Mibefradil was approved for the treatment of chronic stable angina and hypertension on June 16, 1997. More than 25 drugs have been found to be potentially dangerous if used with Posicor®. Due to its complexity of the prescribing information needed to address these interactions, Roche Laboratories withdrew the drug from the US market on June 8, 1998.

contraction (23), which is an intracellular process. The calcium in plasma is present in three forms: bound to plasma proteins, chelated to plasma anions, and the ionized form. Approximately 50% of the calcium is bound to plasma proteins, and albumin is responsible for 80% of the protein binding. An additional 5–10% is chelated to plasma anions such as sulfates and phosphates. The remainder is present as free calcium ions. This ionized fraction is the physiologically active fraction of calcium in plasma.

Calcium plays an important role in vital cell functions. It is involved in the maintenance of hemostasis, a broad array of enzymatic reactions, bone metabolism, electrical activation of various excitable cells, the coupling of electrical activation to cellular secretion, and excitation-contraction of cardiac and vascular smooth muscle. Alexander Sandow first proposed calcium's role in excitation-contraction coupling in 1952 (24). Calcium ions (Ca^{2+}) are important chemical mediators of signal transduction in humans, linking electrical stimuli with physiologic response (25,26). Calcium influx from outside the cell results in formation of a second messenger called inositol triphosphate, which leads to release of intracellular calcium from the sarcoplasmic reticulum (Figure 19–2).

This increased intracellular calcium results in calcium binding to calmodulin and formation of calcium-calmodulin complex, which promotes phosphorylation of myosin by activating myosin light-chain kinase. Phosphorylation of the light chain of myosin allows actin and myosin to bind together and subsequently produce muscle contraction (27).

TABLE 19–2. *Demographic profile of calcium antagonists exposure cases*

Number of exposures		8844
Total cases in either the cardiovascular drug or plant categories		43,856
Age	<6	2304 (26.05%)
	6–19	554 (6.26%)
	>19	5665 (64.05%)
Reason	Unintentional	6831 (77.24%)
	Intentional	1723 (19.48%)
	Other	5 (0.06%)
	Adverse reaction	252 (2.85%)
Outcome	None	3622 (40.95%)
	Minor	801 (9.06%)
	Moderate	883 (9.98%)
	Major	243 (2.75%)
	Death	61 (0.69%)

Modified from Litovitz TL, Klein-Schwartz W, Cobaugh DJ, et al: 1999 Annual Report of the American Association of Poison Control Centers Toxic Exposure Surveillance System. *Am J Emerg Med* 2000;18:517–574.

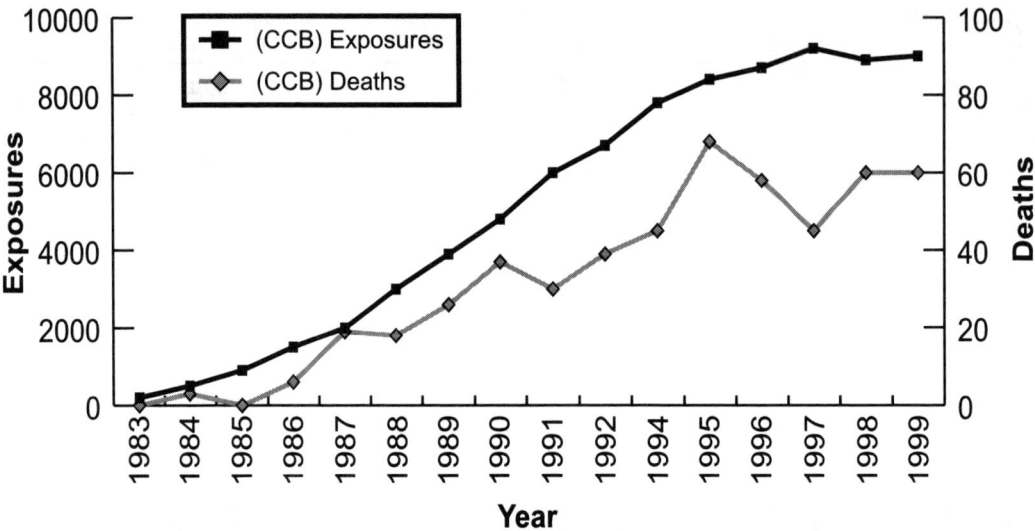

FIG. 19–1. Temporal trends in exposures and deaths from calcium channel blocker overdose as reported by participating centers of the American Association of Poison Control Centers (AAPCC). Data extracted from reports of the AAPCC as published annually in the *American Journal of Emergency Medicine* (1980–2000).

Calcium Channels

Calcium channels are composed of large membrane-spanning glycoproteins containing a pore through which calcium ions selectively pass after the channels open. The driving force for intracellular calcium movement is a large electrochemical gradient across the plasma cell membrane. Because it is a cation, Ca^{2+} carries an electrical charge allowing it to depolarize the negatively charged cytosol after gaining intracellular access (25,26). Calcium ion influx is increased by differences in the concentration gradient across both sides of the cell. The intracellular concentration of calcium in resting cells is less than 4 μg/dL (1 μmol/L), which is approximately 1/1000 the concentration of extracellular calcium (25). Once the channel is opened, calcium diffuses across the plasma cell membrane from an area of high concentration to low concentration, starting the cascade that leads to physiologic response.

Based on their mechanism of action, three types of calcium channels have been described: receptor-operated, leak, and voltage-dependent (28). Voltage-dependent channels are further subdivided into two classes based on high- and low-voltage-dependent depolarization characteristics. Normal resting membrane potential is around −80 mV (29). High-voltage-dependent calcium channels require larger electrical currents (depolarized at membrane potential of −10 to −30 mV) to stimulate channel opening, and they remain open longer than other types of channels.

Several different high-voltage-dependent channels have been identified, including L-, N-, P-, Q-, and R-types. The more recently discovered low-voltage-dependent calcium

TABLE 19–3. *List of calcium channel blockers in the top 200 brand name drugs in 1999*

Rank	Product	Total Rx retail units
5	Norvasc (amlodipine)	24,393,000
35	Cardizem CD (diltiazem)	9,581,000
41	Procardia XL (nifedipine)	8,385,000
52	Adalat CC (nifedipine)	6,866,000
111	Plendil (felodipine)	3,216,000
120	Tiazac (diltiazem)	2,970,000
168	Covera-HS (verapamil)	2,008,000

Modified from *Drug Topics Red Book 1999.* Montvale, NJ: Medical Economics Company, 1999.

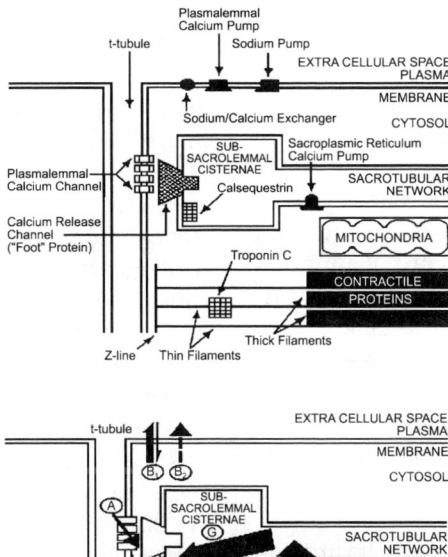

FIG. 19–2. The role of calcium: Schematic diagram showing (*top*) key structures and (*bottom*) major calcium ion (Ca^{2+}) fluxes involved in cardiac excitation-contraction coupling in adult cardiac myocytes. The thickness of the arrows indicates the size of the Ca^{2+} fluxes; their direction describes "energetics" of the Ca^{2+} fluxes (downward arrows describe passive Ca^{2+} fluxes, and upward arrows describe energy-dependent Ca^{2+} transport). Ca^{2+} enters the cell from the extracellular fluid by way of plasma membrane (plasmalemmal) Ca^{2+} channels (*A*); although most of this Ca^{2+} triggers calcium release from the sarcoplasmic reticulum, a small portion directly activates the contractile proteins (*A1*). Ca^{2+} transport back into the extracellular fluid involves two plasma membrane system: Na^+/Ca^{2+} exchange (*B1*) and the plasmalemmal Ca^{2+} pump (*B2*). The sarcoplasmic reticulum membrane regulates two Ca^{2+} fluxes: Ca^{2+} release from the subsarcolemmal cisternae by way of the intracellular Ca^{2+} release channels ("ryanodine receptors") (*C*) and active Ca^{2+} uptake by the Ca^{2+} pump of the sarcotubular network (*D*). Ca^{2+} diffuses within the sarcoplasmic reticulum in a third Ca^{2+} flux (*G*), returning to the subsarcolemmal cisternae, where it is stored in complex with calsequestrin and other Ca^{2+}-binding proteins. Binding of Ca^{2+} to (*E*) and release of Ca^{2+} from (*F*), the high-affinity calcium-binding sites of troponin C, define its affinity for Ca^{2+}. Movements of Ca^{2+} into and out of mitochondria (*H*) buffer cytosolic Ca^{2+} concentration. Reprinted with permission from Katz AM: Calcium channel diversity in the cardiovascular system. *J Am Coll Cardiol* 1996;28:522–529. (Modified from Katz AM: *Physiology of the heart.* 2nd ed. New York: Raven Press, 1992.)

channels have a lower threshold for stimulation than L-type channels and are stimulated in response to slight deviations from resting potential (depolarized at 60 to −40 mV). These channels are referred to as T-type channels (T stands for transient, tiny current) and are inactivated more quickly than other types of calcium channels.

The high-voltage-activated (HVA) channels are heterooligomeric complexes of five proteins from four genes:

- α1 subunit: contains the binding site for all known CCBs, the voltage-sensor, the selectivity filter, and the ion-conducting pore
- β subunit: located intracellularly

- α2δ subunit: a disulfide-linked dimmer
- γ subunit: a transmembrane protein

Figure 19–3 shows the proposed structures of calcium channel subunits (30).

The exact subunit composition of the low-voltage-activated (LVA) channels is unknown. Three α1 subunits have been identified that induce large T-type current after expression in Xenopus oocyte and in human embryonic kidney cells in the absence of additional subunits (31–34). These LVA channels are possibly composed of a single α1 subunit protein that contains a voltage sensor, the selectivity filter, and the ion-conduction pore as well as the binding site for the T-type CCBs such as mibefradil and kurtoxin.

The α1 subunit contains most of the prominent features of the calcium channel complex. It contains the ion-conducting pore, the selectivity filter of the pore, the voltage sensor, and the interaction sites for the β subunits, the βγ subunits of the G proteins, the α2δ subunit, the CCBs, and activators. To date, nine different genes have been identified for the α1 subunit that are homologous to each other and encode proteins of predicted molecular masses of 212–273 kDa. These genes belong to the same multigene family as voltage-activated sodium and potassium channels and share a common ancestral protein with them. Hydrophobicity analysis of the α1 subunits predicts a transmembrane topology with four homologous repeats, each containing five hydrophobic putative α helices and one amphipathic segment.

An early evolutionary event separated the α1 subunits into the electrophysiologically different LVA and HVA calcium channels, which share less than 30% sequence identity. The two LVA genes G and H induce T-type currents in the absence of additional subunits

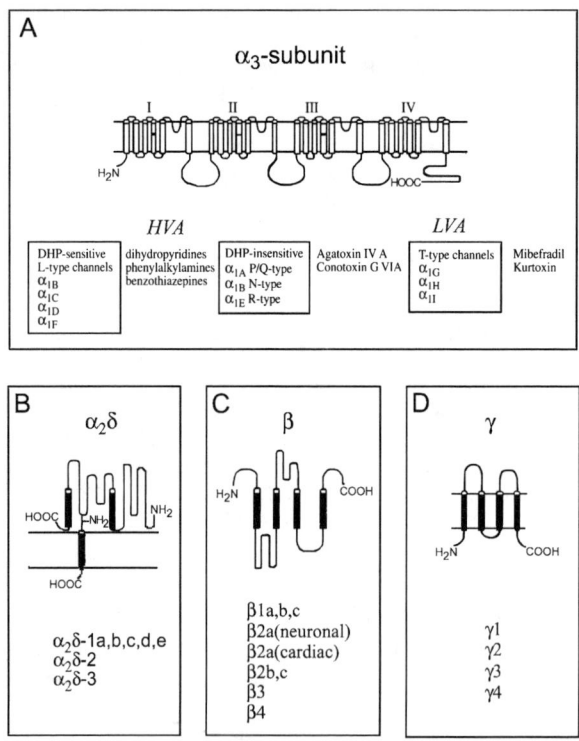

FIG. 19–3. Proposed structures of the calcium channel subunits. (*A*) Membrane topology of the pore-forming α1 subunit, molecular diversity of the α1 genes, and pharmacological properties of different classes. (*B, C,* and *D*) Putative structures and genes of the accessory α2δ, β, and γ subunits. Small letters indicate splice variants. Reprinted with permission from Hoffman F, Lacinová L, Klugbauer N: Voltage-dependent calcium channels: from structure to function. *Rev Physiol Biochem Pharmacol* 1999;139:33–87.

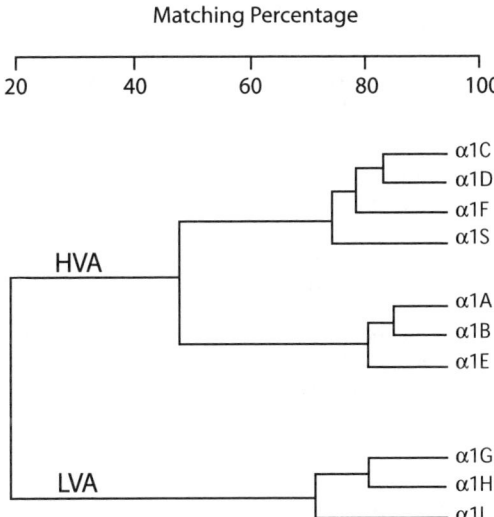

FIG. 19-4. Phylogenetic comparison of the α1 subunits based on the primary structure alignment of the membrane-spanning regions. Matching percentage was calculated using the program CLUSTAL (see Higgins DG, Sharp PM: CLUSTAL: a package for performing multiple sequence alignment on a microcomputer. *Gene* 73:1988;237–244; and Perez-Reyes E: Molecular characterization of a novel family of low-voltage activated, T-type, calcium channels. *J Bioenerg Biomembr* 1998;30:313–318.). Reprinted with permission from Hoffman F, Lacinová L, Klugbauer N: Voltage-dependent calcium channels: from structure to function. *Rev Physiol Biochem Pharmacol* 1999;139:33–87.

(31,33). In 1998, Perez-Reyes et al. identified a third LVA channel, an α1I (32). HVA channels can be divided into two subfamilies, the four (C, D, F, S) dihydropyridine (DHP)-sensitive and the three (A, B, E) DHP-insensitive calcium channels. The A, B, and E genes are expressed almost exclusively in neuronal tissues. Both groups share approximately 50% identical amino acids, whereas the amino acid identity of the individual members of each subfamily is generally over 60% (Figure 19-4).

Based on the biophysical and pharmacological criteria, the native currents of the HVA calcium channels have been divided into five classes:

- L (long-lasting)-type channels
- P (Purkinje)-type channels
- N (neither L or T channel)-type channels
- Q-type channels
- R (remaining)-type channels

L-type currents (α1C, α1D, α1F, α1S) have been found most abundantly in the cardiovascular system. They are also in skeletal muscle, gastrointestinal smooth muscle, uterus, fibroblasts, kidneys, and neuronal and endocrine tissues. The P-, Q-, N-, and R-type channels have been found mainly in neuronal and endocrine tissues.

The P-type current is mediated by α1A channels and is blocked by the funnel web spider *Agelenopsis aperta* toxin ω-agatoxin (<10 nM) and ω-conotoxin MVIIC (5–30 nM) extracted from several marine snails of the genus *Conus* (35–38). The Q-type current is mediated by α1A channels and blocked by ω-conotoxin MVIIC (>100 nM) and ω-agatoxin IVA (>10 nM) (36,39). The N-type current is mediated by α1B channels and is blocked by ω-conotoxin GVIA (100–500 nM), a peptide isolated from the venom of a cone snail and MVIIC (>100 nM) (40).

Three groups of classical CCBs readily block the L-type channels: the dihydropyridines, phenylalkylamines, and benzothiazepines. The L-type CCBs mainly act on cardiovascular channels. These L-type channels have rather diverse functional roles. In neurons, they supply the calcium for activation of small-conductance, calcium-activated K^+ channels (40). They do not have any role in neurotransmitter secretion, a process that is linked to N- and P/Q-type channels in many neuronal cells. In skeletal muscle, L-type channels are vital for excitation-contraction coupling, which does not require influx of calcium through the channel (41). In smooth muscle they play a role in tension development. In the cardiovascular system, these channels are important for electrical impulse generation and for the initiation of atrial and ventricular contraction and contribute to pacemaker potential of the sinoatrial (SA) node and conduction through the atrioventricular (AV) node.

The R-type channels (α1E) have been identified by cloning as a major neuronal calcium channel (42–44). Table 19-4 describes various types of calcium channels and the α1 subunit, location, proposed function, and

TABLE 19–4. Plasma membrane calcium channels

Channel subtype (physiologically defined current)	α1 Subunit	Location	Proposed function	Selective blocker
Voltage-gated Ca^{2+} channels				
L-type	C, D, S, F	Cardiac muscle, smooth muscle	Skeletal excitation-contraction coupling; cardiac excitation-contraction coupling; cardiac pacemaker activity; cardiac atrioventricular conduction; transmitter release from endocrine cells	DHP antagonists
T-type	G	Brain, heart	Growth regulation; cardiac pacemaker activity	Scorpion toxin "kurtoxin"
	H	Kidney, heart, brain		
	I			
P-type	A	Cerebellar Purkinje neurons	Transmitter release from neurons and endocrine cells	ω-Agatoxin and ω-conotoxin MVIIC
Q-type	A	Neurons	Transmitter release from neuronal cells	ω-Conotoxin MVIIC and ω-agatoxin
N-type	B	Neuronal and endocrine tissues	Transmitter release from neuronal cells	ω-Conotoxin GVIA and MVIIC
R-type	E		Transmitter release from neuronal cells	
Intracellular Ca^{2+} release channels				
Ryanodine receptors	—	Skeletal and cardiac muscle	Smooth muscle excitation-contraction coupling; responses to stimuli in nonmotile cells	
Inositol triphosphate receptors	—	Smooth muscle and cardiac muscle	Pharmacomechanical coupling, ? growth regulation, ? diastolic tension	

blocker of each. T channels are found in a variety of tissues including vascular and myocardial smooth muscle. They are less widespread in myocardium than L-type channels, have a lower voltage threshold, and are found in low density in cardiac muscle and in higher density in active vascular smooth muscle cells and nodal cells (45). They do not normally exist in ventricular myocytes except under pathological conditions such as ventricular hypertrophy (45). These channels are also found in SA cells and are thought to have a role in automaticity (25). T-type channels are also found in the neurosecretory cells of the adrenal cortex and medulla and juxtaglomerular cells of the kidney. Figure 19–5 shows the distribution of L- and T-type voltage-dependent calcium channels in humans (46).

PHARMACOLOGY

Although CCBs are classified together, they all differ in their pharmacokinetic properties (47). Table 19–5 describes the pharmacokinetics of the different CCBs. Clinical use of these drugs in different pa-

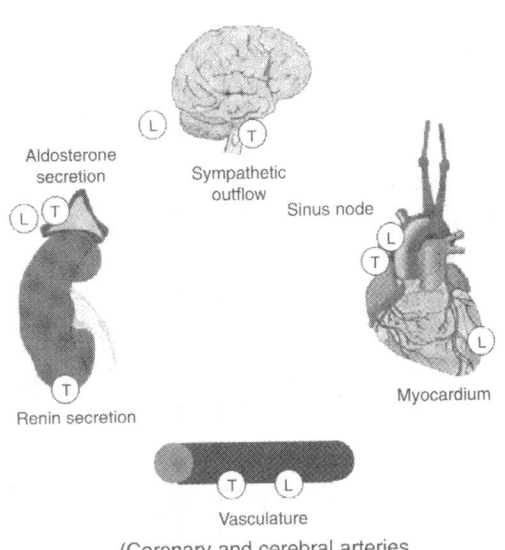

FIG. 19–5. Distribution of L-type and T-type calcium ion channels in the cardiovascular system under normal physiologic conditions. In the myocardium, T-type channels signal remodeling and growth but not contraction. T-type channel expression is associated with hypertrophy and vascular remodeling in the ventricle and aorta. Reprinted with permission of the publisher from Hermsmeyer K, Mishra S, Miyagawa K, et al: Physiologic and pathophysiologic relevance of T-type calcium ion channels: potential indications for calcium antagonists. *Clin Ther* 1997;19[Suppl A];18–26. Copyright 1997 Excerpta Medica Inc.

tients may be influenced by differences in completeness of gastrointestinal absorption, amount of first-pass hepatic metabolism, protein binding, extent of distribution in the body, and the pharmacologic actions of different metabolites (48). The prototype CCBs have short half-lives requiring multiple-dose administration per day, which results in fluctuations in drug concentrations throughout the day. For this reason, various sustained-release delivery systems are available: diffusion type (diltiazem, verapamil), bioerosion (diltiazem, nifedipine, nicardipine), osmosis (verapamil, isradipine, nifedipine), diffusion/erosion (felodipine) (49–51). Nisoldipine was recently approved as a once-daily therapy in a coat-core formulation and verapamil in a daily-onset sustained-release osmotic drug-delivery system (52). All CCBs are available in an oral form, and diltiazem, verapamil, and nicardipine are available in intravenous formulations as well.

After administration of an oral dose, the CCBs, which are metabolized in the liver, show large interindividual variation in circulating plasma concentrations (53–55). In people with hypertension and angina pectoris, wide individual differences also exist in the relationship between plasma concentrations of CCBs and their associated therapeutic effect (54,55). The CCBs are completely and rapidly absorbed, but extensive hepatic first-pass metabolism yields generally low bioavailability. Exceptions to this are nifedipine and amlodipine. Times of peak plasma concentrations for immediate-release formulations are 0.5–3 hours; however, all major classes of CCBs are now available in sustained-release forms with peak plasma concentrations occurring between 6 and 12 hours.

Also, CCBs have extensive volumes of distribution and protein-binding capabilities. The CCBs are extensively metabolized by the liver, with little excretion in the urine. In the case of verapamil, ultimate excretion of the parent compound, as well as the active hepatic metabolite norverapamil, is 75% via the kidneys and 25% via the gastrointestinal tract. Norverapamil is a hepatic metabolite of verapamil, which appears rapidly in the plasma after oral administration of verapamil and in concentrations similar to those of the parent compound; like verapamil, norverapamil undergoes delayed clearance during chronic dosing. Diltiazem is acetylated in the liver to deaceyldiltiazem (with 40% of the activity of the parent compound), which accumulates with chronic therapy. Unlike verapamil, only 35% of diltiazem is excreted by the kidneys (65% by the gastrointestinal tract). The elimination half-lives range from 1–2 hours for nimodipine to 30–50 hours for amlodipine. In liver disease, studies have shown that the half-life of verapamil is increased (56,57). Animal studies involving verapamil (58) demonstrated decreased hepatic blood flow, and in the case of nifedipine (59) clearance of nifedipine is linearly related to hepatic blood flow. The clearance of all the CCBs would be expected to be

TABLE 19–5. Pharmacokinetics of calcium channel blockers

Parameters	Verapamil	Diltiazem, Diltiazem SR	Amlodipine	Felodipine	Isradipine	Nicardipine	Nifedipine, Nifedipine SR	Nimodipine	Nisoldipine	Bepridil
Absorption (%)[a]	90	80–90	ND	~100	90–95	~100	90	ND	ND	~100
Bioavailability (%)[a]	20–35	40–67	64–90	20	15–24	35	45–70 86	13	5	59
Volume of distribution (L/kg)	4.7	5.3	21.4	9.7	69–161	—	0.8–1.4	0.94	2.3	8.0
Onset of action: oral (min)	30[b]	30–60	ND	120–300	120	20	20	ND	ND	60
Time of peak plasma concentrations (h)	1–2.2	2–3 6–11	6–12	2.5–5	1.5	0.5–2	0.5 6	<1	6–12	2–3
Protein binding (%)	83–92	70–80	93	99	95	>95	92–98	95	>99	99
Therapeutic serum concentrations (ng/mL)	80–300	50–200	ND	ND	ND	28–50	25–100	ND	ND	1–2
Metabolite	Norverapamil[c]	Desacetyldiltiazem[e]	90% converted to inactivate	Six inactive	Monoacids and cyclic lactone[f]	Glucuronide conjugates	Acid or lactone[f]	Unknown[g]	5 major urinary metabolites	4-OH-N-phenyl-bepridil
Excreted unchanged in urine (%)	3–4	2–4	10	<0.5	0	<1	1–2	<1	Trace	±
Elimination half-life (h)	3–7[d]	3.5–6 5–7	30–50	11–16	8	2–4	2–5	1–2	7–12	24

SR, sustained release; ND, no data.
[a] Although these agents are well absorbed (80–90%) after oral administration, they are subject to extensive first-pass effects, resulting in an absolute bioavailability that is considerably less.
[b] Peak therapeutic effects occur within 3–5 minutes after intravenous administration.
[c] Pharmacologic activity 20% of verapamil.
[d] 4.5 to 12 hours with multiple dosing; may be prolonged in elderly.
[e] Pharmacologic activity 25% to 50% of diltiazem; plasma concentrations 10–20% of parent drug.
[f] Of six metabolites identified, account for >75%.
[g] Inactive.

TABLE 19–6. Electrophysiology, hemodynamics, and electrocardiogram changes

Parameters		Verapamil	Diltiazem / Diltiazem SR	Amlodipine	Felodipine	Isradipine	Nicardipine	Nifedipine SR	Nimodipine	Nifedipine / Nislodipine	Bepridil
Electrophysiology											
Effective refractory period	Atrium	0	0	0	0	0	0	0	NA	0	↑
	AV node	↑↑	↑	0	0	0	↑↓	±		0	↑
	His-Purkinje	0	0	0	0	0	→	0		0	↓
	Ventricle	0	0	0	0	0	0	0		0	↑↑
	Accessory pathway	±	NA	0	0	ND	0	0		0	
SA node automaticity		↓↓	↓	0	0	0	0	0		0	↓
AV node conduction		↓↓↓	↓↓	0	0	0	0–↑	±		0	↓
Sinus node recovery time		0ᵃ	0ᵃ	0	0	±	0	0		0	ND
ECG changes	Heart rate	↑↓	↓–0	±	±	±	↑	↑	↑	±	↑–0
	QRS complex	0	0	0	0	0	0	0	NA	0	0
	PR interval	↑	↑	0	0	0	ND	0		0	↑↑
	QT interval	ND	ND	0	0	↑	↑	ND		0	↑↑
Hemodynamics											
Myocardial contractility		↓↓	↓	↑	↑	0	0	↓		0	↓
Cardiac output		0–↑	↑	↑	↑	↑↑	↑↑	↑	0	0	
Peripheral vascular resistance		↓↓	↓	↓↓↓	↓↓↓	↓↓↓	↓↓↓	↓↓↓	↓↓	↓↓↓	↓
Coronary blood flow		↑	↑	↑	↑	↑	↑	↑	↑	↑	↑
Myocardial oxygen demand		↓	↓	↓	↓	↓	↓	↓	↓	↓	↓

↑, slight increase; ↑↑, moderate increase; ↑↑↑, pronounced increase; ↓, slight decrease; ↓↓, moderate decrease; ↓↓↓, pronounced decrease; ±, negligible effect; NA, not applicable; ND, no data; SR, sustained release.
ᵃ Prolonged sick sinus syndrome.

decreased in patients with disease states such as congestive heart failure, cardiomyopathy, or hypotensive overdose in which hepatic blood flow is reduced. Various dihydropyridine CCBs such as felodipine, nifedipine, and nisoldipine should not be administered with grapefruit juice because it has been shown to interfere with the drugs' metabolism, resulting in a mean increase in the area under the plasma concentration-time curve of almost twofold and in a mean increase in maximum concentration (C_{max}) of approximately threefold (60).

All CCBs are metabolized to less-active metabolites in the liver by oxidative pathways, mainly by cytochrome P450CYP3A, a subgroup of the cytochrome P450 enzyme family, and to a lesser extent by other members of this enzyme family (61–63). All CCBs, with the exception of diltiazem and nifedipine, are administered as racemic mixtures, with one active and one inactive stereoisomer with respect to blockade of L-type calcium channels (64). The cytochrome P450CYP3A enzymes metabolize each isomer at a different rate, resulting in stereoselective drug clearance (65). Hepatic biotransformation of calcium antagonists such as verapamil may be greater in women than in men (66).

All CCBs differ in their tissue selectivity and therefore they have different electrophysiologic, hemodynamic, and inotropic effects (Table 19–6) (67–73). These differences have important clinical implications. Verapamil and diltiazem are classified as class IV antiarrhythmic agents, whereas dihydropyridines have a negligible effect on cardiac conduction at therapeutic doses. Verapamil and diltiazem are fairly selective for the myocardium, the SA and AV nodes, and the conducting tissues. These nodal tissues possess only calcium channels, which explains why verapamil and diltiazem are effective in treating arrhythmias dependent on nodal conduction. The dihydropyridines are more selective for vascular smooth muscle than cardiac tissue. In the vascular system, arterioles appear to be more sensitive than veins; orthostatic hypotension is not a common side effect. Dihydropyridines are the most potent vasodilators of all CCBs and are associated with the most reflex activation of the sympathetic nervous system.

Important differences in "vascular selectivity" exist among the CCBs. This tissue selectivity can be summarized in a vascular-cardiac selectivity ratio. In general, the dihydropyridines have a greater ratio of vascular smooth muscle effects relative to cardiac effects than do bepridil, diltiazem, and verapamil (Table 19–6) (74). Furthermore, the dihydropyridines may differ in their potency in different vascular beds. For example, nimodipine is claimed to be particularly selective for cerebral blood vessels. All CCBs have negative inotropic effects in vitro (75), but in dihydropyridines these effects are overshadowed by the reflex sympathetic activation and decreased afterload with resultant minimal depression in cardiac function in vivo. Thus, although all CCBs are vasodilators, their in vivo effect on the cardiac tissue, SA and AV node, and arterial beds varies with the individual agent.

The adverse-effect profiles of CCBs are diverse, as are their therapeutic uses. Among the CCBs, the dihydropyridines have the greatest incidence of side effects, mainly related to potent systemic vasodilation leading to headache, dizziness, flushing, and pedal edema. The newer dihydropyridines with longer half-lives and gradual onsets of action have fewer vasodilative side effects such as headaches but have more pedal edema (76).

In the benzothiazepines class, diltiazem has dose-dependent vasodilatory and conduction-related side effects. More common and clinically significant cardiovascular effects that occur relatively frequently with diltiazem include first-degree AV block, sinus bradycardia, exacerbation of congestive heart failure or pulmonary edema, and excessive hypotension.

In the phenylalkylamines class, verapamil has greater affinity for gastrointestinal smooth muscle, and constipation occurs more frequently with verapamil than with other CCBs. The most serious adverse effects associated with verapamil therapy represent extensions of its therapeutic effects on the AV and SA node (sinus bradycardia, reflex sinus tachycardia, AV block of any degree, and hypotension) and its vasodilatory effect on vasculature (dizziness, flushing, and headache). Table 19–7 describes major side effects and their incidence.

CCBs are involved in a significant number of drug interactions. Table 19–8 describes major drug interactions of verapamil, nifedipine, diltiazem, and nicardipine. The classes with significant conduction effects (such as verapamil and diltiazem) interact with β-blockers or digoxin to produce symptomatic bradycardia and AV block, which can also result when used in patients with preexisting SA or AV node dysfunction. Therefore, high-grade AV block, sick sinus syndrome, and sinus bradycardia are relative contraindications to these CCBs. In addition, most CCBs are metabolized by the hepatic cytochrome P450 system and can interact with other drugs undergoing hepatic metabolism. Adverse interactions of mibefradil with other drugs including some hydroxymethyl coenzyme A inhibitors led to several case reports of rhabdomyolysis and voluntary withdrawal of the drug from the market (77,78).

CALCIUM CHANNEL BLOCKER TOXICITY

A Case of Calcium Channel Blocker Poisoning

A 54-year-old black woman with a history of bipolar disorder, hypertension, and "a heart problem" was found unresponsive by her family. When paramedics arrived, the patient was combative, thrashing about, and without a recordable blood pressure. On arrival at the emergency department the patient was agitated and thrashing. Physical examination revealed a resting bradycardia of 40 beats/min, undetectable blood pressure, respiratory rate of 24/min, and temperature of 97.3 °F (36.3 °C). There were bibasilar rales, and neurologically the patient was agitated, following simple commands, and thrashing about but otherwise nonfocal.

Intravenous access was obtained, a monitor was placed, and oxygen was delivered by 100% nonrebreather mask, and the patient was initially treated with a fluid bolus of 500 mL normal saline and initiation of a dopamine infusion. An electrocardiogram at that time revealed an idioventricular rhythm at 50 beats/min, and an external pacemaker was unable to capture the rhythm (Figure 19–6).

Within 15 minutes after arrival the patient's blood pressure rose to 129/100 mmHg, with a heart rate of 54 beats/min. The patient then was able to give a history of left-sided chest pain with radiation to the left arm. Fortunately, in the course of this initial assessment the nurse noticed that the patient's Calan SR 240 mg (verapamil sustained release) prescription was empty and that it had just been filled on the previous day. The diagnosis of acute verapamil poisoning was made. The patient's other medications included Prozac (fluoxetine) and Prilosec (omeprazole).

The patient was emergently intubated and treated with orogastric lavage and activated charcoal, and whole bowel irrigation (WBI) was initiated. The patient's blood pressure remained stable, but the idioventricular rhythm continued. Five hours later, in the intensive care unit, the patient converted to normal sinus rhythm at 70 beats/min. The WBI was stopped for 1 hour during a procedure and the patient went into second-degree heart block, which resolved after the WBI was initiated again.

In this patient a myocardial infarction was ruled out, and she admitted to taking an overdose. She developed atrial fibrillation and flutter and suffered an embolic stroke. She was discharged 2 weeks later with residual expressive aphasia.

Incidence of Calcium Channel Blocker Toxicity

As noted earlier, as the use of CCBs has increased, there has been a concomitant increase in poisoning cases. In 1986, there were just over 1200 exposures and 7 deaths associated with CCB exposures reported to AAPCC. In 1998, these figures increased to 8666 exposures, including 1188 characterized

TABLE 19–7. Calcium channel blockers: clinical indications, dosing information, potential drug interactions, and adverse effects

Drug	Indications	Dose	Drug interactions	Adverse effects	
Verapamil	Angina pectoris: stable, unstable, variant Arrhythmia	80–160 mg PO tid (immediate release) • Oral dosing: Digitalized patients with chronic atrial fibrillation: 240–360 mg/d in 3–4 divided doses; PSVT prophylaxis in nondigitalized patients: 240–480 mg/d in 3–4 divided doses; do not exceed 480 mg/d • Parenteral dosing: SVT: Initial 5–10 mg (0.075–0.15 mg/kg) over 2 min; repeat 10 mg i.v. (0.15 mg/kg) 30 min after initial dose if response is not adequate	β-Blockers: bradycardia, hypotension Digoxin: ↑ serum digoxin concentration Lithium: ↓ serum lithium concentration Theophylline: ↑ theophylline concentration Cyclosporine: ↑ cyclosporine concentration Carbamazepine: ↑ carbamazepine concentration	Dizziness, lightheadedness Headache Constipation Nausea Hypotension Peripheral edema Bradycardia AV block Pulmonary edema Other (shortness of breath, dyspnea, wheezing)	3.5 % 2.2% 7.3% 2.7% 2.5% 2.1% 1.4% 1.0% 1.8% 1.4%
	Hypertension	Immediate release: 80–160 mg PO tid; sustained release 120–480 mg qd			
	Unlabeled use: migraine headache, cardiomyopathy	Prophylaxis of migraine headaches (40–80 mg 3–4 times/d); cluster headaches (180 mg/d)			
Diltiazem	Angina pectoris: stable, variant Atrial fibrillation SVT	Immediate release: 30–90 mg qid; Cardizem CD: 120–480 mg qd Direct i.v. single injections: 0.25 mg/kg over 2 min; a second bolus (0.35 mg/kg over 2 min) can be given 15 min later if needed. Subsequent doses need to be individualized. Start initial infusion at 10 mg/h and increase in 5 mg/h increments up to 15 mg/h as needed. Infusion duration >24 h and rate >15 mg/h are not recommended	β-Blockers: bradycardia, hypotension Cimetidine and ranitidine: ↑ the effect of diltiazem Digoxin: ↑ digoxin concentration Carbamazepine: ↑ carbamazepine concentration Cyclosporine: ↑ cyclosporine concentration	Dizziness, lightheadedness Headache Asthenia Nausea Constipation Peripheral edema AV block Bradycardia Abnormal electrocardiogram Flushing Dermatitis, rash	1.5–7% 2–12% 2.8–5% 1.6–1.9% 1.6% 2.4–9% 0.6–7.6% 1.5–6% 4.1% 1.7–3% 1–1.5%
	Hypertension	Cardizem CD: 180–480 mg/qd; Cardizem SR: 60–180 mg bid; Dilacor XR: 120–540 mg qd			
	Unlabeled use: Raynaud's phenomenon				

Drug	Indication	Dose	Interactions	Adverse effects	%
Amlodipine	Angina pectoris Hypertension	5–10 mg qd In geriatric patients, small frail individuals, and patients with hepatic insufficiency, initial daily doses of 2.5 mg and 5 mg are recommended for hypertension and angina, respectively	β-Blockers: hypotension, bradycardia	Dizziness, lightheadedness Headache Peripheral edema Palpitations Tachycardia Congestive heart failure Fatigue, lethargy Flushing	1.1–3.4% 7.3% 1.8–14.6% 0.7–4.5% 1% 1% 4.5% 0.7–4.5%
Felodipine	Hypertension Unlabeled use: Raynaud's phenomenon, congestive heart failure	5–20 mg qd Daily dose >10 mg is not advised in geriatric patients or those with hepatic impairment	β-Blockers: bradycardia, hypotension Barbiturates: ↓ felodipine effects Erythromycin: ↑ felodipine effects Cimetidine, ranitidine: ↑ felodipine effects Hydantoins: ↓ felodipine effects Carbamazepine: ↑ carbamazapine concentratio	Dizziness, lightheadedness Headache Asthenia Paresthesia Nausea Abdominal pain Peripheral edema Chest pain Palpitations Flushing Muscle cramps Upper respiratory infections Cough Dyspepsia	18.6% 5.8% 4.7% 2.5% 1.9% 1.8% 22.3% 2.1% 1.8% 6.4% 1.9% 5.5% 2.9% 2.3%
Isradipine	Hypertension	2.5–10 mg bid	β-Blockers: bradycardia, hypotension	Dizziness Headache Nausea Abdominal discomfort Edema Chest pain Palpitations Tachycardia Flushing Fatigue Rash	7.3% 13.7% 1.8% 1.7% 7.2% 2.4% 4.0% 1.5% 2.6% 3.9% 1.5%

(Table continued on next page.)

TABLE 19-7. Calcium channel blockers: clinical indications, dosing information, potential drug interactions, and adverse effects (Continued)

Drug	Indications	Dose	Drug interactions	Adverse effects	
Nicardipine	Angina pectoris: stable Hypertension Unlabeled use: congestive heart failure	20–40 mg tid (sustained-release capsules not recommended for angina) Immediate release: 20–40 mg tid; sustained release: 30–60 mg bid	β-Blockers: bradycardia, hypotension Cimetidine, ranitidine: ↑ nicardipine effects Cyclosporine: ↑ cyclosporine concentration	Dizziness Headache Asthenia Somnolence Dry mouth Nausea Abdominal discomfort Peripheral edema Angina Palpitations Tachycardia Flushing	4–6.9% 6.4–8.2% 4.2–5.8% 1.1–1.4% 0.4–1.4% 1.9–2.2% 0.8–1.5% 7.1–8% 5.6% 3.3–4.1% 0.8–3.4% 5.6–9.7%
Nifedipine	Angina pectoris: stable, variant Hypertension Unlabeled use: migraine headache, Raynaud's phenomenon, congestive heart failure, cardiomyopathy	10–60 mg tid; sustained-release tablets: 30–120 mg qd	β-Blockers: bradycardia, hypotension Digoxin: ↑ serum digoxin concentration Cimetidine, ranitidine: ↑ nifedipine effects Quinidine: ↓ quinidine concentration	Dizziness Headache Nervousness, mood change Nausea, heatburn Peripheral edema Palpitations Congestive heart failure Acute myocardial infarction Flushing Weakness Muscle cramps Dyspnea, cough, wheezing Nasal congestion, sore throat	4.1–27.0% 10–23.0% 7.0% 3.3–11.0% 7.0–30.0% 7.0% 2.0–6.7% 4.0–6.7% 3.0–25.0% 12.0% 8.0% 6.0–8.0% 6.0%

Drug	Indication	Dosage	Drug Interactions	Adverse Effects	
Nimodipine	Subarachnoid hemorrhage Unlabeled use: migraine headaches	Initiate within 96 h of event; 60 mg PO every 4 h × 21 d Reduce dose to 30 mg every 4 h in hepatic failure	Hypotensive agents: ↑ hypotension Phenytoin: ↑ phenytoin effects	Hypotension (most common) Edema Headache	5.0%
Nisoldipine	Hypertension	20–60 mg qd	Cimetidine: ↑ nisoldipine effects Quinidine: ↓ effects of nisoldipine	Headache Dizziness Peripheral edema Vasodilation Chest pain Palpitations Nausea Other pharyngitis Sinusitis	22.0% 5.0% 22.0% 4.0% 2.0% 3.0% 2.0% 5.0% 3.0%
Bepridil	Angina pectoris: stable	200–400 mg qd Discontinue if QT interval >0.52 s	β-Blockers: bradycardia, hypotension Antiarrhythmics, tricyclic antidepressants: ↑ QT interval	Dizziness, lightheadedness Nervousness Headache Asthenia Tremor Nausea Abdominal discomfort Dry mouth Bradycardia Tachycardia Peripheral edema Congestive heart failure Torsades de pointes Shortness of breath, dyspnea, wheezing Agranulocytosis	11.6–27% 7.4–11.6% 7.0–13.6% 6.5–14.0% 9.3% 7.26% 7.0% 3.4% 2.0% 2.0% 2.0% 1.0% ↑ QT interval 8.7% Rare

↑, increased; ↓, decreased; PO, by mouth; tid, three times a day; PSVT, paroxysmal supraventricular tachycardia; i.v., intravenous; qd, every day; qid, four times a day; bid, twice a day; SVT, supraventricular tachycardia.

TABLE 19–8. Drug interactions of calcium channel blocking drugs

CCB	Interacting drug	Mechanism	Consequence	Management	Reference
Verapamil	β-Blockers	SA and AV nodal inhibition; myocardial failure	Added nodal and negative inotropic effects	Care during cotherapy, monitor EKG, blood pressure, heart size.	Ellrodt et al. 1985[a] Yeh et al. 1984[b]
	Cimetidine	Hepatic metabolic interaction	Serum verapamil concentration increases	Adjust dose	Peipho et al. 1987[c]
	Digitalis poisoning	Added SA and AV nodal inhibition	Asystole, complete heart block after i.v. verapamil	Avoid i.v. verapamil in digitalis poisoning	Opie 1995[d]
	Digoxin	Decreases digoxin clearance	Increased risk of digoxin toxicity	Halve digoxin dose, monitor serum digoxin concentrations	Pederson 1985[e]
	Disopyramide	Pharmacodynamic	Hypotension, constipation	Check left ventricular function before initiating cotherapy	—
	Flecainide	Added negative inotropic effect	Hypotension	Check left ventricular function, monitor serum flecainide concentrations	—
	Prazosin	Hepatic interaction	Excess hypotension	Check blood pressure during cotherapy	Reid et al. 1985[f]
	Quinidine	Added α-receptor inhibition; verapamil decreases quinidine clearance	Hypotension, increases quinidine concentrations	Monitor quinidine concentrations, check blood pressure	Maisel et al. 1985[g]
	Theophylline	Inhibition of hepatic metabolism	Increases serum theophylline concentrations	Reduce theophylline dose, check concentrations frequently	Hansten and Horn 1986[h]
Diltiazem	β-Blockers	Added SA nodal inhibition; negative inotropism	Bradycardia, hypotension	Check EKG and monitor left ventricular function	Hung et al. 1983[i]
	Cimetidine	Hepatic metabolism interaction	Increases diltiazem concentrations	Reduce diltiazem dose by one-third	Peipho et al. 1987[c]
	Cyclosporine	Hepatic metabolism of cyclosporine inhibited	Increased blood cyclosporine concentrations	Decrease cyclosporine dose	Grino et al. 1986[j]
	Digoxin	Some reduction in digoxin clearance	Only in renal failure	Monitor digoxin concentration	Peipho et al. 1987[c]
	Flecainide	Added negative inotropic effect	Hypotension	Check left ventricular function, check flecainide concentration	—
Nicardipine	Cyclosporine	Hepatic metabolism of cyclosporine inhibited	Increased blood cyclosporine concentrations	Decrease cyclosporine concentration	Bourbigot et al. 1986[k]
	Digoxin	Decreased digoxin clearance	Blood digoxin doubles	Decrease digoxin dose, check digoxin concentrations	Kirch et al. 1984[l]

Nifedipine	β-Blockers	Added negative inotropism	Excess hypotension	Check blood pressure, use test dose of nifedipine	Opie and White 1980[m]
	Cimetidine	Hepatic metabolism interaction	Increased blood nifedipine concentrations	Decrease nifedipine dosage by 40%	Peipho et al. 1987[c]
	Digoxin	Minor or modest changes in digoxin	Increased digoxin concentrations	Check digoxin concentrations	Kleinbloesem et al. 1985[n]
	Prazosin	Prazosin blocks α-reflex to nifedipine	Postural hypotension	Test dose of nifedipine or prazosin	Kiss and Farsang 1989[o]
	Propranolol	Nifedipine and propranolol have opposite effects on blood liver flow	Nifedipine decreases propranolol concentrations; propranolol increases nifedipine concentrations	Readjust propranolol and nifedipine doses if needed	Kleinbloesem et al. 1985[p]
	Quinidine	Nifedipine improves poor left ventricular function, quinidine clearance faster	Decreased quinidine effect	Check quinidine concentrations	Farrington et al. 1984[q]

EKG, electrocardiogram; i.v., intravenous.

[a] Ellrodt AG, Ault MJ, Riedinger MS, et al: Efficacy and safety of sublingual nifedipine in hypertensive emergencies. *Am J Med* 1985;79[Suppl 4A]:19–25.
[b] Yeh R, Gulamhusein SS, Klein GJ: Combined verapamil and propranolol for supraventricular tachycardia. *Am J Cardiol* 1984;53:757–763.
[c] Peipho RW, Culbertson VL, Rhodes RS: Drug interactions with the calcium-entry blockers. *Circulation* 1987;75:181–194.
[d] Opie LH, ed: *Drugs for the heart*, 4th ed. Philadelphia: Saunders,1995.
[e] Pederson KE: Digoxin interactions: the influence of quinidine and verapamil on the pharmacokinetics and receptor binding of digitalis glycosides. *Acta Med Scand* 1985;697(Suppl):12–40.
[f] Reid JL, Meredith PA, Pasanisi F: Clinical pharmacological aspects of calcium antagonists and their therapeutic role in hypertension. *J Cardiovasc Pharmacol* 1985;7[Suppl 4]:S18–S20.
[g] Maisel AS, Motulsky HJ, Insel PA: Hypotension after quinidine plus verapamil: possible additive competition at alpha-adrenergic receptors. *N Engl J Med* 1985;312:167–171.
[h] Hansten PD, Horn JR: Calcium channel blocker-induced drug interactions: evidence for metabolic inhibition. *Drug Interact Newsl* 1986;6:35–40.
[i] Hung J, Lamb IH, Connolly SJ, et al: The effect of diltiazem and propranolol, alone and in combination, on exercise performance and left ventricular function in patients with stable effort angina: a double-blind, randomized, and placebo-controlled study. *Circulation* 1983;68:560–567.
[j] Grino JM, Sabate I, Castelao AM, et al: Influence of diltiazem on cyclosporin clearance. *Lancet* 1986;2:1387–1388.
[k] Bourbigot B, Guiserix J, Airiau J, et al: Nicardipine increases cyclosporin blood levels. *Lancet* 1986;1:1447.
[l] Kirch W, Hutt HJ, Heidemann H, et al: Drug interactions with nitrendipine. *J Cardiovasc Pharmacol* 1984;6:S982–S985.
[m] Opie LH, White DA: Adverse interaction between nifedipine and beta blockade. *Br Med J* 1980;281:1462–1464.
[n] Kleinbloesem CH, van Brumelen P, Hilliers J, et al: Interaction between digoxin and nifedipine at steady-state in patients with atrial fibrillation. In: Kleinbloesem CH, ed: *Nifedipine: clinical pharmacokinetics and haemodynamic effects*. The Hague: Drukkerij JH Pasmans BV, 1985:167–173.
[o] Kiss I, Farsang C: Nifedipine-prazosin interaction in patients with essential hypertension. *Cardiovasc Drugs Ther* 1989;3:413–415.
[p] Kleinbloesem CH, van Brummelen P, Sandberg TH, et al: Kinetic and haemodynamic interactions between nifedipine and propranolol in healthy subjects utilizing controlled rates of drug input. In: Kleinbloesem CH, ed: *Nifedipine: clinical pharmacokinetics and haemodynamic effects*. The Hague: Drukkerij JH Pasmans BV, 1985:151–165.
[q] Farringer JA, McWay-Hess K, Clementi WA: Cimetidine-quinidine interaction. *Clin Pharmacol* 1984;3:81–83.

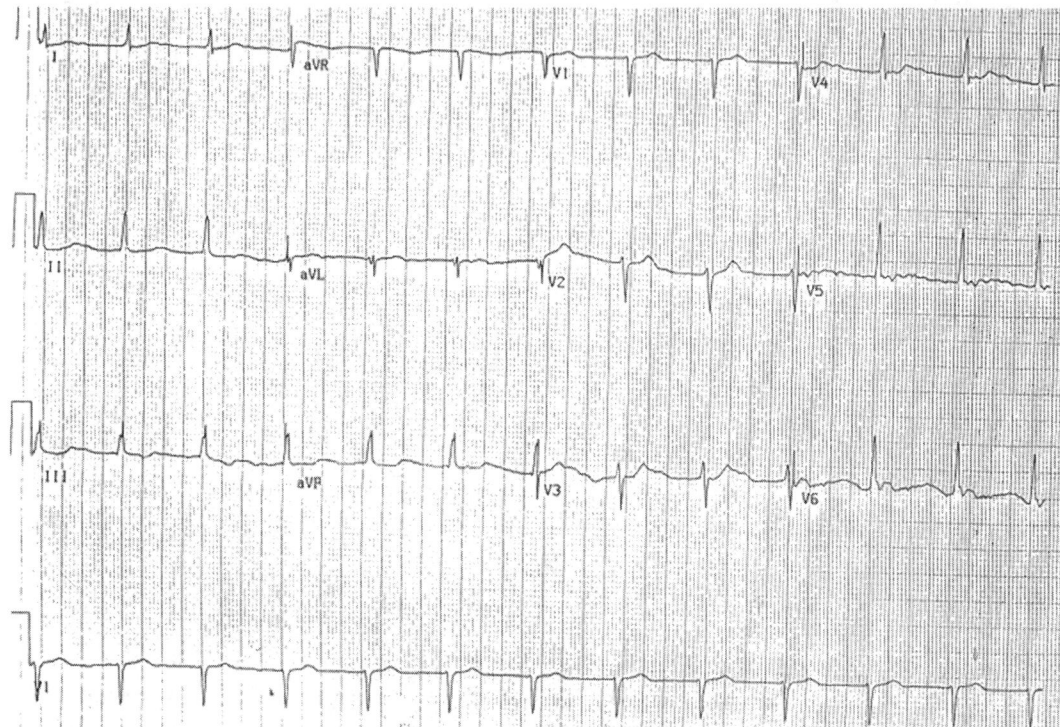

FIG. 19–6. This 12-lead electrocardiogram taken several hours postingestion of Calan SR 240 mg of unknown amount demonstrates accelerated junctional rhythm (no P-waves found) and nonspecific T-wave abnormality.

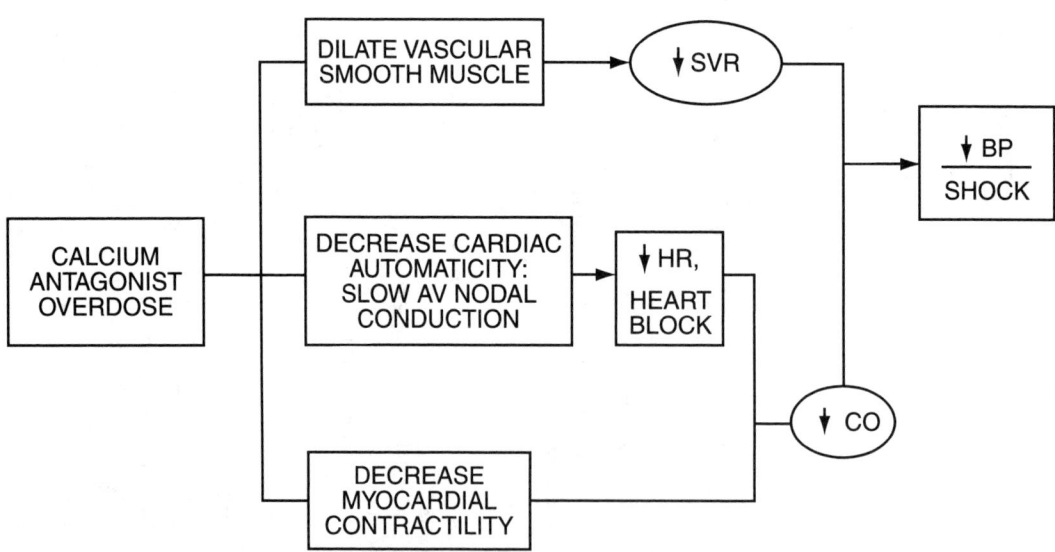

FIG. 19–7. Pathophysiology of cardiovascular consequences of calcium-channel blocker overdose. BP, blood pressure; HR, heart rate; SVR, systemic vascular resistance. Reprinted from permission from Pearigen PD. Calcium channel blocker poisoning. In: Haddad LM, Shannon MS, Winchester JF, eds: *Clinical management of poisoning and drug overdose*, 3rd ed. Philadelphia: WB Saunders, 1998.

by moderate to major toxicity, and 61 deaths, representing nearly 52% of deaths caused by all cardiovascular drugs.

Pathophysiology

The most serious toxic effects of CCBs results from their alteration of transmembrane flow of calcium throughout the cardiovascular system and elsewhere, and are generally an extension of their therapeutic effects on heart rate, conduction, myocardial contractility, and systemic vascular resistance (79). Figure 19–7 illustrates pathophysiology of cardiovascular consequences of CCB overdose. The general pathophysiology of CCB overdose is the same—reduction in cardiac output (decrease in cardiac automaticity and slow AV nodal conduction leads to decrease in heart rate; these factors together with decrease in myocardial contractility lead to reduction in cardiac output) with associated reduction in systemic vascular resistance from dilation of vascular smooth muscle leading to severe hypotension and a state of shock, as occurred in the patient case study.

The severity of poisoning may be affected by several factors including the specific agent involved, dose ingested, product formulation, the presence of co-ingestants, underlying cardiovascular health, and presence of other comorbidities such as age and renal and liver diseases. Comorbidity and age are two factors that adversely affect both morbidity and mortality in CCB poisonings. Elderly patients and those with underlying cardiovascular diseases—for example, congestive heart failure—are more sensitive to the myocardial depressive effects of CCBs (80,81). Even at therapeutic doses these individuals are more frequently prone to developing symptoms of mild hypoperfusion, such as dizziness and fatigue (82–85). The onset of symptoms and duration of toxicity are affected by the product formulation (immediate vs. sustained-release). The toxicity is often present within 2–3 hours of ingestion with regular-release preparations, whereas the onset of toxicity may be delayed for 6–15 hours in sustained-release preparations (86–88). Because the drug's half-life is prolonged in sustained-released preparations, the toxicity may persist for greater than 48 hours (89–91).

Manifestations

The life-threatening toxicity of CCBs is manifested largely within the cardiovascular system and is mainly an extension of its therapeutic effects. Table 19–9 describes clinical manifestations of CCB poisoning. Signs and symptoms of CCB overdose generally begin

TABLE 19-9. *Clinical manifestations of calcium channel blocker poisoning*

Cardiovascular	Central nervous system	Gastrointestinal	Metabolic
• Hypotension • Shock • Bradycardia • Accelerated AV nodal rhythm • Secondary and tertiary AV block with nodal or ventricular escape rhythm • Sinus arrest with AV nodal escape rhythm • Asystole	• Lethargy, slurred speech, confusion • Coma • Respiratory arrest • Seizure	• Nausea, vomiting • Ileus, obstruction • Bowel ischemia, infarction	• Hyperglycemia • Lactic acidosis

Modified from Pearigen PD: Calcium channel blocker poisoning. In: Haddad LM, Shannon MS, Winchester JF, eds: *Clinical management of poisoning and drug overdose*, 3rd ed. Philadelphia: WB Saunders, 1998:1020–1031.

within 1–2 hours of ingestion, but the onset of moderate to severe cardiovascular manifestations can be delayed as long as 24 hours with ingestion of sustained-release preparations (88.92). Myocardial depression and peripheral vasodilation occur, leading to bradycardia and hypotension (93). Bradycardias and conduction defects are among the most frequent findings associated with verapamil or diltiazem overdose, as were observed in the patient case study, and hypotension is present in most significant exposures to any CCB (94,95). Myocardial conduction may be impaired, producing varying degrees of AV conduction abnormalities and junctional or idioventricular rhythms, which were present in the patient case study, and bundle-branch blocks and asystole have been reported (77,90,96–102). Ventricular tachyarrhythmias are not typical of CCB overdose, and their presence suggests underlying cardiac disease or toxic effects of drugs other than CCBs (79).

Some patients with paroxysmal and/or chronic atrial fibrillation or atrial flutter and a coexisting accessory AV pathway (for example, Wolff-Parkinson-White syndrome) have developed increased antegrade conduction across the accessory pathway bypassing the AV node, producing a very rapid ventricular response or ventricular fibrillation after receiving intravenous verapamil (103). Although a risk of this occurring with oral verapamil has not been established, patients receiving oral verapamil may be at risk, and its use in these patients is contraindicated. Treatment is usually cardioversion with a direct current. Intravenous verapamil in combination with β-blockers has been associated with cardiovascular collapse resulting from additive negative effects on heart rate, AV conduction, and/or cardiac contractility. In patients with underlying conduction system disease or in combination with digoxin or β-blockers, the use of oral verapamil or diltiazem in the therapeutic doses may depress the sinus node and AV conduction. Conduction defects are rarely seen with nifedipine overdose.

The most common physical finding in CCB overdose, and which occurred in the patient case study, is hypotension (84,104). Diltiazem commonly produces hypotension from reduced cardiac output, systemic vasodilation, and disturbances in AV conduction (105). Of the CCBs, nifedipine and other dihydropyridines produce the greatest vasodilatory effects. In nifedipine overdose, profound hypotension associated with reflex tachycardia is common (86,106,107). The signs and symptoms represent the degree of cardiovascular compromise and hypoperfusion of the central nervous system. Early or mild symptoms include dizziness, fatigue, and lightheadedness, whereas more severe CCB overdose patients may present with lethargy, altered mental status, syncope, coma, and death (96,101,108–112). Coma most commonly occurs in the setting of severe cardiovascular collapse with profound hypotension. Cases of seizures, however rare, have been reported in overdose due to diltiazem, nifedipine, and verapamil (96,107,113). Cases of strokes from hypotension-induced cerebral hypoperfusion, as described in the patient case, have been reported (114). Severe central nervous system depression is uncommon, and if respiratory depression or coma is present without severe hypotension, consider co-ingestants or other causes of altered mental status (115).

Gastrointestinal symptoms of nausea and vomiting frequently develop in patients with CCB overdose. Verapamil has been associated with paralytic ileus and small bowel obstruction, with and without mechanical obstruction, as well as with fecal impaction (98,116). Cases of mesenteric ischemia and colonic infarction have also been reported in the setting of cardiogenic shock from CCB overdose (117–120).

The effect of therapeutic doses of CCBs on glucose tolerance and insulin release is controversial. However, there have been many reported cases of hyperglycemia in patients with CCB overdose (96,100,102, 108,109,121–126). The concentration of extracellular calcium ions is known to play a crucial role in insulin release from the pancreatic β-cells, the secretory process being triggered by the cytosolic accumulation of calcium (127,128). In vitro, calcium antagonists are known to inhibit glucose and sulfonylurea-induced insulin release by interfering with calcium entry into the islet β-cells (129–131). In CCB overdose, calcium channel selectivity is lost, leading to impairment of calcium influx, and this leads to reduction in insulin release (132). The hyperglycemic effect may be exacerbated in a diabetic patient or if glucagon is used as inotropic therapy (133). Hyperglycemia related to CCB overdose is reversible and usually resolves within 24 hours, but it may necessitate insulin therapy. Metabolic acidosis is also a common finding in severe CCB overdose and is most likely related to decreased tissue perfusion from both CCB-mediated reduction in cardiac output and peripheral vasodilation (122).

Cases of renal failure in the setting of CCB poisoning have been reported (100,102). Nifedipine has been associated with acute, reversible deterioration in renal function in several patients despite the absence of any documented episodes of systemic hypotension (134). Altered renal hemodynamics and possible worsening of left ventricular failure as a result of negative inotropism were implicated as possible explanations.

There have been reports of noncardiogenic pulmonary edema associated with massive CCB poisoning (95,121,135–139). The mechanism by which CCBs contribute to the development of pulmonary edema is not clear. However, experimental studies have demonstrated that these agents can contribute to edema formation either by directly increasing capillary hydrostatic pressure (140,141) or by increasing capillary permeability, either by a direst effect of calcium on cell shape (142) or by a mechanism mediated by the prostaglandins PGE2 and PGF2 (143,144). In addition, massive sympathetic discharge, which has been shown to contribute to "neurogenic" pulmonary edema (145), may occur as a reflex response to the bradycardia and hypotension.

Adverse cutaneous reactions to CCBs are rare, but they have included Stevens-Johnson syndrome, erythema multiforme, exfoliative dermatitis, toxic epidermal necrolysis, cutaneous vasculitis, angioedema, and urticaria (146). The mechanism of adverse reactions reported with CCBs remains unknown. Qualitatively, but not quantitatively, the spectrum of cutaneous adverse reactions appears to be similar among these drugs even though they are chemically dissimilar. The only common feature among these agents is their mode of therapeutic action, that is, calcium channel blockage. Additional studies are necessary to define the exact mechanism of adverse dermatological reactions associated with these antiarrhythmic agents.

Treatment of CCB Overdose

There are no available agonists; therefore, treatment consists of removal of drug by aggressive gastrointestinal decontamination by orogastric lavage and by oral ingestion of activated charcoal. Syrup of ipecac should be avoided because CCB-poisoned patients can rapidly deteriorate and become severely hypotensive. The cardiovascular problems associated with these conditions such as hypotension, left ventricular conduction defects, bradycardia, nodal blocks, and asystole are treated with appropriate measures (Table 19–10). Abnormalities in cardiac rhythm and conduction may be resistant to therapy, and responses to calcium, atropine, or isoproterenol may be insufficient, necessitating temporary transvenous pacing.

CCB Measurement

For diagnosis and confirmation of a diagnosis or for exclusion of intoxication caused by a CCB, validated screening methods have been developed and are available for the main classes, the phenylalkylamines (verapamil), the benzothiazepines (such as diltiazem), and the dihydropyridines (such as amlodipine and nifedipine). Generally, these methods detect metabolites of the CCB in urine and would be performed be-

TABLE 19–10. *Management of calcium channel blocker poisoning*

General management
- Airway, ventilatory, circulatory support
- Activated charcoal (with or without gastric lavage)
- Whole-bowel irrigation in sustained-release ingestion
- Intravenous calcium chloride in critically ill patients
- Admit to intensive care unit and initiate continuous electrocardiogram monitoring
- Seizures: Lorazepam, 1–2 mg/dose i.v.; phenytoin, 15 mg/kg i.v., or phenobarbital, 15 mg/kg i.v., each loaded over 20–30 min

Arrhythmias
- Asymptomatic: Supportive measures and continued electrocardiogram monitoring
- Symptomatic: Atropine, isoproterenol, or cardiac pacing
  - Profound bradycardia: Atropine sulfate (not always effective)
  - Sinus bradycardia: 10% calcium gluconate or calcium chloride SA and AV node block: Isoproterenol or dobutamine
  - Asystole: External cardiac massage and cardiac pacing

Hypotension
- Treat significant bradyarrhythmias as above
- Calcium salts: Calcium chloride, 1 g (10 mL of 10% solution); or calcium gluconate, 3 g (30 mL of 10% solution), i.v. over 5 min; repeat as needed if favorable response
- 0.9% Sodium chloride, 200 mL i.v. every 10 min up to 1–2 L or evidence of pulmonary edema
- Glucagon, 2–5 mg i.v. over 5 min; repeat or initiate continuous infusion
- If patient remains hypotensive, place pulmonary artery catheter and consider administration of the following: norepinephrine, Neo-Synephrine, dopamine, dobutamine, amrinone

Severe left ventricular dysfunction
- 10% calcium gluconate or calcium chloride; norepinephrine or dopamine

Other therapies
- Amrinone[a]
- 4-Aminopyridine[b,c,d]
- Bay k 8644 (a dihydropyridine analog that acts as a calcium entry promoter)[e,f]
- Digoxin[g]
- Hyperinsulinemic euglycemia[h]
- Intra-aortic balloon counterpulsation[i]
- Extracorporeal bypass[j]

i.v., intravenous.

[a] Goenen M, Col J, Compere A, et al: Treatment of severe verapamil poisoning with combined amrinone-isoproterenol therapy. *Am J Cardiol* 1986;58:1142–1143.

[b] Agoston S, Maestrone E, van Hezik EJ, et al: Effective treatment of verapamil intoxication with 4-aminopyridine in the cat. *J Clin Invest* 1984;73:1291–1296.

[c] Gay R, Algeo S, Lee R, et al: Treatment of verapamil toxicity in intact dogs. *J Clin Invest* 1986;77:1805–1811.

[d] ter Wee PM, Hovinga TKK, Uges DRA, et al: 4-Aminopyridine and hemodialysis in the treatment of verapamil intoxication. *Hum Toxicol* 1985;4:327–329.

[e] Korstanje C, Jonkman FAM, Van Kemenade JE, et al: Bay k 8644, a calcium entry promoter, as an antidote in verapamil intoxication in rabbits. *Arch Int Pharmacodyn Ther* 1987;287:109–119.

[f] Pearigen PD, Benowitz NL: Poisoning due to calcium antagonists: experience with verapamil, diltiazem and nifedipine. *Drug Safety* 1991;6:408–430.

[g] Ramo MP, Grupp I, Pesola MK, et al: Cardiac glycosides in the treatment of experimental overdose with calcium-blocking agents. *Res Exp Med (Berl)* 1992;192:335–343.

[h] Kline JA, Lenova E, Raymond RM: Beneficial myocardial metabolic effects of insulin during verapamil toxicity in the anesthetized canine. *Crit Care Med* 1995;23:1251–1263.

[i] Frierson J, Bailly D, Shultz T, et al: Refractory cardiogenic shock and complete heart block after unsuspected verapamil-SR and atenolol overdose. *Clin Cardiol* 1991;14:933–935.

[j] Hendren WC, Schreiber RS, Garretson LK: Extracorporeal bypass for the treatment of verapamil poisoning. *Ann Emerg Med* 1989;18:984–987.

fore quantification of the parent drug in serum. Several methods have been described in the literature to screen for these agents in urine. Among the ones that would be recommended to a laboratory considering setting up would be gas chromatography–mass spectrometry screening methods for basic and neutral drugs that have been shown to be suitable for detection of the phenylalkylamines and the benzothiazepines and their metabolites (147.148) and for the dihydropyridine metabolites (149). For follow-up evaluation and monitoring of CCB poisoning, reliable quantification is achievable using high-performance liquid chromatography methodology (150–152). Nifedipine is very light sensitive such that reliable quantification requires protection of blood samples from light (153).

CURRENT CONTROVERSIES

Since the mid-1990s, there has been an ongoing controversy over the safety of CCBs. There have been reports of increased adverse cardiovascular events (myocardial infarction, arrhythmias, stroke), malignancies, and gastrointestinal and intraoperative bleeding.

Calcium Channel Blockers and Adverse Cardiovascular Effects

Despite CCBs' apparent beneficial properties, controversies exist regarding their ability to reduce cardiac risk or possibly increase risk. The evidence as to whether CCBs increase cardiovascular risk is not conclusive. This section reviews the evidence presented by cohort, case control, and large randomized trials comparing use of CCBs with increased cardiovascular complications.

In four case-controlled studies, two studies observed an increase in risk and two studies did not. In 1995, Aursnes et al. examined the relationship between various drugs used for treating blood pressure and the incidence of acute myocardial infarction involving 95 cases and 329 control subjects (154). The risk of myocardial infarction for drug treatment during the last 5 years vs. non–drug treatment was 0.70 (95% confidence interval 0.42–1.18). The risk for diuretics and β-blockers tested against no treatment was 0.91 (0.52–1.61). The corresponding risk for vasodilating drugs was 0.43 (0.20–0.91). On the other hand, 4 weeks of exposure to α-blockers tested against the other drug treatments indicated an odds ratio of 4.62 (1.01–24.0) for individuals with a history of angina. The authors of this study concluded that their data confirm that treatment with diuretics and β-blockers has only little effect on the incidence of myocardial infarction. As a whole, vasodilators are associated with a significant reduction in this incidence, but α-blockers increase the risk in patients with angina.

Another study conducted in the United Kingdom in 1996 involving 210 cases and 793 control subjects did not demonstrate increased cardiovascular risk in hypertensive patients treated with CCBs (155). The study by Psaty et al. in 1995 included 623 cases and 2032 control subjects from the Group Health Cooperative of Puget Sound with pharmacologically treated hypertension (156). Compared with users of diuretics alone, the adjusted risk ratio of myocardial infarction was increased by approximately 60% among users of CCBs with or without diuretic (risk ratio = 1.62; 95% confidence interval 1.11–2.34). The use of CCBs compared with β-blockers was associated with an approximately 60% increase in the adjusted risk of myocardial infarction (risk ratio = 1.57; 95% confidence interval 1.21–2.04). High doses of β-blockers were associated with a decreased risk of myocardial infarction, while use of high doses of short-acting CCBs were associated with an increased risk.

In another case-controlled study Alderman et al. matched 189 hypertensive patients who had a first cardiovascular event with control subjects (157). Long-acting and short-acting CCBs differed in cardiovascular outcomes. Compared with those on β-blocker monotherapy, patients on long-acting calcium antagonists (n = 136) had no increased risk of a cardiovascular event [adjusted odds ratio 0.76 (95% confidence interval 0.41–1.43)], whereas patients on short-acting CCBs (n = 27) were at significantly greater risk [adjusted odds ratio 3.88 (1.15–13.11), $P = 0.029$].

Several cohort studies examining the relation between use of CCBs and cardiovascular events also demonstrate conflicting evidence. In 1995, Pahor et al. in Rome prospectively evaluated 906 hypertensive patients aged =71 who were treated with a single hypertensive agent (158). During 3538 person-years follow-up, 188 patients died (53 deaths per 1000 person-years). Compared with β-blockers, the relative risks for mortality associated with use of verapamil, diltiazem, nifedipine, and angiotensin-converting enzyme inhibitors were 0.8 (0.4–1.4), 1.3 (0.8–2.1), 1.7 (1.1–2.7), and 0.9 (0.6–1.4), respectively. Compared with β-blockers, short-acting nifedipine was associated with decreased survival in older hypertensive persons (relative risk 1.7; 95% confidence interval 1.1–2.7).

Koenig et al. in 1997 studied the long-term survival, for a median follow-up time of 4.4 years, of 197 nondiabetic patients in the population-based acute myocardial infarction registry in Augsburg, Germany. The patients were 25–74 years old and had survived a first Q-wave acute myocardial infarction for at least 28 days (159). The use of CCBs was not associated with a reduction in total mortality (relative risk 1.23; 95% confidence interval 0.89–1.69). Use of diltiazem was associated with increased risk (relative risk 1.55; 95% confidence interval 1.04–2.32) of death compared with nifedipine (relative risk 1.00; 95% confidence interval 0.68–1.48). Using patients on β-blockers only as a reference, the prescription of CCBs was consistently associated with an increased total mortality (nifedipine, without β-blockers relative risk 1.20; 95% confidence interval 1.12–3.57; diltiazem, without β-blockers relative risk 2.87; 95% confidence interval 1.75–4.70).

In Japan, Ishikawa et al. conducted a cohort study to elucidate the effect of short-acting nifedipine and diltiazem on cardiac events in 1115 patients with healed myocardial infarction (160). The patients included 595 who did not receive a calcium antagonist, 341 who received short-acting nifedipine 30 mg/d, and 179 who received short-acting diltiazem 90 mg/d. Cardiac events occurred in 51 patients (8.6%) in the no-calcium-antagonist group and 54 (10.4%) in the calcium-antagonist group (odds ratio 1.24; 95% confidence interval, 0.83–1.85), demonstrating that the CCBs did not reduce the incidence of cardiac events. This study showed that use of short-acting nifedipine and diltiazem in the post myocardial infarction population was associated with a 24% higher cardiac event rate, but this strong adverse trend was not statistically significant.

In 1998, Michels et al. prospectively explored relative risks of cardiovascular diseases and mortality in 14,617 hypertensive women in the Nurse's Health Study (161). They found significant elevation in relative risk of total myocardial infarction among women who used CCBs compared with those who did not. CCB monodrug users had an age-adjusted relative risk of myocardial infarction of 2.36 (95% confidence interval, 1.43–3.91) compared with those prescribed thiazide diuretics. Women prescribed CCBs had a higher prevalence of ischemic heart disease. Comparing the use of any CCB (monodrug and multidrug users) with that of any other antihypertensive agent, the adjusted relative risk was 1.42 (95% confidence interval 1.01–2.01).

In a cohort of 3539 hypertensive patients from the Framingham Heart Study, there were no differences in mortality between subjects with hypertension using CCBs and those who were not (162). Results were similar among subjects with hypertension with and without coronary heart disease.

Several randomized controlled trials examining the relationship between use of CCBs and cardiovascular events demonstrates conflicting evidence. The GLANT, STONE, and Syst-Eur trials demonstrate safety of CCB use in patients with hypertension, whereas the MIDAS and ABCD trials raise questions about the effect of CCBs on coronary heart disease. The GLANT study evaluated patients with mild to moderate essential hypertension who were mainly treated with an angiotensin-converting enzyme inhibitor (delapril, n = 980) or a calcium antagonist (n = 956) for 12 months (163). There was no significant difference in the incidence of cerebrovascular and cardiovascular events between the two groups, and the results suggested that blood pressure reduction did not necessarily lead to a parallel decrease in cerebrovascular and cardiovascular complications.

In the single-blinded and placebo controlled STONE trial (Shanghai Trial of Nifedipine in Elderly Hypertensives) (164), 1632 mainly Chinese patients aged 60–79 years with systolic blood pressure ≥160 mmHg or diastolic blood pressure ≥96 mmHg were allocated to either nifedipine or placebo after a 4-week placebo. A significant reduction in relative risk was observed for strokes and severe arrhythmia, with an overall decrease from 1.0 to 0.41 (95% confidence interval 0.27–0.61). Nifedipine treatment diminished the number of severe clinical outcomes in elderly hypertensives significantly.

In 1997, in the Syst-Eur trial (165), 4695 patients were randomly assigned to nitrendipine 10–40 mg daily with the possible addition of enalapril 5–20 mg daily and hydrochlorothiazide 12.5–25.0 mg daily, or matching placebos. Among elderly patients with isolated systolic hypertension, antihypertensive drug treatment starting with nitrendipine reduces the rate of cardiovascular complications. Treatment of 1000 patients for 5 years with this type of regimen may prevent 29 strokes or 53 major cardiovascular end points.

The MIDAS trial (Multicenter Isradipine Diuretic Atherosclerosis Study) (166) compared the rate of progression of mean maximum intimal-medial thickness in carotid arteries, using quantitative B-mode ultrasound imaging, during antihypertensive therapy with isradipine vs. hydrochlorothiazide. There was no difference in the rate of progression of mean intimal-medial thickness between isradipine and hydrochlorothiazide over 3 years ($P = 0.68$). There was a higher incidence of major vascular events (such as myocardial infarction, stroke, congestive heart failure, angina, and sudden death) in isradipine (n = 25, 5.65%) compared with hydrochlorothiazide (n = 14; 3.17%) ($P = 0.07$), and a significant increase in major vascular events and procedures (for example, transient ischemic attack, dysrhythmia, aortic valve replacement, and femoral popliteal bypass graft) in isradipine (n = 40, 9.05%) compared with hydrochlorothiazide (n = 23; 5.22%) ($P = 0.02$).

The meta-analysis by Stason and colleagues evaluated event reduction in patients treated with nifedipine alone or in combination with other antihypertensive agents for a median of 8 weeks (167). The results suggested that when nifedipine is used as part of combination therapy for patients with hypertension, cardiovascular events are reduced, although withdrawal rates are higher. Events are not reduced when nifedipine is used as monotherapy.

The ABCD trial (Appropriate Blood Pressure Control in Diabetes) (168) compared nisoldipine with enalapril as a first-line antihypertensive agent in terms of the prevention and progression of complications of diabetes and incidence of myocardial infarction. Nisoldipine was associated with a higher incidence of fatal and nonfatal myocardial infarctions (n = 24) than enalapril (n = 4) (relative risk 9.5; 95% confidence interval 2.7–33.8) in patients with diabetes and hypertension.

In patients with preexisting coronary artery disease, the data on coronary vascular disease risk are also conflicting. CCBs appear to increase cardiovascular event rates in patients with acute coronary syndromes. The following studies demonstrate that nifedipine may increase cardiovascular events in patients with acute coronary syndrome, and data on diltiazem appear to be equivocal.

In 1984, the Danish Study Group evaluated verapamil in acute myocardial infarction in 3498 patients (68). Of 1436 patients with acute myocardial infarction, 717 were treated with verapamil and 719 with placebo. After six-months, 92 patients (12.8%) in the verapamil group and 100 patients (13.9%) in the placebo group were dead (not significant). After 12 months the mortality rates were 15.2% and 16.4%, respectively (not significant). By 6 months after entry a total of 56 reinfarctions were recorded in 50 patients (7%) in the verapamil group and 66 reinfarctions in 60 patients (8.3%) in the placebo group (not significant).

Gibson and colleagues performed a multicenter, double-blind, randomized study to evaluate the effect of diltiazem on reinfarction after a non–Q-wave myocardial infarction (169). This study showed that there was no increase in mortality or reduction in frequency of refractory postinfarction angina in the diltiazem group compared with placebo. They concluded that diltiazem was effective in preventing early reinfarction and severe angina after non–Q-wave infarction and that it was generally safe and well tolerated.

The Trial of Early Nifedipine in Acute Myocardial Infarction (TRENT) study enrolled 4491 patients with chest pain of presumed ischemic origin within 24 hours of symptom onset (170). Patients were assigned to receive nifedipine or placebo. There was no difference in mortality between the two groups in the subgroup taking β-blockers. This trial was terminated early due to the similarity of cardiac events in the two groups.

The Holland Interuniversity Nifedipine/Metoprolol Trial (HINT) (171) group in 1986 conducted a multicenter, double-blind, placebo controlled, randomized trial of nifedipine, metoprolol, and nifedipine and metoprolol combined in a group of 338 patients with unstable angina not pretreated with a β-blocker and of nifedipine in 177 patients pretreated with a β-blocker. This study was terminated early because patients taking nifedipine alone had a high infarction rate (27% vs. 14%). The combination of nifedipine and metoprolol had no advantage over metoprolol alone, although in patients already taking β-blockers nifedipine was slightly beneficial, with a relative risk of 0.68 for recurrent ischemia or myocardial infarction.

The Secondary Prevention Reinfarction Israeli Nifedipine Trial II (SPRINT II) (172) investigators randomly assigned 1358 patients with suspected myocardial infarction within 3 hours of symptom onset to receive nifedipine 60 mg/d or placebo. An excess mortality was observed in the nifedipine group with 105 deaths (15.4%) compared with 90 deaths (13.3%) in the placebo group. This difference was entirely due to excess in early (6-day) mortality with nifedipine. The study was discontinued as soon as this was known.

Furberg and colleagues in 1995 assessed the effect of the dose of nifedipine on increased risk of mortality seen in the randomized secondary-prevention trials by performing a meta-analysis of 16 trials (173). Overall, nifedipine was associated with a significant adverse effect on total mortality (relative risk 1.16, 95% confidence interval 1.01–1.33), and high doses of nifedipine were significantly associated with increased mortality ($P = 0.01$).

In patients with stable angina, there is also controversy about whether CCBs increase cardiovascular event rates. Three studies between 1986 and 1991 involving nicardipine (174,175) and nisoldipine (176) suggested that despite reduction in angina with short-acting dihydropyridines, the incidence of serious adverse cardiovascular events was increased. These data suggest disparity between the ability of short-acting dihydropyridines to relieve angina and their ability to decrease event rate associated with stable coronary disease.

The TIBET (Total Ischaemic Burden European Trial) study group investigated whether total ischemic burden has important prognostic implications in patients with stable angina on a standard antianginal regimen (177). Six hundred eighty-two patients with a diagnosis of chronic stable angina were randomized to receive atenolol, 50 mg twice a day; nifedipine, 20 mg twice a day; or the combination of atenolol and nifedipine, 50 and 20 mg, respectively, twice a day. There was a nonsignificant trend to a lower rate of hard end points (cardiac death, nonfatal myocardial infarction, unstable angina) in the combination therapy group. The nifedipine group did not have a significant increase in cardiovascular event rate.

TABLE 19-11. *Major trials evaluating risk of GI bleeding associated with use of CCBs*

Reference	Patients	Significant results	Interpretation
Pahor et al. 1996[a]	n = 1636. A prospective cohort study conducted from 1985 through 1992 on 1636 hypertensive patients' age =68 y taking β-blockers, ACE inhibitors, or calcium antagonists	Compared with β-blockers (4819 person-years, 65 events), the relative risk for GI bleed associated with ACE inhibitor (772 person-years, 13 events) was 1.23 (95% CI 0.66–2.28) and with calcium antagonist (1510 person-years, 42 events) it was 1.86 (1.22–2.82). The risks for verapamil, diltiazem, and nifedipine did not differ significantly.	Calcium antagonists were associated with an increased risk of GI bleeding in the population studied.
Wagenknecht et al. 1995[b]	n = 149 (entered in study). A R, PC, DB clinical trial of patients undergoing cardiac valve replacement surgery to test whether nimodipine (30 mg started 12 h before surgery and continued for 5 d four times a day) would reduce the combined incidence of neurological, neurophthalmological, and neuropsychological deficits over 6 mo compared with placebo.	From an interim analysis of only 149 patients, an independent monitoring committee recommended suspension of recruitment due to an excess mortality in the patients receiving nimodipine compared with those receiving placebo [8/75 (11%) vs. 1/74 (1%)]. Major bleeding occurred often and as per their definition, 10 (13%) patients were receiving nimodipine compared with 2 (3%) in placebo group. Six of the nine deaths (5 in patients receiving nimodipine and one in a patient receiving placebo) occurred in patients with major bleed.	This was an unexpected finding, observed in a post hoc analysis. This was the first time major surgical bleeding associated with calcium antagonists was reported. The bleeding may be the result of vasodilation promoted by calcium antagonists or by the drug's known antiplatelet action.
Hynynen et al. 1996[c]	n = 120. A prospective randomized trial of patients undergoing coronary artery bypass graft surgery	Of the 81 patients receiving long-term treatment with calcium antagonists, 38 received nifedipine, and 35 diltiazem, while the other 8 were taking verapamil, nisoldipine, or felodipine. No patient received nimodipine. There was no difference in postoperative blood loss found when comparing patients who were on long-term treatment vs. those who were not on calcium antagonists.	The possibility that calcium antagonists increase the tendency to bleeding in patients having cardiac surgery requires further testing

ACE, angiotensin-converting enzyme; GI, gastrointestinal; CI, confidence interval; R, PC, DB, randomized placebo-controlled double-blind.
[a] Pahor M, Guralnik JM, Furberg CD, et al: Risk of gastrointestinal haemorrhage with calcium antagonists in hypertensive persons over 67 years old. *Lancet* 1996;347:1061–1065.
[b] Wagenknecht LE, Furberg CD, Hammon JW, et al: Surgical bleeding: unexpected effect of a calcium antagonist. *BMJ* 1995;310:776–777.
[c] Hynynen M, Kuitunen A, Salmenpera M: Surgical bleeding and calcium antagonists [Letter, Comment]. *BMJ* 1996;312:313.

Rehnqvist et al. in Sweden compared long-term treatment effects of metoprolol or verapamil on combined cardiovascular end points in patients with stable angina (178). They demonstrated that both drugs are well tolerated and showed no difference in the effect on mortality, cardiovascular end points, and measures of quality of life.

In patients post myocardial infarction without left ventricular dysfunction, several trials of secondary prevention demonstrated the safety of CCB use. The efficacy of nifedipine in the prevention of postinfarction morbidity and mortality was evaluated in a double-blind (SPRINT) trial of 2276 survivors of acute myocardial infarction recruited from the cardiac departments of 14 Israeli hospitals (179). Compared with placebo, nifedipine had no effect on cardiac events in survivors of acute myocardial infarction.

The Multicenter Diltiazem Post-Infarction Trial (MDPIT) showed 11% fewer first recurrent cardiac events (death from cardiac cause or nonfatal reinfarction) with diltiazem (202 events) than with placebo (226 events). Several trials showed that long-term treatment with verapamil after an acute myocardial infarction caused a significant reduction in major events (180–182).

There is a tremendous amount of literature available that challenges the safety of CCBs in patients with cardiovascular diseases. Most of the data is available for dihydropyridine CCBs. The following conclusions can be made after evaluation of the literature:

- In patients with acute myocardial infarction, CCBs should be avoided.
- In patients with stable coronary disease (post myocardial infarction or low-risk stable angina), all types of CCBs are safe, and they are not associated with a higher incidence of adverse events.
- In patients with mild to moderate hypertension at low risk for cardiovascular events, long-acting dihydropyridine CCBs do not increase cardiovascular event risk.
- In patients with high risk for coronary events, further studies are necessary to demonstrate the role of CCBs.

CCBs and Gastrointestinal Bleeding

It has been reported in the literature that CCBs interfere with gastrointestinal motility and enhance bleeding by inhibiting platelet aggregation while preventing normal vasoconstrictive response to bleeding (183,184). However, large-scale trials have not shown a significant increase in gastrointestinal bleeding risk. Table 19–11 describes major trials evaluating this risk.

REFERENCES

1. Lindner E: Phenyl-propyl-diphenyl-propyl-amin, eine neue Substanz mit coronargefasserweiternder Wirkung. *Arzneim Forsch* 1960;10:569.
2. Hass H, Hartfelder G: α-Isopropyl-α[(N-methyl-N-bomoveratryl)-γ-amino-propyl]-3,4-dimethoxyphenylactonitril, eine Substanz mit coronargefasserweiternden Eigenschaften. *Arzneimittelforschung* 1962;12:549.
3. Nayler WG: Biochemical aspects of prenylamine action. Preliminary communication. *Biochim Appl* 1967;14 [Suppl 1]:305–321.
4. Fleckenstein A, Doring HJ, Kammermeier H, et al: Influence of prenylamine on the utilization of high energy phosphates in cardiac muscle. *Biochim Appl* 1967;14 [Suppl 1]:323–344.
5. Fleckenstein A: Peter Harris Award lecture. History and prospectus in calcium antagonist research. *J Mol Cell Cardiol* 1990;22:241–251.
6. Fleckenstein A, Tritthart H, Fleckenstein B, et al: A new group of competitive Ca-antagonists (Iproveratril, D 600, prenylamine) with highly potent inhibitory effects on excitation-contraction coupling in mammalian myocardium. *Pflugers Arch* 1969;307:R25.
7. Fleckenstein A, Fleckenstein B, Spah F, et al: Gallopamil (D-600)—ein Kalziumantagonist von hoher Wirkungsstarke und Spezifitat. Effekte auf Myokard und Schrittmacher. In: Kaltenbach M, Hopf R, eds: *Gallopamil: pharmacological and clinical profile of a calcium antagonist*. Berlin: Springer-Verlag, 1984:1–34.
8. Fleckenstein-Grun G, Fleckenstein A: Blockade of the calcium dependent bioelectrical automaticity and electromechanical coupling of smooth muscle cells by gallopamil (D600). In: Kaltenbach M, Hopf R, eds: *Gallopamil: pharmacological and clinical profile of a calcium antagonist*. Berlin: Springer-Verlag, 1984:33–48.
9. Raschack M, Gries J, Buhler V, et al: Studies on the cardiovascular effects of gallopamil. In: Kaltenbach M, Hopf R, eds: *Gallopamil: pharmacological and clinical profile of a calcium antagonist*. Berlin: Springer-Verlag, 1984:72–80.

10. Fleckenstein A: Specific inhibitors and promoters of calcium action in the excitation-contraction coupling of heart muscle and their role in the production or prevention of myocardial lesions. In: Harris P, Opie L, eds: *Calcium and the heart.* [Proceedings of the meeting of the European Section of the International Study Group for Research in Cardiac Metabolism, London, September 6, 1970.] New York: Academic Press, 1971:135–188.
11. Fleckenstein A, Tritthart H, Döring HJ, et al: Bay a 1040—ein hochaktiver Ca+2–antagonistischer Inhibitor der elektromechanischen Koppelungsprozesse im Warmblüter-Myokard. *Arzneimittelforschung* 1972;22:22–33.
12. Fleckenstein A, Fleckenstein-Grün G, Byon YK, et al: Vergleichende Untersuchungen über die Ca+2–antagonistischen Grundwirkungen von Niludipine (Bay a 7168) and Nifedipine (Bay a 1040) auf Myokard, Myometrium and glatt Gefäßmuskulatur. *Arzneimittelforschung* 1979;29:230–246.
13. Takenaka T, Usuda S, Nomura T, et al: Vasodilator profile of a new 1,4-dihydropyridine derivative, 2,6-dimethyl-4(3-nitrophenyl)-1,4-dihydropyridine-3,5-dicarboxylic acid 3-[2-(N-benzyl-N-methyl amino)]-ethyl ester 5-methyl ester hydrochloride (YC-93). *Arzneimittelforschung* 1976;26:2172–2178.
14. Nagao T, Sato M, Nakajima H, et al: Studies on a new 1,5 benzothiazepine derivative (CRD-401). II. Vasodilator actions. *Jpn J Pharmacol* 1972;22:1–10.
15. Nagao T, Narita H, Sato M, et al: Development of diltiazem, a calcium antagonist: coronary vasodilating and antihypertensive actions. *Clin Exp Hypertens [A]* 1982;4:285–296.
16. Sato M, Nagao T, Yamaguchi I, et al: Pharmacological studies on a new 1,5-benzothiazepine derivative (CRD-401). *Arzneimittelforschung* 1971; 21:1338–1343.
17. Glasser SP: The relevance of T-type calcium antagonists: a profile of mibefradil. *J Clin Pharmacol* 1998;38:659–669.
18. Triggle DJ: Pharmacologic and therapeutic differences among calcium channel antagonists: profile of mibefradil, a new calcium channel antagonist. *Am J Cardiol* 1996;78[Suppl A]:7–12.
19. Litovitz TL, Klein-Schwartz W, White S, et al: 1999 Annual Report of the American Association of Poison Control Centers Toxic Exposure Surveillance System. *Am J Emerg Med* 2000;18:517–574.
20. *Drug Topics Red Book 1999.* Montvale, NJ: Medical Economics Company, 1999.
21. Zaloga GP, Chernow B: Divalent cations: calcium, magnesium and phosphorus. In: Chernow B, ed: *The pharmacologic approach to the critically ill patient,* 3rd ed. Baltimore: Williams & Wilkins, 1994:777–804.
22. Marino PL: Calcium and magnesium in serious illness: a practical approach. In: Sivak ED, Higgins TL, Seiver A, eds: *The high risk patient: management of the critically ill.* Baltimore: Williams & Wilkins, 1995:1183–1195.
23. Smith JB: Calcium homeostasis in smooth muscle. *New Horiz* 1996;4:2–18.
24. Sandow A: Excitation-contraction coupling in muscular response. *Yale J Biol Med* 1952;25:176.
25. Katz A: Calcium channel diversity in the cardiovascular system. *J Am Coll Cardiol* 1996;28:522–529.
26. Tsien RW, Ellinor PT, Zhang J-F, et al: Molecular biology of calcium channels and structural determinants of key functions. *J Cardiovasc Pharmacol* 1996;27[Suppl A]:S4–10.
27. Robertson RM, Robertson D: Drugs used for the treatment of myocardial ischemia. In: Molinoff PB, Rudden RW, eds: *Goodman and Gilman's the pharmacological basis of therapeutics,* 9th ed. New York: McGraw-Hill, 1996:759–779.
28. Schwartz A, McKenna E, Vaghy PL: Receptors for calcium antagonists. *Am J Cardiol* 1988;62:3G-6G.
29. Fang LM, Osterrieder W: Potential dependent inhibition of cardiac Ca+2 inward currents by Ro 40-5967 and verapamil: relation to negative inotropy. *Eur J Pharmacol* 1991;196:205–207.
30. Hoffman F, Lacinova L, Klugbauer N: Voltage-dependent calcium channels: from structure to function. *Rev Physiol Biochem Pharmacol* 1999;139:33–87.
31. Perez-Reyes E, Cribbs LL, Daud A, et al: Molecular characterization of a neuronal low-voltage activated T-type calcium channel. *Nature* 1998; 391:896–900.
32. Perez-Reyes E, Cribbs LL, Daud A, et al: Molecular characterization of T-type calcium channels. In: Tsien RW, Clozel JP, Nargeot J, eds: *Low-voltage-activated T-type calcium channels.* Chester, England: Adis International Ltd., 1998:290–307.
33. Cribbs LL, Lee JH, Yang J, et al: Cloning and characterization of α1H from human heart, a member of the T-type Ca^{+2} channel gene family. *Circ Res* 1998;83:103–109.
34. Klugbauer N, Marais E, Lacinová L, et al: A T-type calcium channel from brain. *Pflugers Arch* 1999;437(5):710–715.
35. Mintz IM, Adams ME, Bean BP: P-Type calcium channels in rat central and peripheral neurons. *Neuron* 1992;9:85–95.
36. Sather WA, Tanabe T, Zhang JF, et al: Distinctive biophysical and pharmacological properties of class A (BI) calcium channel α1 subunits. *Neuron* 1993;11:291–303.
37. McDonough SI, Swartz KJ, Mintz IM: Inhibition of calcium channels in rat central and peripheral neurons by omega-conotoxin MVIIC. *J Neurosci* 1996;16:2612–2613.
38. Zhang JF, Randall AD, Ellinor PT, et al: Distinct pharmacology and kinetics of cloned neuronal Ca+2 channels and their possible counterparts in mammalian CNS neurons. *Neuropharmacology* 1993;32:1075–1088.
39. Plummer MR, Logothetis DE, Hess P: Elementary properties and pharmacological sensitivities of calcium channels in mammalian peripheral neurons. *Neuron* 1989;2:1453–1463.
40. Marrion NV, Tavalin SJ: Selective activation of Ca+2 activated K+ channels by co-localized Ca+2 channels in hippocampal neurons. *Nature* 1998;395:900–905.
41. Rios E, Pizarro G, Stefani E: Charge movement and the nature of signal transduction in skeletal muscle excitation-contraction coupling. *Annu Rev Physiol* 1992;54:109–133.
42. Niidome T, Kim MS, Friedrich T, et al: Molecular

cloning and characterization of a novel calcium channel from rabbit brain. *FEBS Lett* 1992:308:7–13.
43. Soong TW, Stea A, Hodson CD, et al: Structure and functional expression of a member of the low voltage-activated calcium channel family. *Science* 1993:260:1133–1136.
44. Schneider T, Wei X, Olcese R, et al: Molecular analysis and functional expression of the human type E neuronal Ca+2 channel α1 subunit. *Receptors Channels* 1994;2:255–270.
45. Hermsmeyer K: Role of T channels in cardiovascular function. *Cardiology* 1998;89[Suppl 1]:2–9.
46. Hermsmeyer K, Mishra S, Miyagawa K, et al: Physiologic and pathophysiologic relevance of T-type calcium ion channels: potential indications for calcium antagonists. *Clin Ther* 1997;19[Suppl A];18–26.
47. Pitt B: Diversity of calcium antagonists. *Clin Ther* 1997;59:617–627.
48. Keefe D, Frishman WH: Clinical pharmacology of the calcium-channel blocking drugs. In: Packer M, Frishman WH, eds: *Calcium channel antagonists in cardiovascular disease*. Norwalk, CT: Appleton-Lang; 1984:3–19.
49. Brogden RN, McTavish D: Nifedipine gastrointestinal therapeutic system (GITS). *Drugs* 1995; 50:495–512.
50. Katz B, Rosenberg A, Frishman WH: Controlled-release drug delivery systems in cardiovascular medicine. *Am Heart J* 1995;129:359–368.
51. Mitchell J, Frishman W, Heiman M: Nislodipine: a new dihydropyridine calcium channel blocker. *J Clin Pharmacol* 1993;33:46–52.
52. White WB: A chronotherapeutic approach to the management of hypertension. *Am J Hypertens* 1996;9:29S–33S.
53. Kates R: Calcium antagonists—pharmacokinetic properties. *Drugs* 1983;25:113–124.
54. Frishman WH, Kirstein E, Klein M, et al. Clinical relevance of verapamil plasma levels in stable angina pectoris. *Am J Cardiol* 1982;50:1180–1184.
55. Frishman WH, Charlap S, Kimmel B, et al: Diltiazem compared to nifedipine and combination treatment in patients with stable angina: effects on angina, exercise tolerance and the ambulatory ECG. *Circulation* 1988;77:774–786.
56. Somgyi A, Albrecht M, Kliems G, et al: Pharmacokinetics, bioavailability and ECG response of verapamil in patients with liver cirrhosis. *Br J Clin Pharmacol* 1981;12:51–60.
57. Woodcock BG, Rietbrock I, Vohringer HF, et al: Verapamil disposition in liver disease and intensive-care patients: Kinetics, clearance, and apparent blood flow relationships. *Clin Pharmacol Ther* 1981;29:27–34.
58. Hamann SR, Blouin RA, Chang SL, et al. Effects of hemodynamic changes on the elimination kinetics of verapamil and nifedipine. *J Pharmacol Exp Ther* 1984;231:301–305.
59. McAllister RG, Hamann SR, Blonin RA: Pharmacokinetics of calcium-entry blockers. *Am J Cardiol* 1985;55:30B–40B.
60. Bailey DG, Arnold JMO, Spence JD: Grapefruit juice and drugs. *Clin Pharmacokinet* 1994;26:91–98.
61. Kroemer HK, Gautier J-C, Beaune P, et al: Identification of P450 enzymes involved in metabolism of verapamil in humans. *Naunyn Schmiedebergs Arch Pharmacol* 1993;348:332–337.
62. Pichard L, Gillett G, Fabre I, et al: Identification of the rabbit and human cytochrome P-450IIIA as the major enzymes involved in the N-demethylation of diltiazem. *Drug Metab Dispos* 1990;18:711–719.
63. Guengerich FP, Brian WR, Iwasaki M, et al: Oxidation of dihydropyridine calcium channel blockers and analogues by human liver cytochrome P-450 IIIA4. *J Med Chem* 1991;34:1838–1844.
64. Abernethy DR, Schwartz JB: Pharmacokinetics of calcium antagonists under development. *Clin Pharmacokinet* 1988;15:1–14.
65. Kroemer HR, Echizen H, Heidemann H, et al: Predictability of the in vivo metabolism of verapamil from in vitro data: contribution of individual metabolic pathways and stereoselective aspects. *J Pharmacol Exp Ther* 1992;260:1052–1057.
66. Schwartz JB, Capili H, Daugherty J: Aging of women alters S-verapamil pharmacokinetics and pharmacodynamics. *Clin Pharmacol Ther* 1994; 55:509–517.
67. Wood A: Calcium antagonists: pharmacologic differences and similarities. *Circulation* 1989; 80[Suppl]:IV184–IV188.
68. The Danish Study Group on Verapamil in Myocardial Infarction: verapamil in acute myocardial infarction. *Eur Heart J* 1984;5(7):516–528.
69. Schwinger R, Böhm M, Erdmann E: Different negative inotropic activity of Ca+2-antagonists in human myocardial tissue. *Klin Wochenschr* 1990;68:797–805.
70. Ferrari R, Cucchini F, Bolognesi R, et al: How do calcium antagonists differ in clinical practice? *Cardiovasc Drugs Ther* 1994;8[Suppl 3]:565–575.
71. Opie L: Calcium channel antagonists in the treatment of coronary artery disease: Fundamental pharmacological properties relevant to clinical use. *Prog Cardiovasc Dis* 1996;38:273–290.
72. Borchard U: Calcium antagonists in comparison: view of the pharmacologist. *J Cardiovasc Pharmacol* 1994;24[Suppl]:S85–S91.
73. Vetroves G: Hemodynamic and electrophysiologic effects of first- and second-generation calcium antagonists. *Am J Cardiol* 1994;73 [Suppl]:34A–38A.
74. Spedding M, Fraser S, Clarke B, et al: Factors modifying the tissue selectivity of calcium-antagonists. *J Neural Transm Suppl* 1990;31:5–6.
75. Peipho RW: Calcium antagonist in patients with congestive heart failure: still a bridge too far. *J Clin Pharmacol* 1995;35:443–453.
76. Triggle D: Calcium-channel antagonists: mechanism of action, vascular selectivity, and clinical relevance. *Cleve Clin J Med* 1992;59:617–627.
77. Anonymous: Roche, FDA announces new drug-interaction warnings for mibefradil. *Am J Health-System Pharmacy* 1998;55:210.
78. Schassmann-Suhijar D, Bullingham R, Gasser R, et al: Rhabdomyolysis due to interaction of simvastatin with mibefradil. *Lancet* 1998;351:1929–1930.
79. Pearigen PD: Calcium channel blocker poisoning. In: Haddad LM, Shannon MS, Winchester JF, eds: Clinical management of poisoning and drug over-

dose, 3rd ed. Philadelphia: WB Saunders, 1998:1020–1031.
80. Clifton DG, Booth DC, Hobbs S, et al: Negative inotropic effect of intravenous nifedipine in coronary artery disease. Relation to plasma levels. *Am Heart J* 1990;119:283–290.
81. Materne P, Legrand V, Vandormael M, et al: Hemodynamic effects of intravenous diltiazem with impaired left ventricular function. *Am J Cardiol* 1984;54:733–737.
82. Hattori VT, Mandel WJ, Peter D: Calcium for myocardial depression from verapamil. *N Engl J Med* 1982;306:238.
83. Hossack KF: Conduction abnormalities due to diltiazem. *N Engl J Med* 1982;307:953–954.
84. Ishikawa T, Imamura T, Koiwaya Y, et al: Atrioventricular dissociation and sinus arrest induced by oral diltiazem. *N Engl J Med* 1983;309:1124–1125.
85. Morris DL, Goldschlager N: Calcium infusion for reversal of adverse effects of intravenous verapamil. *JAMA* 1981;249:3212–3213.
86. Ramoska EA, Spiller HA, Winter M, et al: A one-year evaluation of calcium channel blocker overdose: toxicity and treatment. *Ann Emerg Med* 1993;22:196–200.
87. Spiller HA, Meyers A, Ziemba T, et al: Delayed onset of cardiac arrhythmias from sustained-release verapamil. *Ann Emerg Med* 1991;20:201–203.
88. Tom PA, Morrow CT, Kelen GD: Delayed hypotension after overdose of sustained release verapamil. *J Emerg Med* 1994;12:621–625.
89. Ashraf M, Chaudhary K, Nelson J, et al: Massive overdose of sustained-release verapamil: a case report and review of literature. *Am J Med Sci* 1995;310:258–263.
90. Barrow PM, Houston PL, Wong DT: Overdose of sustained-release verapamil. *Br J Anaesth* 1994;72:361–365.
91. Kozlowski JH, Kozlowski JA, Schuller D: Poisoning with sustained-release verapamil. *Am J Med* 1988;85:127.
92. Rankin RJ, Edwards IR: Overdose of sustained-release verapamil. *N Z Med J* 1990;103:165.
93. Schoffstall JM, Spivey WH, Gambone LM, et al: Effects of calcium channel blocker overdose-induced toxicity in the conscious dog. *Ann Emerg Med* 1991;20:1104–1108.
94. Pearigen PD, Benowitz NL: Poisoning due to calcium antagonists: experience with verapamil, diltiazem and nifedipine. *Drug Safety* 1991;6:408–430.
95. Howarth DM, Dawson AH, Smith AJ, et al: Calcium channel blocking drug overdose: an Australian series. *Hum Exp Toxicol* 1994;13:161–166.
96. Horowitz BZ, Rhee KJ: Massive verapamil ingestion: a report of two cases and a review of literature. *Am J Emerg Med* 1989;7:624–631.
97. Zoghbi W, Schwartz JB: Verapamil overdose: report of a case and review of the literature. *Cardiovasc Rev Rep* 1984;5:356.
98. Fauville JP, Hantson P, Honore P, et al: Severe diltiazem poisoning with intestinal pseudo-obstruction: case report and toxicological data. *J Toxicol Clin Toxicol* 1995;33:273–277.
99. Kuo MJ, Tseng YZ, Chen TF, et al: Verapamil overdose and severe hypocalcemia. *J Toxicol Clin Toxicol* 1992;30:309–311.
100. MacDonald D, Alguire PC: Case report: fatal overdose with sustained-release verapamil. *Am J Med Sci* 1992;303:115–117.
101. Orr GM, Bodansky HJ, Dymond DS, et al: Fatal verapamil overdose. *Lancet* 1982;2:1218–1219.
102. Quezado Z, Lippmann M, Wertheimer J: Severe cardiac, respiratory, and metabolic complications of massive verapamil overdose. *Crit Care Med* 1992;19:436–438.
103. McGovern B, Garan H, Ruskin JN: Precipitation of cardiac arrest in patients with Wolff-Parkinson-White syndrome. *Ann Intern Med* 1986;104:791–794.
104. Ramoska EA, Spiller HA, Myers A: Calcium channel blocker toxicity. *Ann Emerg Med* 1990;19:649–653.
105. Ferner RE, Odemuyima O, Field AB, et al: Pharmacokinetics and toxic effects of diltiazem in massive overdose. *Hum Toxicol* 1989;8:497–499.
106. Welch RD, Todd K: Nifedipine overdose accompanied by ethanol intoxication in a patient with congenital heart disease. *J Emerg Med* 1990;8:169–172.
107. Wells TG, Graham CJ, Moss MM, et al: Nifedipine poisoning in a child. *Pediatrics* 1990;86:91–94.
108. Crump BJ, Holt DW, Vale JA: Lack of response to intravenous calcium in severe verapamil poisoning. *Lancet* 1982;2:939–940.
109. Hendren WC, Schreiber RS, Garretson LK: Extracorporeal bypass for the treatment of verapamil poisoning. *Ann Emerg Med* 1989;18:984–987.
110. Hofer CA, Smith JK, Tenholder MF: Verapamil intoxication: a literature review of overdoses and discussion of therapeutic options. *Am J Med* 1993;95:431–438.
111. Koch AR, Vogelaers DP, Decruyenaere JM, et al: Fatal intoxication with amlodipine. *J Toxicol Clin Toxicol* 1995;33:253–256.
112. Roper TA, Sykes R, Gray C: Fatal diltiazem overdose: report of four cases and review of the literature. *Postgrad Med J* 1993;69:474–476.
113. Malcolm N, Callegari P, Goldberg J, et al: Massive diltiazem overdosage: clinical and pharmacokinetic observations. *Drug Intell Clin Pharm* 1986;20:888.
114. Shah AR, Passalacqua BR: Case report: sustained release verapamil overdose causing stroke: an unusual complication. *Am J Med Sci* 1992;304:357–359.
115. DeRoos F: Calcium channel blockers. In: Goldfrank LR, Flomenbaum NE, Lewin NA, et al, eds: *Goldfrank's toxicologic emergencies*, 6th ed. Stamford, CT: Appleton & Lange, 1998:829–843.
116. Ward DJ, Ward JW, Griffo W, et al: Intravenous calcium for fecal impaction secondary to verapamil. *N Engl J Med* 1982;307:1709–1710.
117. Goglin WK, Elliott BM, Deppe SA: Nifedipine-induced hypotension and mesenteric ischemia. *South Med J* 1989;82:274–275.
118. Gutierrez H, Jorgensen M: Colonic ischemia after verapamil overdose. *Ann Intern Med* 1996;124:535.

119. Sporer KA, Manning JJ: Massive ingestion of sustained-release verapamil with a concretion and bowel infarction. *Ann Emerg Med* 1993;22:603–605.
120. Wax P: Intestinal infarction due to nifedipine overdose. *J Toxicol Clin Toxicol* 1995;33:725–728.
121. Brass BJ, Winchester-Penny S, Lipper BL: Massive verapamil overdose complicated by noncardiogenic pulmonary edema. *Am J Emerg Med* 1996;14:459–461.
122. Enyeart JJ, Price WA, Hoffman DA, et al: Profound hyperglycemia and metabolic acidosis after verapamil overdose. *J Am Coll Cardiol* 1983;2:1228–1231.
123. Goenen M, Col J, Compere A, et al: Treatment of severe verapamil poisoning with combined amrinone-isoproternol therapy. *Am J Cardiol* 1986;58:1142–1143.
124. McMillan R: Management of acute severe verapamil intoxication. *J Emerg Med* 1988;6:193–196.
125. Spurlock BW, Virani NA, Henry CA: Verapamil overdose. *West J Med* 1991;154:208–211.
126. Watling SM, Crain JL, Edwards TD, et al: Verapamil overdose: case report and review of the literature. *Ann Pharmacother* 1992;26:1373–1377.
127. Malaisse WJ: Role of calcium in insulin secretion. *Isr J Med Sci* 1972;8:224–251.
128. Devis G, Somers G, Malaisse WJ: Dynamics of calcium-induced release. *Diabetologia* 1977;13:531–536.
129. Malaisse WJ, Devis G, Pipeleers DG, et al: Calcium antagonists and islet function. IV. Effect of D600. *Diabetologia* 1976;12:77–81.
130. Somers G, Devis G, Van Obberghen E, et al: Calcium antagonists and islet function. II. Interaction of theophylline and verapamil. *Endocrinology* 1977;99:114–124.
131. Levy J, Herchuelz A, Sener A, et al: Inhibition by verapamil of calcium influx in the B-cell [Abstract]. *Diabetes* 1975;24:400.
132. Devis G, Somers G, Van Obberghen E: Calcium antagonists and islet function. I. Inhibition of insulin release by verapamil. *Diabetes* 1975;24:547–551.
133. Thomas SH, Stone K, May WA: Exacerbation of verapamil-induced hyperglycemia with glucagons. *Am J Emerg Med* 1995;13:27–29.
134. Diamond JR, Cheung JY, Fang LST: Nifedipine-induced renal dysfunction: alteration in renal hemodynamics. *Am J Med* 1984;77:905–909.
135. Buckley NA, Whyte IM, Dawson AH: Overdose with calcium channel blockers [Letter]. *BMJ* 1994;308:1639.
136. Gelbke HP, Schlicht HG, Schmidt G: Fatal poisoning with verapamil. *Arch Toxicol* 1980;37:89–94.
137. Herrington DM, Insley BM, Weinman GG: Nifedipine overdose. *Am J Med* 1986;81:344–346.
138. Humbert VH, Munn NJ, Hawkins RF: Noncardiogenic pulmonary edema complicating massive verapamil overdose. *Chest* 1994;105:606–607.
139. Leesar MA, Martyn R, Talley JD, et al: Noncardiogenic pulmonary edema complicating massive verapamil overdose. *Chest* 1994;105:606–607.
140. Gustafsson D: Microvascular mechanisms involved in calcium antagonist edema formation. *J Cardiol Pharm* 1987;10[Suppl 1]:S121–S131.
141. Law RI, Takeda P, Mason DT, et al: The effects of calcium channel blocking agents on cardiovascular function. *Am J Cardiol* 1982;49:547–553.
142. Green K, Cheeks L, Hull DS: Effects of calcium channel blockers on rabbit corneal endothelial function. *Curr Eye Res* 1994;13:401–408.
143. Payne DK, Fuseler JW, Owens MW: Modulation of endothelial cell permeability by lung carcinoma cells. A potential mechanism of malignant pleural effusion formation. *Inflammation* 1994;18:407–417.
144. Fedorak RN, Empey LR, Walker K: Verapamil alters eicosanoid synthesis and accelerates healing during experimental colitis in rats. *Gastroenterology* 1992;102:1229–1235.
145. Simon RP: Neurogenic pulmonary edema. *Neurol Clin* 1993;11:309–323.
146. Stern R, Khalsa JH: Cutaneous adverse reactions associated with calcium channel blockers. *Arch Intern Med* 1989;149:829–832.
147. Maurer HH: Identification of antiarrhythmic drugs and their metabolites in urine. *Arch Toxicol* 1990;64:218–230.
148. Maurer HH: Systematic toxicological analysis of drugs and their metabolites by gas chromatography-mass chromatography. *J Chromatogr* 1992;580:3–41.
149. Maurer HH, Arlt JW: Screening procedure for detection of dihydropyridine calcium channel blocker metabolites in urine as part of a systematic toxicological analysis procedure for acidic compounds by gas chromatography-mass spectrometry after extractive methylation. *J Anal Toxicol* 1999;23:73–80.
150. Koppel D, Wagemann A: Plasma level monitoring of d,l-verapamil and three of its metabolites by reversed-phase HPLC. *J Chromatogr* 1991;570:229–234.
151. Hussain MD, Tam YK, Finegan BA, et al: Simple and sensitive high-performance liquid chromatographic method for the determination of diltiazem and 6 of its metabolites in human plasma. *J Chromatogr* 1992;582:203–210.
152. Soons PA, Schlens JHM, Roosemalen MCM, et al: Analysis of nifedipine and its pyridine metabolites dehydronifedipine in blood and plasma. *J Pharm Biomed Anal* 1991;9:475–484.
153. Tucker FA, Minty PSB, MacGregor GA: Study of nifedipine photodecomposition in plasma and whole blood using capillary gas-liquid chromatography. *J Chromatogr* 1985;342:193–198.
154. Aursnes I, Litleskare I, Froyland H, et al: Association between various drugs used for hypertension and risk of acute myocardial infarction. *Blood Pressure* 1995;4(3):157–163.
155. Jick H, Derby LE, Gurewich V, et al: The risk of myocardial infarction associated with antihypertensive drug treatment in persons with uncomplicated essential hypertension. *Pharmacotherapy* 1996;16:321–326.
156. Psaty BM, Heckbert SR, Koepsell TD, et al: The risk of myocardial infarction associated with antihypertensive drug therapies. *JAMA* 1995;274:620–625.
157. Alderman MH, Cohen H, Roque R, et al: Effect of

long-acting and short-acting calcium antagonists on cardiovascular outcomes in hypertensive patients. *Lancet* 1997;349:594–598.
158. Pahor M, Guralnik JM, Corti MC, et al: Long term survival and use of antihypertensive medications in older persons. *J Am Geriatr Soc* 1995;43:1191–1197.
159. Koenig W, Lowel H, Lewis M, et al: Long-term survival after myocardial infarction: relationship with thrombolysis and discharge medication. Results of the Augsburg Myocardial Infarction Follow-up Study 1985 to 1993. *Eur Heart J* 1996;17:1199–1206.
160. Ishikawa K, Nakai S, Takenaka T, et al: Short-acting nifedipine and diltiazem do not reduce the incidence of cardiac events in patients with healed myocardial infarction. Secondary Prevention Group. *Circulation* 1997; 95:2368–2373.
161. Michels KB, Rosner BA, Manson JE, et al: Prospective study of calcium channel blocker use, cardiovascular disease, and total mortality among hypertensive women: the Nurse's Health Study. *Circulation* 1998;97:1540–1548.
162. Abascal VM, Larson MG, Evans JC, et al: Calcium antagonists and mortality risk in men and women with hypertension in the Framingham Heart Study. *Arch Intern Med* 1998;158:1882–1886.
163. The GLANT Study Group: A 12-month comparison of ACE inhibitor and CA antagonist therapy in mild to moderate essential hypertension—The GLANT study. Study Group on Long-term Antihypertensive Therapy. *Hypertension Res* 1995;18(3):235–244.
164. Gong L, Zhang W, Zhu Y, et al: Shanghai trial of nifedipine in the elderly (STONE). *J Hypertens* 1996;14(10):1237–1245.
165. Staessen JA, Fagard R, Thijs L, et al: Randomized double-blind comparison of placebo and active treatment for older patients with isolated systolic hypertension. The Systolic Hypertension in Europe (Syst-Eur) Trial investigators. *Lancet* 1997;350:1632–1633.
166. Borhani NO, Mercuri M, Borhani PA, et al: Final outcome results of the Multicenter Isradipine Diuretic Atherosclerosis Study (MIDAS). A randomized controlled trial. *JAMA* 1996;276:785–791.
167. Stason WB, Schmidt CH, Niedzwiecki D, et al: Safety of nifedipine in patients with hypertension: a meta-analysis. *Hypertension* 1997;30[1 Pt 1]:7–14.
168. Estacio RO, Jeffers BW, Hiatt WR, et al: The effect of nisoldipine as compared with enalapril on cardiovascular outcomes in patients with non-insulin-dependent diabetes and hypertension. *N Engl J Med* 1998;338:645–652.
169. Gibson RS, Boden WE, Theroux P, et al: Diltiazem and reinfarction in patients with non-Q-wave myocardial infarction. Results of a double-blind, randomized, multicenter trial. *N Engl J Med* 1986;315:423–429.
170. Wilcox RG, Hampton JR, Banks DC, et al. Trial of early nifedipine in acute myocardial infarction: The Trent Study. *Br Med J (Clin Res Ed)* 1986;293:1204–1208.
171. The Holland Interuniversity Nifedipine/Metoprolol Trial (HINNT) Research Group: Early treatment of unstable angina in the coronary care unit: a randomized, double-blind, placebo controlled comparison of recurrent ischaemia in patients treated with nifedipine or metoprolol or both. *Br Heart J* 1986;56(5):400–413.
172. The SPRINT Study Group. The Secondary Prevention Reinfarction Israeli Nifedipine Trial (SPRINT) II. Design, methods, and results. *Eur Heart J* 1988;9:354–364.
173. Furberg CD, Psaty BM, Meyer JV: Nifedipine. Dose-related increase in mortality in patients with coronary heart disease. *Circulation* 1995;94:1326–1331.
174. Scheidt S, LeWinter MM, Hermanovich J, et al: Efficacy and safety of nicardipine for chronic stable angina pectoris: a multicenter randomized trial. *Am J Cardiol* 1986;58:715–721.
175. Gheorghiade M, Weiner DA, Chakko S, et al: Monotherapy of stable angina with nicardipine hydrochloride: double-blind, placebo-controlled, randomized study. *Eur Heart J* 1989;10:695–701.
176. Thadani U, Zellner SR, Glasser S, et al: Double-blind, dose-response, placebo-controlled multicenter study of nisoldipine. A new second-generation calcium channel blocker in angina pectoris. *Circulation* 1991;84:2398–2408.
177. Dargie HJ, Ford I, Fox KM: Total Ischaemic Burden European Trial (TIBET). Effects of ischaemia and treatment with atenolol, nifedipine SR, and their combination on outcome in patients with chronic stable angina. The TIBET Study Group. *Eur Heart J* 1996;17:104–112.
178. Rehnqvist N, Hjemdahl P, Billing E, et al: Effects of metoprolol vs. verapamil in patients with stable angina pectoris. The Angina Prognosis Study in Stockholm (APSIS). *Eur Heart J* 1996;17:76–81.
179. The Israeli Sprint Study Group: Secondary Prevention Reinfarction Israeli Nifedipine Trial (SPRINT). A randomized intervention trial of nifedipine in patients with acute myocardial infarction. *Eur Heart J* 1988;9:354–364.
180. The Danish Study Group on Verapamil in Myocardial Infarction: effect of verapamil on mortality and major events after acute myocardial infarction (The Danish Verapamil Infarction Trial II-DAVIT II). *Am J Cardiol* 1990;66:779–785.
181. Rengo F, Carbonin P, Pahor M, et al: A controlled trial of verapamil in patients after acute myocardial infarction: results of the Calcium Antagonist's Reinfarction Italian Study (CRIS). *Am J Cardiol* 1996;77:365–369.
182. Hansen JF, Hagerup L, Pedersen F, et al: Cardiac event rates after acute myocardial infarction inpatients treated with verapamil and trandolapril versus trandolapril alone. Danish Verapamil Infarction Trial (DAVIT) Study Group. *Am J Cardiol* 1997;79:738–741.
183. Konrad-Dalhoff I, Baunack AR, Ransch KD, et al: Effect of the calcium channel antagonists nifedipine, nitrendipine, nimodipine and nislodipine on oesophageal motility in man. *Eur J Clin Pharmacol* 1991;41:313–316.
184. Pahor M, Guralnik JM, Furberg CD, et al: Risk of gastrointestinal haemorrhage with calcium antagonists in hypertensive persons over 67 years old. *Lancet* 1996;347:1061–1065.

SELF-ASSESSMENT QUESTIONS

1. Which of the following is not a class of calcium channel antagonists (CCBs)?
   a. thienopyridine
   b. phenylalkylamines
   c. benzothiazepines
   d. dihydropyridines
   e. diarylaminopropylethers

2. High-voltage-activated calcium channels are heterooligomeric complexes of five proteins from which of the four gene subunits?
   a. $\alpha 1, \beta, \alpha 2\delta, \gamma$
   b. $\beta 2\alpha, \beta 2\gamma, \delta, \alpha$
   c. $\alpha 1, \beta 2, \gamma 2\delta, \alpha 1\beta$
   d. $\alpha 1, \beta 1\gamma, \delta, \gamma\delta$

3. Which of the following subunit contains most of the prominent features of calcium channel complex–ion conducting pore, the selectivity filter of the pore, the voltage sensor and interaction sites for other subunits, and the CCBs and activators?
   a. β subunit
   b. $\alpha 1$ subunit
   c. $\alpha 2\delta$ subunit
   d. $\gamma 2$ subunit

4. Which of the following CCBs are also classified as antiarrhythmic agents?
   a. amlodipine, bepridil
   b. mibefradil, diltiazem
   c. verapamil, diltiazem
   d. nicardipine, felodipine

5. Which of the following statements about CCB toxicity is not true?
   a. The most common physical finding in CCB overdose is hypotension.
   b. The general pathophysiology of CCB overdose is the reduction in cardiac output with associated increase in systemic vascular resistance from dilation of vascular smooth muscle leading to severe hypotension and state of shock.
   c. The life-threatening toxicity of CCB is manifested largely within the cardiovascular system and is mainly an extension of its therapeutic effects.
   d. Bradycardias and conduction defects are among the most frequent findings associated with verapamil and diltiazem overdose.

6. Which of the following about current controversies in use of CCBs is not true?
   a. In patients with stable coronary disease (post myocardial infarction or low-risk stable angina), all types of CCBs are safe, and they are not associated with a higher incidence of adverse events.
   b. In patients with high risk for coronary events, further studies are necessary to demonstrate the role of CCBs.
   c. In patients with mild to moderate hypertension at low risk for cardiovascular events, long-acting dihydropyridine CCBs do not increase cardiovascular event risk.
   d. In patients with acute myocardial infarction, there is evidence supporting use of CCBs safely.

CHAPTER 20

Biological Monitoring of Chemical Exposure

Lee M. Blum, Ph.D., DABFT, and Edward A. Emmett, M.D., M.S.

LEARNING OBJECTIVES

After completing this chapter, the reader should be able to:

1. Define biological monitoring and biomarkers.
2. Describe the role and uses of biological monitoring.
3. Understand the issues that may affect biological exposure testing.
4. Describe concerns about analytical methods used in biological exposure testing.
5. Characterize the importance of proper specimen collection.
6. List factors that need to be considered when interpreting biological monitoring test results.
7. Explain workplace exposure limits and the limitations in their use.

DEFINITIONS: BIOLOGICAL MONITORING AND BIOMARKERS

Biological monitoring traditionally refers to the measurement of internal exposure to a xenobiotic through the analysis of a biological specimen. The term biological monitoring has sometimes been extended to include the measurement of certain reasonably specific biological tests to measure internal exposure, such as the measurement of cholinesterase activity as an indicator of exposure to cholinesterase-inhibiting pesticides.

A related term is biomarker. The term biomarker (1,2) is used broadly to include three different types of markers:

- "Biomarkers of exposure" are exogenous substances within the system, the interactive product between a xenobiotic compound and endogenous compound(s), or measures of some other event in the biological system related to the exposure.
- "Biomarkers of effect" are indicators of an endogenous component of the biological system, measures of the functional capacity of the system, or indicators of an altered state of the system that are recognized as impairment or disease.
- "Biomarkers of susceptibility" are indicators that the health of the system is especially sensitive to the challenge of exposure to a xenobiotic compound.

Today, with the exception of cholinesterase inhibitors and methemoglobin formation, biomarkers of effect generally offer little advantage over measurement of the chemical itself. Biomarkers of susceptibility have yet to be shown to have practical routine applicability (3).

THE ROLES OF BIOLOGICAL MONITORING AND ENVIRONMENTAL MONITORING

Biological monitoring is generally used to determine whether potentially toxic amounts of chemicals are being absorbed or accumulated in the body from the ambient environment, from food or water sources, or from an occupational setting.

To illustrate the place of biological monitoring, it is instructive to consider a general sequence for the development of disease from one or more toxic amounts of chemicals as follows (4).

There is a source of toxic concentrations of substances such as an industrial process; chemical storage, hazardous waste, or mine site; or contaminated air, water, food, or surfaces in contact with the skin. In these media the substance at issue is measured through an appropriate form of environmental monitoring.

The toxicant is absorbed into the human body through the lungs, skin, and other exposed surfaces or through the digestive tract and then distributed to body tissues and/or metabolized and/or excreted. These processes are often modeled rather than measured directly. The toxicokinetics will likely be different given different routes of absorption.

As a result of the above processes there is a certain internal dose of the toxicant or its derivative at the target organ. The internal dose of the substance or a metabolite in either the target organ or in another surrogate tissue sample is measured through biological monitoring.

As a result of interaction with a receptor in the target organ, biochemical, physiological, or cellular events occur. The agent-receptor interaction (for example adducts between DNA and xenobiotics) may be measured directly; or biochemical, physiologic, or cellular events subsequent to agent-receptor interaction (for example, released inflammatory mediators) may be measured as biomarkers of effect. Other types of measurement of subsequent processes may be available such as measures of repair function (such as sister chromatid exchange).

As a result of genetic, biochemical, and cellular damage, occupational or environmental diseases occur with symptoms, signs, and evidence of disordered pathology or physiology that can ultimately be detected by a physician and enumerated by an epidemiologist.

Both biological and environmental monitoring have important roles in the prevention and assessment of potentially hazardous occupational and environmental chemical exposures.

The relative utility of monitoring the environment or of monitoring internal biological exposure may be examined from the perspectives of controlling exposure, legally enforcing regulatory standards, and protecting health.

Environmental measurements are the most useful for identifying the actual source(s) of emissions and the environmental pathway(s) from the source to the medium of human exposure (air, water, food, surfaces contacting the skin, etc.). For these purposes the main disadvantage of biological monitoring is that the measurements do not readily allow different sources of the xenobiotic in question to be separately evaluated.

Where there are legal issues of causation, biological monitoring can determine that increased absorption has occurred but cannot directly show the source of that exposure.

However, from the perspective of protection of health, measurements of internal biological exposures are generally more germane than environmental measures because the biological measurement confirms that absorption has occurred. With respect to predicting the occurrence of a possible health effect from the exposure, the internal exposure measurement integrates exposures from all sources (occupational, environmental, hobbies, etc.) and all routes of absorption (inhalation, ingestion, percutaneous) as well as the effects of distribution, metabolism, and excretion. In that sense, the measurement of internal biological exposure may reflect the effective dose that causes adverse biological consequences.

USES OF BIOLOGICAL MONITORING

Biologic concentrations from exposure may be used to assess potential toxicity in an individual; for monitoring community levels of exposure to a toxic substance; or for monitoring a worker population occupationally exposed to a toxicant. The principles of the use of biologic concentrations of toxicants in each situation will be briefly discussed.

In an individual case of deliberate or inadvertent exposure, biologic concentrations assist in diagnosis, in determining the prognosis, in monitoring the course of the illness, and in assessing the efficacy of therapy. The utility of the information obtained will depend partially on how well the dose-response or dose-effect relations are known for the agent in question. This information is obtained from the literature, including case reports from previous exposures and poisonings. Attention needs to be given to comparability of information with respect to analytical techniques, to the route of absorption (particularly as it influences half-lives and time to peak absorption), and to the time from exposure to sampling.

Monitoring of population levels of pollutants may be performed on a local, national, or international basis. This type of monitoring may establish the distribution of pollutant uptake and/or demonstrate changes in pollutant levels over time. In some cases historical materials may be examined such as Egyptian mummies or pre-Columbian remains to determine changes following industrialization. An example of a community monitoring program is the Human Pesticide Monitoring Program of the US Environmental Protection Agency. In this program, adipose samples were collected at autopsy or operation and used to determine the levels of such chlorinated hydrocarbons as dichlorodiphenyltrichloroethane (DDT) and its metabolite dichlorodiphenyldichloroethylene (DDE), chlordane, polychlorinated biphenyls (PCBs), and related substances in the US population over time.

In the United States, national estimates of the extent and severity of recent human exposures to lead have been possible as a result of incorporating blood lead measurements in the US National Health and Nutrition Surveys (5). These studies are designed so that results may be extrapolated to the population as a whole.

Biological monitoring of occupationally exposed populations is performed to monitor occupational groups to help ensure that working conditions are safe. Some substances are not suitable for biologic monitoring, including those whose major effect is direct damage to surface tissues of the lungs, eyes, skin, or GI epithelium. Biological monitoring is not usually suitable for assessing occupational exposure to allergens, because of the extraordinarily small amounts of substance that can trigger a reaction in those who are allergic; to teratogens, because the toxic effect can occur from quite transient exposures limited to the time of organogenesis; or to carcinogens, because of the long latency period between exposure and the occurrence of cancer and the at least theoretical possibility that cancer could result from a single chemical exposure.

Biological monitoring is of most use where the toxic effects to be prevented are the end result of chronic low-level exposures. Biological monitoring is particularly useful for agents that have a major component of skin absorption, because this route is relatively poorly assessed using the standard industrial hygiene measurement techniques that are most effective for airborne exposure.

BIOLOGICAL EXPOSURE TESTING

There are many factors that influence the interpretability of biological exposure testing. The effect from an exogenous substance (therapeutic or environmental agent) on a biological system depends on its ability to interact with target molecules within the organism. The extent of that effect is related to the amount of the compound reaching these target molecules. Factors that can affect the concentrations attained at these sites depend on the manner of exposure; the chemical properties of the exposure agent; and biological processes such as absorption, distribution, storage, metabolism, and excretion. These factors influence biological exposure testing and affect the detection and/or the interpretation of the test results.

The manner of exposure to a toxicant contributes to the amount detected in the body. The absorptive properties of an agent and route of exposure influence the extent of its uptake. Typically, gases, vapors of solvents, and aerosols are inhaled via the lungs with absorption through the lungs dependent on the toxicant's solubility in blood. On the other hand, the skin is generally not highly permeable to some agents and can act as a reasonably good barrier from the environment. For a toxicant to penetrate through the skin, it must pass through several layers of dermal cells. Furthermore, the air concentration of the inhaled toxicant, the rate of respiration, and the area of skin contacted also alters the amount absorbed.

The frequency and duration of exposure affects the extent of circulating toxicant concentrations. Chronic exposure to a chemical, where absorption exceeds metabolism and excretion, allows an accumulation of the compound in the body. However, if the exposure to the same compound were of short duration with each exposure allowing for its clearance between exposures, the accumulation hazard would be reduced.

The activity of an individual during an exposure can influence the absorptive rate. During physical activity, normal physiological changes in the body include increased breathing and increased blood flow to the skin, among others. These physiological changes increase the absorption potential of a toxicant through the lungs and skin. Liira et al. demonstrated higher blood concentrations of methyl ethyl ketone after exposure during exercise than after sedentary exposure (6).

Differences in chemical properties of a toxicant can cause changes in its toxicokinetic properties. Mercury is an example of the influence different chemical properties can have on exposure testing. There are three forms of mercury: elemental mercury, inorganic mercury salts, and organic mercury. These three forms have different toxicokinetic profiles. Elemental mercury is well absorbed through the lungs via inhalation or through the skin, yet is poorly absorbed through the GI tract. This form of mercury is distributed primarily to the kidney and the brain and is excreted in the urine and feces. Inorganic mercury compounds are absorbed through the GI tract and through the skin. The kidney is the main site of deposition of inorganic mercury, with elimination occurring through both the urine and feces. Organic mercury is a general term for a diverse group of compounds. Because of its prevalence, the characteristics of methyl mercury are presented. Methyl mercury is absorbed through the GI tract, the lungs, and to a small extent the skin. Because of its lipid solubility, methyl mercury concentrates more in the blood and brain than the inorganic form, with excretion primarily through the bile into the feces (7).

The biological processes on chemicals as they move through the human body influence the outcome of exposure testing. These processes include absorption, distribution, storage, metabolism, and excretion. Each can have an effect on the chemical composition of the toxicant or metabolite and its interaction with biochemical functions (8).

Absorption. The absorption of many industrial toxicants primarily occurs through the lungs via inhalation. However, other common routes of absorption of these compounds are through the skin or the GI tract. The absorption of any toxicant depends on its physicochemical properties and the characteristics of the absorptive site. The large surface area and the thin barrier of the alveolar membrane provide rapid access to the circulating blood for gases and vapors. The transfer of the toxicant from the alveolar space to the blood depends on the partial pressure of the gas in both the alveolar air and the blood, and the solubility of the agent as expressed by its partition coefficient. Solubility also plays a role in the absorption of inhaled particles and droplets. Uptake may be delayed until they break down into smaller particles or until they enter through the cellular lining of the respiratory surface or through the lymphatic system.

The multicellular layers of intact skin provide a formidable barrier against the dermal absorption of many environmental agents. Absorption through the top layer, the epidermis, is proportional to the lipid solubility of the compound, while permeability through the lower dermis layer is more dependent on its water solubility. Dermal absorption can be enhanced through damage to the skin, such as physical abrasions or preexisting dermatitis. Some compounds are more likely to be absorbed through the skin than through the lungs. Workers in an acrylic fiber factory exposed to dimethylformamide were more likely to have absorbed it dermally than from inhalation (9).

Exposure of industrial toxicants through the GI tract typically occur by ingestion of compounds as a result of poor hygiene habits such as eating, drinking, or smoking in areas where manufacturing processes occur. The GI tract is anatomically and physiologically structured for the absorption of essential nutrients. To some extent, many types of toxicants can be absorbed through the GI tract. Substances that are weak organic acids or bases can be absorbed by diffusion in the appropriate area of the GI tract where they exist in the lipid-soluble form. Absorption of some nutrients occurs through specialized transport mechanisms. Many of these same absorptive transport mechanisms are used by some toxicants. For example, cobalt uses the same transport system as iron does (10).

Distribution. After absorption, distribution of the toxicant in the body depends on the blood flow to the various organs and its permeability through the capillary bed and into the particular tissue. Inhalation through the lungs transports the compound to the heart first. Dermal absorption carries the toxicant to the right side of the heart and then to the lungs, where if volatile, it may be partially excreted. Absorption through the GI tract delivers the compound to the liver. Other than in the central nervous system, which is closely monitored, the perfusion rate through the various organs depends on the extent of activity. In a person at rest, the kidney, liver, and GI tract typically are relatively highly perfused. However, this changes to some extent with exercise when the blood is diverted to areas where it is needed, such as the skeletal muscles and the skin.

Storage. Once circulating in the blood, some compounds accumulate in various sites in the body. The sites that accumulate or store toxicants may include adipose, bone, liver, kidney, and plasma protein. This accumulation of chemicals at these sites may serve as a protection mechanism. This mechanism reduces the concentrations of toxicant at the target site. Because the stored compound is in equilibrium with its circulating concentrations, it is slowly released as the chemical is eliminated. This process prolongs the duration of the toxicant in the body.

Metabolism. To facilitate their elimination, many foreign substances undergo biochemical conversion. Most of these reactions are the result of enzymatic activity located primarily in the liver. These reactions include hydrolysis, oxidation, reduction, and conjugation. The activity of these enzymes depends on a variety of factors including age, heredity, nutritional and health status, co-administration of therapeutic drugs and alcohol, and environmental exposures. The products of these chemical reactions are typically less toxic and more water-soluble than their parent compound. There are instances, however, when metabolic intermediates or the final product is more toxic than the parent. The ability to create these products, therefore, could determine the extent of toxicity. For example, methanol itself does not appear to have cytotoxic properties. However, metabolism by dehydration to formaldehyde and then to formic acid results in two highly reactive substances, with formic acid responsible for much of the toxicity. The treatment for methanol poisoning is the administration of ethanol, which competes for the enzyme responsible for methanol metabolism (11).

Excretion. The elimination of toxicants and their metabolites from the body occurs by various means. The kidney probably excretes more chemicals from the body than any other route. Glomerular filtration in the kidneys as well as tubular diffusion and tubular secretion provide an effective means of excreting water-soluble compounds. The liver through the biliary system is another route of elimination leading to excretion in the feces. For some compounds, the enterohepatic circulation prolongs elimination such that the compound is excreted in the bile, then reabsorbed in the intestine. The lungs can excrete toxic gases and volatile compounds as an equilibrium is reached between the alveolar space and the passing blood. Other bodily secretions such as sweat, breast milk, and semen provide additional routes of excretion.

ANALYTICAL METHODS

An understanding of the biological processes and biochemical effects of various exposure agents has led to the development of analytical methods for the determination of these agents and/or their metabolites in biological matrices. These methods need to assist the medical community in assessing and managing exposures to potentially harmful chemicals. Their ability to measure accurately and precisely the exposure determinant is crucial to their success.

The availability of testing is essential for biomonitoring. Exposure testing in hospital or reference laboratories by routine analysis may not always be available for a variety of reasons. Not only is the absence of testing capabilities self-limiting, but so is the timeliness of that availability. The selected determinant to be used for monitoring may not be sufficiently stable to allow for testing after prolonged periods of storage. Methemoglobin concentrations in the blood can be used to monitor excessive exposure to several single aromatic nitro- and amino- compounds such as aniline and nitrobenzene. Methemoglobin is the result of chemical oxidation of the heme iron of hemoglobin from the ferrous to the ferric state, and approximately 1% is present in erythrocytes of normal individuals. The analysis for methemoglobin needs to be performed within a few hours of the blood collection, because of its rapid reduction (12).

Therefore, a delay in analysis or prolonged storage will give misleading analytical results.

Biomonitoring testing in hospital or reference laboratories may be limited for several reasons. The lack of human toxicokinetic or toxicodynamics studies of a specific exposure agent may limit the determination of the analytical specifications including the analyte to be determined (that is, the substance itself, the metabolite, or a biological effect marker), the specimen of choice, and the timing of specimen collection, among others. In addition, research laboratories that investigate human chemical exposures may use synthesized chemicals and specialized equipment not readily available in the hospital or reference laboratory. Furthermore, analytical procedures developed in research laboratories may not be amenable to routine laboratory testing because of lengthy and complex testing procedures. Finally, the cost of testing may be prohibitive because procedures that require extensive extractions and sophisticated instrumentation incur a great expense to the testing laboratory.

Sensitivities and Specificities

Although testing may be available in the hospital or reference laboratories, the analytical methods may not be sensitive enough to distinguish normal background concentrations from concentrations associated with harmful effects, nor may they be specific enough to exclude other possibilities for the result. A substance that is undetected by routine analysis does not exclude the possibility that exposure has not occurred. As newer analytical methods and instrumentation become available, the ability to detect ever-decreasing concentrations of distinct substances improves.

The use of early colorimetric tests could detect only toxic or lethal concentrations of organic and inorganic compounds in biological matrices. As technology improved with various chromatographic techniques and atomic absorption spectrometry, chemical concentrations in the parts per million typically were detected. Now with some of the newer commercially available instruments, including liquid and gas chromatography with tandem mass spectrometry and inductively coupled plasma mass spectrometry, the analysis for organic and inorganic toxicants are more specific in identifying these compounds and can reach sensitivities to subparts per billion.

SAMPLE COLLECTION

Specimen collection is a critical aspect of biomonitoring testing. The biological processing of the toxicant and the considerations to the patient influences the collection of an appropriate specimen. These factors as well as the specimen type and the collection techniques can affect the analysis and the interpretation of the test findings.

Biological Process

Prior knowledge of the toxicokinetics of an agent is important when determining the type of specimen, the analyte to be measured, and the timing of specimen collection. A study monitoring isopropanol exposure found that isopropanol could be detected in the alveolar air of workers but not in their blood or urine. However, acetone, the isopropanol metabolite, was detected in all three specimens during exposure, and the urinary acetone concentration was higher the next morning compared with at the end of the exposure (13). For substances with long biological half-lives in various body compartments, the time of specimen collection may not be critical; however, for compounds that are rapidly eliminated, the timing of specimen collection becomes important.

For some compounds, blood and breath concentrations may change rapidly soon after an exposure, causing markedly variable concentrations if time since exposure is not known. Therefore, for many industrial toxicants, the collection of blood or breath samples is often performed just before the start of the next shift.

The use of urine to assess exposure ideally requires a 24-hour specimen, but this method is inconvenient and impractical. A method called "double-void" sampling is sometimes used when the time between two urine sample collections is known (8). The results of the second sample can be related to previous exposure if the excretion kinetics of the chemical is established. For practical reasons, a random urine is usually collected and a correction for the dilution of the urine is made, either by adjusting to a constant specific gravity or by relating it to the creatinine present in the urine.

The American Conference of Governmental Industrial Hygienists (ACGIH) in establishing their Biological Exposure Indices (BEIs) has determined the sampling time of each analyte based on the length of retention in the body (14). For substances that accumulate, the ACGIH does not require any specific specimen sampling time. Sample timing and the associated recommended collection are described by the ACGIH. A collection before the work shift suggests collection 16 hours after the end of the exposure. Specimens to be collected during the shift are to be performed anytime after 2 hours of exposure. The end-of-shift sampling time recommends collection as soon as possible after the exposure. Specimens to be drawn at the end of the work week are sampled after 4 to 5 consecutive working days with exposure, and for sampling time indicated as discretionary, specimens could be collected at anytime.

Invasiveness and Safety

With the collection of biological samples, invasiveness and safety of the collection procedure to the patient may also determine the type of specimen to be collected. Blood typically reflects what is circulating in the body. However, the collection of blood is invasive and has potential problems such as infection. The handling of blood specimens by phlebotomists and laboratory personnel requires awareness and concern about potentially infectious diseases. The collection of blood also requires precaution and proper disposal of biohazardous waste. The collection of other biological specimens such as urine may not be invasive, but the issue of patient privacy and safety and the safety of collection personnel in the handling of potentially biohazardous material should be of concern. The analysis of alveolar breath offers an alternative as a noninvasive specimen that avoids the handling of potentially infectious waste.

Samples Types

Blood

Because internal concentrations of toxicants are typically correlated with toxicological effects, blood is the apparent specimen of choice for monitoring exposure. The analysis of blood has its advantages in that it gives a direct measurement of internal concentrations. Unfortunately, blood is a complex matrix requiring extensive clean-up procedures before analysis for many exposure agents. Blood is used to measure exposure agents or their metabolites directly, and it is used indirectly to determine exposure by examining blood constituents. For example, organophosphorus pesticides inhibit cholinesterase enzyme activity. Therefore, the measurements of pseudocholinesterase activity in plasma and/or cholinesterase activity in erythrocytes are commonly used as indicators of organophosphorus pesticide exposure.

Urine

Urine has always been a preferred specimen in biological monitoring primarily because of its availability and relative ease in collection. Using the same mechanisms to eliminate normal metabolic end products, the kidney excretes many exogenous chemicals from the body. The urinary concentration of a toxicant depends on the concentration in the blood circulating through the kidneys. It is important to remember, however, that in many cases the exposure agent undergoes extensive metabolism before urinary excretion, and it is these metabolites that may be used as exposure markers.

Urinary concentrations also depend on kidney function. Because kidneys regulate water and solute balance in the body, the urinary amount of the parent chemical or its metabolite may be more concentrated or diluted depending on the intake or loss of water through such activities as drinking, eating, or sweating, among others. In an attempt to compensate for this variability, urinary characteristics such as creatinine concentrations and specific gravity have been used.

The use of these factors, however, does have its limitations. The excretion of creatinine is related to body mass and may be increased by physical activity. The conventional method of using the specific gravity adjustment assumes that a change in the urinary flow preserves the relative ratio between the mass of the toxicant and the mass of the total dissolved solids. However, Vij and Howell demonstrated that excretion rates for different substances vary with changes in urinary flow (15). They suggest a mathematical formula that modifies the specific gravity adjustment using an appropriately weighted exponential factor.

Breath

The analysis of exhaled breath is used to assess the exposure to volatile organic substances. Because the breath coming from the alveolar region of the lung is nearly in equilibrium with the blood leaving the lungs, the concentration of a compound in the alveolar gas can be related to that in the arterial blood by the solubility of the compound in blood. The analysis of exhaled breath can be an indicator of recent exposure or represent a biological response of an individual (16). The collection of breath is noninvasive, avoids potentially infectious waste, and offers a limitless supply of sample. Although breath is a less complex biological matrix than others such as blood, high concentrations of CO_2 and water vapor may cause problems in the analytical determinations. The acceptability of breath analysis as a viable method for biological monitoring has been established by the ACGIH through their publication of BEIs for toxicants such as n-hexane and perchlorethylene (14).

Alternative specimens have been used to monitor exposures. These specimens have included hair, adipose tissue, breast milk, and semen, among others (17–20). Examination of cellular samples may provide a closer look at the cytotoxic effects of these toxins. Chromosomal abnormalities and DNA adducts have been used to indicate exposure (21–24). However, studies using these cytogenic markers should include several end-point measurements combining adduct formation with specific disease-related gene sites to further elucidate health risks.

Specimen Collections

Proper collection and subsequent storage before analysis of specimens to be used for biological monitoring is necessary in ascertaining an analytical result that is truly reflective of the amount of determinant in the specimen. Errors may be associated with specimen collection and storage through evaporation, chemical deterioration, precipitation and adsorption on surfaces, and/or contamination (25). To reduce the potential for loss of volatile compounds through evaporation, fill the container to the top and refrigerate the specimen collected. Although the storage stability of each determinant should be verified, refrigeration or freezing of the specimen may be necessary to reduce deterioration, especially of organic compounds. The acidification of urine specimens decreases the loss of trace elements due to urinary precipitation or surface adsorption to the sides of the specimen container.

Contamination of the specimen is a considerable problem especially in the analysis of trace metals (26). The analysis of blood and urine specimens can be misleading if proper precautions are not taken to minimize sample contamination. Contamination may come from the patients themselves or from the procedures and devices used in the collection. Dust from the workplace on a patient's hands, skin, and clothing are potential sources of contamination. The specimen collection should take place in a clean environment clear of dust of the workplace. The patient should be free of any potential contamination before collection through proper cleansing of the venipuncture site, washing of hands, and showering and changing of clothing, if necessary.

Procedures and devices used in specimen collection are additional sources of contamination. Collection tubes and metal vessels, syringes, powder in protective gloves, certain disinfectants, glass pipettes, and pipette tips may contain varying amounts of trace elements. Several manufacturers offer evacuated blood-collection tubes to be used specifically for trace elemental analysis. These tubes are typically suitable for commonly analyzed elements such as cadmium or lead. However, for less-common elements, either contact the manufacturer for the findings of their evaluation for potential contaminants, or self-verify the absence of the elements to be analyzed before use. Urine should be not be collected in a metal pan or urinal, but in an acid-washed plastic container with a plastic-lined screw cap.

On occasion, circumstances may arise in the collection of specimens outside the realm of workplace urine drug testing in which for medicolegal reasons, verifiable specimens are required. In these cases, a chain of custody would need to be initiated identifying the specimen, the type of specimen, the date and time of collection, and the persons coming in contact with the specimen. Labels and seals should be placed on and over the top of the specimen, respectively, for purposes of identification with the accompanying chain-of-custody form as well as maintaining the integrity of the specimen before analysis.

INTERPRETATION OF BIOLOGICAL MONITORING RESULTS

Several factors may need to be addressed in interpreting biological monitoring results. The interpretation may have to take into account potential analytical variation, interindividual variations, and the comparability of the population from which reference values were obtained. Not all these considerations will apply to any situation, but all can be relevant at times.

Analytical Variation

In comparing values against established reference values, it is important to ascertain whether the reference values were obtained using the same analytic techniques. Often, the analytic methods used in the past were less precise than those now available and were more susceptible to interferences that are no longer relevant. Consequently, the values obtained with older analytic methods were often higher than what would be obtained using current methods.

Interindividual Variability

Sources of individual variability include genetic variations (for example, in the metabolism of a xenobiotic) and physical differences between individuals (such as variations in fat stores among subjects). Diurnal variations occur in the tissue concentrations of some xenobiotics. The uptake of gases and respirable particles depends on the pulmonary ventilation rate and thus will be increased in those who work at higher energy levels. The uptake may vary in those who work different shift patterns and therefore have different exposure durations. Lifestyle, diet, hobby and recreational exposures, and smoking and alcohol habits may also be sources of interindividual variation.

Comparability of Reference Populations

A "normal" population used to establish the reference values will ideally be similar to the exposed population in demographic factors (age and sex), places of residence, and habits that may affect background values. For example, concentrations of cadmium are characteristically around 70% higher in smokers than nonsmokers (27), and concentrations of urinary mercury are higher in those who have amalgam fillings (28).

The way in which values are distributed in a population may be important. Many pollutants such as metals and organic chemicals may be log-normally distributed among the population. This distribution may affect the type of appropriate statistical analysis for a set of results so that the use of geometric means may be more appropriate than arithmetic means.

The demographic characteristics, health status, and disease vulnerability of the population used to determine the reference range may be important.

The reference range for normal healthy working-age adults may not be appropriately safe for potentially susceptible populations such as children, pregnant women, the elderly, and those who are already ill. Those with liver or kidney disease may have impaired distribution, metabolism, or excretion. In addition, diseases of target end organs may render an individual more susceptible to toxic effects involving that organ.

Other Sources of Variability

Depending on the particular chemical toxicant, other factors may be important. Seasonal variations may occur in the concentrations of some metals. For example, blood lead is usually somewhat lower in the summer, probably due to increased excretion in the sweat. Seasonal variation in the proportions of meat, seafood, or other components in the diet affect the concentrations of pollutants for which these foods are an important source.

Interactions among chemicals may be important in several ways. An exogenous agent may affect the distribution, metabolism, or excretion of another substance; may interact at a target organ; or may interfere with the chemical analysis.

CHOOSING A BASE-LINE VALUE

Sometimes, especially in the evaluation of occupational exposures to a chemical, a choice of base-line values is available. Moreover, in some cases, no reference range is available, and a set of comparison reference values will need to be developed. Several types of comparison values may be considered in these instances.

Preplacement group mean values—those measured before a worker takes the job where he or she will be exposed to the chemical in question—may be particularly valuable. Such measurements normally incorporate the usual lifestyle, diet, geographic location, community exposures, and other factors relevant to a group of workers.

An individual ideally may act as his or her own control in any evaluation of dose received from any particular source. Depending on the half-life of the substance under consideration and the circumstances of work, observations can be made before and after shifts for agents with a fairly short half-life. For agents with a longer half-lives comparisons of measurements obtained at the beginning and again at the end of the work week, or after a working period of some weeks compared with observations at the end of a vacation period or plant shutdown, will be appropriate.

If statistical studies are to be performed, comparisons using individuals as their own controls are more statistically powerful, that is, fewer subjects are required to obtain statistically significance, for example, to demonstrate that absorption is occurring in the workplace (29).

BIOLOGICAL WORKPLACE EXPOSURE LIMITS

Biological workplace exposure limits represent acceptable concentrations of a toxicant in a specified matrix of an exposed population. These limits imply a permissible level of exposure under working conditions below which most workers should not experience adverse health effects. Several governmental regulations and standards for workplace health incorporate biological monitoring. In the United States the applicable regulations are from the US Department of Labor under the Occupational Safety and Health Act (OSHA) of 1970. Many other developed countries have analogous regulations including the United Kingdom (30) and Germany (31). In Germany, there are two sets of published reference values, one for noncarcinogenic substances and a separate list of reference values for carcinogens that is derived more stringently on the basis that no safe level can be given for a carcinogenic substance. US OSHA standards as part of an overall medical surveillance program usually incorporate additional provisions covering the specimens to be tested, frequency of required sampling, actions to be taken at varying levels of results, and communication of the results to employees (32).

The World Health Organization has developed biomonitoring action levels based on judgment of acceptable exposures, which are useful as accepted reference values if regulatory standards either do not apply or are considered insufficient under the circumstances. These have been derived either from the extant clinical, epidemiologic, and toxicologic studies on the relationship between a measured concentration of the chemical or metabolite in body fluids and the health outcome or from experience with good working practices (33,34).

Some private bodies have also published recommended exposure levels. One of the most widely quoted may be the list of BEIs published annually by ACGIH (14). Although these do not have the force of law, they may be used for guidance. Curiously, the ACGIH BEIs indicate the concentration of the substance most likely observed in specimens collected in a healthy worker who has an inhalation exposure to the threshold limit value (TLV), an inhalation exposure the ACGIH considers safe for most workers. This paradigm ignores the effects of exposure routes other than airborne and does not purport to address any "susceptible" group of workers. There has been some criticism that the standard-setting process used by ACGIH is not impartial and has been overly influenced by industries that have a vested interest in having higher acceptable limits

(35). Accordingly, these exposure values should be used with caution.

CAUTIONS IN THE USE OF ACCEPTABLE LIMIT VALUES IN BIOLOGICAL MONITORING

Biological exposure results require medical interpretation and understanding beyond the mere comparison with general reference values. For example in the workplace, biological monitoring results measure the cumulative effect of exposure from all routes of absorption, reflect the degree of protection being attained in practice from the use of protective equipment, and reflect the work practices and personal hygiene practices of individual workers. It is common to see quite large variations in biological monitoring values among workers who are doing the same work. By carefully evaluating these situations, differences in habits, in hygiene, or in the ways tasks are performed can frequently be identified that explain these high variations. For example, a welder of small industrial parts who does not wear respiratory protection and who positions his head close to the welding fumes, thereby breathing in more fumes, will likely have higher blood or urinary values of metals from the plume than co-workers. Inconsistency in using protective clothing, poor fit of a respirator, or failure to wash before eating lunch may similarly lead to higher values than found in co-workers.

The use of regulated or privately derived exposure limits requires an appreciation that the limits reflect the state-of-the-art of science and the available knowledge base at the time the limit was established. As a result, exposure limits are subject to change. With the advances in bioanalytical chemistry, molecular biochemistry, and diagnostic health science, exposure limits change, or more-specific determinants may replace nonspecific exposure markers. For example, traditionally, urinary phenol concentrations have been used to monitor benzene exposure. However, the ingestion of certain foods or pharmaceutical agents can contribute to the urinary phenol concentration, thereby complicating the interpretative value of this analyte. The more specific benzene metabolites, t,t-muconic acid and S-phenylmercapturic acid, are now suggested as urinary markers for benzene exposure (36).

Questions regarding specificity of the determinant for an exposure agent and confounders such as environmental or dietary exposures may also limit the use of these values. Furthermore, exposure limits have yet to be established for many workplace chemicals, and there is even less information about acceptable levels, if any, for most chemicals found in the environment. More research and public awareness of issues surrounding the potential health hazards from the ambient and workplace environments can be expected to lead to the development and improvement of more dependable techniques in assessing the absorption and toxicities of these substances.

CASE EXAMPLE OF THE USE OF BIOLOGICAL MONITORING DATA

A 2-year-old child who had moved to live in a 70-year-old inner-city house had a routine well visit to her pediatrician. A blood lead test was performed to screen for lead poisoning.

Comment: Until almost 1960, paint used in houses could contain lead as a pigment. The developing brain in children up to 6 years of age is particularly susceptible to lead poisoning. Effects on the developing brain appear to occur at concentrations lower than those confirmed for other health effects of lead and at concentrations lower than those that cause adverse effects in adults. If children are continually exposed to lead, their blood lead tends to increase during ages 0–2 years and peaks at ages 18–24 months. Because children with mild to moderate elevations of lead are asymptomatic, screening for lead is recommended at the ages of 1 and 2 years to detect and intervene in cases in whom excessive lead absorption is occurring.

The blood lead test results showed a concentration of blood lead of 35 µg/dL (1.69 µmol/L). According to the Centers for Disease Control and Prevention (CDC), this value was considered elevated and to reflect excessive lead absorption. The child was referred to a physician specializing in the environmental effects of lead. The child saw that physician 6 weeks later. Additional medical history at that visit revealed that there was chipped and peeling paint in the house in which the child resided, and that this house was located on a busy street. The child also spent significant time in the houses of two relatives. The child had a strong history of pica and was seen to mouth toys, fingers, and other objects. On one occasion the child was observed to put paint chips into the mouth. A blood sample was taken for lead and was reported at 33 and 31 µg/dL (1.59 and 1.50 µmol/L), respectively, in two separate analyses. The free erythrocyte protoporphyrin (FEP) was also measured and was elevated at 97 µg/dL (1.72 µmol/L).

Because the half-life for blood lead in the case of a naive exposure to lead is approximately 1 month, the repeat blood lead result at about the same concentration as before confirmed that increased absorption of lead was still occurring. The FEP is a biomarker that reflects the blood concentration in the bone marrow at the time the erythrocytes are formed. The FEP concentration remains stable in the erythrocyte over the 120-day lifespan of that cell. Thus, the FEP corresponded to a cumulative blood lead over the 120-day period of just above 25 µg/dL (1.21 µmol/L). The fact that the FEP was not quite as elevated as might be expected, suggested that the elevated lead concentrations were relatively recent. At least if the lead concentration had been in the 33–35 µg/dL (1.59–1.69 µmol/L) range over the entire 120-day period, one would have expected a higher FEP. To interpret the elevated FEP as reflecting an effect of lead, the other causes of an elevated FEP needed to be ruled out, that is, iron deficiency anemia and hereditary protoporphyria.

There were multiple potential sources of lead for this child including several houses that may have contained lead paint and possible lead dust particles condensed from the exhaust of automobiles using leaded gasoline through the 1980s. The blood lead determination does not allow for the distinction among these possible sources. Environmental measurements of potential sources may have helped identify the source in this child's case. The presence of pica is an important factor in lead absorption with children who consume lead containing dust and paint chips. The slightly sweet taste of lead paint can contribute to the hazard.

The family told the physician that they were moving to a different house in 3 days. To decrease lead absorption, the mother was instructed to use dust-control measures with frequent mopping, not to sweep or use high-phosphate detergents, and to reduce the amount of fat in the diet. The child was to return 1 month later for another blood lead determination.

Efforts to reduce the lead exposure were indicated to prevent continuing lead absorption during the susceptible ages of up to 6 years old.

When the blood lead test was repeated the next month, the concentration was 15 µg/dL (0.72 µmol/L).

Tracer studies have shown the half-life of lead in human blood to be around 36 days, the half-life in soft tissues around 40 days, and in bone around 27 years. The 50% reduction in the blood lead concentration in this child over the 1-month period was consistent with the termination of new lead absorption. It appeared that the previous source(s) of lead were not a factor in the child's new environment.

The current CDC interpretation of what constitutes an elevated blood lead concentration and represents increased absorption is 10 µg/dL (0.48 µmol/L) or more in children aged 1–5 years. Recent community monitoring data confirms that blood lead concentrations among urban children have generally declined in the past 20 years. This change has been attributed both to the removal of lead from gasoline and to the success of programs directed at reducing the amount absorbed from lead-based paint found in older houses. The focus on lower levels of lead has implications for the screening tests used. In the past, the FEP was used as a rapid, efficient screen test for possible increased lead absorption. However, the FEP does not give a good reflection of the blood lead at concentrations below 15–20 µg/dL (0.72–0.97 µmol/L); therefore, the use of the FEP as a screening tool is no longer appropriate.

REFERENCES

1. Committee on Biological Markers of the National Research Council: Biological markers in environmental health research. *Environ Health Perspect* 1987;74:3–9.
2. International Programme on Chemical Safety (IPCS): *Biomarkers and risk assessment: concepts and principles.* [Environmental Health Criteria, No 155.] Geneva: World Health Organization, 1993.
3. Aitio A: Biomarkers and their use in occupational medicine. *Scand J Work Environ Health* 1999; 25:521–528.
4. Emmett EA, Foa V: Biochemical and cellular indices of human toxicity. In: Foa V, Emmett EA, Maroni M, et al, eds: *Occupational and environmental chemical hazards: cellular and biochemical indices for monitoring toxicity.* Chichester, England: Ellis Horwood, 1987:32–40.
5. Annest JL, Mahaffey KR: *Blood lead levels for persons aged 6 months-74 years: United States, 1976–1980.* [Report No. 223, Series 11. DHHS Publication No (PHS) 84-1683.] Hyattsville, MD: U.S. Public Health Service, 1994.
6. Liira J, Riihimaki V, Pfaffli P: Kinetics of methyl ethyl ketone in man: absorption, distribution and elimination in inhalation exposure. *Int Arch Occup Environ Health* 1998;60(3):195–200.
7. Berlin M: Mercury. In: Friberg L, Nordberg GF, Vouk VB, eds: *Handbook on the toxicology of metals,* vol. II. Amsterdam: Elsevier, 1990:387–445.
8. Morgan MS: The toxicologic basis for biological monitoring of industrial chemicals. Presented at "Biological monitoring for the detection and quantification of chemical exposures." American Industrial Hygiene Conference and Exposition, Orlando, FL, May 21, 2000.
9. Lauwerys RR, Hoet P: Principal advantages of biological monitoring. In: Lauwerys RR, Hoet P: *Industrial chemical exposure: guidelines for biological monitoring,* 2nd ed. Boca Raton, FL: Lewis Publishers, 1993:5–7.
10. Schade SG, Felsher BF, Glader BE, et al: Effect of cobalt upon iron absorption. *Proc Soc Exp Biol Med* 1970;134:741–743.
11. Kruse JA: Methanol poisoning. *Intensive Care Med* 1992;18: 391–397.
12. Lauwerys RR, Hoet P: Amino and nitro derivatives. In: Lauwerys RR, Hoet P: *Industrial chemical exposure: guidelines for biological monitoring,* 2nd ed. Boca Raton, FL: Lewis Publishers, 1993:199–206.
13. Brugnone F, Perbellini L, Apostoli P, et al: Isopropanol exposure: environmental and biological monitoring in a printing works. *Br J Ind Med* 1983; 40(2):160–168.
14. American Conference of Governmental Industrial Hygienists: *Threshold limit values for chemical substances and physical agents and biological exposure indices.* Cincinnati, OH: American Conference of Governmental Industrial Hygienists, 2000.
15. Vij HS, Howell S: Improving the specific gravity adjustment method for assessing urinary concentrations of toxic substances. *Am Ind Hyg Assoc J* 1998; 58:375–380.
16. Pleil JD, Fisher JW, Lindstrom AB: Trichloroethene levels in human blood and exhaled breath from controlled inhalation exposure. *Environ Health Perspect* 1998;106:573–580.
17. Bonde JP, Joffe M, Danscher G, et al: Objectives, designs and populations of the European Asclepios study on occupational hazards to male reproductive capability. *Scand J Work Environ Health* 1999; 25[Suppl 1]:49–61.
18. Larsen SB, Abell A, Bonde JP: Selection bias in occupational sperm studies. *Am J Epidemiol* 1998; 147(7): 681–685.
19. DeBruin LS, Josephy PD, Pawliszyn JB: Solid-phase microextraction of monocyclic aromatic amines from biological fluids. *Anal Chem* 1998; 70(9):1986–1992.
20. Heinzow BG, McLean A: Critical evaluation of current concepts in exposure assessment. *Clin Chem* 1994;40[7 Pt 2]:1368–1375.
21. Kriek E, Rojas M, Alexandrov K, et al: Polycyclic aromatic hydrocarbons—DNA adducts in humans: relevance as biomarkers for exposure and cancer risk. *Mutat Res* 1998;400(1-2): 215–231.
22. Surralles J, Autio K, Nylund L, et al: Molecular cytogenetic analysis of buccal cells and lymphocytes from benzene-exposed workers. *Carcinogenesis* 1997;18(4):817–823.
23. Peluso M, Amasio E, Bonassi S, et al: Detection of DNA adducts in human nasal mucosa tissue by 32P-postlabeling analysis. *Carcinogenesis* 1997;18 (2):339–344.
24. Stone JG, Jones NJ, McGregor AD, et al: Development of a human biomonitoring assay using buccal mucosa: comparison of smoking related DNA adducts in mucosa versus biopsies. *Cancer Res* 1995;55(6):1267–1270.
25. Aitio A, Jarvisalo J: Biological monitoring of occupational exposure to toxic chemicals. *Ann Clin Lab Sci* 1985;15(2):121–139.
26. Anand VD, White JM, Nino HV: Some aspects of specimen collection and stability in trace element analysis of body fluids. *Clin Chem* 1975;21(4): 595–602.
27. Alessio L, Apostoli P, Ferioli A: Identification of reference values for metals in general population groups. The example of cadmium. *Toxicol Environ Chem* 1990;27:39–48.
28. Jokstad A, Thommassen Y, Bye E, et al: Dental amalgam and mercury. *Pharmacol Toxicol* 1992; 70:308–313.
29. Woolfson RF: *Statistical methods for the analysis of biomedical data.* New York: John Wiley and Sons, 1987:149–156.
30. Health and Safety Executive: *Biological monitoring for chemical exposures in the workplace.* [Guidance Note EH 56.] London: Health and Safety Executive, Her Majesty's Stationery Office, 1992.
31. Deutsche Forschungsgemeinschaft: MAK-und BAT-Werte-Liste, 1996. *Senatskommission zur Prüfung gesundheitsschädlicher Arbeitsstoffe,* 32nd ed. Weinheim, Germany: VCH, 1996.

32. U.S. Department of Labor, Occupational Safety and Health Administration: 29 CFR. Part 1910.1025 (Lead); Part 1910.1027 (Cadmium); and Part 1910.1028 (Benzene). Revised as of July 1, 1998. Washington, DC: U.S. Government Printing Office, 1998.
33. World Health Organization: *Biological monitoring of chemical exposures in the workplace*, vol. 1. Geneva: World Health Organization, 1996.
34. World Health Organization: *Biological monitoring of chemical exposures in the workplace*, vol. 2. Geneva: World Health Organization, 1996.
35. Castleman BI, Ziem GE: Corporate influence on threshold limit values. *Am J Ind Med* 1988;13:531–539.
36. Boogaard PJ, van Sittert NJ: Suitability of S-phenylmercapturic acid and *trans, trans*-muconic acid as biomarkers for exposure to low concentrations of benzene. *Environ Health Perspect* 1996;104[Suppl 6]: 1151–1157.

SELF ASSESSMENT QUESTIONS

1. Which of the following is not a type of biomarker?
   a. biomarker of exposure
   b. biomarker of effect
   c. biomarker of susceptibility
   d. biomarker of sensitivity

2. What is biological monitoring used for in an individual case?
   a. to establish the distribution of pollutant uptake and/or demonstrate pollutant levels over time on a local or national basis
   b. to monitor occupational groups to help ensure that working conditions are safe
   c. to assess the hazard or risk of a potential toxicant in an individual subject
   d. to identify the source of a chemical emission

3. True or false: There is no difference between sedentary or physical work in the absorption potential of a toxicant.
   a. true
   b. false

4. Which of the following is not a type of sample used in biological monitoring?
   a. air samples
   b. breath samples
   c. adipose tissue
   d. breast milk

5. Interpretation of biological monitoring results may need to account for each of the following except which one?
   a. lifestyle, diet, and hobby and recreational exposures
   b. population values obtained from older analytical methods
   c. interactions with exposures to other chemicals or therapeutic agents
   d. type of charcoal filter used in an air sampling collection device

6. True or false: The list of Biological Exposure Indices published by the American Conference of Governmental Industrial Hygienists has the force of law behind them.
   a. true
   b. false

CHAPTER 21

Carbon Monoxide Poisoning

Jo-Anne Vergilio, M.D.

LEARNING OBJECTIVES

After completing this chapter, the reader should be able to:

1. List the major causes of carbon monoxide poisoning in the United States.
2. Describe the pathophysiologic effects of carbon monoxide intoxication.
3. Describe two methodologies by which to quantify carboxyhemoglobin concentrations.
4. List clinical manifestations of carbon monoxide poisoning.
5. Understand current controversies surrounding hyperbaric oxygen therapy.
6. Suggest two methods by which to reduce carbon monoxide poisoning.

INTRODUCTION

Carbon monoxide poisoning has been described throughout history. Since the creation of fire and enclosed spaces, humans have suffered from its effects. Claude Bernard is credited with the first accurate description of carbon monoxide poisoning in 1857 (1), but earlier observations pertaining to the effects of coal fumes have been recorded by others including Aristotle (3rd century B.C.) and Hippocrates (A.D. 500) (2).

Today, carbon monoxide is the most common cause of unintentional poison-related death in the United States. Recent analyses estimate 40,000 visits to emergency departments for carbon monoxide poisoning each year (3). Carbon monoxide gives no indication of its presence, and the symptoms of carbon monoxide poisoning are nonspecific. Laboratory testing performed routinely in the emergency room or physician's office that is directed at evaluating oxygenation, as opposed to carbon monoxide or carboxyhemoglobin concentrations, may mislead the unsuspecting, and therefore carbon monoxide poisoning may go undetected.

Case Study

FP was a 71-year-old man with a history of coronary artery disease who, after waking one cold winter morning, went to examine his furnace in the basement. While downstairs he acutely developed headache, nausea, dizziness, and near-syncope with nonradiating substernal chest pain. He called to his wife, who promptly contacted 911.

On arrival in the emergency department, the patient was noted to be tachycardic and tachypneic with a 100% non-rebreather face mask in place. Physical examination revealed the patient's skin to be pink, though his distal extremities were cool to touch. Pulse oximetry identified an O_2 saturation of 98%, and an arterial blood gas demonstrated pH 7.35, pCO_2 36 mmHg, pO_2 306 mmHg, HCO_3 20 mmHg, O_{2sat} 99%, and O_2Hb saturation of 67% with a carboxyhemoglobin concentration of 32%.

Upon further inquiry, the patient reported that he had cleaned his gas furnace with a wire brush the preceding week. The patient's wife, who had no significant past medical history, confided that she too had experienced headache and nausea while in the house, but she had attributed these symptoms to recent painting and other home renovations. The wife was immediately tested and found to have a carboxyhemoglobin concentration of 28% even though her only symptoms were intermittent headache and nausea.

Both individuals were maintained on 100% oxygen until their carboxyhemoglobin concentrations normalized. The husband, however, developed an ataxic gait 2 days after therapy, prompting his transfer to an out-of-state hospital for hyperbaric oxygen therapy. Despite administration of hyperbaric oxygen, the husband's symptoms did not improve at 2 months follow-up.

Discussion

This actual case illustrates just some of the issues relevant to the development, clinical and laboratory diagnosis, and management of carbon monoxide poisoning. These aspects include the periodicity of unintentional carbon monoxide poisoning, the causes of carbon monoxide intoxication in the United States, the nonspecific symptoms of carbon monoxide poisoning that may confound its diagnosis, the importance of clinical history-taking in identifying victims, the inconsistency between carboxyhemoglobin concentrations and symptomatology, and the potential neurologic sequelae of carbon monoxide intoxication.

CARBON MONOXIDE—INSIDE AND OUT

Carbon monoxide is a colorless, odorless, tasteless gas that possesses a density of 0.968 and is chemically stable at ambient temperature and pressure. It is ubiquitous in the environment, with an atmospheric concentration less than 0.001% (4). Natural sources of carbon monoxide include forest fires, volcanic activity, natural gases emitted from mines, marine algae, and human activities and metabolism (5).

Carbon monoxide is produced endogenously during the breakdown of hemoglobin and other heme-containing components in the body (6). Heme is converted from its protoporphyrin ring form to the linear tetrapyrrole biliverdin, with the release of both carbon monoxide and iron. This endogenous production yields blood carboxyhemoglobin concentrations of 0.4–0.7%.

Conditions that increase endogenous concentrations of carboxyhemoglobin include pregnancy and hemolytic anemia. During pregnancy, elevated carboxyhemoglobin concentrations are attributed to both an increase in circulating erythrocyte mass by the mother as well as endogenous carbon monoxide production by the fetus (7). With hemolysis, there is increased erythrocyte turnover and hence increased erythrocyte degradation with a compensatory increase in erythrocyte production.

Approximately 40% of global carbon monoxide production is secondary to these natural sources, whereas approximately 60% is generated by manmade pollution (8). Carbon monoxide is a byproduct of hydrocarbon combustion, such as from the burning of coal, petroleum, wood, and plant matter. Major exogenous sources of carbon monoxide include the combustion of fuels for transportation and heating as well as for industrial processing and refuse removal. Tobacco smoke is another significant source of environmental carbon monoxide, and smoking affects endogenous carboxyhemoglobin concentrations in humans. Chronic smokers have elevated carboxyhemoglobin concentrations that generally range from 5% to 10% depending upon the amount of tobacco consumption. Routine passive inhalation by nonsmokers can also elevate base-line carboxyhemoglobin concentrations.

Given the continuous production of carbon monoxide by both manmade and natural sources, various carbon monoxide "sinks" exist to scavenge and stabilize concentrations in the environment. These sinks include removal by both plants and microorganisms as well as by the atmospheric oxidation of carbon monoxide to carbon dioxide by ultraviolet solar radiation (5,8).

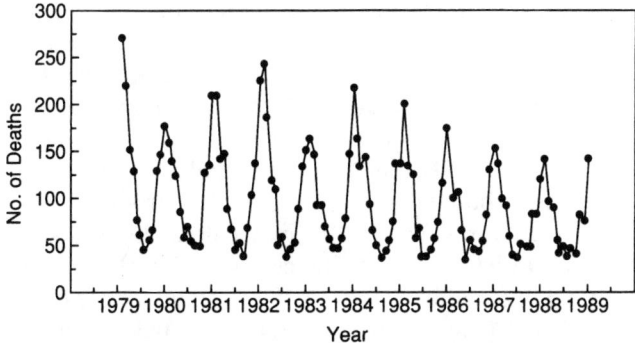

FIG. 21-1. Unintentional carbon monoxide–related deaths by month, 1979 through 1988. From Cobb NC, Etzel RA: Unintentional carbon monoxide-related death in the United States, 1979 through 1988. *JAMA* 1991;266(5):659–663.

CARBON MONOXIDE POISONING—CAUSE AND EFFECT

During the years 1979–1988, the Center for Environmental Health and Injury Control identified >53,100 deaths related to carbon monoxide poisoning in the United States (9). Of these, ~25,900 were secondary to suicide, ~15,500 were related to severe burns or house fires, ~11,500 were unintentional, and ~200 were the result of homicide. Carbon monoxide is responsible for more unintentional poisoning deaths in the United States than any other single agent.

These unintentional carbon monoxide–related fatalities demonstrate a cyclical periodicity, with the highest incidences occurring during winter months in the northern United States, as demonstrated in our case history (Figures 21–1 and 21–2) (9); such trends offer clues to causality. Motor vehicles, heaters, and appliances that use carbon-based fuels constitute the major sources of exposure. Of the ~11,500 unintentional deaths reported above, 57% were attributed to motor vehicle exhaust, with 83% of these associated with stationary automobiles (9).

Mechanisms of motor vehicle–related exposure include malfunctioning exhaust systems, operation of motor vehicles in en-

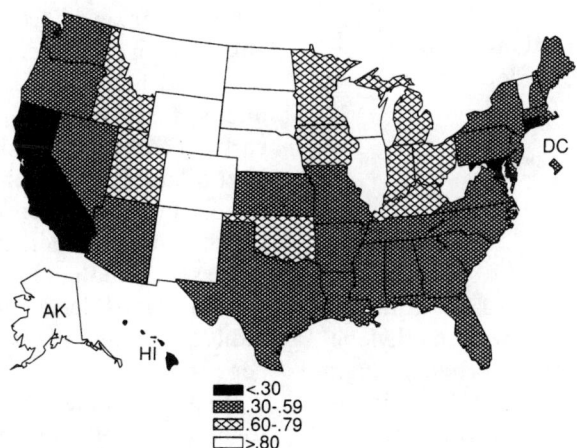

FIG. 21-2. Unintentional carbon monoxide–related deaths by state, 1979 through 1988 (age-adjusted rates per 100,000 resident population). From Cobb NC, Etzel RA: Unintentional carbon monoxide-related death in the United States, 1979 through 1988. *JAMA* 1991;266(5):659–663.

closed spaces with inadequate ventilation, and use of fuel-burning heaters in passenger compartments of automobiles and recreational campers (10). Snow-obstructed vehicle exhaust systems (11) and recreational gasoline-powered boats with enclosed cabins (12) have created dangerous carbon monoxide–rich environments. Children traveling in the rear of pickup trucks that have leaking exhaust systems or rear-exiting (rather than side-exiting) tailpipes have also fallen victim to carbon monoxide poisoning (13). Liquid petroleum gas-powered forklifts used in industry (14) and propane-powered ice-resurfacing machines used in indoor ice arenas (15) are additional culprits.

Vehicles, however, are not the only cause for concern. Domestic stoves and fireplaces are also major sources of carbon monoxide poisoning. Charcoal grills and propane gas stoves have been used as sources of warmth both in homes during power outages and in camping enclosures with adverse, and sometimes fatal, consequences (16,17). Fireplaces and furnaces may malfunction or may be tampered with inadvertently, as was the case with our patient.

Pathophysiology of Carbon Monoxide Poisoning

Carbon monoxide is generally absorbed in the body via inhalation and is eliminated essentially unchanged in the lungs. The absorption of carbon monoxide depends on four basic factors:

- Carbon monoxide concentration
- Minute ventilation
- Activity level
- Duration of exposure

One particular exception to this route of entry relates to exposure with methylene chloride, a chemical that is absorbed through both the skin and lungs in vapor form and is then transported to the liver, where carbon monoxide is produced as an end product of its metabolism (18).

Once absorbed, carbon monoxide possesses an affinity for hemoglobin that is 200–250 times that of oxygen (19), and in binding and forming carboxyhemoglobin, carbon monoxide exerts its devastating effects. Inhaled carbon monoxide undergoes competitive reversible binding to the heme sites of hemoglobin. When two of the four sites are bound by carbon monoxide, the hemoglobin molecule undergoes a change in its allosteric configuration such that bound oxygen molecules can no longer be easily released under normal partial pressures of oxygen (20). As a consequence, the oxyhemoglobin dissociation curve is shifted to the left (Figure 21–3), resulting in impaired oxygen release to tissues and consequent tissue hypoxia (21). Via this mechanism, carbon monoxide interferes with tissue oxygenation in two separate respects:

- In occupying the oxygen-binding sites of hemoglobin, carbon monoxide directly lowers the capacity of blood to transport oxygen.
- In binding hemoglobin, carbon monoxide hinders oxygen release.

The resultant tissue hypoxia manifests initially as shortness of breath with a compensatory increase in minute ventilation and, consequently, an increased work of breathing. As a result, more carbon monoxide is inspired, which further exacerbates the increased oxygen demands.

Carbon monoxide exposure with its resultant hypoxia affects humans and animals alike. Given their increased basal metabolic rates and hence increased oxygen requirements, children and animals usually succumb early to carbon monoxide exposure. In fact, in the latter part of the 19th century, coal miners brought caged canaries into underground mines as "detectors." Canaries, being more sensitive than humans to the effects of carbon monoxide, would collapse, serving as indicators of an otherwise undetectable danger (7).

Pregnant women are similarly susceptible to carbon monoxide exposure given their increased metabolic requirements; however, the effects of hypoxia are twofold because carbon monoxide affects both mother and fetus. Fetal carboxyhemoglobin concentra-

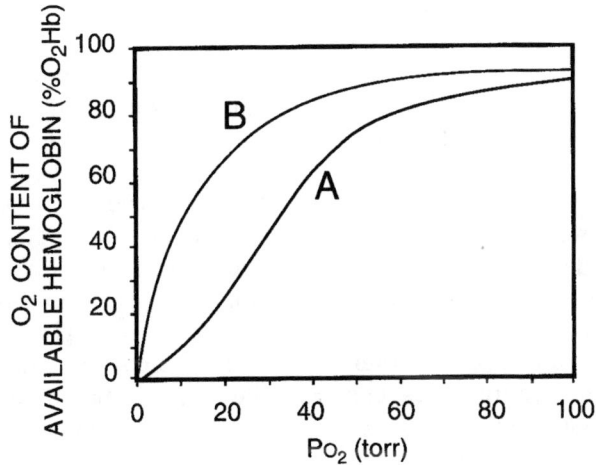

FIG. 21–3. The effect of carboxyhemoglobin (COHb) on the dissociation curve of oxyhemoglobin. (line A) Normal oxygen dissociation curve with 80 g/L of hemoglobin in the blood (mimicking anemia). (line B) Oxygen dissociation curve with 50% COHb and 160 g/L of hemoglobin in the blood (normal hemoglobin concentration). Though the oxygen-carrying capacity is the same in both cases, when COHb is present, oxygen dissociates from hemoglobin at lower values of PO_2. From Klaassen CD: Nonmetallic environmental toxicants. In: Hardman JGG, Limbird LL, eds: *Goodman and Gilman's the therapeutic pharmacological basis of therapeutics.* New York: McGraw-Hill, Health Professional Divisions, 1996:1676.

tions depend on maternal carboxyhemoglobin concentrations as well as endogenous fetal production. Uptake and elimination of carbon monoxide occur more slowly in the fetus than in the mother, in part because of delayed transfer through the placenta (8). Once absorbed through the placenta, fetal hemoglobin has an even higher affinity for carbon monoxide than adult hemoglobin, making its release all the more difficult. As a result, fetal steady-state concentrations can occur as long as 40 hours after maternal steady states are reached and the final carboxyhemoglobin concentrations in the fetus may exceed, by 10–15%, those of the mother (7).

Although the hypoxic effects of carbon monoxide are well understood, other mechanisms of injury that implicate carbon monoxide as a direct intracellular toxin have been proposed and are being investigated in various experimental models. In these studies, carbon monoxide has been shown to bind other heme-containing compounds including myoglobin, cytochrome c oxidase, and cytochrome P450 (2). Binding to myoglobin may interfere with oxygen transfer to cardiac and skeletal muscle and therefore compromise muscle function. Binding to cytochrome c oxidase, a terminal enzyme in oxidative cellular respiration, may interfere with energy production, and binding to the cytochrome P450 enzyme system may interfere with hepatic metabolism. Carbon monoxide poisoning is also believed to be associated with a reperfusion-type injury in the brain whereby toxic oxygen radicals produced after reoxygenation contribute to lipid and protein injury within the central nervous system (22,23). These investigations suggest that carbon monoxide exerts a multiplicity of effects.

Clinical Manifestations of Carbon Monoxide Poisoning

The clinical manifestations of acute carbon monoxide poisoning are nonspecific, and individuals exposed to carbon monoxide at the same time and in the same manner may present with differing complaints. Malaise, fatigue, headache, nausea, and vomiting are not uncommon and, given the increased incidence of carbon monoxide exposure during winter months, may mimic a nonspecific viral illness.

Tachycardia and tachypnea result as compensatory mechanisms of cellular hypoxia. In individuals with underlying cardiac and/or pulmonary disease, deterioration in function may be observed with decreased exercise tolerance, angina, arrhythmia, and/or pulmonary edema (24). In our case history, the husband, who had underlying coronary artery disease, presented with substernal chest pain and a near-syncopal episode while his wife, who had no significant past medical history, presented with only headache and nausea. Cardiac manifestations may contribute to misdiagnosis because an underlying etiology for either new-onset or recurrent angina, arrhythmias, or myocardial infarction may not be sought by clinicians. With acute severe poisoning, cardiac compromise with secondary hypotension and/or arrhythmia may ultimately be the cause of an individual's demise.

Other common manifestations of carbon monoxide poisoning involve the central nervous system. Individuals may notice impaired mental function, and they may present with agitation, inability to concentrate, and changes in personality. With worsening hypoxia, individuals may develop cerebral vasodilation with cerebral edema, syncope, seizures, coma, and eventual death.

Cherry red skin and retinal hemorrhages are features classically associated with carbon monoxide exposure; however, their occurrence is rare. Interestingly, our patient did present with a reddish hue to his skin, though this may not be observed in most poisoning victims.

CLINICAL AND LABORATORY EVALUATION

Because the symptoms of carbon monoxide exposure are nonspecific, clinicians must exercise a high index of suspicion when evaluating patients and must also attempt to elicit clues from the clinical history. Such clues may include symptoms affecting multiple individuals at home, at school, or in a workplace. The symptoms may be intermittent, developing only in one particular environment, or they may persist or worsen in individuals who have been medically evaluated and discharged back to the toxic source, only to return again for medical assessment. In the case provided, the husband relayed the relevant history of cleaning his furnace only after the carboxyhemoglobin concentration was attained and further questions were asked of him; likewise, his wife did not come forward with her symptoms until the diagnosis had already been reached.

Pulse Oximetry

Pulse oximetry, a standard technique routinely used in ambulances, medical offices, and emergency rooms to evaluate oxygen saturation, may further confound the clinical picture. Pulse oximeters falsely elevate oxygen saturation in the presence of carboxyhemoglobin (25,26) and therefore may not accurately indicate an individual's true state of hypoxic compromise. Pulse oximeters are easy to use, generate a result in seconds, and are a noninvasive means of assessment because they require only a finger or earlobe probe that attaches to a portable instrument.

The methodology of pulse oximetry is based on two principles: the differential absorption characteristics of oxyhemoglobin and deoxyhemoglobin at two wavelengths of light [660 nm (red) and 940 nm (infrared)] and the pulsatile nature of arterial blood flow. One side of the pulse oximeter probe emits light while the other side of the probe serves as a photodetector. Received light is divided into two separate components, one that occurs with systole and corresponds to pulsatile flow and another that occurs with diastole, is of a constant intensity, and reflects "background" tissues such as muscle, fat, bone, skin pigmentation, and venous blood (Figure 21–4A) (26). The pulse oximeter determines an absorption ratio by dividing the pulsatile absorption by the nonpulsatile absorption for each wavelength of

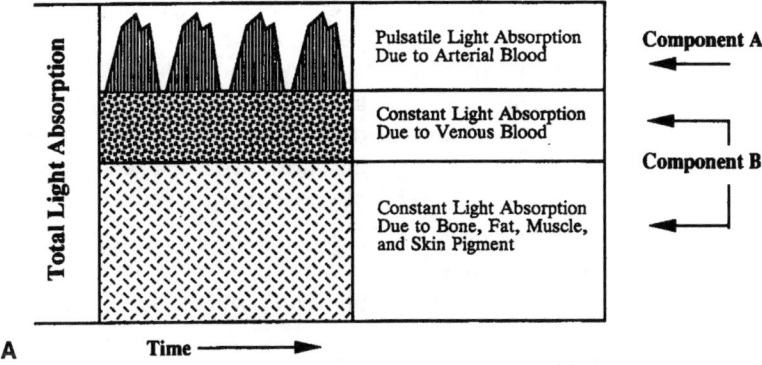

FIG. 21–4. The diagram shows the components of light absorption as a function of time. The equations show how to calculate related parameters. HB, hemoglobin; HbO_2, oxyhemoglobin; COHb, carboxyhemoglobin; MetHb, methemoglobin. From Mengelkoch LJ, Martin D, Lawler J: A review of the principles of pulse oximetry and accuracy of pulse oximeter estimates during exercise. *Phys Ther* 1994;74(1):40–49.

light (Figure 21–4B). From this, an algorithm is used to calculate oxygen saturation that estimates the arterial oxygen saturation of available hemoglobin (Figure 21–4C).

Though pulse oximeters can only distinguish oxyhemoglobin from deoxyhemoglobin, other forms of hemoglobin are present in normal individuals but in low concentration (for example, methemoglobin, a form of hemoglobin that can no longer bind oxygen, and carboxyhemoglobin). A more accurate assessment of oxygen saturation estimates the arterial oxygen saturation of total hemoglobin (Figure 21–4D); however, because concentrations of methemoglobin and carboxyhemoglobin are low in normal individuals, this contribution is minor. In individuals with high concentrations of either methemoglobin or carboxyhemoglobin, this contribution is great, and estimates of oxygen saturation by pulse oximetry are prone to error.

The absorption of carboxyhemoglobin at a wavelength of 660 nm is similar to that of oxyhemoglobin. As a result, carboxyhemoglobin appears similar to oxyhemoglobin by photodetection and leads to the overestimation of oxygen saturation.

In the case presented, pulse oximetry calculated an oxygen saturation of 98% while CO-oximetry determined the oxygen saturation of hemoglobin to be only 67% with a carboxyhemoglobin concentration of 32%. Confounding the clinical picture is the arterial blood gas determination of pO_2 306 mmHg and O_{2sat} of 99%; however, these values were obtained after therapy was initiated with 100% oxygen. Certainly, the institution of oxygen therapy should never be delayed by the collection of blood samples, but these details must be factored into data analysis.

Gas Chromatography

Gas chromatography is the most accurate and precise method to quantify carboxyhemoglobin concentrations and is considered to be the reference method of detection. Using this methodology, carbon monoxide must first be released from hemoglobin before it can be quantified. Blood is treated with potassium ferricyanide, a reagent that oxidizes iron in the heme moieties of hemoglobin. This oxidative process converts carboxyhemoglobin to methemoglobin with release of carbon monoxide into the gaseous phase (27). The carbon monoxide is then measured by gas chromatography using a molecular sieve column and a thermal conductivity detector. Lower levels of detection can be achieved by using a reducing catalyst between the column and detector (27). The catalyst converts carbon monoxide into methane that can then be detected by flame ionization. Attachment of a heated mercuric oxide reaction chamber to the column with an ultraviolet detector allows for even lower levels of detection (27). With this technique, carbon monoxide elutes from the column and reacts with mercuric oxide to form mercury gas, a compound that absorbs at 254 nm.

Gas chromatography is both accurate and precise and can detect low concentrations of carboxyhemoglobin (<1.5%). In addition, unlike other methodologies, gas chromatography is not generally susceptible to interfering substances. Operation is straightforward, but does require skilled technicians.

Spectrophotometry (CO-Oximetry)

Spectrophotometric quantification is based on the differences in spectral absorption exhibited by oxyhemoglobin and carboxyhemoglobin. Absorption measurements are performed on blood samples using a range of different wavelengths, and these measurements are then compared electronically to compute oxyhemoglobin and carboxyhemoglobin concentrations.

Both oxyhemoglobin and carboxyhemoglobin exhibit two separate peaks in absorption. The absorption maxima for oxyhemoglobin are at 576–578 nm and 540–542 nm, whereas for carboxyhemoglobin, the absorption maxima are at 568–572 nm and 538–540 nm (Figure 21–5A) (27). When a dilute sample of blood is treated with sodium hydrosulfite, oxyhemoglobin is converted to deoxyhemoglobin while carboxyhemoglobin is unaffected. The spectral absorption pattern for deoxyhemoglobin consists of a single peak at approximately 555 nm. Therefore, under deoxygenating conditions, the absorp-

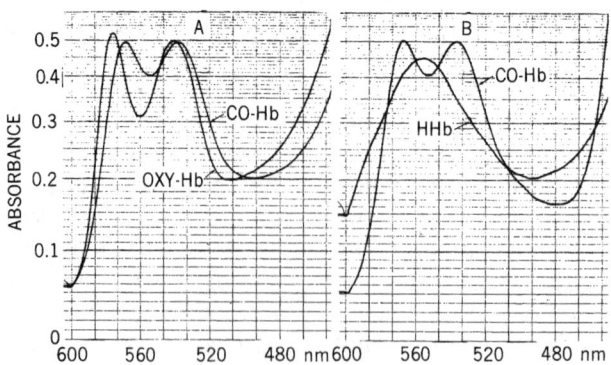

FIG. 21–5. Spectral absorption. (*A*) 100% oxyhemoglobin (OXY-Hb) and 100% carboxyhemoglobin (CO-Hb). (*B*) 100% oxyhemoglobin (now HHb) and 100% carboxyhemoglobin after treatment of both with sodium dithionite. From Porter WH: Clinical toxicology. In: Burtis CA, Ashwood ER, eds: *Tietz textbook of clinical chemistry.* Philadelphia: WB Saunders, 1999:917–921.

tion pattern for oxyhemoglobin changes, whereas the pattern for carboxyhemoglobin does not (Figure 21–5B) (27). By quantifying the different absorption peaks, oxyhemoglobin and carboxyhemoglobin concentrations can be determined.

Spectrophotometric analysis correlates well with gas chromatographic analysis at carboxyhemoglobin concentrations greater than 2–3%, but not at concentrations below this threshold (28,29). Spectrophotometry is therefore most useful for detecting measurements related to exogenous carbon monoxide exposure as opposed to endogenous carbon monoxide production. Spectrophotometry allows for rapid and easy analysis, making it the method of choice in most routine laboratory practices when low sensitivity is not generally a necessity.

Caveats

Though the measurement of carboxyhemoglobin is critical to diagnosing carbon monoxide poisoning, carboxyhemoglobin concentrations have not been shown to correlate accurately with clinical manifestations or to be reliable indicators of either the severity of exposure or the expected prognosis (30).

TREATMENT

Once carbon monoxide poisoning has been confirmed, 100% oxygen should be instituted immediately and delivered via either a continuous positive airway pressure mask or endotracheal intubation. Oxygen therapy increases the oxygen content of the blood by maximizing the fraction dissolved in plasma; as a result, sufficient oxygen is supplied directly to tissues to meet oxygen requirements, and the rate of elimination of carbon monoxide is increased as well. The half-life of elimination of carbon monoxide on room air with normal breathing is 4–6 hours; this decreases to 40–80 minutes with the administration of 100% oxygen and is further reduced to 15–30 minutes with the administration of hyperbaric oxygen (100% oxygen at 2.5 atmospheres of absolute pressure) (31).

The general recommendations for treatment of carbon monoxide intoxication are at least 6 hours of 100% oxygen therapy with simultaneous cardiac monitoring using a 12-lead electrocardiogram (32). Other basic laboratory tests including arterial blood gas, electrolyte panel, blood urea nitrogen, creatinine, and glucose should also be performed. Though severe lactic acidosis secondary to hypoxia can precipitate fatal arrhythmias, treatment of mild acidosis with sodium bicarbonate further shifts the oxyhemoglobin dissociation curve to the left and is to be discouraged, because it only compromises already impaired oxygen delivery to tissues (32). When the cause of carbon monoxide poisoning relates to smoke inhalation, the presence of other toxic compounds such as cyanide should also be assessed. Once the patient has stabilized after initial therapy, a neuropsychologic examination should be administered. Patients are asked to perform various tasks, some of which relate to attention and concentration, number processing, memorization, and rapid fine hand movements (33).

The presence of loss of consciousness, impaired neuropsychologic assessment, abnormal neurologic findings, or cardiovascular symptoms are considered by some to be indications for hyperbaric oxygen therapy (32); however, the utility of hyperbaric oxygen therapy in the treatment of carbon monoxide poisoning is one of the major controversies today.

Proposed Mechanisms of Hyperbaric Oxygen Therapy

Though hyperbaric oxygen therapy increases the amount of dissolved oxygen in the blood and expedites the rate of elimination of carbon monoxide from the body, its effects on associated morbidity and mortality are less well defined. Hyperbaric oxygen therapy is believed by some to minimize the adverse effects of ischemic-reperfusion injury. In this injury process, an acute interruption of blood flow initially causes metabolic dysfunction with accompanying cellular damage and potential cell death. Once blood flow is restored, polymorphonuclear leukocytes are attracted to sites of injury, but these further compromise perfusion by causing stasis and occlusion of the microvasculature. Leukocytes also release oxygen radicals that can damage cellular components via lipid peroxidation (22). Carbon monoxide poisoning is not considered to be a classical model of ischemic-reperfusion injury; however, analogies have been drawn particularly with respect to its effects in the central nervous system.

During carbon monoxide poisoning, the brain undergoes initial intracellular hypoxia with subsequent transient ischemia secondary to hypotension (22). Carbon monoxide is then removed with restoration of normal blood flow and re-oxygenation. Animal models of carbon monoxide poisoning have demonstrated sequestration of polymorphonuclear leukocytes within microvasculature that is associated first with increased production of xanthine oxidase and then of toxic oxygen radicals with subsequent evidence of lipid peroxidation (23,34). These findings are similar to those seen in ischemic-reperfusion injury, as described above. Treatment with hyperbaric oxygen inhibited neutrophil accumulation and its consequences in these animal models (35).

These studies have offered potential mechanisms, as well as a potential therapy, for the neurologic sequelae observed with carbon monoxide intoxication.

SHORT- AND LONG-TERM SEQUELAE OF CARBON MONOXIDE POISONING

The most severe consequences of carbon monoxide poisoning are coma and death, but other more subtle neurologic and behavioral effects have been described with numerous conflicting reports in the literature (36). Generally speaking, carbon monoxide poisoning is believed to be associated with both short- and long-term neurologic sequelae. A high percentage of survivors of carbon monoxide poisoning experience some permanent physical, cognitive, emotional, and/or psychological impairment (37).

Persistent neurologic sequelae are deficits that occur with, or immediately after, exposure and may improve over time, but usually never with complete resolution (2). These changes include depression, anxiety, personality changes, and emotional lability.

Delayed neurologic sequelae (DNS) are symptoms that can occur from 2 days to 2 months after exposure and usually follow a period of lucidity. Symptoms include mental deterioration, disorientation, bradykinesia, gait disturbances, cogwheel rigidity, aphasia, incontinence, and apathy (30). In a study reported by Choi (38), increased age and the presence (and extended duration) of unconsciousness at the time of carbon monoxide exposure appeared to be risk factors associated with the development of DNS. Furthermore, patients with DNS demonstrated abnormalities on computed tomographic imaging of the brain that included low-density lesions in the globus pallidus and decreased density in the white matter of the cerebral cortex.

In the case presented, the husband developed an ataxic gait 2 days after his acute deterioration, prompting his transfer to an institution that offered hyperbaric oxygen therapy. By this time, his carboxyhemoglobin concentration had already normalized, so hyperbaric oxygen therapy was instituted for its potential neurologic benefits. A 2-month follow-up assessment identified no improvement in the patients neurologic symptoms.

HYPERBARIC OXYGEN THERAPY THROUGH THE AGES

Despite the proposed mechanisms of hyperbaric oxygen therapy, there are numerous contradictory reports within the literature with little definitive confirmatory evidence of its benefits (2,33,39–48). Limitations to all of these studies have been cited and include:

- Failure to blind patients and examiners to treatment protocols

- Administration of inadequate doses of hyperbaric oxygen
- Extended delays (>6 hours) from the cessation of poisoning to the administration of therapy
- Use of "soft" outcome measures (such as telephone interviews of both patients and their physicians) as opposed to neuropsychological testing
- Inadequate or poorly characterized controls in cases in whom neuropsychological testing was performed
- Poor follow-up
- Inadequate or absent statistical analysis

The administration of hyperbaric oxygen therapy is not a minor undertaking and is not a procedure without risk. Limited facilities are available across the United States, often necessitating transfer of patients over extended distances.

Transport incurs both additional expenses and additional risks to the patient. Patients that are critically ill must be monitored during transport as well as during the administration of therapy. Depending on the hyperbaric facility, the patient may be isolated in a chamber with difficult access to health-care personnel. Risks of hyperbaric oxygen therapy include tympanic membrane rupture, reversible myopia (secondary to toxic effects of oxygen on the lens), pneumothorax, seizure, and, rarely, fire (49,50).

Even among hyperbaric oxygen facilities, there is little universal agreement in the selection criteria used to identify patients deemed appropriate for therapy (51). Most facilities treat carbon monoxide–poisoned patients who have demonstrated loss of consciousness, ischemic changes on electrocardiogram, focal neurologic deficits, abnormal neuropsychologic testing, or coma, regardless of their carboxyhemoglobin concentration. A small majority of facilities implement a minimum carboxyhemoglobin concentration as the only criterion used to assess asymptomatic patients. In general, there is better agreement regarding the most severely poisoned patients as opposed to those more mildly affected.

There are presently three prospective, randomized, double-blinded, controlled clinical trials to investigate the benefits and risks of hyperbaric oxygen therapy for acute carbon monoxide poisoning; however, preliminary data are inconclusive (32,39,40, 47,48). Particular groups for which no definitive studies have been performed include pregnant women and children, for whom the institution of therapy is equally controversial.

CARBON MONOXIDE POISONING— ITS PREVENTION

Given the high incidence of carbon monoxide poisoning and its potential consequences, numerous efforts have been made to minimize exposure. These efforts include a form of air pollution control that attempts to regulate the amount of carbon monoxide in ambient air as well as in occupational settings. In the outdoor environment, air quality cannot exceed 9 ppm carbon monoxide over an 8-hour period, whereas in an occupational environment, air quality cannot exceed 25–50 ppm carbon monoxide over the same time period (52). In order to identify these limits, carbon monoxide detectors have been developed.

Carbon monoxide detectors use colorimetric, metal oxide semiconductor, and electrochemical sensor equipment (53). They can be powered by alternating current or battery, and they can produce a visual or audible alarm. Carbon monoxide detectors are generally sensitive to temperature, humidity, and the presence of other interfering substances, and they can become unstable with time (52,53). Such monitoring devices have now been introduced into residences in an analogous manner to smoke detectors.

Though carbon monoxide monitors can help to identify exposures, they are no substitute for education of the general public or the health-care community with respect to the sources, risks, and modes of protection against carbon monoxide poisoning.

REFERENCES

1. Bernard C: Leçons sur les effets des substances toxiques et medicamenteuses. Paris: JB Baillere et fils, 1857.
2. Weaver LK: Carbon monoxide poisoning. *Crit Care Clin* 1999;15(2): 297–317.
3. Hampson NB: Emergency department visits for carbon monoxide poisoning in the Pacific Northwest. *J Emerg Med* 1998;16(5):695–698.
4. Ernst A, Zibrak JD: Carbon monoxide poisoning. *N Engl J Med* 1998;339(22):1603–1608.
5. Jaffe LS: Sources, characteristics, and fate of atmospheric carbon monoxide. *Ann N Y Acad Sci* 1970; 174(1):76–87.
6. White P: Carbon monoxide production and heme catabolism. *Ann N Y Acad Sci* 1970; 174(1):23–31.
7. Longo LD: The biological effects of carbon monoxide on the pregnant woman, fetus, and newborn infant. *Am J Obstet Gynecol* 1977;129:69–103.
8. Vreman HJ, Wong RJ, Stevenson DK, et al: Carbon monoxide in breath, blood, and other tissues. In: Penney DG, ed: *Carbon monoxide toxicity*. Boca Raton, FL: CRC Press, 2000:19–60.
9. Cobb NC, Etzel RA: Unintentional carbon monoxide-related death in the United States, 1979 through 1988. *JAMA* 1991;266(5):659–663.
10. Centers for Disease Control and Prevention: Deaths from motor-vehicle-related unintentional carbon monoxide poisoning—Colorado, 1996, New Mexico, 1980–1995, and United States, 1979–1992. *MMWR Morb Mortal Wkly Rep* 1996;45(47):1029–1032.
11. Centers for Disease Control and Prevention: Carbon monoxide poisonings associated with snow-obstructed vehicle exhaust systems—Philadelphia and New York City, January 1996. *MMWR Morb Mortal Wkly Rep* 1996;45(1):1–3.
12. Silvers SM, Hampson NB: Carbon monoxide poisoning among recreational boaters. *JAMA* 1995; 274(20)1614–1616.
13. Hampson NB, Norkool DM: Carbon monoxide poisoning in children riding in the back of pickup trucks. *JAMA* 1992;267(4):538–540.
14. Centers for Disease Control and Prevention: Carbon monoxide poisoning associated with use of LPG-powered (propane) forklifts in industrial settings—Iowa, 1998. *MMWR Morb Mortal Wkly Rep* 1999;48(49):1121–1124.
15. Centers for Disease Control and Prevention: Carbon monoxide poisoning at an indoor ice arena and bingo hall—Seattle, 1996. *MMWR Morb Mortal Wkly Rep* 1996;45(13):265–267.
16. Hampson NB, Kramer CL, Dunford RG, et al: Carbon monoxide poisoning from indoor burning of charcoal briquets. *JAMA* 1994;271(1):52–53.
17. Centers for Disease Control and Prevention: Carbon monoxide poisoning deaths associated with camping—Georgia, March 1999. *MMWR Morb Mortal Wkly Rep* 1999;48(32):705–706.
18. Stewart RD, Hake CL: Paint-remover hazard. *JAMA* 1976:235:398–401.
19. Allen TA, Root WS: Partition of carbon monoxide and oxygen between air and whole blood of rats, dogs and men as affected by plasma pH. *J Appl Physiol* 1957;10:186–209.
20. Weaver LK: Carbon monoxide poisoning. *Crit Care Clin* 1999;15(2):297–317.
21. Roughton FJW, Darling RC: The effect of carbon monoxide on the oxyhemoglobin dissociation curve. *Am J Physiol* 1944;141:17–31.
22. Buras J: Basic mechanisms of hyperbaric oxygen in the treatment of ischemia-reperfusion injury. *Int Anesthesiol Clin* 2000;38(1):91–109.
23. Thom SR: Carbon monoxide-mediated brain lipid peroxidation in the rat. *J Appl Physiol* 1990;68:997–1003.
24. Allred EN, Bleecker ER, Chaitman BR, et al: Short-term effects of carbon monoxide exposure on the exercise performance of subjects with coronary artery disease. *N Engl J Med* 1989:321:1426–1432.
25. Bowes WA, Corke BC, Hulka J 3rd: Pulse oximetry: a review of the theory, accuracy, and clinical applications. *Obstet Gynecol* 1989;74[3, part 2]:541–546.
26. Mengelkoch LJ, Martin D, Lawler J: A review of the principles of pulse oximetry and accuracy of pulse oximeter estimates during exercise. *Phys Ther* 1994;74(1):40–49.
27. Porter WH: Clinical toxicology. In: Burtis CA, Ashwood ER. eds: *Tietz textbook of clinical chemistry*. Philadelphia: WB Saunders, 1999:917–921.
28. Mahoney JJ, Vreeman HJ, Stevenson DK, et al: Measurement of carboxyhemoglobin and total hemoglobin by five specialized spectrophotometers (CO-oximeters) in comparison with reference methods. *Clin Chem* 1993;39(8):1693–1700.
29. Vreman HJ, Mahoney JJ, Van Kessel AL, et al: Carboxyhemoglobin as measured by gas chromatography and with the IL 282 and 482 CO-Oximeters. *Clin Chem* 1988;34(12):2562–2566.
30. Hardy KR, Thom SR: Pathophysiology and treatment of carbon monoxide poisoning. *Clin Toxicol* 1994;32(6):613–629.
31. Pace N, Strajman E, Walker EL, et al: Acceleration of carbon monoxide elimination in man by high pressure oxygen. *Science* 1950;111:652–654.
32. White S: Update on the clinical treatment of carbon monoxide poisoning. In: Penney DG, ed: *Carbon monoxide toxicity*. Boca Raton, FL: CRC Press, 2000:274–279.
33. Seger D, Welch L: Carbon monoxide controversies: neuropsychologic testing, mechanism of toxicity, and hyperbaric oxygen. *Ann Emerg Med* 1994;24: 242–248.
34. Thom S: Leukocytes in carbon monoxide-mediated brain oxidative injury. *Toxicol Appl Pharmacol* 1993;123:234–237.
35. Thom S: Functional inhibition of leukocyte B2 integrins by hyperbaric oxygen in carbon monoxide-mediated brain injury to rats. *Toxicol Appl Pharmacol* 1993;123:248–256.
36. Benignus VA: Behavioral effects of carbon monoxide exposure: results and mechanisms. In: Penny DG, ed: *Carbon monoxide*. Boca Raton, FL: CRC Press, 1996:211–238.
37. Helffenstein DA: Neuropsychological evaluation of the carbon monoxide-poisoned patient. In: Penny DG, ed: *Carbon monoxide toxicity*. Boca Raton, FL: CRC Press, 2000:439–440.
38. Choi S: Delayed neurologic sequelae in carbon

monoxide intoxication. *Arch Neurol* 1983;40:433–435.
39. Scheinkestel CD, Jones K, Cooper DJ, et al: Interim analysis—controlled clinical trial of hyperbaric oxygen in acute carbon monoxide (CO) poisoning [Abstract]. *Undersea Hyperb Med* 1996;23 [Suppl]:7.
40. Mathieu D, Wattel F, Mathieu-Nolf M, et al: Randomized prospective study comparing the effect of HBO versus 12 hours NBO in non-comatose CO poisoned patients: results of the interim analysis. *Undersea Hyperb Med* 1996;23[Suppl]:7–8.
41. Olson KR: Hyperbaric oxygen for carbon monoxide poisoning: does it really work? [Editorial]. *Ann Emerg Med* 1995;25(4):535–537.
42. Ducasse JL, Celsis P, Marc-Vergnes JP: Non-comatose patients with acute carbon monoxide poisoning: hyperbaric or normobaric oxygenation? *Undersea Hyperb Med* 1995;22(1):9–15.
43. Thom SR, Taber RL, Mendiguren II, et al: Delayed neuropsychologic sequelae after carbon monoxide poisoning: prevention by treatment with hyperbaric oxygen. *Ann Emerg Med* 1995;25(4):474–483.
44. Raphael J-C, Elkharrat D, Jars-Guincestre MC, et al: Trial of normobaric and hyperbaric oxygen for acute carbon monoxide intoxication. *Lancet* 1989; 2:414–419.
45. Myers RA, Snyder SK, Emhoff TA: Subacute sequelae of carbon monoxide poisoning. *Ann Emerg Med* 1985;14:1163–1167.
46. Tibbles PM, Perrotta PL: Treatment of carbon monoxide poisoning: a critical review of human outcome studies comparing normobaric oxygen with hyperbaric oxygen. *Ann Emerg Med* 1994;24(2)269–276.
47. Weaver LK, Hopkins RO, Larson-Lohr V, et al: Double-blind, controlled, prospective, randomized clinical trial (RCT) in patients with acute carbon monoxide (CO) poisoning: outcome of patients treated with normobaric oxygen or hyperbaric oxygen (HBO_2)—an interim report. *Undersea Hyperbar Med* 1995;22[Suppl]:14.
48. Weaver LK, Hopkins RO, Howe S, et al: Outcome at 6 and 12 months following acute CO poisoning. *Undersea Hyperbar Med* 1996;23[Suppl]:9–10.
49. Tibbles PM, Edelsberg JS: Medical progress: hyperbaric-oxygen therapy. *N Engl J Med* 1996;334(25): 1642–1648.
50. Sheridan RL, Shank ES: Hyperbaric oxygen treatment: a brief overview of a controversial topic. *J Trauma* 1999;47(2):426–435.
51. Hampson, NB, Dunford RG, Kramer CC, et al: Selection criteria utilized for hyperbaric oxygen treatment of carbon monoxide poisoning. *J Emerg Med* 1995;13(2):227–231.
52. Raub JA, Mathieu-Nolf M, Hampson NB, et al: Carbon monoxide poisoning—a public health perspective. *Toxicology* 2000;145:1–14.
53. Kwor R: Carbon monoxide detectors. In: Penney DG, ed: *Carbon monoxide toxicity*. Boca Raton, FL: CRC Press, 2000:61–82.

SELF-ASSESSMENT QUESTIONS

1. List two major causes of carbon monoxide poisoning in the United States.

2. List two mechanisms by which carbon monoxide causes hypoxia.

3. Describe two methodologies by which to quantify carboxyhemoglobin concentrations.

4. Describe the potential sequelae of carbon monoxide poisoning.

5. List some of the present controversies related to the use of hyperbaric oxygen therapy.

6. Suggest two methods by which to reduce carbon monoxide poisoning.

CHAPTER 22

Organophosphate and Carbamate Pesticide Poisoning

Tai C. Kwong, Ph.D., DABCC

LEARNING OBJECTIVES

After completing this chapter, the reader should be able to:

1. Define pesticide poisoning and describe the extent of the problem in the United States and worldwide.
2. Describe the classification and basic chemistry of pesticides and their relative toxicity.
3. Describe the toxicokinetics and toxicodynamics of pesticide poisoning.
4. Describe the mechanism of toxicity and treatment rationale.
5. Describe the mechanism of action of oxime antidotes.
6. Describe the role of the clinical toxicology laboratory in pesticide testing.

INTRODUCTION

The organophosphates and carbamates are effective pesticides used in large quantities around the world. Poisoning by these toxic chemicals due to occupational or accidental exposure is thus a serious global public health problem (1). Toxic exposures are most common among agricultural workers, manufacturing workers, and small children (2). Their extensive use poses serious challenges to countries where safety measures to minimize toxic exposures are less developed or not rigorously enforced. For example, 1000 deaths per year associated with poisoning involving these pesticides were reported to occur in Sri Lanka, a country of 15 million inhabitants, and the World Health Organization has estimated that 1 million serious accidental poisonings and 2 million suicide attempts involving them occur each year worldwide (1).

In the United States, 13,348 toxic organophosphate exposures were reported in 1999, one-quarter of which were managed at health-care facilities, and there were 5 fatal outcomes (3). There were no fatalities associated with the 4194 reported cases of carbamate poisoning.

The organophosphates and carbamates are closely related compounds, being esters of phosphoric acid, phosphothioic acid, or carbamic acid. They differ in the structures of the side chains, which can be quite variable; there are more than 200 organophosphates and 25 carbamates formulated into thousands of products (Figures 22–1 and 22–2). The side chains determine the toxicokinetics and toxicodynamics of pesticide poisoning (4). Therefore, identification of the active ingredient is important and can be accomplished by examination of the product label, the Material Safety Data Sheet, a trade reference such as the *Farm Chemical Handbook* (published annually by Meister Publishing, Willoughby, OH), or an online database of pesticides available in the United States (http://www.cdpr.ca.gov).

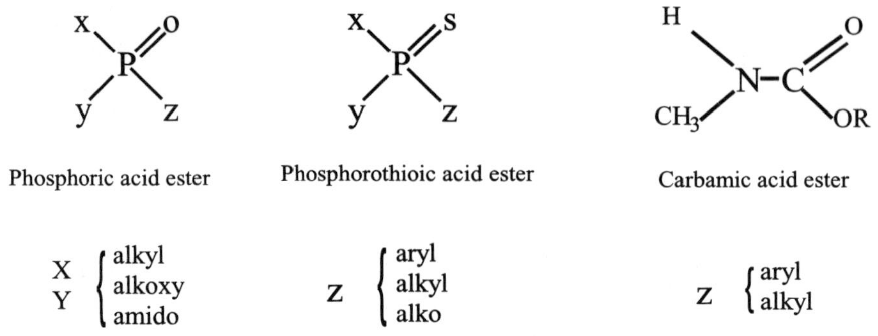

FIG. 22–1. Organophosphorus and carbamate esters.

Phosphoric acid ester

Phosphorothioic acid ester

Carbamic acid ester

X, Y { alkyl, alkoxy, amido }

Z { aryl, alkyl, alko }

Z { aryl, alkyl }

Case Study

A previously healthy 4-year-old was brought to a local hospital because of nausea, vomiting, and increased work of breathing. The patient's landlord had brought over milk-white "roach poison" in a milk jug that was accidentally left in the refrigerator. The patient thought it was milk and ingested an unknown quantity. The "poison" was later identified as diazinon. The patient was given 0.4 mg of atropine and transferred from the local hospital to a tertiary care center. On admission, the patient was awake but had little spontaneous movement. The patient also presented with pinpoint pupils, diaphoresis, salivation, rhinorrhea, bronchorrhea, gross motor fasciculation, particularly of the lower extremities, and pulmonary edema. The patient was intubated and given atropine doses at 0.02 mg/kg and started on pralidoxime with a loading dose of 25 mg/kg over 30 min, followed by a maintenance dose of 10 mg/kg per hour. The patient showed marked decrease in

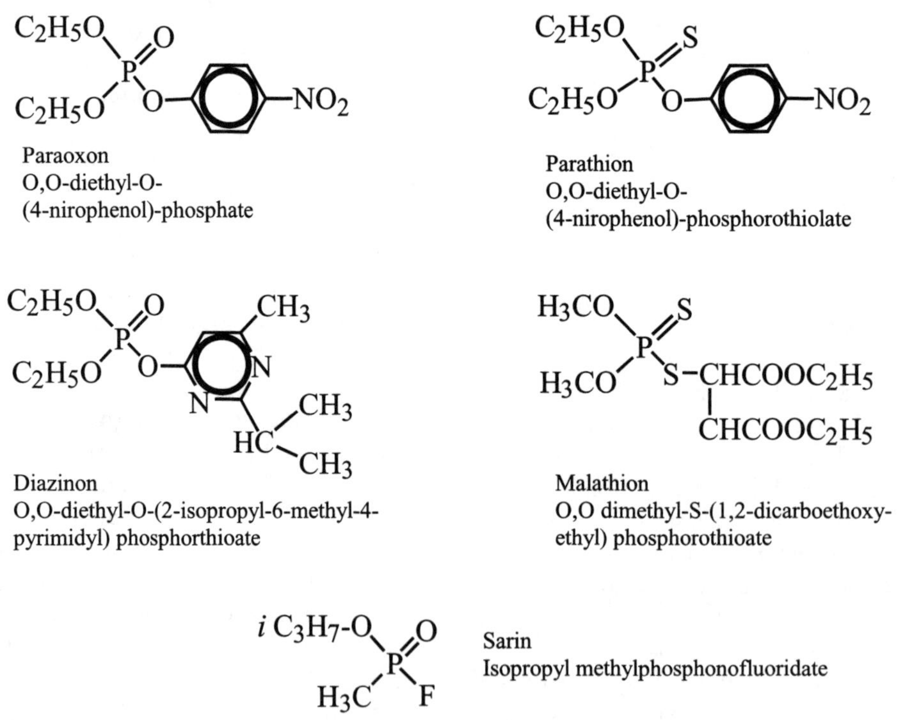

Paraoxon
O,O-diethyl-O-(4-nirophenol)-phosphate

Parathion
O,O-diethyl-O-(4-nirophenol)-phosphorothiolate

Diazinon
O,O-diethyl-O-(2-isopropyl-6-methyl-4-pyrimidyl) phosphorthioate

Malathion
O,O dimethyl-S-(1,2-dicarboethoxy-ethyl) phosphorothioate

Sarin
Isopropyl methylphosphonofluoridate

FIG. 22–2. Examples of anticholinesterase organophosphates.

TABLE 22–1. *Times of intervention and cholinesterase activity results for case study*

	Time (h)	Erythrocyte AChE (U/L; reference range 11,000–15,000)	Plasma ChE (U/L; reference range 3200–6600)
Ingestion	0		
	5.5	4250	–
Atropine, 2-PAM	6		
	19	11,800	832
Extubated	36		
	116	10,290	694

2-PAM, pralidoxime.

pulmonary secretions and oxygen requirement, but still had muscle fasciculation after 6 hours and miosis after 18 hours. Thirty hours after therapy the patient was extubated and discharged from the pediatric intensive care unit. Table 22–1 shows the times of intervention and cholinesterase (ChE) activity results.

TOXICOKINETICS

Toxic exposures to these pesticides can be due to rapid absorption by all routes—respiratory, gastrointestinal, ocular, and dermal. The onset of symptoms is most rapid after inhalation (5). Dermal absorption is slower but can result in severe toxicity if exposure is prolonged and can be enhanced if the agent is lipophilic and helped by the solvent and emulsifier used in the formulation. Oral ingestion is often accidental by children (as exemplified by the 4-year-old in the case presented), but is usually associated with suicide attempts by adults. The pesticides distribute rapidly and accumulate in fat, liver, and kidney.

The phosphothioates (P=S) are more lipophilic than the phosphates (P=O) and are stored extensively in fat. The phosphothioates have to be bioactivated by conversion to their active phosphate (P=O) analogs by oxidative desulfuration mediated by cytochrome P450. This is the reason for delayed appearance of toxic symptoms for the phosphothioate pesticides. Except for the fat-soluble organophosphates (such as fenthion) or those requiring metabolic activation (such as parathion), most patients become symptomatic within 12 hours of exposure (5,6). The patient in the case study ingested diazinon, a phosphoric, not a thiophosphoric, ester, and rapidly exhibited signs of acute organophosphate toxicity in <5 hours after ingestion. Elimination is slow because of extensive fat storage; for the more lipophilic phosphothioates, it can take many days.

PHARMACODYNAMICS

The organophosphates and carbamates are powerful inhibitors of carboxylic ester hydrolases, including acetylcholinesterase (AChE, found in nervous tissues and erythrocytes) and butyrylcholinesterase (plasma or pseudocholinesterase). The toxic effects are almost entirely due to the inhibition of AChE in the nervous system, which leads to an accumulation of the neurotransmitter acetylcholine at synapses and myoneural junctions. The continued stimulation and eventual paralysis of the acetylcholine receptors account for the clinical signs and symptoms of organophosphate poisoning, including muscarinic, nicotinic, and central nervous system effects (2,6). The most common manifestations of muscarinic effects of the parasympathetic nervous system—salivation, lacrimation, urination, diarrhea, gastrointestinal distress, and emesis—can be remembered with the help of the acronym "SLUDGE." Sweating is due to inhibition at the sympathetic postganglionic sites. Other frequently seen muscarinic effects include bradycardia, bronchorrhea, and miosis. Nicotinic effects at the somatic nerve endings include muscle fasciculation, weakness, and paralysis and at the autonomic synapses include hypertension, tachycardia, and dilated pupils (2,6). Central nervous system manifestations include restlessness, headache, drowsiness, confusion, slurred speech, emotional lability, psychosis, ataxia, tremor, delirium, and seizure.

The young patient in the case study experienced all the classic muscarinic and nicotinic symptoms associated with organophosphate poisoning. Many other young patients, however, may have different clinical presentations—the "classic" muscarinic symptoms typically associated with adults may be absent and the primary toxic manifestations are those of the central nervous system (7).

MECHANISM OF TOXICITY

AChE is one of the most efficient enzymes, being capable of an extremely rapid rate of hydrolysis of acetylcholine and regeneration of the active enzyme. Acetylcholine attaches itself to the hydroxyl group of serine residue 203 at the active center of AChE to form an enzyme intermediate (8). Breakdown of this intermediate regenerates an active enzyme and, in the process, acetylcholine is hydrolyzed (Figure 22–3A).

The molecular mechanism of organophosphate and carbamate toxicity lies in the phosphorylation or carbamylation of the same serine residue and the formation of an organophosphoro- or carbamyl-intermediate with AChE (Figures 22–3B and 22–3C). The phosphorylated or carbamylated enzyme is much more stable and has a lower rate of hydrolysis and regeneration of the active enzyme. For some phosphorylated enzymes, the regeneration rate can be so slow that the phosphorylated AChE is essentially inactive (4,8).

Moreover, some of these phosphorylated enzymes might lose an alkyl group over the next 24–48 hours before active enzymes can be regenerated. This process is referred to as "aging" (Figure 22–3B). An aged enzyme is permanently phosphorylated and cannot be

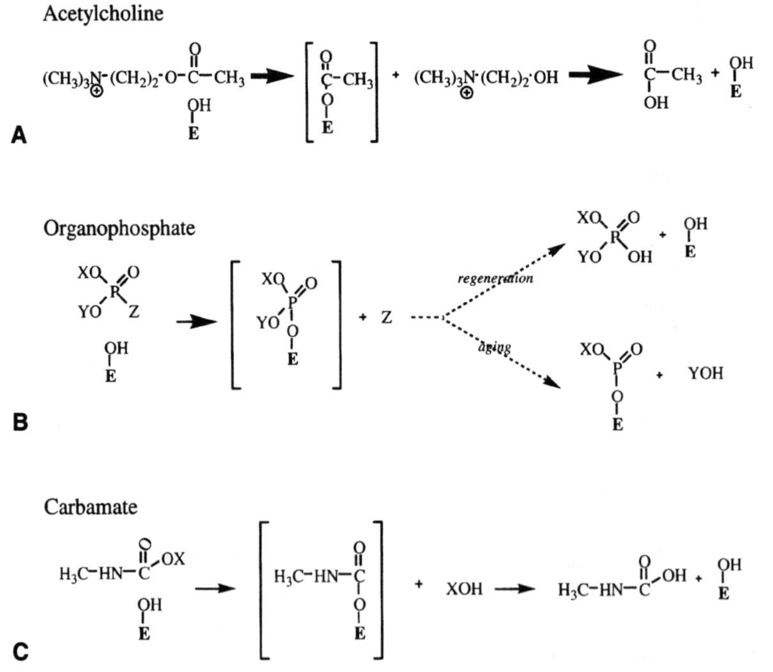

FIG. 22–3. Acetylcholinesterase hydrolysis of acetylcholine and pesticide inhibition of acetylcholinesterase.

regenerated by spontaneous hydrolysis or an oxime antidote (5). The powerful toxic action of a nerve agent such as sarin is due to its very rapid rate of aging, rendering the AChE permanently inactivated very quickly (8). The hydrolysis of the carbamyl-AChE intermediate to regenerate an active enzyme is more rapid. Therefore, carbamate poisoning is not as severe and is self-limiting. Oxime antidote therapy is usually not needed; atropine treatment in such cases is adequate (6).

In general, the chemical structure of the organophosphate (mostly the side groups) determines the affinity to AChE, time for hydrolysis and regeneration of the enzyme, and therefore the toxicity and time of onset of symptoms (4,5).

TREATMENT

The standard treatment of AChE pesticide poisoning entails a two-pronged approach (6,9):

- Atropine, a competitive acetylcholine antagonist, to reverse the biochemical abnormalities at the synapses due to excess acetylcholine
- Nucleophilic oxime, pralidoxime or obidoxime, to regenerate AChE

Pralidoxime (2-hydroxyiminemethyl-1-methyl pyridinium chloride) or Protopam is the only currently available regenerating agent in the United States; obidoxime is available in Europe and other parts of the world (Figure 22–4). Pralidoxime works by its nucleophilic attack on the phosphate moiety of the phosphorylated enzyme, resulting in phosphate removal and regeneration of an active AChE (Figure 22–5). An aged AChE cannot be reactivated by pralidoxime. Therefore, pralidoxime therapy should be initiated as soon as possible after the exposure. The structure of the organophosphate side chain is an important determinant of the rates of enzyme aging and pralidoxime reactivation. For example, the rate of reactivation by pralidoxime was significantly decreased with time for the fast-aging fenitrothion and methylparathion, with no reactivation occurring at 48 hours. In contrast, effective reactivation was still apparent at 48 hours for ethylparathion (4,9).

Pralidoxime works synergistically with atropine in reversing the toxicity of organophosphate. Its action is most dramatic at nicotinic sites, whereas atropine works most effectively with muscarinic sites. The pralidoxime adult dose is 1–2 g given intravenously (i.v.) over a short infusion period of 30

Pralidoxime Chloride
(Protopam)

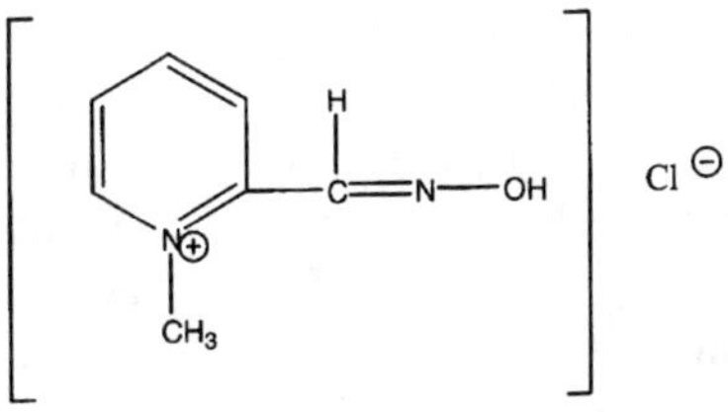

Obidoxime Chloride
(Toxigonin)

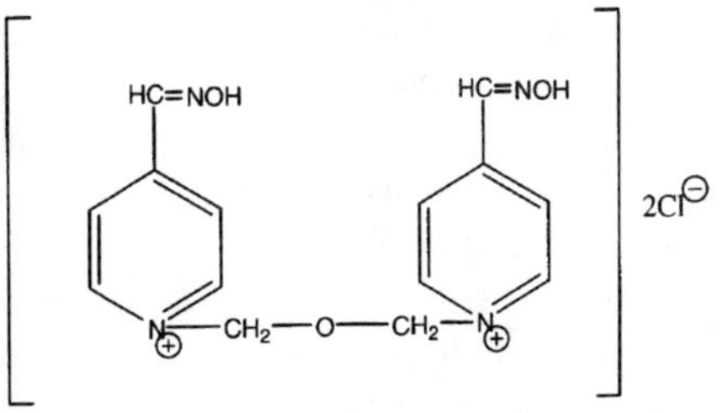

FIG. 22–4. Chemical structures of nucleophilic oximes.

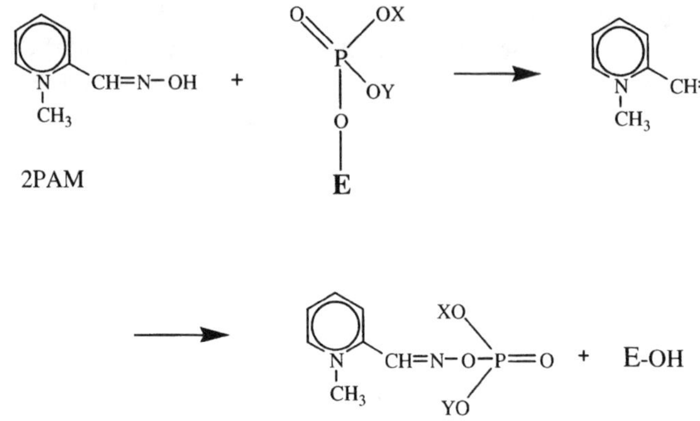

FIG. 22–5. Pralidoxime reactivation of phosphorylated acetylcholinesterase.

minutes and the pediatric dose is 20–40 mg/kg to a maximum of 1 g/dose. Pralidoxime can be given repeatedly, generally 1 g every 6–12 hour for 48 hours (9). The therapeutic endpoint for atropine and pralidoxime is the reversal of muscarinic and nicotinic manifestations. Therapeutic drug monitoring of pralidoxime has not been advocated.

Pralidoxime is excreted renally, with 80% of the dose recovered unchanged in the urine. It has been characterized by a two-compartment model and a steady-state volume of distribution of about 0.8 kg/L in volunteers. Thiamine co-administration has been reported to increase pralidoxime volume of distribution and peak concentration, and to decrease renal clearance, probably due to competition for renal secretion (9). Animal data suggest that the effective plasma concentration is greater than 4 µg/mL (10). In a human study, a loading dose of 4 mg/kg over 15 min followed by a maintenance dose of 3.2 mg/kg per hour for 3.75 hours was able to maintain therapeutic concentrations for 257 minutes, compared with 118 minutes after a short infusion regimen (15 min) of the same dose. Adverse effects, consisting of blurred vision, dizziness, and increased diastolic blood pressure were associated with high peak pralidoxime plasma concentrations after short i.v. infusions (11).

Carbamate poisoning is less severe than that caused by organophosphate exposure. This less severity is due to the more rapid hydrolysis of the carbamyl-AChE intermediate to regenerate an active enzyme. The clinical course of carbamate poisoning is therefore more self-limiting. Oxime antidote therapy is usually not needed for these patients; atropine treatment is adequate (6).

DIAGNOSTIC TESTING, ERYTHROCYTE AND PLASMA CHOLINESTERASE

The targets of organophosphates and carbamate toxicity are the cholinesterases (ChE). Hence the measurement of the erythrocyte AChE or plasma ChE activity is used for the diagnosis and monitoring of poisoning (12). Erythrocyte ChE (E.C. 3.1.1.7 acetylcholine acetylhydrolase) is also known as acetylcholinesterase (AChE) or true cholinesterase. Because it is also found in nervous tissues and skeletal muscle, erythrocyte AChE activity reflects peripheral tissue, muscle, and brain AChE activity and is therefore a more accurate assessment of toxicity than plasma ChE. Erythrocyte AChE activity, typically derived from subtracting plasma ChE activity from whole blood lysate total ChE activity, is generally not readily available in clinical laboratories.

Plasma ChE (E.C. 3.1.1.8 acylcholine acylhydrolase) is also known as pseudocholinesterase, butyrylcholinesterase, and benzoylcholinesterase. It is a liver protein found in plasma, heart, and brain. Its endogenous function is unknown. Plasma ChE is the cholinesterase measured by hospital clinical

$$\text{Butyrylthiocholine} \xrightarrow{H_2O \quad H^+} \text{Butyrate + Thiocholine}$$

$$\text{Thiocholine + DTNB*} \rightleftharpoons \text{Mixed disulphide + 5-mercapto-2-nitro-benzoic acid}$$
$$(410 \text{ nm})$$

*4,4-dithio-2-nitro-benzoic acid (Ellman's Reagent)
Stable at room temperature
Do Not use NaF as anticoagulant

FIG. 22–6. Plasma cholinesterase assay.

laboratories. A typical modern clinical laboratory assay is based on Ellman's reaction—hydrolysis of a thiocholine derivative (such as butyrylcholine) and the reaction of the released thiocholine with a dye to yield a colored product that is monitored photometrically (13,14) (Figure 22–6).

After exposure to a toxic dose of anticholinesterase pesticide, plasma ChE activity is rapidly reduced. First toxic symptoms can appear at 40–50% inhibition and serious neuromuscular effects at 80%. Some organophosphates affect plasma ChE more than they do erythrocyte AChE (such as diazinon, as seen with the patient in the case study); others (such as parathion and nerve gas) inhibit AChE more (15). Untreated patients may see a gradual return to normal activity in 4–6 weeks. Small repeated exposures may gradually reduce ChE activity to very low levels while the patient exhibits minimal symptoms. Thus, very low ChE activity does not always correlate with toxic symptoms. Therefore, almost as important as the degree of inhibition is the velocity at which inhibition occurs. The decrease in erythrocyte AChE and its subsequent normalization is less rapid than that of plasma ChE. Establishment of plateau activity is when the patient's ChE activity is back at base line and, if it is an occupational exposure, when the person can return to work.

Laboratory identification of the organophosphate after acute toxic exposure is not available for diagnostic or management purposes because the target of organophosphate poisoning, the ChEs, can be measured directly. Identification of the organophosphate involved will not change the management of these patients. Moreover, most clinical laboratories lack the advanced analytical capability necessary for organophosphate analysis.

Nevertheless, exposure to organophosphate, particularly for occupational monitoring, can be determined from the presence of urinary alkyl phosphate metabolites. The six primary alkyl phosphate metabolites are dimethyl phosphate, diethyl phosphate, dimethyl thio phosphate, diethyl thio phosphate, dimethyl dithio phosphate, and diethyl dithio phosphate. Exposure to most organophosphates in use will result in the presence of one or more of these metabolites in urine (16,17). The profile of alkyl phosphate metabolites in urine can point to the identity of the parent organophosphate in many instances. Identification by metabolite analysis will not be possible if exposure has been to a mixture of organophosphates. Analytical methods suitable for surveillance or monitoring require detection limits at 10 μg/L or less. Typically, these are methods based on solid-phase extraction, chemical derivation (such as pentofluorobenzylbromide), and gas chromatography using flame-ionization detection (17).

DIAGNOSTIC PROBLEMS OF CHOLINESTERASES

Testing for ChE activity is not available in most hospital-based laboratories. Diagnosis of anticholinesterase pesticide poisoning is, therefore, usually confirmed retrospectively. Additionally, the base-line value of AChE or plasma ChE of an individual is generally not known, and with reference values having a 300% spread, a patient's ChE activity can suffer a 50% reduction and still fall within the reference range. Therefore, an isolated value may neither confirm nor exclude exposure unless it is markedly abnormal. Diagnosis and monitoring are best served by serial determinations (12). Because plasma ChE is a liver protein, it is reduced in malnutrition and liver disease states such as hepatitis, cirrhosis, and neoplasm. Certain drugs such as succinylcholine and lidocaine can depress plasma ChE activity. Individuals carrying the genetic variant $E^a_1 E^a_1$ have reduced plasma ChE concentrations (14). For erythrocyte AChE, low enzyme activity is associated with factors that shorten the circulatory life of erythrocytes, such as hemoglobinopathies.

Although most ChE activity assay kits are based on Ellman's reaction, depending on the modification adopted, different assays (kits) can give very disparate results. A variety of substrates, substrate concentrations, pH, and other assay conditions are used and accounts for the variations. Moreover, a substrate chosen for its specificity for plasma ChE may not be optimal for erythrocyte AChE (15).

NEUROLOGICAL SEQUELAE

In most instances, prompt intervention can reverse the acute toxicity of organophosphate poisoning. Two neurological sequelae after severe intoxication, however, have been described—organophosphate-induced delayed neuropathy (OPIDM) and intermediate syndrome.

OPIDN occurs 1–3 weeks after exposure and significant ChE inhibition (6,18). In such cases the patient usually presents initially with weakness of the lower extremities, progressing to the upper extremities, and ataxia, and eventually paralysis will set in. Patients suffering from mild cases will recover over the next several months; more serious cases will have persistent symptoms.

OPIDN is caused by organophosphates that have high affinity for and inhibit an enzyme, neuropathy target esterase (NTE). All organophosphates in use can cause OPIDN after massive exposure, but none will at concentrations that do not inhibit ChE activity. Inhibition of NTE, however, is not obligatory for OPIDN development. Therefore, the role of NTE in the initiation of OPIDN is not known. NTE inhibition is useful for screening organophosphates for their potential to cause OPIDN.

Intermediate syndrome was first described in 1987 in Sri Lanka (19). Onset of symptoms usually occurs 24–96 hours after resolution of a severe cholinergic crisis and before the development of OPIDN. Initial presentation includes proximal limb weakness, neck flexion weakness, and decrease in reflexes, and the usual cause of death is acute respiratory paralysis. Although atropine and oxime antidote treatments are ineffective, patients can recover if they have not suffered hypoxic brain damage. Intermediate syndrome is associated with certain organophosphates, for example, parathion, methylparathion, diazinon, malathion, and fenthion. The exact cause of the intermediate syndrome is not known, but it has been speculated that it could be due to inadequate oxime therapy during redistribution of tissue organophosphate after initial successful oxime therapy.

REFERENCES

1. Jeyaratnam J: Acute pesticide poisoning: a major global health problem. *World Health Stat Q* 1990; 43:139–144.
2. O'Malley M: Clinical evaluation of pesticide exposure and poisoning. *Lancet* 1997;349:1161–1166.
3. Litovitz TL, Klein-Schwartz W, White S, et al: 1999 Annual Report of the American Association of Poison Control Centers Toxic Exposure Surveillance System. *Am J Emerg Med* 2000;18:517–574.
4. Moretto A: Experimental and clinical toxicology of anticholinesterase agents. *Toxicol Lett* 1998;102–103:509–513.
5. Vale JA: Toxicokinetic and toxicodynamic aspects of organophosphorus (OP) insecticide poisoning. *Toxicol Lett* 1998;102–103:649–652.
6. Aaron CK, Howland MA: Insecticides: organophosphates and carbamates. In: Goldfrank LR, Flomenbaum NE, Lewin NA, et al, eds: *Goldfrank's toxicologic emergencies*, 6th ed. Stamford, CT: Appleton and Lange, 1998:1429–1449.

7. Lifshitz M, Shahak E, Sofer S: Carbamate and organophosphate poisoning in young children. *Pediatr Emerg Care* 1999;15:102–103.
8. Taylor P: Anticholinesterase agents. In: Hardman JE, Limbird LE, eds: *Goodman & Gilman's the pharmacological basis of therapeutics*, 9th ed. New York: McGraw-Hill, 1996:161–176.
9. Howland MA, Aaron CK: Pralidoxime. In: Goldfrank LR, Flomenbaum NE, Lewin NA, et al, eds: *Goldfrank's toxicologic emergencies*, 6th ed. Stamford, CT: Appleton and Lange, 1998:1445–1449.
10. Bokonjic D, Jovanovic D, Jokanovic M, et al: Protective effects of oximes HI-6 and PAM-2 applied by osmotic minipumps in quinalphos-poisoned rats. *Arch Int Pharmacodyn Ther* 1987;288:309–318.
11. Medicis JJ, Stork CM, Howland MA, et al: Pharmacokinetics following a loading plus a continuous infusion of pralidoxime compared with the traditional short infusion regimen in human volunteers. *Clin Toxicol* 1996;34:289–295.
12. Coye MJ, Barnett PG, Midtling JE, et al: Clinical confirmation of organophosphate poisoning by serial cholinesterase analyses. *Arch Intern Med* 1987; 147:438–442.
13. Ellman GL, Courtney KD, Andres VJ: A new and rapid colorimetric determination of acetylcholinesterase activity. *Biochem Pharmacol* 1961;7:88–95.
14. Moss DW, Henderson AR: Clinical enzymology. In: Burtis CA, Ashwood ER, eds: *Tietz textbook of clinical chemistry*, 3rd ed. Philadelphia: WB Saunders, 1999:708–711.
15. Wilson BW, Sanborn JR, O'Malley MA, et al: Monitoring the pesticide-exposed worker. *Occup Med* 1997;12:347–363.
16. Davies JE, Peterson JC: Surveillance of occupational, accidental, and incidental exposure to organophosphate pesticides using urine alkyl phosphate and phenolic metabolite measurement. *Ann N Y Acad Sci* 1997;837:257–268.
17. Moate TF, Lu C, Fenske RA, et al: Improved cleanup and determination of dialkyl phosphates in the urine of children exposed to organophosphorus insecticides. *J Anal Toxicol* 1999;23:230–236.
18. Abou-Donia MB, Lapadula DM: Mechanism of organophosphorous ester-induced delayed neurotoxicity. *Annu Rev Pharmacol Toxicol* 1990;30:405–440.
19. Senanayake N, Karalliedde L: Neurotoxic effects of organophosphorus insecticides. *N Engl J Med* 1987; 316:761–763.

SELF-ASSESSMENT QUESTIONS

1. Which route leads to the most rapid absorption of organophosphates?
   a. inhalation
   b. oral
   c. dermal
   d. ocular

2. Which are clinical manifestations of organophosphates poisoning?
   a. nicotinic effects
   b. muscarinic effects
   c. central nervous system effects
   d. all of the above

3. Which of the following statements about organophosphate toxicity is incorrect?
   a. It causes reversible phosphorylation of acetylcholinesterase and a much reduced rate of spontaneous regeneration.
   b. It causes permanent phosphorylation of acetylcholinesterase.
   c. It causes nucleophilic attack on acetylcholinesterase, causing rapid degradation of the enzyme.
   d. Nervous tissue esterase is inhibited.

4. Which of the following statements is incorrect?
   a. Nervous tissue and plasma cholinesterases are the same.
   b. Erythrocyte acetylcholinesterase is not a liver protein.
   c. Erythrocyte acetylcholinesterase is a better marker for acute organophosphate intoxication than plasma cholinesterase.
   d. Diagnosis of organophosphate poisoning is best confirmed with serial determinations of cholinesterase activity.

5. Which of the following statements is incorrect?
   a. A malnourished patient has reduced plasma cholinesterase activity.
   b. Infants have cholinesterase activity lower than that of adults.
   c. Patients with the $E^u E^u$ genotype for plasma cholinesterase have reduced enzyme activity.
   d. Sodium fluoride is an inhibitor of plasma acetylcholinesterase.

CHAPTER 23

Lead Testing

Petrie M. Rainey, M.D., Ph.D., DABMT

LEARNING OBJECTIVES

After completing this chapter, the reader should be able to:

1. List indications for lead testing.
2. List sources of preanalytical error in lead testing.
3. Describe the pros and cons of lead testing methods.
4. Describe current controversies in lead testing.
5. Suggest future directions of lead testing.
6. Reduce personal exposure to lead.

INTRODUCTION

Lead poisoning has been recognized as a significant medical problem for millennia (see Table 23–1). Ironically, interest in lead poisoning has recently reached a peak, even as average blood lead concentrations in the United States have fallen to their lowest values in decades. Acute, symptomatic lead poisoning has become extremely uncommon. The American Association of Poison Control Centers has reported only two lead-related deaths in the past 10 years, both associated with ingestion of lead-contaminated moonshine whiskey (1). The increased interest has resulted from the recognition that lead may have harmful effects at low concentrations that were previously common and considered safe. The most intensely investigated effects have been the intellectual, developmental, and behavioral effects in young children of blood lead concentrations as low as 10 µg/dL (0.48 µmol/L). The deficits produced, while real, have been too small to demonstrate in individuals. Therefore, the definition of lead poisoning has become one based entirely on the measured blood lead concentration. Laboratories have experienced greatly increased lead-testing volumes. They have also sometimes become victims themselves, not of lead poisoning itself, but of the fear of lead poisoning, which may become the most harmful side effect of a moderately elevated blood lead concentration.

Case Study

PB had a fingerstick lead test done at his 2-year checkup by Dr. Bills, his pediatrician. Three weeks later, the state lab reported a concentration of 57 µg/dL [2.75 µmol/L; 94% of children 1–2 years old have blood lead <10 µg/dL (<0.48 µmol/L)]. Alarmed, Dr. Bills called PB's home immediately, but unsuccessfully. He called again from his own home that evening, reaching a babysitter, who provided the number of the restaurant where PB's parents were having an anniversary dinner. Dr. Bills instructed them to bring the child to the emergency room immediately for a retest.

Your lab receives PB's venous blood specimen at 2 a.m. on Saturday. At 8 a.m., PB's father (a lawyer) calls for the results. The clerk informs him that the next lead run is on Monday. He demands the supervisor, who points out that Centers for Disease Control and Pre-

TABLE 23–1. *Lead poisoning through the ages*

~5000 B.C.	Discovery of cuppelation
? B.C.	"They [Pharaoh's army] sank as lead in the mighty waters." Moses, Exodus 15:10
370 B.C.	Hippocrates reports first case of occupational lead poisoning
A.D. 476	Speculative cause of the fall of the Roman empire (decreased fertility leads to extinction of the ruling class)
1817	"If we were to judge of the interest excited by any medical subject by the number of writings to which it has given birth, we could not but regard the poisoning by lead as the most important to be known." Orfila
1904	Linkage of pediatric lead poisoning to ingestion of leaded paint
1922	Lead paint banned in Queensland, Australia
1970	Lead paint banned for residential use in the United States
1976	Reduction of lead in gasoline begins
1986	Lead solder banned from plumbing
1991	CDC issues recommendations for universal pediatric lead screening
1997	CDC revises guidelines to target selected populations for screening

vention (CDC) guidelines suggest follow-up in 48 hours. The father points out that his child has been poisoned for >3 weeks. You get paged and call in a technologist for a special run. The results are 13 µg/dL (0.63 µmol/L), 9 µg/dL (0.43 µmol/L), and 11 µg/dL (0.53 µmol/L) on three repeats.

PB's father calls that afternoon for the result and is told that results can only be released to the doctor. He asks for the supervisor, who tells him that the doctor is needed to explain elevated concentrations. You are paged again when the father becomes abusive and begins to talk about lawsuits.

Discussion

This hypothetical case is based on several actual cases and illustrates some practical problems associated with pediatric lead screening. These include delays in reporting results, difficulties in following up elevated screening results, false-positives due to redistribution of lead during the time between screening and confirmatory testing, limited assay precision at the 10 µg/dL (0.48 µmol/L) concentration that currently defines "poisoning," and the serious emotional stress that parents may feel when told that their child is poisoned.

The most likely cause for the elevated screening concentration was contamination of the specimen by skin-associated lead. It is also possible, but much less likely, that PB had an acute exposure just before the screening test, and that most of the lead measured in his blood subsequently redistributed to soft tissue and bone during the interval between the screening and confirmatory tests (resulting in a false false-positive). Children with lead-contaminated dust on their skin usually have managed to get lead into their bodies as well. Moreover, children with acute exposures rarely have had only one exposure. In most cases in which the screening concentration is substantially greater than the confirmatory concentration, the latter still exceeds 10 µg/dL (0.48 µmol/L).

Because PB's blood lead concentration was minimally elevated, no formal environmental assessment was undertaken. A follow-up venous concentration measured 3 months later was 8 µg/dL (0.39 µmol/L).

LEAD POISONING IS ELEMENTARY

Lead is a soft, blue-white metal that is malleable, ductile, and a poor conductor of electricity. It is an element with four stable isotopes (Pb 204, 206, 207, and 208), and is found naturally in various minerals, in which it exhibits valences of 2 or 4. The metallic form is readily produced by reductive processes. Because lead is an element, it is not biodegradable.

Lead is a cumulative poison, mainly deposited as the divalent ion in bone, where it is relatively harmless, but serves as a reservoir to maintain tissue concentrations. In the tissues, lead binds to sulfhydryl groups and inhibits the activity of a variety of enzymes. Heme synthesis is decreased by inhibition of δ-aminolevulinic acid (δ-ALA) dehydratase (porphobilinogen synthetase) and by impairment of iron incorporation into protoporphyrin. At higher lead concentrations, renal function is impaired by mechanisms that remain unclear. Saturnine gout (alchemists associated lead with the planet Saturn) results from inhibition of uric acid clearance. In addition, some researchers speculate that lead may alter neurological function by affecting calcium-dependent signal transduction. This may be a mechanism underlying the behavioral and mild hypertensive effects seen at relatively low blood lead concentrations.

Only 5–10% of ingested lead is absorbed in adults. However, children have increased needs for the divalent cations iron and calcium and may absorb as much as 50% of ingested divalent lead. Diets poor in iron and calcium promote increased absorption. When inhaled in forms that reach the alveoli (for example, lead vapor), lead is efficiently absorbed by both adults and children, with bioavailability of 50–70%.

Absorbed lead is initially found in the blood compartment, primarily bound to red cells. For this reason, blood lead concentrations are the most responsive to recent exposures. Lead then redistributes to the soft tissues. The apparent elimination half-life of lead from blood is 30–50 days in adults (Figure 23–1). This probably reflects both elimination (urinary, fecal, shed skin, etc.) and redistribution to bone. Soft-tissue half-life is 30–60 days.

Lead is initially deposited primarily on bone surfaces and in trabecular bone, where it has a half-life of 3–5 years. The elimination half-life from cortical bone is very long, on the order of 10–20 or more years. Bone half-lives are probably shorter in children as a result of the extensive bone remodeling that accompanies growth. When current lead intake is lower than previous exposure (as is the case for those who lived during the era of leaded gasoline), lead released from bone may contribute substantially to the circulating lead pool. Increased bone resorption, such as occurs with pregnancy, lactation, or immobilization (for example, after a fracture), may lead to increased blood lead concentrations.

LEAD POISONING TODAY

Symptomatic lead poisoning is uncommon today. Instead of being diagnosed by signs or symptoms, lead poisoning is defined by blood

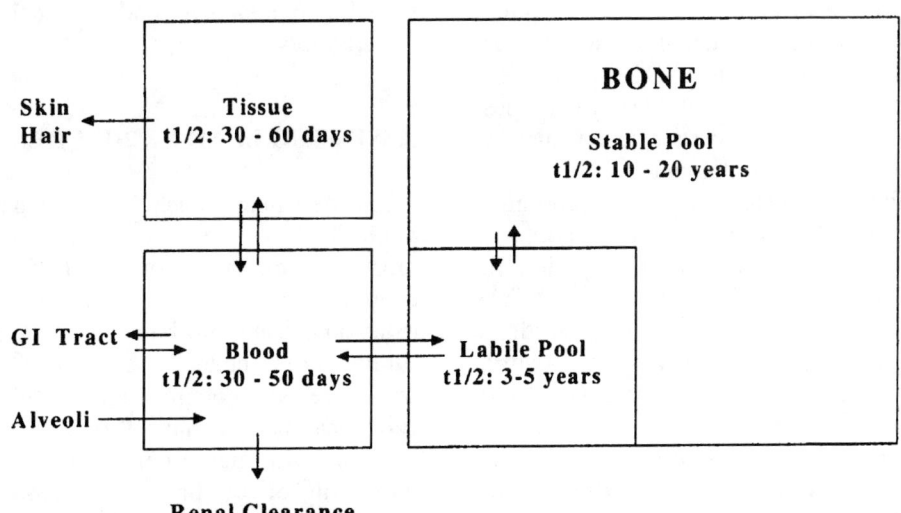

FIG. 23–1. Lead toxicokinetics. GI, gastrointestinal; t1/2, half-life.

lead concentrations. The definitions of lead poisoning differ substantially for the two populations at greatest risk, namely, young children and occupationally exposed adults.

Young children are especially at risk for lead poisoning. In its recommendations for preventing lead poisoning in young children, the CDC defines lead poisoning in children as any confirmed blood lead concentration of ≥10 µg/dL (≥0.48 µmol/L) (2). The primary source of excessive exposure for young children is not from eating lead paint chips, as is commonly believed, but from chronic ingestion and inhalation of lead-contaminated dust. Being low to the ground, where dust accumulates, and engaging in extensive hand-to-mouth activities greatly increase ingestion of lead-containing dust. Ingested lead is more efficiently absorbed in young children, especially those who are iron- or calcium-deficient. Once absorbed, lead more readily crosses the incomplete blood-brain barrier and has greater effects on the still-developing nervous system.

House dust produced by the weathering and wearing of exposed lead paint is the most common source of lead exposure in children, but not the only one. Lead in the soil and in the water supply also contribute to the overall lead burden (3). Children may also be exposed to lead brought home by parents who are lead workers, or to lead in second-hand cigarette smoke (a typical cigarette contains ~1 µg of lead). There are numerous other less common sources, including lead solder in cans and utensils originating outside the United States, improperly glazed ceramics, folk remedies, and some calcium supplements (bone meal or dolomite).

Significant adult lead poisoning is confined almost entirely to those who are exposed in their occupations or hobbies. Occupations at particularly high risk include battery manufacturing and recycling, smelting, welding, printing, glass making, and stained glass work. Painters whose jobs include removal of old paint are also at high risk. The Occupational Safety and Health Administration (OSHA) requires that workers with blood lead concentrations >60 µg/dL be removed from active exposure, as must workers with lead concentrations repeatedly >50 µg/dL. Exposure may not resume until two successive concentrations have been <40 µg/dL (4).

This rule should not be taken to indicate that lead concentrations <40 µg/dL (<1.93 µmol/L) pose no risks to adults, but rather that the risks of higher lead concentrations are unacceptable. The World Health Organization (WHO) has set a limit of 30 µg/dL as the maximum acceptable concentration for adults.

Although symptomatic lead poisoning is uncommon, it is useful to recognize the signs and symptoms, because not all poisonings are detected by pediatric and occupational screening programs. Acute lead poisoning typically presents with severe abdominal pain and mental confusion or deterioration in adults. Acute encephalopathy is typically seen only in children, who may present with persistent vomiting, ataxia, seizures, and coma. In chronic lead poisoning, adults will experience abdominal pain, fatigue, neuropsychiatric symptoms, and constipation, as well as mild hypertension. More serious exposures may present with peripheral neuropathy, especially wrist drop, and saturnine gout. Anemia is relatively uncommon in adults. Although not a typical presenting symptom, decreased fertility is also a result of chronic lead poisoning. In children, signs and symptoms of chronic lead poisoning include anemia, abdominal pain, developmental delay, hyperactivity, and other behavioral disturbances.

ASSESSING LEAD EXPOSURE

A variety of approaches have been used to assess lead exposure. By far the most common is the direct measurement of the lead content of various bodily specimens including blood, urine, and hair. The in vivo measurement of bone lead content by X-ray fluorescence is becoming more widely used. Biological markers have also been used. At one time, the measurement of zinc protoporphyrin or erythrocyte protoporphyrin was the favored screening test for pediatric lead poisoning. δ-ALA, which can increase as the result of inhibition of δ-ALA dehydratase, has also been proposed, particularly for assessing occupational exposure.

The measurement of lead in whole blood specimens is the most widely used method today for assessing lead exposure. The correlation of blood lead concentrations with biological effects is also the best understood. Because the blood compartment is the first one traversed by absorbed lead, blood lead best reflects recent exposures. In the absence of acute exposure, blood lead largely reflects the overall lead burden in the body. A disadvantage of blood lead measurement stems from the fact that ~99% of blood lead is associated with the erythrocytes. In persons with comparable lead burdens, the blood lead concentration will be largely proportional to the hematocrit. The use of serum or plasma lead concentrations can avoid the hematocrit bias, but lead concentrations in this compartment are too low to be accurately measured by any widely available technique.

Urine lead concentrations also reflect a composite of recent exposure and total body burden, but are somewhat less affected by recent exposure than blood lead concentrations. Urine lead may be preferable for occupational monitoring because specimens may be obtained noninvasively. A disadvantage of urine samples is that lead concentration is flow dependent. To adjust for this, either 24-hour urinary lead elimination should be measured or, if spot urines are used, the lead concentration should be normalized using the urine creatinine concentration.

Urine lead testing was also previously used in the lead mobilization test. This test (which is currently out of favor) evaluated the increase in urinary lead excretion in response to a test dose of a chelating agent. This was suggested as an indicator of children with blood lead concentrations <45 μg/dL (<2.17 μmol/L) who would benefit from chelation therapy. Recent findings suggest that chelation therapy is no more beneficial than simple removal from exposure in children with blood lead concentrations <45 μg/dL (<2.17 μmol/L) and have led the American Academy of Pediatrics to declare the lead mobilization test obsolete (5).

Hair has also been used to assess lead exposure. Hair testing offers the advantage of being noninvasive and may also allow the time of an acute exposure to be estimated by measurement of lead in serial segments. A major disadvantage of measuring lead in hair is the problem of external contamination. Poor interindividual correlation with blood concentrations and lack of well-accepted normative data make interpretation difficult.

Capillary blood lead samples have become the preferred specimens for pediatric lead screening because they are much easier to obtain from young children than venous specimens. Previously, there was some concern that contamination problems would be common with this approach. However, several studies have now shown excellent correlation with venous blood concentrations, provided that careful hand washing and collection techniques are followed. Nonetheless, false-positives do occur with this technique. It is recommended that all positive results [blood lead ≥10 μg/dL (≥0.48 μmol/L)] be confirmed with a venous specimen.

Although the most common cause for falsely elevated results is contamination of the specimen by environmental lead on the subject's hand, most subjects with lead on their hands also have excessive lead in their blood and are still "positive" under the screening definition. Contamination of the collection area can lead to falsely elevated concentrations in children with low true values. Every effort should be made to avoid false-positive results because of the significant anxiety that parents may experience during the interval between the false-positive result and the negative confirmatory test. Because a substantial number of children have lead concentrations in the region of 10 μg/dL (0.48 μmol/L), laboratory imprecision may also account for a number of false positives.

An interesting phenomenon is the false false positive. This occurs when there is a significant interval between the screening test and the confirmatory test. Because the half-life of lead in blood is probably <30 days in children, sporadic exposure may result in a true-positive concentration at the time of screening that has declined to a true-negative concentration by the time of confirmation.

LEAD-TESTING METHODS

Numerous methods have been developed for measuring blood lead concentrations or other indicators of lead exposure, as reviewed below. Currently, electrothermal atomic absorption spectroscopy is the most popular technique for laboratories with larger testing volumes, whereas anodic stripping voltammetry is more widely used in lower volume settings. Properly performed, both have adequate precision to meet current regulatory requirements [that is, to successfully pass proficiency testing with an error of no more than 4 µg/dL or 10%, whichever is greater, as mandated by the Clinical Laboratory Improvement Amendments (CLIA 88)]. In many states, low-cost lead testing may be provided by the state health department laboratory, although turn-around times may be long (≥ 1 week). The introduction of inexpensive point-of-care testing devices promises to revolutionize lead screening.

Atomic Absorption

Atomic absorption methods have long been a mainstay for the measurement of heavy metals, including lead. The basic technique involves atomization of lead at high temperature, either in a flame or in a furnace. At elevated temperature, some portion of the lead will exist in a nonionized atomic form [Pb(0)]. This atomic lead will absorb light at several characteristic wavelengths. Light at these wavelengths may be generated in a lamp that uses the discharge from the vaporized metal. Subsequently, one characteristic wavelength (usually 283.3 nm) is selected by a monochromator. When this light is passed through the atomized lead, light will be absorbed in proportion to the lead concentration. Because light intensity may also be attenuated by nonspecific scattering, techniques must be used to correct for this effect.

Flame Atomic Absorption

In the past, lead was commonly measured by flame atomic absorption techniques. A widely used technique involved extraction of the lead by complexation with either ammonium pyrrolidine dithiocarbamate or diethyldithiocarbamate dissolved in methyl isobutyl ketone. This served both to concentrate the lead and to separate it from potentially interfering substances. The extracts burned cleanly in the flame, thereby minimizing background scattering. A more direct method involved use of the Delves cup (6). In this technique, a dry specimen was introduced directly into the flame. Background could be minimized by ashing techniques before analysis. However, the most common use of this technique in lead testing involved direct analysis of dried blood spots on filter paper. This allowed extremely rapid testing because no preliminary sample preparation was required. However, background scattering was high, leading to poor precision and necessitating multiple determinations on the same specimen. Although this technique is still in use in a few laboratories, concerns about precision and increased likelihood of specimen contamination have resulted in recommendations against its use. This controversy may soon fade, because equipment for Delves cup analyses is no longer being made.

Electrothermal Atomic Absorption

The development of flameless atomic absorption techniques (7,8) represented a major advance in lead analysis. Rather than using a flame, lead is vaporized in a hollow graphite tube heated by an electric current (that is, electrothermal). In addition to minimizing fluctuations inherent in a flame, the confinement of the vapor to the volume of the tube allowed use of much smaller samples. And stepwise heating protocols reduced smoke formation and resulting nonspecific light scattering.

In the traditional design, electrodes surround either end of the graphite tube, providing longitudinal current flow. The heat sink provided by the electrodes results in more rapid heating of the middle of the tube than the ends. This problem may be overcome through the use of current flow perpendicular to the axis of the tube, although this design is technically more difficult. Temperature uniformity may be enhanced by the use of a stabilized temperature platform. This is a platform within the graphite tube on which the specimen is placed. The platform itself is heated not by current flow but by radiant heat from the surrounding graphite tube. This results in more homogenous temperature development and a very short duration for the vaporization phase.

Another advantage of the graphite furnace is that heating may be done in distinct steps. The tube is initially heated gently (100–200 °C) to dry the specimen. A pyrolysis, or charring, step occurs at a temperature just below that at which lead begins to vaporize. This step removes much of the volatile organic material that would otherwise create a light-scattering smoke. To allow the charring step to occur at as high a temperature as possible, modifiers that decrease the volatility of the lead are often added to the specimen. After charring, a cool-down step may be introduced. Then the temperature is rapidly increased to one at which lead fully and rapidly vaporizes (≥ 1500 °C), and the measurement is made during this step. A final temperature increase cleans the furnace for the next specimen (2400–2500 °C).

Even with the above modifications, vaporization of the lead occurs over a finite time interval. To capture the maximum signal and to avoid variations caused by different vaporization profiles, the light attenuation signal is integrated over an interval wider than the vaporization time. This allows the use of aqueous calibrators instead of matrix-matched calibrators. Although the vaporization profiles are different for aqueous and whole blood specimens, the integration step results in comparable signals from both types of specimens.

Although the pyrolysis step removes much of the background due to light scatter by nonlead "smoke," a background correction is still necessary. One approach has been to use light at a second wavelength not absorbed by lead (for example, from a deuterium lamp) as a measure of nonspecific light scattering. A more sophisticated approach uses the Zeeman effect: Application of a strong magnetic field changes the energy level of the orbitals in the lead atom and shifts the wavelength of the absorption peak. Turning the magnetic field on allows measurement of the light scattering alone using the primary wavelength. By rapidly turning on and off the magnetic field, both total absorption and background absorption tracings can be generated and the background subtracted from the total absorption to give the lead-specific absorption.

Anodic Stripping Techniques

Another approach to lead testing involves anodic stripping methods (9). These techniques have the advantage relative to atomic absorption techniques of requiring much less expensive equipment. A disadvantage is the requirement for specimen pretreatment by ashing or by treatment with a decomplexation agent before analysis. The initial step in anodic stripping methods involves removing and concentrating the lead from the specimen by reductive plating onto an electrode, most typically to form an amalgam in a thin film of mercury on a carbon electrode. The lead is then reoxidized and measured during the oxidation step. In anodic stripping voltammetry, an increasing voltage is applied to the electrode after the plating step. Lead reoxidizes at a characteristic voltage, and the integrated current flow during the reoxidation is a measure of the amount of lead present.

In anodic stripping potentiometry, a constant current is applied to the electrode. In the absence of oxidizable species, this results in an increasing voltage. When the oxidation potential of lead is reached, the current is then carried by the oxidizing lead, resulting in a plateau in the voltage. The current flow resulting from lead oxidation, conveniently measured as the duration of the plateau, is a measure of the lead present.

Inductively Coupled Plasma Mass Spectroscopy

Inductively coupled plasma mass spectroscopy (10) is a rapid and accurate method for measurement of heavy metals with low reagent costs. However, the required instrumentation is quite expensive and can generally be justified only for laboratories with a high volume of heavy metal testing. The initial step is atomization of the specimen in a plasma (an ionized gas) generated from a carrier gas by electrical induction. Introduction of specimen into the high-temperature plasma results in breakdown to basic atomic and ionic species. The plasma is then sampled into a mass spectrometer that has been programmed to detect ions at the characteristic mass-to-charge (m/z) ratios of the natural isotopes of the metal being measured. Because multiple ions may be determined in rapid sequence, this technique allows the simultaneous quantification of multiple heavy metals in a specimen. In addition, the ratio of the lead isotopes in the sample may be determined and can be used as a signature, suggesting the possible source of the lead.

For greater accuracy, isotope dilution mass spectroscopy may be used (11). In the case of lead, a known amount of the rare isotope Pb^{204} is introduced into the specimen as an internal standard. The more common lead isotopes are then measured based on their ratio to Pb^{204}.

Neutron Activation Analysis

In neutron activation analysis, the specimen is bombarded with a high flux of neutrons. Nuclear capture of some of these neutrons results in the generation of radioactive isotopes of the element to be analyzed. The type and energy of the resulting emitted radioactivity serve to identify the source, and the amount provides quantification. Because this technique requires a high-intensity neutron source, its use is primarily limited to research applications.

X-Ray Fluorescence

X-ray fluorescence is similar to light fluorescence, except that the excitation and emission wavelengths lie within the X-ray region of the electromagnetic spectrum. Because of the ability of X-rays to readily penetrate tissue, this technique can be used to measure lead concentration in vivo. In particular, it can be used to assess the lead content of bones (12). It is relatively inconvenient in that measurement times are long (≥ 30 minutes). Nonetheless, it is being increasingly used as a way of assessing total body lead burden, particularly in a research setting. Two types of X-rays are used. L X-rays penetrate only the top 1–2 mm of bone, yielding values that correspond with recently deposited (and possibly chelatable) lead. K X-rays penetrate 2–4 cm, with results that are thought to correspond with whole body burden.

Biological Markers

Zinc protoporphyrin (ZPP) is an alternate end product of the heme biosynthetic pathway that is formed when iron is not sufficiently available. (Heme is iron protoporphyrin.) Concentrations of ZPP are increased in erythrocytes formed at elevated blood lead concentrations. This process has often been attributed to lead inhibition of ferrochelatase, the enzyme that inserts iron into protoporphyrin. However, ZPP is also formed by the action of ferrochelatase, suggesting that some other mechanism is operative (13). Because ZPP is fluorescent, it may be measured directly in intact erythrocytes. This allows an extremely rapid measurement and explains the popularity that ZPP measurements once had as a screening test for pediatric lead exposure. ZPP is no longer recommended as the primary screening test because ZPP concentrations consistently exceed the upper limit of normal only at lead concentrations >35 µg/dL (1.69 µmol/L).

ZPP also becomes elevated when the final step of heme synthesis is slowed by iron deficiency.

δ-ALA is a heme precursor that undergoes buildup after lead inhibition of the zinc metalloenzyme δ-ALA dehydratase (porphobilinogen synthetase). Because δ-ALA is excreted in urine and can be measured noninvasively, it has been suggested as a marker for occupational lead exposure.

CURRENT CONTROVERSIES

Pediatric Lead Poisoning—Epidemic by Edict?

In 1991, the CDC issued new guidelines for the prevention of pediatric lead poisoning. These guidelines contain several recommendations (Table 23–2), among which are a call for universal lead screening of children <6 years old and a tiered definition of lead poisoning that specifies varying interventions based on blood lead concentration (Table 23–3) (2). Although these were only recommendations, they became the basis of regulations in several states.

Objections were raised to the recommendation for universal screening on the assertion that pediatric lead poisoning was uncommon in many areas, particularly areas with few older houses. At the time, there were relatively few hard data to support or deny those assertions.

Objections were also raised to the poisoning threshold being set too low at 10 μg/dL. It was noted that in the era of leaded gasoline, the average lead concentration in children was 16 μg/dL (0.77 μmol/L), with little evidence of significant harm. It was further argued that there was no evidence to support the specific choice of 10 μg/dL, and that using this definition was directly harmful in producing unnecessary worry in parents of children with lead concentrations between 10 and 20 μg/dL (0.48 and 0.97 μmol/L), which many felt to pose an insignificant risk.

Proponents of the limit argued that lead has no beneficial function and that lead had been shown to impair development and intelligence in children with no apparent lower concentration threshold. The most desirable lead concentration would be zero, but 10 μg/dL (0.48 μmol/L) represented an achievable target.

At the time of the recommendations, the impairment of intelligence was generally accepted, but the extent of the impairment remained debatable. An important issue that was largely overlooked was that there was no evidence at the time that reducing modestly elevated lead concentrations in any way reversed any effects that might have been produced. The extent of the disagreement was reflected in the observation that only ~50% of pediatricians adhered to the CDC guidelines.

Since the time of these guidelines, new data have clarified some of the previous uncertainties. Broad-based surveys revealed that the average blood concentration in children had fallen dramatically before issuance of the guidelines and fell further thereafter, to 2.7 μg/dL (0.13 μmol/L) in 1991–94 (14,15) and to 2.0 μg/dL (0.10 μmol/L) in 1999 (16). Pediatric lead poisoning was indeed largely confined to specific areas and populations. Studies were done that suggested that reduction in lead concentrations did result in improved intellectual functioning and developmental catch-up (17). The extent of the effect of low lead concentrations was better quantified, with a meta-analysis of a large number of studies suggesting an average loss of 1–3 intelligence quotient (IQ) points with an increase of blood lead concentration from 10 to 20 μg/dL (0.48 to 0.97 μmol/L) (18). Other studies suggested that modestly elevated lead concentrations had more pronounced effects on behavior than on IQ, par-

TABLE 23–2. *1991 CDC childhood lead poisoning guidelines: major recommendations*

- Primary prevention of lead poisoning
- Universal lead screening for children <6 years old
- Screening by direct measurement of blood lead
- Multiple action levels
- Oral chelation therapy with succimer

TABLE 23-3. *1991 CDC childhood lead poisoning guidelines: action concentrations*

<10 μg/dL	Not considered to indicate lead poisoning
10–14 μg/dL	A high prevalence of children with concentrations ≥10 μg/dL should trigger community-wide lead poisoning prevention. Individual children should be retested more frequently.
15–19 μg/dL	Parents should receive education on preventing lead poisoning and on nutrition. Children should be retested more frequently; if concentrations persist, sources of lead exposure should be investigated.
20–44 μg/dL	Full medical evaluation is indicated. Sources of lead exposure should be removed from the child's environment. Drug therapy may be indicated.
45–69 μg/dL	Chelation therapy is indicated. The child should be removed to a lead-free environment and not returned until sources of lead exposure have been eliminated.
≥70 μg/dL	A medical emergency requiring immediate hospitalization and chelation therapy.

ticularly in terms of incidence of aggressive behavior, poor classroom performance, and attention-deficit disorder.

In late 1997, the CDC issued revised guidelines calling for continued universal pediatric lead screening in communities where past screening suggested that ≥12% of children had blood lead concentrations ≥10 μg/dL or where ≥27% of houses were built before 1950. In areas with lower demographic risk, selective screening was recommended, targeting only children with specific personal risk factors. Other recommendations were essentially unchanged (19). Shortly thereafter, the American Academy of Pediatrics issued guidelines that differed only slightly from the CDC recommendations (20).

Does Chelation Help Asymptomatic Poisoning?

There is a consensus that chelation therapy is indicated in all symptomatic lead-poisoned patients, as well as those with blood lead concentrations ≥45 μg/dL (2.17 μmol/L) (2,5,19,20). This consensus presumably reflects the observations that symptoms largely result from the action of lead in the soft tissues and that chelation rapidly reduces soft-tissue lead concentrations. However, considerable controversy exists as to whether chelation is appropriate in asymptomatic children with blood lead concentrations between 25 and 45 μg/dL (1.21 and 2.17 μmol/L). In the past, a lead mobilization test was recommended to identify children who could theoretically benefit from chelation therapy. A test dose of ethylenediamenetetraacetic acid (EDTA) was given, and the amount of lead excreted in the urine in response to this test dose was measured. Children who mobilized significant amounts of lead were considered likely to benefit from chelation therapy, although no outcome data were ever collected to support this theory.

Recent studies have tested the effectiveness of chelation in children with blood lead concentrations in the 25–45 μg/dL (1.21–2.17 μmol/L) range. These studies compared the effect of chelation, coupled with removal from ongoing lead exposure, with the effect of removal from exposure only. Although chelation produced a more rapid decline in blood lead, there was no difference in the amount of decline after several months. This was true even when the chelated group was chosen based on a positive lead mobilization test (21,22). The American Academy of Pediatrics currently does not recommend chelation for children with lead concentrations <45 μg/dL (5).

LEAD TESTING IN THE 21ST CENTURY

Several trends in lead testing may have a significant effect on the amount of lead testing being done in clinical laboratories. The first is the CDC's decision to recommend targeted rather than universal pediatric lead screening in low-risk communities. In areas where there are few children with blood lead concentrations ≥10 μg/dL (≥0.48 μmol/L), universal lead screening is not cost effective. Therefore, the CDC has proposed to focus and intensify efforts in areas and populations that continue to have a significant number of children with elevated blood lead concentrations. The recommendations direct public health officials to determine regions in which universal pediatric lead screening should continue. In addition, screening is recommended for children living in low-prevalence areas who have personal risk factors for increased lead exposure (19). Laboratories serving regions with a low prevalence of elevated blood lead concentrations may have experienced a substantial drop in their testing volume.

A second major trend will be the introduction of lead screening with point-of-care testing devices. This approach will offer significant advantages over lead screening using central laboratories. The most significant will be the ability to obtain the screening result while the child is still in the clinic, thereby allowing the confirmatory venous specimen to be drawn at the same visit. (The confirmatory specimen should be tested at the point of care to rule out false-positives due to contamination of the capillary screening specimen, and to initiate immediate interventions after verification of substantially elevated results. It may additionally be sent to a central laboratory for a definitive measurement with higher-precision methodology.)

It is likely that several instruments may soon be available for point-of-care lead screening. When the CDC issued its 1991 guidelines, it also made available grants for the development of portable, low-cost lead-screening devices. Devices under development use several techniques including anodic stripping, electrothermal atomic absorption spectroscopy, plasma atomic emission spectroscopy, and a novel approach using a membrane with lead-specific binding sites. One instrument has already received Food and Drug Administration approval and is commercially available. It uses anodic stripping with a disposable colloidal gold electrode mounted on a plastic strip and can give results in 5 minutes. Thus, even laboratories serving high-risk populations may see their screening testing volume move to the clinic.

Another trend is the effort to identify and abate potentially hazardous housing prospectively. Currently, most hazards are identified only after a child has developed an elevated lead concentration. Efforts are underway to develop models that can predict the risk of developing elevated blood lead concentrations from measurements of the lead content of surface dust collected by wipe testing (23). Should these efforts be successful, the bulk of lead testing may move to environmental laboratories.

A trend that may have a future effect on laboratories is the current effort to identify genetic determinants of susceptibility to lead poisoning. It is important to recognize that exposure effects identified in a population may present in individuals with greater or lesser intensity than the population average (24,25). This variability undoubtedly includes a component of genetic susceptibility. For example, persons homozygous or heterozygous for the variant δ-ALA dehydratase gene, $ALAD_2$, tend to have higher blood lead concentrations, due to the increased affinity for lead of the variant gene product (δ-ALA dehydratase is the major lead-binding protein in erythrocytes). But they have lower zinc protoporphyrin concentrations, suggesting that the increased binding of lead decreases the toxicologically active fraction and results in a net protective effect (26). In the future, laboratories may include phenotypic or genotypic susceptibility markers in a lead screening panel.

LEAD TESTING AND YOU

Guidelines have been issued for appropriate lead concentrations in children and occupationally exposed adults, but what about adults without occupational exposure? Should OSHA guidelines be considered appropriate for them as well? Certainly not. OSHA concentrations represent not an absence of risk but rather absence of an unacceptable risk. In persons not occupationally exposed, there are no benefits to offset possi-

TABLE 23–4. Blood lead concentrations in the United States, 1991–1994

Age (y)	Geometric mean (μg/dL)	% with <10 μg/dL
1–2	3.1	94.1
3–5	2.5	96.5
6–11	1.9	98.0
12–19	1.5	99.2
20–49	2.1	98.5
50–69	3.1	97.1
≥70	3.4	95.4
Total	2.3	97.8

To convert lead values in μg/dL to μmol/L, multiply by 0.0483. Data from Centers for Disease Control and Prevention: Update: blood lead levels—United States, 1991-94. *MMWR Morb Mortal Wkly Rep* 1997;46:141–146.

ble risks. Most adults can and should have much lower concentrations. Lead has no known function in the body. Moreover, there is no apparent threshold below which lead has no effect. It is prudent to minimize exposure regardless of your blood lead concentration. Thanks to the removal of lead from gasoline, most adults who are not occupationally exposed currently have lead concentrations <10 μg/dL (<0.48 μmol/L) (Table 23–4).

The most common source of lead exposure for adults as well as children is lead-contaminated dust. The best way to avoid exposure is to live in a new house with no lead in the paint, no lead in the pipes, and no lead in the yard. However, lead exposure can be kept at an acceptable level even in older houses with all of these lead sources. Removal of lead paint can be very effective but carries an extremely high cost and may worsen the problem if done incorrectly. Studies have shown that dust control is effective in lowering lead concentrations in children with high concentrations [>20 μg/dL (>0.97 μmol/L)]. It is less effective in children with lower concentrations, whose exposure may be through multiple sources.

The basic principles of dust control are to cover it up or wash it away. Covering lead paint with wallpaper or a coating of lead-free paint is effective in areas free of friction and abrasion. Dust created by abrasion of lead paint on moving surfaces can be readily removed by wet cleaning with high-phosphate detergents focused on areas where such dust is likely to collect (window sills and wells, and floors, especially around windows and doors). Encouraging children to wash and dry their hands and faces before eating can also substantially reduce ingested lead burden. For houses with external lead paint, soil removal has been shown to lack substantial effectiveness. Planting shrubbery to discourage play along exterior walls may be equally effective and is much less expensive.

Another source of lead exposure is drinking water. Lead is primarily introduced into the water within the house as a result of contact with leaded solder in the plumbing. The amount of lead leached from the solder increases with water temperature and with contact time. Lead exposure from this source can be minimized by using only cold water for cooking and drinking and by flushing standing water from the pipes before using water in the morning. The use of tobacco and alcohol has been associated with increased blood lead concentrations that correlate with amount of use. Some hobbies also increase lead exposure, including visiting shooting ranges.

Some nutritional interventions may help reduce lead exposure. Diets high in calcium and iron reduce lead absorption, particularly in children. (Note: a high iron intake may be harmful for adults.) Low-fat diets also decrease absorption of ingested lead. Serum concentrations of vitamin C have recently been shown to correlate with lower blood lead concentrations. There is also limited data to suggest that vitamin C may promote lead excretion (23).

REFERENCES

1. Litovitz TL, Klein-Schwarz W, White S, et al: 1999 annual report of the American Association of Poison Control Centers Toxic Exposure Surveillance System. *Am J Emerg Med* 2000;18:517–574.
2. Centers for Disease Control: Preventing lead poisoning in young children. Atlanta: United States Department of Health and Human Services, Public Health Service, 1991. (Available from Publication Activities, National Center for Environmental Health and Injury Control, Mailstop F-29, CDC, 1600 Clifton Road, NE, Atlanta, GA 30333.)

3. Lanphear BP, Matte TD, Rogers J, et al: The contribution of lead-contaminated house dust and residential soil to children's blood lead levels. A pooled analysis of 12 epidemiologic studies. *Environ Res* 1998;79:51–68.
4. Rempel D: The lead-exposed worker. *JAMA* 1989;262:532–534.
5. American Academy of Pediatrics Committee on Drugs: Treatment guidelines for lead exposure in children. *Pediatrics* 1995;96:155–160.
6. Delves HT: A microsampling method for rapid determination of lead in blood by atomic absorption spectroscopy. *Analyst* 1970;95:431–438.
7. Jackson KW, ed: Electrothermal atomization for analytical atomic spectrometry. New York: John Wiley, 1999.
8. Parsons PJ, Slavin W: A rapid Zeeman graphite furnace atomic absorption spectrometric method for determination of lead in blood. *Spectrochim Acta Part B* 1993;48:925–939.
9. Roda SM, Greenland RD, Bornschein RL, et al: Anodic stripping voltammetry procedure modified for improved accuracy of blood lead analysis. *Clin Chem* 1988;34:563–537.
10. Nuttall KL, Gordon WH, Ash KO: Inductively coupled plasma mass spectrometry for trace element analysis in the clinical laboratory. *Ann Clin Lab Sci* 1995;25:264–271.
11. Paschal DC, Caldwell KL, Ting BG: Determination of lead in whole blood using inductively coupled argon plasma mass spectrometry with isotope dilution. *J Anal Atom Spectrom* 1995;10:367–370.
12. Todd AC, Chettle DR: In vivo X-ray fluorescence of lead in bone: review and current issues. *Environ Health Perspect* 1994;102:172–177.
13. Labbe RF, Vreman HJ, Stevenson DK: Zinc protoporphyrin: a metabolite with a mission. *Clin Chem* 1999;45:2060–2072.
14. Centers for Disease Control and Prevention: Update: blood lead concentrations—United States, 1991–94. *MMWR Morb Mortal Wkly Rep* 1997;46: 141–146.
15. Pirkle JL, Brody DJ, Gunter EW, et al: The decline in blood lead concentrations in the United States. The National Health and Nutrition Examination Surveys (NHANES). *JAMA* 1994;272:284–291.
16. Centers for Disease Control and Prevention: Blood lead levels in young children—United States and selected states—1996–1999. *MMWR Morb Mortal Wkly Rep* 2000;49:1133–1137.
17. Ruff HA, Bijur PE, Markowitz M, et al: Declining blood lead concentrations and cognitive changes in moderately lead-poisoned children. *JAMA* 1993; 269:1641–1646.
18. Pocock SJ, Smith M, Baghurst P: Environmental lead and children's intelligence: a systematic review of the epidemiological evidence. *BMJ* 1994;309: 1189–1197.
19. Centers for Disease Control: Screening young children for lead poisoning: guidance for state and local public health officials. Bethesda, MD: US Department of Health and Human Services, Public Health Service, 1997. (Document can be obtained by calling the toll free number 888-232-6789.)
20. American Academy of Pediatrics Committee on Environmental Health: Screening elevated blood lead levels. *Pediatrics* 1998;101:1072–1078.
21. Markowitz ME, Bijur PE, Ruff H, et al: Effects of calcium disodium versenate ($CaNa_2EDTA$) chelation in moderate childhood lead poisoning. *Pediatrics* 1993;92:265–271.
22. O'Connor ME, Rich D: Children with moderately elevated lead levels: is chelation with DMSA helpful? *Clin Pediatr* 1999;38:325–331.
23. Matte TD: Reducing blood lead levels: benefits and strategies. *JAMA* 1999;281:2340–2342.
24. Ruff HA: Population-based data and the development of individual children: the case of low to moderate lead levels and intelligence. *J Dev Behav Pediatr* 1999;20:42–49.
25. Tong S, McMichael AJ, Baghurst PA: Interactions between environmental lead exposure and sociodemographic factors on cognitive development. *Arch Environ Health* 2000;55:330–335.
26. Onalaja AO, Claudio L. Genetic susceptibility to lead poisoning. *Environ Health Perspect* 2000;108 [Suppl 1]:23–28.

SELF-ASSESSMENT QUESTIONS

1. What are three factors that increase the risk of lead poisoning in children?

2. What are three causes of false-positive lead results?

3. List three distinct strategies for assessing lead exposure.

4. List three different methods for measuring blood lead concentration.

5. List three trends in pediatric lead screening that will affect laboratories.

CHAPTER 24

Arsenic, Mercury, and Cadmium

Gabor Komaromy-Hiller, Ph.D., DABCC

LEARNING OBJECTIVES

After completing this chapter, the reader should be able to:

1. Describe the clinical symptoms of acute and chronic arsenic, mercury, and cadmium poisoning.
2. List possible biomarkers of arsenic, mercury, and cadmium exposures based on each toxin's biochemistry.
3. Assess the importance and toxicity of the various forms of arsenic and mercury.
4. Select specimen type to assess arsenic, mercury, and cadmium exposure based on clinical history.
5. Evaluate preanalytical issues, especially contamination.
6. Critically evaluate analytical methods for measuring total arsenic, mercury, and cadmium, and set up analytical methods for speciation of arsenic and mercury.

ARSENIC

Elemental arsenic is a gray, shiny, brittle metalloid. When exposed to air its surface turns black because of arsenic trioxide formation (As_2O_3). In nature, to a small extent it can be found in its elemental form, but mostly as the arsenide of true metals. Smelting of ores of such metals produces arsenic trioxide. Arsenic compounds have been described and used since antiquity. However, metallic arsenic was not produced until the Middle Ages. Most arsenic compounds are toxic, and in fact arsenic has had a bad historical reputation since the Renaissance era for its use as an easy way of eliminating political rivals. Inducible tolerance, which may protect against acute arsenic poisoning, can develop through chronic low-dose intake of arsenic. Arsenic has also been used as an antisyphilitic agent.

Today, the largest source of human arsenic exposure is pesticide exposure. Other uses of arsenic and arsenic compounds include pharmaceuticals, in the glass and ceramic industry, and in metallurgy. The proposed acceptable limit of arsenic in the drinking water in the United States is 5 µg/L (66.5 nmol/L) (1). Seafood can contain 2–22 mg/kg arsenic, mostly in an organic form, which is considered nontoxic. The dietary intake of arsenic for an average adult is 25–30 µg/kg per day.

In 1999 there were 1338 reported cases of arsenic exposure, 335 due to pesticides and 1003 due to other arsenic compounds. Most of those exposed to nonpesticide arsenic compounds were adults (68%), and 13% were children <6 years of age (13%). Two exposures resulted in death, and 47 (4.7%) were intentional exposures.

The picture is different for arsenic pesticide exposure, for which 18% of the affected individuals were adults and 74% were children <6 years of age. No deaths were reported due to arsenic pesticide exposure, and 3.3% of the cases were intentional exposures (2). The overwhelming majority of children (<19 years old) accidentally exposed to arsenic pesticides (81%) involved pesticides stored in a garage or storage shed accessible to children.

Case Study

A 50-year-old man was admitted to the hospital with dry and itchy skin, nausea and vomiting, diarrhea, general tiredness, and a weight loss of 25 pounds (11.3 kg) in the past couple of months. Clinical history revealed an idiopathic cyclic neutropenia and hyperpigmentation, which had developed 2 years earlier. In the previous 3 months he had been admitted to another hospital for depression and an apparent Guillain-Barré syndrome with severe dysesthesias.

In the past he had been admitted to the hospital several times with the main complaint of nausea and vomiting. Examinations showed hyperpigmentation, but a normal abdominal exam. The neurological examination revealed decreased muscle strength in the lower extremities. Leukocyte counts were repeatedly low [$1.4-4 \times 10^3/\mu L$ ($1.4-4 \times 10^9/L$)] with elevated eosinophils (6–13%). Blood urea nitrogen (BUN) values at the first two admissions were 35 mg/dL (12.5 mmol/L), but at later admissions they were 40 and 55 mg/dL (14.3 and 19.6 mmol/L). A similar trend was observed for serum aspartate aminotransferase (AST), which was normal initially, but with time increased to 30 U/L (normal <20 U/L). His general condition usually improved within days for the first two hospital admissions and he was discharged. At later admissions he improved after weeks or months.

At his last admission his blood pressure was low (70/50 mmHg), his liver function tests were highly abnormal [AST 1520 U/L, bilirubin 4 mg/dL (68 μmol/L), alkaline phosphatase (ALP) 550 U/L, and lactate dehydrogenase (LDH) 840 U/L], and significant impairment of kidney function followed [creatinine = 2.8 mg/dL (247.5 μmol/L)]. In the hospital he developed a bleeding diathesis (elevated prothrombin time, partial thromboplastin time, and thrombocytopenia). Four days later he died. The day before his death a urine specimen was sent to the laboratory for arsenic analysis, where arsenic poisoning was confirmed [urinary As = 500 μg/L (6.65 μmol/L)].

General Symptoms of Arsenic Toxicity

The case study highlighted the major presenting signs of chronic arsenic poisoning. Recurring abdominal symptoms of nausea and vomiting, unexplained sudden weight loss, peripheral neuropathy affecting the lower extremities, progressively deteriorating liver function tests, jaundice, kidney failure, leukopenia, and thrombocytopenia are all difficult to evaluate one by one but in this context point to chronic arsenic exposure. In general, symptoms of arsenic poisoning depend on the route of exposure (ingestion vs. inhalation), and vary for acute and chronic exposure. Table 24–1 summarizes the differences between acute and chronic arsenic exposure (3,4).

Briefly, acute arsenic toxicity presents usually 30 minutes after ingestion with gastrointestinal symptoms, including constriction of the throat, gastric pain, vomiting, and diarrhea. Muscle cramps, hypertension, and tachycardia follow. Multiorgan failure may occur. Chronic arsenic toxicity usually presents with dermatological, neurological, hematological, and gastrointestinal symptoms, including hyperkeratosis and numbness and tingling in the extremities. Peripheral vascular disorders, leukopenia, anemia, and electrocardiogram abnormalities may develop. Acute exposure to arsine gas, which is the most toxic form of arsenic and was used in the past as a military poison, usually presents with nonspecific symptoms. In 1–2 days the developing triad of abdominal pain, hematuria, and jaundice is fairly characteristic of arsine gas poisoning.

Arsenic Exposure

The primary routes of arsenic exposure are inhalation, ingestion, and dermal contact. The highest number of exposure cases oc-

TABLE 24–1. Symptoms of acute and chronic arsenic poisoning

Target system	Acute	Chronic
Skin	Delayed hair loss; Mee's lines (2–3 weeks after exposure)	Melanosis, facial edema, hyperkeratosis, cutaneous cancers, hyperpigmentation
Neurologic	Hyperpyrexia, delirium, convulsions, tremor, coma; peripheral neuropathy, motor deficits	Encephalopathy, polyneuropathy, tremor, axonal degeneration, muscle wasting (lower extremities are most severely affected)
GI tract	Abdominal pain, dysphagia, vomiting, diarrhea	Nausea, vomiting, diarrhea, anorexia, weight loss
Liver	Fatty infiltration	Hepatomegaly, jaundice, cirrhosis
Kidney	Tubular and glomerular damage, oliguria, uremia	Nephritic findings, renal failure
Hematologic		Anemia, leukopenia, thrombocytopenia, impaired folate metabolism, basophilic stippling
Cardiovascular	Conduction blocks, QT prolongation, T-wave changes, ventricular tachycardia, fibrillation	Myocarditis, pericarditis, blackfoot disease

GI, gastrointestinal.

curs during manufacturing and processing operations of arsenic and its compounds. Higher than average exposure may occur in smelters, and during pesticide application and wood preservation. Indirect exposure may occur when arsenic-containing pesticides or arsenic from mineral deposits leach into drinking water. The proposed Environmental Protection Agency (EPA) maximum limit for arsenic in drinking water is 5 µg/L (1). Because many uses of arsenical pesticides are banned in the United States, and arsenic production was discontinued, the number of cases of workplace exposure has significantly reduced since the early 1980s. The maximum permissible exposure limit set by the Occupational Safety and Health Administration (OSHA) for workplace airborne arsenic is 10 µg/m³.

Direct consumption of arsenic and arsenic compounds occurs through our food supply, which contributes approximately 50 µg (0.66 µmol) of arsenic per day. Trace concentrations of arsenic can be found in livestock and at much higher concentrations in seafood. Arsenic is also part of some medications, and those can be a source of exposure for a limited portion of the population.

Biochemistry and Metabolism

Arsenic and arsenic compounds are absorbed easily through the gut. The trivalent form is more lipid soluble at physiological pH and is absorbed better through dermal contact. Exposure to arsine gas results in pulmonary absorption. After absorption, arsenic is present in the blood bound to proteins. It is thought that the cellular uptake of the trivalent arsenic species, arsenite, is several fold greater than that of arsenate (5). Unlike arsenate, arsenite is uncharged at physiological pH, and moreover, arsenate uptake is kinetically slowed down by phosphate uptake. Presumably, arsenate and phosphate share a common transporter (6). Redistribution occurs to the liver, lungs, intestinal wall, and spleen. Small amounts can penetrate the blood-brain barrier. Arsenic can replace phosphorus in bone, and there it can persist for years. Arsenic is also deposited in keratin-rich tissues such as hair and nails.

Measurement of Arsenic Toxicity

Arsenic toxicity results from three different mechanisms. Arsenic strongly binds to vicinal thiol groups. Dihydrolipoic acid, which is a cofactor of pyruvate dehydrogenase, is an endogenous compound with vicinal thiol groups that shows high affinity to arsenic. Arsenic binding to this cofactor results in disruption of the Krebs cycle in mitochondria (7). Also, arsenate competes with phosphate for reaction with adenosine diphosphate (ADP). The product ADP-arsenic has lower energy than adenosine triphosphate, which would form normally. Arsenic also binds to sulfhydryl groups of other proteins such as glutathione reductase, DNA ligase II, and glucocorticoid receptors, resulting in loss of enzyme and steroid-binding activity (8,9).

Today, chronic arsenic exposure is more of a concern because of its carcinogenic effect. Although its carcinogenic effect is not confirmed by animal studies, arsenic is on the list of carcinogenic substances. It is thought that arsenite, As(III), itself is not mutagenic but affects the mutagenicity of other carcinogens by blocking DNA repair (10). Arsenite forms a complex through vicinal thiol groups with the enzyme DNA ligase II, and it inhibits the DNA excision and repair mechanisms. It is also speculated that some genotoxic effects of arsenic can be caused by dimethylarsine (DMA), its peroxy radical, or other peroxy radicals resulting from glutathione and lipoic acid depletion (11).

Inorganic arsenic species are the more toxic forms of arsenic. Trivalent arsenic, As(III), is thought to be more toxic than the pentavalent form, As(V), an observation that is explained by its greater cellular uptake as noted earlier. The methylation of arsenite, As(III), involves S-adenosyl methionine (SAM), and glutathione:

$$H_3AsO_3 + SAM \rightarrow H_2AsO_3CH_3$$

$$H_2AsO_3CH_3 + GSH \rightarrow H_2AsO_2CH_3 \text{ (MMA)}$$

$$H_2AsO_2CH_3 + SAM \rightarrow HAsO_2(CH_3)_2 \text{ (DMA)}$$

Arsenite is oxidized and methylated in the liver to the less toxic forms of monomethylarsine (MMA) and DMA. Arsenate is first reduced to arsenite and then follows the same metabolic pattern (10). Because of these detoxification steps, As(III) and As(V) are the predominant forms in urine shortly (<10 hours) after ingestion. Their concentrations return to normal in 20–30 hours. MMA and DMA become the predominant urinary species only 40–60 hours after the exposure, and return to baseline in 6–20 days. Blood half-life of the inorganic species is 4–6 hours and of the methylated arsenic is 20–30 hours (8).

Laboratory Assessment of Arsenic Exposure

Direct measurements of arsenic in whole blood, urine, hair, or fingernail are used almost exclusively to assess arsenic exposure. However, it has been reported that arsenic can interfere with porphyrin metabolism (12). Several enzymes, including aminolevulinate synthase, porphobilinogen deaminase, uroporphyrinogen III synthase, uroporphyrinogen decarboxylase, coproporphyrinogen oxidase, ferrochelatase, and heme oxygenase are affected, and therefore the decreased enzyme activity or modified porphyrin excretion pattern can be used as an early indicator of chronic arsenic exposure.

Serum is the least useful specimen for identifying arsenic exposure, because arsenic rapidly disappears to the phosphate pool (<4 hours). Whole blood specimens are useful for identifying acute exposures. Urine is a useful specimen for assessing arsenic exposure because it is easy to collect and analyze and in general is the sample of choice. However, it is not useful for long-term (>1 week) exposures. Because arsenic has a high affinity for keratin, fingernails or hair can be used for assessing long-term (6 months to 1 year) exposure (13).

Because of the greatly different toxicity of the various forms of arsenic (Table 24–2), total arsenic determination is only partially useful for the assessment of arsenic exposure. Several methods have been reported for arsenic speciation (14–21). It is used to distinguish between toxic inorganic and nontoxic organic forms. Methods that are based on the different reactivity of inorganic and organic arsenic to form arsine (AsH_3) are not very accurate; however, they are fairly simple and inexpensive. A method using a simple disposable cartridge to separate inorganic and organic arsenic has also been reported recently (15). Methods based on high-performance liquid chromatography (HPLC) provide the most accurate determination of the different arsenic species. Detection methods require fairly complex systems because arsenic species are not electrochemically active or ultraviolet (UV) absorbent (16–21). This complexity of instrumentation limits their practical utility. Liquid chromatography–mass spectrometry (LC/MS) may provide an easy way to speciate arsenic (20,21), though only large reference laboratories can afford such instrumentation.

For the determination of total arsenic, atomic absorption spectroscopy (AAS) or inductively coupled plasma mass spectrometry (ICP-MS) is used most frequently. AAS has the advantage of simple and inexpensive instrumentation. On the other hand, ICP-MS has a much higher throughput and multi-element detection capability that can be used to analyze several toxic metals with only one injection (for example, As, Hg, and Pb). For both AAS and ICP-MS, an aliquot of the sample (whole blood or urine) is mixed with a dilution liquid, followed by injection into the analyzer. A well-known interfering species in ICP-MS determination is $^{40}Ar^{35}Cl$. Several strategies have been described to correct for this interference (22).

As with any other trace metal, specimen contamination is a major problem. This is especially true for urine specimens, which are the most easily contaminated of all sample types. Dust is the major source of contamination. Elevated values of urine arsenic should be interpreted with caution and reviewed in correlation with the clinical symptoms.

Seafood consumption before testing can easily elevate urine arsenic levels to 350 μg/d (4.66 μmol/d), well into the toxic range. Dietary evaluation is necessary before sample collection. Seafood consumption should be avoided for at least 10 days before collection of the urine specimens (23).

Interpretation of Laboratory Results

Arsenic poisoning can occur with some very nonspecific symptoms, and as noted, the laboratory plays an important role in the diagnosis. Elevated arsenic results, however, should be interpreted with caution because contamination is quite possible. Moreover, seafood consumption prior to sample collection can greatly elevate arsenic concentrations. In such cases arsenic speciation can confirm the presence of nontoxic arsenic species.

Treatment

Treatment consists of removal of the patient from the source of exposure, supportive measures, and chelation therapy (3). For arsenic ingestion lavage with activated charcoal may be applied; however, activated charcoal does not adsorb significant quantities of arsenic. In cases of severe poisoning a Foley catheter and a venous catheter should be placed, and a high urine output needs to

TABLE 24–2. *Various chemical forms of arsenic*

Chemical name	Formula
Arsenious acid	As_4O_6
Arsenic pentoxide	As_2O_5
Arsenic acid	H_3AsO_4
Arsenous acid	H_3AsO_3
Monomethyl arsenic (MMA)	$CH_3H_2AsO_2$
Dimethyl arsine (DMA)	$(CH_3)_2HAsO_2$
Arsenocholine	$(CH_3)_3As^+CH_2CH_2OH$
Arsenobetaine	$(CH_3)_3As^+CH_2COOH$

be maintained (1–2 mL · kg⁻¹ · h⁻¹). Alkalinizing the urine helps to prevent the deposition of erythrocyte breakdown products in renal tubules.

For chelation British antilewisite (BAL) has been used for many years but is no longer recommended. 2,3-Dimercaptosuccinic acid (succimer or DMSA) and 2,3-dimercaptopropane-1-sulfonate (DMPS) are more efficient than BAL. Chelation should be initiated for symptomatic patients and if urine arsenic concentrations exceed 200 µg/L (2.66 µmol/L).

In patients involved in arsine gas poisoning, chelation therapy may not be effective in preventing hemolysis. Exchange transfusions and hemodialysis should be considered for arsine-induced renal failure (3).

MERCURY

Elemental mercury is a heavy, silvery, slightly volatile liquid at room temperature, hence its historical name "quicksilver." It forms alloys with most metals except iron. In organic or inorganic compounds mercury has two oxidation states, monovalent mercurous ion and divalent mercuric ion. The monovalent form can undergo a disproportionation reaction by which elemental and divalent mercury form, producing toxic effects. Divalent mercury also forms organometallic compounds, some of which are highly toxic.

Mercury compounds have been found to have medicinal (calomel, mercurial salts, elemental mercury) and fungicidal (alkylmercurials) uses. Phenyl mercury compounds were also widely used in the paper industry and in latex paint manufacturing (24). Mercurial compounds have been phased out from medicinal and agricultural usage; however, mercury amalgam for tooth filling is still widely used.

In 1999 there were 4148 reported cases of mercury exposure due to mercurial compounds (3861), mercury-containing antiseptics (267), mercury fungicides (11), or mercury oxide used in batteries (9). Most of those exposed to mercury fungicides or antiseptics were children <6 years of age (8 of 11 and 211 of 267, respectively). About 25% of individuals exposed to other mercury compounds were children <6 years of age. Two hundred sixty cases (6.3%) were reported as intentional exposure, and only 1 case of 4148 resulted in death (2).

Recently, a major issue regarding the neurotoxic effect of methylmercury on the developing fetus surfaced (25). In 1997, the EPA prepared a study report to Congress on mercury, its distribution, environmental fate, and public health effect (26). Also concerns were raised about commercially available fish that have mercury concentrations in excess of the Food and Drug Administration (FDA) action limit of 1 ppm (27). Mercury concentrations exceeding 1 ppm (1 mg/L, 4.99 µmol/L) were measured in fish along the Western Gulf and Pacific coasts (28).

Case Study

A 48-year-old woman was admitted to the hospital on January 20, 1997. Her clinical history in the previous 5 days included progressive deterioration in balance, gait, and speech. She had lost 15 pounds (7 kg) during the prior 2 months and had suffered periods of nausea, diarrhea, and abdominal discomfort.

The patient looked thin but healthy, and was concerned about her neurologic problems. Computed tomography (CT) and magnetic resonance imaging (MRI) looked normal apart from a probable meningioma, 1 cm in diameter. The cerebrospinal fluid appeared clear, with no detectable cells, and with a total protein concentration of 42 mg/dL. Because mercury exposure was suspected, a urine and whole blood specimen were sent for laboratory analysis.

In the following days the patient experienced tingling in her fingers, flashes of light in both eyes, soft background noise in both ears, and progressive difficulty with speech, walking, hearing, and vision. Laboratory results of mercury analysis were 4000 µg/L (19,960 nmol/L) in blood [toxic concentration >200 (>998 nmol/L)], and 234 µg/L (1.17 µmol/L) in urine [toxic concentration

>50 µg/L (>250 nmol/L)]. Chelation therapy with oral succimer was initiated.

The progression of neurological deterioration continued, and vitamin E was added to the regimen as an antioxidant. Repeated CT and MRI scans were normal, and there was no evidence of occipital or cerebral damage.

On February 6, 1997, the patient became unresponsive to all visual, verbal, and light-touch stimuli. Despite the aggressive chelation regimen urinary excretion of mercury declined rapidly. Her neurological status continued deteriorating, and the patient's condition deteriorated to a vegetative state with spontaneous episodes of agitation and crying. The source of mercury exposure was tracked down to a chemistry laboratory incident involving dimethylmercury. The patient died, 298 days after the exposure, on June 8, 1997 (29,30).

As illustrated in this case report, mercury toxicity appears with neurotoxic as well as some rather general symptoms. The damage involves the cerebral cortex, and involves extensive neuronal death and loss. Although chelation therapy was initiated about 5 months after the suspected exposure, it was of little use because mercury had already deposited in the cerebral cortex. The fact that the urinary excretion of mercury decreased shows a deteriorating kidney function. The lag phase between the exposure and toxic symptoms can be months long and can delay clinical intervention.

General Symptoms of Mercury Toxicity

Symptoms of mercury toxicity depend on the type of exposure (for example, inhalation, ingestion, or dermal contact) and on the chemical form of mercury. The symptoms of acute and chronic mercury exposure are summarized in Table 24–3.

Generally, as the case report highlighted, symptoms of mercury poisoning include mental status changes, tremor, ataxia, slurred speech (for organic mercury poisoning), chorea or athetosis, abnormal reflexes, sensory or motor loss, metallic taste in the mouth, blue line along the gums, skin and lens discoloration, cough, peripheral neuropathy, and renal failure. The neuropathy and renal failure can develop later regardless of the source

TABLE 24–3. Symptoms of acute and chronic mercury poisoning

Target system	Acute	Chronic
Skin	Embolization, abscess formation (after injection of Hg compounds); blue-black pigmentation, urticaria, eczema, burns, exfoliation	Mercurialentis, band-shaped corneal opacities, acrodynia
Neurologic	Irritability, fatigue, insomnia, hearing loss, constricted visual fields, headache, confusion, lethargy, tremor, seizures, dysarthria, ataxia	Tremors, ataxia of lower limbs, salivation, erethism, delirium, manic-depressive psychosis, shyness
GI tract	Metallic taste, nausea, vomiting, anorexia, colitis, hematemesis, hematochezia, tenesmus, bloody diarrhea	Loose teeth, blue gums
Pulmonary	Hemoptysis, cyanosis, pneumonitis, respiratory distress, pulmonary hemorrhage, edema	
Hematologic	Thrombocytopenia, anemia	
Fetal and teratogenic		Fetus: severe brain damage, mental retardation, spasticity, cerebral palsy, ataxia, tremors, hearing deficiency, seizures, small size, anemia

GI, gastrointestinal.

of exposure. Inhalation of mercury vapor presents with chest tightness, fever, general weakness, and gastrointestinal upset, followed by renal failure and peripheral neuropathy. Inorganic mercury can cause hematemesis, hematochezia, oral burning, renal failure, and peripheral neuropathy. As mentioned earlier, organic mercury poisoning presents with gastrointestinal upset, mental status changes, brain damage (in utero), renal failure, and peripheral neuropathy (3,31).

Mercury Exposure

The EPA-established benchmark dose of mercury is 0.1 µg/kg body weight per day, which can come from various sources (26). This rate of intake corresponds to a "no adverse effect level" of <5 µg/L (<25 nmol/L) in whole blood. Dietary intake (notably from seafood) can be significant, and chronic low dose intake of methylmercury is a major concern for people who eat seafood more than once a week. Mercury leaching from dental amalgams seems not to pose a significant threat to the general public (32), although amalgam-based filling materials are continuously being phased out. The use of mercury in folk medicine or its use for ritualistic purposes involves only a limited number of patients (33). There are no prevalence data for the use of mercury in folk medicine. However, it was noted that mercury is still used in some areas to control head lice or is used in skin-lightening creams. Further uses of mercury included rituals such as burning mercury on candles; mixing it with ammonia, camphor, or soap; adding it to bath water; and even using it as a dietary supplement (33).

Exposure to mercury via leaching from dental amalgams can contribute to the body's mercury burden. It was reported that up to 100 µg/d of mercury could be released from mercury amalgams with gum chewing. Under normal circumstances, and if there is no other source of mercury exposure, the amount of mercury released from dental amalgams appears to be insignificant (34,35).

Furthermore, there is no evidence that chelation therapy, to reduce mercury burden due to leaching from dental amalgams, would provide beneficial health effects (33).

As in the case of arsenic, it was noted that methylmercury can induce porphyria. It is known that mercury interferes with the enzymes participating in heme biosynthesis, and can cause increased elimination of porphyrins in the feces. Some investigators have suggested that monitoring porphyrin excretion patterns can serve as an early biomarker for mercury exposure (36).

Biochemistry of Mercury Toxicity

Elemental mercury is very poorly (~0.01%) absorbed from the gastrointestinal tract (37). However, if it is trapped and is bioconverted to ionized or methylmercury, it is absorbed and can cause chronic mercury poisoning. Inhalation is the major route of elemental mercury exposure. It has been estimated that 75–85% of inhaled mercury vapor is absorbed by the body (38). Elemental mercury is oxidized to divalent mercury (Hg^{2+}) in a catalase- and peroxide-mediated reaction. Inhibitors of catalase can considerably slow down this reaction. Elemental mercury readily crosses the blood-brain barrier and the placenta, whereas oxidized mercuric ion does not. Mercuric ion detected in these compartments is probably an oxidation product, and because it cannot cross these barriers can accumulate in the brain and in the developing fetus.

About 20% of ingested inorganic mercury is absorbed, depending on intestinal pH, the type of mercuric salt, age, diet, biliary secretion, and excretion in the feces (39). It is rapidly redistributed to the brain and other nervous tissues. Ionized mercury is concentrated in the kidneys. The major excretion route is through the feces, and with time, renal excretion becomes important. The elimination half-life of inorganic mercury is about 20–65 days (40). The major elimination routes are feces and, with time, the kidney.

Methylmercury is absorbed completely in

the gut and is accumulated in lipid-rich tissues such as the central nervous system. Absorption through inhalation or dermal contact was also observed. Blood concentrations of methylmercury follow organ concentrations initially, but with time some organs such as brain, muscle, and kidney start to accumulate mercury. The elimination half-life of methylmercury is between 35 and 189 days (41).

Methylmercury crosses the placenta and can cause neurotoxicity in the developing fetus. We have incomplete knowledge about the dose-effect relationship of methylmercury concentration in cord blood. Methylmercury concentration in cord blood is about fivefold that in maternal hair (42). Furthermore, various studies have shown that methylmercury concentration in hair is about 250 times that of whole blood (43–46). Therefore, we can speculate that the concentration of methylmercury in cord blood is about a thousandfold that in maternal blood. The major route of metabolism is demethylation; however, this is a slow process. Alkylmercurial compounds with long carbon chains appear to be metabolized faster to mercuric ion (47).

There are three different mechanisms of mercury toxicity. Inorganic mercury (Hg^{2+}) binds to sulfur-containing proteins, distorting their tertiary structure, which leads to loss or a decrease of biological activity. Secondly, as the tertiary structure changes the protein may become immunogenic and elicit an antibody response (48,49). Alkyl mercury is highly hydrophobic and binds to proteins in lipid-rich tissues. Myelin is particularly susceptible to disruption by alkyl mercury (23).

Of the several nutrients that affect mercury toxicity, selenium is the most widely studied. Selenium probably affects mercury toxicity at various levels. Mercury decreases the activity of several enzymes in the synthesis of glutathione, leading to decreased glutathione concentration and antioxidant activity and the subsequent increase of free radicals. Selenium prevents the depression of the enzymes in the glutathione (GSH) synthesis pathway caused by mercury. Also sodium selenite decreases the amount of inorganic mercury bound to renal metallothionein. In general, selenium delays the onset of inorganic and alkylmercury toxicity or reduces the severity of the toxic effects. In this regard there seems to be a difference between the various forms of selenium. Selenite is more effective than selenate and selenocystine, whereas selenomethionine seems to have no protective effects (50,51).

The toxicity of the various forms of mercury differs widely. In general, elemental mercury is considered nontoxic when ingested because of its low absorption rate. When inhaled and absorbed elemental mercury is oxidized to divalent mercuric ion and becomes toxic. Inorganic mercury is toxic. However, its hydrophilic nature inhibits passage across the blood-brain barrier or the placenta. Its major target organ is the kidney. Methylation of inorganic mercury to methylmercury occurs in microorganisms through different mechanisms (52,53). Extreme hazards are associated with dimethylmercury as emphasized by the recent Dartmouth Medical Center incident (29,30). Also, this species, covalently attached probably to a thiol-containing carrier molecule, can cross both the blood-brain barrier and the placenta and cause irreversible neurological damage, especially in the developing fetus.

Laboratory Assessment of Mercury Exposure

Because of its relatively long elimination half-life, whole blood can be used to assess mercury exposure. Plasma is not recommended because mercury concentration in plasma is about one-twentieth that in erythrocytes. Urine can be used to follow the effectiveness of chelation therapy, and it is also thought to indicate better the mercury burden of the kidneys. It should be noted that organic mercury is mainly eliminated through the feces. To assess long-term exposure hair specimens are recommended.

Methods recommended for total mercury

analysis include neutron activation analysis (NAA), AAS, gas chromatography with electron-capture detection (GC-ECD), X-ray fluorescence (XRF), inductively coupled plasma atomic emission spectrophotometry (ICP-AES) and ICP-MS. With the most sensitive methods of AAS and ICP-MS, detection limits are in the 1 ppb (1 µg/L, 4.99 nmol/L) range.

AAS using cold vapor atomization is the most widely used analytical method measurement of total mercury concentration. The assay is very specific, and if the strong oxidizing agent is omitted from the procedure, inorganic mercury can be analyzed specifically. If ICP-MS instrumentation is used for mercury analysis, mercury should preferably be run on a dedicated system, because it tends to adsorb to the transfer tubing of the autosampler and nebulizer. This adsorption can be avoided by adding small quantities of gold to every solution involved in mercury analysis.

Several methods are available for mercury speciation (54,55). In assessing mercury poisoning in the clinical setting the three key determinants are the form of mercury, the route of exposure, and the dose (50). Other information, such as selenium status, may be important as well.

In terms of preanalytical issues, seafood consumption should be assessed and eliminated for at least 2 months before sample collection. Apart from that, contamination, especially for hair specimens, can be a problem. The use of mercury to control for lice significantly elevates mercury concentration found in hair; however, this mercury is only adsorbed to the outside of the hair and does not reflect mercury body burden (33). Rigorous wash of the collected hair sample is necessary to remove this contamination. Specimen refrigeration may be necessary during transportation and storage, because elemental mercury and some organic mercury species are volatile at room temperature. Because specimen contamination is an issue, elevated mercury concentrations are preferably confirmed from a follow-up specimen. This is especially true when the laboratory findings do not correlate with the clinical symptoms.

Treatment

Treatment includes removal from the source of exposure, supportive care and decontamination, and chelation therapy. When mercury poisoning is suspected chelation therapy should be initiated. However, chelation therapy does not reverse the developed symptoms of chronic mercury poisoning. BAL was used in the past, but now it is contraindicated because it actually raises the mercury concentration in the brain. The first choice of chelator is DMPS, but N-acetyl-dl-penicillamine and d-penicillamine can be used, too. In severe cases hemodialysis should be started after chelation. If elemental mercury is trapped in the intestine, surgical removal is necessary (3).

CADMIUM

Elemental cadmium is a silver-white metal that can be easily cut with a knife. On exposure to moist air it is slowly oxidized to cadmium oxide (CdO). In nature it is found in zinc ores and also as the sulfide (CdS). Cadmium and its salts are highly toxic. Cadmium is ubiquitous and all rocks and soils contain cadmium.

Today, cadmium and its compounds are used in electroplating, batteries, pigments, and plastic production as a stabilizer. Cadmium salts have limited use as fungicides for golf courses and home lawns. Major industrial releases of cadmium occur in waste water, landfills, and during the smelting of ores. The EPA standard for maximum concentration of cadmium in drinking water is 5 ppb (µg/L). The EPA does not allow cadmium in pesticides. The maximum allowed limit of cadmium in food coloring is set by the FDA to 15 ppm (mg/L).

In 1999 there were 95 reported cases of cadmium exposure, 10 due to nickel-cadmium batteries and 85 due to other cadmium compounds. Most of the exposed individuals (59 out of 95) were adults. None of the incidents resulted in death, and there were no intentional exposures (2).

Case Study

A 40-year-old male was admitted to the hospital with accidental cadmium fume inhalation. Two hours after the incident he had experienced coughing and a metallic taste in his mouth, and he was taken to the hospital. There he experienced rapidly decreasing shortness of breath, chest pain, and flu-like symptoms. He also complained of weakness, headache, chills, and muscular pains. His body temperature was 101.3 °F (38.5 °C). The following day he developed acute pulmonary edema, which worsened over the next few days. Five days after admission his shortness of breath began to resolve, and a week after admission he was released from the hospital.

The patient's whole blood cadmium result was 432 µg/L (3843.5 nmol/L), well into the toxic range, in a sample obtained upon admission. However, at this point his urinary cadmium was normal, 5 µg/L (44.5 nmol/L). Subsequent specimens obtained over the following days showed an increased urinary excretion of cadmium, up to 1200 µg/L (10.7 µmol/L). Further laboratory evaluations revealed general proteinuria, glucosuria, and an increasing concentration of β_2-microglobulin in urine, 250 µg/g creatinine (251 µmol/mol creatinine) to 1634 µg/g creatinine (1643 µmol/mol creatinine) [normal is <300 µg/g creatinine (<302 µmol/mol creatinine)].

As this case illustrates, acute cadmium exposure presents with general symptoms of metal toxicity such as metallic taste in mouth, weakness, headache, muscle pains, and fever. The primary target organ of cadmium is the kidney, and shortly after exposure several biomarker proteins can be detected in the urine. Depending on the cadmium load kidney failure can ensue. In this case although some tubular damage occurred, as was indicated by the increasing concentration of β_2-microglobulin in urine, it did not progress to complete renal failure.

General Symptoms of Cadmium Toxicity

Acute cadmium poisoning can occur through ingestion and inhalation. The usual symptoms after cadmium ingestion present rapidly, within about 30 minutes after the ingestion. These symptoms include increased salivation, choking, nausea and vomiting, abdominal pain, and diarrhea. Because cadmium attacks the renal system, kidney dysfunction follows with glucosuria, proteinuria, and aminoaciduria. Renal and hepatic failure can occur.

Inhalation seems to be a more important route of acute cadmium poisoning. It usually presents with throat dryness, cough, headache, vomiting, chest pain, restlessness, and irritability. Pulmonary involvement includes pneumonitis, edema, and bronchopneumonia. Permanent lung damage and cardiovascular collapse may occur (56).

Chronic cadmium exposure may result in decreased renal functions, renal damage including proteinuria, Fanconi-like renal dysfunction, severe osteomalacia, and "itai-itai" disease. Based on various reports from the 1980s, cadmium has been classified as a human carcinogen (57–59). The primary sites of cancer associated with cadmium exposure are the lung and the prostate.

Table 24–4 summarizes the differences of the presentation of acute and chronic cadmium poisoning.

Cadmium Exposure

Exposure to cadmium may occur through eating food, drinking water, smoking tobacco, or inhaling cadmium-containing dust and fumes. Ingestion may occur from food improperly stored in a cadmium-plated container. Inhalation may occur in occupational settings. It has been estimated that average cadmium concentrations in cigarettes range from 1000 ppb (1000 ng/g dry weight, 8.9 nmol/g dry weight) to 3000 ppb (3000 ng/g dry weight, 26.7 nmol/g dry weight), and smokers absorb about 1–4 µg/d from cigarettes. From food an average person's cadmium burden is 1–3 µg/d, and from drinking water 2 µg/d (60). OSHA mandates monitoring of cadmium-exposed workers at regular intervals (61). Whole blood and urinary cadmium, to estimate current exposure and cad-

TABLE 24–4. *Symptoms of acute and chronic cadmium poisoning*

Target system	Acute	Chronic
GI tract	Nausea, vomiting, salivation, diarrhea, abdominal pain, metallic taste, constriction of the throat	
Pulmonary	Chest pain, shortness of breath, pulmonary edema, pneumonitis, bronchopneumonia	Lung cancer
Neurologic	Fever, sweating, muscle pain, headache, restlessness, irritability	
Renal	Glucosuria, proteinuria, kidney failure	Decreased renal function, renal damage, proteinuria, Fanconi-like renal dysfunction
Bone		Severe osteomalacia, "itai-itai"

GI, gastrointestinal.

mium body burden, as well as urinary β_2-microglobulin, reflecting tubular damage, are monitored. Occupational action levels are summarized in Table 24–5. Removing workers from the source of exposure is mandatory at action level C.

The biomarkers for cadmium exposure reflect the tubular damage caused by cadmium. About 0.02% of the cadmium body burden is excreted daily into urine. Four threshold cadmium concentrations were identified that correspond to various degrees of kidney damage. At low cadmium excretion (<4 μg Cd/g creatinine, <4 nmol Cd/mmol creatinine) sialic acid and 6-ketoprostaglandin $F_{1\beta}$ appear in the urine. Between 4 and 7 μg Cd/g creatinine (4 and 7 nmol Cd/mmol creatinine), N-acetyl-β-D-glucosaminidase, intestinal alkaline phosphatase, albumin, and transferrin have been detected (62). Up to about 10 μg Cd/g creatinine (10 nmol Cd/mmol creatinine) an additional two protein biomarkers appear, β_2-microglobulin and Tamm-Horsfall glycoprotein (63–65). Higher cadmium excretion rates indicate irreversible glomerular damage, and tissue nonspecific alkaline phosphatase, brush border antigen HF5, retinol-binding protein, and glycosaminoglycans are detectable in the urine (63–65).

Biochemistry of Cadmium Toxicity

The fraction of orally ingested cadmium absorbed is about 2–7%. It appears that cadmium absorption is affected by iron status, because up to 20% of ingested cadmium can be absorbed when iron stores are low. The pulmonary absorption rate is much greater,

TABLE 24–5. *OSHA action levels for workers exposed to cadmium*

Laboratory assays	Results	Action
Level A		
Cadmium, whole blood	≤5 μg/L	Monitor annually, biennial medical examination, no action required
Cadmium, urine	≤3 μg/g creatinine	
β_2-Microglobulin, urine	≤300 μg/g creatinine	
Level B		
Cadmium, whole blood	5–10 μg/L	Monitor semiannually, annual medical examination, discretional removal from source of exposure
Cadmium, urine	3–7 μg/g creatinine	
β_2-Microglobulin, urine	300–750 μg/g creatinine	
Level C		
Cadmium, whole blood	>10 μg/L	Monitor quarterly, semiannual medical examination, mandatory removal from source of exposure
Cadmium, urine	>7 μg/g creatinine	
β_2-Microglobulin, urine	>750 μg/g creatinine	

25–50% for cadmium oxide. However, the absorption rate can vary depending on the inhaled cadmium species (56).

Most cadmium in the circulation (90%) is bound to erythrocytes, and therefore, whole blood is the specimen of choice for assessment of cadmium exposure. Cadmium is a cumulative element, and it accumulates in three target organs: liver, kidney, and bone. The biological half-life of cadmium is extremely long, 15–20 years (65,66).

Cadmium can interfere with normal calcium metabolism of the bone (66). It is deposited in the osteoid tissues and disrupts calcification, decalcification, and bone remodeling. It also interferes with zinc metabolism because it avidly binds to metallothionein (67). Metallothionein is a group of low-molecular-weight proteins that may be involved in the cellular defense against toxic metal stress. Metallothionein I and II are expressed in all tissues, whereas metallothionein III is expressed in brain and is rich in zinc. Because cadmium cannot cross the blood-brain barrier, neurotoxicity induced by cadmium may be an indirect effect through disturbed zinc metabolism. Selenium, similar to its effect on mercury toxicity, can abolish the toxic effects of cadmium. Cadmium, with its high affinity for sulfhydryl groups, preferably binds to proteins containing –SH moieties, such as the selenium-containing protein glutathione peroxidase, which plays a key part in preventing oxidative damage. Upon cadmium binding, glutathione peroxidase becomes inactive. Synthesis of new glutathione peroxidase molecules requires adequate selenium supplies. Therefore, selenium, through the de novo synthesis of glutathione peroxidase, can prevent or diminish the toxic effects of cadmium (68).

Laboratory Assessment of Cadmium Exposure

Medical surveillance regulated by OSHA is aimed at accomplishing three goals. First is to identify workers who are at risk for adverse effects associated with chronic cadmium exposure. This goal is accomplished by regular monitoring of employees. Second is to prevent cadmium-induced diseases, mainly to prevent kidney damage and lung cancer. Third is to minimize existing disease related to cadmium exposure.

Laboratories providing cadmium analysis service to occupational clinics are mandated to participate in an interlaboratory comparison program. The Centre de toxicologie du Québec informed participating laboratories in November 2000 that it will discontinue its cadmium proficiency testing program and instead supply the College of American Pathologists (CAP) with proficiency-testing materials. Interested laboratories therefore should contact CAP about participating in its cadmium interlaboratory comparison program.

Urine specimens are used to assess the cadmium body burden. Urinary excretion increases with age and is well correlated with the concentration of cadmium in the kidney. Once the body becomes saturated with cadmium, it is rapidly excreted in urine, and under these circumstances urinary cadmium reflects recent exposure. Also, with the progression of renal tubular damage, more and more cadmium is spilled into urine because of its loss from the kidney depot (56).

Cadmium has a long biological half-life (15–20 years). Bone, kidney, and liver serve as "deep" compartments, while blood acts as a "peripheral" compartment. Cadmium half-life in the blood is approximately 2–3 months; therefore, whole blood is useful to monitor both cadmium body burden and recent exposure. Cadmium concentration in blood is also influenced by smoking status (see above). Nonsmokers usually have half the whole blood cadmium observed in smokers.

Hair is not a recommended specimen. Although cadmium accumulates in hair, it is such an ubiquitous element that it is very difficult to distinguish endogenous cadmium from contamination. Therefore, hair is not recommended for cadmium monitoring or assessment of cadmium exposure because contamination-free specimens cannot be guaranteed. Elevated concentrations in any specimen type should be confirmed by biomarker evaluation or with a second collection.

TABLE 24–6. *OSHA recommendation on quality objectives for the cadmium medical monitoring program*

Analyte concentration pool	Limit of detection	Precision (%CV)	Accuracy
Cadmium in blood	0.5 µg/L		±1 µg/L or 15% of the assigned value
≤2 µg/L		40	
>2 µg/L		20	
Cadmium in urine	0.5 µg/g creatinine		±1 µg/L or 15% of the assigned value
≤2 µg/L		40	
>2µg/L		20	

Total cadmium can be measured by flame or graphite furnace atomic absorption spectroscopy (FAAS, GFAAS), NAA, ICP-AES, or ICP-MS. The advantage of FAAS and GFAAS over NAA, ICP-AES, and ICP-MS methods is the less expensive instrumentation. However, the operating cost for GFAAS is higher than for the other methods. The advantage of ICP-AES and ICP-MS methods is their tremendous throughput. Detection limits achieved by any of these methods are adequate [<1 µg/L (<8.9 nmol/L)] for biological monitoring of cadmium in body fluids (69–71). Table 24–6 summarizes the OSHA recommendation for quality objectives for methods used to measure cadmium in biological fluids (61).

Speciation of cadmium is mostly of concern for environmental monitoring (60) and has not been investigated in a clinical setting. However, some authors have recommended measuring urinary metallothionein concentration along with cadmium, as a better indicator of body burden (67).

Treatment

Therapy includes removal of the patient from the source of exposure and supportive therapy. Chelation therapy is contraindicated because it exposes the kidney to large quantities of nephrotoxic cadmium.

REFERENCES

1. Environmental Protection Agency (EPA): National primary drinking water regulations; arsenic and clarifications to compliance and new source contaminants monitoring. *Fed Regist* 2000;65(121): 38887–38983.
2. Litovitz TL, Klein-Schwartz W, White S, et al: 1999 annual report of the American Association of Poison Control Centers Toxic Exposure Surveillance System. *Am J Emerg Med* 2000;18(5):517–574.
3. Graeme KA, Pollack CV Jr: Heavy metal toxicity, part I: arsenic and mercury. *J Emerg Med* 1998; 16:45–56.
4. Gorby MS: Arsenic poisoning. *West J Med* 1988; 149:308–315.
5. Lerman SA, Clarkson TW, Gerson RJ: Arsenic uptake and metabolism by liver cells is dependent on arsenic oxidation state. *Chem Biol Interact* 1983; 45:401–406.
6. Murer H, Markovich D, Biber J: Renal and small intestinal sodium-dependent symporters of phosphate and sulfate. *J Exp Biol* 1994;196:167–181.
7. Aposhian HV: Biochemical toxicology of arsenic. *Rev Biochem Toxicol* 1989;10:265–299.
8. Crecelius EA: Changes in the chemical speciation of arsenic following ingestion by man. *Environ Health Perspect* 1997;19:147–150.
9. Lopez S, Miyashita Y, Simons SS: Structurally based, selective interaction of arsenite with steroid receptors. *J Biol Chem* 1990;265:16039–16042.
10. Rossman TG: Molecular and genetic toxicology of arsenic. In: Rose J, ed: *Environmental toxicology*. Amsterdam: Gordon and Breach Science Publishers, 1998:171–187.
11. Snow ET: Metal carcinogenesis: mechanistic considerations. *Pharmacol Ther* 1992;53:31–65.
12. García-Vargas GG, Hernández-Zavala A: Urinary porphyrins and heme biosynthetic enzyme activities measured by HPLC in arsenic toxicity. *Biomed Chromatogr* 1996;10:278–284.
13. Poklis A, Saady JJ: Arsenic poisoning: acute or chronic, suicide or murder? *Am J Forensic Med Pathol* 1990;11:226–232.
14. Benramdane L, Bressolle F, Vallon JJ: Arsenic speciation in humans and food products: a review. *J Chromatogr Sci* 1999;37:330–344.
15. Nixon DE, Moyer TP: Arsenic analysis II: rapid separation and analysis of inorganic arsenic plus metabolites and arsenobetaine from urine. *Clin Chem* 1992;38:2479–2483.
16. Del Razo LM, Aguilar C, Sierra-Santoyo A, et al: Interference in the quantitation of methylated arsenic species in human urine. *J Anal Toxicol* 1999; 23:103–107.
17. Le XC, Ma M: Short-column liquid chromatography with hidride generation atomic fluorescence

detection for the speciation of arsenic. *Anal Chem* 1998;70:1926–1933.
18. Pretty JR, Blubaugh EA, Caruso JA: Determination of arsenic(III) and selenium(IV) using an on-line anodic stripping voltammetry flow cell with detection by inductively coupled plasma atomic emission spectrometry and inductively coupled plasma mass spectrometry. *Anal Chem* 1993;65:3396–3403.
19. Pergantis SA, Heithmar EM, Hinners TA: Speciation of arsenic animal feed additives by microbore high-performance liquid chromatography with inductively coupled plasma mass spectrometry. *Analyst* 1997;122:1063–1068.
20. Corr JJ, Larsen EH: Arsenic speciation by liquid chromatography coupled with ionspray tandem mass spectrometry. *J Anal Atom Spectrom* 1996;11:1215–1224.
21. Corr JJ: Measurement of molecular species of arsenic and tin using elemental and molecular dual mode analysis by ionspray mass spectrometry. *J Anal Atom Spectrom* 1997;12:537–546.
22. Kershisnik MM, Kalamegham R, Ash KO, et al: Using 16O35Cl to correct for chloride interference improves accuracy of urine arsenic determinations by inductively coupled plasma mass spectrometry. *Clin Chem* 1992;38:2197–2202.
23. Moyer TP: Toxic metals. In: Burtis CA, Ashwood EA, eds: *Tietz textbook of clinical chemistry*, 3rd ed. Philadelphia, WB Saunders Company, 1999:982–998.
24. Clarckson TW: The toxicology of mercury. *Crit Rev Clin Lab Sci* 1997;34(3):369–403.
25. Mercury Policy Project & California Communities Against Toxics: The one that got away: FDA fails to protect the public from high mercury levels in seafood. April 2000. Available on the Internet at http://www.mercurypolicy.org/exposure/documents/one_that_got_away.pdf.
26. United States Environmental Protection Agency: Mercury Study Report to Congress. [EPA-452/R-97-003.] Washington, DC: Environmental Protection Agency, 1997.
27. Food and Drug Administration: *Fed Regist* 1979;44:3990,3992.
28. O'Connor TP, Beliaeff B: Recent trends in coastal environmental quality: results from the Mussel Watch Project. 1986 to 1993. Silver Spring, MD: U.S. Department of Commerce, National Oceanic and Atmospheric Administration, National Ocean Service, Office of Ocean Resources, Conservation and Assessment, 1995.
29. Nierenberg DW, Nordgren RE, Chang MB, et al: Delayed cerebellar disease and death after accidental exposure to dimethylmercury. *N Engl J Med* 1998;338:1672–1676.
30. Kulig K: A tragic reminder about organic mercury. *N Engl J Med* 1998;338:1692–1694.
31. O'Carroll RE, Masterton G, Dougall N, et al: The neuropsychiatric sequelae of mercury poisoning. The Mad Hatter's disease revisited. *Br J Psychiatry* 1995;167:95–98.
32. Oskarsson A, Schütz A, Skerfving S, et al: Total and inorganic mercury in breast milk and blood in relation to fish consumption and amalgam fillings in lactating women. *Arch Environ Health* 1996;51(3):234–241.
33. Risher JF, De Rosa CT, Jones DE, et al: Summary report for the expert panel review of the toxicological profile for mercury. *Toxicol Ind Health* 1999;15:483–516.
34. Hahn LJ, Kloiber R, Leininger RW, et al: Whole-body imaging of the distribution of mercury released from dental fillings into monkey tissues. *FASEB J* 1990;4:3256–3260.
35. Hanson M, Pleva J: The dental amalgam issue. A review. *Experientia* 1991;47:9–22.
36. Leonzio C, Fossi MC, Casini S: Porphyrins as biomarkers of methylmercury and PCB exposure in experimental quail. *Bull Environ Contam Toxicol* 1996;56:244–250.
37. Bornmann G, Henke G, Alfes H, et al: Über die enterale Resorption von metallischem Quecksilber. [Intestinal absorption of metallic mercury.] *Arch Toxikol* 1970;26(3):203–209.
38. Oikawa K, Saito H, Kufune I, et al: Mercury absorption by inhaling through the nose and expiring through the mouth at various concentrations. *Chemosphere* 1982;11(9):943–951.
39. Nielsen JB: Toxicokinetics of mercuric-chloride and methylmercuric chloride in mice. *J Toxicol Environ Health* 1992;37(1):85–122.
40. Hall LL, Allen PV, Fisher HL, et al: The kinetics of intravenously-administered inorganic mercury in humans. In: Subramanian KMS, Wastney ME, eds: *Kinetic models of trace elements and mineral metabolism during development*. Boca Raton, FL: CRC Press, 1994:1–21.
41. Al-Shahristani H, Shihab KM: Variation of biological half-life of methyl mercury in man. *Arch Environ Health* 1974;28:342–344.
42. Grandjean P, Budtz-Jørgensen E, White RF, et al: Methylmercury exposure biomarkers as indicators of neurotoxicity in children aged 7 years. *Am J Epidemiol* 1999;150(3):301–305.
43. Gearhart J, Clewell H, Crump K, et al: Pharmacokinetic dose estimates of mercury in children and dose response curves of performance tests in a large epidemiological study. *Water Air Soil Pollut* 1995;80:49–58.
44. Sherlock JC, Quinn MJ: Underestimation of dose-response relationship with particular reference to the relationship between the dietary intake of mercury and its concentration in blood. *Hum Toxicol* 1988;7(2):129–132.
45. Rice G, Swartout J, Mahaffey K, et al: Derivation of U.S. EPA's oral reference dose (RFD) for methylmercury. *Drug Chem Toxicol* 2000;23(1):41–54.
46. Swartout J, Rice G: Uncertainty analysis of the estimated ingestion rates used to derive the methylmercury reference dose. *Drug Chem Toxicol* 2000;23(1):293–306.
47. Berlin M: Mercury. In: Friberg L, Nordberg GR, Vouk VB, eds: *Handbook on the toxicology of metals*, 2nd ed. New York: Elsevier Press, 1986.
48. Hultman P, Enström S: Mercury induced antinuclear antibodies in mice: characterization and correlation with renal immune complex deposits. *Clin Exp Immunol* 1988;71:269–274.
49. Nielsen JB, Hultman P: Experimental studies on genetically determined susceptibility to mercury-induced autoimmune response. *Ren Fail* 1999;21(3-4):343–348.
50. Goyer RA: Toxic and essential metal interactions. *Annu Rev Nutr* 1997;17:37–50.
51. Cuvin-Aralar MLA, Furness RW: Mercury and se-

lenium interaction: a review. *Ecotoxicol Environ Saf* 1991;21:348–364.
52. Wood JM, Kennedy FS, Rosen CG: Synthesis of methyl-mercury compounds by extracts of a methanogenic bacterium. *Nature* 1968;220(163): 173–174.
53. Jernelov A: A new biochemical pathway for the methylation of mercury and some ecological implications. In: Miller MW, Clarkson TW, eds: Mercury, mercurials and mercaptans. Springfield, IL: Charles C Thomas, 1973:315–323.
54. Engqvist A, Colmsjö A, Skare I: Speciation of mercury excreted in feces from individuals with amalgam fillings. *Arch Environ Health* 1998;53: 205–213.
55. Liang L, Bloom NS, Horvat M: Simultaneous determination of mercury speciation in biological materials by GC/CVAFS after ethylation and room-temperature precollection. *Clin Chem* 1994;40:602–607.
56. Oberdörster G: Pulmonary deposition, clearance and effects of inhaled soluble and insoluble cadmium compounds. *IARC Sci Publ* 1992;118:189–204.
57. Thun MJ, Schnoor TM, Smith AB, et al: Mortality among a cohort of U.S. cadmium production workers—an update. *J Natl Cancer Inst* 1985;74(2):325–333.
58. Takenaka S, Oldiges H, Konig H, et al: Carcinogenicity of cadmium chloride aerosols in Wistar rats. *J Natl Cancer Inst* 1983;70:367–373.
59. Cadmium and cadmium compounds. *IARC Monogr Eval Carcinog Risks Hum* 1993;58:119–237.
60. Robards K, Worsfold P: Cadmium: toxicology and analysis. *Analyst* 1991;116:549–568.
61. OSHA, Department of Labor: Occupational exposure to cadmium. 29CFR1910.1027.
62. Buchet JP, Lauwerys R, Roels H, et al: Renal effects of cadmium body burden of the general population. *Lancet* 1990;336:699–702.
63. Roels HA, Hoet P, Lison D: Usefulness of biomarkers of exposure to inorganic mercury, lead, or cadmium in controlling occupational and environmental risks of nephrotoxicity. *Ren Fail* 1999; 21(3-4):251–262.
64. Fels LM: Risk assessment of nephrotoxicity of cadmium. *Ren Fail* 1999;21(3-4):275–281.
65. Lauwerys RR, Bernard AM, Roels HA, et al: Cadmium: exposure markers as predictors of nephrotoxic effects. *Clin Chem* 1994;40:1391–1394.
66. Järup L, Berglund M, Elinder CG, et al: Health effects of cadmium exposure—a review of the literature and a risk estimate. *Scand J Work Environ Health* 1998;24[Suppl 1]:1–51.
67. Jin T, Lu J, Nordberg M: Toxicokinetics and biochemistry of cadmium with special emphasis on the role of metallothionein. *Neurotoxicology* 1998;19: 529–536.
68. Ogjanovic B, Zikic RV, Stajn A, et al: The effects of selenium on the antioxidant defense system in the liver of rats exposed to cadmium. *Physiol Res* 1995; 44:293–300.
69. Pruszkowska E, Carnick G, Slavin W: Direct determination of cadmium in urine with use of a stabilized temperature platform furnace and Zeeman background correction. *Clin Chem* 1983;29:477–480.
70. Stoeppler K, Brandt M: Contributions to automated trace analysis. Part V. Determination of cadmium in whole blood and urine by electrothermal atomic absorption spectrophotometry. *Fresenius Z Anal Chem* 1980;300:372–380.
71. Roberts C, Clark J: Improved determination of cadmium in blood and plasma by flameless atomic absorption spectroscopy. *Bull Environ Contam Toxicol* 1986;36(4):496–499.

SELF-ASSESSMENT QUESTIONS

1. What is the specimen of choice for assessing chronic arsenic exposure?

2. What is the importance of arsenic speciation?

3. What role does selenium play in mercury toxicity?

4. Does mercury exposure from dental amalgams present an additional body burden or intoxication?

5. What is the target organ of cadmium, and what biomarkers can be used to assess chronic cadmium exposure?

CHAPTER 25

Acute Iron Poisoning

Fred M. Henretig, M.D., Henry R. Drott, Ph.D., and Kevin C. Osterhoudt, M.D.

LEARNING OBJECTIVES

After completing this chapter, the reader should be able to:

1. Describe the epidemiology of childhood iron poisoning.
2. Discuss the pathophysiology of iron toxicity.
3. Discuss the clinical picture of acute iron poisoning.
4. Describe the current state-of-the art analytic techniques in use for the evaluation of iron overdose.
5. Discuss the current treatment modalities in use for iron poisoning.

EPIDEMIOLOGY

Acute iron poisoning is one of the leading causes of pediatric overdose mortality in the United States. Less morbid exposures are very common, with ~20,000 reports to poison control centers each year, mostly in young children (<6 years) (1). Pediatric formulations of multivitamin and iron combinations are often marketed to look like candy or cartoon figures. Adult-intended iron tablets, unfortunately, have also been historically produced to look like candy and are often small, smooth-coated, and easy for toddlers to swallow intact. Frequently prescribed to expectant or postpartum women, they are common around the homes of young children.

It is important to note that iron supplements are typically prescribed to be taken three times per day for several months and may not be thought of as a "serious" medication by parents. Thus, the usual precautions regarding access to the medication or use of child-resistant caps are often less vigilant than optimal. Intentional adult overdoses of iron also occur, particularly in pregnant women. Since 1997 the United States Food and Drug Administration has required that a warning label appear on pharmaceutical iron products, and that products containing more than 30 mg of elemental iron per dose need to be distributed in unit dose packaging (2) (for example, blister packages—Figure 25–1).

ILLUSTRATIVE CASE

Case Description

A previously well 2-year-old boy was brought to the emergency department (ED) with vomiting of 4 hours duration, followed by progressive lethargy. In the ED he was mottled, cyanotic, and minimally responsive to pain. Vital signs included temperature = 100.8 °F (38.2 °C), heart rate = 180/min, blood pressure = 60 mmHg/palpation, and respiration rate = 44/min.

The general examination was notable for a pale, clammy child with prolonged capillary refill, stupor without focal neurologic signs, and a supple neck. The child's clothing was stained by copious green vomitus.

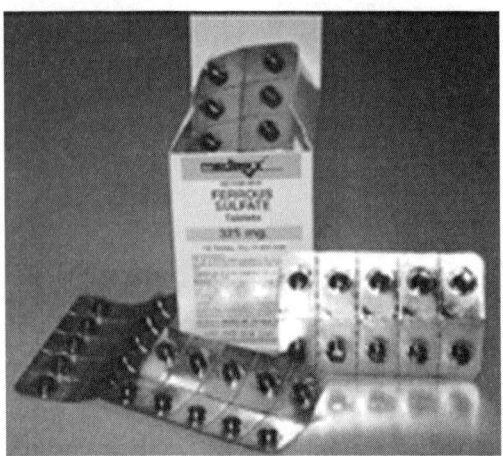

FIG. 25–1. Ferrous sulfate tablets in unit dose packaging.

Differential Diagnosis

The differential diagnosis of this child's presentation is that of acute onset of vomiting followed by altered mental status and hypotension. Several medical, surgical, and toxicologic conditions should be considered. Medical diseases presenting in this manner might include diabetic ketoacidosis (though no history of the classic prodrome of polyuria, polydipsia, and polyphagia was offered), Reye's syndrome (although the child had not recently had varicella or influenza or therapeutic salicylate use), and a fulminant sepsis or meningitis syndrome (this child did have a low-grade fever that was later ascribed to coincident viral illness). Surgical conditions might include gastrointestinal (GI) trauma with perforation (no history of trauma existed, but occult trauma such as child abuse must always be kept in mind), appendicitis with rupture (no history of prodromal abdominal pain and anorexia was offered; though appendicitis, while uncommon, is often misdiagnosed in this age group), and intussusception (similarly to iron poisoning, this malady often presents with altered mentation and bloody stools).

Toxic causes of this syndrome are particularly compelling in the context of the pica-prone-aged toddler (1–4 years) with acute onset of a multisystem disorder with associated alteration in mental and cardiovascular status. Several important considerations include:

- Salicylate intoxication, which may manifest with vomiting, altered mental status (more typically agitation and combativeness, with progression to seizures), and often concomitant hyperventilation, fever, and tinnitus, with shock as a late or preterminal manifestation
- Theophylline, with vomiting, tachycardia, agitation, and tremor and hypotension, seizures, and dysrhythmias in severe cases
- Digoxin, with vomiting, visual changes, bradycardia, and dysrhythmias

Other less likely considerations include miscellaneous drugs such as β-blockers, calcium antagonists, cyclic antidepressants, phenothiazines, and clonidine as causes of hypotension, but usually without prominent vomiting. In the presented case, however, the abrupt onset of vomiting (often with cramping abdominal pain and diarrhea) followed within a few hours by hypotension and lethargy in a toddler is particularly characteristic of acute iron poisoning.

PHARMACOLOGY AND TOXICOLOGY

The body iron content is 3–5 g in adults, 70% as the ferrous form in hemoglobin and myoglobin, and 25% in the ferric state in transferrin, ferritin, and hemosiderin. Iron is absorbed in the duodenum and upper jejunum. Plasma iron is bound in the ferric state to transferrin with typically 30–40% saturation. Iron excretion is very minimal, limited primarily to GI erythrocyte loss and mucosal cell exfoliation (accounting for ~1 mg/d in males) plus menstrual blood loss in women (adding up to 2 mg/d).

Iron toxicity relates to the ingested dose of elemental iron. The most commonly available preparations include ferrous sulfate (20% elemental iron), ferrous fumarate (33% elemental iron), and ferrous gluconate (12% elemental iron). In evaluating the potential for toxicity when the ingested preparation and number of tablets are known, these percentages can be used to estimate the total ingested dose of elemental iron and then can be compared with the child's weight. Overdoses of <20 mg/kg of elemental iron are generally nontoxic, ingestions of 20–60 mg/kg typically result in moderate toxicity with predominantly GI symptoms, and ingestion of >60 mg/kg may lead to threateningly toxic manifestations including shock, acidosis, hepatotoxicity, coagulopathy, and acute respiratory distress syndrome (ARDS). Overdoses of >200–250 mg/kg elemental iron are potentially lethal.

PATHOPHYSIOLOGY

Iron toxicity is manifested by several predictable multiorgan system disturbances. GI toxicity includes direct mucosal corrosion with vomiting, diarrhea, crampy abdominal pain, hematemesis, and melena. The cardiovascular system (CV) is affected by free iron damage to small vessels with consequent postarteriolar dilation, venous pooling, and capillary leak leading to decreased venous return and shock; in addition, some direct cardiomyopathic toxicity is described. Metabolic aberrations include metabolic acidosis with a positive anion gap that is multifactorial. Processes that contribute to metabolic acidosis after iron overdose include (3):

- Mitochondrial dysfunction (lipid peroxidation of mitochondrial membranes)
- In situ conversion of iron from ferrous to ferric state (which releases protons)
- Interference with Krebs cycle enzymes
- Lactic acidosis from the above-noted CV disturbances

Hepatotoxicity occurs as free iron accumulates in the liver, with resultant cloudy swelling, followed by periportal necrosis 3–4 days after ingestion. Fulminant hepatic failure may occur but is rare. The hematologic system is affected with coagulation disturbances including inhibition of thrombin synthesis and decreased thrombin-induced conversion of fibrinogen to fibrin. Central nervous system (CNS) dysfunction is usually related to systemic effects of volume depletion and poor perfusion, though cerebral edema may occur rarely. The lungs may manifest ARDS in severe cases, thought to represent alveolar membrane damage from iron-generated free radicals.

CLINICAL PRESENTATION

The clinical presentation of acute iron poisoning characteristically manifests in four distinct stages (4) (Table 25–1). These include an initial Stage I (30 minutes to 6 hours) of predominantly GI effects including vomiting, diarrhea, and acute GI hemorrhage. In severe

TABLE 25–1. *Clinical stages of acute iron poisoning*

Stage	Time postingestion		Signs and symptoms
I (early acute)	0–6 h	GI:	nausea, vomiting, diarrhea, abdominal pain, melena, hematemesis
		CNS:	lethargy, coma
		CV:	pallor, tachycardia, hypotension
		Other:	hyperglycemia
II (quiescent)	6–24 h	CNS:	intermittent lethargy
III (recurrent)	12–48 h	GI:	hematemesis, melena
		CNS:	lethargy, coma
		CV:	cyanosis, cardiovascular collapse, pulmonary edema
		Other:	metabolic acidosis, leukocytosis, coagulation defects, liver damage, oliguria
IV (late)	4–6 wk	GI:	gastric scarring, pyloric obstruction

Adapted with permission from Henretig FM, Temple AR: Acute iron poisoning in children. *Emerg Med Clin North Am* 1984;2:121–132.

exposures, this stage may also manifest CNS and CV effects (such as lethargy and hypotension). Stage II follows (6–24 hours) and is classically described as "quiescent," with modest improvement of GI, CNS, and CV symptoms, particularly after supportive care with adequate fluid resuscitation. This phase may not occur in severe cases. Stage III (12–48 hours) finds a recurrence of severe systemic effects with GI symptoms, shock, acidosis, coagulopathy, and rarely ARDS. A late Stage IV (4–6 weeks after recovery) is described with evolution of gastric or intestinal strictures, presenting with signs of gastric outlet or proximal intestinal obstruction.

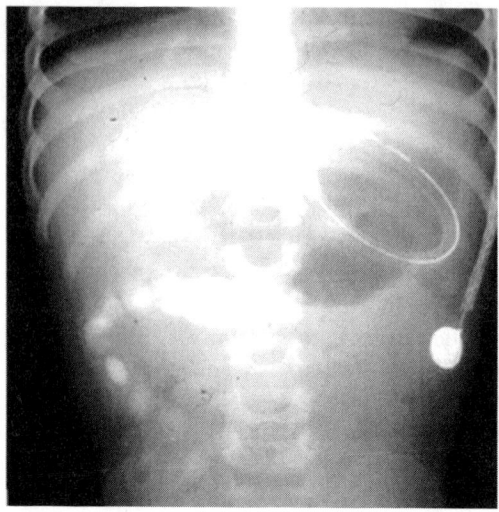

FIG. 25–2. Plain abdominal radiograph displaying intestinal iron pills.

EVALUATION

The history of total iron amount (as noted above) and formulation (serious effects are rare with ingestion of pediatric multivitamin-iron preparations) is important in gauging prognosis. Associated features of early symptom onset, home remedies or efforts at GI decontamination, medical history, suicidal ideation, and/or psychiatric history in intentional overdoses are to be noted.

The physical examination focuses on evidence of abnormal vital signs, particularly heart rate, blood pressure, and capillary refill. CNS depression and signs of GI injury may be noted. Signs of hepatotoxicity and pulmonary toxicity are usually delayed several days and are not apparent at initial presentation.

Plain radiographs of the abdomen may reveal iron pills (Figure 25–2). This finding may confirm the diagnosis and, if present, may indicate repeat films to evaluate the efficacy of subsequent GI decontamination efforts. Pediatric multivitamin with iron products are not typically radiopaque.

Laboratory studies, particularly the serum iron concentration, may help in prognosis and therapeutic decision making. The serum iron 4–6 hours after ingestion is most predictive of the clinical course, with concentrations >500 μg/dL (>89.5 μmol/L) usually toxic, whereas those of 350–500 μg/dL (62.7–89.5 μmol/L) in asymptomatic patients are usually predictive of a benign outcome (see further discussion of analytic considerations below). Serum iron determinations obtained later than 6 hours postingestion are less valid indicators, and concentrations may become undetectable by 24 hours owing to distribution of iron to tissues.

Comparison of the total iron-binding capacity (TIBC) to serum iron was formerly thought to be of prognostic significance (serum iron greater than TIBC was thought to correlate with severe clinical effects). However, the unreliability of the TIBC in the context of acute iron overdose, and the lack of its availability in most hospitals on a stat basis, have made reliance on this test obsolete (3,5).

Several additional nonspecific studies may correlate with a toxic clinical course, including elevated serum glucose [>150 mg/dL (>8.3 mmol/L)] and leukocytosis (>15,000/mm^3), but are not sensitive or specific enough to have clinical utility (6,7). In the absence of an available serum iron concentration, the presence of a wide anion gap metabolic acidosis may be the best laboratory predictor of toxicity. Seriously ill patients obviously need close monitoring of fluid and electrolyte, hematologic, hepatic, pulmonary, and renal function and coagulation status. Critically ill patients may require typed and cross-matched blood for transfusion and frequent monitoring of arterial blood gases and pH. An algorithm for the evaluation of a patient with acute iron poisoning is presented in Figure 25–3.

TREATMENT

The initial approach to acutely iron-poisoned patients involves consideration of options for GI decontamination. Compared with many common overdoses, the choices in this context are problematic. Iron tablets are difficult to pull through a pediatric lavage tube, and iron does not bind well to activated charcoal. Early use of syrup of ipecac may be warranted in some patients, but it complicates later evaluation of the child for evidence of GI symptoms, which are an important prognostic indicator in mildly affected children. Ipecac might be considered for pediatric iron overdoses in the range of >40–60 mg/kg that come to medical attention within 30 minutes of ingestion and who have not exhibited spontaneous vomiting. Patients with overt symptoms on presentation may warrant an attempt at gastric lavage followed by catharsis. Special lavage solutions were formerly recommended in an effort to complex insoluble iron salts in the stomach (for example, phosphates and bicarbonates), but are associated with adverse effects themselves, are hard to prepare in an emergency, and in general have lacked efficacy in specific trials.

Currently, for all but the most trivial exposures, catharsis with a polyethylene glycol-balanced electrolyte formulation (such as Golytely®) until all pills and pill fragments have passed might be considered the decontamination procedure of choice (8,9). A rare patient may require endoscopy or even gastrotomy for embedded pills with rising iron concentrations or evolving signs of an acute abdomen. Possible future approaches undergoing clinical trials now include the intragastric use of magnesium hydroxide or

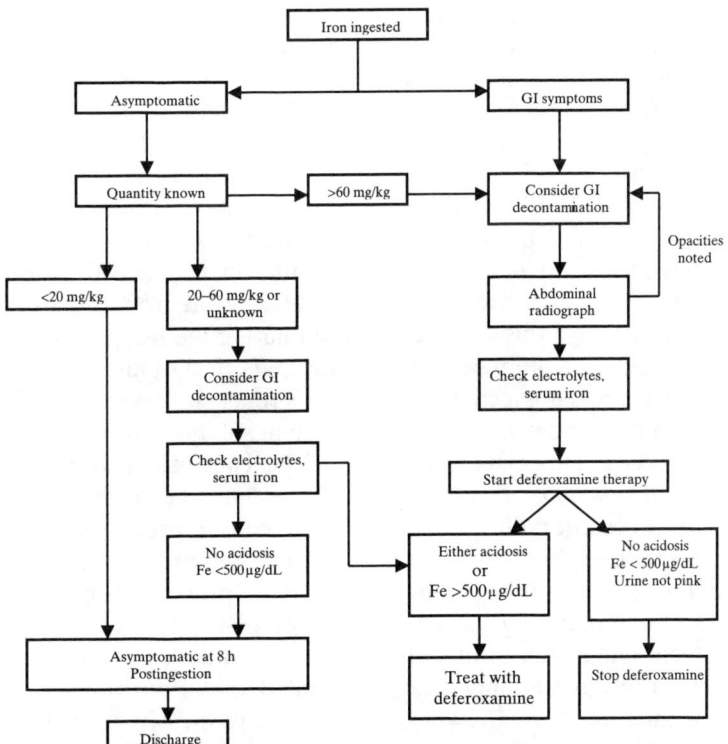

FIG. 25–3. Algorithm for evaluation of patients after acute iron ingestion.

deferoxamine (Desferal, DFO) complexed to activated charcoal.

Chelation therapy is the mainstay of modern treatment for the systemic effects of severe iron poisoning. The specific iron chelator DFO binds iron in a 1:1 molar ratio (100 mg of DFO binds 9 mg of iron) and hastens its excretion in the urine, where its iron complex (feroxamine) confers a brownish-orange coloration. Its volume of distribution (V_d) is that of total body water, 0.6 L/kg, and hence is likely distributed intracellularly and binds both circulating free iron and intracellular iron, though not that in hemoglobin or transferrin (10). Chelation therapy with DFO is indicated for all ill patients, for those with serum iron concentrations >500 μg/dL (>89.5 μmol/L), and for those with significant exposure history with a positive radiograph.

The usual recommended dose for moderate to severely symptomatic cases is 15 mg/kg per hour by continuous intravenous infusion. A higher dose may be considered (for example, 25–30 mg/kg per hour) for the initial treatment of the critically ill patient. It is rarely necessary, and of questionable safety, to continue therapy for >24 hours. The indications to discontinue treatment include resolution of clinical effects, serum iron <150 μg/dL (<26.9 μmol/L), lack of iron pills on abdominal X-ray, return of normal urine color, and resolution of acidosis. Complications of DFO therapy are rare, but have included infection (especially *Yersinia enterocolitica* sepsis) and ARDS with prolonged (>24 hours) infusions (though this has been difficult to distinguish from iron-induced toxicity itself) (11). Pregnancy is not a contraindication for the usual treatment guidelines, because DFO is not believed to cross the placenta significantly, and most morbidity to the fetus in maternal iron overdose is due to complications in the mother herself.

Supportive care for the iron-poisoned patient includes careful monitoring of vital signs, fluid and electrolyte status, urine output, need for blood transfusions, and renal, pulmonary, and hepatic function. The value of extracorporeal techniques of enhanced elimination is negligible. The iron-DFO complex is removed by hemodialysis if acute renal failure should occur.

ANALYTIC CONSIDERATIONS

In the first volume of the series *Advances in Clinical Chemistry*, published some 40 years ago, the first chapter was entitled "Plasma Iron" (12). The immediate tack was to set the course to the universally accepted use of photometric methods. The fundamental steps involved were separation of the iron from its natural "siderophilin" (known today as transferrin), followed by reaction with a suitable chromogenic substance. It should go without saying that an optically clear solution was and still is desirable, although not as absolute as today for selected methods. The avoidance of the common spectral interference from hyperbilirubinemia and hyperlipidemia prevalent in many specimens is still a major focus of analytical procedures. Hemolysis in this test has a twofold adverse effect: hemoglobin iron being measured and optical interference.

The evolution of the extraction of iron from the matrix occurred in three phases. The steps were use of trichloroacetic acid (TCA) solely, use of prior treatment with hydrochloric acid (HCl), and inclusion of a reducing substance. Although TCA was known to extract iron from the plasma, the failure to extract the iron completely was a significant problem for the laboratory. Occlusion of the iron in the precipitate results in diluting the sample, which causes the undesired effect of insufficient color. Hot TCA and repeated extraction all increased the iron in solution but at an analytical price.

The classical analytical method allowed the serum to be incubated with 2N (2 mol/L) HCl before precipitation of the proteins, with TCA to release the iron from the iron-binding proteins. Subsequent observation found that the inclusion of thioglycollic acid facilitated the extraction of iron. It was concluded that iron in the ferric state has a greater propensity to be trapped in the precipitated proteins or phospholipids through absorption or entrapment. Because iron in many complexes is trivalent, the reduction to the ferrous state helps in the decomposition of the complex.

The next step in most procedures is separation. Filtration, centrifugation, or organic extractions were once common modalities of physically separating the iron in a protein-free solution from the precipitated proteins. The migration from a sample requiring physical separation to a sample with iron available for reaction with a chromogenic agent in the presence of solubilized precipitated proteins has reached maturity. Most analytical platforms and assays have perfected methodologies designed as "direct" measurement.

The list of chromogens is large owing to the analytical pressure for greater spectral sensitivity and selectivity. Although thiocyanate is one of the oldest color reagents for iron, the phenanthrolines and the newer heterocyclic compounds containing two or more nitrogen-containing ring structures form well-defined stable complexes with ferrous iron. The molar absorptivity for the common chromogens in 1977 exceeded 30,000 (13). One of the author's (H.R.D.) initial introductions to pediatric clinical chemistry was to develop a micromethod for serum iron, especially for toxicology purposes. Spectral sensitivity for most clinical applications has been achieved; selectivity probability could be further refined. Other metal ions, such as copper and zinc, undergo chemical reaction with the chromogens. Normal plasma contains copper and zinc in trace amounts. However, these ions can be present in higher concentrations due to either extraneous or endogenous factors and react with the chromogen, resulting in misleading data.

Similarly, an extensive number of reducing agents are available. Both inorganic (sodium dithionite) and organic (thioglycollic acid) compounds have been used to reduce iron to the ferrous state. Many combinations of chromogen, extraction, and reducing agents have been tested and will continue to be developed and evaluated. The need for either a chromogen or a system that can overcome the effect of DFO on the assay still exists.

Most published reports on iron ingestion in the United States today deal with pediatric patients, in spite of societal and governmental efforts to dispense the iron preparations in child-resistant containers. The first requirement in a pediatric setting is that the blood volume requirement for this test should be minimized and consistent with other routine clinical analyses such as glucose. The phlebotomy skills must be refined to properly collect the blood without hemolysis. At our institution, a venipuncture using a 25- or 23-gauge needle is routinely performed with minimal hemolysis and limited discomfort to most children. The pneumatic tube delivery system of today can speed the delivery of sample to the laboratory. The use of plastic in the blood-collection materials and the gel-containing tubes forming a physical separation barrier of the erythrocytes from the plasma during centrifugation are advances that have improved the quality of the specimen. The analytical platforms permit the delivery of an analytical result easily within an hour or less. Speed of service is a key factor for the toxicology service to develop and maintain. High-quality reagents, good manufacturing practices, and optimized methodologies are factors that consistently ensure reliable results.

In the case of acute iron toxicity, the known antidote, Desferol, is administered to patients who ingest iron-containing products or who are on chronic transfusion protocols. Because the affinity constant of DFO for ferric iron (10^{31}) is greater than that of transferrin, colorimetric procedures are unreliable after administration of this drug. DFO has a serum elimination half-life of 50 minutes, so specimens should be collected at least 4 hours after the last administration of DFO to minimize assay interference. The toxicologist should look for other avenues to address the problem of analytical interference. For example, the patient who has overdosed with digoxin is treated with digoxin-immune Fab (Digibind), a known interferant with the routine immunoassays for digoxin. This interference has been resolved with a "free" digoxin assay. One approach to the laboratory dilemma is use of atomic absorption spectroscopy, shown to accurately measure iron in the presence of DFO (14). Other pretreatment strategies have been reported for DFO-containing specimens followed by analysis on commercial clinical analyzers.

It is well documented that iron concentrations have a diurnal variation. This diurnal variation can be a complicating factor in the differential diagnosis of iron-deficient anemia. However, this physiological function is not an important factor in acute iron toxicity.

Recent reviews have questioned the reliability of the classical TIBC test in assessing the care of the acute iron poisoning. It is clear that neither a "single test" nor a "single determination" provides a clear and simple approach to the management of a complex clinical presentation (15). Multiple laboratory findings coupled with clinical presentations and radiological evidence are useful screening and management tools used by the physician caring for the potentially intoxicated patient.

REFERENCES

1. Litovitz TL, Klein-Schwartz W, White S, et al: 1999 Annual Report of the American Association of Poison Control Centers Toxic Exposure Surveillance System. *Am J Emerg Med* 2000;18:517–574.
2. Morris CC: Pediatric iron poisonings in the United States. *South Med J* 2000;93:352–358.
3. Siff JE, Meldon SW, Tomassoni AJ: Usefulness of the total iron binding capacity in the evaluation and treatment of acute iron overdose. *Ann Emerg Med* 1999;33:73–76.
4. Mills KC, Curry SC: Acute iron poisoning. *Emerg Med Clin North Am* 1994;12:397–413.
5. Thompson DF: Reassessment of measuring total iron binding capacity in acute iron overdose. *Ann Pharmacother* 1994;28:63–66.
6. Lacouture PG, Wason S, Temple AR, et al: Emergency assessment of severity in iron overdose by clinical and laboratory methods. *J Pediatr* 1981;99:89–91.
7. Palatnick W, Tenenbein M: Leukocytosis, hyperglycemia, vomiting, and positive X-rays are not indicators of severity of iron overdose in adults. *Am J Emerg Med* 1996;14:454–455.
8. Perrone J: Iron. In: Goldfrank LR, Flomenbaum NE, Lewin NA, et al, eds: *Goldfrank's toxicologic emergencies*, 6th ed. Stamford, CT: Appleton and Lange, 1998:618–627.
9. Tenenbein M: Whole bowel irrigation in iron poisoning. *J Pediatr* 1987;111:142–145.
10. Tenenbein M: Benefits of parenteral deferoxamine for acute iron poisoning. *J Toxicol Clin Toxicol* 1996;34:485–489.
11. Howland MA: Risks of parenteral deferoxamine for acute iron poisoning. *J Toxicol Clin Toxicol* 1996;34:491–497.
12. Ramsay WNM: *Advances in clinical chemistry*, vol. 1. New York: Academic Press, 1958;1:1–39.
13. Fairbanks VF, Klee GG: Biochemical aspects of hematology. In: Burtis CA, Ashwood ER, eds: *Tietz textbook of clinical chemistry*, 3rd ed. Philadelphia: WB Saunders, 1999:1701–1703.
14. Helfer RE, Rodgerson DO: The effect of deferoxamine on the determination of serum iron and iron-binding capacity. *J Pediatr* 1966;68:804–806.
15. Roberts WL, Smith PT, Martin WJ, et al: Performance characteristics of three serum iron and total iron binding capacity methods in acute iron overdose. *Am J Clin Pathol* 1999;112:657–664.

SELF-ASSESSMENT QUESTIONS

1. What medication represents the most common fatal pharmaceutical ingestion in toddlers?

2. Why is the peripartum period a particularly dangerous time for unintentional iron ingestions in young children?

3. How does iron toxicity result in metabolic acidosis?

4. What are the principal clinical features of acute iron poisoning?

5. Is measuring iron-binding capacity useful in the laboratory evaluation of acute iron poisoning? Why or why not?

6. Which chelating medication is currently the drug of choice for iron poisoning, and what are the typical indications for its use?

CHAPTER 26

Alternative Samples: Oral Fluid (Saliva), Sweat, Hair, and Meconium*

Edward J. Cone, Ph.D.

LEARNING OBJECTIVES

After completing this chapter, the reader should be able to:

1. Identify appropriate alternative specimens suitable for use in new drug-testing applications.
2. Describe chemical and biologic factors that influence the deposition of drugs and metabolites in alternative matrices.
3. Delineate the time course of appearance and disappearance of drugs in alternative matrices.
4. List advantages and disadvantages encountered when using new alternative matrices for drug testing.

INTRODUCTION

Illicit drug administration is often perceived by society to be risky or antisocial. Such behavior can lead to many unfavorable outcomes for the individual and for society at large. The frequency of illicit drug use within various populations is a subject of much speculation and study. Drug-policy decisions and intervention efforts aimed at reducing illicit drug usage are often predicated on drug-use measurements obtained through self-reports of drug-use history. In addition, clinicians and health-care specialists continually seek accurate means of diagnosis and quantification of drug exposure.

Objective measures of drug use, such as urinalysis, provide a means of determining whether drug use has occurred. The technology of urinalysis has progressed rapidly over the past two decades because of widespread implementation of drug-testing programs by the federal government, the military, and private industry. The need for reliable, inexpensive urine-based drug tests led to significant efforts in research and commercial development of such tests. At the same time, research has progressed on the evaluation of other biological fluids and tissues as useful matrices for drug detection.

Currently, there is growing interest in the use of alternate body fluids and tissues such as oral fluid (saliva), sweat, hair, and meconium for the diagnosis of drug use. Indeed, the Substance Abuse and Mental Health Services Administration presently has a program underway for review of alternate matrices (oral fluid, sweat, and hair) for drug testing. In addition, draft guidelines have been developed for use of alternate matrices in federal workplace drug-testing programs. Approval of these guidelines will create standardized screening and confirmation cutoffs for oral fluid, sweat, and hair in a similar manner as has been established for urine testing. The following discussion provides an

---

*Parts of this work were published in Cone EJ: New developments in biological measures of drug prevalence. NIDA Res Monogr 1997;167:108–129.

overview of drug-testing principles and is followed by a discussion of the potential uses and limitations of urine, saliva, sweat, and hair testing for drugs of abuse as objective measures of drug exposure.

DRUG-TESTING PRINCIPLES

The usefulness of a drug test resides in its ability to accurately detect the presence of parent drug or metabolite in a biological fluid or tissue after human drug administration. This ability has been referred to as the "validity" of the test system (1). This definition reflects both chemical factors that influence test outcome such as sensitivity (the least amount of detectable drug), specificity (how selective the assay is for the drug), and accuracy and pharmacologic considerations including dose, time of drug administration, and route of drug administration. Individual differences in rate of absorption, metabolism, and excretion also are pharmacologic variables that may influence test outcome. With the proliferation of forensic drug testing, this definition of validity has been extended to include confirmation of initial test results by a different chemical method, such as gas chromatography–mass spectrometry (GC/MS). When there is a possibility of litigation, it is extremely important to use assay methods that are highly accurate, reliable, and specific for the analyte of interest.

A plethora of commercial assays and published methodology may be used for drug testing. For the most part, these methods can be grouped into two categories: screening assays and confirmation assays. These assays can be adapted for measurement of drugs in other body fluids and tissues, but must be properly validated before use. Generally, screening assays [immunoassays and thin-layer chromatography (TLC)] are commercial-based tests that are inexpensive and simple to perform. However, few forensic laboratories now use TLC. The labor-intensive nature and subjective endpoint of TLC has led most laboratories to switch to immunologically based methods.

In contrast, confirmation assays (GC/MS) are more expensive and more labor intensive, but specificity is usually higher than with screening tests. Immunoassay-based screening tests may cross-react with a variety of similar chemical substances. For example, most commercial immunoassays for opiates give positive test results for specimens containing either morphine or codeine. In this case, a more specific methodology is needed to distinguish between these two drugs. Often, the less expensive screening tests are used to eliminate specimens containing no drug or drug below the cutoff concentration. The more expensive, labor-intensive tests are used for absolute drug identification and accurate quantification.

URINE

Urine is produced continuously by the kidney as an ultrafiltrate of blood. During urine production, the kidney reabsorbs essential substances and excess water, and waste products such as urea, organic substances, and inorganic substances are eliminated from the body. The daily amount and composition of urine vary widely depending on many factors such as fluid intake, diet, health, drug effects, and environmental conditions. The volume of urine produced by a healthy adult in a 24-hour period ranges from 1 to 2 L, but normal values outside these limits are frequently encountered. Creatinine, a byproduct of muscle metabolism, is present at a relatively constant concentration in blood and is excreted in urine. Consequently, the average 24-hour output of creatinine in urine also is relatively constant. Comparing creatinine concentration in urine with blood provides a means of assessing renal function. For most people, urine creatinine concentrations exceed 20 mg/dL (1.768 mmol/L), although concentrations <20 mg/dL (1.768 mmol/L) are occasionally encountered.

Urine specimens with creatinine concentrations <20 mg/dL (1.768 mmol/L) can be produced by excessive water intake. Drug users who are being urine tested sometimes attempt evasion by drinking large amounts of water or herbal teas in an attempt to dilute drug concentrations below cutoff concentrations (2). Consequently, many laboratories test both for creatinine and specific gravity. A specimen that contains <20 mg/dL (1.768 mmol/L) creatinine and for which the specific gravity is <1.003 is reported as a "Dilute Specimen." Specimens that do not exhibit clinical signs or characteristics associated with normal human urine [creatinine ≤ 5 mg/dL (0.442 mmol/L) and specific gravity ≤1.001 or ≥1.020] are reported as "Substituted." Medical Review Officers (MRO) who review results with abnormally low creatinine concentrations may request retesting of the subject for drugs. Drug-creatinine ratios can be evaluated for evidence of attempted dilution of urine (3). A highly dilute specimen might test negative, but evaluation of the drug-creatinine ratio could provide convincing evidence that the sample would have been positive if normal water intake had occurred. Unfortunately, creatinine normalization procedures are not normally considered by MROs in their evaluation of negative specimens.

Adulteration of urine specimens also has become commonplace. In attempts to foil the efforts of drug testing, many household products have been used as urine additives including salt, vinegar, bleach, soap, and Visine®. In addition, numerous commercial adulteration products are now advertised on the Internet. Products that may affect results such as Urinaid (glutaraldehyde), Klear (potassium nitrite), Whizzies (sodium nitrite), and Urine Luck (pyridinium chlorochromate) can be purchased. As a result, many laboratories now perform specimen validity tests for creatinine concentration, specific gravity, pH, nitrite concentration, pyridine (as a marker for pyridinium chlorochromate), glutaraldehyde, bleach, and soap. Detection of these substances at higher than normal physiological concentration can result in a report of "Adulterated Specimen" to the MRO.

When a drug is administered by the intravenous or smoking route, absorption is nearly instantaneous, and drug excretion in urine begins almost immediately. Absorption is slower when a drug is administered by the oral route, and excretion in urine may be delayed for several hours. Normally, specimens voided within 6 hours after drug administration contain the highest concentration of parent drug and metabolites. Because drug excretion in urine normally occurs at an exponential rate, most of the drug dose of many illicit drugs is eliminated within 48 hours after administration.

Detection times for drugs of abuse vary according to dose, frequency of administration, cutoff concentration, and numerous other factors. Despite wide variance, it is helpful to know average detection times when interpreting urine test data. A list of average detection times and commonly used urine cutoff concentrations is shown in Table 26–1.

ORAL FLUID (SALIVA)

Saliva is secreted primarily by three glands, the parotid, submandibular, and sublingual glands. Secretions from serous and mucosal cells in these glands form saliva. Serous cells secrete watery fluid containing electrolytes and amylases and mucous cells produce mucins (mucoproteins and mucopolysaccharides). Recently, the more correct term "oral fluid," which is a composite of all the fluids and substances in the mouth, has been used interchangeably with "saliva"; however, the bulk of earlier literature utilizes the term "saliva." The flow of saliva depends on neurotransmitter stimulation and can vary widely from zero flow to rates as high as 10 mL/min. The pH of saliva generally is slightly acidic but increases with saliva flow rate from a low of pH 5.5 to pH 7.9. Saliva composition also depends on flow, but generally consists of ~90% water, with the remainder being electrolytes (sodium, calcium, bicarbonates, magnesium, etc.), amylase, organics (glucose, urea, lipids), proteins (low concentrations), and hormones (cortisol, testosterone, estrogens, progesterone).

Drugs may enter saliva by passive diffusion from blood, by ultrafiltration, and by active secretion. Of these processes, passive diffusion represents the most important

TABLE 26–1. Typical screening and confirmation cutoff concentrations and detection times for drugs of abuse in urine

Drug	Screening cutoff concentrations [ng/ml (μmol/L)]	Analyte tested in confirmation	Confirmation cutoff concentrations [ng/ml (μmol/L)]	Urine detection time
Amphetamine	1000 (7.40)	Amphetamine	500 (3.70)	2–4 days
Barbiturates	200 (0.84) (secobarbital calibrator)	Amobarbital, secobarbital, other barbiturates	200 (0.84)	2–4 days for short acting; ≤30 days for long acting
Benzodiazepines	300 (1.05), 200 (0.70) (oxazepam calibrator)	Oxazepam, diazepam, other benzodiazepines	200 (0.70)	≤30 days
Cocaine	300 (1.04)	Benzoylecgonine	150 (0.52)	1–3 days
Opiates	2000 (7.01), 300 (1.05) (morphine calibrator)	Morphine, codeine	2000 (7.01), 300 (1.05)	1–3 days
Marijuana	None	6-Acetylmorphine	10 (0.03)	<12 hours
	50 (0.15), 20 (0.06)	11-nor-9-Carboxy-Δ9-tetrahydrocannabinol	15 (0.04)	1–3 days for casual use; ≤30 days for chronic use
Methadone	300 (0.97)	Methadone	300 (0.97)	2–4 days
Methamphetamine	1000 (6.70)	Methamphetamine, amphetamine	500 (3.35), 200 (1.34)	2–4 days
Phencyclidine	25 (0.10)	Phencyclidine	25 (0.10)	2–7 days for casual use; ≤30 days for chronic use

route of entry for most drugs, with the possible exception of ethanol, a molecule small enough to enter by ultrafiltration. Several reports and reviews have appeared on the occurrence of drugs of abuse in saliva (4-8).

Saliva offers several advantages and some disadvantages compared with urine testing for drugs. The major advantages of saliva as a test medium include its ready accessibility for collection, a less objectionable nature (compared with urine), presence of parent drug in higher abundance than metabolites, and high correlation of saliva drug concentration with the free fraction of drug in blood (Table 26-2). Significant correlations of saliva drug concentrations with plasma also have been reported for cocaine. Cone et al. (9) reported finding significant correlations of saliva cocaine concentrations with plasma concentrations and with responses on self-rating scales for drug sensation, psychotomimetic effects and feelings of rush, and heart rate.

Despite the numerous advantages of saliva, there are some disadvantages. The use of saliva drug concentrations to predict blood concentrations is limited because of the possibility of contamination of saliva from drug use by the oral, smoked, and intranasal routes of drug administration. When drugs are administered by these routes, contamination of the oral cavity and saliva can greatly distort saliva-plasma ratios, thereby distorting useful pharmacokinetic relationships. A recent study demonstrated that both intranasal and smoked cocaine distorted saliva-plasma ratios immediately after drug administration, but contamination cleared rapidly and saliva specimens that were obtained 2 hours after dosing were free of contamination (10). Despite this minor limitation, saliva measurements can be used as evidence of recent drug use, even in some situations in which oral contamination is likely to have occurred, for example, marijuana smoking.

The short time course for detectability of drugs in saliva prevents this biological fluid from being used to detect long-term historical drug use. At the same time, this feature of saliva makes it useful for detection of very recent drug use. Most drugs disappear from saliva and blood within 12–24 hours after administration. There is often a temporal relationship between the disappearance of drugs in saliva and the duration of pharmacologic effects. Consequently, saliva is useful in the detection of recent drug use in automobile drivers and accident victims and for testing employees before engaging in safety-sensitive activities (11).

SWEAT

Sweat is a watery fluid produced primarily by eccrine glands; these glands are distributed widely across the skin surface of humans. The primary purpose of sweat production is heat regulation; consequently, the amount of sweat produced is highly dependent on environmental conditions. Sweat consists mostly of water (99%), with the greatest concentrated solute being sodium chloride. Routine sweat collection has proved difficult in the past because of large variations in the rate of sweat production and the lack of collection devices that are suitable for collection of this type of biological fluid.

A variety of drugs of abuse have been identified in sweat including amphetamine, cocaine, ethanol, methadone, methamphetamine, morphine, nicotine, and phencyclidine. The mechanism for drug entry into sweat is unclear, but it most likely occurs by passive diffusion from blood to the sweat gland. An alternate mechanism could involve drug diffusion through the stratum corneum to the skin surface, where drug would be dissolved in sweat.

A sweat collection device has been developed for the collection of sweat for drug monitoring. The device resembles a Band-Aid® bandage that is applied to the skin and can be worn for several days to several weeks. The "sweat patch" consists of an adhesive layer on a thin transparent film of surgical dressing to which a rectangular absorbent cellulose pad (14 cm^2) is attached. Sweat is absorbed and concentrated on the cellulose pad. The transparent film allows oxygen, carbon dioxide, and water vapor to escape, but prevents loss of drug constituents excreted in sweat. Over several days of wear, sweat saturates the pad

TABLE 26-2. *Comparison of usefulness of urine, saliva (oral fluid), sweat, and hair as a biological matrix for drug detection*

Biological matrix	Drug detection time	Major advantages	Major disadvantages	Primary use
Urine	2–4 days	Mature technology; on-site methods available; established cutoffs	Only detects recent use; specimen adulteration possible; invasive collection procedure	Detection of recent drug use
Saliva (oral fluid)	12–24 hours	Easily obtainable; samples "free" drug fraction; commercial screening methods available and FDA approved	Short detection time; oral drug contamination; collection methods influence pH and saliva-plasma ratios; only detects recent use	On-site testing; workplace testing; linking positive drug test to behavior and performance
Sweat	1–2 weeks	Cumulative measure of drug use; FDA-approved collection and screening methods	Potential for environmental contamination	Detection of drug use during period wearing patch
Hair	Months	Detects long-term or chronic drug use; similar sample can be recollected	Potential for environmental contamination and color bias	Detection of drug use in recent past (1–3 months)

FDA, Food and Drug Administration

and diffuses out through the membrane; drug, if present, slowly concentrates on the pad. The patch is then removed, and the absorbent pad is detached from the device and analyzed for drug content.

Sweat testing for cocaine with the sweat patch was evaluated by Cone et al. (12). Cocaine was administered in doses of 1–25 mg by the intravenous route to four cocaine-experienced, drug-free subjects. Sweat patches were worn for 24–48 hours after drug administration. After removal, the patches were extracted and analyzed for cocaine and metabolites by GC/MS. The primary analyte excreted in sweat was parent cocaine, followed by ecgonine methyl ester and benzoylecgonine. Generally, there appeared to be a dose-concentration relationship; however, there was wide intersubject variability. Limited data were also collected in the same study on the excretion of heroin in sweat. Like cocaine, parent heroin was excreted in sweat along with metabolites that consisted of 6-acetylmorphine and morphine. Drug appeared in sweat as early as 1 hour after drug administration and peaked in concentration within 24 hours.

Advantages of the sweat patch for drug monitoring include high subject acceptability of wearing the patch, a low incidence of allergic reactions to the patch adhesive, and the ability to monitor drug intake for a period of 1–2 weeks with a single patch. In addition, the patch appears to be relatively tamper-proof, that is, the patch adhesive is specially formulated so that the patch can only be applied once and cannot be removed and successfully reapplied to the skin surface.

Disadvantages of the sweat patch include high intersubject variability, the possibility of environmental contamination of the patch before application or after removal, and a risk of accidental removal during a monitoring period. During patch application, extreme care must be taken to cleanse the skin surface before placement of the patch and also to avoid contamination of the cellulose pad during handling. Similar care must be taken when removing the patch and handling for analysis.

Recent studies have demonstrated the utility of the sweat patch for monitoring subjects in treatment programs (13–16). Moderate to good agreement of sweat-patch results with urine-test results has generally been found for methadone, morphine, and cocaine. There are few side effects from wearing sweat patches, and use in a busy outpatient clinic setting was judged to be practical (13).

HAIR

Hair is composed primarily of a fibrous network of keratin strands that are intertwined to form elongated strands. The strands are stabilized by interlinking disulfide and hydrogen bonds that give hair a semicrystalline structure. The inner structure of hair is protected by a layer of cuticle cells that restrict or retard entry of environmental pollutants. As hair ages, the cuticle deteriorates from exposure to ultraviolet radiation, chemicals, and mechanical stresses. Head hair grows at an average rate of 1.3 cm/mo, although there is some variation according to sex, age, and ethnicity (17). Collection of hair for testing is most often performed by cutting locks of hair near the scalp surface at the vertex of the head. During collection, the root and tip of the hair lock are identified for later use. Other types of hair such as pubic, axillary, and arm hair have also been used for drug testing.

Baumgartner et al. (18) reported the first evidence of drug in human hair by analyzing head hair of cocaine abusers by radioimmunoassay for benzoylecgonine, the major metabolite of cocaine. Many other reports have subsequently appeared regarding the presence of drugs in hair. Drug representatives from almost all classes of abused drugs have now been detected in hair (19,20). Presently, research on hair testing for drugs of abuse is performed in many laboratories around the world. In the United States, a limited number of laboratories offer commercial hair testing services for drugs of abuse.

When hair is analyzed for drugs of abuse, the parent drug is often present in greater abundance than its corresponding polar metabolites. For example, the major analyte found in hair of cocaine users is parent cocaine. Benzoylecgonine, the primary urinary metabolite, is present in hair in amounts varying from trace concentrations to approximately one-third of parent cocaine (21). Heroin is found in hair in varying amounts together with 6-acetylmorphine and morphine (22). 6-Acetylmorphine is usually found in greatest abundance in hair; in contrast, conjugated morphine is the major metabolite found in urine.

Although the technology of hair testing has progressed rapidly over the past two decades, several highly controversial aspects of hair testing remain unresolved. It remains unclear how drugs enter hair. The most likely drug entry routes involve diffusion from blood into the hair follicle and hair cells with subsequent binding to hair cell components, drug entry from sweat that bathes hair follicles and hair strands, and drug entry from oily secretions (sebum) in the hair follicle and onto the skin surface. Finally, drug entry into hair can also occur from the external environment.

Possible drug entry from sweat and from the environment is particularly troubling because this suggests that false-positives may be possible if an individual's hair absorbs drugs from the environment or from another person's drug-laden sweat (23). Numerous procedures have been devised by laboratories in an attempt to overcome this problem either by cleansing the hair of contamination by washing before analysis or by monitoring for metabolites that prove active drug use occurred (such as cocaethylene). Another controversial issue in hair analysis is the interpretation of dose and time relationships. Although it has been generally assumed that segmental analysis of hair provides a record of drug usage, studies with labeled cocaine have not supported this interpretation (24). At best, only limited dose and time relationships have been found (25). Other controversial issues that remain unresolved are the possibility of color bias in hair testing (26,27), appropriate means of differentiating drug users' hair from environmentally contaminated hair, appropriate applications of hair testing, the feasibility of hair testing for marijuana usage, and performance differences between different laboratories (28).

Despite the controversial nature of some aspects of hair testing, this technique is being used on an increasingly broad scale in a variety of circumstances. One of the most promising applications of hair testing appears to be its use in prevalence studies. The time record of drug use available from hair is considerably longer than any other biological specimen currently used for drug testing. Self-reported drug use over a period of several months can be compared with hair test results from a hair strand (~3.9 cm length) representative of the same time period. It is expected that this type of comparison will be more effective than urine testing because urine provides a historical record of only 2–4 days under most circumstances. Indeed, Mieczkowski et al. (29) compared self-reported cocaine use with hair and urine analysis in a group of 256 arrestees and found that hair analysis was far more effective than either urinalysis or self-report in detecting drug use. Of the 256 interviewed, 8.5% of the arrestees reported cocaine use within the past 30 days and 21.8% had positive urine tests, whereas 55% had positive hair tests.

Generally, hair analysis provides a longer estimate of drug use than either self-report measures or urinalysis. The wider window of detection is a clear advantage of hair testing as a drug-use prevalence measure. Other advantages include ease of obtaining, storing, and shipping specimens; ability to obtain a second sample for reanalysis; low potential for evasion or manipulation of test results; and low risk of disease transmission in the handling of samples. A potential disadvantage of hair analysis would be its inability to detect recent drug usage because of slow growth rate; however, this has not been thoroughly investigated. Mounting evidence points to the likelihood that drug excretion in sweat is an important route of drug entry into hair (30). This allows the possibility of drug appearing in hair within hours of drug administration. Another consideration regarding the use of hair analysis is the limited number of laboratories offering commercial hair-testing services. Clearly, as demand for hair testing services grows, commercial development also will proceed in simultaneous fashion. In addition, as more attention is focused on this area of drug testing, many of the early controversies may be resolved by additional research.

MECONIUM

According to a 1993 National Pregnancy and Health Survey, the incidence of pregnant women in the general population who abuse cocaine is ~1.1% (31). In high-risk urban populations, prevalence estimates have ranged as high as 44%. Concurrent use of cocaine, ethanol, and other chemical substances is common among women of childbearing age who use cocaine. Meconium begins to be formed by the fetus between the 12th and 16th week of gestation and accumulates until after birth. It is a complex mixture of epithelial cells, squamous cells, residual amniotic fluid, water, and many other products. Drugs of abuse such as cocaine and morphine become sequestered in meconium and remain until birth.

The analysis of meconium for cocaine and metabolites has proved to be a reliable method for the detection of fetal cocaine exposure. Meconium is readily available as a specimen and is easily collected from diapers of newborns. Specimens must be homogenized before analysis. Screening methods generally use immunoassay and are followed by confirmation by GC/MS (32,33). Better sensitivity and a larger gestational window of detection have been demonstrated for meconium testing than for neonatal urine testing. Most drugs of abuse have now been detected in meconium. Advantages of meconium include its ease of collection, its noninvasiveness, and its wide window of drug detection (last two trimesters). Disadvantages include the complexity and nonhomogeneity of the specimen.

SUMMARY

Interest has grown steadily over the past decade in the use of alternate specimens for drug testing. Saliva testing offers different information than urinalysis regarding recency of drug use. The detection times for drugs in saliva are similar to or longer than those for blood (4–24 hours). Consequently, saliva testing may offer the possibility of revealing current drug use that affects an individual's performance in complex psychomotor tasks such as driving and operating heavy equipment. Sweat testing has become feasible through the development of a new sweat patch device designed to collect nonvolatile drugs of abuse from human skin. The device is applied like a Band-Aid® bandage to the skin. Substances with the volatility of water or greater leave the device through a membrane barrier. Less volatile substances such as drugs are concentrated on an absorption pad inside the patch. Subjects can wear the patch for up to several weeks, followed by removal, storage, and analysis of the contents of the absorption pad. Preliminary studies with the sweat patch indicate that it may be useful for detection of single and multiple episodes of drug use over 1–2 weeks. Currently, its usefulness as a quantitative measure of drug use is being evaluated. Hair testing appears to offer the possibility of monitoring drug use over an extended period of time that depends on the length of an individual's hair. Drugs are sequestered in hair and remain bound for an extensive period of time. Because hair grows at an average rate of 1.0–1.5 centimeters per month, analysis of segments of hair for drug content can reveal historical drug use dating back months to years. Recent prevalence studies have indicated that substantially higher drug use rates are generally revealed by hair analysis than by urinalysis or self-report.

How each of the new drug detection technologies will be used in the future for detecting and quantifying drug exposure is uncertain; however, it is clear that even greater reliance will be placed on alternative specimens as complementary or independent methods to urine testing. The technological base and general understanding of the usefulness of saliva, sweat, and hair for drug detection are evolving rapidly. The use of different biological specimens offers uniquely different information about the extent, frequency, and effect of drug use in selected populations.

REFERENCES

1. Gorodetzky CW: Detection of drugs of abuse in biological fluids. In: Martin WR, ed. *Handbook of experimental pharmacology.* Berlin, Germany: Springer-Verlag, 1977:319–409.
2. Cone EJ, Lange R, Darwin WD: In vivo adulteration: excess fluid ingestion causes false-negative marijuana and cocaine urine test results. *J Anal Toxicol* 1998;22:460–473.
3. Huestis MA, Cone EJ: Differentiating new marijuana use from residual drug excretion in occasional marijuana users. *J Anal Toxicol* 1998;22:445–454.
4. Cone EJ, Jenkins AJ: Saliva drug analysis. In: Wong SHY, Sunshine I, eds. *Handbook of analytical therapeutic drug monitoring and toxicology.* New York: CRC Press, 1997:303–333.
5. Kidwell DA, Holland JC, Athanaselis S: Testing for drugs of abuse in saliva and sweat. *J Chromatogr B* 1998;713:111–135.
6. Samyn N, Verstraete A, van Haeren C, et al: Analysis of drugs of abuse in saliva. *Forensic Sci Rev* 1999;11:1–19.
7. Huestis MA, Cone EJ: Alternative testing matrices. In: Karch S, ed: *Drug abuse handbook.* Boca Raton, FL: CRC Press, 1998:799–857.
8. Liu H, Delgado MR: Therapeutic drug concentration monitoring using saliva samples. *Clin Pharmacokinet* 1999;36:453–470.
9. Cone EJ, Kumor K, Thompson LK, et al: Correlation of saliva cocaine levels with plasma levels and with pharmacologic effects after intravenous cocaine administration in human subjects. *J Anal Toxicol* 1988;12:200–206.
10. Cone EJ, Oyler J, Darwin WD: Cocaine disposition in saliva following intravenous, intranasal, and smoked administration. *J Anal Toxicol* 1997;21:465–475.
11. Skopp G, Potsch L: Perspiration versus saliva—basic aspects concerning their use in roadside drug testing. *Int J Legal Med* 1999;112:213–221.
12. Cone EJ, Hillsgrove MJ, Jenkins AJ, et al: Sweat testing for heroin, cocaine, and metabolites. *J Anal Toxicol* 1994;18:298–305.
13. Taylor JR, Watson ID, Tames FJ, et al: Detection of drug use in a methadone maintenance clinic: sweat patches versus urine testing. *Addiction* 1998;93:847–853.
14. Kintz P, Brenneisen R, Bundeli P, et al: Sweat testing for heroin and metabolites in a heroin maintenance program. *Clin Chem* 1997;43:736–739.
15. Preston KL, Huestis MA, Wong CJ, et al: Monitoring cocaine use in substance–abuse-treatment patients by sweat and urine testing. *J Anal Toxicol* 1999;23:313–322.
16. Huestis MA, Cone EJ, Wong CJ, et al: Monitoring opiate use in substance abuse treatment patients with sweat and urine drug testing. *J Anal Toxicol* 2000;24:509–521.
17. Saitoh M, Uzuka M, Sakamoto M. Rate of hair growth. In; Montagna W, Dobson RL, eds: *Advances in biology of skin, vol. IX. Hair growth.* Oxford: Pergamon, 1969:183–201.
18. Baumgartner WA, Black CT, Jones PF, et al: Radioimmunoassay of cocaine in hair: concise communication. *J Nucl Med* 1982;23:790–792.
19. Gaillard Y, Pepin G: Testing hair for pharmaceuticals. *J Chromatogr B* 1999;733:231–246.
20. Nakahara Y: Hair analysis for abused and therapeutic drugs. *J Chromatogr B* 1999;733:161–180.
21. Cone EJ, Yousefnejad D, Darwin WD, et al: Testing human hair for drugs of abuse. II. Identification of unique cocaine metabolites in hair of drug abusers and evaluation of decontamination procedures. *J Anal Toxicol* 1991;15:250–255.
22. Cone EJ, Welch P, Mitchell JM, et al: Forensic drug testing for opiates: I. Detection of 6-acetylmorphine in urine as an indicator of recent heroin exposure; drug and assay considerations and detection times. *J Anal Toxicol* 1991;15:1–7.
23. Smith FP, Kidwell DA: Cocaine in hair, skin swabs, and urine of cocaine users' children. *Forensic Sci Int* 1996;83:179–189.
24. Henderson GL, Harkey MR, Zhou C, et al: Incorporation of isotopically labeled cocaine and metabolites into human hair: 1. Dose-response relationships. *J Anal Toxicol* 1996;20:1–12.
25. Pragst F, Rothe M, Spiegel K, et al: Illegal and therapeutic drug concentrations in hair segments—a timetable of drug exposure. *Forensic Sci Rev* 1998;10:81–111.
26. Kidwell DA, Lee EH, DeLauder SF: Evidence for bias in hair testing and procedures to correct bias. *Forensic Sci Int* 2000;107:39–61.
27. Kronstrand R, Forstberg-Peterson S, Kagedal B: Codeine concentration in hair after oral administration is dependent on melanin content. *Clin Chem* 1999;45:1485–1494.
28. Wennig R: Potential problems with the interpretation of hair analysis results. *Forensic Sci Int* 2000:107:5–12.
29. Mieczkowski T, Barzelay D, Gropper B, et al: Concordance of three measures of cocaine use in an arrestee population: hair, urine, and self-report. *J Psychoactive Drugs* 1991;23:241–249.
30. Joseph RE Jr, Hold KM, Wilkins DG, et al: Drug testing with alternative matrices II. mechanisms of cocaine and codeine disposition in hair. *J Anal Toxicol* 1999;23:396–408.
31. Moore C, Negrusz A: Drugs of abuse in meconium. *Forensic Sci Rev* 1995;7:103–118.
32. Kadehjian L. Drug testing of meconium: determination of prenatal drug exposure. In: Wong SHY, Sunshine I, eds: *Handbook of analytical therapeutic drug monitoring and toxicology.* New York: CRC Press, 1997:265–279.
33. Moore C, Negrusz A, Lewis D: Determination of drugs of abuse in meconium. *J Chromatogr B* 1998;713:137–146.

SELF-ASSESSMENT QUESTIONS

1. What are the relative detection times for cocaine appearing in saliva, sweat, and hair?

2. Which alternative specimen provides drug-testing results most likely to correlate with impairment?

3. What are the principal advantages of sweat and hair for drug testing?

4. Meconium drug testing provides information about in utero drug exposure over what period of pregnancy?

CHAPTER 27

Advanced Analytical Techniques*

Larry D. Bowers, Ph.D., DABCC, FACB

LEARNING OBJECTIVES

After completing this chapter, the reader should be able to:

1. List five types of liquid introduction interfaces for mass spectrometry (MS) and an advantage and limitation of each.
2. List five types of mass analyzers and an advantage and limitation of each with respect to toxicology.
3. Describe the difference between mass scans and selected ion monitoring in the quadrupole mass spectrometer.
4. Describe the function of tuning a mass spectrometer and its effects on analytical results.
5. Define the following terms associated with ion-trap mass spectrometers: mass instability scan, axial modulation, automatic gain control, and selected ion storage.
6. Describe the difference between MS-MS in space and MS-MS in time.
7. List two types of MS-MS analyzers.
8. Define "characteristic ion" as it applies to identification criteria by selected ion monitoring.
9. List three factors that affect the quality of a library spectrum match.
10. Describe the effects of ion suppression on electron ionization and electrospray ion sources.
11. Discuss the advantages and disadvantages of the use of electrospray high-performance liquid chromatography–MS for quantitative and qualitative analysis.
12. Compare the operation of electrospray and atmospheric pressure chemical ionization for quantitative analysis of drugs and provide three examples of each.

INTRODUCTION

Advanced analytical techniques are used in toxicology and therapeutic drug monitoring to increase the information available about the identity of an unknown compound, to improve the specificity of detection, or to improve limits of detection for trace concentration compounds. The analytical problems in toxicology can be divided into two categories:

- Threshold determinations, in which the measurements are made relative to an administratively or scientifically based threshold concentration
- Nonthreshold determinations, in which any signal that meets acceptance criteria is to be reported

Interpretation of the results of an analysis in either category are affected by whether the analysis is quantitative or qualitative, whether identification of the compound is required, and whether the analyte concentration is high or near the limit of detection

---

*Reprinted with permission from Bowers LD: Advanced analytical techniques. In: Shaw LM, Kwong TC, Orsulak PJ, et al, eds: *Contemporary practice in clinical toxicology.* Washington, DC: American Association for Clinical Chemistry, 2000:24-1–24-16.

of the technique. There are four main areas of concern:

- Quantitative results when the signal-to-noise ratio is large (= 50)
- Quantitative results when the signal-to-noise ratio is close to the limits of detection
- Identification when the signal-to-noise ratio is large (= 50)
- Identification when the signal-to-noise ratio is close to the limits of detection

Matrix considerations can also affect the selection of the analytical technique and ancillary parts of the procedure such as sample cleanup. To ensure that the data obtained are accurate, the laboratory should verify that the methodology is fit for the analytical purpose and that the instrumentation is properly used, maintained, and validated.

This chapter will focus on these considerations when gas chromatography (GC) and high-performance liquid chromatography (HPLC) are coupled to either single mass spectrometry (MS) or tandem MS (MS-MS) techniques. Although there are other advanced analytical methods used in toxicology, these have been considered the gold standard for confirmatory analyses and have increasing potential for use as reference methods for therapeutic drug monitoring.

INSTRUMENTAL CONSIDERATIONS

Chromatographic Systems

Both GC and HPLC are mature analytical techniques. A detailed discussion of the principles of GC and HPLC are beyond the scope of this chapter, but has been done elsewhere, including the interfacing of the techniques to MS (1–4). In GC, the degree of separation between the different components in a mixture is affected by the chemical nature of the analyte or derivative, the mobile-phase flow rate, the liquid stationary phase used, and the temperature and/or rate of temperature change. The strength of capillary GC is its inherent peak-resolving capacity, which is due to the narrow chromatographic peaks obtained. The sample-injection system remains one of the problematic areas of GC analysis. Maintenance of the GC system is an important consideration for the production of valid analytical data (2).

In HPLC, the separation is determined by the chromatographic selectivity of the column–mobile phase system. Analytical HPLC can be performed in reverse-phase, normal-phase, or ion-exchange mode. The majority of separations have been done in reverse-phase mode, although method development using other systems has increased in recent years (3). The strength of HPLC is its ability to separate a wide variety of compounds that cannot be analyzed by GC, including peptides and proteins. Nevertheless, if a separation of small molecules can be readily performed by GC, it remains the method of choice (1). The weakness of HPLC is the relatively low resolving capacity of the technique. The high degree of selectivity of the mobile phase–stationary phase system is offset by the relatively broad chromatographic peaks.

The recently developed capillary electrochromatography (CEC) and capillary electrophoresis (CE) have potential importance in analytical toxicology. CE is a separation technique, but not a chromatographic technique because it does not involve partition between a stationary and a mobile phase. CE can resolve large numbers of compounds based on differences in their charge and electrophoretic mobility (5). The technique is particularly well suited to the separation of macromolecules. Separation of neutral molecules has been reported using micellar additives to the supporting electrolyte (micellar electrokinetic chromatography), although this approach has numerous limitations. The supporting electrolyte moves through a capillary tube and past the detector by means of electro-osmotic flow. Typical column dimensions are 0.32 mm i.d. × 25 cm. Because of the extremely narrow chromatographic bands (due to eluent volumes in nanoliters), in-column detection is most prevalent using ultraviolet absorbance, diode-array detectors, fluorescence, and laser-induced fluorescence. MS detectors have been used. The major limitation for trace toxicological work is the very small sample capacity. Thus, while impressive limits of detection are reported (for example, in picograms), the fact that only nanoliters of sample can be injected makes the limits of detection for the injection solution on the order of micrograms per milliliter.

CEC is a hybrid of HPLC and CE (6). The driving force to percolate the mobile phase through the column is electroosmosis. Thus, there is no pressure drop, and extremely small particles for stationary-phase support can be used. This, combined with the plug flow profile maintained with electromotive flow, results in peak capacities and widths approaching those of capillary GC. Interestingly, because the electrical field strengths are in the kilovolts per centimeter range, separation occurs due to both partition between phases and electrophoretic mobility. The limitations for CEC at present are the limited sample capacity (submicroliter injections), variability in column performance, and the lack of availability of commercial instrumentation.

Chromatographic–Mass Spectrometric Systems

Interfaces and Ion Sources

Many of the recent advances in the application of MS to toxicological problems have been the result of improved methods of introducing both gas and liquid sample streams into the mass analyzer. For the purposes of this chapter, the interface and ion source will be considered together. The purpose of the interface is to facilitate the transfer of analyte from the chromatographic system to the mass spectrometer. This is more easily accomplished for a continuous stream of gas than for a comparable stream of liquid, due to the volume of gas produced when the respective phase is expanded into the vacuum required for operation of the mass analyzer. An ideal source of ions for MS has the following characteristics:

- Stable, intense ion current
- High ionization efficiency
- Production of ions characteristic of the analyte structure
- Molecular ion
- Controllable, reproducible formation of fragment ions
- Compatibility with chromatographic systems
- Independence of matrix effects

The ideal ion source does not exist. Recognition of the limitations of the interface and ion source for chromatographic–mass spectrometric systems is important for their effective application.

Continuous-Flow Gas Introduction Interfaces

GC is well suited for interfacing with a mass spectrometer. With the advent of capillary GC with gas flow rates on the order of a milliliter per minute, direct interface of the column into the ion source has been routinely achieved. With the availability of electronic flow control, constant flow conditions provide more consistent ion source performance over the range of temperatures used in a temperature-programmed GC run. Despite the general perception that the GC/MS interface is free of interference, ion suppression has been reported in GC/MS when co-eluting substances are present in sufficiently high concentrations (7).

Either electron ionization (EI) or chemical ionization (CI) of the GC effluent achieves a continuous production of ions. In EI, energetic cation radicals are produced from the interaction of the analyte molecules with an electron beam generated by the filament in the ion source. The radical cations fragment to form a cation and a neutral. One or more electronic lenses focus the cation beam produced into the mass analyzer. The lenses also serve the function of providing the ions with a relatively homogenous momentum or velocity, which is important for separation of the ions in the mass analyzer.

In CI mode, an ionized reagent gas reacts with the GC effluent in a specially designed ion source to provide a less-energetic CI of the sample (8). The most common reagent gases are methane, ammonia, and isobutane. The most common type of CI reactions resulting in positive ions is proton transfer. There is usually little fragmentation in CI compared with EI. This limited fragmentation can decrease the amount of identification information present in the spectrum. Negative ions can be produced either by electron capture of thermalized electrons or reaction with proton-abstracting reagents such as O^-. The former has become a method of choice for molecules containing electronegative atoms, such as the halogens contained in the benzodiazepine drug class (Table 27–1). Negative ion chemical ionization (NICI) should be clearly indicated, because by convention CI is used for positive ions.

Continuous-Flow Liquid Introduction Interfaces

Interfacing a continuous liquid stream with the vacuum system of a mass spectrometer has been more difficult. All liquid-phase introduction interfaces must accomplish four things:

- Nebulization. This consists of dispersing the liquid into fine droplets so that evaporation of the bulk solvent is enhanced.
- Transportation. The molecules must be transferred from the near-atmospheric pressure region to the vacuum surrounding the mass analyzer.
- Ionization. Formation of ions in the near-atmospheric pressure region results in CI with the formation of even-electron ions.
- Desolvation. The tightly bound water surrounding the ion must be eliminated; otherwise, the mass spectral data would reflect cluster mass and not analyte mass. The desolvation requires significant energy and is usually accomplished by countercurrent flow of gas or elevated temperature in a narrow tube.

In practice, multiple operations may occur together (for example, nebulization and ionization). Although several interfaces have been described, the electrospray (ES) and atmospheric pressure chemical ionization (APCI) interface approaches have become most popular for HPLC-MS (4). The most common liquid interface techniques are summarized in Table 27–2.

In ES ionization, a high voltage is placed onto the liquid contained in a very small diameter tube. The voltage itself is sufficient to nebulize the liquid, but only at flow rates below 10 μL/min. Above this flow rate, a

TABLE 27–1. Summary of gas-phase ionization techniques

Method	Ionizing agent	Pressure	Characteristics	Applications
Electron ionization	Electrons (~70 eV)	~10^{-5} Torr	Reproducible spectra, extensive fragmentation	Innumerable
Chemical ionization	Gaseous ions (>100-fold excess of reagent gas)	~ 1 Torr	Molecular ions and controllable fragmentation	Sensitive analyses
Chemical ionization (negative ion)	Thermalized electrons	~ 1 Torr	High sensitivity due to 100-fold increase in electron mobility relative to chemical reagent ions	Benzodiazepines; halogen derivatives

TABLE 27-2. *Liquid-MS interfacing techniques*

Interface	Nebulizing source	Ionizing agent	Comments
Atmospheric pressure chemical ionization (APCI)	Pneumatic	CI reagent ion formed from corona discharge	Robust interface; broad range of compound applicability
Electrospray	Electrostatic	Ion desorption	Flow rate limited to 10 µL/min if no nebulization assist; difference in efficiency with solvent composition; multiply charged ions; ion suppression
Thermospray	Thermal	CI in gas phase at near-atmospheric pressure	Very temperature-dependent spectra; can be subject to solvent noise; sometimes requires discharge for ion formation
Particle beam (MAGIC)	Pneumatic	EI, CI at 10^{-5} Torr	Limited to relatively volatile (nonpolar) compounds; not sensitive; library-searchable spectra
Moving belt	Thermal	EI, CI at 10^{-5} Torr	Limited to volatile solvents; excellent enrichment; mechanically complex; carryover problems
Direct liquid injection (DLI)	Supersonic expansion	EI, CI at 10^{-5} Torr	Limited to very low flow rates; inefficient; mechanically simple

method of assisting nebulization, such as pneumatic or ultrasonic devices, must be used. The ES is a concentration-dependent detection technique, so the response is optimal at low flow rates; nebulization assistance basically provides a gas phase splitting of the liquid flow. Ionization with ES is thought to occur on the highly charged surface of the microdroplets that form as a result of evaporation and Raleigh disruption. The best response is obtained from molecules that are charged in the liquid state. The production of multiply charged ions has allowed the detection of high-molecular-weight species such as polymers and proteins. For low-molecular-weight species, adducts of the molecule with protium, ammonium, sodium, and potassium can be observed. Because the charge on the droplets is limited, ion suppression has been reported for analyses in which the ionization of co-eluting substances overwhelms the ionization of the analyte of interest. This can be severe enough to result in lack of detection of the analyte.

In APCI ionization, ionization is thought to occur in the gas phase after neutral molecules "evaporate" from the surface of the liquid droplets. The molecules interact with a rich plasma of protons to produce protonated species. Thus, APCI is similar to CI in GC. In general, the energy of ionization for APCI is greater than for ES, and frequently some fragmentation is observed. Because the energy required to "evaporate" a neutral molecule from a liquid surface is much less than that required for a charged molecule, APCI is more effective for neutral molecules and is seldom observed for multiply charged species in solution. Due to the external source of ionization, APCI is much less subject to ion suppression than ES. APCI is a mass flow–dependent (milligrams per minute) detection technique rather than a concentration-dependent technique. This trait suggests that constant flow is required for a concentration-dependent response.

Mass Analyzers

The mass analyzer is the device that separates ion fragments formed in the source according to their mass-to-charge (m/z) ratio. The most frequently used mass analyzer is the quadrupole mass spectrometer, although there has been increased interest recently in another analyzer based on the same operating principles, the ion trap. The advantages of these analyzers is their relatively low cost, small size, and relative ease of operation under computer control. Recently there has been increased interest in other mass analyzers such as magnetic-electrical sector analyzers and time-of-flight (TOF) analyzers. Sector analyzers have excellent ion transmission and high mass-resolution capabilities and thus can provide increased selectivity and sensitivity in an analytical scheme. TOF mass analyzers depend on the time required for an ion to move from the source region to the detector and thus can provide extremely fast scan times. These are well suited to fast GC, in which peak widths are on the order of tenths of seconds. TOF analyzers can also be configured to provide relatively high mass resolution, making them a potential alternative for sector instruments in some applications. The advantages and disadvantages of a variety of mass analyzers are shown in Table 27–3.

Quadrupole Mass Spectrometers

The separation of ions in a quadrupole mass analyzer occurs by varying the radio frequency (RF) and direct current (DC) voltages on an opposing set of quadrupole

TABLE 27–3. Advantages and disadvantages of mass analyzers in toxicology

Mass Analyzer	Advantages	Disadvantages
Magnetic-electric sector	High mass resolution Highest sensitivity and lowest limit of detection Capable of high-energy MS-MS linked scans	Relatively low scan speed Large physical size Expensive Limited mass range in SIM modes
Quadrupole mass spectrometer	High scan speed Inexpensive Small physical size Capable of low-energy MS-MS scans and SRM (MS-MS in space)	Limited (unit) mass resolution Mass-dependent ion transmission (required tune) Low-energy collision-induced dissociation MS-MS spectra depend on collision cell design, energy, collision gas, pressure, and other factors
Ion trap (Paul trap)	High scan speed Inexpensive Small physical size Unique chemical ionization modes Capable of low-energy MS-MS scans and SRM (MS-MS in time)	Limited dynamic range (space charge effect) Ion-molecule interactions Unfamiliar ion manipulation procedures Low-energy collision-induced dissociation MS-MS spectra depend on collision cell design, energy, collision gas, pressure, and other factors
Time of flight	Extremely high scan speed High ion transmission Highest mass range	Requires pulsed ion source or ion beam switching Fast detectors required may limit dynamic range
Ion cyclotron (FTMS) (Penning trap)	Extremely high mass resolution Nondestructive ion measurement Extensive ion manipulation possibilities	Limited dynamic range (space charge effect) Requires very high vacuum Ion-molecule interactions Requires superconducting magnet Requires extensive operator expertise

SRM, selected reaction monitoring.

rods. The alternating fields created by the changing voltages cause the ions to spiral down the center of the quadrupole rods according to their m/z ratio. As shown in Figure 27–1, a set of RF and DC voltages roughly described by a triangle permits stable trajectory of a particular m/z ratio ion through the mass analyzer. Operated in the spectral scan mode, the voltage applied to the rods is a linear ramp that bisects the upper portion of the stability triangle. Quadrupole mass spectrometers must be tuned, which entails adjusting the voltages and the slope of the scan line to achieve transmission of the correct m/z ion and with an acceptable relative intensity.

In the selected ion monitoring (SIM) mode, a specific DC and RF voltage corresponding to the apex of the triangle is applied to ensure the passage of essentially a single m/z ratio ion. In SIM, a dynamic tuning procedure should be followed. In this approach, data are acquired every 0.1 atomic mass unit across the GC peak around the nominal mass peak. The maximum peak area response defines the mass setting that should be used for the SIM analysis.

Quadrupole Ion Traps (Paul Trap)

Paul and Steinwedel described the ion trap in 1953 at the same time as the quadrupole mass spectrometer. Analytical interest in the ion trap has only emerged in the last decade. Although the ion trap is operated with oscillating electrical fields like the quadrupole mass spectrometer, there are significant differences in the performance of the instruments. These differences result in some concern among toxicologists that the results obtained from the ion trap are less reliable. The ions formed in the trap are retained by imposition of an RF-only potential. Note in Figure 27–1 that when the DC voltage is zero, all ions have stable trajectories. The ions can be selectively ejected from the end caps of the trap by ramping only the DC voltage. When the edge of the stability triangle is reached (see point B in Figure 27–1), the ion of that m/z ratio is selectively ejected from the ion trap. This mode of operation is called mass selective ejection or a mass instability scan. The fact that all ions are stored in the trap creates some potential problems in that the ion cloud in the trap can shield ions of a particular m/z from the applied field (space charging), and mass misassignment may occur. A variety of computer-controlled functions, such as automatic gain control, has been added to improve mass peak shape and system performance (9,10).

There is no analog of SIM on the ion trap. Because all ions are stored in the ion trap under full scan conditions, an improvement in limit of detection should be achieved with selective storage of the ions of interest and ejection of those ions arising from matrix. Several such selective ion storage techniques have been described (10). Bowers and Borts (11) showed that the improvement in limits of detection is related to the specific storage technique used and the proximity of the interfering ion to the ion of interest. Interestingly, two groups have shown that for drug testing, the full scan GC/MS mode for an ion

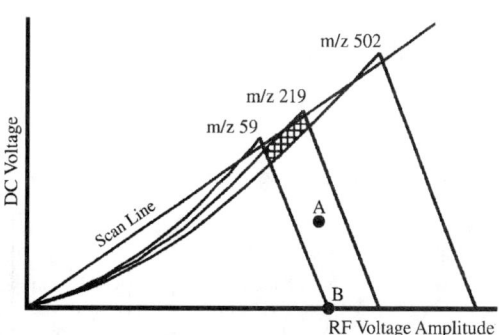

FIG. 27–1. Stability diagram for operation of quadrupole mass spectrometers and ion traps. When the RF frequency and analyzer dimensions are fixed, the stability diagram triangles can be plotted for individual m/z ions. Areas of non-overlap between the stability "triangles" for different m/z ions (the hatched area) are regions where only one ion has a stable trajectory and will be detected. At point A, both m/z 219 and 502 would be detected, whereas m/z 59 would be ejected. At point B, m/z 59 would be ejected using an RF-only mode. In the ion trap, ions are ejected and detected when their trajectory is unstable.

trap is comparable in sensitivity to the SIM mode for quadrupole mass spectrometers (12,13). Given the differences in acquisition mode, this is not surprising.

Ionization can be achieved inside the trap (internal) with a gated electron beam that passes through the trap. Alternatively, the analyte can be ionized in an external source; the ions can subsequently be injected into the trap for mass analysis. These two approaches may vary significantly in performance. Because neutral molecules do not enter the ion trap in the external ionization mode, the potential for charge transfer reactions during the ion storage period is greatly diminished. The result of charge transfer reactions is sometimes observed as an unusually high relative abundance of protonated molecular ion in the mass spectrum. This can cause poor library matches in library searching schemes if spectra acquired with an ion trap are not included in the library. A similar problem arose when the quadrupole mass spectrometer was introduced: Library spectra were predominantly acquired on sector instruments. It remains to be seen whether external or internal ionization will dominate in GC/MS applications. External ionization must be used to interface the ion trap to HPLC.

CI on the ion trap is also somewhat different from its counterpart on the quadrupole mass spectrometer. Ion traps are advantageous in that reagent ions can be generated and stored in the trap, allowing much longer reaction times and, consequently, the use of single reagent species and "liquid" reagent gases such as acetonitrile. Ion traps with internal ionization cannot perform negative ionization because thermalized electrons cannot be stored in the trap. Although it is theoretically possible to perform NICI with reagents to advantage due to the longer reaction times, commercial traps do not have the requisite software to store negative ion reagent gases.

Tandem Mass Spectrometry

The technique of MS-MS offers the potential for increased specificity in analysis. After selection of a precursor ion from a first mass analyzer, the ion undergoes a second fragmentation after collision with a neutral gas molecule. Subsequent to this collision-induced dissociation (CID), the product ion fragments are separated in a second mass analysis step. Both high-energy (kilovolts) and low-energy (volts) collision energy experiments can be done. A multiple sector instrument must be used for kilovolt collision

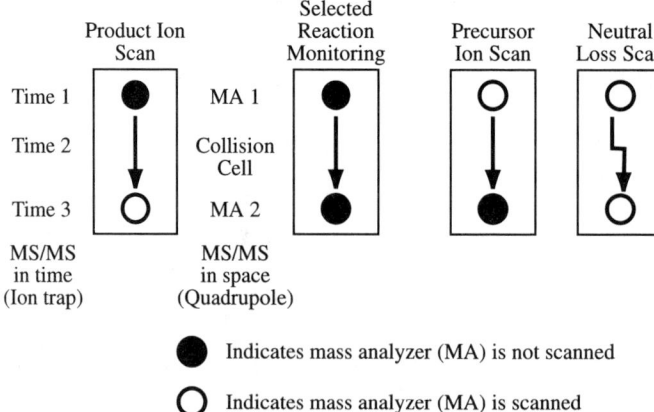

FIG. 27–2. Schematic diagram of the operational modes of tandem mass spectrometry. A closed circle indicates that a single ion is selected in the mass analyzer, and an open cirlce indicates that all ions are scanned from the analyzer. A straight arrow indicates that the product ions are directly transmitted to the second analyzer. The offset arrow indicates that a fixed mass difference is maintained between the two mass analyzers.

energies, and the collision gas is usually helium. Tandem quadrupole mass spectrometers can be used for low-energy collision work, and the collision gas is usually a larger target such as argon. The fragmentation observed is a function of the collision energy, the collision gas thickness, and the collision gas used. The techniques and applications of MS-MS have been reviewed (14,15).

The various MS-MS operational modes are summarized in Figure 27-2. The product ion scan is the most common qualitative MS-MS technique. As shown, the first mass analyzer isolates a single precursor ion and the second mass analyzer is scanned to identify the product ion masses. Selected reaction monitoring is the MS-MS analog of SIM, in which the mass of the precursor ion(s) and product ion(s) are known, and the mass analyzers are stepped from mass to mass to optimize the data acquisition time. In the precursor scan mode, the mass of the product ion is fixed, and the first mass analyzer is scanned to determine the mass of the precursor ion(s). This can be useful to determine metabolites present in a sample if a characteristic product ion is formed. Finally, the mass difference between the two mass analyzers can be maintained so that any precursor ions giving rise to a common neutral fragment loss will be detected.

The implementation of MS-MS on the ion trap is somewhat different than the MS-MS in space approach described above in which the ions occupy different regions of the mass spectrometer during each mass analysis and collision. Because the ions are stored in the ion trap, conditions can be established in which a precursor ion is selected, and in the same space but at a different time, undergoes CID and mass analysis of the product ions. This approach has been called MS-MS in time. The advantage of this approach is the relatively low cost of adding the MS-MS function, because it is primarily software driven. The disadvantage is that only the product ion scan function can be performed.

MS-MS is particularly well suited to situations where primarily molecular ions or molecular ion adducts are formed during ionization. Thus, GC/MS-MS with CI and HPLC-MS-MS are ideally matched.

IDENTIFICATION AND QUANTIFICATION ISSUES IN GC/MS

As mentioned above, the primary rationale for the use of techniques such as GC/MS and HPLC-MS is to improve the selectivity of quantification and the confidence that the analyte is correctly identified. The quantitative aspects of GC/MS are well established and will not be further discussed here. Acceptable quantitative data can be obtained from total ion current data, extracted ion profile data, or SIM data. The determining factor is primarily the concentration of analyte, with SIM being the most specific and having the lowest limits of detection. The use of GC/MS-MS as a routine analytical tool has been well established, and again the degree of improvement of quantification depends on the background interference and the specificity of the precursor-product ion pair.

For identification, it is generally accepted that a full mass spectral scan provides the most complete information when the concentration of analyte is large enough to obtain a good spectrum. As an alternative to a full scan, a scan of the region of the mass spectrum that contains characteristic ions from the compound of interest can be used. A recent proposed guideline for confirmation of drugs by GC/MS addresses many of these issues (16). Information theory has also been applied to qualitative GC/MS data to assess the accuracy of the technique in identification (17,18).

Computer-assisted library searches are frequently used as an aid in identifying an unknown compound. In most cases for an unknown, a reverse search is performed. In this approach, the degree of inclusion of the library spectrum in the unknown spectrum is assessed, with larger match factors indicating more precise inclusion. The effect of the presence of additional mass fragments in the spectrum on the library match factor is thus minimized. In general, the quality of the spectrum obtained from the unknown compound is the determining factor in any library match. The quality of the mass spectrum decreases as the amount of compound supplied to the mass spectrometer is decreased. Other factors that affect the library match factor include the presence of a spectrum of the correct analyte in the library collection, the quality of the library spectrum (including the instrument type used to collect the spectrum), the use of complete vs. condensed spectra, instrumental factors (including the tune and source temperature), and the search algorithm itself. There are two commercially available library search algorithms: probability-based matching (PBM) (19) and dot product (NIST) (20). Recently, computer-aided chromatographic peak deconvolution software based on the mass spectral properties of the peak has become available (21). The effect of this advance on the quality of the match factor, and its admissibility as evidence has not been determined.

Sphon originally described the use of SIM for compound identification (22). Using diethylstilbestrol as a model compound, three characteristic ions with appropriate relative abundances were shown to have a reliability of about 1 in 30,000. Subsequent experiments have shown that using ±20% around the observed relative abundance, a SIM analysis with three ions was able to select diethylstilbestrol from a library of over 270,000 compounds (23). It should be pointed out that these experiments used only the MS data, and that the additional information content of the capillary GC retention time should significantly increase the reliability of the identification. For toxicological identification, the information content of a chromatographic method is related to the number of peaks that can be resolved in a given period of time or peak capacity of the method. De Ruig et al. (24) suggest uncertainty of 1:50–1:200 in capillary GC. An ad hoc committee of forensic scientists summarized identification using SIM analysis by stating (25):

> ... the SIM mode generally is more sensitive and less affected by potential interferences from coeluting compounds. The specificity (certainty of identification) of a SIM assay depends on many factors including: the number of ions monitored, the uniqueness of the monitored ions, the selectivity of the extraction procedure, the type of derivative, the efficiency of the chromatographic separation, and the selectivity of the method of ionization. A well-designed SIM assay can provide a very reliable method of identification. However, it may be difficult to evaluate the reliability of a SIM assay without personal experience with the method or access to data from the analysis of a substantial number of specimens.

There remain some areas in which consensus has not been attained, such as the approach to be taken for chromatographic peaks experiencing overload conditions and the associated retention time shifts.

IDENTIFICATION AND QUANTIFICATION ISSUES IN HPLC-MS AND HPLC-MS-MS

The use of HPLC-MS and HPLC-MS-MS has increased exponentially over the past decade. This is primarily due to the availability of reliable interface techniques and the recognition that many analytes are not amenable to GC analysis due to their lability, polarity, or molecular weight. The analysis of immunosuppressive drugs is an excellent example of the development of HPLC-MS and HPLC-MS-MS routine and reference methods. The quantitative nature of these methods is well established, and relatively sensitive methods can be developed.

There are several factors that differentiate HPLC-MS techniques from GC/MS techniques at this stage of their development. First, whether the ES or APCI interface technique is selected, the mild CI frequently results in little fragmentation. When fragmentation does occur, the energy retained by the molecule in transitioning the interface is different for each instrumental interface design, which has made the development of a universal library difficult, if not impossible. In cases in which fragmentation does not occur, either CID in the interface region or in a collision cell (MS-MS) has been used to improve spectral information content. Again, these approaches are instrument dependent and have made the development of compound libraries difficult. Thus, while it is possible to create libraries in a laboratory, the sharing of these libraries and the fine-tuning of a computer-assisted spectral matching algorithm have yet to be accomplished. In addition, the computer control of the interface and collision cell conditions make it possible to alter the kind of data acquired in each scan across a chromatographic peak. The ultimate benefit of these approaches is not fully determined.

For trace analysis using ES ionization, the potential for ion suppression presents a serious limitation (26,27). Ion suppression may be sample-specific and even compound-specific within a given sample (27). To eliminate the potential for ion suppression to provide false-negative or inaccurate quantitative results, one should use a co-eluting internal standard for each compound. Alternatively, specific and more extensive clean-up methods can be used. The effect of this problem on drug screening using HPLC-MS has not been determined.

REFERENCES

1. Ullman MD, Bowers LD, Burtis CA: Chromatography/mass spectrometry. In: Burtis CA, Ashwood ER, eds: *Teitz textbook of clinical chemistry*, 3rd ed. Philadelphia: WB Saunders, 1999:164–204.
2. Rood D: *A practical guide to the care, maintenance, and troubleshooting of capillary gas chromatography systems.* Heidelberg, Germany: Hüthig, 1991.
3. Snyder LR, Kirland JJ, Glajch JL: *Practical HPLC method development*, 2nd ed. New York: Wiley & Sons, 1997.
4. Bowers LD: Coupled mass spectrometric-chromatographic systems. In: Wong SHY, Sunshine I, eds: *Handbook of analytical therapeutic drug monitoring and toxicology.* New York: CRC Press, 1996:173–199.
5. Thormann W: Drug monitoring by capillary electrophoresis. In: Wong SHY, Sunshine I, eds: *Handbook of analytical therapeutic drug monitoring and toxicology.* New York: CRC Press, 1996:1–19.
6. Colón LA, Guo Y, Fermier A: Capillary electrochromatography. *Anal Chem* 1997;69:461A–467A.
7. Wu AH, Ostheimer D, Cremese M: Characterization of drug interferences caused by co-elution of substances in gas chromatography/mass spectrometry confirmation of targeted drugs in full-scan and selected-ion monitoring modes. *Clin Chem* 1994;40:216–220.
8. Harrison AG: *Chemical ionization mass spectrometry*, 2nd ed. Boca Raton, FL: CRC Press, 1992.
9. March RE: An introduction to quadrupole ion trap mass spectrometry [Tutorial]. *J Mass Spectrom* 1997;32:351–369.
10. March RE, Todd JFJ, eds: *Practical aspects of ion trap mass spectrometry, volume III: chemical, environmental and biomedical applications.* New York: CRC Press, 1995.
11. Bowers LD, Borts DJ: Evaluation of selected-ion storage ion-trap mass spectrometry for detecting urinary anabolic agents. *Clin Chem* 1997;43:1033–1039.
12. Wu AH, Onigbinde TA, Wong SS, et al: Evaluation of full-scanning GC/ion trap MS analysis of NIDA drugs-of-abuse urine testing in urine. *J Anal Tox* 1992;16:202–206.
13. Fitzgerald RL, O'Neal CL, Hart BJ, et al: Comparison of an ion-trap and a quadrupole mass spectrometer using diazepam as a model compound. *J Anal Toxicol* 1997;21:445–450.
14. de Hoffmann E: Tandem mass spectrometry: a primer [Tutorial]. *J Mass Spectrom* 1996;31:129–137.
15. Busch KL, Glish GL, McLuckey SA: *Mass spectrometry/mass spectrometry: techniques and applications of tandem mass spectrometry.* New York: VCH Publishers, 1988.
16. National Committee for Clinical Laboratory Standards: *Gas chromatography/mass spectrometry (GC/MS) confirmation of drugs: proposed guideline.* [NCCLS document C43-P.] Wayne, PA: National Committee for Clinical Laboratory Standards, 2000.
17. Cleij P, Dijkstra A: Information theory applied to qualitative analysis. *Fr J Anal Chem* 1979;298:97–109.
18. Thieme D, Müller RK: Information theory and systematic toxicological analysis in "general unknown" poisoning cases. *Fr J Anal Chem* 1997;358:785–792.
19. McLafferty FW, Hertel RH, Villwock RD: Probability based matching of mass spectra. *Org Mass Spectrom* 1974;9:690–702.
20. Stein SE, Scott DR: Optimization and testing of mass spectral library searching algorithms for compound identification. *J Am Soc Mass Spectrom* 1994;5:859–866.
21. Stein SE: An integrated method for spectrum extraction and compound identification from gas chromatography/mass spectrometry data. *J Am Soc Mass Spectrom* 1999;10:770–781.
22. Sphon JA: Use of mass spectrometry for confirmation of animal drug residues. *J Assoc Off Anal Chem* 1978;61:1247–1252.
23. Baldwin R, Bethem RA, Boyd RK, et al: Limits to confirmation, quantitation, and detection. *J Am Soc Mass Spectrom* 1997;8:1180–1190.
24. de Ruig WG, Dijkstra G, Stephany RW: Chemometric criteria for assessing the certainty of qualitative analytical methods. *Anal Chim Acta* 1989;223:277–282.
25. Wu N, Chen NB, Cody JT, et al: Report of 1988 Ad Hoc Committee on Forensic GC/MS: Recommended Guidelines for Forensic GC/MS Proce-

dures in Toxicology Laboratories Associated with Offices of Medical Examiner and/or Coroners. *J Forensic Sci* 1990;35:236–242.
26. Matuszewski BK, Constanzer ML, Chavez-Eng CM: Matrix effect in quantitative LC/MS/MS analyses of biological fluids: A method for determination of finasteride in human plasma at picogram per milliliter concentrations. *Anal Chem* 1998; 70:882–890.
27. Borts DJ, Bowers LD: Direct detection of testosterone and epitestosterone glucuronide and sulfate by HPLC/MS/MS. *J Mass Spectrom* 2000;35:50–61.

SUGGESTED READING

1. Boyd, RK, Henion JD, Alexander M, et al: Mass spectrometry and good laboratory practice. *J Am Soc Mass Spectrom* 1996;7:211–218.
2. McLafferty FW, Turecek F: *Interpretation of mass spectra,* 4th ed. Mill Valley: University Science Books, 1993.

SELF-ASSESSMENT QUESTIONS

1. Increasing the temperature program rate in a gas chromatography (GC) method does which of the following?
   a. decreases retention time
   b. decreases peak width
   c. decreases peak resolution
   d. decreases column life time due to increased column bleed

2. Which of the following changes in GC column parameters decrease retention under isothermal conditions?
   a. a decrease in diameter (film thickness constant)
   b. a decrease in film thickness
   c. a decrease in length
   d. a decrease in flow rate

3. In reverse-phase high-performance liquid chromatography (HPLC), an increase of 10% in the organic modifier (for example, methanol) content causes a decrease in which one?
   a. retention time by a factor of two
   b. capacity by a factor of two
   c. pressure by a factor of two
   d. resolution by a factor of two

4. In reverse-phase HPLC, which of the following increases resolution under isocratic conditions?
   a. an increase in column length
   b. an increase in particle diameter
   c. an increase in temperature
   d. an increase in mobile-phase velocity

5. Indicate the base peak (a) and the parent ion (b) in the following spectrum.

6. Does an extracted ion chromatogram (or profile) usually give a limit of detection that is lower than, the same as, or higher than a selected ion monitoring chromatogram?
   a. lower
   b. the same
   c. higher

7. Which of the following has the most information content?
   a. HPLC
   b. thin-layer chromatography
   c. capillary gas chromatography
   d. packed-column gas chromatography

8. True or false: Tandem mass spectrometry (MS-MS) in space can be accomplished with an ion trap mass analyzer.
   a. true
   b. false

9. True or false: Any GC/MS assay that can be performed on a sector mass spectrometer operated in the high mass resolution mode can be performed on a quadrupole mass spectrometer.
   a. true
   b. false

10. What is the most popular interface for HPLC-MS?
    a. particle beam
    b. electrospray
    c. atmospheric pressure chemical ionization
    d. thermospray

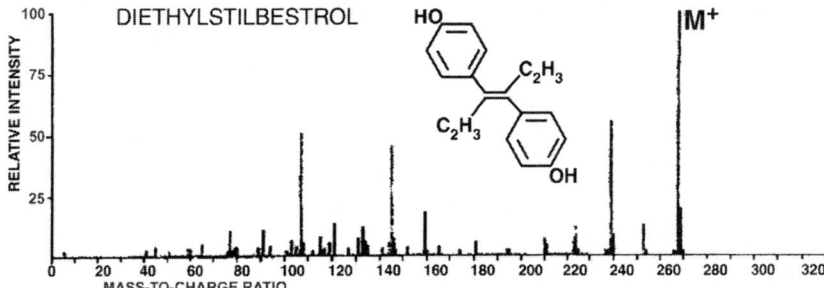

11. Which of the following is the operation mode of ion trap used in most currently available commercial instrumentation?
   a. mass selective storage
   b. mass selective detection
   c. mass selective ejection
   d. mass selective destruction

12. For a quadrupole mass spectrometer, the purpose of tuning is to establish which?
   a. the accuracy of the mass axis calibration
   b. the relative abundance of the signal in various regions of the mass axis
   c. the width of the mass peaks
   d. a and b
   e. a, b, and c

13. True or false: A library match quality index of 0.80 rules out the possibility of a match between the unknown and library spectrum.
   a. true
   b. false

14. True or false: Ion suppression can only occur in HPLC-MS interfaces.
   a. true
   b. false

15. Which of the following is true about a characteristic ion?
   a. It arises from the core structure of the molecule.
   b. It arises from the derivatization reagent.
   c. It can only be the molecular ion.
   d. It is observed only in selected ion monitoring.

16. True or false: Selected reaction monitoring requires the presence of three characteristic ions for acceptance as an identification.
   a. true
   b. false

17. True or false: Ion suppression can be corrected by use of an internal standard with a different retention time than the analyte.
   a. true
   b. false

CHAPTER 28

Fundamentals of Pharmacogenetics

*Mark W. Linder, Ph.D., FACB, DABCC,
and Roland Valdes Jr., Ph.D., FACB*

LEARNING OBJECTIVES

After completing this chapter, the reader should be able to:

1. Define the two components of pharmacology, pharmacokinetics and pharmacodynamics, and discuss the role of drug-metabolizing enzymes and cellular receptors in each.
2. Define terms and nomenclature common to genetics and pharmacology as they relate to pharmacogenetics and discuss the relation between pharmacogenetics and pharmacokinetics and pharmacodynamics.
3. Outline the molecular mechanisms of genetic polymorphism and discuss the most common genetic polymorphisms of cytochrome P4502D6 associated with the extremes of drug metabolism.
4. Discuss the important variables that must be considered when implementing a pharmacogenetic diagnostic procedure.
5. Understand how pharmacogenetic information can be applied to clinical and forensic toxicology and support of pharmacotherapy.

INTRODUCTION

The science of pharmacogenetics links differences in gene structure (polymorphism) with pharmacologic differences in drug-action and carcinogen disposition. It is now well established that many of the genes that encode proteins dictating the pharmacology of xenobiotics display genetic polymorphism. These polymorphisms in turn alter the functionality of the protein product and lead to dramatic phenotypic differences in response to medicines or susceptibility to carcinogenesis. As depicted by the diagram in Figure 28–1 (1), pharmacogenetics forms the third and base leg of the pharmacology triangle as the molecular basis supporting pharmacokinetics and pharmacodynamics.

Through the application of pharmacogenetic information, extreme phenotypic differences in the pharmacokinetics or pharmacodynamics of drugs or the risk of exposure-linked cancer can be predicted. In addition, pharmacogenetic information reveals molecular mechanisms of drug action and carcinogenesis. As a result, pharmacogenetics can be applied to clinical pharmacology and toxicology, forensic toxicology, cancer-risk stratification, and pharmaceutical design and lead to a better understanding of the pharmacology of drugs and xenobiotics.

OVERVIEW OF PHARMACOLOGY

Pharmacology is traditionally divided into two mechanistic components, pharmacokinetics and pharmacodynamics (Figure 28–2).

- Pharmacokinetics relates to the effect of the body on the drug and principally includes drug dose, bioavailability, distri-

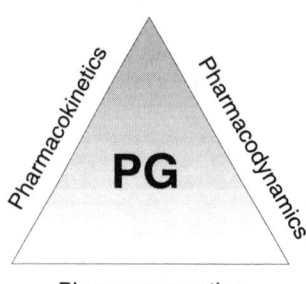

FIG. 28–1. Pharmacogenetics (PG) establishes the functional basis of pharmacokinetics and pharmacodynamics.

bution, and clearance. These pharmacokinetic variables control the systemic drug concentration.
- Pharmacodynamics relates to the effect of the drug on the body mediated through cellular receptors and is driven by the systemic drug concentration.

Depending on the drug concentration, the response may be a subtherapeutic effect, the desired therapeutic effect, or a toxic effect, hence the importance of attaining the target therapeutic interval for optimized therapy.

Drug Metabolism and Genetic Polymorphism

The enzymes involved in drug metabolism drive pharmacokinetics, whereas pharmacodynamics is driven by the change in cellular function resulting from drug interaction with specific receptor proteins.

The pharmacokinetics of a drug in balance with the dose rate maintains the steady-state systemic drug concentrations that drive the pharmacodynamic response. Principally, the average steady-state concentration (C_{aveSS}) of drug is maintained by dose rate, bioavailability of the drug, and clearance of the drug (Equation 1) (2).

$$C_{aveSS} = \frac{\text{dose rate} \times \text{bioavailability}}{\text{clearance}} \quad (1)$$

The activity of drug-metabolizing enzymes is the principal determinant of the drug extraction ratio (ER), which is a measure of the fraction of drug cleared from circulation as it passes through whichever organ is involved in the metabolism of the drug. The extraction ratio is a component of both bioavailability (f) and clearance (Cl) (2).

$$\text{Bioavailability (f)} = (\% \text{ absorption}) \times (1 - \text{ER}) \quad (2)$$

where % absorption is the fraction of the dose absorbed

$$\text{Clearance (Cl)} = q \times \text{ER} \quad (3)$$

where q is the blood flow to the organ

From Equations 2 and 3 it becomes evident that as the extraction ratio decreases, f increases and Cl decreases. This concept of how

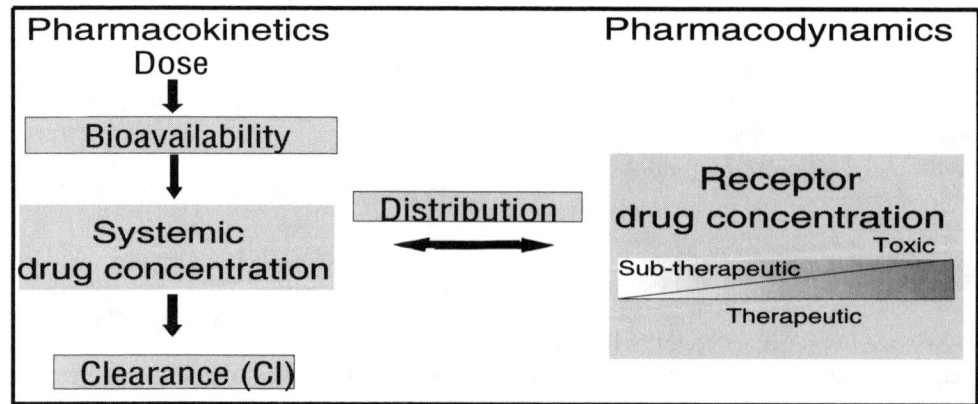

FIG. 28–2. Pharmacology overview.

changes in the activity of drug-metabolizing enzymes alter the steady-state concentration of drug at a constant dose rate may be best illustrated by a hypothetical example. The drug desipramine (an antidepressant) is avidly metabolized by a specific hepatic enzyme, cytochrome P4502D6. Thus, this drug demonstrates a relatively large extraction ratio ranging from 0.5 to 0.75 in most subjects (2). If, for example, an individual with 100% absorption of the drug and a hepatic extraction ratio of 0.6 is administered a standard dose of desipramine of 150 mg/d, the resulting steady-state concentration of desipramine in serum is on the order of 115 ng/mL (431 nmol/L). If however, a second individual is administered the same dose, but has 50% of the drug-metabolizing activity or a 50% decrease in the ER, using equations 1, 2, and 3 the combined effect of a decreased ER on bioavailability and clearance can be predicted and in this scenario would result in a steady-state drug concentration of 405 ng/mL (1519 nmol/L), a fourfold increase in the steady-state drug concentration. Thus, a 50% decrease in ER can lead to more than a 400% increase in the resulting steady-state drug concentration under identical dosing conditions.

Drug metabolism is traditionally divided into two categories based on the type of reaction catalyzed by the enzyme. Phase I enzymes catalyze oxidative metabolism of principally hydrophobic substrates. Phase II enzymes are conjugative and act to conjugate polar moieties such as acetate, sulfate, glutathione, and glucuronic acid to products of phase I enzymes or substrates with appropriate functional groups necessary for the conjugation to occur. Table 28–1 lists a sample of various phase I and phase II enzymes that are known to exhibit genetic polymorphism.

To illustrate several points regarding genetic polymorphism of drug-metabolizing enzymes and the prevalent nomenclature, we will use a phase I enzyme that has been well characterized as an example. Cytochrome P4502D6 (debrisoquine hydroxylase) is a member of the supergene family of heme-thiolate proteins, designated cytochrome P450 due to a characteristic absorption spectrum of the protein products, which peaks at a wavelength of 450 nm.

TABLE 28–1. *Polymorphic drug-metabolizing enzymes*

Phase I
- Cytochrome P450s
- Monoamine oxidase
- Alcohol dehydrogenase
- Aldehyde dehydrogenase
- Dopamine β-hydroxylase

Phase II
- *N*-Acetyl transferases
- Thiopurine methyltransferase (TPMT)
- Sulphotransferase

This supergene family is subdivided into several families. The family to which the gene belongs is designated by a number (Figure 28–3). Following the family designation is a capital letter that designates the gene subfamily, and the specific isoenzyme product is then designated by a number (3). To distinguish between alleles of a given gene, the isoenzyme designation is followed by an asterisk and a number from 1 through the number of alleles identified for that gene. The example shown is the nomenclature for the *4 allele of cytochrome P4502D6. In addition, alleles may include additional structural variations that may or may not alter the function of the protein product. These additional variations of a specific allele are indicated by a capital letter after the allele designation (not shown). Nomenclature systems for the cytochromes P450 (3) and polymorphic alleles (4) have been published.

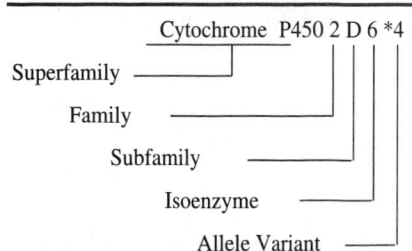

FIG. 28–3. Nomenclature system for designating enzymes and allozymes of cytochrome P450.

REVIEW OF GENE STRUCTURE

A gene, in the most fundamental sense, is a linear sequence of nucleotides. You will recall that nucleotides are joined to one another in sequence via a phosphodiester bond between the 5' and 3' carbons of the deoxyribose moiety of the nucleotide. This structural element provides a basis for structural orientation. The deoxyribonucleic acid is a double-stranded molecule with antiparallel polarity. The coding strand (the strand that is transcribed into RNA) is referred to as the sense strand and is conventionally depicted in the 5' to 3' orientation. The antisense strand is thus complementary (A-T, G-C) in sequence to the sense strand and is depicted by convention in the 3' to 5' orientation.

The linear sequence of a gene includes a minimum of four structural domains. The most 5' domain is the regulatory region, which includes the structural attributes (nucleotide sequence) required for transcriptional control of gene expression including cell-type or tissue-specific regulation and the regulatory response to intra- or intercellular signals. This domain terminates at the nucleotide where the RNA polymerase initiates transcription. This nucleotide is the +1 position of the gene. Nucleotides 5' to the start site of transcription are numbered with sequential negative numbers, and conversely, nucleotides 3' to the start site of transcription are numbered with positive numbers. Downstream or 3' to the +1 nucleotide, the general structure of a gene includes the nucleotide sequence domains (exons), which ultimately direct the amino-acid structure of the protein gene product and may be interrupted by nucleotide sequence domains (introns), which in some genes play a variety of roles including fine-tuning transcriptional regulation.

Introduction to Genetic Variation

With the general structure of genes in mind we can now turn to mechanisms of variation in gene structure called genetic polymorphism. Genetic polymorphism occurs in the form of gross structural changes including complete gene deletion, gene duplication, and genetic translocation, in which portions of similar genes are combined, creating a new gene hybrid (see Table 28–2). By far the most common form of genetic polymorphism is the single-nucleotide polymorphism (SNP), in which the nucleotide sequence at one specific position is changed, inserted, or deleted. Each of these changes in the gene structure introduces a variant form of the gene into the population gene pool and is designated an allele of the original gene. (Some standards suggest that a gene variant can be classified as an allele when the frequency of the variation is >1%). Thus an allele is an inherited gene, present in each nucleated cell of the body. Due to the diploid structure of the human genome, each cell carries two copies of each gene. Two copies of the same allele yields a homozygous genotype and any combination of two different alleles yields a heterozygous genotype.

The various types of genetic polymorphism can be generally classified by their resulting influence on protein expression or phenotype. Genetic polymorphism resulting in gene deletion invariably leads to loss of function and no production of the gene product. In contrast, gene duplication and multiduplication most commonly leads to increased expression of the gene product and a hyperactivity phenotype. An exception is duplication of an allele that includes additional

TABLE 28–2. *Molecular mechanisms of genetic polymorphism*

Single nucleotide
- Coding region (amino-acid substitution)
- Noncoding region
  - Regulatory sequences: over- or underproduction of active protein
  - Intron: inappropriate mRNA splicing, truncated protein product, diminished or absent enzymatic activity

Gene rearrangements
- Gene deletion
- Gene duplication
- Genetic recombination
- Nucleotide repeats

structural variation leading to loss of function. Genetic translocation typically yields a nonfunctional gene. SNPs can result in a variety of changes in the expressed protein function depending upon where the polymorphism occurs in the overall gene structure. SNPs in the 5' regulatory domain may influence gene regulation (5). SNPs in the coding exons only influence function if there is a resulting amino-acid change that alters the protein function. SNPs within the intron regions are typically silent unless the SNP alters a nucleotide critical for splicing the RNA during maturation, in which case the SNP typically leads to loss of or decrease in protein function.

PHARMACOGENETICS: LINKING GENETIC POLYMORPHISM WITH DRUG METABOLISM PHENOTYPE

Now we can begin to make the link between genetic polymorphism and drug metabolism phenotype. The genes encoding drug-metabolizing enzymes reside on somatic chromosomes, and thus there are two copies of each gene that are inherited in a non-sex-linked Mendelian fashion. Homozygous individuals have two genes that are indistinguishable and may be active or inactive. Heterozygous individuals have more than one allelic form of the gene and may be in the context of one active allele or doubly heterozygous in that they are heterozygous for more than one variant allele. When one allele of a gene predominates in the expressed trait, the allele is said to be dominant, and the one that does not dictate the phenotype is said to be recessive. Both dominant and recessive alleles of drug-metabolizing enzymes have been described and will be illustrated with examples later in the discussion.

In addition, for most examples thus far in pharmacogenetics, the phenotype is responsive to gene dose, and activity either increases or decreases as a function of the number of alleles encoding active protein. An excellent example of the influence of genetic polymorphism on drug metabolism is again debrisoquine hydroxylase (CYP2D6). The drug-metabolism phenotype of this enzyme is defined as the debrisoquine metabolic ratio, which is the ratio of unchanged debrisoquine to 4-hydroxydebrisoquine recovered in the urine after a standard dose of debrisoquine. Deficiency in debrisoquine hydroxylase activity is most prevalent among subjects of Caucasian descent. The dramatic range of debrisoquine metabolic ratios within the Caucasian population is illustrated in Figure 28–4 (6). When the debrisoquine metabolic ratio is very low (for example <0.2), almost no unchanged debrisoquine is recovered in the urine sample and the majority of drug is in the metabolized form, indicating extremely rapid or ultrarapid metabolism (UM). In contrast, when the majority of drug recovered is in the unchanged state, the metabolic ratio is very high (>12.6, Figure 28–4) indicating poor metabolism (PM).

Metabolic ratios intermediate to the UM and PM phenotypes make up the extensive metabolizer (EM) phenotype, which is characteristic of most of the population. Heterozygous individuals are commonly grouped with the homozygous EMs. However, it is apparent that these heterozygous EMs have partially impaired metabolic activity and have been designated as intermediate metabolizers (IM). The clinical significance of heterozygosity for one inactive allele has not been clearly demonstrated, and thus for practical purposes homozygous and heterozygous individuals are both considered EM.

The relationship between the debrisoquine metabolic ratio and the *CYP2D6* genotype has been extensively characterized (3,7) and was elegantly illustrated by the work of Agundez et al. (8) illustrated in Figure 28–5. Individuals homozygous for two inactive alleles show the least metabolic capacity (PM) as indicated by an excessively high ratio of unchanged drug to metabolite recovered in the urine. Heterozygous individuals with only one active *CYP2D6* allele demonstrate metabolic activity intermediate between PMs and subjects homozygous for two active *CYP2D6* alleles (EMs). In contrast, approximately 5–

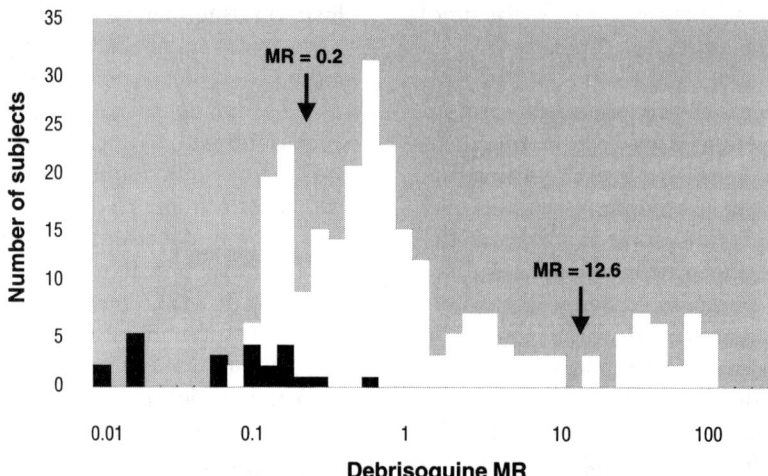

FIG. 28–4. Population distribution of debrisoquine 4-hydroxylase metabolic ratios (MR) and the corresponding phenotypic designations, ultrarapid metabolizer (MR < 0.2), extensive metabolizer (MR 0.2–12.6), and poor metabolizer (MR > 12.6). Black bars indicate subjects with duplicated or multiduplicated *CYP2D6* alleles.

10% of subjects have a duplication of the *CYP2D6* on one chromosome. These individuals illustrate the highest metabolic capacity and constitute the UM phenotype. Thus, there is a relationship between the *CYP2D6* genotype and the resulting drug metabolism phenotype.

Cytochrome P4502D6 Genotyping: Strategies and Methods

From the perspective of the clinical or forensic laboratory, application of this information requires an understanding of the genetic diversity that contributes to common

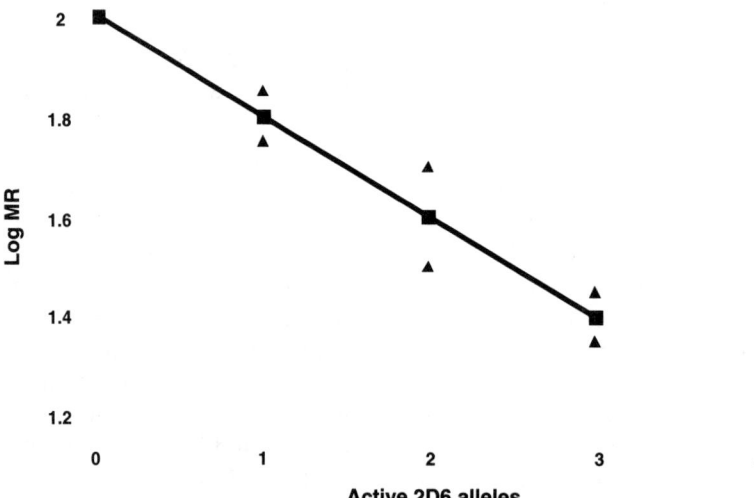

FIG. 28–5. Logarithm of metabolic ratio defined as the percentage of unmetabolized debrisoquine relative to 4-hydroxydebrisoquine recovered in urine from unrelated subjects with zero, one, two, or three functional *CYP2D6* alleles. Adapted with permission from Agundez JAG, Ledesma MC, Ladero JM, et al: Prevalence of *CYP2D6* gene duplication and its repercussion on the oxidative phenotype in a white population. *Clin Pharmacol Ther* 1995;57:265–269.

metabolic phenotypes. Again, an example is *CYP2D6*. The structure, function, and prevalence of *CYP2D6* alleles have recently been reviewed (5). Most alleles identified are rare and constitute <1% of all variant *CYP2D6* alleles. The most common *CYP2D6* alleles that are associated with the PM phenotype are listed in Table 28–3 along with the structural change that defines the allele, the effect on the protein gene product, and the prevalence in mostly Caucasian populations. Note that most of these alleles are the consequence of single-nucleotide polymorphisms that lead to either lack of a protein product or production of an inactive enzyme.

The most prevalent alleles are the *CYP2D6*4* (18–24%) followed by *CYP2D6*3* (1.2–3%) and the gene deletion *CYP2D6*5* (1.2–2.9%). The remaining alleles are less common but are present in certain populations to the extent that they must be included in genotyping strategies to achieve >95% confidence in properly predicting the poor metabolizer phenotype.

Several methods have been described to screen for these individual alleles, and other methods have been described that are designed to detect multiple alleles through a sequential methodologic strategy (7,9). These methods have evolved based on the understanding of the *CYP2D6* gene structure and the relative nucleotide positions that account for the polymorphic alleles. A map illustrating the relative location of the various *CYP2D6* nucleotide variations is shown in Figure 28–6.

The method of Chen et al. (9), which allows for the detection of the most common PM alleles (*3,*4,*6, and *7), is illustrated in Figure 28–7. By the application of nested and heminested allele-specific polymerase chain reaction (PCR), an initial 1818–base pair (bp) region of the *CYP2D6* gene is amplified and used as a template for subsequent PCR reactions. This preamplification step is necessary due to the presence of highly homologous *CYP2D* pseudogenes that must be differentiated from the *CYP2D6* gene. Using the 1818-bp product from the first round of amplification, allele-specific oligonucleotide primers differing only in the identity of the 3′ base, which is specific for either the native (*1) or variant allele of interest, are used in parallel amplification reactions. A PCR product is only produced when the allele-specific primer forms a stable heteroduplex with the 1818-bp template. Thus, a PCR product is only produced from alleles that are complementary to the allele-specific primers.

An example result of analysis for the *3 and *4 alleles is shown in Figure 28–8. The allele-specific primers used for preamplification of the 1818-bp product and for determination of the *3 and *4 primers are listed in Table 28–4. The *4 is differentiated from the *1 allele by parallel amplifications using primers RB7 and RB8-B. Primer RB-7 is complementary at the 3′ end with the native *1 allele. Primer RB8-B is complementary at the 3′ end with the *4 allele. Thus, subject 1 is determined to be *1*4 based on the fact that both RB7 and RB-8B primers yield a PCR product. In contrast, subject 2 in Figure 28–7 is determined to be *1*1 based on the fact that only the RB7 primer generated a PCR product. In an analogous fashion, the *3 allele is differentiated from the *1 allele using allele-specific primers FA5 and FA-6A. The combined interpretation of all four reactions leads to the conclusion that the genotype of subject 1 is *1*4 and subject 2 is *1*1. Neither subject yielded a product using FA5-A and thus neither is a carrier of the *3 allele. Two additional PCR reactions can be run using the 1818-bp product to identify the *6 and *7 alleles (not shown).

The gene-deletion allele *CYP2D6*5* cannot be directly assessed by the method of Chen et al. As previously discussed, the *CYP2D6* PM alleles are considered recessive, and thus the phenotype of subjects who are heterozygous for the *5 allele depends on the nature of the complementary allele. In the allele-specific methods, an individual who is heterozygous *1*5 will yield the result *1*1, and a subject who is heterozygous *3*5 will appear to be *3*3. In each scenario, the phenotype will be accurately predicted as EM in the first case and PM in the case of the *3*5 heterozygote. In addition, the allele-specific methods do assist in ruling out the *5 allele. For example, subject 1, whose genotype was

TABLE 28-3. Description of CYP2D6 alleles

Allele	Functional nucleotide changes	Structural effect	Activity	Phenotype	Allele frequency
CYP2D6*1	None	None	Normal	EM	0.364 (0.337–0.392)
CYP2D6*1X2	Gene duplication	None	Increased	UM	0.0051 (0.0019–0.033)
CYP2D6*2	$C_{2938}T$, $G_{4268}C$	$Arg_{296}Cys$, $Ser_{486}Thr$	Decreased	EM	0.324 (0.298–0.352)
CYP2D6*2XN[a]	Gene duplication	$Arg_{296}Cys$, $Ser_{486}Thr$	Increased	UM	0.0134 (0.008–0.022)[b]
CYP2D6*3	A_{2637} deletion	Frameshift	None	PM	0.0204 (0.0131–0.0302)
CYP2D6*4	$G_{1934}A$	Splicing defect	None	PM	0.207 (0.184–0.231)
CYP2D6*4X2	$G_{1934}A$, gene duplication	Splicing defect	None	PM	0.0008 (0.0000–0.0047)
CYP2D6*5	Gene deletion	CYP2D6 deleted	None	PM	0.0195 (0.0124–0.0292)
CYP2D6*6	T_{1795} deletion	Frameshift	None	PM	0.0093 (0.0047–0.0166)
CYP2D6*7	$A_{3023}C$	$His_{324}Pro$	None	PM	0.0008 (0.0000–0.0047)
CYP2D6*8	$G_{1846}T$	Stop codon	None	PM	0.0000 (0.0000–0.0031)
CYP2D6*9	A_{2701}-A_{2703} or G_{2702}-A_{2704} del	Lys_{281} deleted	Decreased	EM	0.0178 (0.0111–0.0271)
CYP2D6*10	$C_{188}T$	$Pro_{34}Ser$, $Ser_{486}Thr$	Decreased	EM	0.0153 (0.0091–0.0240)
CYP2D6*11	$G_{971}C$	Splicing defect	None	PM	0.0000 (0.0000–0.0031)
CYP2D6*12	$G_{212}A$	$Gly_{42}Arg$	None	PM	0.0000 (0.0000–0.0031)
CYP2D6*13	CYP2D6/CYP2D7 hybrid	Frameshift	None	PM	0.0000 (0.0000–0.0031)
CYP2D6*14	$G_{1846}A$	$Gly_{169}Arg$	None	PM	0.0000 (0.0000–0.0031)
CYP2D6*15	T_{226} insertion	Frameshift	None	PM	0.0008 (0.0000–0.0047)
CYP2D6*16	CYP2D7P/CYP2D6 hybrid	Frameshift	None	PM	0.0008 (0.0000–0.0047)

[a] n = 2, 3, 4, 5, or 13.
[b] Frequency for n = 2.
Modified from Linder MW, Valdes R Jr: Pharmacogenetics in the practice of laboratory medicine. *Mol Diag* 1999;4:365–379.

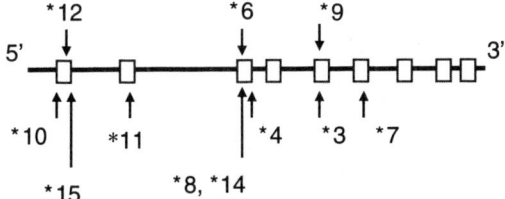

FIG. 28–6. Map illustrating the relative position of the structural changes that define some of the most common *CYP2D6* PM alleles. Open boxes represent exons 1–9. Position and designation of the allele sequence changes are indicated by an asterisk and number.

determined to be *1*4, could not have the gene deletion because both alleles *1 and *4 were accounted for by the PCR results.

The caveat of the *5 deletion is when a subject is homozygous for the gene deletion *5*5. This subject will not produce an 1818-bp amplification product due to homozygous deletion of the gene. Within the Caucasian population the estimated incidence of subjects of the *5*5 genotype is <1 in 330. To verify the gene deletion, a robust PCR method has been developed by Steen et al. (10).

Based on the effect and prevalence of the various alleles shown in Table 28–3, it becomes evident that multiple genotypes (allele combinations) can yield a common phenotype. There are five major alleles that can combine to give 14 genotypes associated with the PM phenotype. The prevalence of these genotypes account for ~97.5% of poor metabolizers. In contrast, nearly 100% of UMs can be identified by observation of the *CYP2D6* duplication allele (11). Based on what is currently understood concerning genetic polymorphism of the *CYP2D6* system, a genotyping strategy that includes testing for the *3, *4, *5, *6, *7, and *15 PM alleles in combination with the test for the duplication allele provides for >97.5% reliability in predicting the PM or UM phenotype (12).

Role of the Clinical or Forensic Laboratory

For pharmacogenetic testing to contribute to improved therapeutic management or to be used as a basis for accounting for toxicologic or forensic discrepancies between apparent drug dose and the measured drug concentrations, these tests must be made available by laboratories. Several considerations must be addressed when establishing these services. Initially for each enzyme of interest, a genetic profiling strategy must be designed to maximize sensitivity and specificity for predicting phenotype. For example, profiling for the five most common CYP2D6

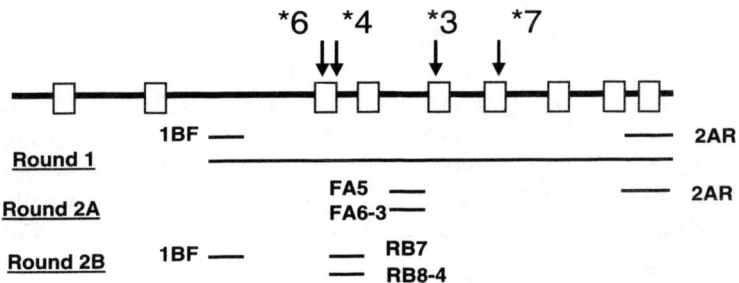

FIG. 28–7. Schematic diagram illustrating the allele-specific PCR-based method for discrimination of four *CYP2D6* PM alleles. An initial amplification is carried out using *CYP2D6*-specific primers 1BF and 2AR to amplify an 1818-bp region of *CYP2D6*, which spans the nucleotide positions defining the *3, *4, *6, and *7 alleles. Eight separate allele-specific PCR reactions are then carried out to test for the specific sequence variations consistent with these alleles. Shown are the PCR primer combinations used to distinguish the *3 [2AR in combination with FA5 (*1-specific) or FA6-A (*3-specific)] and *4 [1BF in combination with RB7 (*1-specific) or RB8-B (*4-specific)] alleles from the common active *1 allele. From Chen S, Chou W-H, Blouin RA, et al: The cytochrome P450 2D6 (CYP2D6) enzyme polymorphism: screening costs and influence on clinical outcomes in psychiatry. *Clin Pharmacol Ther* 1996;60:522–534.

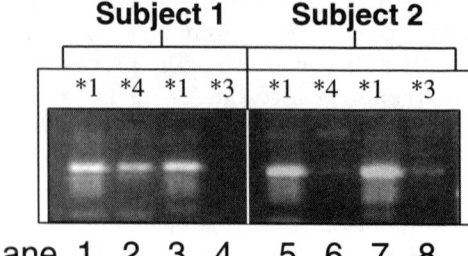

FIG. 28–8. Allele-specific PCR genotyping results for the *CYP2D6*3* and **4* alleles. PCR products for two individuals, subject 1 and subject 2, generated from allele-specific amplification reactions for the **3* and **4* alleles were electrophoresed on a 2% agarose gel and visualized by ethidium bromide staining and ultraviolet trans-illumination. Primer combinations: lanes 1 and 5, 1BF with RB7; lanes 2 and 6, 1BF with RB8-B; lanes 3 and 7, 2AR with FA5; and lanes 3 and 8, 2AR with FA6-A.

alleles shown in Table 28–3 has been reported to provide >95% sensitivity for predicting the debrisoquine hydroxylase PM phenotype (13).

Secondly, methods must be developed and improved to reduce testing cost and technical difficulty. Recent advances in the automation of DNA extraction and sequence determination are expected to significantly hasten the availability of cost-effective pharmacogenetic profiling strategies.

Individual laboratories will also be faced with the decision of what enzyme or receptor polymorphisms are important at their individual institutions, which is contingent on the needs of certain clinical subspecialties. Validation of these methods will require cooperation between laboratories to exchange samples for reference materials, quality control, and proficiency testing.

In addition to analytical demands placed on the clinical laboratory, application and interpretive guidelines must be communicated through the service laboratory. Many questions will need to be addressed such as: When should testing be performed? How should drug dosage be adjusted based on the pharmacogenetic genotype? Which drugs are affected by a specific genetic polymorphism? And can an alternative therapeutic be considered based on its metabolic route or the role of metabolism in overall clearance of the drug?

As to when testing should be performed, there are two schools of thought. One recommends that testing be considered before initiation of therapy to identify individuals who would demonstrate dramatic pharmacokinetic differences from the norm and, based on published case reports, the dosage regimen can be adjusted to accommodate the unique pharmacokinetics of the individual (14). This approach is of particular importance when the severity of inappropriate dosing is unacceptable. A second approach is to genotype only individuals who fail therapy in an attempt to define the basis for therapeutic failure. These questions require future careful outcomes-related considerations. To address the question of how dosing of a particular medication should be adjusted in light of the pharmacogenetic information, guidelines can be formulated by review of case reports in the literature and consultation with clinical pharmacologists. The following cases from the literature provide preliminary guidelines for certain substrates of cytochrome P4502D6 and other drug-metabolizing enzymes that exhibit genetic polymorphism.

PHARMACOGENETICS IN PSYCHIATRY

The first example is a 59-year-old woman (15) treated with imipramine for depression. Treatment was begun at the lower end of the standard dosing interval and was followed with 7 weeks of unsuccessful treatment including increasing doses of imipramine (75–150 mg/d). As a result of suspected toxic side effects, the blood concentration of imipramine and its pharmacologically active metabolite desipramine were measured. The concentration of imipramine [125 ng/mL (446 nmol/L)] was consistent with the dosage regimen; however, the concentration of desipramine (1730 ng/mL), which is a substrate of CYP2D6, was well above the accepted threshold for life-threatening toxicity (1000 ng/mL).

Treatment was immediately discontinued for 4 weeks and resumed at a dose of 25 mg/d. The patient exhibited marked improvement and the concentrations of imipramine and desipramine were again measured and found to be slightly lower than the typical therapeutic interval (25 ng of imipramine/mL + 135 ng of desipramine/mL, combined therapeutic range = 180–350).

This case illustrates several points of the poor metabolizer phenotype. First, although treatment was initiated at the lower end of the standard dosage interval, the resulting steady-state concentration of drug was dangerously high and not verified by therapeutic drug monitoring until after toxicity developed. Second, due in part to the delayed pharmacodynamics of this drug, the patient endured more than 12 weeks of depression and drug toxicity before the selection of an acceptable dosing protocol. If the metabolism phenotype of this patient had been determined before initiation of therapy, poor metabolism could have been anticipated, the appropriate dosing could have been established much sooner, and the patient could have been successfully treated within 4 weeks. Thus, pharmacogenetic profiling could have avoided more than 8 weeks of failed therapy and a potentially fatal overdose.

Depression is a major illness that affects both quality of life and occupational productivity. This disease can be managed through pharmacologic means. However, therapeutic failure due to genetic variability is not uncommon and leads to unnecessary patient morbidity and frustration. This example illustrates the need to reduce the dose of imipramine in poor metabolizers approximately sixfold to attain typical therapeutic concentrations and appropriate response (13).

In another case (13), a patient was treated with clomipramine for agoraphobia with a relatively high dose of 150 mg/d (typical dose 25–150 mg/d). Therapeutic drug monitoring revealed subtherapeutic concentrations of clomipramine and its pharmacologically active metabolite. This patient was eventually stabilized on 300 mg/d, which is 2- to 12-fold the standard dose. This patient displays the UM phenotype, and pharmacogenetic analysis revealed *CYP2D6* gene duplication.

A decision of what drugs will be affected by a genetic polymorphism is not entirely based on the metabolism of the drug. Brosen and Gram suggest criteria for evaluating the need of pharmacogenetic analysis for management of therapeutics (16). To illustrate, consider three drugs commonly used in the treatment of depression, desipramine, fluoxetine, and sertraline. Desipramine is metabolized solely by CYP2D6 and has a narrow therapeutic index. The pharmacokinetic variation of this drug resulting from pharmacogenetic differences is substantial and clearly alters the response to desipramine (17). In contrast, fluoxetine, which is also principally metabolized by cytochrome P4502D6 (18), has a relatively broad therapeutic index, and the pharmacokinetic variation resulting from genetic polymorphism of *CYP2D6* is not currently considered clinically important.

This understanding may change, however, in light of recent reports of a fatal fluoxetine overdose in a child. This case report emphasizes the importance of pharmacogenetic information in resolving unusual cases of drug overdose. This child was determined to have CYP2D6 deficiency based on genotyping, and failure to metabolize fluoxetine was considered an important component in understanding the basis for the unfortunate fatal outcome (19). In addition, CYP2D6 poor metabolizers convert less of the fluoxetine dose to the nor-fluoxetine metabolite that is suggested to play an important role in certain aspects of antidepressant response (20). Finally, sertraline, also an antidepressant, is not a substrate for CYP2D6 and thus is clearly not subject to the polymorphism of *CYP2D6*.

Again, this can be illustrated by a case in the literature. A 41-year-old woman was treated >10 years with nortriptyline for management of recurrent depression. This patient required a three- to fivefold increased dose of nortriptyline to achieve therapeutic concentrations. However, depression persisted, and extremely high concentrations of hydroxylated metabolites of nortriptyline were measured in the patient's blood. This patient's phenotype was determined to be an ultrarapid metabolizer, which was corroborated by genetic analysis that revealed *CYP2D6* gene duplication. This patient was switched to a selective serotonin uptake inhibitor, which is not a CYP2D6 substrate, with markedly improved response.

At this time, pharmacogenetics has the greatest potential for influence in the areas of psychiatry, hematology, and oncology. We have thus far addressed the role of cytochrome P4502D6 in the metabolism of a variety of therapeutics used in psychiatry. In the following sections we will outline the role of pharmacogenetics in other areas of medicine.

PHARMACOGENETICS IN ONCOLOGY

Thiopurines including mercaptopurine, azathioprine, and thioguanine are used to treat many diseases including cancer, autoimmune hepatitis, and myasthenia gravis and they are also used as immunosuppressants after organ transplantation. Thiopurines are catabolized through the action of thiopurine methyltransferase (TPMT). Individuals with diminished TMPT activity are poor metabolizers of thiopurines and thus accumulate 5- to 25-fold increased concentrations of toxic 6-thioguanine nucleotides and may suffer from severe myelosuppression, which can compromise the success of life-saving and costly transplant surgeries (21).

The thiopurine methyltransferase phenotype is defined by erythrocyte 6-mercaptopurine methylation (22). There is a trimodal distribution of TPMT activity leading to a 40-fold range of activity among healthy populations (23). Approximately 90% of subjects have high activity, ~10% have intermediate activity, and ~0.3% demonstrate deficiency in TPMT activity. These phenotypic groups are accounted for in part by genotypic differences of the *TPMT* gene. Extremely elevated TPMT activity has also been reported; however, a genetic basis for this phenotype has not been elucidated.

Genetic Polymorphism of TPMT

Nine nonfunctional *TPMT* alleles are associated with inheritance of low enzymatic activity TPMT*2–*8 (24,25), with *TPMT*1* representing the most common active allele. Subjects with high activity are believed to be homozygous for the common active *TPMT* allele, intermediate subjects heterozygous with one *TPMT* allele variant, and low-activity subjects homozygous with two variant *TPMT* alleles. The most common *TPMT* allele variants in order of decreasing frequency are *TPMT*3A*, *TPMT*3C*, and *TPMT*2* (26).

Due to the potential threat of fatal hematopoietic thiopurine toxicity resulting from reduced TPMT activity, it is recommended that a testing strategy with the highest level of sensitivity be adopted. Although marked differences in the prevalence of the individual PM alleles exist between populations of differing ethnic origin, the identity of the alleles is constant, allowing for a uniform pharmacogenetic screening approach for all populations studied to date. The *TPMT*3A* allele includes two nucleotide sequence variants, G_{460}-A and A_{719}-G. Each of these nucleotide variants have been shown independently to result in a dramatic decrease in TPMT expression and catalytic activity because these variants are found in isolation within the context of the rare *TPMT*3B* and *TPMT*3C* alleles, respectively. When analysis is performed for both variants (for example, the *TPMT*3A* allele), the *TPMT*3B* and *TPMT*3C* alleles will be revealed in the process. For example, an individual heterozygous for the *TPMT*3A* and *TPMT*3B* alleles will demonstrate heterozygosity for the A_{719}-G variant and homozygosity for the G_{460}-A variant. In contrast, an individual heterozygous for the *TPMT*3B* and *TPMT*3C* alleles will demonstrate heterozygosity for both variants. This individual cannot be differentiated from individuals heterozygous for *TPMT*1*3A*. However, subjects heterozygous for both variants can be presumed to be heterozygous for *TPMT*1*3A* based on the extremely low frequency of *TPMT* 3B*3C* heterozygotes, which is estimated at <1 in 100,000.

Interpretation of *TPMT* Genotyping

The clinical use of the thiopurine antimetabolites is an important component of successful therapy of several diseases. The use of these compounds can be directed based on the knowledge of the individual's TPMT activity determined either directly or inferred from the pharmacogenetic analysis. TPMT deficiency occurs as a result of homozygosity for two variant *TMPT* alleles and is associated with grossly elevated thioguanine nucleotide concentrations and severe hematopoietic toxicity in individuals treated with standard dosages. In contrast, very high TPMT activity is associated with a higher risk of therapeutic failure and disease relapse. Based on pharmacogenetic analysis, >95% of TPMT-deficient individuals can be correctly identified (26). Identification of patients with TPMT deficiency can be applied to the successful treatment of these individuals with thiopurine medications by reducing the dose to 6–10% of the conventional dosage (27–29). Individuals with high TPMT activity appear to benefit from doses near the upper limit of the dosage generally recommended (30).

It should be noted that genetic polymorphism accounts for only two-thirds of the total variance in erythrocyte TPMT activity, suggesting the possibility of additional genetic factors beyond those currently described that might regulate TPMT activity (31).

The genetic basis for deficiency in TPMT activity has been well characterized, and through genotyping techniques most poor metabolizers can be identified before initiation of thiopurine therapy (32). The dramatic differences in the dose required for essential immunosuppression between individuals with varying TMPT activity may be as great as 8- to 15-fold (33). Thus, characterization of the TMPT activity based on genotypic approaches identifies individuals who require much lower doses, thus avoiding toxicity and failed therapy.

PHARMACOGENETICS IN HEMATOLOGY

Warfarin is the most commonly used therapeutic for control of coagulation. There is a poor relation between the warfarin dose and the therapeutic response measured as the international normalized ratio (INR), which is a ratio of the time required for the patient's blood to coagulate relative to a standardized coagulation time. This poor relation leads to difficulty in achieving the desired therapeutic response with the drug and often prolongs hospital stays as the dose is titered to achieve the target INR (34). The (S)-enantiomer of warfarin is responsible for most of the pharmacologic effect. This enantiomer is metabolized principally by the cytochrome P4502C9 isoenzyme (35).

Effectiveness of anticoagulant therapy is routinely monitored as the INR. Epidemiologic evidence has established that the warfarin dose required to achieve a given target INR may vary by as much as 120-fold between individuals (36,37). These distributions typically demonstrate that ~7% of subjects require dosages of <15 mg warfarin/wk and that ~4% of patients require dosages of >60 mg/wk. This degree of variation in response presents a dangerous and costly dilemma for both patients requiring maintenance anticoagulation therapy and their physicians (38).

Proposed Mechanism of Warfarin Hypersensitivity

A high affinity hepatic (S)-warfarin 7-hydroxylase ascribed to cytochrome P4502C9 is responsible for both regioselective and stereoselective metabolism of (S)-warfarin (40). The cytochrome P4502C9 enzyme is encoded by the *CYP2C9* gene. This gene exhibits structural polymorphism leading to two gene variants (alleles) that encode qualitatively different proteins with differing catalytic activities with respect to the 7-hydroxylation of (S)-warfarin. The allele expressing the wild-type protein is designated *CYP2C9*1*. This gene sequence is the reference sequence to which all variants are compared. The *CYP2C9*2* allele is defined by a $C_{416} \to T$ nucleotide substitution resulting in substitution of cysteine for arginine at amino-acid position 144. The *CYP2C9*2* protein product expressed in vitro demonstrates ~12% of the wild-type protein activity (39). The *CYP2C9*3* allele is defined by an $A_{1061} \to C$ nucleotide substitution resulting in substitution of leucine for isoleucine at amino-acid position 359. The *CYP2C9*3* protein product expressed in vitro demonstrates ~5% of the wild-type protein activity (Table 28–3; 39,40).

The influence of these *CYP2C9* alleles on warfarin dose requirements is currently a subject of great interest. Multiple studies have clearly demonstrated increased frequency of *CYP2C9* allele variants in patients stabilized on low-dose warfarin therapy (41–43) and have established relations between genetic deficiency in CYP2C9 with increased likelihood of extremely elevated INRs and major bleeding complications compared with the general clinic population. Thus, these data clearly implicate genetic polymorphism of *CYP2C9* in the etiology of warfarin hypersensitivity. However, among those subjects requiring the lowest warfarin dosages, 20% of these were homozygous for the common active *CYP2C9*1* allele, raising the possibility that additional unidentified *CYP2C9* allele variants are present within the population.

In summary, the studies reported to date strongly support a role for *CYP2C9* polymorphism in the etiology of warfarin hypersensitivity. The limited information available on human subjects and the in vitro activity assessment of the CYP2C9*2 and CYP2C9*3 variant proteins suggests that these alleles will demonstrate nonequivalent effects on warfarin metabolism and dose requirement. Therefore, more in-depth studies that include identification of all relevant *CYP2C9* alleles and measurement of plasma (S,R)-warfarin concentrations and ratios are needed to strengthen the relationship between *CYP2C9* genetics and warfarin maintenance dose requirements.

This enzyme is subject to genetic polymorphism (44) in which 36% of subjects are heterozygous for an inactivating *CYP2C9* allele and ~4% are homozygous for the variant allele. In a population of subjects treated with warfarin to maintain the INR between 2.0 and 4.0, heterozygous subjects were found to require one-half to one-fifth as much drug as homozygous common individuals (44).

In conclusion, understanding pharmacogenetics requires a general understanding of pharmacology and genetics and a review of definitions and nomenclature common to genetics and pharmacology. Due to the fundamental nature of pharmacogenetic information, extreme phenotypic differences in the pharmacokinetics or pharmacodynamics of drugs or the risk of exposure-linked cancer can be predicted. In addition, pharmacogenetic information reveals molecular mechanisms of drug action and carcinogenesis. As a result, pharmacogenetics can be applied to all aspects of drug response and toxicology. Pharmacogenetics represents an exciting opportunity for laboratory medicine to draw closer to the goal of individualized pharmacotherapy.

REFERENCES

1. Linder MW, Valdes R Jr: Pharmacogenetics: fundamentals and applications. *Ther Drug Monit* 1999; 20:9–17.
2. Schumacher GE: Introduction to therapeutic drug monitoring. In: Schumacher GE, ed: *Therapeutic drug monitoring*. Norwalk, CT: Appleton & Lang, 1995.
3. Nelson DR, Kamataki T, Waxman DJ, et al: The P450 superfamily: update on new sequences, gene mapping, accession numbers, early trivial names of enzymes and nomenclature. *DNA Cell Biol* 1993; 12:1–51.
4. Daly AK, Brokmoller J, Broly F, et al: Nomenclature for human *CYP2D6* alleles. *Pharmacogenetics* 1996;6:193–201.
5. Hayashi S, Watanabe J, Kawajiri K: Genetic polymorphisms in the 5'-flanking region change transcriptional regulation of the human cytochrome P450IIE1 gene. *J Biochem* 1991;110:559–565.
6. Dahl M-L, Johansson I, Bertilsson L, et al: Ultrarapid hydroxylation of debrisoquine in a Swedish population. Analysis of the molecular genetic basis. *J Pharmacol Exp Ther* 1995;274:516–520.

7. Sachse C, Brockmoller J, Bauer S, et al: Cytochrome P450 2D6 variants in a Caucasian population: allele frequencies and phenotypic consequences. *Am J Hum Genet* 1997;60:284–295.
8. Agundez JAG, Ledesma MC, Ladero JM, et al: Prevalence of *CYP2D6* gene duplication and its repercussion on the oxidative phenotype in a white population. *Clin Pharmacol Ther* 1995;57:265–269.
9. Chen S, Chou W-H, Blouin RA, et al: The cytochrome P450 2D6 (CYP2D6) enzyme polymorphism: screening costs and influence on clinical outcomes in psychiatry. *Clin Pharmacol Ther* 1996;60: 522–534.
10. Steen VM, Andreassen OA, Daly AK, et al: Detection of the poor metabolizer-associated *CYP2D6(D)* gene deletion allele by long-PCR technology. *Pharmacogenetics* 1995;5:215–223.
11. Lovlie R, Daly AK, Molven A, et al: Ultrarapid metabolizers of debrisoquine: characterization and PCR-based detection of alleles with duplication of the *CYP2D6* gene. *FEBS Lett* 1996;392:30–34.
12. Linder MW, Valdes R Jr: Pharmacogenetics in the practice of laboratory medicine. *Mol Diag* 1999; 4:365–379.
13. Linder MW, Prough RA, Valdes R Jr: Pharmacogenetics: a laboratory tool for optimizing therapeutic efficiency. *Clin Chem* 1997;43:254–266.
14. Bertilsson L, Aberg-Wistedt A, Gustafsson LL, et al: Extremely rapid hydroxylation of debrisoquine: a case report with implication for treatment with nortriptyline and other tricyclic antidepressants. *Ther Drug Monit* 1985;7:478–480.
15. Balant-Gorgia AE, Balant LP, Garrone G: High blood concentrations of imipramine or clomipramine and therapeutic failure: a case report study using drug monitoring data. *Ther Drug Monit* 1989;11:415–420.
16. Brosen K, Gram LF: Clinical significance of the sparteine/debrisoquine oxidation polymorphism. *Eur J Clin Pharmacol* 1989;36:537–547.
17. Brosen K, Klysner R, Gram LF, et al: Steady-state concentrations of imipramine and its metabolites in relation to the sparteine/debrisoquine polymorphism. *Eur J Clin Pharmacol* 1986;30:679–684.
18. Hamelin BA, Turgeon TJ, Vallee F, et al: The disposition of fluoxetine but not sertraline is altered in poor metabolizers of debrisoquin. *Clin Pharmacol Ther* 1996;60:512–521.
19. Sallee FR, DeVave L, Ferrell RF: Fluoxetine-related death in a child with cytochrome P-450 2D6 genetic deficiency. *J Child Adolesc Psychopharm* 2000;10:27–34.
20. Goodnick PJ: Pharmacokinetics of second generation antidepressant: fluoxetine. *Psychopharmacol Bull* 1991;27:503–512.
21. Schutz E, Gummert J, Armstrong VW, et al: Azathioprine pharmacogenetics: the relationship between 6-thioguanine nucleotides and thiopurine methyltransferase in patients after heart and kidney transplantation. *Eur J Clin Chem Clin Biochem* 1996;34:199–205.
22. Weinshilboum RM, Raymond FA, Pazmino PA: Human erythrocyte thiopurine methyltransferase: radiochemical microassay and biochemical properties. *Clin Chim Acta* 1978;85:323–333.
23. Szumlanski CL, Honchel R, Scott MC, et al: Human liver thiopurine methyltransferase pharmacogenetics: biochemical properties, liver-erythrocyte correlation and presence of isozymes. *Pharmacogenetics* 1992;2:148–159.
24. Lee D, Szumlanski C, Houtman J, et al: Thiopurine methyltransferase pharmacogenetics. Cloning of human liver cDNA and a processed pseudogene on human chromosome 18q21.1. *Drug Metab Dispos* 1995;23:398–405.
25. Otterness D, Szumlanski C, Lennard L, et al: Human thiopurine methyltransferase pharmacogenetics: gene sequence polymorphisms. *Clin Pharmacol Ther* 1997;62:60–73.
26. Yates CR, Krynetski EY, Loennechen T, et al: Molecular diagnosis of thiopurine S-methyltransferase deficiency: genetic basis for azathioprine and mercaptopurine intolerance. *Ann Intern Med* 1997; 126:608–614.
27. Evans WE, Horner M, Chu YQ, et al: Altered mercaptopurine metabolism, toxic effects, and dosage requirement in a thiopurine methyltransferase-deficient child with acute lymphocytic leukemia. *J Pediatr* 1991;119:985–989.
28. McLeod HL, Miller DR, Evans WE: Azathioprine-induced myelosuppression in thiopurine methyltransferase deficient heart transplant recipient [Letter; Comment]. *Lancet* 1993;341:1151.
29. Lennard L, Gibson BE, Nicole T, et al: Congenital thiopurine methyltransferase deficiency and 6-mercaptopurine toxicity during treatment for acute lymphoblastic leukaemia. *Arch Dis Child* 1993; 69:577–579.
30. Chocair PR, Duley JA, Simmonds HA, et al: The importance of thiopurine methyltransferase activity for the use of azathioprine in transplant recipients. *Transplantation* 1992;53:1051–1056.
31. Vuchetich JP, Weinshilboum RM, Price RA: Segregation analysis of human red blood cell thiopurine methyltransferase activity. *Genet Epidemiol* 1995; 12:1–11.
32. Otterness D, Szumlanski C, Lennard L, et al: Human thiopurine methyltransferase pharmacogenetics: gene sequence polymorphisms. *Clin Pharmacol Ther* 1997;62:60–73.
33. Yates CR, Krynetski EY, Loennechen T, et al: Molecular diagnosis of thiopurine S-methyltransferase deficiency: genetic basis for azathioprine and mercaptopurine intolerance. *Ann Intern Med* 1997;126: 608–614.
34. Doecke CJ, Cosh DG, Gallus AS: Standardized initial warfarin treatment: evaluation of initial treatment response and maintenance dose prediction by randomized trial, and risk factors for an excessive warfarin response. *Aust N Z J Med* 1991;21:319–324.
35. Rettie AE, Korzekwa KR, Kunze KL, et al: Hydroxylation of warfarin by human cDNA expressed cytochrome P-450: a role for P-4502C9 in the etiology of (S)-warfarin-drug interactions. *Chem Res Toxicol* 1992;5;54–59.
36. James AH, Britt RP, Raskino CL, et al: Factors affecting the maintenance dose of warfarin. *J Clin Pathol* 1992;45:704–706.
37. Hallak HO, Wedlund PJ, Modi MW, et al: High clearance of (S)-warfarin in a warfarin resistant subject. *Br J Clin Pharmacol* 1993;35:327–330.
38. Doecke CJ, Cosh DG, Gallus AS: Standardized initial warfarin treatment: evaluation of initial treatment response and maintenance dose prediction by randomized trial and risk factors for an excessive

warfarin response. *Aust N Z J Med* 1991;21:319–324.

39. Sullivan-Klose T, Ghanayem BI, Bell DA, et al: The role of the *CYP2C9*-Leu^{359} allelic variant in the tolbutamide polymorphism. *Pharmacogenetics* 1996;6:341–349.
40. Haining RL, Hunter AP, Veronese ME, et al: Allelic variants of human cytochrome P-450 2C9: Baculovirus-mediated expression purification, structural characterization, substrate stereoselectivity and prochiral selectivity of the wild-type and I359L mutant forms. *Arch Biochem Biophys* 1996;333:447–458.
41. Furuya H, Fernandez-Salguero P, Gregory W, et al: Genetic polymorphism of CYP2C9 and its effect on warfarin maintenance dose requirement in patients undergoing anticoagulation therapy. *Pharmacogenetics* 1995;5:389–392.
42. Aithal GP, Day CP, Kesteven JL, et al: Association of polymorphisms in cytochrome P450 CYP2C9 with warfarin dose requirement and risk of bleeding complications. *Lancet* 1999;353:717–719.
43. Steward DJ, Haining RL, Henne KR, et al: Genetic association between sensitivity to warfarin and expression of CYP2C9*3. *Pharmacogenetics* 1997;7:361–367.
44. Rettie AE, Wienkers LC, Gonzalez FJ, et al: Impaired (S)-warfarin metabolism catalyzed by the R144C allelic variant of CYP2C9. *Pharmacogenetics* 1994;4:39–42.

SELF-ASSESSMENT QUESTIONS

1. What is the functional entity governing the pharmacokinetic variables of bioavailability and clearance?
   a. drug administration rate
   b. cellular receptors
   c. drug-metabolizing enzymes
   d. blood flow

2. What is pharmacogenetics?
   a. pharmacology directed towards genetic mechanisms of cell function
   b. administration of nucleic acid–based pharmaceuticals
   c. the effect of differences in gene structure on pharmacology
   d. genetic variation resulting from pharmacotherapy

3. What are the three general types of genetic polymorphism and what effect do these have on the drug metabolism phenotype? (UM, ultrarapid metabolism; PM, poor metabolism; EM, extensive metabolizer; IM, intermediate metabolizer)
   a. SNP: EM or PM
      gene deletion: PM
      gene duplication: UM
   b. SNP: PM
      gene deletion: UM
      gene duplication: EM
   c. SNP: UM
      gene deletion: PM
      gene duplication: EM
   d. SNP: UM
      gene deletion: PM
      gene duplication: PM

4. What is the expected phenotype of an individual heterozygous for the *1 allele and an allele with duplication of the CYP2D6*4 allele?
   a. UM
   b. PM
   c. EM

For questions 5–8, use these PCR results:

Representation of PCR results for the *3 and *4 alleles of CYP2D6.

	FA5	FA6-A	RB7	RB8-B
a.	X		X	X
b.	X	X	X	X
c.	X	X		
d.			X	X

5. Choose the PCR results consistent with the CYP2D6*1*3 genotype.
   a.
   b.
   c.
   d.
   e. none

6. Choose the results consistent with the CYP2D6*3*4 genotype.
   a.
   b.
   c.
   d.
   e. none

7. Choose the results consistent with the CYP2D6*4*5 genotype.
   a.
   b.
   c.
   d.
   e. none

8. Choose the results consistent with the CYP2D6*4*4 genotype.
   a.
   b.
   c.
   d.
   e. none

9. What is the toxic consequence of thiopurine methyltransferase deficiency?
   a. therapeutic failure due to lack of thio-guanine nucleotides
   b. accumulation of thio-guanine nucleotides
   c. decreased elimination of thiopurines
   d. increased elimination of thio-guanine nucleotides

10. What enzyme polymorphism is associated with decreased (S)-warfarin metabolism?
    a. N-acetyl transferase
    b. cytochrome P4502D6
    c. thiopurine methyltransferase
    d. cytochrome P4502C9

CHAPTER 29

The Toxicology Laboratory

K. Michael Parker, Ph.D., DABCC,
and Tai C. Kwong, Ph.D.. DABCC

LEARNING OBJECTIVES

After completing this chapter, the reader should be able to:

1. Describe two levels of toxicology testing used to support the diagnosis and treatment of impaired and overdose patients.
2. Describe the process of developing an essential drug screen and the drug assays that can be included in the screen.
3. Describe the chromatographic techniques that can be used for advanced toxicology testing, and list examples of drugs that can be identified with these procedures.

INTRODUCTION

"Laboratory services must be available within a reasonable period of time for the provision of appropriate diagnostic tests for individuals who require these services." This guideline, approved by the American College of Emergency Physicians' Board of Directors and published as a Policy Statement, indicates the critical need for laboratory support of patients presenting for emergency medical care (1). The policy provides a list of suggested laboratory tests for hospitals with 24-hour emergency departments. In addition to the routine laboratory tests, the guideline calls for toxicology screening and drug concentrations.

The challenge for today's laboratory is to decide how much toxicology testing is appropriate. There are three questions that must be considered:

- What testing is necessary for patient care?
- What technological capabilities are required?
- What testing can be financially justified?

This chapter will consider each of these questions. Background information related to toxicology testing and a more thorough description of methods may be obtained from recent textbooks (2–5).

WHAT TESTING IS NECESSARY FOR PATIENT CARE?

Clinical laboratories routinely provide several tests that physicians need when a poisoning or drug overdose is suspected (Table 29–1). Blood gases, electrolytes, glucose, urea nitrogen, liver function tests, and osmolality are general tests that should be obtained for any potentially toxic patient. These tests not only assess metabolic and organ function, but also allow the determination of the anion gap and osmolal gap. Depending on the situation, other routine laboratory tests may be useful (such as renal tests, complete blood count, and urinalysis).

In addition to routine chemistry, hematology, and urinalysis tests, is more specific toxicology testing medically necessary? Should

TABLE 29–1. *General tests*

- Blood gases
- Electrolytes
- Glucose
- Urea nitrogen
- Osmolality
- Hepatic tests
- Renal tests
- Complete blood count
- Urinalysis

the laboratory offer additional testing to identify and/or quantify a suspected toxin?

Opponents of drug screening point out that most toxicologic diagnoses and treatment decisions are made on the basis of clinical information. Numerous studies (6–11) have attempted to demonstrate the value of toxicology results for patient care. Even though situations occur in which unexpected substances are identified or clinical assessment is inaccurate, these studies have shown that drug-screening results influence clinical judgment in a minority of cases. However, the effect of even rare findings in comprehensive outcome studies remains to be demonstrated.

Proponents of toxicology testing usually counter with arguments based on the following.

- Knowledge of the specific agent present and, in some cases, the concentration of the toxin in blood can affect the diagnosis and treatment of the patient. Patient histories are often very helpful but are not always accurate, especially when the patient has taken an illicit substance. Signs and symptoms at presentation may provide essential clues that indicate the nature of the poisoning; however, patients who have ingested one of several drugs (for example, acetaminophen and iron) may present to the emergency department when toxicity is occurring but is not clinically apparent. Patients may present either before or after apparent symptoms develop; therefore, it is useful to determine what substances are actually present.
- A drug may be identified by toxicology testing for which a specific antidote is available or extracorporeal removal is required. Quantification initially and after therapy may be helpful to determine the adequacy of treatment.
- Multiple drugs may be identified by toxicology testing. Seldom is a single drug ingested. History and clinical assessment may not adequately indicate all substances present; however, knowledge of an unsuspected substance may be helpful in the treatment of the patient.
- "Negative" results from toxicology testing may lead the clinician to consider other etiologies to explain the patient's condition. The signs and symptoms of a substance poisoning are not unique. When a drug cannot be detected, other causes for coma, seizures, psychosis, etc., should be investigated. Additional procedures [for example, computed tomography scan or lumbar tap] or consultation may be appropriate in these situations.
- Quantification of a toxin may affect the approach to therapy. Obviously, low concentrations of acetaminophen and methanol, along with other drugs, are treated differently from high concentrations.

If one concludes that toxicology testing can be beneficial, then how much and what testing should be done? Toxicology testing can have diagnostic, management, and/or prognostic value (12), as illustrated in Figure 29–1. Therefore, the decision as to which tests to provide must take into consideration the purpose of the testing.

For the initial identification of a substance, a test should have a high analytical accuracy.

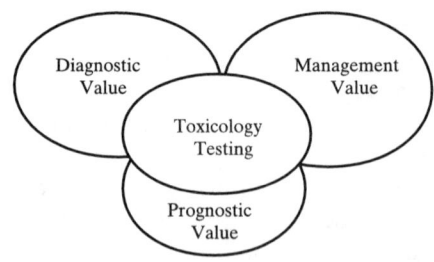

FIG. 29–1. Value of toxicology testing.

Most analytical methods can detect substances at concentrations encountered in overdoses. Analytical false-positives result from nonspecificity in testing procedures. If false-positive results are immediately acted on without elimination through confirmatory testing, patient care can be compromised. Clinical false-positives occur when a substance is detected that is not contributing to the clinical situation. It remains the task of the clinician to determine the contribution of detected substances to the patient's medical condition. Clinical false-negatives occur because those using the results do not understand the limitations of the drug screen. If a report stating "no drugs detected" is considered negative for all relevant drugs, then the possibility of drug toxicity may be too quickly dismissed. To reduce clinical false-negatives it is imperative that the medical staff understand what substances can be detected in the toxicology testing offered by the laboratory.

Analytical procedures used to make serial quantitative measurements to follow treatment should be accurate but also must possess a high degree of precision because multiple results will be compared. Laboratories should strive to develop and use methods with coefficients of variation of less than 8%. This precision is achievable with many quantitative drug assays and is sufficient to detect most medically significant changes in serial results. When this cannot be achieved in a single analytical run, the mean from replicate measurements may provide the desired precision. The medical staff should understand the contribution of imprecision to serial results in order to identify significant changes in patient values.

Essential Drug Screen

In recent years, the concept of a set of essential toxicology assays targeted for selective drugs has emerged as the choice for routine testing preferred over the general drug screen. The assays comprising the "essential drug screen" identify a limited number of drugs for which urine and serum testing have the greatest immediate impact on diagnostic and management decisions. Due to the essential nature of this screen, a panel composed of these assays should be available for emergency patient care 24 hours a day and 7 days a week. Typically, results from the panel should be available within 60 minutes—a time frame useful for patient care.

Bailey (13) compared results obtained by comprehensive and limited drug screens in an emergency and hospital population. Over a 6-month period, 1734 consecutive cases were tested using both comprehensive and limited drug screens. Results from the limited screen were available within 60 minutes; results from the comprehensive screen were not available until much later. In 71% of the urines screened, comparable results were obtained from the two approaches. As would be expected, in most cases in which discrepant results were obtained, additional substances were identified with the comprehensive drug screen. No attempt was made in the study to determine whether the additional findings obtained with the comprehensive approach added value to patient care.

Warner (14) described a process used to define the composition of a limited drug screen for emergency toxicology based on clinical utility. Only drugs or drug metabolites that can be quickly and accurately identified and fulfill one or more of the following criteria were included:

- An antidote or specific treatment for the drug is available.
- Delayed symptoms requiring recognition and treatment are associated with the drug.
- The drug is commonly encountered in the patient population and detection is helpful in delivering appropriate patient care.

Through a joint effort of the laboratory and clinical staffs, drug assays were selected for the stat emergency drug screen. This entire panel can be completed within 45 minutes using assays that are available around the clock each day. Note that the composition of the emergency drug screen does not contain some substances routinely included in workplace drug screens and some point-of-care testing (POCT) devices (such as phencyclidine).

The selective drug screens used by Bailey and Warner differ in composition, but both were developed in cooperation with their clinical staffs to provide rapid results with high clinical utility. Input from the medical staff is necessary when identifying the assays to be included in the limited drug screen. Not only will this process lead to the most appropriate drug screen for the patient population, but it also allows the clinician staff to understand the limitations of the testing panel. In both reports results from the panels were available within 1 hour for use in patient care. Such a response time for the laboratory is much faster than is usually experienced with traditional toxicology testing. One of us (K.M.P.) has used this approach in our laboratory to develop the limited drug screen shown in Table 29–2. Results from this panel are available within 1 hour. Table 29–3 lists the candidate drug assays that should be considered in designing the emergency drug screen.

Advanced Toxicology Testing

Obviously, patients can be encountered in emergency situations who have been exposed to substances not identified with the limited drug screen (such as clonidine overdose). More extensive toxicology testing may be useful in the care of these patients. Parker et al. (15) reported results obtained by comprehensive testing in a pediatric population. Others (16,17) have reported results from comprehensive testing of adult patients in the setting of emergency care. However, testing using more advanced toxicology procedures requires additional time beyond 1 hour to complete. Typically, at least 4 hours are necessary to complete testing for uncomplicated cases and additional time for complex ones. Acute care of these patients is supportive and in most cases adequate despite the lack of comprehensive drug-testing results. Advanced toxicology testing may contribute helpful information in a difficult case and may provide data useful for follow-up care.

The situations requiring advanced toxicology testing are less frequently encountered than those in which the limited screen is useful. Most laboratories will not find it cost effective or of sufficient clinical utility to obtain the equipment required for this level of testing. Consequently, a hospital laboratory may determine that patient care needs are adequately met by outsourcing more ad-

TABLE 29–3. Tests appropriate for an essential toxicology screen

Urine assays (qualitative)
- Cocaine metabolites
- Opiates
- Amphetamines
- Barbiturates
- Benzodiazepines
- Propoxyphene
- Tricyclic antidepressants

Serum assays (quantitative)
- Acetaminophen
- Salicylates
- Ethanol
- Lithium
- Digoxin
- Iron
- Carbamazepine
- Phenytoin
- Theophylline
- Phenobarbital (if urine barbiturates are positive)

TABLE 29–2. Limited drug screen used at OU Medical Center[a]

Urine testing
- Cocaine metabolites
- Opiates
- Benzodiazepines
- Amphetamines
- Cannabinoid metabolites
- Barbiturates
- Ethanol

Serum testing
- Salicylates
- Acetaminophen
- Ethanol
- Tricyclic antidepressants

[a] OU Medical Center is the medical center campus and hospital of The University of Oklahoma.

vanced toxicology testing. A single toxicology laboratory providing advanced testing may serve multiple hospitals. If a hospital laboratory chooses to outsource specimens for advanced testing, a process should be in place to preserve and save sufficient quantities of specimens from the acute episode in case further testing is needed. If these specimens are not available for later testing valuable information may be lost.

WHAT TECHNOLOGICAL CAPABILITIES ARE REQUIRED?

The selective drug screen should be considered as the essential toxicology service required to support an emergency department. It can be provided by any clinical laboratory, including those that do not have advanced instrumentation. Such testing is feasible in any laboratory supporting emergency medical care because the drugs can be detected using assays that are either available as reagent kits (such as urine immunoassays for drugs of abuse) or simple chemical tests (such as color tests) that can be performed around the clock each day. The quantitative serum tests are either routine therapeutic drug monitoring assays (such as digoxin) or clinical chemistry tests (such as lithium) that are available in most, if not all, clinical laboratories. Follow-up confirmation for most positive results should be done to eliminate false-positives and to prevent adverse consequences to the patient of an incorrect result. Whether confirmation is performed in the laboratory or outsourced to another laboratory will depend on the technology available. Laboratories that offer only the limited screen should arrange with another laboratory to provide confirmatory testing for drugs for which positive results have implications beyond the acute episode of care.

Essential Drug Screen

Simple chemical (spot) tests can be useful to rule out certain toxins and require no additional technology. These tests may require some familiarity but are easy to perform directly on the specimen. They should be used to provide a rapid preliminary result. The following toxins can be identified by spot tests:
- Acetaminophen
- Salicylates
- Phenothiazines
- Ethchlorvynol

Spot tests, while adequately sensitive, particularly in overdoses, are not specific. Moreover, a very high drug concentration may give a much darker color, and a combination of drugs (each producing a different color) may produce a final color that is not characteristic of a single component. A positive spot test is best interpreted as the presence of drug(s) and should be followed by additional testing. For example, a positive spot test for salicylates or acetaminophen should prompt the quantification of these substances in serum.

POCT devices have become available to provide immunoassay-based methodology near the patient (18). It is our experience that such methodology can be helpful, but that results should be verified or confirmed by another assay. Valentine and Komoroski (19) reported problems with a visual panel detection device. Because most immunoassays are relatively nonspecific, verification or confirmation is recommended, especially when results indicate an illicit drug or results do not correlate with the patient's clinical presentation. Results obtained with these screening tests may be sufficient for medical purposes when interpreted in relation to clinical information; however, in the absence of clinical data or for forensic purposes they should be clearly regarded as presumptive results. The laboratory report should clearly indicate that the results are preliminary, unconfirmed findings. Whenever possible, there should be follow-up confirmation for most positive results to eliminate false-positives and to prevent adverse consequences to the patient of an incorrect result. Further testing by another laboratory for confirmation may be necessary in situations in which more certainty is required. Specimen identity, that is, chain of custody, will also be an issue in this latter situation.

For assessing performance of POCT assays we recommend that two levels of control materials (one positive and one negative) be performed every 24 hours or when the assays are run. Positive controls should be selected with analyte concentrations close to the cutoff values used to distinguish positive from negative results.

Quantification of drug concentrations should be performed for substances with an established relationship between concentration and clinical effect (20). Drug concentrations are usually carried out using serum, although plasma or whole blood may be acceptable for some measurements. Serum, obtained without the use of separator material, is the preferred specimen for quantitative immunoassays and lithium measurements (21). The presence of ethanol can be identified in urine, but all quantifications must be performed on blood, breath, serum, or plasma. (Urine ethanol concentrations are not useful in determining blood ethanol concentrations due to wide variations in the urine-blood ethanol ratio.) Because the enzymatic methods for ethanol use serum or plasma, either of these specimens should be used. Whole blood ethanol measurements are more appropriately performed by gas chromatography (GC) at the next level of toxicology testing.

Advanced Toxicology Service

Over the past three decades the advent of new technology has expanded the menu of drugs that can be detected. With comprehensive drug testing often more than 500 substances (drugs and/or metabolites) can be identified.

This level of testing requires technology that may not be readily available in most laboratories. Equipment and instrumentation for liquid chromatography (LC), GC, and thin-layer chromatography (TLC) greatly expand the toxicology screening capabilities of a laboratory. Specific compounds in several drug groups can be identified. In addition, this technology can be used to verify screening results obtained with immunoassay or spot testing. This form of confirmation (LC, GC, or TLC) may not be sufficient for forensic services but is adequate for testing in the medical setting. Table 29–4 illustrates some of the drugs that can be identified through more extensive testing.

Capabilities to perform more advanced toxicology testing also expand the specimen types that may be tested. Immunoassay drug-screening methods were designed for detection of compounds in urine; however, they have been applied to the detection of drugs in other specimens. A commonly used alternative specimen is meconium (22). Drug

TABLE 29–4. *Drugs identified by extensive toxicology testing*

Gas chromatography (qualitative or quantitative)
- Methanol
- Isopropanol
- Ethylene glycol
- Basic drugs (sedatives, hypnotics, opioids)

Liquid chromatography
- Amines (identify specific compound)
- Opiates (identify specific compound)
- Tricyclic antidepressants (identify specific compound)
- Phenothiazines (identify specific compound)
- H_2-receptor blockers (identify specific compound)
- Benzodiazepines (identify specific compound)
- Others [for example, the Remedi® HS System (Bio-Rad) is cleared by the Food and Drug Administration to identify >900 drugs and metabolites]

Thin-layer chromatography
- Tricyclic antidepressants (identify specific compound)
- Opiates (identify specific compound)
- Carbamates (meprobamate, carisoprodol)
- Analgesics (acetaminophen, propoxyphene, tramadol)
- Quinine and quinidine
- Amines (identify specific compound; distinguish amphetamine and methamphetamine from ephedrine, pseudoephedrine, diphenhydramine, and phentermine)
- Others (for example, the Toxi-Lab® Drug Compendium contains many other drugs and metabolites that can be identified by TLC)

extraction from meconium is relatively straightforward and can be done without acquiring additional equipment or expertise. However, these methods should be used with caution with alternative specimens. Urine metabolites for which the immunoassays were designed may not be the most significant compounds or the most useful substances to measure in another specimen type (23). With appropriate extraction techniques and chromatographic methods, drug metabolites found in the alternative specimen matrix can be identified.

Different detectors are required for GC depending on the substance to be detected. To identify and quantify the alcohols, the GC system should be equipped with a flame ionization detector. Identification of basic drugs is improved with a nitrogen-phosphorus detector.

Substances are identified with GC by their retention times. Therefore, an internal standard should be used for both qualitative and quantitative measurements as a monitor for retention time. Retention time and relative mobility are important parameters for LC and TLC, respectively. In addition, other parameters can be used to identify compounds. Information obtained by scanning or diode-array spectrophotometry can be used along with retention time in LC. In TLC, spot color, after chemical development of the chromatogram, provides important information in addition to relative mobility.

Quality-control materials for the broad range of drugs identified by chromatographic methods are not routinely available in general-purpose laboratories. Controls that contain multiple drugs must be obtained or prepared and should be analyzed daily to evaluate analytical performance. The control materials should contain drugs selected for their different polarities and other properties in order to serve as effective indicators of extraction and chromatography performances.

GC and LC systems are readily available at modest prices. The skills required to operate these are beyond those often available in the routine laboratory. Special training of laboratory personnel is required. TLC is inexpensive to set up, but may be somewhat expensive to operate if materials are purchased as kits. Results from chromatographic systems are quite dependent on the skills of the operator. We have found that extensive training and experience are required to maximize the benefits of these systems. A significant volume of testing is necessary to retain the expertise. Laboratories in which testing volume is low often find that quality of the drug screen is low due to infrequent use.

Laboratories that have the instrumentation and expertise to perform testing with mass spectrometry can further expand their testing capabilities. This requires the addition of gas chromatography–mass spectrometry (GC/MS) to the technology available in the laboratory. This instrument provides a wide range of capabilities for performing forensic confirmations, quantifying specific drugs, and identifying a much larger group of substances. By building an in-house library, one can identify >500 drugs and metabolites by GC/MS. Table 29–5 gives a list of some

TABLE 29–5. *Drugs identified by GC/MS*

- Amobarbital
- Atropine
- Betaxolol
- Butabarbital
- Carisoprodol
- Chlordiazepoxide
- Chloroquine
- Chlorpheniramine
- Clonidine
- Clozapine
- Dextromethorphan
- Diazepam
- Doxylamine
- Fentanyl
- Fluoxetine
- Fluvoxamine
- Haloperidol
- Hydroxyzine
- Ibuprofen
- Lorazepam
- Meprobamate
- Metoprolol
- Pentobarbital
- Phenobarbital
- Propoxyphene
- Secobarbital
- Selegiline
- Sertraline
- Tacrine

drugs for which GC/MS is particularly good for identification in an underivatized urine extract.

Multidrug control material should be analyzed routinely when the GC/MS system is being used to screen for a variety of drugs. Developing a comprehensive toxicology laboratory usually requires that personnel be specifically employed for this function. We have found that the GC/MS system performs more reliably with less downtime if use of the instrument is restricted to a few, more experienced personnel.

Recent developments in the interfacing of a LC to a mass spectrometer and in the technology for ionization have resulted in liquid chromatography–mass spectrometry (LC/MS) instruments that are robust and sufficiently economical to deploy in an advanced toxicology testing setting. The LC/MS is particularly suitable for analysis of biological specimens for drugs and metabolites (24).

WHAT TESTING CAN BE FINANCIALLY JUSTIFIED?

Each laboratory operates in an environment of decreasing reimbursements and increasing costs. The revenue obtained for clinical toxicology testing is set by payors (frequently at levels insufficient to cover the usual cost of testing), leaving laboratories very little opportunity to adjust how much they are paid for services; therefore, profitability and in many cases survival depend largely on finding ways to reduce costs.

Payors base reimbursement for laboratory testing on the American Medical Association's Current Procedural Terminology (CPT) codes. Chemistry codes 82000–84999 or therapeutic drug assay codes 80150–80299 are appropriate for quantification of drugs in a toxicology screen. The specific code is selected for each drug quantitated. The options are limited for coding qualitative drug screens and are based on the procedures used rather than the drug identified. CPT 80100 is used for a drug screen when a chromatographic method is used; CPT 80101 is appropriate when an immunoassay technique is used. Either code can be used multiple times in a single case; however, 80100 and 80101 cannot be used together. CPT 80102 is designated for each confirmatory procedure; however, confirmation must be linked to a positive screening result. These options for reimbursement can accommodate limited drug screening but are not supportive of comprehensive drug screening in which both chromatographic and immunoassay procedures are applied concurrently.

In most laboratories, labor contributes 50–60% of the total cost of operation. Testing that can be provided using automated equipment can effectively reduce the dollars spent on personnel. The cost of instrumentation for drug assays is greatly reduced when the same system is used for general chemistry testing; however, the purchase of capital equipment dedicated solely to toxicology testing is more difficult to justify. Chromatography equipment is generally not available through reagent purchase plans. Dollars must be spent upfront to acquire this technology. Consequently, only laboratories that have a significant volume of drug testing are likely to be able to justify the expense.

SUMMARY

As recommended by the American College of Emergency Physicians, laboratories serving emergency departments should provide rapid toxicology screening and drug concentrations in addition to routine chemistry, hematology, and urinalysis tests. Three important issues must be considered when

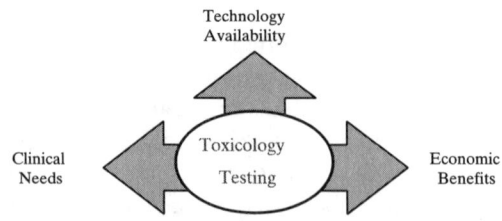

FIG. 29–2. Considerations for toxicology testing.

deciding what and how much toxicology testing should be done (Figure 29–2).

We have described the concept of a limited, targeted, and rapid drug screen (essential drug screen) that should be the basic toxicology service provided by a clinical laboratory in support of an emergency department. This drug screen is within the capabilities of most laboratories. In consultation with the physicians providing emergency care and in consideration of the patient population, the specific composition of the drug screen and the list of drug to be quantified can be determined. This level of testing is justifiable both clinically and economically. POCT devices may be a part of this approach; however, the cost of these devices can significantly increase the cost of toxicology testing and restrict the flexibility required to tailor the menu to the clinical needs.

We have also described a more comprehensive toxicology service (advanced toxicology testing) that requires dedicated chromatographic equipment. This capability may not be appropriate for every laboratory, but may be extremely useful in some situations. The expanded services of a comprehensive toxicology laboratory are essential to confirm screening results and to assist with cases in which the ingested substance can only be detected by chromatography procedures.

Many laboratories may choose to outsource requests for comprehensive toxicology testing; however, metropolitan areas are best served when these results can be obtained as soon as possible, preferably within 6–12 hours. A minimum of one comprehensive toxicology laboratory is an important resource in the community to meet this service goal. For laboratories that offer only the essential drug screen and outsource more extensive toxicology testing, we recommend that specimens from the first encounter with the patient be saved and properly preserved should additional testing be needed. A delay in collecting the required specimens may diminish the value of subsequent toxicology testing.

The approach to providing appropriate clinical toxicology testing described in this chapter is similar to the selective and progressive strategy recently recommended by the Alberta Medical Association (http://www.amda.ab.ca). The approach is also consistent with the draft recommendations for the use of laboratory tests to support the impaired and overdosed patient from the emergency department currently promulgated by an expert panel of the National Academy of Clinical Biochemistry (http://www.nacb.org).

REFERENCES

1. Emergency care guidelines. *Ann Emerg Med* 1997;29:564–571.
2. Diagnostic procedures. In: Ellenhorn MJ, Schonwald S, Ordog G, et al: *Ellenhorn's medical toxicology: diagnosis and treatment of human poisoning*, 2nd ed. Baltimore: Williams and Wilkins, 1997:47–65.
3. Osterloh JD: Laboratory diagnoses and drug screening. In: Haddad LM, Shannon MW, Winchester JF, eds. *Clinical management of poisoning and drug overdose*, 3rd ed. Philadelphia: WB Saunders, 1998:44–59.
4. Weisman RS, Howland MA, Verebey K: The toxicology laboratory. In: Goldfrank LR, Flomenbaum NE, Levin NA, et al, eds: *Goldfrank's toxicologic emergencies*, 5th ed. East Norwalk, CT: Appleton & Lange, 1994:99–108.
5. Kwong TC: Toxicology. In: McClatchy KP, ed: *Clinical laboratory medicine*. Baltimore: Williams and Wilkins, 1994:445–467.
6. Wiltbank TB, Sine HE, Brody BB: Are emergency toxicology measurements really used? *Clin Chem* 1974;20:116–118.
7. Helliwell M, Hampel G, Sinclair E: Value of emergency toxicological investigations in differential diagnosis of coma. *Br Med J* 1979;2:819–821.
8. Bury RW, Mashford ML: Use of a drug-screening service in an inner-city teaching hospital. *Med J Aust* 1981;1:132–133.
9. Kellermann AL, Fihn SD, Logerfro JP, et al: Impact of drug screening in suspected overdose. *Ann Emerg Med* 1987;16:1206–1216.
10. Kellermann AL, Fihn SD, Logerfro JP, et al: Utilization and yield of drug screening in the emergency department. *Am J Emerg Med* 1988; 6:14–20.
11. Brett AS: Implications of discordance between clinical impression and toxicology analysis in drug overdoes. *Arch Intern Med* 1988;148:437–441.
12. Osterloh JD: Utility and reliability of emergency toxicologic testing. *Emerg Med Clin North Am* 1990;8:693–723.
13. Bailey DN: Results of limited versus comprehensive toxicology screening in a university medical center. *Am J Clin Pathol* 1996;105:572–575.
14. Warner A: Cost-effective toxicology testing. *Therapeutic Drug Monitoring and Toxicology In-Service Training and Continuing Education* 1996;17:35–43.
15. Parker KM, White BN, Beattie DJ, et al: Compre-

hensive drug screening for a pediatric population. *Clin Chem* 1988;34:748–750.
16. Bailey DN: Comprehensive toxicology screening in patients admitted to a university trauma center. *J Anal Toxicol* 1986;10:147–149.
17. Taylor RL, Cohan SL, White JD: Comprehensive toxicology screening in the emergency department: an aid to clinical diagnosis. *Am J Emerg Med* 1985;3:507–511.
18. Wu AHB: Near-patient and point-of-care testing for alcohol and drugs of abuse. *Therapeutic Drug Monitoring and Toxicology In-Service Training and Continuing Education* 1995;16:227–235.
19. Valentine JL, Komoroski EM: Use of a visual panel detection method for drugs of abuse: clinical and laboratory experience with children and adolescents. *J Pediatr* 1995;126:135–140.
20. Stone JA: Appropriate use of serum drug concentration measurement in overdose. *Therapeutic Drug Monitoring and Toxicology In-Service Training and Continuing Education* 1996;17:205–214.
21. Warner A: Therapeutic drug monitoring guidelines for laboratories. In: Warner A, Annesley T, eds: *Guidelines for therapeutic drug monitoring services*. Washington, DC: National Academy of Clinical Biochemistry, 1999:3–9.
22. Kwong TC, Ryan RM: Detection of intrauterine illicit drug exposure by newborn drug testing. *Clin Chem* 1997;43:235–242.
23. Steele BW, Bandstra ES, Wu N-C: m-Hydroxybenzoylecgonine: an important contributor to the immunoreactivity in assays for benzoylecgonine in meconium. *J Anal Toxicol* 1993;17:348–352.
24. Fitzgerald RL, Rivera JD, Herold DA: Broad spectrum drug identification directly from urine using liquid chromatography-tandem mass spectrometry. *Clin Chem* 1999;45:1224–1234.

SELF-ASSESSMENT QUESTIONS

1. For the essential drug screen, which statement is not true?
   a. All assays for the screen should be performed on a urine specimen.
   b. All assays included in the screen should be available within 1 hour.
   c. The composition of the screen should be developed in consultation with the emergency physicians the laboratory serves.
   d. The required assays should be available in most hospital laboratories.

2. Advanced toxicology testing is based on the use of which of the following for drug identification?
   a. chemical spot tests
   b. chromatography procedures
   c. immunoassay methods
   d. enzyme measurements

3. Select the true statement.
   a. Reagent costs are usually the major component of the total cost of toxicology testing.
   b. There are no established limits for the amount a laboratory can charge for toxicology testing.
   c. Both statements are true.
   d. Neither statement is true.

4. What kind of value does toxicology testing have for patients presenting for emergency medical care?
   a. diagnostic
   b. management
   c. prognostic
   d. all of the above

5. Which is true for quantitative drug measurements?
   a. They correlate the concentration of the drug with the clinical effect.
   b. They should be performed on serum rather than urine.
   c. Both are true.
   d. Neither is true.

APPENDIX A

Answers to Self-Assessment Questions

Chapter 1. Epidemiology of Poisoning

1. d
2. d
3. c
4. c
5. d

Chapter 2. Toxicokinetics: Principles and Practical Applications

1. First calculate the amount of drug in the body.
   C_p = 22 mg/L
   Amount in body = $C_p \times V_d$
   Mean $V_{d\ Theo}$ = 0.5 L/kg
   Therefore, amount in body = 11 mg/kg.
   But 110 mg/22 kg = 5 mg/kg.
   Possibilities:
   - Lab error (line draw, skin contamination, mixed samples, interference, initial concentration not zero)
   - Dosing error—too much given (nursing, pharmacy, physician), amount or volume, concentration, number of doses
   - Weight incorrect (few 1-year-olds weigh 22 kg!)
   - Initial concentration not zero (history of no theophylline in question, possible use of over-the-counter preparations containing theophylline)
   - A combination of above

2 and 3. Concentrations drawn during the distribution phase are extremely difficult to interpret without sophisticated pharmacokinetic computer programs. The result cannot be compared with published "therapeutic ranges" [0.5–2 ng/mL (0.6–2.6 nmol/L)], which refer to predose, steady-state concentrations.

The fact that these patients are asymptomatic is additional evidence that this concentration is not toxic.

Additional data on the method and site of sampling, presence of potential interacting substances, electrolytes (especially potassium), pH, calcium and magnesium, electrocardiogram, and clinical condition must all be known to assess such a result.

Concentrations must be done at least 6–8 hours after ingestion or when clinical symptoms appear. Concentrations drawn sooner after ingestion may be very high but not indicative of toxicity.

Chapter 3. Pharmacokinetics

1. C_{peak} = dose/[$V_d \times$ weight (kg)]
2. acute-on-chronic > chronic > acute
3. $K_e = [\ln(C_{p1}/C_{p2})]/t$
4. $t_{1/2} = 0.693/K_e$
5. Steady-state volume of distribution is greater than central compartment.

Chapter 4. Clinical Approach to the Poisoned Patient

1. c
2. d

3. b

4. a

Chapter 5. Marijuana

1. e
2. c, d
3. d
4. c
5. c

Chapter 6. Opioids

1. c
2. b
3. e
4. b
5. a
6. d

Chapter 7. Cocaine

1. e benzoylecgonine
   c cocaine
   b ecgonine methyl ester
   a ethylcocaine
   g methylecgonidine
   d m-hydroxybenzoylecgonine
   f norcocaine
2. inhibits amine pump, blocks sodium channel conductance
3. cortical arousal: pupillary dilation; sympathomimetic: euphoria; local anesthetic: increased blood pressure
4. a
5. Rush: After a dose of cocaine, brain concentrations of norepinephrine and dopamine increase. Crash: Rush is followed by a marked reduction to below-normal norepinephrine and dopamine concentrations.
6. b

Chapter 8. Amphetamines

1. phenyl-2-propanone, ephedrine
2. the presence of a halogen (Br>I>F) at the para-position on the benzyl ring
3. formation of permanent diastereomers with a chiral derivatizing agent, formation of temporary diastereomers with a chiral column
4. the use of selegiline (deprenyl), clobenzorex, benzphetamine, famprofazone, fenproporex, or mefenorex
5. c
6. d
7. d-Methamphetamine is primarily a central nervous system (CNS) stimulant; l-methamphetamine has much less CNS activity and is used primarily as a peripheral vasoconstrictor in nasal decongestant inhalers.

Chapter 9. γ-Hydroxybutyrate

1. d
2. a
3. c
4. b
5. e
6. a
7. b
8. e
9. c
10. a

Chapter 10. Point-of-Care Testing for Drugs of Abuse

1. Cocaine, opiates, and phencyclidine are among the most dangerous drugs that can contribute significant acute clinical problems and would warrant emergency testing. The benzodiazepines and barbiturates can also cause significant sedation and may warrant emergency therapeutic measures.
2. Quick turn-around time (TAT) for results is the major advantage for POCT testing in either the emergency department or workplace drug-testing environment. For the emergency department, rapid TAT enables quicker triage of intoxicated or

overdosed patient to the appropriate level of care. For workplace drug testing, on-site testing enables the program to receive negative results immediately, enabling the subject to return to work.

3. Although pilot programs are underway, the Substance Abuse and Mental Health Services Administration currently does not permit on-site drug testing with POCT devices.

4. There are subtle differences in the cross-reactivities of POCT devices. Some are targeted to methamphetamine but not both amphetamine and methamphetamine. Some devices detect free cocaine in addition to the benzoylecgonine metabolite. For the Triage system, the assay is more sensitive to glucuronides than laboratory-based assays.

5. In a negative-indicating POCT devices, a sample containing the targeted drug is indicated by the absence of a line in the test area, which is unlike most POCT devices for other analytes, for example, pregnancy or cardiac marker tests, in which a line is produced in the presence of the analyte.

Chapter 11. Urine Adulteration before Testing for Drugs of Abuse

1. c
2. d
3. b
4. b
5. a
6. d

Chapter 12. Volatile Alcohols: Ethanol, Methanol, and Isopropanol

Set 1

1. The concentration of ethanol does not sufficiently explain the mental status in this patient. Altered mental status in a chronic alcoholic may be due to other conditions, including hypoglycemia, electrolyte derangement, ingestion of other toxins, ethanol withdrawal, head trauma (acute or chronic), infection, hepatic encephalopathy, uremia, Wernicke-Korsakoff syndrome, and postictal state (after stroke or seizure). Breath alcohol was used as a rapid estimation of the blood ethanol concentration in this patient and an assessment of the level of intoxication at the point of care.

2. The estimated blood ethanol concentration by breath testing was lower than, but consistent with, the serum ethanol concentration measured by the laboratory. The higher serum ethanol concentration reflects the greater water content of serum compared with whole blood. Laboratory-based testing ruled out hypoglycemia, other metabolic abnormalities, and volatile alcohol intoxication. The mildly elevated amylase activity, with a normal lipase and abdominal examination, suggested a parotid source of the hyperamylasemia rather than acute pancreatitis (chapter 12 reference 15). Ultimately, the history, physical examination, and imaging led to a diagnosis of head injury in this case, and prophylactic treatment with antiseizure medication was initiated on admission.

3. Yes, the ethanol concentration in this case accounts for the osmolal gap within 5 mOsm/kg.

Formulae for Question 3:

Calculated osmolality in mOsm/kg H_2O =
[sodium in mEq/L $\times$ 2] +
[glucose in mg/dL $\div$ 18] +
[BUN in mg/dL $\div$ 2.8]

Calculated osmolal gap = calculated osmolality − measured osmolality

Ethanol osmolal equivalent
= ethanol in mg/dL $\times$ 0.23 mOsm/kg

Calculations for Question 3:

Calculated osmolality = (138 $\times$ 2) + (121 $\div$ 18) + (17 $\div$ 2.8) = 289 mOsm/kg

Osmolar gap = 310 mOsm/kg − 289 mOsm/kg = 21 mOsm/kg

Ethanol osmolal equivalent
= 69 mg/dL ethanol × 0.23
= 16 mOsm/kg

where 0.23 is the mOsm/kg equivalent of 1 mg/dL ethanol

4. Thiamine is a cofactor required for acetylcoenzyme entry into the Krebs cycle during the final stage of ethanol metabolism to carbon dioxide and water. Multivitamins and thiamine were also given because of nutritional deficiencies in alcoholic patients. Thiamine deficiency is associated with development of encephalopathy in chronic alcoholics.

Set 2

1. The estimated oral dose of ethanol is 49 g.
   Formula for Question 1:

   $$D (g) = F \times Vol (mL) \times 0.8 (g/mL)$$

   where
   D = grams of ethanol in the drink
   F = fraction of ethanol (% vol./vol.)
   Vol = volume in milliliters (1 ounce = ~30 mL)
   0.8 = approximate specific gravity of ethanol (actual specific gravity = 0.789 at 20 °C).

   Calculation for Question 1:

   Ethanol in mouthwash = 0.185 × (11 oz × 30 mL/oz) × 0.8 g/mL = 49 g

2. The ethanol content of 11 ounces of mouthwash is equivalent to three to four alcoholic beverages.
   Formula for Question 2:

   $$D (g) = F \times Vol (mL) \times 0.8 (g/mL)$$

   where
   D = grams of ethanol in the drink
   F = fraction of ethanol (% vol./vol.)
   Vol = volume in milliliters (1 ounce = ~30 mL)
   0.8 = approximate specific gravity of ethanol (actual specific gravity = 0.789 at 20 °C).

   Calculations for Question 2:

   Ethanol content of 1.25 oz whiskey (43%)
   = 0.43 × (1.25 oz × 30 mL/oz) × 0.8 g/mL
   = 13 g

   Ethanol content of 6 oz of table wine (9%)
   = 0.09 × (6 oz × 30 mL/oz) × 0.8 g/mL
   = 13 g

   Ethanol content of 12 oz of beer (4%)
   = 0.04 × (12 oz × 30 mL/oz) × 0.8 g/mL
   = 13 g

3. The dose of ethanol per kilogram of body weight is 3.7 g/kg. This is within the lethal range. The reported lethal dose of ethanol in children [3 g/kg (65 mmol/L)] is lower than the threshold for adults [5–6 g/kg (109–130 mmol/L)]. Based on the ethanol concentration at the time of admission, the estimated amount of ethanol in the child's body was 28 g (see answer to question 8) or 2.2 g/kg.

   Calculation for Question 3:

   Dose of ethanol per kilogram = 49 g (from answer 5) ÷ 13.1 kg = 3.7 g/kg

4. 28.4 g
   Formula for Question 4:

   $$E_t (g) = C_t (g/L) \times V_d (L/kg) \times W (kg)$$

   where
   E_t = ethanol in body at time t
   C_t = blood ethanol concentration at time t
   V_d = volume of distribution
   W = body weight

   Calculation for Question 4:

   3.10 g/L × 0.70 L/kg × 13.1 kg = 28.4 g

Set 3

1. The osmolar gap is 139 mOsm/kg. Methanol accounts for only a portion of the osmolar gap in this case.
   Formulae for Question 1:

   Calculated osmolality in mOsm/kg H_2O
   = [sodium in mEq/L × 2] +
   [glucose in mg/dL ÷ 18] +
   [BUN in mg/dL ÷ 2.8]

   Calculated osmolal gap = calculated osmolality − measured osmolality

   Methanol osmolal equivalent = ethanol in mg/dL × 0.34 mOsm/kg

   Calculations for Question 1:

   Measured osmolality = 465 mOsm/kg

   Calculated osmolality = [2 × 154] + [225 ÷ 18] + [14 ÷ 2.8] = 326 mOsm/kg

   Osmolal gap
   = 465 mOsm/kg −
   326 mOsm/kg
   = 139 mOsm/kg

   Methanol osmolal equivalent = 260 mg/dL methanol × 0.34 mOsm = 88 mOsm/kg

   where 0.34 is the mOsm/kg equivalent of 1 mg/dL methanol

2. Formate exhibited a shorter half-life than methanol and was more effectively removed by hemodialysis.
3. Ethanol treatment, when used without hemodialysis, significantly increases the half-life of methanol.
4. In this patient, during dialysis, the half-life of methanol was only 3.5 hours. When ethanol treatment is used without hemodialysis, the half-life of methanol increases to 30.3–52 hours (chapter 12 reference 30).
5. Acidosis in this patient was severe. The cause was metabolic production of formate. After metabolism occurs, the methanol concentration alone may not be a reliable diagnostic index of methanol intoxication, and the degree of acidosis is a better reflection of toxicity from metabolic production of formate. In this case, both the extreme degree of metabolic acidosis and the initial methanol concentration indicate a life-threatening intoxication. The methanol concentration was in the range in which fatalities have occurred in untreated patients. The osmolal gap was also elevated but overestimated the amount of methanol (see answer to question #1, Set 4), consistent with other cases reported in the literature (chapter 12 reference 33).

Set 4

1. The CNS depression produced by isopropanol poisoning is generally considered to be approximately twice the severity of an equivalent amount of ethanol (Ellenhorn MJ: Alcohols and glycols. In: Ellenhorn MJ, ed: *Ellenhorn's medical toxicology*. Baltimore: Williams & Wilkins, 1997:1127–1165). Acetone also contributes to the depression.
2. Elimination of both toxic agents follows first-order kinetics, but the half-life of acetone is three times as long as that of isopropanol. With a delayed patient presentation, acetone may be the predominant finding in blood.
3. Co-ingestion of ethanol prolongs isopropanol elimination and reduces the rate of metabolism to acetone. Co-ingestion of ethanol can affect the relative amount of isopropanol and acetone because isopropanol and ethanol compete for alcohol dehydrogenase. As the pharmacokinetic data in this case show, the concentration of both isopropanol and acetone vary significantly during the postingestion course.

Chapter 13. Glycols

1. e
2. b
3. c
4. b
5. a

6. e
7. a

Chapter 14. Psychotropic Agents: The Benzodiazepines

1. treatment of insomnia, treatment of anxiety, prophylactic treatment for epileptic seizures, muscle relaxant
2. may cause or aggravate clinical depression, may induce drug dependence, may cause disinhibition, occasionally may result in paradoxical violent reactions
3. The most highly prescribed benzodiazepines in the United States are alprazolam, lorazepam, clonazepam, diazepam, and temazepam. The Internet Web site http://www.rxlist.com/top200.htm lists the 200 most highly prescribed drugs by number in US prescriptions dispensed each year. In 1999, benzodiazepines listed on the top 200 list in a study sample of 2.8×10^9 prescriptions were alprazolam (#45) marketed by Greenstone, alprazolam (#72) sold by Geneva, lorazepam (#78) available from Mylan, clonazepam (#86) marketed by Teva Pharm, diazepam (#127) sold by Mylan, temazepam (#159) from Mylan, lorazepam (#161) marketed by ESI Lederle, alprazolam (#178) sold by Purepac, and lorazepam (#199) from Purepac.
4. high mean drug dose, prolonged duration of treatment (>90 days), dependent personality, history of drug and/or alcohol abuse, use of benzodiazepines with short elimination half-lives
5. cross-reactivity of kit antibodies for flunitrazepam metabolites when used for screening, choice of biological specimen (blood, serum, or urine), selection of confirmation method [high-performance liquid chromatography (HPLC) or gas chromatography/mass spectrometry (GC/MS)], screening and confirmation cutoff values, time interval since last use of drug or exposure
6. **Advantages of immunoassays**: objective measurement, ease of automation, consistent application of a defined cutoff value

Limitations of immunoassays: Antibodies found in various vendors' products have different cross-reactivity to various benzodiazepine metabolites, and normal cutoff values (200–300 ng/mL) are too high to detect some low-dose benzodiazepines. Also, optimal cross-reactivity to several benzodiazepines is only possible with enzymatic hydrolysis of urine specimens before the screening test. Most diagnostic companies have not incorporated β-glucuronidase hydrolysis enzyme into their urine benzodiazepine screening kits.

Advantages of chromatographic assays: One can detect specific drugs and/or drug metabolites; one can use several different analytical approaches for chromatographic analysis including gas chromatography, HPLC, GC/MS, and liquid chromatography–mass spectrometry.

Limitations of chromatographic methods: These analytical procedures are generally more technically complex, have a longer turn-around time, and have a higher instrumentation cost. Chromatographic methods allow one to use >1 cutoff value for the benzodiazepines. A confirmation cutoff value of 100–200 ng/mL (used for higher-dose benzodiazepines such as diazepam or oxazepam) should be lowered to 25–50 ng/mL to detect lower-dose drugs such as alprazolam and triazolam (by measuring their α-hydroxy metabolites). Analytical run times for benzodiazepine analysis by GC/MS is long (often up to 20–25 minutes) compared with other analyses, so throughput is relatively low.

Chapter 15. Antidepressant Drugs

1. d
2. e
3. b
4. b

5. e
6. e
7. d

Chapter 16. Agents for the Treatment of Bipolar Disorder

1. a
2. a
3. a
4. d
5. c

Chapter 17. Antiepileptic Drugs

1. e
2. b
3. d
4. b
5. a
6. c
7. b

Chapter 18. Digoxin and Other Cardiac Glycosides

1. c
2. d
3. a
4. d
5. b
6. d
7. e
8. a. Over the years, the clinical value of digoxin in the treatment of heart failure in patients with normal sinus rhythm has been controversial. Part of the controversy is because the design of clinical studies has in general been poor, leading to hard-to-interpret conclusions. Recently, three prospective randomized controlled studies (PROVED, RADIANCE, and DIG) have shown that the major effect of digoxin together with either diuretic therapy or both diuretic and angiotensin-converting enzyme (ACE) inhibitor therapy is on quality of life, including improvement in the signs and symptoms of congestive heart failure, increased exercise tolerance, improvement in left ventricular ejection fraction, less need for other medications, and reduction in the rate of hospitalization. There has not been an effect of digoxin on the overall rate of death in these studies.

b. The development of new, improved methods for digoxin measurement continues to evolve. There is a need to define the most reliable and accurate method for measurement of digoxin and to evaluate this in prospective therapeutic drug monitoring studies.

c. Although there is a generally accepted therapeutic range for digoxin in clinical practice, reevaluation of this range is important for two reasons. The first is that today's most frequently used immunoassays are less subject to interferences, particularly digoxin-like immunoreactive factors, than the earlier commonly used procedures. Thus, measured digoxin concentrations are generally lower than previously. The second reason is that digoxin is often given with an ACE inhibitor (to reduce the afterload) and a diuretic, thus providing a different pharmacologic milieu than previously. It is possible that a lower therapeutic range will be required at least for some patients than was previously the case.

Chapter 19. Calcium Channel Blockers: An Overview

1. a
2. a
3. b
4. c
5. b
6. d

Chapter 20. Biological Monitoring of Chemical Exposure

1. d
2. c
3. b
4. a
5. d
6. b

Chapter 21. Carbon Monoxide Poisoning

1. motor vehicles and heaters
2. decreased capacity of blood to transport oxygen and impaired oxygen release by hemoglobin
3. gas chromatography and CO-oximetry
4. personality changes, emotional lability, gait disturbances, and aphasia
5. no proven benefits to hyperbaric oxygen therapy, many limitations in historical studies, and numerous potential risks of therapy
6. education of the general population and health-care providers; installation of carbon monoxide detectors in business and residential settings

Chapter 22. Organophosphate and Carbamate Pesticide Poisoning

1. a
2. d
3. c
4. a
5. c

Chapter 23. Lead Testing

1. increased ingestion of lead-contaminated dust as a result of proximity to the floor and hand-to-mouth activities, increased absorption of ingested lead, developing nervous system has incomplete blood-brain barrier and is more susceptible to damage
2. analytical imprecision, specimen contamination, decrease in blood lead concentration between screening test and confirmatory test due to redistribution and elimination after an acute exposure
3. direct measurement of lead concentration in blood or urine (or hair); in vivo measurement of lead content of bone using X-ray fluorescence; measurement of a biological marker, for example, zinc protoporphyrin or δ-aminolevulinic acid
4. atomic absorption spectrophotometry, anodic stripping techniques, inductively coupled plasma mass spectroscopy
5. targeted screening, rather than universal screening, pediatric lead screening using point-of-care devices, identifying and abating hazardous housing prospectively using measurement of lead in household dust

Chapter 24. Arsenic, Mercury, and Cadmium

1. The primary choice is urine because it is easy to collect and easy to handle in the clinical laboratory. Of course, contamination is a greater concern for urine specimens than for venous collection. Urine arsenic values can provide an assessment of arsenic exposure during the previous 2–3 days. Hair and nail are also frequently used specimens to assess long-term (previous 6–12 months) exposure to arsenic; however, specimen contamination is of greatest concern.
2. Arsenic exists in several inorganic and organic forms. In general, inorganic arsenic species are more toxic than organic species. Most elevated urine arsenic concentrations are due to nontoxic organic species such as arsenobetaine and arsenocholine. The major source of these compounds is seafood, and arsenic speciation should be used to confirm the source of elevated arsenic.
3. The interaction between mercury and selenium in the body is complex and concentration dependent. In general, selenium displays an overall protective effect against mercury toxicity. This protective effect may be due to redistribution of mercury in the presence of selenium,

competition for binding sites, formation of mercury-selenium complexes, and conversion of mercury to less toxic forms. However, a high concentration of selenium enhances mercury toxicity.

4. Public concern is high regarding mercury exposure from dental amalgams. Mercury slowly leaches from amalgams used in dental fillings; however, the amount ingested represents an additional body burden rather than intoxication. In recent literature there are no clear findings showing adverse effects of mercury released from dental amalgam fillings.

5. The major target organ is kidney. Metallothionein in urine is such an early indicator of cadmium body burden that its presence precedes renal tubular damage. Several other markers can be used to assess the tubular damage caused by cadmium. The ones that indicate irreversible kidney damage include β_2-microglobulin, retinol-binding protein, and glycosaminoglycans.

Chapter 25. Acute Iron Poisoning

1. Iron preparations represent the most common fatal ingestion in toddlers.

2. The peripartum period is a particularly dangerous time for unintentional iron ingestion in young children because of a combination of parental stress and distraction, the universal prescription of iron supplements to pregnant and postpartum women, toddler impulsivity heightened by the arrival of a new sibling and parental attention to the baby, use of iron over a long time three times a day, and consequent failure to consider it a true medication (with potentially dangerous side effects).

3. Iron toxicity results in metabolic acidosis because of the following: decreased intravascular volume from vomiting, gastrointestinal (GI) blood loss, and capillary leak; direct effects of iron on release of H^+ ions; and interference with Krebs cycle enzymes and mitochondrial oxidative metabolism.

4. The principal clinical features of acute iron poisoning are GI effects including vomiting, diarrhea, and GI bleeding; cardiovascular dysfunction that may progress to shock; metabolic acidosis; hepatic toxicity that may progress to fulminant necrosis; and pulmonary injury including an acute respiratory distress syndrome–like picture.

5. No, measuring the iron-binding capacity is not useful in the laboratory evaluation of acute iron poisoning because it is usually difficult to obtain, often falsely elevated, and does not add much to the clinical picture and serum iron concentration.

6. Deferoxamine is the chelating medication that is currently the drug of choice for iron poisoning. It is indicated for moderate to severe poisoning clinically and/or for serum iron concentrations >500 µg/dL (>89.5 µmol/L).

Chapter 26. Alternative Samples: Oral Fluid (Saliva), Sweat, Hair, and Meconium

1. saliva, 1 day; sweat, 1 week; hair, 2–3 months

2. saliva

3. Sweat provides the convenience of weekly monitoring periods and is approved by the Food and Drug Administration; hair provides a long-term (as long as 3 months) monitoring period, and additional samples can be obtained (within a few days) that are similar to the original collection.

4. the last two trimesters of pregnancy

Chapter 27. Advanced Analytical Techniques

1. a, b, c
2. b, c
3. b
4. a
5. base peak and parent ion = mass-to-charge ratio 268
6. c
7. c

8. b
9. b
10. b
11. c
12. e
13. b
14. b
15. a
16. b
17. b

Chapter 28. Fundamentals of Pharmacogenetics

1. c
2. c
3. a
4. c
5. e
6. b
7. c
8. c
9. b
10. d

Chapter 29. The Toxicology Laboratory

1. a
2. b
3. d
4. d
5. c

APPENDIX B

Various Toxins Associated with Vital Sign Abnormalities

Francis J. De Roos, M.D.

TEMPERATURE

Hyperthermia: increased temperature

- Cocaine and amphetamines
- Theophylline and caffeine
- Thyroid hormone
- Antihistamines
- Phencyclidine
- Phenothiazines and neuroleptics (neuroleptic malignant syndrome)
- Salicylates
- Dinitrophenol
- Serotonin reuptake inhibitors (serotonin syndrome)

Hypothermia: decreased temperature

- Sedative-hypnotics including ethanol, benzodiazepines, and barbiturates
- Hypoglycemic agents
- Opioids
- Clonidine, guanabenz
- Carbon monoxide
- Phenothiazines

HEART RATE

Tachycardia: increased heart rate

- Cocaine and amphetamines
- Theophylline and caffeine
- Ethanol and other sedative-hypnotic withdrawal
- Tricyclic antidepressants
- Antihistamines
- Phenothiazines
- Iron
- Hallucinogens (bad trip)

Bradycardia: decreased heart rate

- Calcium channel antagonists
- β-Adrenergic antagonists
- Cardiac glycosides

Dysrhythmia: irregular heart beat

- Antidysryhthmic drugs
- Clonidine, guanabenz
- Cholinergics including carbamates and organophosphates
- Phenylpropanolamine

BLOOD PRESSURE

Hypertension: increased blood pressure

- Cocaine and amphetamines
- Theophylline (early)
- Phenylpropanolamine
- Ergot alkaloids
- Lead
- Monoamine oxidase inhibitor (drug interaction)

Hypotension: decreased blood pressure

- Theophylline (late)
- Tricyclic antidepressants
- Calcium channel antagonists

- β-Adrenergic antagonists
- Cardiac glycosides
- Clonidine, guanabenz
- Iron
- Sedative-hypnotics including ethanol, benzodiazepines, and barbiturates
- Nitrates and nitroprusside
- Sedative-hypnotics
- Disulfiram reaction
- Monoamine oxidase inhibitor (overdose)

RESPIRATORY RATE

Tachypnea: increased respiratory rate

- Salicylates
- Cocaine and amphetamines
- Theophylline and caffeine
- Cyanide and hydrogen sulfide
- Sodium monofluoroacetate
- Ethylene glycol
- Phenformin and metformin
- Carbon monoxide
- Methemoglobin inducers
- Simple asphyxiants
- Hydrocarbons
- Paraquat

- Dinitrophenol
- Thyroid hormone

Bradypnea: decreased respiratory rate

- Opioids
- Sedative-hypnotics, including ethanol, benzodiazepines, and barbiturates
- Clonidine, guanabenz
- Botulism

PUPILLARY RESPONSES

Miosis: small pupils

- Opioids
- Cholinergics (organophosphate, carbamates)
- Nicotine
- Clonidine, guanabenz
- Phenothiazines

Mydriasis: dilated pupils

- Anticholinergics
- Sympathomimetics
- Methanol
- Sedative-hypnotic withdrawal

APPENDIX C

Classic Toxidromes: Clinical Manifestations and Agents Commonly Involved

Francis J. De Roos, M.D.

Toxidrome	Clinical manifestations	Agents commonly involved
Opioid	Hypopnea or bradypnea, lethargy, obtundation, miosis (pinpoint pupils), hypothermia	Opioids, clonidine and guanabenz, phenothiazines, hypoglycemic agents
Sympathomimetic	Hyperthermia, tachycardia, hypertension, agitation, delirium, seizures, mydriasis, diaphoresis, increased peristalsis (bowel sounds)	Cocaine, amphetamines, theophylline, caffeine, salicylates, monoamine oxidase inhibitors
Anticholinergic	"Hot as Hades, mad as a hatter, blind as a bat, dry as a bone, and red as a beet"; hyperthermia, tachycardia, hypertension, agitation, delirium, seizures, mydriasis, decreased peristalsis (bowel sounds), dry, flushed skin	Diphenhydramine, hydroxyzine, tricyclic antidepressants, antipsychotics, benztropine, many plants [Jimson weed (*Datura* genus), deadly nightshade (*Atropa belladonna*), henbane (*Hyoscyamus niger*)], hyoscyamine, scopolomine, atropine
Cholinergic	Bradycardia (muscarinic), tachycardia (nicotinic), hypertension (nicotinic), miosis, "SLUDGE" (salivation, lacrimation, urination, diarrhea, gastrointestinal upset, and emesis)	Organophosphates, carbamates, physostigmine, pilocarpine, betel nut (*Areca catechu*), mushrooms (*Clitocybe dealbata, C. illudens, Inocybe lacera*)
Sedative-hypnotic	Hypothermia, bradypnea or hypopnea, rarely hypotension, lethargy, stupor, obtundation	Ethanol, benzodiazepines, barbiturates, zolpidem, ethchlorvynol, meprobamate, chloral hydrate, glutethimide

APPENDIX D

Concentrations of Compounds that Produce Positive Results

Barbarajean Magnani, Ph.D., M.D.

TABLE D–1. Amphetamine-methamphetamine immunoassay: urine concentrations of compounds that produce positive results

Drug	Abbott FPIA[a] (ng/mL)	Abbott RDS[b] amph. (ng/mL)	Abbott RDS[c] meth. (ng/mL)	BIOSITE Triage[d] (ng/mL)	Microgenics Corp. CEDIA[e,f] (ng/mL)	Roche Abuscreen ONLINE[g] (ng/mL)	Roche Abuscreen ONTRAK[h] (ng/mL)	Dade Behring (Syva) EMIT II Plus[i,j] (ng/mL)
d-Amphetamine (Dexedrine)	1000[k]	1000[l]	50,000	650	1000	1000[k]	500	1070
l-Amphetamine (Cydril)	3000	7500	10,000	40,000	40,000	24,700	25,000	7660
dl-Amphetamine	1000	1500	—	—	1500	1650	1000	1680
4-Chloroamphetamine	1000	—	—	1500	—	—	—	3000
Ephedrine	—	—	25,000	—	—	—	—	—
l-Ephedrine	—	—	—	—	250,000	see m	—	218,000
Fenfluramine (Pondimin)	50,000	—	—	10,000	—	—	—	47,000
p-Hydroxyamphetamine	10,000	—	—	1500	—	9288	1000	—
p-Hydroxymethamphetamine	—	—	—	2000	—	476,000	—	—
d-Methamphetamine (Desoxyn)	1000	—	—	650	1000[k]	287,000	—	1000[k]
l-Methamphetamine	8000	90,000	1000	30,000	8000	>750,000	—	2420
dl-Methamphetamine	3000	—	—	—	1000	580,000	—	1310
4-Methyl-2,5-dimethoxy-amphetamine (DOM)	100,000	—	—	—	—	—	—	—
3,4-Methylenedioxy-amphetamine (MDA)	3000	3000	50,000	1200	50,000	2665	2000	2130
3,4-Methylenedioxy-N-ethylamphetamine (MDEA) (MDE)	8000	50,000	—	2500	—	—	—	15,000
3,4-Methylenedioxy-methamphetamine (MDMA) (Ecstasy)[n]	3000	—	500	2000	1500	697,000[o]	—	9140
Methylphenidate (Ritalin)	—	—	—	—	—	>750,000	—	—
Phendimetrazine (Bontril)	—	—	—	—	—	—	—	—

Compound								
Phenmetrazine	100,000	—	—	—	—	—	6000	—
Phenethylamine (PEA)	100,000	45,000	—	—	—	—	—	—
β-Phenethylamine	—	—	750,000	—	74,632	—	100,000	—
Phentermine	10,000	—	—	50,000	>750,000	—	—	2000
Phenylpropanolamine (PPA)	—	100,000	—	—	—	100,000	100,000	330,000
d-Phenylpropanolamine	—	90,000	—	—	>500,000	—	—	—
l-Phenylpropanolamine	—	—	—	—	92,451	—	—	—
d,l-Phenylpropanolamine	—	—	—	500,000	134,800	—	—	—
Propylhexidrine	10,000	—	7500	—	331,000	—	—	—
Pseudoephedrine	—	—	1000	—	—	—	—	742,000
d-Pseudoephedrine	—	—	—	160,000	see m	—	—	—
Tyramine HCl	100,000	—	—	—	487,000	200,000	—	217,000

amph., amphetamine; meth., methamphetamine; —, No data available, or reported not to cross-react above the sensitivity of the assay.

[a] Abbott Laboratories. Package insert. 66-8410/R2; November, 1997.

[b] Abbott Laboratories. (American BioMedica Corp.) Rapid Drug Screen Amphetamine Package Insert. Revision D. September 1998. (Point of Care Device).

[c] Abbott Laboratories. (American BioMedica Corp.) Rapid Drug Screen Methamphetamine Package Insert. Revision D. September 1998. (Point of Care Device).

[d] BIOSITE Triage. Specificity tables. Rev. J—12/22/97. (Point of Care Device).

[e] Microgenics (CEDIA, Roche Diagnostics). Package insert. December 1999.

[f] CEDIA DAU amphetamine assay provides a choice of two cutoff concentrations: 500 and 1000 ng/mL.

[g] Roche Abuscreen ONLINE. Package insert. Rev. No. 011649300; 1999.

[h] Roche Abuscreen ONTRAK TesTcup-er. Order No.11-1771-8. November1998; ONTRAK TesTcup-5 M2K. Order No. 11-1850-1. April 1999; ONTRAK TesTcup. Order No. 47518. Art. No. 07-6422-1. December1998 (Point of Care Device).

[i] Dade Behring (Syva) EMIT II Plus Monoclonal. Package insert. 9C022UL.2SW; January 2000.

[j] Specimens from patients taking chlorpromazine (Thorazine) may produce positive results with this assay.

[k] Cutoff calibrator concentration.

[l] Combinations of amphetamine and methamphetamine that will produce a positive reaction: amph./meth. 500/500 and amph./meth. 200/500.

[m] The following compounds were tested at a concentration of 100,000 ng/mL in pooled human urine and were shown to have a cross-reactivity of <0.2%: d-ephedrine, dl-ephedrine, l-ephedrine, d-pseudoephedrine, l-pseudoephedrine, l-norpseudoephedrine.

[n] Only 36% of respondents using microparticle immunoassays and 11% of respondents using enzyme immunoassays reported a positive result for the Amphetamine Group when challenged with a sample containing 2000 ng/mL of MDMA. Over 99% of those testing with either FPIA or CEDIA reported positive findings, while 87% detected the drug using CMI (Triage). Poklis A: American Association for Clinical Chemistry/College of American Pathologists, Urine Drug Testing (Screening) Survey Set UDS-C, Final Critique 1999, Northfield, IL.

[o] Approximate cross-reactivity of MDMA to cutoff calibrator concentration is 0.1%.

TABLE D–2. Barbiturates immunoassay: urine concentrations of compounds that produce positive results

Drug	Abbott FPIA[a] (ng/mL)	Abbott RDS[b] (ng/mL)	BIOSITE Triage[c] (ng/mL)	Roche Boehringer Mannheim Corporation CEDIA[d,e] (ng/mL)	Roche Abuscreen ONLINE[f] (ng/mL)	Roche Abuscreen ONTRAK[g] (ng/mL)	Dade Behring (Syva) EMIT II Plus[h,e] 200 ng/mL cutoff cal. (ng/mL)	Dade Behring (Syva) EMIT II Plus[h,e] 300 ng/mL cutoff cal. (ng/mL)
Allobarbital (Diadol)	400	300	300	—	230	200	213	430
Alphenal (Phenallylmal)	200	—	300	—	—	—	278	708
Amobarbital (Amytal)	700	300	300	207	699	500	202	655
Aprobarbital (Somnipron)	200	150	300	195	221	200	204	367
Barbital (Barbitone)	2000	1000	500	1000	985	1000	918	3163
Barbitulic acid	—	—	—	—	—	—	—	—
Barbituric acid	—	—	—	—	>500,000	>100,000	—	—
1,3-Dimethylbarbituric acid	—	—	—	—	>500,000	>100,000	—	—
5-Ethyl-5-(4-hydroxyphenyl)-barbituric acid or (p-hydroxyphenobarbital)	2000	—	300	—	540	1000	707	3435
Brallobarbital	200	—	—	—	—	—	—	—
Butabarbital (Butisol)	200	150	300	198	312	400	195	410
Butalbital (Sandoptal, Plexonal)	200	—	300	213	342	200	198	355
Butalbutal	—	300	—	—	—	—	—	—

Compound								
Butethal	—	250	300	—	—	—	—	—
Butobarbital	400	—	—	—	—	—	265	744
Cyclobarbital	—	—	300	—	—	—	—	—
Cyclopentobarbital (Cyclopal)	200	—	300	190	170	200	216	379
Diphenylhydantoin	—	—	—	—	428,000	—	—	—
Glutethimide (Doriden)	10,000	—	2500	—	75,694	—	—	—
Hexobarbital (Hexanal)	100,000	—	25,000	—	692,000	>100,000	—	—
Mephobarbital (Mebaral)	—	—	2500	—	58,636	>100,000	—	—
Metharbital	1,000,000	—	3000	—	—	—	—	—
Methohexital (Brevital)	1,000,000	—	250,000	—	—	—	—	—
Pentobarbital (Nembutal)	200	250	300	270	425	400	210	390
Phenobarbital	200	1000	400	195	505	500	648	2835
Secobarbital (Seconal)	200[i]	300	300	200[i]	200[i]	200[i]	200[i]	300[i]
Talbutal (Lotusate)	200	150	300	130	—	—	166	275
Thiopental (Pentothal)	2000	—	25,000	—	—	—	18,646	65,507

cal., calibrator concentration; —, No data available, or reported not to cross-react above the sensitivity of the assay.

[a] Abbott Laboratories. Package insert. 66-8379/R2; October 1997.
[b] Abbott Laboratories. (American BioMedica Corp.) Rapid Drug Screen. Package Insert. Revision D. September 1998. (Point of Care Device).
[c] BIOSITE Triage. Specificity tables. Rev. J—12/22/97. (Point of Care Device).
[d] Roche Boehringer Mannheim. Package insert. 98-692-2. September 1997.
[e] Can be used at either 200 or 300 ng/mL cutoff calibrator concentration.
[f] Roche Abuscreen ONLINE. Package insert. Revision No.011649500; 1999.
[g] Roche Abuscreen ONTRAK TesTcup-er. Package insert. Order No.11-1771-8; November 1998. (Point of Care Device).
[h] Dade Behring (Syva) EMIT II Plus. Package insert. 9D022UL.2SW; January 2000.
[i] Cutoff calibrator concentration.

TABLE D-3. Benzodiazepine immunoassay: urine concentrations of compounds that produce positive results

Drug	Abbott FPIA[a] (ng/mL)	Abbott RDS[b] (ng/mL)	BIOSITE Triage[c] (ng/mL)	Roche Boehringer Mannheim Corp. CEDIA[d,e] 300 ng/mL cutoff cal. (ng/mL)	Roche Boehringer Mannheim Corp. CEDIA[d,e] 200 ng/mL cutoff cal. (ng/mL)	Roche Abuscreen ONLINE[f,g] (ng/mL)	Roche Abuscreen ONTRAK[h] (ng/mL)	Dade Behring (Syva) EMIT II Plus[i,e] 300 ng/mL cutoff cal. (ng/mL)	Dade Behring (Syva) EMIT II Plus[i,e] 200 ng/mL cutoff cal. (ng/mL)
Alprazolam (Xanax)	200	300	450	138	100	119	900	101	77
Alprazolam glucuronide	—	—	—	—	200	—	—	—	—
α-Hydroxyalprazolam	—	—	400	163	115	113	125	108	85
4-Hydroxyalprazolam	—	—	500	—	—	128	150	—	—
Bromazepam (Durazanil)	200	300	400	300	190	154	75	653	426
Chlordiazepoxide (Librium)	800	—	1250	2083	1200	291	900	1521	695
Desmethylchlordiazepoxide	—	—	—	—	—	273	—	—	—
Norchlordiazepoxide	800	—	1500	—	—	—	—	2268	1075
Clobazam (Frisium)	1000	300	700	400	300	—	—	205	138
Clonazepam (Klonopin)	200	300	350	188	225	270	—	1064	502
NH$_2$-Clonazepam	—	—	—	—	200	—	—	—	—
7-NH$_2$-Clonazepam	—	—	—	—	—	—	—	4553	1619
Clorazepate (Tranxene)	—	—	5000	325	300	109	900	see j	see j
Clotiazepam	—	—	—	—	—	—	—	420	250
Delorazepam	—	—	350	150	100	—	—	—	—
Demoxepam	400	—	2000	1900	1000	162	400	1090	557
Diazepam (Valium)	200	300	350	110	125	82	—	101	80
Desmethyldiazepam	—	300	—	—	—	—	—	111	87
Estazolam (Eurodin)	100	—	300	125	95	—	—	114	88
Flunitrazepam (Rohypnol)	200	—	350	188	175	292	—	336	194
7-Aminoflunitrazepam	—	—	2000	—	200	—	100	552	295
Desmethylflunitrazepam	—	—	—	—	—	278	700	—	—
3-Hydroxyflunitrazepam	—	—	—	—	—	424	1000	—	—
Flurazepam (Dalmane)	200	300	450	150	100	170	600	140	112
N-Desalkylflurazepam	200	—	300	138	115	175	400	160	115
Didesethylflurazepam	—	—	—	—	—	146	150	—	—
1-N-Hydroxyethylflurazepam	200	—	350	200	200	121	150	136	107
Halazepam	—	—	750	—	—	—	—	148	113
Ketazolam	—	—	—	—	—	—	—	129	100

Compound							
Lorazepam (Ativan)	200	550	208	175	337	400	1159	603
Lorazepam glucuronide	300	400	10,000	400	—	—	—	—
Lormetazepam (Ergocalm)	—	600	163	150	—	—	300[k]	200[k]
Medazepam (Nobrium)	1000	3500	200	150	91	—	165	125
Desmethylmedazepam	200	—	—	—	238	—	—	—
Midazolam (Versed)	—	900	—	—	132	—	130	97
Nimetazepam	200	—	—	—	—	—	—	—
Nitrazepam (Insomin)	200	750	300[k]	200[k]	155	—	296	176
7-Aminonitrazepam	—	—	—	250	297	125	—	—
Nordiazepam (Nordaz)	200[k]	300	150	120	100[k]	900	—	—
Nordiazepam glucuronide	—	300	—	—	—	—	—	—
Oxaprozin[l] (Daypro)	1,000,000	35,000	10,000	10,000	—	—	—	—
Oxazepam (Serax)	200	700	275	165	138	200	247	163
Oxazepam glucuronide	—	800	10,000	800	—	—	—	—
N-Methyloxazepam	—	—	—	—	121	—	—	—
Pinazepam	—	—	—	—	117	—	—	—
Prazepam (Centrax)	200	1300	150	160	139	—	124	103
Temazepam (Restoril)	200	550	175	180	—	500	165	122
Temazepam glucuronide	—	500	10,000	750	—	—	—	—
Tetrazepam	—	—	—	—	—	—	186	131
Triazolam (Halcion)	200	400	138	90	128	4500	139	102
Alphahydroxytriazolam	—	700	150	125	118	200	185	132
4-Hydroxytriazolam	—	1000	—	—	278	400	—	—

cal., calibrator concentration; —, No data available, or reported not to cross-react above the sensitivity of the assay.

[a] Abbott Laboratories. Package insert. 66-9510/R4; February 1998.
[b] Abbott Laboratories.(American BioMedica Corp.) Rapid Drug Screen. Package Insert. Revision D. September 1998. (Point of Care Device).
[c] BIOSITE Triage. Specificity tables. Rev. J—12/22/97. (Point of Care Device).
[d] Roche Boehringer Mannheim. Package insert. September 1997.
[e] Can be used at 300 or 200 ng/mL cutoff calibrator concentration. (200 ng/mL cutoff reflects CEDIA DAU High Sensitivity Assay to detect glucuronide conjugates.)
[f] Roche Abuscreen ONLINE. Package insert. Revision No. 0116497OO; 1999.
[g] Can be used at 300, 200, or 100 ng/mL cutoff calibrator concentration.
[h] Roche Abuscreen ONTRAK TesTcup-er. Package insert. Order No.11-1771-8; November 1998. (Point of Care Device).
[i] Dade Behring (Syva) EMIT II Plus. Package insert. 9F022UL.25W; January 2000.
[j] Clorazepate degrades rapidly in stomach acid to nordiazepam. Nordiazepam hydroxylates to oxazepam, which is detected by the assay at 163 ng/mL and 247 ng/mL.
[k] Cutoff calibrator concentration.
[l] Cross-reactivity to 2500 ng/mL of oxaprozin was approximately 79% for respondents using a microparticle point-of-care device and 75% using a KIMS (kinetic interation of microparticles in solution) assay; FPIA showed a concentration cutoff response with oxaprozin (positives were reported with a benzodiazepine cutoff ≤100 ng/mL, but none at a cutoff value of ≥150 ng/mL of benzodiazepine.) Poklis A: American Association for Clinical Chemistry/College of American Pathologists, Urine Drug Testing (Screening) Survey Set UDS-B, Final Critique 1999, Northfield, IL.

TABLE D–4. Cannabinoids immunoassay: urine concentrations of compounds that produce positive results

Drug	Abbott FPIA[a] (ng/mL)	Abbott RDS[b] (ng/mL)	BIOSITE Triage[c] (ng/mL)	Microgenics Corp. CEDIA[d,e] (ng/mL)	Roche Abuscreen ONLINE[f] (ng/mL)	Roche Abuscreen ONTRAK[g] (ng/mL)	Dade Behring (Syva) EMIT II Plus[h] 20 ng/mL cutoff cal. (ng/mL)	Dade Behring (Syva) EMIT II Plus[h] 50 ng/mL cutoff cal. (ng/mL)	Dade Behring (Syva) EMIT II Plus[h] 100 ng/mL cutoff cal. (ng/mL)
11-Nor-Δ8-THC-9-Carboxylic acid	25	75	40	40	—	—	—	—	—
11-Nor-Δ9-THC-9-Carboxylic acid	50[i]	50	3000	50[i]	50[i]	50[i]	20[i]	50[i]	100[i]
11-Nor-9- carboxy-Δ9-THC glucuronide	—	—	100	62	—	—	79	95	328
11-OH-Δ8-THC	—	—	—	—	—	—	43	67	129
11-OH-Δ9-THC	25	5000	150	125	314	1000	42	77	124
8-α-OH-Δ9-THC	—	—	—	1000	258	400	—	—	—
8-β-OH-Δ9-THC	—	—	—	500	457	250	26	68	146
8-β-11-diOH-Δ9-THC	—	—	—	—	—	—	24	58	109
Cannabinol	25	5000	2000	1000	2638	>100,000	—	—	—
11-Hydroxy-cannabinol	—	—	—	—	1018	2000	—	—	—
Cannabidiol	—	—	—	1000	50,000	>100,000	—	—	—
Cannabinol, Δ8-	—	—	2000	—	—	—	—	—	—
Cannabinol, Δ9-	—	—	1000	—	—	—	—	—	—
Chloroquine	—	—	—	—	—	50,000	—	—	—
Δ9-THC	—	5000	—	500	488	30,000	—	—	—
Δ8-THC	—	5000	—	—	—	—	—	—	—

cal., calibrator concentration; —, No data available, or reported not to cross-react above the sensitivity of the assay; THC, tetrahydrocannabinol.

[a] Abbott Laboratories. Package insert. 66-8389/R3; October 1997.
[b] Abbott Laboratories. (American BioMedica Corp.) Rapid Drug Screen. Package Insert. Revision D. September 1998. (Point of Care Device).
[c] BIOSITE Triage. Specificity tables. Rev. J—12/22/97. (Point of Care Device).
[d] Microgenics (Roche) CEDIA. Package insert. 98-697-2 August 1999; can also use a 20 ng/mL or 100 ng/mL cutoff.
[e] A metabolite of the anti-HIV drug Sustiva™ (formerly known as DMP 266) causes false-positive results in the CEDIA THC assay.
[f] Roche Abuscreen ONLINE. Package insert. Art. No.07-3464-0; March 1996.
[g] Roche Abuscreen ONTRAK TesTcup-5 M2K. Package insert. Order No.11-1850-1; April 1999 and ONTRAK TesTcup Order No.45518 Art. No. 07-6422-1; December 1998. (Point of Care Device).
[h] Dade Behring (Syva) EMIT II Plus. Package insert. 9N022UL.2SW; January 2000.
[i] Cutoff calibrator concentration.

TABLE D-5. Cocaine metabolite immunoassay: urine concentrations of compounds that produce positive results

Drug	Abbott FPIA[a] (ng/mL)	Abbott RDS[b] (ng/mL)	BIOSITE Triage[c] (ng/mL)	Microgenics Corp. (Roche) CEDIA[d,e] (ng/mL)	Roche Abuscreen ONLINE[f] (ng/mL)	Roche Abuscreen ONTRAK[g] (ng/mL)	Dade Behring (Syva) EMIT II Plus[h] 300 ng/mL cutoff cal. (ng/mL)	Dade Behring (Syva) EMIT II Plus[h] 150 ng/mL cutoff cal. (ng/mL)
Cocaine	10,000	800	550	312	21,200	75,000	64,000	26,000
Ecgonine	10,000	—	—	10,000	19,700	>100,000	28,000	12,000
Benzoylecgonine	300[i]	225	300	300[i]	300[i]	300[i]	300[i]	150[i]
Benzoylnorecgonine	—	—	5000	—	—	—	—	—
Cocaethylene	—	300	100,000	312	—	—	—	—
Propylbenzoylecgonine	—	—	40,000	—	—	—	—	—
Ecgonine methyl ester	—	—	75,000	10,000	326,000	>100,000	—	—

cal., calibrator concentration; —, No data available, or reported not to cross-react above the sensitivity of the assay.
[a] Abbott Laboratories. Package insert. 66-8394/R4; October 1997.
[b] Abbott Laboratories. (American BioMedica Corp.) Rapid Drug Screen. Package Insert. Revision D. September 1998. (Point of Care Device).
[c] BIOSITE Triage. Specificity tables. Rev. J—12/22/97. (Point of Care Device).
[d] Microgenics (Roche). Package insert. 98-694-3. December 1997.
[e] Can also be used at a 150 ng/mL calibrator cutoff.
[f] Roche Abuscreen ONLINE. Package insert. Revision # 01164990; 1999.
[g] Roche Abuscreen ONTRAK TesTcup-er. Order. No.11-1771-8. November 1998; ONTRAK TesTcup-5 M2K. Order No. 11-1850-1. April 1999; ONTRAK TesTcup Order No. 47518. Art. No. 07-6422-1 December 1998. (Point of Care Device).
[h] Dade Behring (Syva) EMIT II Plus. Package insert. 9H022UL.2SW; January 2000.
[i] Cutoff calibrator concentration.

TABLE D–6. *Methadone immunoassay: urine concentrations of compounds that produce positive results*

Drug	Abbott FPIA[a] (ng/mL)	BIOSITE Triage[b] (ng/mL)	Boehringer Mannheim Corp. Roche CEDIA[c] (ng/mL)	Roche Abuscreen ONLINE[d] (ng/mL)	Dade Behring (Syva) EMIT II Plus[e,f] 300 ng/mL cutoff cal. (ng/mL)	Dade Behring (Syva) EMIT II Plus[e,f] 150 ng/mL cutoff cal. (ng/mL)
d-β-Acetylmethadol	1000	—	—	—	—	—
l-β-Acetylmethadol	1000	—	—	—	—	—
l-α-Acetylmethadol (LAAM)	4000	—	20,000	448	—	—
l-α-Acetyl-N-normethadol (nor-LAAM)	4000	—	—	—	—	—
Chlorpheniramine (Chlor-Trimeton)	—	100,000	—	91,000	—	—
Chlorphenoxamine	100,000	—	—	—	—	—
Chlorpromazine (Thorazine)	—	60,000	—	250,000	—	—
2-Ethylidine-1,5-dimethyl-3,3-diphenyl pyrroline (EDDP)	—	—	500,000	273,000	—	—
2-ethyl-5-methyl-3,3-diphenyl pyrroline (EMDP)	—	—	100,000	333,000	—	—
Mesoridazine	100,000	20,000	—	—	—	—
α-Methadol	—	—	33,333	—	—	—
d-α-Methadol	4000	—	—	—	—	—
l-α-Methadol	500	—	—	—	—	—
Methadol	—	—	25,000	250	—	—
Methadone (Dolophine)	250[g]	300	300[g]	300[g]	300[g]	150[g]
d-Methadone	—	600	—	—	—	—
l-Methadone (Polamidon)	—	300	—	—	—	—
OH-Methadone	—	—	—	557	—	—
Promethazine	—	—	—	12,000	—	—
Thioridazine	50,000	50,000	—	—	—	—

cal., calibrator concentration; —, No data available, or reported not to cross-react above the sensitivity of the assay.

[a] Abbott Laboratories. Package insert. 66-9482/R2; September 1997.
[b] BIOSITE Triage. Specificity tables. Rev. J—12/22/97. (Point of Care Device).
[c] Boehringer Mannheim (Roche). Package insert. 10000531-2. September 1997.
[d] Roche Abuscreen ONLINE. Package insert. Art. No. 07-3481-0. March 1996.
[e] Dade Behring (Syva) EMIT II Plus. Package insert. 9E022UL.2SW; January 2000.
[f] Can be used at a 150 ng/mL calibrator cutoff.
[g] Cutoff calibrator concentration.

TABLE D-7. Opiates immunoassay: urine concentrations of compounds that produce positive results

Drug	Abbott FPIA[a] (ng/mL)	Abbott RDS[b] (ng/mL)	BIOSITE Triage[c] (ng/mL)	Roche CEDIA DAU[d] (ng/mL)	Roche CEDIA DAU 2K[e] (ng/mL)	Roche Abuscreen ONLINE[f] (ng/mL)	Roche Abuscreen ONTRAK[g] (ng/mL)	Roche Abuscreen ONTRAK 2K[h] (ng/mL)	Dade Behring (Syva) EMIT II Plus[i] 300 ng/mL cutoff cal. (ng/mL)	Dade Behring (Syva) EMIT II Plus[i] 2000 ng/mL cutoff cal. (ng/mL)
Codeine	50	300	300	300	1600	225	300	2000	247	1406
6-Acetyl-codeine	—	—	400	—	—	—	—	—	—	—
Diacetylmorphine (heroin)	200	400	400	300	2000	—	—	—	—	—
Dextromethorphan	100,000	—	1,000,000	—	—	—	—	—	—	—
Dextrorphan	100,000	—	—	—	—	—	—	—	—	—
Dihydrocodeine (Drocode)	200	—	300	300	2000	317	400	2000	291	1872
Dihydromorphine	100	—	—	—	—	371	400	2500	—	—
Ethylmorphine (Dionin)	200	400	400	—	—	265	300	2500	—	—
Hydrocodone (Vicodin)	100	5000	300	300	2000	479	500	6000	364	2690
Hydromorphone (Dilaudid)	100	—	500	300	2000	620	700	6000	498	5349
Levallorphan (Lorfan)	100,000	—	15,000	—	—	—	—	—	>7500[j]	>120,000[j]
Levorphanol (Dromoran)	100	—	1500	—	—	—	—	—	2752	36,939
Meperidine (Demerol)	250,000	—	75,000	150,000	1,000,000	30,508	25,000	>600,000	>50,000[k]	>800,000
Morphine	300[l]	300	300	300[l]	2000[l]	300[l]	300[l]	2000[l]	300[l]	2000[l]
6-Monoacetylmorphine	50	400	400	300	2500	311	500	2500	1088	9593
Morphine-3-glucuronide	—	300	—	300	2000	480	350	3500	626	6167
Morphine-3β-D-glucuronide	100	—	490	—	—	—	—	—	—	—
Morphine-6-glucuronide	—	—	—	300	3500	—	—	—	—	—
Nalorphine (Lethidrone)	1000	—	2500	—	—	—	—	—	9862	>100,000
Naloxone	100,000	—	—	—	—	—	30,000	—	828,139	>3,500,000
Naltrexone	100,000	—	—	—	—	—	70,000	—	—	—

(Continued on next page)

TABLE D-7. Opiates immunoassay: urine concentrations of compounds that produce positive results *(Continued)*

Drug	Abbott FPIA[a] (ng/mL)	Abbott RDS[b] (ng/mL)	BIOSITE Triage[c] (ng/mL)	Roche CEDIA DAU[d] (ng/mL)	Roche CEDIA DAU 2K[e] (ng/mL)	Roche Abuscreen ONLINE[f] (ng/mL)	Roche Abuscreen ONTRAK[g] (ng/mL)	Roche Abuscreen ONTRAK 2K[h] (ng/mL)	Dade Behring (Syva) EMIT II Plus[i] 300 ng/mL cutoff cal. (ng/mL)	Dade Behring (Syva) EMIT II Plus[i] 2000 ng/mL cutoff cal. (ng/mL)
N-Norcodeine	1000	—	20,000	—	—	11,744	25,000	>600,000	—	—
N-Normorphine	10,000	—	25,000	—	—	—	—	—	—	—
Noroxymorphine	100,000	—	—	—	—	—	—	—	—	—
Ofloxacin (Floxin)	see *m*	see *m*	see *m*	see *m*	see *m*	see *m*	see *m*	see *m*	see *m*	see *m*
Oxycodone (OxyContin) (Percodan, Percocet)	1000	—	20,000	10,000	64,000	23,166	12,000	>600,000	5388	104,587
Oxymorphone (Numorphan)	200	—	40,000	20,000	100,000	—	—	—	>20,000	>500,000
Rifampicin (Rifampin)	see *n*	see *n*	see *n*	see *n*	see *n*	see *n*	see *n*	see *n*	see *n*	see *n*
Thebaine (Paramorphine)	100	6500	2000	—	—	351	1000	7000	—	—

cal., calibrator concentration; —, No data available, or reported not to cross-react above the sensitivity of the assay.

[a] Abbott Laboratories. Package insert. 69-2120/R6; November 1998.
[b] Abbott Laboratories. (American BioMedica Corp.) Rapid Drug Screen. Package Insert. Revision D. September 1998. (Point of Care Device).
[c] BIOSITE Triage. Specificity tables. Rev. J—12/22/97. (Point of Care Device).
[d] Roche Boehringer Mannheim CEDIA DAU. Package insert. May 1996.
[e] Microgenics (Roche) CEDIA DAU Opiate 2K: Package Insert, Catalog No. 1815296. October 1998.
[f] Roche Abuscreen ONLINE. Package insert. Art. No.07-3472-1; March 1996.
[g] Roche Abuscreen ONTRAK TesTcup-er. Order. No.11-1771-8. November 1998. ONTRAK TesTcup Order No. 47518. Art. No. 07-6422-1 December 1998. (Point-of-Care Device).
[h] Roche Abuscreen ONTRAK TesTcup-5 M2K. Order No. 11-1850-1. April 1999. (Point of Care Device).
[i] Dade Behring (Syva) EMIT II Plus. Package insert. 9B322UL.1SW; February 2000.
[j] Therapeutic or toxic urinary concentrations of levallorphan and nalorphine are not reported in the literature.
[k] Meperidine urinary concentrations of 150,000 ng/mL have been measured in cases of fatal meperidine overdosage.
[l] Cutoff calibrator concentration.
[m] Cross-reactivity to 250 µg/mL of ofloxacin was approximately 86% for respondents using a CMI (Triage) method, 50% using an EMIT enzyme immunoassay, 70% using CEDIA, 1.7% using FPIA, and 1.4% using MIA (KIMS, kinetic interaction of microparticles in solution); American Association for Clinical Chemistry/College of American Pathologists, Urine Drug Testing (Screening) Survey Set UDS-C 1999, Northfield, IL.
[n] Cross-reactivity to 10 µg/mL of rifampicin was approximately 91% for respondents using a microparticle point-of-care device and 96% using a KIMS assay; Poklis A: American Association for Clinical Chemistry/College of American Pathologists, Urine Drug Testing (Screening) Survey set UDS-A, Final Critique 1999, Northfield, IL.

TABLE D–8. Phencyclidine (PCP) immunoassay: urine concentrations of compounds that produce positive results

Drug	Abbott FPIA[a] (ng/mL)	Abbott RDS[b] (ng/mL)	BIOSITE Triage[c] (ng/mL)	Roche Boehringer Mannheim Corp. CEDIA[d] (ng/mL)	Roche Abuscreen ONLINE[e] (ng/mL)	Roche Abuscreen ONTRAK[f] (ng/mL)	Dade Behring (Syva) EMIT II Plus[g] (ng/mL)
4-OH pip Phencyclidine (PCP metabolite)	1000	—	—	—	—	—	—
4-OH-Phencyclidine	—	—	10,000	—	—	—	—
N,N-diethyl-1-phenyl-cyclohexalamine (NNdiethyl PCA)	—	—	—	100	—	—	234
N-Ethyl-1-phenylcyclohexylamine (PCE)	—	—	500	1000	—	—	—
1-(4-hydroxypiperidino)phenylcyclohexane	—	—	—	—	—	—	237
1-(4-Methoxyphenyl)-4-piperidinocyclohexane	—	30	—	—	—	—	—
1-[1-(2-Thienyl)-cyclo-hexyl]morpholine (TCM)	—	—	5000	1000	—	—	80
1-[1-(2-Thienyl)-cyclohexyl]piperidine (TCP)	—	30	25	25	31	50	24
1-[1-(2-Thienyl)-cyclohexyl]pyrrolidine (TCPY)	—	—	—	—	—	—	50
Cyclazocine	100,000	—	—	—	—	—	—
Dextromethorphan (Romilar)	100,000[h]	see i	500,000	—	272,000[j]	—	12,000[k]
Dextrorphan (dextromethorphan metabolite)	1,000,000	—	—	—	—	—	20,000
Dicyclomine	1,000,000	—	—	—	—	—	—
Diphenhydramine	—	—	500,000	—	—	—	—
Fencamfamine	1,000,000	—	—	—	—	—	—
Ketamine	—	—	—	—	—	300,000	—
Mesoridazine	—	—	400,000	—	—	—	28,000
Phencyclidine (Sernylan) (PCP)	25[l]	25	25	25[l]	25[l]	25	25[l]
1-Phenylcyclohexylamine (PCA)	—	—	—	10,000	—	—	—
1-(1-Phenylcyclohexyl)morpholine (PCM)	—	—	—	1000	—	—	41
1-(1-Phenylcyclohexyl)pyrrolidine (PCPy) (PHP)	—	—	50	100	—	50	54
4-Phenyl-4-Piperidinocyclohexanol	—	—	—	5000	—	300	32
1-(1-Phenylcyclohexyl)-4-hydroxypiperidine (PCHP)	—	—	—	100	—	200	—

(Continued on next page)

TABLE D-8. Phencyclidine (PCP) immunoassay: urine concentrations of compounds that produce positive results *(Continued)*

Drug	Abbott FPIA[a] (ng/mL)	Abbott RDS[b] (ng/mL)	BIOSITE Triage[c] (ng/mL)	Roche Boehringer Mannheim Corp. CEDIA[d] (ng/mL)	Roche Abuscreen ONLINE[e] (ng/mL)	Roche Abuscreen ONTRAK[f] (ng/mL)	Dade Behring (Syva) EMIT II Plus[g] (ng/mL)
Phenyltoloxamine	1,000,000	—	—	—	—	—	—
Picenadol	1,000,000	—	—	—	—	—	—
Prolintane	100,000	—	—	—	—	—	—
Thioridazine	1,000,000	—	—	—	—	—	—
Tramadol	1,000,000	—	—	—	—	—	—

—, No data available, or reported not to cross-react above the sensitivity of the assay.
[a] Abbott Laboratories. Package insert. 69-3416/R3; July 1999.
[b] Abbott Laboratories. (American BioMedica Corp.) Rapid Drug Screen. Package Insert. Revision D. September 1998. (Point of Care Device).
[c] BIOSITE Triage. Specificity tables. Rev. J—12/22/97. (Point of Care Device).
[d] Roche Boehringer Mannheim. Package insert. December 1997.
[e] Roche Abuscreen ONLINE. Package insert. Art. No.07-3469-3; March 1996.
[f] Roche Abuscreen ONTRAK TesTcup-5 M2K. Order No. 11-1850-1. April 1999. (Point of Care Device).
[g] Dade Behring (Syva) EMIT II Plus. Package insert 9J022UL.2SW; January 2000.
[h] Individuals taking cold medications containing dextromethorphan may produce concentration results greater than the sensitivity (10.00 ng/mL) or factory set cutoff (25 ng/mL) of the assay. This result will occur when a particular unidentified metabolite(s) of dextromethorphan is present in the urine. The presence and concentration of this metabolite(s) is related to cold medication dosage concentrations, frequency of dosage, and/or individual differences in metabolism.
[i] RDS for PCP also detects high concentrations of the cough suppressant dextromethorphan. In young children, dextromethorphan overdoses may produce positive results with RDS. However, adults ingesting therapeutic dosages of dextromethorphan should not produce a positive result with RDS.
[j] Approximate cross-reactivity is 0.01%.
[k] A positive result from an individual taking preparations containing dextromethorphan should be interpreted with caution and confirmed by another method.
[l] Cutoff calibrator concentration.

TABLE D–9. *Barbiturate immunoassay: serum concentrations of compounds that produce positive results*

Drug	Abbot FPIA[a] (μg/mL)	Syva EMIT[b] (μg/mL)
Allobarbital	3.0	—
Alphenal	3.0	—
Amobarbital (Amytal)	3.0	15.0
Aprobarbital	3.0	—
Barbital	30	—
Brallobarbital	3.0	—
Butabarbital (Butisol)	3.0	7.0
Butalbital	3.0	8.0
Butobarbital	3.0	—
Cyclopentobarbital	3.0	—
Pentobarbital (Nembutal)	3.0	6.0
Phenobarbital	3.0	12–25
Phenytoin	250	—
Secobarbital (Seconal)	3.0[c]	3.0[c]
Talbutal	3.0	5.0
Thiamylal	7.5	—
Thiopental	120	—
Vinylbital	3.0	—

—, No data available.
[a] Abbott Laboratories. Package insert. 66-9128/R3; June 1997.
[b] Syva EMIT tox. Package insert. 7D164UL.7S; September 1998.
[c] Cutoff calibrator concentration.

TABLE D-10. *Benzodiazepine immunoassay: serum concentrations of compounds that produce positive results*

Drug	Abbot FPIA[a] (ng/mL)	Syva EMIT II[b] (ng/mL)
Alprazolam (Xanax)	25	400
Bromazepam (Durazanil)	300	1500
Chlordiazepoxide (Librium)	300	5000
Norchlordiazepoxide	75	—
Demoxepam	75	3000
Clobazam (Frisium)	100	—
Clonazepam (Klonopin)	75	2000
7-amino-Clonazepam	75	—
Diazepam (Valium)	75	300[c]
N-Desmethyldiazepam	—	1000
Nordiazepam	50[c,d]	—
Estazolam (Eurodin)	1000	—
Flunitrazepam (Rohypnol)	75	550
Norflunitrazepam	75	—
Flurazepam (Dalmane)	75	3000
Desalkylflurazepam	75	1000
1-N-Hydroxyethylflurazepam	75	—
Glutethimide	200,000	—
Halazepam	75	—
Lorazepam (Ativan)	75	3000
Medazepam (Nobrium)	75	1000
Midazolam (Versed)	100	—
Nimetazepam	75	—
Nitrazepam (Insomin)	75	1000
7-amino-Nitrazepam	75	—
Oxazepam (Serax)	75	1000
Prazepam (Centrax)	25	1000
Sertraline (Zoloft)	100,000	—
Temazepam (Restoril)	75	1000
Thenyldiamine	100,000	—
Triazolam (Halcion)	25	550
Tripelennamine (PBZ)	20,000	—

—, No data available or reported not to cross-react above the sensitivity of the assay.

[a] Abbott Laboratories. Package insert. 66-7177/R3; February 1996.
[b] Syva EMIT. Package insert. 7B164UL.7S; July 1998.
[c] Cutoff calibrator concentration.
[d] Concentration of the lowest calibrator.

TABLE D–11. *Tricyclic antidepressant immunoassay: serum concentrations of compounds that produce positive results*

	Abbot FPIA[a] (ng/mL)	Syva EMIT[b] (ng/mL)
Drug		
Amitriptyline (Elavil)	100	200–400
Desipramine (Norpramin, Pertofrane)	100	200–400
Imipramine (Tofranil)	75[c]	200–400
Nortriptyline (Aventyl, Pamelor)	100	300[c]
Metabolites		
10-Hydroxyamitriptyline	300	1250
2-Hydroxydesipramine	300	1250
2-Hydroxyimipramine	300	750
10-Hydroxynortriptyline	300	1750
Imipramine-N-oxide	75	—
Other tricyclic antidepressants		
Clomipramine (Anafranil)	100	500
Dothiepin (Prothiaden)	100	500
Doxepin (Adapin, Sinequan)	100	500
Nordoxepin	500	—
Protriptyline (Vivactil)	100	500
Trimipramine (Surmontil)	100	600
Other compounds with similar structure or used concurrently that may produce a positive result		
Carbamazepine (Tegretol)	20,000	Negative at 100,000
Cyclobenzaprine[d] (Flexeril)	300	—
Perphenazine (Trilafon)	300	Negative at 350
Promethazine (Phenazine)	300	Negative at 500

—, No data available.

[a] Abbott Laboratories. Package insert. 69-3236/R3; June 1999.

[b] Behring Diagnostics, Syva EMIT. Package insert. 7C074UL.9B; March 1997.

[c] Cutoff calibrator concentration.

[d] TDx FPIA serum and enzyme-multiplied immuno technique (EMIT) serum immunoassays have both been shown to produce a positive tricyclic assay result with cyclobenzaprine. Polkis A: American Association for Clinical Chemistry/College of American Pathologists Urine Drug Testing (Screening) Survey Set UDS-B: Final Critique, 1998, Northfield, IL.

APPENDIX E

Methods for Salicylate and Acetaminophen Measurement

Tai C. Kwong, Ph.D., DABCC

Knowing the salicylate or acetaminophen serum concentration is critical to the diagnosis and management of patients intoxicated with salicylate or acetaminophen. Therefore, quantification of salicylate and acetaminophen concentrations in serum or plasma should be available in every hospital clinical laboratory at all times. Rapid, easy-to-perform, and accurate assays are available. For patients with unsuspected overdoses, an effective urine drug screen can detect the presence of these two analgesics in urine. In the absence of chromatography-based broad-spectrum urine drug screens, simple qualitative screening tests can quickly detect the presence of salicylate and acetaminophen. All positive urine screening results should trigger quantification of drug concentrations in serum samples.

SALICYLATE METHODS

Most salicylate assays in use are colorimetric assays based on the reaction of salicylic acid with ferric ion in acid medium to give a purple complex (1,2). This simple and rapid reaction, however, is far from specific for salicylate. The metabolites of salicylate (salicyluric acid, the phenolic and acyl glucuronides, gentistic acid), ketone bodies and other β-ketoacids, alcohols, catechols such as tyrosine and the catecholamines, and oxalate, which is used in some blood-collection tubes, all react with ferric ion to give the color reaction. Ketone bodies and catecholamines are frequently elevated in disease states, and the colorimetric method can yield falsely high salicylate concentrations in these patients (2,3).

Trinder's method is the classic colorimetric assay (4). Trinder's reagent contains mercuric chloride and hydrochloric acid to precipitate serum proteins and to reduce serum and plasma blank values to less than 20 mg/L, which are of no clinical significance. Most of the current colorimetric assays are direct reading methods without protein precipitation (Boehringer Mannheim, Indianapolis; Dade ACA, Dimension, Deerfield, IL; Sigma, St. Louis). The elimination of a protein-precipitation step is at the expense of slightly higher serum blank values (>20 mg/L). Despite the lack of specificity of the ferric nitrate reaction, the colorimetric method for salicylate has proven over the years to be acceptable for clinical use (1).

A different approach to the measurement of salicylate is the use of the enzyme salicylate hydrolase (EC 1.14.13.1) purified from *Pseudomonas cepacia* (5–9). This enzyme, in the presence of reduced nicotinamide adenine dinucleotide (NADH) or reduced nicotinamide adenine dinucleotide phosphate (NADPH), converts salicylate to catechol (Figure E–1).

Salicylate concentration can be determined by monitoring, photometrically, the consumption of NADH (5) or NAD(P)H (6)

FIG. E–1. Salicylate hydrolase assay for salicylate measurement.

(Beckman Synchron RGT, Brea, CA; Dade Paramax System) or the formation of catechol (8,9) (Vitros, Johnson & Johnson, Rochester, NY). Ketone bodies and other compounds that are known to interfere with the colorimetric assays do not affect the enzymatic methods. Neither do structurally related compounds such as benzoate, p-OH-benzoate, and p-aminosalicylate (9). Several drugs (acetylcysteine, ascorbic acid, cysteamine, L-dopa, methyldopa, primidone) have been shown to interfere with the assay, but only at the high drug concentration of 1 g/L and not at realistic therapeutic concentrations (9). Salicylate hydrolase has demonstrable reactivity with the metabolites, salicyluric acid and gentistic acid (9), but the extent of interference is clinically insignificant (10). Of greater concern is that the enzyme has a low level of inherent NADH oxidase activity that also consumes NADH. Therefore, assays that monitor the consumption of NADH should have a reagent blank. The presence of NADH oxidase activity also limits the shelf life of a reagent consisting of enzyme and coenzyme. A drawback common to dehydrogenase-based enzyme assays is the spurious results generated by elevated endogenous lactate dehydrogenase activity and high pyruvate concentration (8).

The alternate approach to monitoring the enzyme assay is to measure the formation of catechol, the other reaction product. Catechol can react with a dye to yield a color product for photometric measurement (9). This design offers significant improvement in specificity by circumventing the adverse effects of inherent NAD(P)H oxidase activity and the presence of high dehydrogenase enzyme activity in some samples. This assay system allows the inclusion of an NAD(P)H regenerating system in the assay reagent, thus prolonging the shelf life of the reagent.

The fluorescence polarization immunoassay is rapid and easy to perform. Diflunisal is a serious interference with this assay; therapeutic concentrations of 113 and 74 mg/L (0.45 and 0.68 mmol/L) were measured as apparent salicylate concentrations of 260 and 170 mg/L (18.8 and 12.3 mmol/L), respectively (11). An extensive list of structurally related compounds has been evaluated for potential interference of this assay (12).

Salicylate is not detected by any of the chromatographic drug screens; it is poorly recovered during the extraction process because of its acidity. Its detection in urine is usually achieved by a colorimetric spot test using ferric chloride or Trinder's reagent. This test is very sensitive—it can be positive after the ingestion of one 325-mg aspirin tablet—but nonspecific because phenothiazines and acetoacetate can give false-positive results. Moreover, sodium azide at a bactericidal concentration of 1 mg/mL used in some commercial control material interferes with Trinder's method.

ACETAMINOPHEN METHODS

The acetaminophen concentration of a sample collected at least 4 hours postingestion predicts the severity of hepatotoxicity. Interpretation is aided by a nomogram developed from a database of acetaminophen overdose cases and the relationship between postingestion serum acetaminophen concentrations and clinical outcome (13). Most of the analytical data were obtained when the only routine laboratory tests for acetaminophen were colorimetric assays, of which the

$$\underset{\text{Acetaminophen}}{\underset{\text{NHCOCH}_3}{\text{OH}}} + H_2O \xrightarrow{\text{aryl acylamide amidohydrolase}} \underset{p\text{-aminophenol}}{\underset{\text{NH}_2}{\text{OH}}} + CH_3COOH$$

FIG. E–2. Enzymatic assay for acetaminophen measurement.

most widely used was the method of Glynn and Kendal (14). Their assay is based on mild ring-nitration of acetaminophen with nitrous acid to yield 2-nitro-4-acetamidophenol, which, at alkaline pH, forms a colored substance with strong absorbance at 430 nm. A significant fraction of acetaminophen in plasma is conjugated as glucuronide or sulfate metabolites. The mild acidic conditions for nitration do not cause hydrolysis of the conjugated metabolites (glucuronide and sulfate). Therefore, this method measures only unconjugated acetaminophen (the "parent" drug) and not total acetaminophen. It is thus important to recognize that it is the unconjugated acetaminophen concentration that predicts toxicity and that the use of the nomogram requires unconjugated, not total, acetaminophen. Any assay (none in use, one hopes) incorporating an acid hydrolysis step to convert acetaminophen to p-aminophenol for subsequent color development will also hydrolyze all conjugated metabolites to acetaminophen and is therefore not an acceptable assay (15,16). Applying total, instead of unconjugated, acetaminophen concentration to the nomogram will yield grossly misleading prognostic information.

Although chromatographic assays are available to measure serum acetaminophen concentrations, they are not widely used in clinical laboratories because of the rapid analysis and reporting required. Rapid, accurate, and easy assays in use are the immunoassays (enzyme-multiplied immuno technique and fluorescence polarization immunoassay) and enzyme assays. The enzyme assays are based on aryl acylamide amidohydrolase converting (unconjugated) acetaminophen to p-aminophenol, which is then oxidized or reacted with a dye to form a color (Figure E–2).

In chromatography-based urine drug screen, acetaminophen is easily detected by the TLC-kit (Toxi-Lab, Irvine, CA). Its presence can be masked by other early-eluting substances in a basic extraction of the urine if analysis is based on gas chromatography or high-performance liquid chromatography. It can be detected easily if neutral or acidic extracts are analyzed.

REFERENCES

1. Kwong TC: Salicylate measurement: clinical usefulness and methodology. *CRC Crit Rev Clin Lab Sci* 1987;25:137–159.
2. Stewart MJ, Watson ID: Analytical reviews in clinical chemistry: methods for the estimation of salicylate and paracetamol in serum, plasma and urine. *Ann Clin Biochem* 1987;24:552–565.
3. Kang ES, Todd TA, Capaci MT, et al: Measurement of true salicylate concentrations in serum from patients with Reye's syndrome. *Clin Chem* 1983;29:1012–1014.
4. Trinder P: Rapid determination salicylate in biological fluids. *Biochem J* 1954;57;301–303.
5. You K, Bittikofer JA: Quantification of salicylate in serum by use of salicylate hydrolase. *Clin Chem* 1984;30:1549–1551.
6. Longenecker RW, Trafton JE, Edwards RB: A tableted enzyme reagent for salicylate for use in a discrete multiwavelength system (Paramax®). *Clin Chem* 1984;30:1369–1371.
7. Hammond PM, Ramsay JR, Price CP, et al: A simple colorimetric assay to determine salicylate ingestion utilizing salicylate monooxygenase. *Ann N Y Acad Sci* 1987;50:1288–1291.
8. Chubb SAP, Campbell RS, Ramsay JR, et al: An enzyme mediated, colorimetric method for measurement of salicylate. *Clin Chim Acta* 1986;155:209–220.
9. Morris HC, Overton PD, Ramsay JR, et al: Development and validation of an automated, enzyme-mediated colorimetric assay of salicylate in serum. *Clin Chem* 1990;36:131–135.
10. Badcock NR, Penna AC, Everett DS, et al: Aspirin

metabolites causing misinterpretation of paracetamol results. *Ann Clin Biochem* 1984;21:527–530.
11. Sarma L, Wong SH, DellaFera S: Diflunisal significantly interferes with salicylate measurements by FPIA-TDx and UV-VIS aca methods. *Clin Chem* 1985;31:1922–1923.
12. Koel M, Nebinger P: Specificity data of the salicylate assay by fluorescent polarization immunoassay. *J Anal Toxicol* 1989;13:358–360.
13. Rumack BH, Matthew H: Acetaminophen poisoning and toxicity. *Pediatrics* 1975;55:871–876.
14. Glynn JP, Kendal SE: Paracetamol measurement. *Lancet* 1975;1:1177–1178.
15. Porter WH: In acetaminophen assay, only unconjugated drug should be measured. *Clin Chem* 1984;30:1884–1885.
16. Steward MJ, Adrieaenssens PI, Jarvie DR, et al: Inappropriate methods for the emergency determination of plasma paracetamol. *Ann Clin Biochem* 1979;16:89–95.

APPENDIX F

Selected Book Reviews

1. Adams RM: *Occupational skin disease*, 3rd ed. Philadelphia: WB Saunders, 1999. 792 pages.

   A compendium of the various occupational and environmentally induced skin diseases and their causes. Chapters address the major types of chemicals and products that cause damage to human skin.

2. Baselt RC: *Disposition of toxic drugs and chemicals in man*, 5th ed. Foster City, CA: Chemical Toxicology Institute, 2000. 919 pages.

   This text provides a single source of essential information on the drugs and chemicals most frequently encountered in episodes of human poisoning. Systematic presentation of the referenced data includes usage, blood concentrations, metabolism and elimination, toxicity, analysis, and references. Body fluid and tissue concentrations of agents are summarized from literature reports of fatal poisonings. The textbook is valuable in both clinical and postmortem toxicology practice.

3. Blau K, Halket JM: *Handbook of derivatives for chromatography*, 2nd ed. New York, Wiley & Sons, 1993. 576 pages.

   The chemist's reference book for derivatization chemistry. Somewhat difficult to use because the book is organized by derivatization reactions, but many references are included.

4. Brooks SM, Gochfeld M, Jackson RJ, et al: *Environmental medicine*. St. Louis: CV Mosby, Inc., 1994. 780 pages.

   Textbook dealing with the principles of environmental medicine and toxicology as well as the resultant clinical diseases and the major environmental exposure sources.

5. Bryson PD: *Comprehensive review in toxicology for emergency clinicians*, 3rd ed. Washington, DC: Taylor & Francis, 1996. 848 pages.

   This medical toxicology textbook is written for emergency clinicians but may also be used as a reference for the laboratory-based toxicologist.

6. Dolan JW, Snyder LR: *Troubleshooting LC systems*, 2nd ed. Totowa, NJ: Humana Press, 1989. 523 pages.

   An excellent text on basic LC systems, as well as a systematic approach to determining and fixing problems in the HPLC separation system.

7. Drug Abuse Warning Network (DAWN): Annual medical examiner data and annual emergency department data.

   The medical examiner series presents data and analysis on drug-induced or drug-related deaths reported by medical examiners participating in DAWN. The emergency department series presents information on drug abuse–related emergency department episodes collected through DAWN. Both reports are available on the Internet at http://www.samhsa.gov/oas.dawn.htm or by mail from National Clearinghouse for Alcohol and Drug Information (NCADI), P.O. Box 2345, Rockville, MD 20847-2345.

8. Ellenhorn MJ, Schonwald S, Ordog G, et al: *Ellenhorn's medical toxicology: diagnosis and treatment of human poisoning*, 2nd ed. Baltimore: Williams and Wilkins, 1997. 2047 pages.

   This second edition, written by Dr. Matthew Ellenhorn and published posthumously, is a complete rewrite of the 1988 edition and is organized into major sections on principles of poison management, individual drugs, intoxicants in the home, chemical poisons, and natural toxins. The text focuses on new toxicology information gained over the last 9 years, covering an impressive number of drugs and other chemical poisons. Ellenhorn's text is a valuable reference for any laboratory involved in therapeutic, emergency, abuse, or medicolegal toxicology.

9. Fenton J: *The laboratory and the poisoned patient*. Washington, DC: AACC Press, 1998. 352 pages.

   This text provides a guide to the interpretation of toxicology laboratory data for a wide range of drugs. Information on pharmacokinetics, mechanism, toxicity, interpretive criteria, methods, and other test findings for each agent.

10. Goldfrank LR, Flomenbaum NE, Lewin NA, et al: *Goldfrank's toxicologic emergencies*, 6th ed. Norwalk, CT: Appleton & Lange, 1998. 1917 pages.

    Goldfrank's textbook on medical toxicology focuses on case studies for individual intoxicants and is also a very useful reference for clinical toxicologists. A time-honored source of information in connection with nearly any type of intoxication with recommended practical approaches for diagnosis and management including detailed discussions of standard-of-practice antidotal therapy.

11. Haddad LM, Shannon MW, Winchester JF: *Clinical management of poisoning and drug overdose*, 3rd ed. Philadelphia: WB Saunders, 1998. 1257 pages.

    This is an excellent text to turn to when faced with a question about the mechanism of action, the clinical manifestations, or the recommended therapy for a specific toxic agent. The new edition consists of 106 chapters written by 120 contributing authors. A book with this complexity usually involves a long lead time for final production. Nevertheless, this work is surprisingly up to date. For example, the clinical characteristics of olanzepine toxicity are presented well, especially given that this drug appeared on the U.S. market during 1997, the year immediately preceding that of publication. The completely rewritten chapter "Laboratory Diagnoses and Drug Screening" is must reading for anyone involved in clinical toxicology.

12. Harber P, Shenker MB, Balmes JR: *Occupational and environmental respiratory disease*. St. Louis: CV Mosby, Inc., 1996. 1038 pages.

    Comprehensive multiauthor text addressing all aspects of human respiratory disease caused by environmental agents.

13. Hardman JG, Limbird LE, Molinoff PB, et al: *Goodman & Gilman's the pharmacological basis of therapeutics*, 9th ed. New York: McGraw-Hill, 1996. 1905 pages.

    The 9th edition of Goodman and Gilman's remains an authoritative description of the current pharmacological basis of therapeutics.

14. Klaasen CD, ed: *Casarett and Doull's toxicology: the basic science of poisons*, 5th ed. New York: McGraw-Hill, 1996. 1111 pages.

    This classic text presents the fundamental aspects of toxicology on a broad range of topics. The 5th edition is similar in framework to the previous four editions, but it has undergone significant changes. The prior edition consisted of five sections, but the new edition has seven plus an appendix. The seven sections are General Principles of Toxicology, Disposition of Toxicants, Nonorgan-Directed Toxicity, Target Organ Toxicity, Toxic Agents, Environmental Toxicology, and Applications of Toxicology. The appendix presents recommended limits for exposure to chemicals. Three new chapters have been added: "Toxicokinetics," "Mechanisms of Toxicity," and "Toxic Responses of the Endocrine System."

15. Knapp DR: *Handbook of analytical derivatization reactions.* New York, Wiley-Interscience, 1979. 768 pages.

    Excellent resource for derivatization methods, although the source is somewhat dated. Organized by type of compound derivatized, with many examples for each reaction.

16. Litovitz TL, Klein-Schwartz W, White S, et al: 1999 annual report of the American Association of Poison Control Centers Toxic Exposure Surveillance System. *Am J Emerg Med* 2000;18:517–574.

    This 1998 database report includes >2.2 million human exposure cases reported by 67 participating poison centers. The report is published annually in the *American Journal of Emergency Medicine.*

17. McEvoy GK: 97 *AHFS drug information.* Bethesda, MD: American Society Health-System Pharmacists, Inc., 1997. 2955 pages.

    AHFS Drug Information is a monograph source of drug information on every single-drug entity available in the United States.

18. McLafferty FW, Turecek F: *Interpretation of mass spectra,* 4th ed. Mill Valley, CA: University Science Books, 1996. 371 pages.

    Extensive coverage of mass spectral interpretation, including many problems with solutions. The textbook is widely used in courses on interpretation of mass spectra.

19. Message GM. *Practical aspects of gas chromatography/mass spectrometry.* New York: Wiley & Sons, 1984. 368 pages.

    An excellent, but dated [for example, packed-column gas chromatography (GC)], practical text on GC–mass spectrometry systems. There are practical sections on source cleaning, vacuum pump maintenance, and system troubleshooting. Good coverage of the basics of mass spectrometers.

20. Moffat AC: *Clark's isolation and identification of drugs.* London: The Pharmaceutical Press, 1986. 1223 pages.

    This book is divided into four main sections: Part 1 contains descriptions of analytical techniques and specific types of toxicology testing. Part 2 contains monographs on >1300 drugs and related substances. Part 3 contains 66 indexes of analytical data for chromatography, spectrophotometry, and mass spectrometry. Part 4 consists of descriptions of reagents used in the analytical procedures described in parts 1 and 2. This is a comprehensive and very useful reference book for toxicologists and clinical chemists.

21. National Household Survey on Drug Abuse, SAMHSA Series H-3.

    This database report and interpretation presents the results from the 1998 National Household Survey on Drug Abuse, a sampling of >18,000 persons. The survey, conducted annually by SAMHSA, is available on the Internet at http://www.DrugAbuseStatistics.SAMHSA.gov or by mail from National Clearinghouse for Alcohol and Drug Information (NCADI), P.O. Box 2345, Rockville, MD 20847-2345.

22. Niesink RJM, de Vries J, Hollinger MA: *Toxicology, principles and applications.*

New York: CRC Press and Open University of the Netherlands, 1996. 1284 pages.

The intent of this textbook is to present fundamental principles of toxicology oriented toward molecular and organ systems. Medical and clinical toxicology is covered in a single chapter as are the other areas of applied toxicology.

23. Rood D: *A practical guide to the care, maintenance, and troubleshooting of capillary gas chromatographic systems,* 3rd ed. Weinheim, Germany: Wiley-VCH, 1999. 344 pages.

Good coverage of the basics of capillary GC written as a practical guide. Maintenance of injectors and detectors extensively discussed. Good section on troubleshooting systems.

24. Rosenstock L, Cullen MR. Textbook of clinical occupational and environmental medicine. Philadelphia: WB Saunders, 1994. 909 pages.

A comprehensive text on occupational and environmental diseases. Diseases are organized by organ system. There are several chapters on the effects of classes of compounds. General chapters address principles of occupational and environmental medicine.

25. Willoughby R, Sheehan E, Mitrovich S: *A global view of LC/MS. How to solve your most challenging analytical problems.* Pittsburgh: Global View Publishing, 1998. 554 pages.

An excellent practical guide for getting started in the acquisition and use of high-performance liquid chromatography–mass spectrometry (HPLC-MS). The book is divided into four parts. Part I consists of three chapters that provide enough information about HPLC-MS systems to permit the reader to select the type of HPLC-MS system suited to their analytical needs. Part II contains four chapters that deal with justification and acquisition of alternative HPLC-MS systems and planning for a successful HPLC-MS laboratory. There are five chapters in Part III dealing with practical aspects of sample analysis, problem-solving, and methods development. Finally, Part IV provides five appendices. Appendix A is full of important resources including (1) reference books on all aspects of HPLC-MS; (2) a list of excellent reviews of the most important areas of HPLC-MS including fundamentals and technology overviews, applications overviews, and instrument availability; (3) a list of online services; (4) journals that relate to HPLC-MS; (5) a list of academic centers specializing in HPLC-MS, and contact information. Appendices B, C, and D are equally full of invaluable information and illustrations on practical information on HPLC-MS; interfaces and how they work; and mass analyzers and how they work. Appendix E provides a brief review of the "Pioneers in LC/MS" oral session at the 45th American Society for Mass Spectrometry Conference on Mass Spectrometry and Allied Topics. This invaluable resource is well referenced throughout.

APPENDIX G

Abbreviations Used

5-HT	serotonin	ARDS	acute respiratory distress syndrome
6-AM	6-acetylmorphine	AST	aspartate aminotransferase
11-OH-THC	11-hydroxy-tetrahydrocannabinol	ATP	adenosine triphosphate
AAPCC	American Association of Poison Control Centers	AUC	area under the plasma concentration time curve
AAS	atomic absorption spectrometry	AV	atrioventricular
		b.i.d.	twice a day
ABCD	Appropriate Blood Pressure Control in Diabetes	BAC	blood alcohol concentration
		BD	1,4-butanediol
ACE	angiotensin-converting enzyme	BEI	Biological Exposure Indices
ACGIH	American Conference of Governmental Industrial Hygienists	bp	base pair
		BUN	blood urea nitrogen
AChE	acetylcholinesterase	C_{aveSS}	average steady-state concentration
ADH	alcohol dehydrogenase	C_{max}	maximum concentration
ADP	adenosine diphosphate	C_{min}	minimum (trough) concentration
AF	atrial fibrillation		
AGC	automatic gain control	C_p	serum concentration at any time; the maximal concentration
AIDS	acquired immunodeficiency syndrome		
δ-ALA	δ-aminolevulinic acid	C_{peak}	highest observed serum concentration
ALDH	aldehyde dehydrogenase	cAMP	cyclic adenosine monophosphate
ALP	alkaline phosphatase		
ALT	alanine aminotransferase	CAP	College of American Pathologists
amu	atomic mass unit		
APCI	atmospheric pressure chemical ionization	CB1	central cannabinoid receptor

CB2	peripheral cannabinoid receptor	DEA	Drug Enforcement Agency
CBC	complete blood count	DFO	deferoxamine
CCB	calcium channel blocker	DHD	carbamazepine-10, 11-*trans*-dehydrodiol
CDC	Centers for Disease Control and Prevention	DHP	dihydropyridine
		DIG	Digitalis Investigation Group
CDTA	Chemical Diversion and Trafficking Act of 1988	DLIF	digoxin-like immunoreactive factors
CE	capillary electrophoresis	DMA	dimethylarsine
CEC	capillary electrochromatography	DMPS	2,3-dimercaptopropane-1-sulfonate
CEDIA	cloned enzyme donor immunoassay	DMSA	2,3-dimercaptosuccinic acid
		DNS	delayed neurologic sequelae
CERM	Centre Européan de Recherché Mauvernay	DOT	Department of Transportation
ChE	cholinesterase		
CHF	congestive heart failure	DPC	Diagnostic Products Corporation
CI	chemical ionization		
CID	collision-induced dissociation	DSM-III-R	*Diagnostic and Statistical Manual of Mental Disorders III-R*
Cl	chloride clearance		
CLIA 88	Clinical Laboratories Improvement Act of 1988	DUF	National Institute of Justice Drug Use Forecasting
CNS	central nervous system	DUID	driving under the influence of drugs
CPR	chlorophenolred		
CPRG	chlorophenolred-β-galactoside	EA	enzyme acceptor
		EC	EMIT-dau amphetamine class assay
CPT	Current Procedural Terminology	ECG	electrocardiogram
CT	computed tomography	ED	emergency department; enzyme donor
CV	cardiovascular system		
CYP	cytochrome P450	ED_{50}	effective therapeutic dose
DA	dopamine	EDTA	ethylenediaminetetraacetic acid
DAWN	Drug Abuse Warning Network		
		EI	electron ionization
DC	direct current	EIA	enzyme immunoassay
DDT	dichlorodiphenyl-trichloroethane	EII	EMIT-II amphetamine-methamphetamine assay
		EKG	electrocardiogram

ELISA	enzyme-linked immunosorbent assay	GFAAS	graphite furnace atomic absorption spectroscopy
EM	EMIT-dau monoclonal amphetamine-methamphetamine assay; extensive metabolizer	GHB	γ-hydroxybutyrate
		GI	gastrointestinal
		HCO_3	bicarbonate
EMIT	c9	HDL	high-density lipoprotein
EMIT	enzyme-multiplied immuno technique	HFBA	heptafluorobutyric anhydride
EPA	Environmental Protection Agency	HHS	Health and Human Services
		HI	hydriodic acid
EPH	ephedrine	HINT	Holland Interuniversity Nifedipine/Metoprolol Trial
ER	drug extraction ratio		
ES	electrospray	HIV	human immunodeficiency virus
f	bioavailability		
FAAS	flame furnace atomic absorption spectroscopy	HPLC	high-performance liquid chromatography
Fab	fragment of immunoglobulin G involved in antigen binding	HPPH	5-p-hydroxyphenyl-5-phenylhydantoin
		HVA	high-voltage-activated
Fc	crystallizable fragment of immunoglobulin	i.d.	internal diameter
		i.v.	intravenous; intravenously
FDA	Food and Drug Administration	ICP-AES	inductively coupled plasma atomic emission spectrophotometry
FEP	free erythrocyte protoporphyrin		
		ICP-MS	inductively coupled plasma mass spectrometry
FPIA	fluorescence polarization immunoassay		
		IgG	immunoglobulin G
GABA	γ-aminobutyric acid	IM	intermediate metabolizer; intramuscular
GBL	γ-butyrolactone		
GBL	γ-hydroxybutyrolactone	IN	intranasal; intranasally
GC	gas chromatography; gas-liquid chromatography	INR	international normalized ratio
GC/MS	gas chromatography–mass spectrometry	IP3	inositol triphosphate
		IQ	intelligence quotient
GC-ECD	gas chromatography with electron capture detection	K_a	the rate of drug absorption
		K_e	the rate of elimination
GCS	Glasgow Coma Scale	Ka	absorption rate
GDH	glycerol dehydrogenase		

Ke	population value for the elimination rate of the toxin	MJ	marijuana
		MMA	monomethylarsine
KIMS	kinetic interaction of microparticles in solution; microparticle immunoassay	MRI	magnetic resonance imaging
		MRO	medical review officers
λ_{max}	maximal wavelength	MS	mass spectrometry
LC	liquid chromatography	MS-MS	tandem mass spectrometry
LC/MS	liquid chromatography–mass spectrometry	NAA	neutron activation analysis
		NAC	N-acetylcysteine
LD_{50}	lethal dose	NAD(P)H	the reduced form of nicotinamide adenine dinucleotide or the reduced form of nicotinamide adenine dinucleotide phosphate
LDH	lactate dehydrogenase		
LFT	liver function test		
LLE	liquid-liquid extraction		
LOD	limit of detection		
L-TPC	N-trifluoroacetyl-L-prolyl chloride	NAD+	the oxidized form of nicotinamide-adenine denucleotide
LVA	low-voltage-activated		
LVEF	left ventricular ejection fraction	NADH	the reduced form of nicotinamide-adenine dinucleotide phosphate
m/z	mass-to-charge ratio		
MAO	monoamine oxidase	NADPH	the reduced form of nicotinamide-adenine dinucleotide phosphate
MAOI	monoamine oxidase inhibitors		
mAU	milliabsorbance units	NCADA	National Center for Alcohol and Drug Abuse Interventions
MDA	methylenedioxy-amphetamine		
		NE	norepinephrine
MDMA	methylenedioxy-methamphetamine	NHTSA	National Highway Traffic Safety Administration
MDPIT	Multicenter Diltiazem Post-Infarction Trial	NICI	negative ion chemical ionization
MEKC	micellary electrokinetic chromatography	NIST	dot product library search algorithm; National Institute of Science and Technology
MEOS	microsomal ethanol-oxidizing system		
		NMDA	N-methyl-D-aspartate
MHD	the 10-hydroxy metabolite of oxcarbazepine	NTE	neuropathy target esterase
		NYHA	New York Heart Association
MIDAS	Multicenter Isradipine Diuretic Atherosclerosis Study	OPIDN	organophosphate-induced delayed neuropathy

OSHA	Occupational Safety and Health Act; Occupational Safety and Health Administration		Health Services Administration
		SGOT	serum glutamic-oxaloacetic transaminase
OTC	over the counter	SGPT	serum glutamate-pyruvate transaminase
P.O.	by mouth		
P2P	phenyl-2-propanone	SIM	selected ion monitoring
PBM	probability-based matching library search algorithm	SM	by smoking; smoking
		SNP	single-nucleotide polymorphism
PCB	polychlorinated biphenyls		
PCP	phencyclidine	SPE	solid-phase extraction
PCR	polymerase chain reaction	SPRINT II	Secondary Prevention Reinfarction Israeli Nifedipine Trial II
PE	pseudoephedrine		
PFPA	pentafluoropropionic anhydride	SRM	selected reaction monitoring
		SSI	succinic acid semialdehyde
PM	poor metabolism	STONE	Shanghai Trial of Nifedipine in Elderly Hypertensives
PO	orally		
POC	point of care	$t_{1/2}$	half-life, the time it takes for the amount in the body to drop by half
POCT	point-of-care testing		
PPA	phenylpropanolamine		
ppb	parts per billion	TAT	turn-around time
ppm	part per million	TCA	trichloroacetic acid; tricyclic antidepressant
PROVED	Prospective Randomized Study of Ventricular Failure and the Efficacy of Digoxin		
		TDM	therapeutic drug monitoring
		THC	Δ^9-tetrahydrocannabinol
R_f	rate of flow	THCCOOH	11-nor-9-carboxy-Δ^9-tetrahydrocannabinol
RADIANCE	Randomized Assessment of the Effect of Digoxin on Inhibitors of the Angiotensin-Converting Enzyme Study		
		TIBC	total iron-binding capacity
		TIBET	Total Ischaemic Burden European Trial
		TLC	thin-layer chromatography
RALES	Randomized Aldactone Evaluation Study	TOF	time of flight
		TPMT	thiopurine methyltransferase
RF	radio frequency		
RIA	radioimmunoassay	TRENT	Trial of Early Nifedipine in Acute Myocardial Infarction
SA	sinoatrial		
SAM	S-adenosine methionine	TSH	thyroid-stimulating hormone
SAMHSA	Substance Abuse and Mental	UK	United Kingdom

UM	ultrarapid metabolism	VPA	valproic acid
US	United States	WBC	leukocyte count
UV	ultraviolet	WBI	whole bowel irrigation
V_d	apparent volume of distribution; volume of distribution	WHO	World Health Organization
		wt./vol.	percentage weight in volume
V_{dss}	volume of distribution at steady state	XRF	X-ray fluorescence
		ZPP	zinc protoporphyrin
vol./vol.	percent volume in volume		

Index

Index entries (page numbers) set in italics refer to figures and/or tables.

A

Abbott FPIA assay. *See* Fluorescence polarization immunoassay (FPIA)
Abbott RDS assay. *See* RDS immunoassay
Abbreviations used, table of, 507–512
Abortifacients, 37
Absorption, drug
    calculations, 20–22
    cocaine, 101
    digoxin, 272
    ethanol, 180, 180–181
    γ-hydroxybutyrate (GHB), 133
    of industrial toxicants, 332–333
    isopropanol, 187
    lead, 370–371
    lead poisoning, 371
    opiates and opioids, 74, 76
    peak, 13
    pesticides, 361
    pharmacokinetics of, 11–12
    phenytoin, 250
    THC, 47–48
Abstinence syndrome. *See* Withdrawal syndrome
Abuscreen immunoassays. *See* On-Line systems immunoassay; ONTRAK immunoassay
Abuse, substance, 3–4
Accidents, traffic, marijuana use and, 57–60
Acetaminophen, 500–501
    in opiate preparations, 88–89, 90, 92
    sustained-release, 37
    tests for, *147*, *460*, 500–501, *501*
    toxicity, 19–20
        antidote, 7, 20, 38–39
        case report, 24, 38–39
        death rate, 1
        ethanol use and, 179
        hepatotoxicity and, 24
        overdose nomogram, *21*, 21–22, 38–39, 500–501
        in pregnancy, 38–39
        Rumack-Matthew nomogram, *21*, 21–22, 38–39
        symptoms, 7, 24
        treatment of, *146*
Acetone, *147*, *186*, 187–188, *204*
Acetonemia, 188
Acetylcholinesterase (AChE) inhibition, by pesticides, 361–363, 364–366
N-Acetylcysteine, 7, 38–39
Acetyldigoxins, 269
6-Acetylmorphine (6-AM), *80*, 87, *151*
Acidosis, 35
    in isopropanol poisoning, 188
    metabolic, 34–37
    MUDPILES, *30*, 30
    tests for, *35*
    toxins inducing, *30*
Acids, in foods, urine adulteration by, 160
Acronyms used, table of, 507–512
Acute-on-chronic ingestion, 25
Acylcholine acylhydrolase, 364–365
Adalat. *See* Nifedipine
Addiction, opiates and opioids, 77–78, *79*
Additive effects. *See* Drug interactions
β-Adrenergic antagonists
    calcium channel blocker interactions, *306–311*, 317–318, 320
    toxicity, delayed, 37
    vital signs, effect on, 477–478
α-Adrenergic antagonists, calcium channel blocker interactions, 317
Adulterants, 166
    commercially available, 162–164, *163*
    detection of, 166–168
    foods, 160–161
    mechanism of action, 164
    tests, effect on, 162–166, *163*
    trade names, *163*
    in vitro, 161–164, *163*
    in vivo, 159–161

Adulteration, of urine. *See* Urine adulteration
ADx test, 149
ω-Agatoxin, 299, *300*
Age factor, in poisoning, 3
agelenopsis aperta, 299
Alcohol dehydrogenase, 181, *202*
Alcoholism
  abuse and dependency in US, *177*
  antidepressant clearance increased, 232
  encephalopathy in, 470
  toxicity, case report, 174–175
  toxicokinetics, 179–180
Alcohol(s). *See also* Ethanol; Isopropanol; Methanol
  death rate, 2
  volatile, 173–188
Aldehyde dehydrogenase, *181*
Alfentanil, 75, 89
Alkylephoxysulfonate, 162
Alleles, 443, *444*, 448–449
Allobarbital (Diadol), *484*, *495*
Alphaprodine, 75
Alphenal (Phenallymal), *484*, *495*
Alprazolam, *496*
  chemical structure, *212*
  digoxin drug interaction, *284*
  imipramine drug interaction, 232
  metabolism and metabolites, *216*
  sales rank in US, *212*
  testing and test results, *151*, *486*
  toxicity
    death, 214
    rat LD$_{50}$, 214
American Academy of Pediatrics, 378
American Association of Poison Control Centers (AAPCC)
  calcium channel blocker poisonings, 293, *296*
  cardiac glycoside poisonings, 265
  cocaine exposures and poisonings, 98
  γ-hydroxybutyrate (GHB) exposures, 128
  isopropanol exposures, 187
  lead poisoning, 369
  pediatric cases reported, 1999, *2*
  pesticide poisoning, 8
  poisonings and deaths reported, 1999, 1, *2*
  top 15 substances of exposure, *2*
  volatile alcohol exposures, 173
American College of Emergency Physicians, 462
American Conference of Governmental Industrial Hygienists, 335, 339–340
American Medical Association, 462
Amicron Centrifree micropartition system, 254
Aminogluthemide, *284*
Aminoglycosides, *284*
δ-Aminolevulinic acid, 371, 372–373, 377
4-Aminopyridine, *316*

Aminosalicylic acid, *284*
Amiodarone, 283, *283*, *284*
Amitriptyline, *224*, 225–226, 228–229
Amlodipine
  chemical structure, *294*
  dosage, *307*
  metabolism and metabolites, *302*
  pharmacokinetics of, 301, *302*
  pharmacology of, *303*
  sales rank in US, 293, *296*
  therapeutic uses, *307*
  toxicity, symptoms, *307*
Ammonia, *163*
Amobarbital (Amytal), *151*, *461*, *484*, *495*
Amoxapine, *224*
Amphetamine(s). *See also* Methamphetamine
  chemical structure, *118*
  drugs converted to, 125
  enantiomers, 123–124
  history, 113–114
  metabolism and metabolites, *118*
  pharmacokinetics of, 116–117
  Schedule of Controlled Substances, 114, 115
  synthesis of, 114
  testing and test results
    extensive testing, *460*
    gas chromatography/mass spectrometry (GC/MS), *123*, 123–125
    immunoassay cross-reactivity, *122*, *151*, *482–483*
    immunoassay sensitivity and specificity, *152*
    thin layer chromatography, 117, 119–123
    urine adulteration, 160, *163*
    urine concentrations, *482–483*
    urine screening cut-offs, *414*
  therapeutic uses, 115
  toxicity
    opiate interactions, 82
    respiration effects, *30*
    toxidrome, 479
    vital signs, effect on, 477–478
Amrinone, *316*
Amytal. *See* Amobarbital (Amytal)
Analgesics
  death rate, 2
  opiates and opioids, 73–94
  tests for, *460*
  toxicity, abuse-related emergency room visits, *146*
Anandamide, 45
Anesthetics, 82, 98
Angiotensin-converting enzyme inhibitors, clinical trials, 318, 319
Anhydroecognine methyl ester, 102

Anion gap metabolic acidosis, 35–37, 202–203, *204*
Anodic stripping techniques, 375–376, 379
Antacids, 283, *284*
Antibiotics, 283, *284*, *492*
Anticholinergics
    digoxin drug interaction, *284*
    toxicity, 33
        symptoms, 7
        toxidrome, 479
        vital signs, effect on, 478
Anticholinergic toxidrome, 33–34
Anticonvulsants
    in bipolar disorder, 239, 240–245
    death rate, *2*
Antidepressants
    drug interactions, 231–232, *309*
    monitoring therapeutic ranges, *224*, 224–225, 227–232
    pharmacogenetics, 446–447
    pharmacokinetics of, 231–232
    therapeutic ranges, *224*
    toxicity
        case report, 225–227
        death, 223
        symptoms, 231–232, *232*
        toxidrome, 479
        treatment of, *146*
        vital signs, effect on, 477–478
    tricyclic
        anticholinergic toxidrome, 34
        in bipolar disorder, 239
        death rate, 1, *2*
        drug interactions, *309*
        history and overview, 223–225
        opiate interactions, 82
        tests for, *460*
        toxicity, 225–227, *232*
        toxidrome, 479
        vital signs, effect on, 477–478
Antidysrhythmic agents, 477
Antiepileptics, 247–259
Antifreeze poisoning, 198
Antihistamines, *2*, 34, *284*, 477
Antineoplastics, *284*
Antipsychotics, 34, 479
Aprobarbital (Somnipron), *484*, *495*
Arachidonylethanolamide, 45–46
Areca catechu (betel nut), 479
Arsenic
    epidemiology of, 383–384
    metabolism and metabolites, 385, *386*
    in pesticides, 383–384
    pharmacokinetics of, 385
    routes of exposure, 384–385
    testing and test results, 386–387
    toxicity
        case report, 384
        death, 384
        symptoms, 384, *385*
        treatment of, 387–388
Arsine, 385
Ascorbate, *163*
Asphyxiation, 30, 478
Aspirin
    in opiate preparations, 89, 90, 92
    testing and test results
        ferric chloride test, 31
        Trinder reaction, 31
    toxicity
        anion gap metabolic acidosis, *204*
        antidote, 23
        death rate, 1
        symptoms, 23
Astemizole, 293
Atenolol, 320
Ativan. *See* Lorazepam (Ativan)
Atmospheric pressure chemical ionization, 427, *427*
Atomic absorption, 374–375, 392
Atrial fibrillation, 278
Atropa belladonna (nightshade), 34, 479
Atropine
    anticholinergic toxidrome, 34, 479
    for GHB toxicity, 136
    for pesticide toxicity, 8, *361*, 363
    tests for, *461*
Autoinduction, 242
Automotive products, *2*

B

B&O Supprettes®, 88
Barbital (Barbitone), *151*, *484*, *495*
Barbiturates
    testing and test results
        blood serum tests, *495*
        immunoassay cross-reactivity, *151*, *484–485*, *495*
        immunoassay sensitivity and specificity, *152*
        immunoassays for, *484–485*
        urine screening cut-offs, *414*
    toxicity
        calcium channel blocker interactions, *307*
        digoxin drug interactions, *284*
        opiate drug interactions, 82
        toxidrome, 479
        vital signs, effect on, 477–478
    urine adulterants, *163*
Base-line values, in biological monitoring, 338–339

Bay a1040, 292
Bay a7168, 292
Bay k8644, *316*
Beckman Synchron RGT test, for salicylates, 500
Belladonna, 88
Benadryl®. *See* Dyphenhydramine
Benzalkonium chloride, 165
Benzocaine, in cocaine preparations, 98
Benzodiazepines
    generic manufacturers, *212*
    history and overview, 211
    metabolism and metabolites, *215*
    pharmacokinetics of, *215*
    testing and test results, *151*, *152*, 217–219, *460*
        immunoassay cross-reactivity, *485–486*
        immunoassays for, 214–217
        ion monitoring, 218, *219*
        point-of-care testing, 149
    therapeutic uses, 212–213
    toxicity
        abuse-related emergency room visits, *146*
        death, 214
        digoxin drug interaction, *284*
        suicide attempts, 213–214
        toxidrome, 479
        treatment of, *146*
        vital signs, effect on, 477–478
    urine adulterants, *163*
Benzoic acid, *118*
Benzomorphinans, 74
Benzoylecognine
    assays for cocaine, 105–107, 108
    cocaine metabolite, 101, 102, 103
    hair analysis, 418
    stability in vitro, 108
    testing and test results, 149, *489*
Benzoylnorecognine, *489*
Benzphetamine, *123*, 125
Benztropine, 479
Bepridil
    chemical structure, *295*
    dosage, *309*
    history of, 292
    metabolism and metabolites, *302*
    pharmacokinetics of, *302*
    pharmacology of, *303*
    therapeutic uses, *309*
    toxicity
        digoxin drug interaction, *284*
        symptoms, *309*
Bernard, Claude, 345
Betaxolol, *461*
Betel nut (*Areca catechu*), 479
Beverages. *See* Foods and beverages
    alcoholic. *see* Ethanol
Bezoars, 12
Bicarbonate, *163*

Biguanide, 36
Biological Exposure Indices, 335
Biological monitoring
    base-line values, 338–339
    carbon monoxide, 355
    case report, 340–342
    lead and lead poisoning, 372–374
    lead poisoning, 372–374
    methemoglobin testing, 334
    reasons for, 331–332
    role of, 331
    sample collection, 335–337
    testing and test results, 331–338
    variability in tests, 337–338
    workplace exposure limits, 339–340
Biomarkers, 329, 376–377
BIOSITE Triage immunoassay. *See* Triage DOA point-of-care test device
Bipolar disorder
    anticonvulsant treatment of, 240–245
    epidemiology of, 237–238
    lithium treatment of, 238
Birth control pills, as abortifacient, 37
Bites, 2
Bleach, *163*
Bleeding, gastrointestinal, calcium channel blockers and, *321*, 322
Blood gas concentrations, anion gap metabolic acidosis, 35
Blood pressure, toxin identification and, 477–478
Blood tests
    antidepressants, tricyclic, *497*
    barbiturates, *495*
    benzodiazepines, *496*
    biological monitoring, 336
    cocaine, 105, 107
    lead exposure, 373
    necessary, *456*, *458*
    opiates and opioids, 86–87
    sample collection, 407
    THC, 52–53, 55–56
Boehringer Mannheim (Roche) immunoassays. *See* CEDIA® immunoassay
Bontril (phendimetrazine), *482*
Book reviews, 503–506
Botulism, 478
Bradycardia, 29, 133, 477
Bradypnea, 478
Brallobarbital, *484*, *495*
Breath analysis
    biological monitoring, 336–337
    ethanol, 182–183, *183*, 191–195
Brevital, *485*
Bromazepam (Durazanil), *216*, *486*, *496*
Buprenex®, 92
Buprenorphine
    chemical structure, *76*

clinical pharmacology and pharmacokinetics, 79
metabolism and metabolites, *80*, 92
for opiate detoxification, 83
Schedule of Controlled Substances, 75
therapeutic uses, 93
volume of distribution, 77
Butabarbital (Butisol), *151, 461, 484, 495*
Butalbital, *151, 484, 495*
Butalbutal, *484*
1,4-Butanediol (BD), 127, *130*, 131, *132*
Butethal, *485*
Butisol, *151, 461, 484, 495*
Butobarbital, *485, 495*
Butorphanol
    chemical structure, 76
    clinical pharmacology and pharmacokinetics, 79
    metabolism and metabolites, *80*, 92–93
    Schedule of Controlled Substances, 75
    therapeutic uses, 92–93

C

Cadmium
    epidemiology of, 392
    routes of exposure, 393–394
    testing and test results, 395–396
    toxicity
        biochemistry of, 394–395
        case report, 393
        symptoms, 393, *394*
        treatment of, 396
Caffeine
    in cocaine preparations, 98
    in opiate preparations, 89, 92
    toxicity
        respiration effects, *30*
        toxidrome, 479
        vital signs, effect on, 477–478
    urine adulteration by, 160
Calan®, *294*, 305
Calcium
    lead poisoning treatment, 380
    role of, 293, 295, *297*
Calcium channel blockers, *300*
    clinical trials, 317–322, *321*
    dosage, *306–309*
    history and overview, 291–293
    metabolism and metabolites, 301–304, *302–303*
    pharmacokinetics of, 300–305
    pharmacology of, 292, 293–300, *300*
    sustained-release, 301
    testing and test results, 315, 317
    therapeutic uses, *306–309*
    toxicity
        antidote, *316*
        cardiovascular events increased, 317–320, 322
        clinical trials, 317–320, 322
        death, *295, 296*, 320
        digoxin drug interaction, *284*
        drug interactions, *284, 306–309*
        epidemiology of, *295*, 305, 313
        pathophysiology of, *312*–313
        symptoms, 292–293, 304–305, *306–309, 313–315*
        treatment of, 314, 315, *316*
    trade names, *294–295*
    vital signs, effect on, 477
    volume of distribution, 301
Calcium channels, 296–300, *298, 301*
Calcium oxalate crystalluria, 199, 205
Cannabinoid receptors, 45–46
Cannabinol and cannabinoids
    chemistry and nomenclature, 45–46
    immunoassay cross-reactivity, *488*
    metabolism and metabolites, *44*
    screening tests for, 52–53
    therapeutic uses, 47
Cannabis, history and overview, 43–44
Cannabis seed oil, 161
Cannibis sativa, 43
Capillary electrochromatography (CEC), 424–425
Captopril, *284*
Carbamate pesticides, 8
    esters, *360*
    testing and test results, 364–365
    toxicity
        epidemiology of, 359
        mechanism of, 362–363
        pharmacology of, 361–362
        SLUDGE syndrome, 361–362
        toxicokinetics, 361
        toxidrome, 479
        treatment of, 364
Carbamazepine
    autoinduction of, 242
    in bipolar disorder, 240–241
    calcium channel blocker interactions, *306–307*
    for epilepsy, 247–248
    metabolism and metabolites, *252*, 252–253
    pharmacokinetics of, 242, 255
    pharmacology of, 249
    test schedule for, *244*
    therapeutic ranges, 249
    toxicity, 243–244, 253
        anticholinergic toxidrome, 34
        drug interactions, 253, 258–259, *259*
        opiate interactions, 82
Carbamazepine epoxide, 248, *249*, 255, 258, *259*

Carbon monoxide
  history and overview, 345–346
  poisoning prevention, 355
  sources, 346–348
  testing and test results, 350–353
  toxicity
    case report, 345–346
    causes of, 347–348
    death, 347
    death rate, 1
    epidemiology of, 345, 347, *347*
    hypoxia, *30*
    ischemic-reperfusion injury, 353–354
    long-term effects, 354
    pathophysiology of, 348–349
    symptoms, 345, 349–350
    treatment of, 353–355
    vital signs, effect on, 477–478
Carboxyhemoglobin, 345–346
trans-Carboxypentazocine, *80*
Cardene®. *See* Nicardipine
Cardiac glycosides, 263–288, 477–478. *See also* Digoxin
Cardiovascular drugs. *See also* Calcium channel blockers; Digoxin
  cardiac glycosides, 263–288, *264, 268*
  death rate, *2*
Cardizem, Cardizem CD. *See* Diltiazem
Carfentanil, 89
Carisoprodol, *460, 461*
Catecholamines, *283*
CEDIA® immunoassay
  adulterant effects, 165
  amphetamines, 120, 121, *122, 482–483*
  barbiturates, *484–485*
  benzodiazepines, *486–487*
  cannabinoids, *488*
  cocaine, *489*
  false positives, *162*
  methadone, *490*
  opiates and opioids, 85, *491–492, 493–494*
Cedilanid, *268*
Centers for Disease Control (CDC), 341, 377–378, *377–378*
Centrax. *See* Prazepam (Centrax)
Chelation treatment, 378, 406
Chemical Diversion and Trafficking Act (1988), 115
Chemical exposure. *See* Biological monitoring
Chemical ionization, *426*
Chemicals, *2*
Children
  benzodiazepine overdose in, 213
  diazinon poisoning, 360–361
  digoxin poisoning, 266–267
  fluoxetine fatality, 447
  iron poisoning, 401–402, 407
  lead poisoning, 340–342, 369–370, 377–378
  mouthwash poisoning, 175
  poisonings in, 1–2
  theophylline poisoning, 17–19
Chloral hydrate, *147*, 479
Chlordiazepoxide (Librium), *496*
  chemical structure, *212*
  history and overview, 211
  metabolism and metabolites, *215*
  pharmacokinetics of, *215*
  testing and test results, *151, 461, 486*
Chloroquine, *461*
Chlorpheniramine, *461, 490*
Chlorphenoxamine, *490*
Chlorpromazine, 241, *490*
Cholestyramine, 283, *284*
Cholinergics, 477–478, 479
Cholinesterase inhibitors, 8
Cholinesterases, 364–366, *365*
Chromatography
  cocaine, 105–107
  derivatization, 106–107
  instrumentation, 424–425
  opiates and opioids, 86
Cimetidine, 232, *306–311*
Cisapride, 293
Cleaning substances, death rate, *2*
Clearance, 438–439
Clinical trials, 276–277, 317–322, *321*
Clitocybe sp. (mushrooms), 479
Clobazam (Frisium), 213, *486, 496*
Clobenzorex, 125
Clomipramine, *224*, 447
Clonazepam (Klonopin)
  chemical structure, *212*
  immunoassay cross-reactivity, *496*
  sales rank in US, *212*
  testing and test results, *151*, 213
  therapeutic uses, 213
  toxicity, 239
Clonidine, *461*, 477–478, 479
Clorazepate (Tranxene), *486*
Clozapine, *224, 461*
Cobras Mira analyzer, 207
Coca-Cola, 97–98
Cocaethylene, *489*
Cocaine
  absorption, 101
  chemistry and structure of, 99, *99*
  effects of, 100–101
  elimination, 103
  metabolism and metabolites, *99*, 101, 101–102, 103–104
  pharmacokinetics of, 101–103
  pharmacology of, 99–101, 468

routes of administration, 101
testing and test results, 105–107, *151*, *152*, *414*, 417
   hair specimens, 417–418
   immunoassay cross-reactivity, *489*
   meconium, 419
   stability in vitro, 105, 108
   urine tests, *489*
therapeutic uses, 98, 99–100
toxicity
   abuse-related emergency room visits, *146*
   behavioral, 98
   case report, 98
   death, 98
   impairment, 109
   respiration effects, *30*
   symptoms, 100, 109
   toxidrome, 479
   treatment of, *146*
   vital signs, effect on, 477–478
urine adulterants, *163*
Codeine
   chemical structure, 76
   clinical pharmacology and pharmacokinetics, 79
   metabolism and metabolites, *80*, 88–89
   in opium, 88
   Schedule of Controlled Substances, 75
   testing and test results, 87, *151*, *491*
   Tylenol® with codeine, 88–89
   urine concentrations, 87
   volume of distribution, 77
Cognition, marijuana effect on, 60–65
Colchicines, 37
Colestipol, 283, *284*
Colorimetry test, for salicylates, 499
Committee on Safety of Medicines, 213
Concentration (C), drug, 11–12, 13, 438
   calculations, 17–19, 20–22
   over- and underpredictions, 22–24
Congestive heart failure, 270, 272, 276–277
ω-Conotoxins, 299, *300*
Contraceptives, 257
Controlled Substance Abuse Act, 74, 75, 98
Conus sp., 299
Cosmetics, 2
Cough and cold preparations, 2
Covera-HS®, *294*, *296*
Crack cocaine, 98
CRD-401. *See* Diltiazem
Creatinine urine levels, *159*, 167, 412–413
Cross tolerance, opiates and opioids, 78
Current Procedural Terminology (CPT) codes, 462
Cyanide, 32, 478
Cyclazocine, *493*

Cyclic antidepressants. *See* Antidepressants
Cyclobarbital, *485*
Cyclohexanol opiates, 74
Cyclopal, *485*, *495*
Cyclopentobarbital (Cyclopal), *485*, *495*
Cyclosporine, *283*, *284*, 293
   calcium channel blocker interactions, *306*, *308*, *309*
Cydril, *482*
Cymarin, *268*
Cytochrome P450, 439
Cytochrome P4502C9, 449–450
Cytochrome P4502D6, *439*, 441–446, *442*, 443, *444*, *445*

D

D-600 (niludipine), 292
Dacarel. *See* Nicardipine
Dade Behring (Syva) EMIT immunoassays. *See* EMIT® assays
Dade Paramax System, for salicylate testing, 500
Dalmane. *See* Flurazepam
Darvon®, 92
Date-rape drug. *See* Flunitrazepam
Datura stramonium (jimson weed), 34, 479
DAU immunoassays. *See* CEDIA® immunoassay
Daypro®, *162*, 212, 216, *487*
Debrisoquine hydroxylase, 439, 441–446, *442*, *444*, *445*
Delapril, 319
Delirium, from cocaine intoxication, 109
Delta (δ) opiate receptors, 78
Demerol. *See* Meperidine
Dengel, Ferdinand, 292
Dental amalgam, 390
Department of Health and Human Services (HHS), 121
Department of Labor, 339
Department of Transportation (DOT, 158
Dependence
   ethanol, *177*, 178
   opiates and opioids, 77–78
L-Deprenyl, 116
Depression, clinical approach to, 37–38
Desferal®, 406, 407
Desferoxamine (DFO), 406, 407
Desipramine, *147*, 224, 228
Deslanoside, 263, *269*
Desmethylclomipramine, *224*
Desmethyldinortramadol, *80*
Desmethyldoxepin, *224*
Desmethylnortramadol, *80*
Desmethylsertraline, *224*
Desmethylvenlafaxine, *224*

Desoxyn®, 115, *482*
Despropionylfentanyl, *80*, 89
Detergents, 162, *163*
Detoxification, 83
Dextroamphetamine (Dexedrine®), 82, 115, *482*
Dextromethorphan, 7, 19–20, *461*, *491*, *493*
Dextropropoxyphene, 75
Dextrorphan, *491*, *493*
Diabetes, 206, 319
Diadol, *484*
Diagnosis, clinical approach to, 27–34
Diazepam (Valium)
    chemical structure, *212*, *215*
    for GHB-induced seizures, 135
    history, 211
    immunoassay cross-reactivity, *496*
    metabolism and metabolites, *215*
    pharmacokinetics of, *215*
    sales rank in US, *212*
    testing and test results, *151*, *461*
    toxicity, death, 214
Diazinon, toxicity, 360–361, *361*
Dicyclomine, *493*
Diet, 380
Diethylene glycol, 197–198, *198*
Diethyl ether, *204*
Digibind®, 285, 407
Digitalis, 263–264
Digitalis lanata, 263, *268*
Digitalis purpurea, 263, *268*
Digitoxigenin, *268*
Digitoxin
    chemical structure, *269*
    history and overview, 263–264
    metabolism and metabolites, *277*
    pharmacokinetics of, 275, *276*
    sources, *268*
Digoxin
    in calcium channel blocker overdose, *316*
    chemical structure, *269*, 269
    distribution phase, 12
    history and overview, 263–266
    metabolism and metabolites, *268*, 275, *277*
    monitoring, 278
    pharmacokinetics of, 272–275, *274*, *275*, *276*
    pharmacology of, 270–272, *271*, *273*
    testing and test results, 278–281, *282*
        cross-reactivity in assays, 280
    therapeutic ranges, 278, 283
    therapeutic uses, 275–278
    toxicity
        antidote, 12
        calcium channel blocker interactions, *306*, *308*, *310–311*
        case reports, 266–267
        drug interactions, 282–283, *283*, *284*
        in elderly, 2
        predisposing factors, *283*
        symptoms, 266–267, 281–284, *285*
        treatment of, 284–287, *286*, 407
    Digoxin Fab antibodies, 12, 280–281, 285–286, *287*, 407
Digoxin-like immunoreactive factors (DLIFs), 279–280
Dihydrocodeine
    chemical structure, *76*
    clinical pharmacology and pharmacokinetics, *79*
    metabolites, *80*
    Schedule of Controlled Substances, *75*
    Synalgos®-DC, 89
    testing and test results, *491*
Dihydrolipoic acid, 386
Dihydromorphine, *80*
Dihydronorcodeine, *80*
Dilacor®, *294*
Dilaudid®. *See* Hydromorphone
Diltiazem, 293
    chemical structure, *294*
    clinical trials, 318, 320
    dosage, *306*
    history and overview, 292
    metabolism and metabolites, 301, *302*
    pharmacokinetics of, *302*
    pharmacology of, *303*
    sales rank in US, 293, *296*
    sustained-release, 301
    therapeutic uses, *306*
    toxicity
        digoxin drug interaction, *284*
        symptoms, *304*, *306*, 314
Dilution, in urine adulteration, 159–160, 167–168, 412–413
Dinitrophenol, *30*, 477–478
Dinorpropoxyphene, *80*
Dinortramadol, *80*
Dionin. *See* Dihydrocodeine
Diphenhydramine, *162*, *460*, 479, *493*
Diphenoxylate, 284
Diphenylhydantoin, *485*
Dispyramide, *310*
Distribution, drug, 13
    digoxin, 272
    ethanol, 180–181, *181*
    of industrial toxicants, 333
    lead, 371
    opiates and opioids, 76–77
    pharmacokinetics of, 12, 13
    phenytoin, 249–250
    THC, 48
Disulfiram reaction, 478
Diuretics, 160, *283*, 317
DNA testing, of urine, 169
Dolophine®. *See* Methadone
Dopamine (DA), 100, 115

Doriden, *485*
Doxepin, *224*, 230
Doxylamine, 19–20, *461*
DPC® immunoassays, *122*
Drano®, *163*
Drinking water, as source of lead, 380
Drinks. *See* Foods and beverages
Driving tests, marijuana effect on, 66–67
Dronabinol (Marinol®), 47
Drug Abuse Warning Network (DAWN)
    function of, 3, 145
    γ-hydroxybutyrate (GHB), 128, *128*
    marijuana and THC, 46
    opiate and opioid analgesics, 73
Drug Detector swipe pad, 153
Drug Enforcement Agency (DEA), 114–115
Drug interactions
    antidepressants, 231–232
    calcium channel blockers, *310–311*
    carbamazepine, 253
    digoxin, 282–283, *283*, *284*
    opiates and opioids, 82
    phenytoin, 251–252
    valproic acid, 254
Drug of abuse profile, 83–84
Drugs
    death rate, 2
    emergency room visits, 3
    illegal
        cocaine, 97–109
        death rate, 2
        γ-hydroxybutyrate (GHB), 127–140
        incidence of use, 3
        marijuana, 43–67
        methamphetamine, 114
    over-the-counter
        poisonings from, 1–2
        toxicity, 7
    pediatric cases, rate of, 2
    screening tests for, 457–458, *458*
    sustained-release, 20–22, 36
Drug-Skreen® II system, 119
Drug testing. *See* Tests and testing
DuPont aca system, 206–207
Durazanil, *486*, *496*
Dust control, 380
Dynacirc®, *294*
Dyphenhydramine, 34
Dysrhythmia, 477

E

Ecognine, 102, *489*
Ecognine methyl ester
    cocaine metabolite, 101, 102
    stability in vitro, 108
    testing and test results, *151*, *489*

Ecstasy (MDMA)
    gas chromatography/mass spectrometry (GC/MS), *123*
    immunoassay cross-reactivity, 119, 120, *122*, 124
    immunoassay urine levels, *482*
EDDP, *80*
EDTA, 378
Efavirenz, *162*
Elderly
    calcium channel blocker poisonings, 313
    digoxin poisoning, 267
    monitoring antidepressant levels, 231
    nortriptyline pharmacokinetics, 231
    renal dysfunction and poisonings, 2
Electrocardiography, *312*
Electrochemical oxidation, 192
Electrolytes, 35, 405
Electron ionization, *426*
Electrospray ionization, 426–427, *427*
Electrothermal atomic absorption, 374–375
Elimination, drug
    calculations, 19
    ethylene glycol, 202, *203*
    hepatotoxicity and, 24
    of industrial toxicants, 333–334
    opiates and opioids, 77–78
    pharmacokinetics of, 13–14
    THC, 49
    zero-order, 14
ELISA
    amphetamines, 120–121
    opiates and opioids, 84, 85
Elixir Sulfanilamide, 197
EMDP, *80*
Emergency departments, point-of-care testing, 145–146
EMIT® assays
    adulterant effects, *163*
    amphetamines, 119, 120, *122*, 124, *482–483*
    antidepressants, tricyclic, *497*
    barbiturates, *484–485*, *495*
    benzodiazepines, 217, *486–487*, *496*
    cannabinoids, *488*
    cocaine, *489*
    cross-reactivity of, *122*, *151*
    false positives, *162*
    methadone, *490*
    opiates and opioids, 84, *491–492*, *493–494*
    salicylate interference, 160
    sensitivity and specificity of, 149, *152*
Enalapril, 319
Enantiomers, amphetamine, 123–124
Environmental monitoring, 329–342
Environmental Protection Agency, 385, 390
Enzymes, polymorphic metabolic, *439*

Ephedrine
  in cocaine preparations, 98
  gas chromatography/mass spectrometry (GC/MS), *123*
  immunoassay cross-reactivity, *122*
  immunoassay for, *482*
  in methamphetamine synthesis, 114–115
  tests for, *460*
  toxicity, symptoms, 5
Epidemiology
  abuse-related emergency room visits, *146*
  arsenic poisoning, 383–384
  bipolar disorder, 237–238
  cadmium poisoning, 392
  calcium channel blocker poisonings, 293, *295*, 305, 313
  carbon monoxide poisoning, 345, 347, *347*
  cardiac glycoside poisonings, 264–266, *265*
  epilepsy, 247
  ethanol use, 176, *177*
  GHB use, 131
  iron poisoning, 401
  mercury poisoning, 388
  opiate and opioid use, 73
  pesticide poisoning, 359
  of poisoning, 1–8
  THC impairment studies, 57–67, *58–59*
Epilepsy, 213, 247–248
Epinephrine, toxicity, and ethylene glycol, 199
4,5-Epoxymorphinans, 74
Ergocalm, *487*
Ergot alkaloids, 34, 88, 477, 479
Erythrocyte protoporphyrin, *363*, 372
Erythromycin, 242, *284*, 307
Erythroxylon coca, 97
Escopon®, 88
Esmolol, *284*
Estazolam (Eurodin), *216*, *486*, *496*
1,2-Ethanediol. See Ethylene glycol
Ethanol
  cocaine and, 101
  epidemiology of use, 176
  for ethylene glycol poisoning, 36, 198
  in household products, *176*
  metabolism and metabolites, *181*, 181–182
  for methanol poisoning, 36
  monitoring, in methanol poisoning, *184*
  pharmacokinetics of, 180–182
  pharmacology of, 178–179
  sources, 175–176
  testing and test results, 182–183, *183*, 190–195
  THC impairment studies, *58–59*
  toxicity, 178–180
    abuse-related emergency room visits, *146*
    acute, 178, *179*
    case report, 174–175, 211–212
    chronic, 179–180, *180*
    death rate, 176, *178*
    osmolal gap, *204*, 469–470
    symptoms, 19–20, 173–175, *174*, 178–180
    toxidrome, 479
    vital signs, effect on, 477–478
  toxicokinetics, 180–182
  use rate, 173
  volume of distribution, 180–181, *181*
Ethchlorvynol, *147*, 479
Ethinyl estradiol, 257
2-Ethyl-5-methyl-3,3-diphenylpyrroline (EDMP), *80*
Ethylcocaine, 101
Ethylenediaminetetraacetic acid (EDTA), 378
Ethylene glycol, *30*, 35–36
  characteristics of, 200
  chemical structure, *198*, *202*
  Elixir Sulfanilamide, 197
  metabolism and metabolites, *198*, *202*
  pharmacokinetics of, 200–201, *201*, *202*, *203*
  tests for, *38*, 205–207, *460*
  toxicity
    anion gap metabolic acidosis, *204*
    case report, 198, 199, *199*
    laboratory results of, *203*
    osmolal gap, *204*
    symptoms, 198, *199*, 199–201, *201*
    toxicokinetics, *201*, 201–202
    treatment of, 198
    vital signs, effect on, 478
Ethyl ether, toxicity, osmolal gap, *204*
2-Ethylidene-1,5-dimethyl-3,3-diphenylpyrrolidone (EDDP), *80*
Ethylmorphine, 75
Eurodin, *216*, *486*, *496*
Excited delirium, 109
Exposure, to poisons, 1–4
Extracted ion profile (SIM) data, 431–432
Eyes, pupillary responses, 477
EZ-Screen point-of-care test, 149, *152*

F

False false positive, in lead testing, 373–374
Famprofazone, 125
Federal Register, 115
Felbamate, 257–259, *259*
Felodipine
  chemical structure, *294*
  dosage, *307*
  metabolism and metabolites, *302*
  pharmacokinetics of, *302*
  pharmacology of, *303*
  sales rank in US, *296*
  sustained-release, 301

therapeutic uses, *307*
toxicity
    digoxin drug interaction, *284*
    symptoms, *307*
Fencamfamine, *493*
Fenethylline, 125
Fenfluramine, *482*
Fenproporex, 125
Fentanyl
    abuse of, 93
    chemical structure, 78
    clinical pharmacology and pharmacokinetics, 79
    immunoassay specificity, 85
    metabolism and metabolites, *80*, 89
    Schedule of Controlled Substances, 75
    tests for, *461*
    therapeutic uses, 73
    trade names, 89
Ferndex®, 115
Ferrous sulfate, *402*
Fetal alcohol syndrome, 180
FirstCare point-of-care test, 149, *151*
Flame atomic absorption, 374
Flecainide, *284*, *310*
Fleckenstein, Albrecht, 291–292
Floxin, *492*
Flumazenil, 135
Flunitrazepam
    abuse of, 6
    metabolism and metabolites, *216*
    sexual assault and, 137
    testing and test results
        immunoassay cross-reactivity, *486*, *496*
        immunoassays for, 217
    toxicity, death, 214
Fluorescein, 161
Fluorescence polarization immunoassay (FPIA)
    amphetamines, *482–483*
    antidepressants, tricyclic, *497*
    barbiturates, *484–485*, *495*
    benzodiazepines, *486–487*, *496*
    cannabinoids, *488*
    cocaine, *489*
    methadone, *490*
    opiates and opioids, 84, *491–492*
    phencyclidine (PCP), *493–494*
    salicylates, *500*
Fluoxetine, *224*, 231, *447*, *461*
Flurazepam (Dalmane)
    immunoassay cross-reactivity, *486*, *496*
    metabolism and metabolites, *215*
    pharmacokinetics of, *215*
    toxicity, death, 214
Fluvoxamine, *461*

Fomepizole, 36
Food and Drug Administration, 127–128, 153, 213, 379, 401
Food poisoning, 2
Foods and beverages. *See also* Ethanol
    arsenic in, 387
    cadmium limits, 392
    mercury in seafood, 388, 392
    mushrooms, toxic, 479
    opiates and opioids, 83–84
    toxidrome, 479
    urine adulteration by, 160–161, *163*
Foreign bodies, pediatric cases, rate of, 2
Forensics, amphetamines, 122–123
Formic acid, *184*
Fosphenytoin, 252
Foxglove, 263–264
Free-basing, 98–99
Free erythrocyte protoporphyrin (FEP), 341–342, 372
Frisium, 213, *486*, *496*
Frontline point-of-care test, 149
Fuel cell test equipment, 192
Furosemide, 160, 266

G

GABA transaminase, *130*
Gallopamil, 292
γ aminobutyric acid (GABA), 5, *130*
γ-Butyrolactone (GBL), 127, *130*, 131
    chemical and street names, *132*
    metabolism and metabolites, 133
    toxicity, symptoms, 134
γ-Hydroxybutyrate (GHB)
    abuse of, 127–128
    blood and urine concentrations, *138*
    chemical and street names, *132*
    demographics of use, 131
    endogenous, 137–139
    metabolism and metabolites, *130*, 137–139
    pharmacokinetics of, 131–133, 136
    pharmacology of, 131
    Schedule of Controlled Substances, 128
    stability in vitro, 136–137
    succinic semialdehyde dehydrogenase deficiency, 139
    synthesis of, 5, 129
    testing and test results, 136–137, *138*
    therapeutic uses, 127
    toxicity
        case report, 4–5, 129
        death, 128
        drug interactions, 133
        symptoms, 5, 129, 133–135, *134*, 137, *138*
        toxicokinetics, 131–133

γ-Hydroxybutyrate (GHB) *(continued)*
    treatment of, 129, 135–136
    withdrawal syndrome, 135
γ-Hydroxybutyrolactone (GBL), 5
Gas chromatography/mass spectrometry (GC/MS)
    adulterant effects, 166
    amphetamines, 121–123, *122*, 124
    benzodiazepines, 218
    carbon monoxide poisoning, 352
    cocaine, 106–107
    drugs identified, *460*, *461*
    ethanol, 182, *183*
    ethylene glycol, 205–206
    ibuprofen interference, 160
    identification and quantification issues, 431–432
    instrumentation and methods, 424, 425–430
    interfaces, 425–430
    isopropanol, 188
    methanol, 185–186
    opiates and opioids, 86
Gases and fumes, 2
Gas-phase ionization techniques, 426, *426*
Gastrointestinal effects, calcium channel blockers, *321*, 322
Gastrointestinal preparations, 2
Genes and genetic variation, 438–441
Genetic testing, for adulterants, 169
Genotyping, 442–445, 448–449
Geographic factor, in poisoning, 3
Glipizide, 37
Glucuronide, *251*
Glucuronide conjugates, 77, *80*, *491*
Glutaraldehyde, 168
Glutethimide, 479, 485, 496
Glyburide, 37
Glycerol, *198*, *204*
Glycoaldehyde, *202*
Glycolic acid, *202*
Glycols, 197–207
Glycosides, cardiac, 263–288, 477–478. *See also* Digoxin
Glyoxylic acid, *202*
Golden seal (adulterant), *163*
Golytely®, 405
Grapefruit juice, 304
Graphite furnace, 374–375
Guanabenz, 477–478, 479

H

Hair specimens
    cocaine in, 104
    lead testing, 373
    testing and test results, *416*, 417–419

Halazepam, 486, 496
Halcion®, 213, *216*, 487
Halcion® (triazolam), 496
Hallucinogens, 477
Haloperidol, *461*
Harrison Narcotic Act, 98
Hashish, 43, 44
Hash oil, 44
Health foods, 4–5
Heart disease
    antidepressants and, 231
    cardiac glycosides and, 275–278
    increased, calcium channel blockers and, 317–320, 322
Heart rate, 29, 477
Heavy metals, *147*. *See also* Iron; Lead and lead poisoning
    arsenic, 383–388
    cadmium, 392–396
    mercury, 388–392
Hematology, pharmacokinetics and, 449–450
Hemodialysis, 36
Hemoglobin, carbon monoxide effects, 346, 348, 350–352, *351*
Hemolytic anemia, 346
Hemoperfusion, 40
Hemp seed, 160
Henbane (*Hyoscyamus niger*), 34, 479
Henry's law, 190
Hepatotoxicity, elimination rate and, 24
Herbals, 5–6
Heroin
    clinical pharmacology and pharmacokinetics, 79
    metabolism and metabolites, *80*, 89–90
    Schedule of Controlled Substances, 75
    scopolamine tainted, 4
    testing and test results, 87, 418, *491*
    therapeutic uses, 89–90
    toxicity
        abuse-related emergency room visits, *146*
        case report, 4
        symptoms, 4
    use and abuse of, 73
    volume of distribution, 77
Hexobarbital (Hexanal), 485
High-performance liquid chromatography (HPLC)
    benzodiazepines, 217–218
    instrumentation and methods, 424
    mass spectroscopy, ionization techniques, 427
    opiates and opioids, 86
Hippuric acid, *118*
Histamine antagonists, *460*
Homatropine, 34
Home drug testing, 152–153

Hormones and hormone antagonists, death rate, 2
Hospitals, laboratory testing in, 37
Household products
  antifreeze, 198
  cleaners, 161–162, *163*
  ethanol in, *176*
  isopropanol in, 187
  methanol in, 183
  mouthwash, 175
  nickel-cadmium batteries, 392
  urine adulterants, 159–164, 413
Human Pesticide Monitoring Program, 331
Hydantoins, *284, 307–308*
Hydrocarbons, 2. *See also* Volatile substances
Hydrocarbons, toxicity, *30*, 478
Hydrochlorothiazide, 319
6-Hydrocodol, *80*
Hydrocodone
  chemical structure, 76
  clinical pharmacology and pharmacokinetics, 79
  immunoassay cross-reactivity, *491*
  metabolism and metabolites, *80*, 90
  Schedule of Controlled Substances, 75
  therapeutic uses, 90
Hydrocortisone, 231
Hydrogen sulfide, *30*, 478
6-Hydromorphol, *80*
Hydromorphone
  chemical structure, 76
  clinical pharmacology and pharmacokinetics, 79
  metabolism and metabolites, *80*, 90
  Schedule of Controlled Substances, 75
  testing and test results, *151, 491*
  therapeutic uses, 73, 90
8-Hydroxyamoxapine, *224*
Hydroxyamphetamine, *118*
Hydroxybutorphanol, *80*
Hydroxychloroquine, *284*
Hydroxyfentanyl, *80*, 89
Hydroxymethamphetamine, *118*
Hydroxynorephedrine, *118*
Hydroxynorfentanyl, *80*, 89
Hydroxypentazocine, *80*
Hydroxyzine, 34, *461*, 479
Hyoscyamine, 479
Hyoscyamus niger (henbane), 34, 479
Hyperbaric oxygen therapy, 353–355
Hyperkalemia, 287
Hyperpnea, 30
Hypertension, 477
Hyperthermia, 477
Hypoglycemics, *284*, 477–478, 479
Hyponatremia, 257

Hypotension, 477–478
Hypothermia, 477
Hypoxia, *30*, 348

I

Ibuprofen, 160, *284, 461*
Imipramine
  alprazolam drug interaction, 232
  dosage, 230
  pharmacokinetics of, 230–231
  poor metabolizers, 446–447
  spot test, *147*
  therapeutic ranges, *224*, 228
Immunoassays
  adulterant effects, *163*, 164–165
  amphetamines, 119–121, *122*, 124, *482–483*
  antidepressants, tricyclic, *497*
  barbiturates, *484–485, 495*
  benzodiazepines, 214–217, *486–487, 496*
  cannabinoid, *488*
  cocaine, 105, *489*
  cross-reactivity in, *482–494*
  digoxin serum concentrations, 278–279
  methadone, *490*
  opiates and opioids, 84–86, *491–492*
  phencyclidine (PCP), *493–494*
  point-of-care testing, *482–497*
Impairment, by cannabinoids and THC, 57–67
Indomethacin, *284*
Inductively coupled plasma mass spectroscopy, 376
Industrial chemicals, 330–335
Information processing, marijuana effect on, 60–65
Infrared spectrometry, 192–193
inocybe lacera (mushroom), 479
Inositol triphosphate receptors, *300*
Insomin, *151, 216, 487, 496*
Insomnia, 212–213
Instrumentation, for testing, 424–425, 460–462, 503–506
Interfaces, chromatography/mass spectrometry, 425–430, *426, 427*
Intoxilyzer test equipment, 193
Ion cyclotron mass analyzer, *428*
Ionization techniques, 425–427, *426, 427*
Ion monitoring, 218, *219*
Ion trap (Paul trap) mass analyzer, *428*, 429–430
Ipecac, 405
Iproveratril. *See* Verapamil
Iron
  cadmium storage and, 394
  pharmacology of, 402–403
  testing and test results, 404–405, 406–408
  toxicity

Iron *(continued)*
    anion gap metabolic acidosis, *204*
    case report, 401–402
    epidemiology of, 401
    pathophysiology of, 403
    respiration effects, *30*
    stages of clinical presentation, *403*–404
    symptoms, 401–402, 403, *403*
    toxicokinetics, 402–403
    treatment of, *405*, 405–406
    vital signs, effect on, 477–478
Ischemic-reperfusion injury, 353–354
Isoniazid, *30*, 36, *204*
Isopropanol
    metabolism and metabolites, 187, *187*
    tests for, 188, *460*
    toxicity, *186*
        case report, 186–187
        osmolal gap, *204*
        symptoms, *174*, 186–187
        toxicokinetics, 187–188
    volume of distribution, 187
Isoptin®, *294*
Isradipine
    chemical structure, *294*
    clinical trials, 319
    dosage, *307*
    metabolism and metabolites, *302*
    pharmacokinetics of, *302*
    pharmacology of, *303*
    sustained-release, 301
    therapeutic uses, *307*
    toxicity, symptoms, *307*
Itraconazole, *284*

J

Jimson weed (*Datura stramonium*), 34, 479
Johnson & Johnson Vitros system, for salicylate measurement, 500
Joy® dish detergent, 162, *163*

K

Kaolin-pectin (Kaopectate®), 283, *284*
Kappa (κ) opiate receptors, 78
Ketamine, *493*
Ketazolam, *486*
Ketoacidosis, *204*
KIMS® immunoassay. *See* On-Line systems immunoassay
Kinetic interaction of microparticles in solution (KIMS), 120
Klear adulterant, 162, 413
Klonopin. *See* Clonazepam
Kurtoxin, *300*

L

Laboratories
    in hospitals, 38
    instrumentation, for testing, 424–425, 460–462, 503–506
    rapid testing rationale, *38*
    role of, 445–446, 455
    tests, necessary, 455–459, *456*
Lamotrigine, 248, *249*
Lanatosides, *268*, *269*, 269
Lanoxicaps®, 272
Lanoxin® tablets, 272
Laudanum, 88
Lead and lead poisoning
    base-line values, 341
    blood concentrations, *380*
    CDC guidelines, 377–378, *377–378*
    cutoff values, 372
    exposure, assessment of, 372–374
    history and overview, 369–370, *370*
    poisonings from, 340–342
    sources, 372, 380
    testing and test results, 341–342, 372–379
    toxicity
        blood concentrations, *378*
        chelation treatment, 378
        toxicokinetics, 370–371, *371*
        treatment of, 341, 378, 380
        vital signs, effect on, 477
    universal screening, 379
Lennox-Gastaut syndrome, 257
Lethnidrone (nalorphine), *75*, *491*
Levo-α-acetylmethadol (LAAM), *75*
Levo-Dromoran®, 90
Levomethorphan, *75*
Levonorgestrel, 257
Levorphanol, *75*
    chemical structure, *76*
    clinical pharmacology and pharmacokinetics, *79*
    metabolism and metabolites, *80*, 90
    as opioid metabolite, *80*
    therapeutic uses, 90
Librium®. *See* Chlordiazepoxide
Lidocaine, in cocaine preparations, 98
Lime, *163*
Liquid chromatography (LC), 107, 218–219, *460*
Liquid-liquid extraction (LLE), opiates and opioids, 86
Lithium, 238–240
    calcium channel blocker interactions, *306*
    test schedule for, *244*
    therapeutic ranges, 240
    toxicity, case report, 239–240
Local anesthetics, 98
Lofentanil, 89

Lorazepam (Ativan)
  chemical structure, *212*
  for GHB toxicity, 135
  immunoassay cross-reactivity, *496*
  metabolism and metabolites, *216*
  sales rank in US, *212*
  testing and test results, *151*, *461*, *486*
Lormetazepam (Ergocalm), *487*
Lotusate (talbutal), *485*, *495*
Lysergic acid diethylamide (LSD), abuse-related emergency room visits, *146*

M

Magnetic electron mass analyzers, *428*
Manic depression, 237–238
Mannitol, *204*
Maprotiline, *224*
Marijuana. *See also* Δ⁹-Tetrahydrocannabinol (THC)
  effects of
    behavioral, 46
    impairment, 57–67
    performance, 60–65
  epidemiology studies, 57–60
  history and overview, 43–45
  testing and test results, *414*
  toxicity, treatment of, *146*
Marinol® (dronabinol), 47
Mary Jane's Super Clean 13 adulterant, 162
Mass analyzers, 427–428, *428*
MDA, *122*, *123*
MDMA (Ecstasy)
  gas chromatography/mass spectrometry (GC/MS), *123*
  immunoassay cross-reactivity, 119, 120, *122*, 124
  immunoassay urine levels, *482*
Meconium, 104, 419
Medazepam (Nobrium), *487*, *496*
Memory, marijuana effect on, 60–65
Meperidine
  chemical structure, *78*
  clinical pharmacology and pharmacokinetics, 79
  metabolism and metabolites, *80*, 91
  Schedule of Controlled Substances, 75
  testing and test results, *491*
  therapeutic uses, 73, 90–91
Meperidinic acid, *80*
Mephobarbital (Mebaral), *485*
Meprobamate, *460*, *461*, 479
Mercury
  epidemiology of, 388
  forms, 332
  levels of, 388
  pharmacokinetics of, 390–391
  sources, 390
  testing and test results, 391–392
  toxicity, 332
    biochemistry of, 390–391
    case report, 388–389
    death, 389
    symptoms, 388–390, *389*
    treatment of, 392
Mesoridazine, *490*, *493*
Metabolic acidosis, 34–37, 204
Metabolic ratio, 441–442, *442*
Metabolism and metabolites
  arsenic, 385, 386
  genetic polymorphism and, 438–439, *439*
  of industrial toxicants, 333
  opiates and opioids, 77
  pharmacogenetics and, 441–446
  phases of, 439
  phenytoin, 250–251
  THC, 48–49
Methadol, *80*, *490*
Methadone, *490*
  chemical structure, 77
  clinical pharmacology and pharmacokinetics, 79
  immunoassay cross-reactivity, *490*
  immunoassay specificity, 85
  metabolism and metabolites, *80*, 91
  for opiate detoxification, 83, 91
  Schedule of Controlled Substances, 75
  testing and test results, *414*, *490*
  toxicity
    antidote, 33
    symptoms, 32–33
Methamphetamine, 113. *See also* Amphetamine(s)
  chemical structure, *118*
  drugs converted to, 125
  elimination, 117
  enantiomers, 123–124
  gas chromatography/mass spectrometry (GC/MS), *123*
  metabolism and metabolites, *118*
  pharmacokinetics of, 116–117
  pharmacology of, 115–116, 468
  routes of administration, 117
  Schedule of Controlled Substances, 114
  synthesis of, 114
  testing and test results
    extensive testing, *460*
    immunoassay cross-reactivity, *122*, *482–483*
    immunoassays for, 149, *151*
    urine concentrations, *414*, *482–483*
  therapeutic uses, 115
  toxicity, case report, 116
  toxicity, symptoms, 116–117

Methanol
    hospital laboratory testing, 38
    metabolism and metabolites, 185, *185*
    testing and test results, *460*
    toxicity, 183–186
        acidosis, *30*, 35–36
        anion gap metabolic acidosis, *204*
        case report, 183–184
        osmolal gap, *204*
        symptoms, *174*, 183–184
        toxicokinetics, 185
        treatment of, 184
        vital signs, effect on, 478
    volume of distribution, 185
Metharbital, *485*
Methemoglobin inducers, *30*, 478
Methemoglobin testing, 334
Methohexital (Brevital), *485*
Methotrexate, as abortifacient, 37
Methylbenzoylecgonine. *See* Cocaine
Methylecognine, 102
3,4-Methylenedioxy-methamphetamine. *See* Ecstasy (MDMA)
Methylergonovine, as abortifacient, 37
Methylmercury, 389, 390–391
Methylphenidate (Ritalin), 231, *482*
4-Methylpyrazole, 36
Metoclopramide, *284*
Metoprolol, 320, *461*
Metronidazole, 161, 164–165
Mibefradil, 293, *295*
Microgenics CEDIA immunoassay. *See* CEDIA® immunoassay
Micro-Plate ELISA assay, 121, *122*
Midazolam (Versed), *216*, *487*, *496*
Miosis, 477
Misoprostol, as abortifacient, 37
Monarch 2000 centrifugal analyzer, 207
Monitoring
    antidepressant therapeutic ranges, *224*, 224–225, 227–232
    antimanic drugs, *242*
    biological, of chemical exposure, 329–342
    cadmium, *395–396*
    cardiac glycosides, 278
    environmental, 330
Monoamine oxidase (MAO) inhibitors
    in bipolar disorder, 239
    opiate interactions, 82
    toxicity delayed, 37
    toxidrome, 479
    vital signs, effect on, 477
Morphinans, 74
Morphine
    chemical structure, *76*
    clinical pharmacology and pharmacokinetics, *79*
    metabolism and metabolites, *80*, 88–89, 91
    as opioid metabolite, *80*
    in opium, 88
    Schedule of Controlled Substances, *75*
    testing and test results, 87, *151*, *491*
    therapeutic uses, 73, 91
    toxicity, abuse-related emergency room visits, *146*
    urine concentrations, 87
Morphine glucuronides, *80*
Mouthwash poisoning, 175
Mucomyst, 20
MUDPILES, *30*, 30, 35
Mu (μ) opiate receptors, 78
Muscarinic effects, of pesticides, 361–362
Mushrooms, 479
Mydriasis, 477

N

Nalbuphine
    chemical structure, *76*
    clinical pharmacology and pharmacokinetics, *79*
    metabolism and metabolites, *80*, 93
    therapeutic uses, 93
Nalorphine (Lethnidrone), *75*, *491*
Naloxone, 33, 135, *491*
Naltrexone, *491*
Narcotics, Schedule of Controlled Substances for, *75*
National Center for Alcohol and Drug Abuse Intervention (NCADI), 3
National Health and Nutrition Survey, 331
National Highway Traffic Safety Administration, 193, 194
National Hospital Ambulatory Medical Care Survey, 145
National Household Survey on Drug Abuse, 3–4, 73, 176
National Institute of Justice Drug Use Forecasting (DUF), 4
National Institute on Drug Abuse (NIDA), 46, 73
Nayler, Winfred, 291–292
Nembutal (pentobarbital), *485*, *495*
Neomycin, *284*
Neonates, cocaine in, 104
Neostigmine, 135
Nerdipine. *See* Nicardipine
Nerium oleander, 265
Nerve gas, 365
Neuroleptics, 223–233, 238–240, 477
Neuropathy, organophosphate-induced delayed (OPIDN), 366
Neurotransmitters, cocaine effect on, 100
Neutron activation analysis, 376
Nicardipine, 292

chemical structure, *294*
clinical trials, 320
dosage, *308*
metabolism and metabolites, *302*
pharmacokinetics of, *302*
pharmacology of, *303*
sustained-release, 301
therapeutic uses, *308*
toxicity, symptoms, *308*
Nickel-cadmium batteries, 392
Nicodel. *See* Nicardipine
Nicotine, 478
Nicotinic effects, of pesticides, 362
Niemann, Albert, 97
Nifedipine
chemical structure, *294*
clinical trials, 318, 319, 320
dosage, *308*
history and overview, 292
metabolism and metabolites, *302*
pharmacokinetics of, *302*
pharmacology of, *303*
sales rank in US, 293, *296*
sustained-release, 301
therapeutic uses, *308*
toxicity
digoxin drug interaction, *284*
symptoms, *308*, 315
Nightshade (*Atropa belladonna*), 34, 479
Niludipine, 292
Nimetazepam, *487*, *496*
Nimodipine
chemical structure, *294*
dosage, *309*
metabolism and metabolites, *302*
pharmacokinetics of, 301, *302*
pharmacology of, *303*
therapeutic uses, *309*
toxicity, symptoms, *309*
Nimotop®, *294*
Nisoldipine
chemical structure, *294*
clinical trials, 319, 320
dosage, *309*
metabolism and metabolites, *302*
pharmacokinetics of, *302*
pharmacology of, *303*
therapeutic uses, *309*
toxicity, symptoms, *309*
Nitrates, 478
Nitrazepam (Insomin), *151*, *216*, *487*, *496*
Nitrendipine, clinical trials, 319
Nitrites, 168
Nitroprusside, 478
Nobrium, *487*, *496*
Nomograms, *21*, 22, 38–39, 500–501
Norbuprenorphine, *80*

Norbutorphanol, *80*
Norcocaine, 101
Norcodeine, *80*, 88–89, *492*
Nordaz, *215*, *487*
Nordiazepam (Nordaz), *215*, *487*
Norephedrine, *118*
Norepinephrine (NE), 100, 115
Norfentanyl, *80*, 89
Norfluoxetine, *224*
Norhydrocodone, *80*
Norlevorphanol, 75, *80*
Normeperidine, *80*
Normeperidinic acid, *80*
Normethadol, *80*
Normethadone, 75
Normorphine, *80*, *492*
Nornalbuphine, *80*
Noroxycodone, *80*
Noroxymorphine, *492*
Norpropoxyphene, *80*
Nortramadol, *80*
Nortriptyline, *224*
pharmacogenetics and, 447–448
pharmacokinetics of, 231
therapeutic ranges, 228
toxicity, case report, 226–227
Norvasc®, 293, *294*, *296*
Norverapamil, 301
Noscapine, in opium, 88
Nubain®, 93
Numorphan®. *See* Oxymorphone
Nyquil®, 19–20, 22, 24

O

Obidoxime, *363*
Occupational hazards, diagnosis and, 35
Occupational Safety and Health Administration
arsenic exposure limit, 385
cadmium monitoring, 395–*396*
lead concentration limits, 372, 379–380
role of, 2
surveillance and monitoring, 339
Ofloxacin (Floxin), *492*
Omeprazole, *284*
Omnopon®, 88
Oncology, pharmacogenetics and, 448–449
Ondansetron, 40, 136
ONLINE® KIMS assay, 120, *122*
On-Line systems immunoassay, 217
amphetamines, *482–483*
barbiturates, *484–485*
benzodiazepines, *486–487*
cannabinoids, *488*
cocaine, *489*
methadone, *490*
opiates and opioids, 85, *491–492*, *493–494*

ONTRAK immunoassay, 147
  amphetamines, *152, 482–483*
  barbiturates, *152, 484–485*
  benzodiazepines, *152, 486–487*
  cannabinoids, *488*
  cocaine, *152, 489*
  Δ9-Tetrahydrocannabinol (THC), *152*
  opiates and opioids, *152, 491–492, 493–494*
Opiate and opioid analgesics, 73–94. *See also* Poppy seeds
  addiction potential, *79*
  case report, 83–84
  chemical structures, 74, 76
  clinical pharmacology, *79*
  definitions, 73–74
  distribution phase, 76
  drug interactions, 82
  elimination, 77–78
  metabolism and metabolites, 77, *80*
  pharmacokinetics of, 74–78, *79*
  precautions and contraindications, 81–82
  receptors, 78
  routes of administration, 74, 76, *79*
  Schedule of Controlled Substances, *75*
  testing and test results, 84–88
    extensive testing, *460*
    immunoassay cross-reactivity, *151, 491–492*
    immunoassay sensitivity and specificity, *152*
    urine adulterants, *163*
    urine screening cut-offs, *414*
  therapeutic uses, 81–82
  toxicity
    detoxification, 83
    symptoms, 82–83
    toxidrome, *479*
    treatment of, 82–83, *82–83, 146*
    vital signs, effect on, *477–478*
  trade names, 88–94
  volume of distribution, 78, *79*
  withdrawal syndrome, 78, 81, *81*
Opium, *75, 79*, 88
Opium derivatives, Schedule of Controlled Substances, *75*
Opium Exclusion Act, 88
Oral fluid. *See* Saliva
Organophosphate-induced delayed neuropathy (OPDIN), 366
Organophosphate pesticides
  chemical structures, *360*
  esters, *360*
  testing and test results, 364–365
  toxicity
    antidote, 8
    diazinon poisoning, 360–361
    epidemiology of, 359
    mechanism of, 362–363
    neurologic sequelae (neuropathy), 366
    pharmacology of, 361–362
    SLUDGE syndrome, 29, 361–362
    symptoms, 8, 29, 360–361
    toxicokinetics, 361
    toxidrome, *479*
    treatment of, 363–364
Osmolal gap, 35, *204*
  calculations, *471*
  ethanol, 182
  ethylene glycol, *203*, 203–204
  methanol, 186
  tests for, *35*
Ouabain, *268*
Oxalic acid, *202*
Oxaprozin (Daypro®), *162*, 212, *215*, 216
Oxazepam (Serax), *496*
  chemical structure, *162, 215*
  metabolism and metabolites, *216*
  testing and test results, *487*
  toxicity, death, 214
Oxcarbazepine, 255–257
Oxidizing adulterants, 168
Oxycodone
  chemical structure, *76*
  clinical pharmacology and pharmacokinetics, *79*
  immunoassay cross-reactivity, *492*
  metabolism and metabolites, *80, 92*
  Schedule of Controlled Substances, *75*
  testing and test results, *151*
  therapeutic uses, 73, 91–92
  volume of distribution, 77
OxyContin®. *See* Oxycodone
Oxydess II®, 115
Oxygen, for carbon monoxide poisoning, 353–355
6-β-Oxymorphol, *80*
6-α-Oxymorphol, *80*
Oxymorphone
  chemical structure, *76*
  clinical pharmacology and pharmacokinetics, *79*
  immunoassay cross-reactivity, *492*
  metabolism and metabolites, *80, 92*
  as opioid metabolite, *80*
  Schedule of Controlled Substances, *75*
  therapeutic uses, 92

P

Paint, lead in, 340–341, *370*
Pantopon®, 88
Papaverine, in opium, 88
Paraldehyde, *30*, 36, *204*
Paramax System, for salicylate testing, 500
Paramorphine, *75*, 88, *492*

Paraoxon, *360*
Paraquat, *30, 147,* 478
Parathion, *360,* 365
Paregoric, 88
Paroxetine, *224*
Passive inhalation, of THC, 57
Patient history, diagnosis and, 35
Patients, clinical approach to, 27–40, *33*
Paul trap mass analyzer, *428,* 429–430
PBZ (tripelennamine), *496*
Penicillamine, *284*
Penning trap mass analyzer, *428*
Pentazocine
  chemical structure, 77
  clinical pharmacology and pharmacokinetics, 79
  metabolism and metabolites, *80,* 93
  Schedule of Controlled Substances, 75
  therapeutic uses, 93
Pentobarbital (Nembutal), *151, 461, 485, 495*
Pentothal (thiopental), *485, 495*
Pepto-Bismol, 19–20
Percocet®. *See* Oxycodone
Percodan®. *See* Oxycodone
Perdipine. *See* Nicardipine
Performance studies, marijuana effect on, 60–67, *61–63, 66–67*
Peroxide, *163*
Perphenazine, 225–227
Pesticides and insecticides. *See also* Carbamate pesticides; Organophosphate pesticides
  arsenic, 383–384
  death rate, *2*
  history and overview, 359–361
  monitoring programs, 331
  pediatric cases, rate of, *2*
  toxicity, symptoms, 8
Pharmacodynamics, 77–78, 438
Pharmacogenetics
  γ-hydroxybutyrate (GHB) metabolism, 139
  metabolism and, 139, 438–446
  in oncology, 448–449
  overview, 437
  psychiatry and, 446–448
  succinic semialdehyde dehydrogenase deficiency, 139
Pharmacokinetics, 11–14
  calculations, 17–22
  hematology and, 449–450
  metabolism and genetic polymorphism, 438–439
  overview, 437–438
  principles of, 13–14
Pharmacology, 437–439
Phenallymal, *484, 495*
Phenanthrenes, 74

Phencyclidine (PCP)
  false positives, *162*
  immunoassay cross-reactivity, *493–494*
  testing and test results, 149, *151, 152, 414, 493–494*
  toxicity
    abuse-related emergency room visits, *146*
    treatment of, *146*
    vital signs, effect on, 477
  urine adulterants, *163*
Phendimetrazine (Bontril), *482*
Phenethylamine, *123,* 483
Phenformin, *204,* 478
Phenmetrazine, *483*
Phenobarbital
  drug interactions, 258, *259*
  for epilepsy, 247
  immunoassay cross-reactivity, *485, 495*
  opiate interactions, 82
  testing and test results, *151, 461*
Phenothiazines
  spot test, *147*
  tests for, *460*
  toxidrome, 479
  vital signs, effect on, 477–478
Phentermine
  gas chromatography/mass spectrometry (GC/MS), *123*
  testing and test results
    difficulties, 123
    extensive testing, *460*
    immunoassay cross-reactivity, 119, *122,* 483
Phenylacetone, *118*
Phenylephrine, gas chromatography/mass spectrometry (GC/MS), *123*
Phenylheptylamine opiates, 74
Phenylpiperidine opiates, 74
Phenylpropanolamine
  in cocaine preparations, 98
  gas chromatography/mass spectrometry (GC/MS), *123*
  immunoassay cross-reactivity, *122*
  immunoassay for, *483*
  toxicity, symptoms, 7
  vital signs, effect on, 477
Phenyltoloxamine, *494*
Phenytoin
  calcium channel blocker interactions, *309*
  chemical structure, *251*
  drug interactions, 251–252, 258–259, *259*
  history and overview, 247
  immunoassay cross-reactivity, *495*
  metabolism and metabolites, 250–251, *251*
  pharmacokinetics of, 249–252
  tests for, 254
  therapeutic ranges, 249
  toxicity, 248

pH, of urine, 165, 168
Phosphate, *163*
Photometry, 406
Physical examination, *33*
Physostigmine, 135, 479
Picenadol, *494*
Pilocarpine, 479
Pinazepam, *487*
Plants, *2*, 28, 34
Plasma, 55–56, 102–103
Plendil®, *294, 296*
Plexonal, *484*
Point-of-care testing
    adulterant effects, 165–166
    advantages, 468–469
    essential drug screen, 459–460
    ethanol, 190–195
    immunoassay, drugs detected by, *151*
    test types, 147–153
Poison Control Centers, 13
Poisondex database, 28
Poisoning(s)
    accidental, 1–2
    in children, 1–2, 19–20
    clinical approach to, 27–40
    diagnosis of, 27–34
    epidemiology of, 1–8
    exposures, 3–4
    reported, 1
    toxin identification, 25
Polamidon. *See* Methadone
Polymorphism, genetic, 438–446, 448–449
Pondimin, *482*
Poppy seeds, 83–84, 87, 161
Posicor®, 293, *294*
Potassium cyanide, 32
Potassium nitrite, 162–163
Pralidoxime, 8, *360*, 363, 363–364, *364*
Prazepam (Centrax)
    metabolism and metabolites, *215*
    pharmacokinetics of, *215*
    testing and test results, *487, 496*
    toxicity, death, 214
Prazosin, *310–311*
Pregnancy
    abortifacients, 37
    acetaminophen overdose in, 38–39
    carbon monoxide poisoning, 348–349
    carboxyhemoglobin increase in, 346, 348–349
    clinical approach to, 37
    cocaine use, 419
    ethanol use in, 180
    opiate and opioid use, 83–84
    valproic acid in, 244
Prenylamine, 291–292

Primatene® Mist, 199
Procaine, in cocaine preparations, 98
Procardia, Procardia XL. *See* Nifedipine
Prochlorperazine, for theophylline toxicity, 40
Prolintane, *494*
Promethazine, *490*
Propafenone, *284*
Propantheline, *284*
Propoxyphene
    chemical structure, 77
    clinical pharmacology and pharmacokinetics, *79*
    immunoassay specificity, 85
    metabolism and metabolites, 80, 92
    for opiate detoxification, 83
    tests for, *460, 461*
    therapeutic uses, 92
Propranolol, *309*
Propylbenzoylecognine, *489*
Propylene glycol, 197
    characteristics of, 200
    chemical structure, *198*
    metabolism and metabolites, *198*
Propylhexedrine, *483*
Protein A, 165
Protopam, *363*
Protriptyline, *224*
Pseudoephedrine
    gas chromatography/mass spectrometry (GC/MS), *123*
    immunoassay cross-reactivity, *122*
    tests for, *460, 483*
    toxicity, symptoms, 19–20
Psychiatric patients
    bipolar disorder, 237–238
    clinical approach to, 37–38
    depression, 37–38
    schizophrenia, 37–38, 238
Psychiatry, pharmacogenetics and, 446–448
Psychomotor skills, marijuana effect on, 60–67
Pulse oximetry, 350–352, *351*
Pupillary responses, 477

Q

Quadrupole mass spectrometer, *428*, 428–429
Quality control and assurance, for breath alcohol testing, 193–194
Quinidine, 232, *283*, *284*, *308–311*, *460*
Quinine, 37, *284*, *460*

R

Race, antidepressants and, 231
Radiography, *404*
Radioimmunoassay (RIA)
    adulterant effects, *163*
    amphetamines, 120–121
    opiates and opioids, 84

Ranitidine, 232, *306–308*
Rapid Drug Screen point-of-care test, 149
RDS immunoassay
 amphetamines, *482–483*
 barbiturates, *484–485*
 benzodiazepines, *486–487*
 cannabinoids, *488*
 cocaine, *489*
 methamphetamine, *482–483*
 opiates and opioids, *491–492*
 phencyclidine (PCP), *493–494*
Receptor binding assays, 279
Receptors, cannabinoid, 45–46
Regional Poison Centers, 13
Renal failure, *204*
Reserpine, *283*
Respiration
 depression of, 82
 effort increased, *30*
 toxin effects on, *30*, 478
Restoril. *See* Temazepam (Restoril)
Rifampicin (Rifampin), *284*, *492*
Ritalin (methylphenidate), 231, *482*
Roche Abuscreen immunoassays. *See* On-Line systems immunoassay; ONTRAK immunoassay
Roche Boehringer Mannheim immunoassays. *See* On-Line systems immunoassay; ONTRAK immunoassay
Rohypnol®. *See* Flunitrazepam
Romilar. *See* Dextromethorphan
Roxanol™, 91
RS-69216. *See* Nicardipine
Rumack-Matthew nomogram, 20, *21*, 22, 38–39
Ryanodine receptors, *300*

S

Salicylate hydrolase test, 499, *500*
Salicylates
 adulterant effects, 164–165
 testing and test results
  colorimetry, 499
  photometric tests, 499–500
  salicylate hydrolase, 499, *500*
  spot test, *147*
  test interference by, 160
 toxicity
  acidosis, *30*
  death rate, 1
  respiration effects, *30*
  toxidrome, 479
  treatment of, *146*
  vital signs, effect on, 477–478

Saliva
 cocaine in, 104
 ethanol analysis, 195
 point-of-care testing, 152
 testing and test results, 413, 415, *416*, 419
 THC in, 52, 55–56
Salt, *163*
Samples, alternative, 411–420
Sandoptal, *484*
Schedule of Controlled Substances, 75, 113, 114, 128
Schizophrenia, 37–38, 238
Scopolamine
 heroin-tainted, 4
 toxicity, 34, 479
Screening. *See* Tests and testing
Seasonal factor, in poisoning, 3
Secobarbital (Seconal), *461*, *485*, *495*
Second-hand smoke, THC, 57
Sedatives/hypnotics. *See also* Barbiturates
 death rate, 2
 opiate interactions, 82
 toxidrome, 479
 vital signs, effect on, 477–478
Selegiline, 116, 125, *461*
Selenium, 391
Semylan®. *See* Phencyclidine (PCP)
Serax. *See* Oxazepam (Serax)
Serotonin (5-HT), 100, 115
Serotonin reuptake inhibitors, 477
Sertraline (Zoloft®), *224*, *461*, *496*
Serum tests. *See* Blood tests
Sexual assault, drug use and, 137
SIM data, 431–432
Sinsemilla, 43, 44
SLUDGE syndrome, 29, 361–362
Smoking, 232, 393
Soap, *163*
Sodium hydroxide, 5
Sodium monofluoroacetates, *30*, 478
Solid-phase extraction (SPE), 86, 105–106
Solvents, abuse of, 6–7
Somnipron, *484*, *495*
Specific gravity, of urine, 167
Spectrophotometry, carbon monoxide poisoning, *352*, 352–353
Spot tests, 147–149, *148*, 459
SR141716 (cannabinoid receptor antagonist), 45–46
Stadol®, 92–93
Steroids, *283*
Stimulants, *2*, 30
Storage, of industrial toxicants, 333
Strophanthins, *268*
Strophantus, *268*
Sublimaze®, 89

Substance abuse, 3–4
Substance Abuse and Mental Health Services Administration (SAMHSA), 146, 158, *159*
Substitution, in urine adulteration, 158–159
Succinic acid, *130*
Succinic semialdehyde, *130*
Succinic semialdehyde dehydrogenase, *130*, 137
Succinic semialdehyde dehydrogenase deficiency, 139
Sucralfate, *284*
Sufentanil, 89
Sular®, *294*
Sulfanilamide, Elixir, 197
Sulfasalazine, *284*
Sulfentanil, 75
Sustained-release preparations, 20–22, 36, 37
Sweat
  cocaine in, 104
  patch, for testing, 415, 417
  testing and test results, 415–417, *416*, 419
Sympathomimetics, 113, 117–119, 478, 479. *See also* Amphetamine(s)
Synalgos®-DC, 89
Synchron RGT test, for salicylates, 500
Syva EMIT immunoassays. *See* EMIT® assays

T

Tachycardia, toxins inducing, 29, 477
Tachypnea, 30, 478
Tacrine, *461*
Talbutal (Lotusate), *485*, *495*
Talwin®, 93
Tandem mass spectrometry, *430*, 430–431
TDx/FLx/AxSYM assay, 120, *122*, 124, *162*
Temazepam (Restoril), *487*, *496*
  chemical structure, *212*, 215
  metabolism and metabolites, *216*
  sales rank in US, *212*
  testing and test results, *151*
  toxicity, death, 214
Temperature, toxin identification and, 477
Terfenadine, 293
TesTcup point-of-care test, 149
Tests and testing
  acetaminophen, 500–501, *501*
  acidosis, *35*
  adulteration, 146–147
  advanced, *460*, 460–462
  advantages/disadvantages, *416*
  alternative samples, 411–420
  amphetamines, 117, 119–125, *482–483*
  arsenic, 386–387
  biological, of chemical exposure, 331–338
  blood. *see* Blood tests
  book reviews, 503–506

breath alcohol devices, 192–194
cadmium, 395–396
calcium channel blockers, 315, 317
carbamazepine, 255
cocaine, 105–107
cost-benefit analysis, 462
creatinine levels, 161
cross-reactivity in, 85, 161
Δ^9-Tetrahydrocannabinol (THC), 50–57
digoxin serum concentrations, 278–281
ethanol, 182–183, *183*, 190–195
ethylene glycol, 205–207
false positives, 161
γ-hydroxybutyrate (GHB), 136–137
home testing, 152–153
identification and quantification issues, 431–432
iron poisoning, 406–408
isopropanol, 188
laboratories and, 455–463
lead exposure, 372–379
methamphetamine, *482–483*
methanol, 185–186
necessary, 455–459, *456*
opiates and opioids, 84–86, *85*
osmolal gap, *35*
over- and underpredictions, 22–24
performance, marijuana effect on, 57–67
pesticide poisoning, 364–365
phenytoin, 254, 255
plasma cholinesterase assay, *365*, 365–366
point-of-care testing, 139, 147–153, 190–195
principles of, 412
rationale for rapid, *38*
results confirmation, 53–54, *85*
results interpretation, 54–57
salicylates, 499–500
screening, essential, 457–458, *458*, 459–460
spot tests, 147, 459
techniques, analytical, 423–433
universal screening for lead, 379
urine. *see* Urine tests
valproic acid, 255
value of, *456*, 456–457
in workplace, 146–147, 153–154
Tetracycline, *284*
Δ^9-Tetrahydrocannabinol (THC), 74–78, *163*
  chemical stability, 45
  chemical structure, *44*
  distribution phase, 48
  elimination, 49
  epidemiology of, 57–67
  metabolism and metabolites, *44*, 48–49, *49*, 52
  pharmacokinetics of, 46, 47–52
  pharmacology of, 45–46
  plasma concentrations, *49*, 49–50

stability in vitro, 56–57
structure-activity relationships, 45
testing and test results
  immunoassay cross-reactivity, *151, 488*
  immunoassay sensitivity and specificity, *152*
  impairment studies, *58–59, 61–63, 66–67*
  screening, confirmation, and interpretation, 50–57
toxicity
  abuse-related emergency room visits, *146*
  chronic effects, 46–47
  symptoms, 46–47
urine adulterants, *163*
urine concentrations, 49–50
Tetrahydrozoline, 413
  toxicity, symptoms, 7, 29–30
  urine adulteration by, *163*
Tetrazepam, *487*
THC. See Δ^9-Tetrahydrocannabinol (THC)
Thebaine (Paramorphine), *75*, 88, *492*
Thenyldiamine, *496*
Theobromine, 160
Theo-Dur®, 19–20, 199
Theophylline, 22–24
  calcium channel blocker interactions, *306, 309*
  sustained-release, 20–22, 199
  toxicity
    case report, 39–40
    and ethylene glycol, 199
    pharmacokinetic calculations, 17–19
    respiration effects, *30*
    symptoms, 17–18, 19–20, 39–40
    toxidrome, 479
    treatment of, 40
    vital signs, effect on, *477–478*
  urine adulteration by, 160
  volume of distribution, 11
Thermospray ionization, *427*
Thiamine, 470
Thiamylal, *495*
Thienylcyclohexylpiperidine, 149
Thin layer chromatography (TLC), 86, 106, 117, 119, *460*
Thiopental (Pentothal), *485, 495*
Thiopurine methyltransferase (TPMT), 448–449
Thioridazine, *490, 494*
Thorazine (chlorpromazine), 241, *490*
Threshold limit values, on toxicants, 339–340
Thyroid hormone, *30*, 477–478
Tiazac®, *294, 296*
Time of flight mass analyzer, *428*
Tinctures, of opium, 88
TLC-kit, for acetaminophen testing. *See* Toxi-Lab® TLC systems
Toads, source of cardiac glycosides, 263, *268*

Tolbutamide, *284*
Tolerance
  benzodiazepines, 213
  ethanol, 178
  opiates and opioids, 77–78
Topical preparations, 2
Toxic delirium, 229
Toxicokinetics, 11–14
Toxidromes, 32
  anticholinergics, 33, 479
  cholinergics, 479
  MUDPILES, *30*, 30, 35
  opiates and opioids, 479
  opiate withdrawal, 78, 81, *81*
  organophosphate poisoning, 29
  sedatives/hypnotics, 479
  SLUDGE, 29
  sympathomimetics, 479
  toxicity, 479
Toxigonin, *363*
TOXI-GRAM®, 117, 119
Toxi-Lab® TLC systems, 86, 117, 119, 124, 501
Tramadol
  chemical structure, 77
  clinical pharmacology and pharmacokinetics, 79
  immunoassay cross-reactivity, *494*
  metabolism and metabolites, *80*, 92–93
  tests for, *460*
  therapeutic uses, 92–93
Tranquilizers, opiate interactions, 82
Tranxene, *486*
Trazodone, *224*
Treatment plan, 27–40
Triage DOA point-of-care test device
  amphetamines, *151, 482–483*
  barbiturates, *151, 484–485*
  benzodiazepines, *151, 486–487*
  cannabinoids, *488*
  cocaine, *151, 489*
  cross-reactivity in, *151*
  drugs of abuse, 149
  methadone, *490*
  opiates and opioids, *151, 491–492*
  phencyclidine (PCP), *151, 493–494*
  schematic of, *150*
  sensitivity and specificity of, *152*
  THC, *151*
Triazolam (Halcion®)
  immunoassay cross-reactivity, *496*
  metabolism and metabolites, *216*
  removal from market, 213
  testing and test results, *487*
Trichloroethane, *204*
Tricyclic antidepressants. *See* Antidepressants
Trimipramine, *224*, 230

Tripelennamine (PBZ), *496*
Tylenol®. *See* Acetaminophen
Tylenol® with codeine, 88–89
Tyramine HCl, *483*

U

Ultram®, 92–93
Uremia, *204*
UrinAid, 413
UrinAid adulterant, 162
Urine adulteration, 146–147, 153–154, 412–413
    case report, 154
    methods and practices, 158–166
    prevalence of, 157–158
    regulations, 158
    in vitro adulterants, 161–164
    in vivo adulterants, 159–161
Urine, appearance of, 166–167
Urine Luck adulterant, 413
Urine tests
    adulterant effects, 162–166
    advantages/disadvantages, *416*
    amphetamines, *414*, *482–483*
    appearance and color adulterants, *164*
    barbiturates, *414*, *484–485*
    benzodiazepines, *414*, *486–487*
    biological monitoring, 336
    cannabinoids, *488*
    cocaine, 103, 105–107, 107–108, *414*, *489*
    codeine, 87
    creatinine levels, *159*
    genetic (DNA), 169
    immunoassays, *482–494*
    lead exposure, 373
    marijuana, *414*
    methadone, *414*, *490*
    methamphetamine, 117, *414*, *482–483*
    morphine, 87
    necessary, *456*, *458*
    opiates and opioids, 87, *414*, *491–492*
    pH and adulterant effects, 165
    phencyclidine (PCP), *414*, *493–494*
    principles of, 411–412
    radioimmunoassay POCT tests, 147, 149, *482–483*
    screening and confirmation, drugs of abuse, *414*
    spot tests, 147, *148*
    THC, 49–50, 52, 54–55
    for urine dilution, 167–168

V

Valium®. *See* Diazepam (Valium)
Valproic acid
    in bipolar disorder, 241
    drug interactions, 254, 258, *259*
    metabolism and metabolites, 253
    pharmacokinetics of, 255
    pharmacology of, 249
    test schedule for, *244*
    therapeutic ranges, 244–245, *249*
    toxicity, 244, 248, 249, 253–254
Vanish®, *163*
Vascor®, *294*
Vasonase. *See* Nicardipine
Venlafaxine, *224*
Venoms, 2
Verapamil
    chemical structure, *294*
    clinical trials, 318, 319–320
    dosage, *306*
    electrocardiography, *312*
    history and overview, 291
    metabolism and metabolites, 301, *302*
    pharmacokinetics of, *302*
    pharmacology of, *303*
    sales rank in US, *296*
    sustained-release, 301
    therapeutic uses, *306*
    toxicity
        case report, 305
        digoxin drug interaction, 283, *283*, *284*
        symptoms, 304–305, *306*, 314
Verelan®, *294*
Vicodin. *See* Hydrocodone
Vin Mariani, 97
Vinylbital, *495*
Visine®. *See* Tetrahydrozoline
Vital signs, toxin identification and, 28–32, 477–478
Vitamins, 1–2, 380. *See also* Iron
Vitros system for salicylate measurement, 500
Volatile substances, 6–7, *147*. *See also* Alcohol(s); Hydrocarbons
Volume of distribution (V_d)
    defined, 11
    opiates and opioids, 77, *79*

W

Warfarin, 82, 449–450
Water
    arsenic levels, 383–384, 385
    lead, source of, 380
Weight loss, 232
Whizzies adulterant, 413
Withdrawal syndrome, *81*
    clonazepam, 239
    γ-hydroxybutyrate (GHB), 135
    opiates and opioids, 78, 81
    vital signs, effect on, 477–478

Withering, William, 263
Wohler, Carl, 97
Workplace
    arsenic levels, 385
    cadmium monitoring, 395
    drug testing in, 146–147, 153–154
    environmental monitoring, 331
    exposure limits to toxicants, 339–340
    lead poisoning, 372
World Health Organization, 339
Worthington Diagnostics ultrafree system, 254

X

Xanax. *See* Alprazolam
Xanthines, 160

X-ray fluorescence, 376
Xyrem®, 127, 128

Y

YC-93. *See* Nicardipine

Z

Zero-order elimination, 14
Zinc protoporphyrin (ZPP), 372, 376–377
Zoloft® (sertraline), *224*, *461*, *496*
Zolpidem, 479